Biochemistry and Physiology of Nutrition

VOLUME II

Biochemistry and Physiology of Nutrition

VOLUME II

Edited by

GEOFFREY H. BOURNE

Department of Histology
London Hospital Medical College
London, England

and

GEORGE W. KIDDER

Biological Laboratory
Amherst College
Amherst, Massachusetts

ACADEMIC PRESS INC., *Publishers*

NEW YORK · 1953

LIBRARY OF CONGRESS CATALOG CARD NUMBER: 52-13364

PRINTED IN THE UNITED STATES OF AMERICA

CONTRIBUTORS TO VOLUME II

GEOFFREY H. BOURNE, *Department of Histology, London Hospital Medical College, London, England*

D. P. CUTHBERTSON, *Rowett Research Institute, Aberdeen, Scotland*

GRACE A. GOLDSMITH, *Tulane University, School of Medicine, New Orleans, Louisiana*

B. L. HORECKER, *National Institute for Arthritis and Metabolic Diseases, Bethesda, Maryland*

EUGENE P. KENNEDY, *Ben May Laboratory for Cancer Research and Department of Biochemistry, University of Chicago, Chicago, Illinois*

GEORGE W. KIDDER, *Biological Laboratory, Amherst College, Amherst, Massachusetts*

H. F. W. KIRKPATRICK, *Department of Clinical Investigation, The London Clinic, London, England*

A. KORNBERG, *National Institute for Arthritis and Metabolic Diseases, Bethesda, Maryland*

G. DAVID NOVELLI,* *Biochemical Research Laboratory and Robert W. Lovett Memorial Laboratory, Massachusetts General Hospital, Boston, Massachusetts*

A. T. PHILLIPSON, *Rowett Research Institute, Aberdeen, Scotland*

J. DOUGLAS ROBERTSON, *Department of Clinical Investigation, The London Clinic, London, England*

MORRIS SOODAK, *Biochemical Research Laboratory and Robert W. Lovett Memorial Laboratory, Massachusetts General Hospital, Boston, Massachusetts*

E. J. UNDERWOOD, *Institute of Agriculture, University of Western Australia, Crawley, Western Australia*

WARREN H. YUDKIN, *Department of Chemistry, Northwestern University, Evanston, Illinois*

*Present address: *Department of Microbiology, School of Medicine, Western Reserve University, Cleveland, Ohio.*

Contents

The Nutrition of Invertebrate Animals

By George W. Kidder, *Biological Laboratory, Amherst College, Amherst, Massachusetts*

Energetics and Metabolic Function

By Eugene P. Kennedy, *Ben May Laboratory for Cancer Research and Department of Biochemistry, University of Chicago, Chicago, Illinois*

Hydrolytic and Phosphorolytic Enzymes

By Warren H. Yudkin, *Department of Chemistry, Northwestern University, Evanston, Illinois*

The Respiratory Enzymes

By B. L. Horecker and A. Kornberg, *National Institute for Arthritis and Metabolic Diseases, Bethesda, Maryland*

Coenzymes

By G. David Novelli and Morris Soodak, *Biochemical Research Laboratory and Robert W. Lovett Memorial Laboratory, Massachusetts General Hospital, Boston, Massachusetts*

Iron Metabolism

By H. F. W. Kirkpatrick, *Department of Clinical Investigation, The London Clinic, London, England*

Calcium and Phosphorus Metabolism

By H. F. W. Kirkpatrick and J. Douglas Robertson, *Department of Clinical Investigation, The London Clinic, London, England*

Trace Elements

By E. J. Underwood, *Institute of Agriculture, University of Western Australia, Crawley, Western Australia*

Application to Human Nutrition

By Grace A. Goldsmith, *Tulane University, School of Medicine, New Orelans, Louisiana*

Contents of Volume I

CHAPTER 12

The Intracellular Localization by Histochemical Methods of Enzymes and Vitamins

GEOFFREY H. BOURNE

Department of Histology, London Hospital Medical College, London, England

I. Introduction

Although much is known about the function of enzymes and vitamins in the animal body, little is known of their site of action in the cell. This is a subject which now is attracting the attention of many workers. Therefore a general account of what is already known on this subject would appear to be of value at the present time.

Many of the methods employed to demonstrate enzymes and vitamins in cells involve the fixation of the cell by various means and the use of chemical reactions at very micro levels. It is difficult to be sure, therefore, that the final result obtained represents the exact location of the enzyme or the vitamin in the living cell. Most workers on this subject are aware of these difficulties, however, and take steps to minimize them.

Although this lack of certainty does not mean that all the described localizations are incorrect, it does mean that for the time being claims for intracellular distribution of these substances should be treated with caution.

II. Enzymes

1. Oxidative and Respiratory Enzymes

a. Cytochrome Oxidase. As long ago as 1885 Ehrlich[1] showed that if he injected *p*-phenylenediamine and α-naphthol into animals he could observe a macroscopical bluing of certain organs, which he claimed was due to oxidative processes. He did not make microscopical examinations of his tissues. Winckler[2] was the first to do this in his studies on blood, and Schultze[3] was the first person to apply the technique to histological sections. The enzyme responsible for the development of the blue reaction from the *p*-phenylenediamine and α-naphthol (Nadi reagent) has been described as "indophenol oxidase" or "Nadi oxidase." The blue reaction so formed is due to the oxidative formation of an indophenol when aminodimethylanaline is used with the α-naphthol or to the formation of a diimine when phenylenediamine is used. It is probable that there is considerable diffusion of the colored substance in the cell during and after its formation, which makes the precise location of the enzyme responsible a matter of some doubt. It is now fairly certain that the enzyme responsible for this reaction is, in fact, "cytochrome oxidase."[4] It is probably also identical with Warburg's "respiratory ferment." Although much of this early work on cytochrome oxidase was carried out by means of the color reactions mentioned above, Keilin[5, 6, 7] had been studying the occurrence of cytochrome first in insect wing muscles and later in a variety of other cells by the process of histospectroscopy and had found that in the living cell it was being continually oxidized and reduced. His experiments then suggested that the

[1] P. Ehrlich, Das Sauerstoffbeedürfniss der Organismen, Berlin, 1885.
[2] W. Winckler, *Folia Haematol.* **3,** 323 (1907).
[3] J. Schultze, *Münch. med. Wochschr.* **56,** 167 (1909).
[4] D. Keilin and E. F. Hartree, *Proc. Roy. Soc.* (*London*) **B 125,** 171 (1938).
[5] D. Keilin, *Proc. Roy. Soc.* (*London*) **B 98,** 313 (1925).
[6] D. Keilin, *Proc. Roy. Soc.* (*London*) **B 110,** 129 (1926).
[7] D. Keilin, *Proc. Roy. Soc.* (*London*) **B 104,** 206 (1929).

enzyme responsible for the oxidation of reduced cytochrome might be "indophenol oxidase." As mentioned above, Keilin and Hartree[4] in 1938 confirmed that this enzyme was identical with cytochrome oxidase. Further observations on the intracellular localization of cytochrome oxidase and other respiratory enzymes will be given later in this chapter. Lison[8] has stated that the Nadi reaction is of interest in hematology because the progressive development of blood cells in the marrow runs parallel with the increase in number of indophenol blue granules. Although the development of a blue color has been regarded for some time as being due to "oxidase" activity, there is evidence that it may really be due to the presence of fatty acid peroxides in the granules giving the reaction.

Two other enzymes concerned directly with respiration in the cell have also been demonstrated histochemically. These are aldolase (Allen and Bourne[9]) and succinic dehydrogenase (Rutenberg, Gofstein, and Seligman[10]).

b. Aldolase. The Allen and Bourne technique depended upon the following reactions: (1) Aldolase catalyzed the formation of triose phosphate from hexose diphosphate. (2) The phosphate was liberated from the triose phosphate in the presence of hydroxyl ions. (3) The phosphate was precipitated as magnesium ammonium phosphate which was rendered visible with cobalt chloride and ammonium sulfide as in the Gomori phosphatase reaction. The rapid breakdown of triose phosphate as it was formed was prevented by the presence of iodoacetate, and possible interference by alkaline phosphatase in the reaction was prevented by the addition of fluoride. A considerable positive reaction was obtained with skeletal muscle, smooth muscle, and heart muscle and with liver. All positive cells showed the reaction to be diffused throughout the cytoplasm. These results may be of significance from the point of view of the mechanism of respiration of the cell and will be referred to again later.

c. Succinic Dehydrogenase. If fresh tissue is incubated in a suitably buffered succinate substrate in the presence of ditetrazolium chloride or other tetrazolium salt, the latter accepts the hydrogen split from the succinate by the enzyme. This changes it from a faintly yellow-colored water-soluble substance into a water-insoluble blue diformazan pigment which is precipitated at the site of action of the enzyme (see Rutenberg *et al.*[10]). The reaction appears to be localized in the cytoplasm of a variety of cells (e.g., kidney, liver, muscle; see plate I) and is present mainly as granules in the cytoplasm. There does not appear to be a reaction in the nucleus. In fresh preparations of skeletal muscle the granular distribution of enzymes

[8] L. Lison, Histochemie Animale, Gauthier-Villars, Paris, (1936).

[9] R. L. Allen and G. H. Bourne, *J. Exptl. Biol.* **20**, 62 (1943).

[10] A. M. Rutenberg, R. Gofstein, and A. M. Seligman, *Cancer Research* **10**, 113 (1951).

is very similar to that of sarcosomes, which are modified mitochondria. The reaction in the published photographs of Seligman and Rutenberg's preparations also resembles mitochondria. This may, in fact, be the case, since extracted mitochondria have been shown to contain most of the succinic dehydrogenase activity of the cell. Malaty and Bourne[11] have also found that the sarcosomes of skeletal and heart muscle give a strong positive reaction for succinic dehydrogenase. This is in keeping with the finding of similar activity in sarcosomes isolated by centrifugation from muscle homogenates. Malaty and Bourne also found it difficult to get any results with this method in animals which had been anesthetized with chloroform, ether, or phenobarbitones. The best result was obtained with the tissues of animals which had been killed by a blow on the head.

d. Peroxidases. Peroxidases were first demonstrated by Adler and Adler[12] using benzidine, and Kreibich[13] was the first to apply the reaction histologically. It has since had a wide application in hematological studies because the granules of the polymorph leucocytes give a positive reaction. This enzyme is also present in the mammary gland. The distribution of peroxidase in the eggs in Ascidians and in certain invertebrate eggs (Ries[14, 15] and Reverberi and Pitotti[16]) is mainly in the cytoplasm.

e. Dopa Oxidase. Another oxidative enzyme which has been demonstrated histochemically is dopa oxidase. This was first described by Bloch[17] in the skin of mammals. He found that, after a time, fixed frozen sections of skin treated with dopa (dioxyphenylalanine) showed a dark color in many of the basal cells. The cells which were particularly active in this respect are those which have been called dendritic cells. Bloch believed that only those cells which he called melanoblasts (that is, those capable of producing pigmentation), and not melanophores (cells that store pigment), contained the enzyme. Many workers now believe that the dendritic cells are the only cells in the skin which form pigment. Bloch and his co-workers believed that dopa, a precursor of melanin, normally circulated in the blood and that the melanoblasts, by virtue of the enzyme (dopa oxidase) they contained, were able to convert it into melanin. Dopa, however, is not the only substrate of dopa oxidase. It has now been shown that hydroxytyramine, 6-hydroxyphenylpyruvic acid, and 3,4-dihydroxyphenylpyruvic acid can also act as substrates. According to Zimmermann and

[11] A. H. Malaty and G. H. Bourne, *Nature*, **171**, 295 (1953).
[12] R. Adler and O. Adler, Hoppe Seyler's Zeit. **41**, 38 (1906).
[13] W. Kreibich, *Wien. klin. Wochschr.* **23**, 1443 (1910).
[14] H. Ries, *Z. Zellforsch. u. mikroskop. Anat.* **23**, 63 (1945).
[15] H. Ries, *Pubs. staz. zool. Napoli* **16**, 363 (1937).
[16] G. Reverberi and M. Pitotti, *Pubs. staz. zool. Napoli* **18**, 250 (1940).
[17] Bloch, B., *in* Jadassohn, Handbuch der Haut und Geschlechskrankheiten, Springer, Berlin, 1927, Vol. I, p. 434.

Cornbleet[18] melanin is produced by dendritic cells which are derived from the neural crest; in the negro skin, these cells give a positive dopa oxidase reaction from the third month of fetal life. Subsequently the melanin is transferred to the cells of the Malpighian layer of the dermis.

The dopa reaction has also been studied in the embryonic eye. Miescher,[19] for example, found that in the eye a positive dopa reaction began to appear in the pigment epithelium of the retina at the end of the third day. This is just the time pigment formation begins. After about the twentieth day, when the full load of pigment had been produced in the cells, the dopa reaction became negative. Herrmann and Boss[20] found an association of the pigment granules of the ciliary body not only with dopa oxidase but also with cytochrome oxidase and succinic dehydrogenase.

The papers refer mainly to the localization of the enzyme in various histological or embryological regions; only a few papers made any attempt to describe the intracellular localization of the enzyme. It appears to be localized mainly in the cytoplasm and to be of two types, granular and diffuse, usually found together. There is some doubt about the dopa oxidase reaction, since no attempt seems to have been made to ensure that the reaction observed is due entirely to enzyme activity.

2. Hydrolytic Enzymes

a. Phosphatases. Gomori[21] in Chicago and Takamatsu[22] in Japan simultaneously published identical methods for demonstrating alkaline phosphatase in tissues, and since that date an immense amount of work has been carried out on the histo- and cytochemistry of this enzyme. The method depends upon the fact that the enzyme will survive fixation of the tissue in alcohol or in acetone and that it will also survive embedding of the tissue in hot paraffin and its subsequent mounting on a slide and deparaffinization. Tissues treated in this way are then incubated in a mixture of sodium β-glycerophosphate and calcium chloride together with a suitable buffer. The free calcium ions of this solution trap *in situ* any phosphate split off the glycerophosphate by enzyme action, and an insoluble precipitate of calcium phosphate is deposited at the site of this activity. This is then rendered microscopically visible by treatment of the sections with a solution of soluble cobalt salt which transforms the calcium phosphate to cobalt phosphate. Further treatment with ammonium sulfide produces a black cobalt sulfide which is easily visible. Considerable attention has been paid

[18] A. A. Zimmermann and T. Cornbleet, *J. Investigative Dermatol.* **11,** 383 (1948).
[19] F. Miescher, *Arch. mikroskop. Anat. Entwicklungsmech.* **97,** 326 (1923).
[20] H. Herrmann and M. B. Boss, *J. Cellular Comp. Physiol.* **26,** 131 (1945).
[21] G. Gomori, *J. Cellular Comp. Physiol.* **17,** 71 (1939).
[22] H. Takamatsu, *Trans. Soc. Japan. Path.* **21,** 492 (1939).

to the precision of the localization of the enzyme, as there seems to be ample opportunity for both the enzyme and the reaction products to wander from their original sites during the various manipulations. There is certainly evidence that something diffuses (Martin and Jacoby[23]), and various suggestions have been put forward as to what this substance is. Danielli[24] has brought forward evidence that it is an activator of the enzyme which diffuses, and Leduc and Dempsey[25] that it is the calcium phosphate which does so. Moe[26] has shown that the distribution of the positive phosphatase reaction in cells depends a great deal on the time of incubation. With extended periods of incubation, parts of the cell (e.g., brush borders) which at an earlier stage were positive may become negative. Moe states that this makes it very difficult to decide just where the reaction is localized. Doyle[27] has discussed a variety of factors which influence the reliability of the reaction, and so has Cleland.[28]

There is no space for further discussion here, but this problem of diffusion applies to all histochemical procedures, whether location of enzymes or location of other substances. Nevertheless, with regard to alkaline phosphatase and probably many other enzymes, results, if they are interpreted at all, should be interpreted with an awareness that diffusion may and probably does occur. A fairly detailed account of the intracellular distribution of this enzyme was given by Bourne[29] in 1943, but in the light of more recent developments some of the results must be treated with reserve Nevertheless, they are summarized here as a matter of interest. As might be expected, cells of different organs show not only different amounts of phosphatase activity but also a different distribution. In the alimentary tract, for example, the epithelial cells of the stomach and the colon appeared to contain practically no enzyme; on the other hand, all the epithelial cells of the small intestine and the rectum contained appreciable amounts.

In the small intestine the brush borders of the cells were particularly rich in the enzyme, but it is of interest that it was the outer and inner edges of the border which were positive, and the center portion showed very little activity (see Moe's results).

What could be interpreted as amorphous granules of enzyme were also distributed through the cytoplasm of some cells, and in certain cases the enzyme appeared to be in the Golgi region. A similar association of alkaline phosphatase with the Golgi region has been found by Bourne (1943)[29] in

[23] B. G. Martin, *Biochim. et Biophys. Acta* **3,** 117 (1949).
[24] J. F. Danielli, *J. Exptl. Biol.* **22,** 110 (1946).
[25] E. H. Leduc and E. W. Dempsey, *J. Anat.* **85,** 305 (1951).
[26] H. Moe, *Anat. Record* **112,** 217 (1952).
[27] W. L. Doyle, *Intern. Rev. Cytol.* **2,** (in press) (1953).
[28] K. W. Cleland, *Proc. Linnaean Soc. N.S. Wales* **175,** 35 (1948).
[29] G. H. Bourne, *Quart. J. Exptl. Physiol.* **32,** 1 (1943).

the cells of the mantle (which secretes the shell) of certain mollusks, and in intestinal cells by other authors.

In the kidney it was again the brush border which reacted most strongly (particularly in the cells of the proximal convoluted tubules), and certain cells in the placenta also contained the enzyme. Bladder and urinary epithelium gave a diffuse reaction. Lymphoid tissue also showed a diffuse reaction, and over most of the body the capillary endothelium gave a uniformly positive reaction for the enzyme.

The nuclei of many tissues were strongly positive. In the pituitary gland, for example, two types of nuclear distribution were seen, i.e., one type in which the nucleus showed uniformly dark diffuse reaction, and a second in which the nucleus appeared to contain numerous dark-staining granules. Although it is possible that this difference could be correlated with specific types of cell, no certain evidence of this has been obtained.

A positive nuclear reaction which occurs close to a region of high phosphatase activity is suspect because of the possibility that diffusion phenomena are giving a spurious reaction. But it can safely be said that some nuclei probably do contain alkaline glycerophosphatase.

In their recent investigations on the enzymes of isolated cell nuclei, Allfrey and his colleagues[30] have rightly pointed out that homogenates of tissues prepared in the usual way with aqueous media (e.g., citrate or sucrose) are likely to result either in extraction of enzymes or other components from the nuclei, or in diffusion of substances into the nuclei from the homogenate. To overcome this difficulty, they froze, then lyophilized, tissues. They ground the tissue into a dry powder in a ball mill with petroleum ether and then carried out a series of sedimentations and flotations in mixtures of cyclohexane and carbon tetrachloride of different specific gravities. Nuclei obtained in this way appeared to be free of cytoplasmic contaminants and diffusion artifacts, and their enzymic and general composition could be estimated directly. The results obtained were as follows: crystalline hemoglobin was obtained from the nuclei of fowl and goose erythrocytes, but myoglobin was absent from heart muscle nuclei. Eighty per cent of the arginase activity and 30% of the catalase activity of liver cells was found to be in the nuclei. There appeared to be no significant quantities of amylase, lipase, uricase, or alkaline glycerophosphatase, but it is still possible that they may occur in the nuclei of other tissues. Arginase activity, high in mammalian liver and fowl kidney, was found by the authors to be present in liver nuclei but not in the nuclei of the kidney cells. Similar results were found for catalase. Alkaline phosphatase, adenosinetriphosphatase, the nucleotide-specific phosphatases, and β-glucuronidase appeared to occur only in low concentrations in nuclei. Esterase, on the

[30] V. Allfrey, H. Stern, and A. E. Mirsky, *Nature* **169**, 128 (1952).

other hand, appeared to be present in all the nuclei examined. Enzymes such as adenosine deaminase, nucleoside phosphorylase, and guanase were present in high proportion in many nuclei.

Nuclear composition appeared to vary with the age and the physiological state of the cell. In particular, there was a difference between fetal and adult tissues.

The fact that nuclei do contain enzymes likely to be involved in nucleoside metabolism (nucleoside phosphorylase, adenosine deaminase, and guanase), together with the fact that nucleoside phosphorylase catalyzes the formation of nicotinamide ribosides, leads the authors to point out the possible role of the nucleus in controlling the formation of di- and triphosphopyridine nucleotide coenzymes.

Vincent[31] has also shown that isolated nucleoli show only acid phosphatase activity despite the fact that they often give a strong positive reaction with the alkaline phosphatase technique.

Phosphatase is known always to be present when bone is being formed. It is not surprising, therefore, that osteoblasts possess high phosphatase activity. It appears to be present in both nucleus and cytoplasm and to be largely diffused. Phosphatase also appears at sites of absorption (e.g., brush borders of kidney tubule cells and alimentary tract cells), and some authors have related its presence there to glucose absorption. This may or may not be so, but it is of interest to note that wherever transfer of solutes occurs, in invertebrates as well as in vertebrates, there is a concentration of phosphatase at the transfer surface. This is found also in the placenta, the cells of the outer border of the syncytial trophoblast containing the enzyme in the human, the guinea pig, the rat, and the cat placentae. Phosphatase is also located in sites which suggest that it plays some part in protein and nucleic acid metabolism (see review by Bradfield[32]).

Dempsey and Deane[33] investigated the distribution of enzymes capable of splitting ribosenucleic acid, adenylic acid, glucose-1-phosphate, fructose diphosphate, glycerophosphate, and lecithin in the duodenum. They found that maximal activity of these enzymes occurred at different pH's and suggested that this was evidence that they were actually separate enzymes. Histochemically all these enzymes, with the exception of lecithin phosphatase, had an identical distribution in the brush border and in the Golgi region. They pointed out that the localization of these enzymes near the absorbing borders of the duodenal cells might be concerned with absorption of nutrients. For example, glycerophosphate is concerned in both breakdown and resynthesis of neutral fat during absorption, and glucose-1-phosphatase is probably concerned in carbohydrate absorption. The known presence of

[31] W. S. Vincent, *Proc. Natl. Acad. Sci. U.S.* **38,** 139 (1952).
[32] J. R. G. Bradfield, *Biol. Revs. Cambridge Phil. Soc.* **25,** 113 (1950).
[33] E. W. Dempsey and H. W. Deane, *J. Cellular Comp. Physiol.* **27,** 159 (1946).

phospholipins in the cell membrane, the Golgi zone, and the nuclear membrane probably explains the presence in these sites of lecithin phosphatase which may be concerned with the permeability of the cell membrane Reis,[34, 35, 36] in a series of papers, established the existence of a separate enzyme that splits muscle adenylic acid (5-nucleotidase), and Pearse and Reis[37] have described its histochemical distribution. At a pH of 7.5 they obtained specific activity by this enzyme and found histochemically that it was particularly well localized in the media of arteries, including the aorta, and also in the posterior pituitary, thyroid, and nerve. But, as they point out, diffusion of the reaction product is greater at pH 7.5 than at pH 9, and their method cannot therefore be used for intracellular localization of the enzyme.

Baradi and Bourne[38] in their work on the localization of the enzymes in the gustatory regions of rabbits have obtained evidence for the existence of separate enzymes which split (at a pH of about 9.0) 5-nucleotide, 3-nucleotide, ribonucleic acid, hexose diphosphate, adenosinetriphosphate, and glycerophosphate. The distribution of these enzymes in the cells in the gustatory regions varies considerably, and they are inhibited by different substances. They are mostly in the form of cytoplasmic granules of various sizes whch could not be directly identified with any cell component. Newman *et al.*[39] also brought forward evidence for the separate identity of a number of phosphate-splitting enzymes.

Naidoo and Pratt have demonstrated the existence in brain cells of an enzyme which splits vitamin B_1 pyrophosphate (see plate II, *A* and *B*).

Gomori[40] demonstrated an enzyme present in granular form in the cytoplasm of cells of liver, adrenal cortex, small intestine, bronchi, pancreatic islets (β cells), and lachrymal gland, which splits *p*-chloranilidophosphoric acid. He called this enzyme phosphamidase. It is said to split a N=P bond. He found it to be present at a much higher concentration in the grey matter of the central nervous sytem and in malignant tumors.

An enzyme which splits α-glycerophosphate at a pH of about 5.0 has also been studied, although not as extensively as alkaline phosphatase. It has been found that this enzyme is present in cytoplasmic granules in fibroblasts of a variety of mammals. The enzyme is widely distributed in the nervous system. Shimizu[41] found the reaction most intense in the axone

[34] J. L. Reis, *Enzymologia* **2,** 183 (1937).
[35] J. L. Reis, *Bull. soc. chim. biol.* **22,** 36 (1940).
[36] J. L. Reis, *Biochem. J.* **48,** 548 (1951).
[37] A. G. E. Pearse and J. L. Reis, *Biochem. J.* **50,** 534 (1952).
[38] A. F. Baradi and G. H. Bourne, *Nature* **168,** 977 (1951).
[39] W. Newman, I. Feigin, A. Wolf, and E. A. Kabat, *Am. J. Path.* **26,** 257 (1950).
[40] G. Gomori, *Proc. Soc. Exptl. Biol. Med.* **69,** 407 (1949).
[41] N. Shimizu, Y. Handa, H. Jiro, and T. Kumamato, *Proc. Soc. Exptl. Biol. Med.* **75,** 696 (1950).

and in the Golgi region. Smith[42] found that the activity near the nucleus was primarily perinuclear and that it was also great in the axon hillock and in the axon. If the axon of a cell was cut, the activity increased and spread over the whole cell. Shimizu, on the other hand, found a decrease of acid phosphatase activity in nerve cells in experimental beriberi. In many preparations the acid phosphatase appears diffused in the cytoplasm. Moog and Steinbach,[43] however, have produced evidence that in many cases acid phosphatase and alkaline phosphatase activity of cells is associated with the large granule (mitochondral) fraction. Gomori[44] has also indicated that acid phosphatase activity may occur in nuclei. Naidoo and Pratt[45] showed that acid glycerophosphatase was present in the nuclei of Purkinje cells of the cerebellum of the white rat and that the nucleoli stained more intensely than any other (see plate II, *E* and *F*). The fibers of the basket cells were also deeply stained. Similarly, they found that the nuclei of the ependymal cells lining the ventricles were strongly positive. With adenosinetriphosphate as a substrate at a pH of 6.5, these authors again found strong nuclear staining in muscle, blood vessels, ependymal walls, and astrocytes, and in the nucleoli of neurones (see plate II, *D* and *I*). The nuclei and nucleoli of the Purkinje cells also gave a very strong positive reaction. With adenosinemonophosphate as a substrate at the same pH, the only cellular constituents to stain were the nucleoli of the Purkinje cells. The nerve fibers of most parts of the brain, however, gave a very strong reaction (see plate II, *C* and *G*). Many other valuable papers on acid phosphatase and on the alkaline phosphatases have also been published, but it is obviously impossible in the space available to do more than mention a few representatives of this enormous literature.

b. Lipase. A method for demonstrating lipase histochemically has been described by Gomori.[46, 47] He used palmitic and stearic esters of hexitans which are known commercially as "Tweens." The lipase splits off the palmitate or stearate, which is trapped by Ca ions and precipitated as insoluble calcium palmitate or stearate. These are converted into lead palmitate or stearate by immersion in lead nitrate, and the lead is rendered microscopically visible by changing it into lead sulfide with the aid of ammonium sulfide. The enzyme was found to be widely distributed in various organs including liver, pancreas, lung, kidney, testis, adrenal, stomach, and small intestine (see plate III, *C*). In the liver it is found mostly in the central part of the lobules. Gomori has pointed out that

[42] W. K. Smith, *Anat. Record* **102,** 523 (1948).
[43] F. Moog and H. B. Steinbach, *J. Cellular Comp. Physiol.* **28,** 209 (1946).
[44] G. Gomori, *Arch. Path.* **32,** 189 (1941).
[45] D. Naidoo and O. E. Pratt, *J. Neurol. Neurosurg. Psychiat.* **14,** 287 (1951).
[46] G. Gomori, *Proc. Soc. Exptl. Biol. Med.* **58,** 362 (1945).
[47] G. Gomori, *Arch. Path.* **41,** 121 (1946).

lipase is nearly always found in the cell in the form of granules, sometimes as coarse granules. Baradi and Bourne have found (unpublished work) that in many cases in the gustatory regions of the tongue these granules correspond with the mitochondria. Gomori has shown that in the pancreas the zymogen granules, and in the epididymus the brush border of the cells, frequently gave a very strong reaction. Liver lipase was usually in the form of coarse granules. It is of interest that Gomori has pointed out that the female genital tract has not been found to contain lipase in any of the species of animals examined.

c. Simple esterase. Nachlas and Seligman[48] published a technique for the demonstration of simple esterase. This is an enzyme which splits compounds such as naphthyl acetate and other fatty acid esters of naphthyl (see plate II, *H*, and plate III, *A*, *B*, *D*, and *E*). The technique involves the hydrolysis of naphthyl acetate. The β-naphthol produced by the hydrolytic effect of the esterase becomes coupled in the substrate mixture with a diazonium compound to form an insoluble purplish-red pigment (an azo dye) which is precipitated at the site of enzyme activity. Nachlas and Seligman found that this enzyme was widely distributed in the body and that in rat and dog it was greatest in amount in lung, liver, pancreas, and kidney. Baradi and Bourne[38] found that a strong reaction for this enzyme was given by olfactory mucosa cells and by taste bud cells. In these cells the enzyme was mainly in the form of granules of various size which by their intense reaction indicated an appreciable amount of the enzyme. Each of these granules was surrounded by a diffuse halo, and it seems that either the enzyme or the reaction product can diffuse slightly during the process of demonstration. Nachlas and Seligman picture esterase in the form of granules in both epithelial and parietal cells of rabbit stomach. In their paper Nachlas and Seligman question whether the "lipase" demonstrated by Gomori with his Tween technique is true lipase and suggest that the "esterase" which they have demonstrated appears to occupy similar sites to those which Gomori's lipase occupies. Baradi and Bourne[38] have found that the distribution of these two enzymes in gustatory and olfactory regions of the rabbit is completely different.

d. Cholinesterase. Another "esterase" which has been demonstrated histochemically is cholinesterase (see plate III, *I*). Koelle and Friedenwald[49] and Gomori[50] have published techniques for this enzyme, and it has been shown that both non-specific and specific cholinesterases can be demonstrated in tissues. In Gomori's method the substrate is lauroyl, myristoyl, or related cholines; in Koelle's method it is acetyl thiocholine iodide. The relationship between the enzymes as demonstrated by the two methods

[48] M. M. Nachlas and A. M. Seligman, *J. Natl. Cancer Inst.* **9,** 415 (1949).
[49] G. B. Koelle and J. S. Friedenwald, *Proc. Soc. Exptl. Biol. Med.* **70,** 617 (1949).
[50] G. Gomori, *Proc. Soc. Exptl. Biol. Med.* **68,** 354 (1948).

and the problem of the demonstration of esterases as a whole has been discussed by Gomori.[51] In some nerve cells cholinesterase has been found to be present in the nuclei. In Koelle's method the enzyme often appears in granular form, but its distribution has not been identified with any formed cytoplasmic components. There appears to be some variation in intensity of the histochemical reaction of this enzyme in different species. One of the anomalous facts about the technique is that the enzyme from the electric organ of the electric eel does not split the substrate (e.g., myristoyl choline) used in Gomori's method.

Barnett and Seligman[52] and Holt[53] have described a method of demonstrating esterases in tissues by the production of indigo. In this technique indoxyl acetate is used as a substrate. At the pH optimum for esterase activity the indoxyl is rapidly oxidized to indigo blue and precipitated from solution at the sites of enzyme activity. Barnett and Seligman said that the reaction was granular and cytoplasmic. The granules were fairly small in liver and kidney and large in nervous tissue. In anterior horn cells they were present in the cytoplasm, axon and dendrites. Granules were also found in polymorph leucocytes and striated muscle. The substrate was found to be hydrolyzable by lipase, non-specific esterase and choline esterase.

f. Glucuronidase. Friedenwald and Becker[54] have demonstrated glucuronidases in fresh frozen sections. The substrate used was 8-hydroxyquinoline glucuronic acid. When this compound was split and the 8-hydroxyquinoline combined with iron, it was precipitated at the site of enzyme activity and converted into Prussian blue or combined with a diazo substance to form an insoluble azo dye. The enzyme is said by the authors to be cytoplasmic in distribution.

3. Phosphorylase

Phosphorylase activity has been demonstrated by Cobb[55] by the production and subseauent demonstration of glycogen from glucose-1-phosphate. This formed the basis of a thesis in the University of Illinois which the author has not seen therefore intra-cellular detail cannot be described.

4. Differential Centrifugation of Cell Homogenates for Determining Intracellular Distribution of Enzymes

We have dealt with the direct demonstration of enzymes in cells, but the development of the technique of differential centrifugation of cell homog-

[51] G. Gomori, *Intern. Rev. Cytol.* **1,** 323 (1952).
[52] R. J. Barnett and A. M. Seligman, *Science* **114,** 579, (1951).
[53] S. J. Holt, *Nature* **169,** 271 (1952).
[54] J. S. Friedenwald and B. Becker, *J. Cellular Comp. Physiol.* **31,** 303 (1948).
[55] J. D. Cobb, Thesis, University of Illinois, 1948.

enates has also provided much interesting information on the distribution of enzymes within the cell. However, some care must also be used in interpreting the results of these methods. One does not know whether the effects of homogenizing cells may result in the adsorption of enzymes onto the surfaces of nuclei, mitochondria, etc., or, conversely, in the leaching out of enzymes from the formed elements into the supernatent.

a. Nuclei. The isolation of cell components from cells was first carried out by Miescher in 1871.[56] Miescher first digested pus with pepsin and HCl and then centrifuged out the nuclei. He was also able to obtain the nuclei of fish spermatozoa. Miescher washed the spermatozoa with distilled water until the cytoplasmic portions were removed, but his preparations were imperfect because a quantity of cytoplasmic material continued to adhere to the nuclei. A more modern technique was used by Zittle and Zitin[57] to obtain sperm nuclei. They subjected spermatozoa to the effect of sonic vibrations and literally shook the nuclei free of their cytoplasmic encumbrances. Nuclei have also been obtained from bird erythrocytes, from *Arbacia* eggs, and from a variety of tissue homogenates. In the latter the nuclei come down at a relatively low centrifugal speed and can easily be separated from the supernatant. The isolation of nuclei in this way enabled relatively large amounts of nuclear substance to be obtained for chemical analysis (see Dounce[58, 59]). About 20% of nuclear material isolated in this way was desoxyribosenucleic acid (see plate III, *G* and *H*). There was no glycogen, and up to 10% of lipid. Dounce lists the enzymes which have been found in nuclei. There was no succinic dehydrogenase, but half of the cytochrome oxidase activity was said to be present in them. Since other work shows that the major portion of the cytochrome oxidase activity is found in mitochondria, it seems that we have here an example of the possible adsorption of the enzyme from the homogenates into one or other or both of these structures. All of the amine oxidase and arginase and uricase activities of the cell appear to be in the nuclear fraction, and so does up to 80% of the catalase fraction. The nuclei also contain half the enolase and esterase and nearly half the aldolase activity of the cell. Allen and Bourne[9] found aldolase activity evenly distributed through both nuclei and cytoplasm. These results apply to rat liver cell nuclei. Rat kidney cell nuclei, however, appear to contain no arginase. It is of interest that chicken erythrocyte nuclei contain all the acid phosphatase activity of the cell and beef pancreas nuclei contain half the esterase and half the lipase of pancreatic homogenate. Dounce has discussed the results for cytochrome

[56] K. Miescher, *Hoppe-Seyler's Med. Chem. Untersuch.* **4,** 441 (1871).
[57] C. A. Zittle and B. J. Zitin, *J. Biol. Chem.* **144,** 99 (1942).
[58] A. L. Dounce, *J. Biol. Chem.* **147,** 685 (1943).
[59] A. L. Dounce *in* Sumner and Myrbäck, The Enzymes, Academic Press. New York, 1950, Vol. I,Part 1, p. 215.

oxidase and succinic dehydrogenase. He points out that the acidulated homogenate which he used for his work makes it difficult for him to compare the activity of the nuclei with that of a fresh homogenate. If an attempt is made to allow for this difference, the cytochrome oxidase activity in the nuclei works out at about 10 to 15 % of that of fresh liver. Dounce, however, contents himself with claiming that his work shows that the nucleus does contain some cytochrome oxidase activity. The possibility still exists, and Dounce acknowledges it, that whole mitochondria might stick onto the nuclear membrane from the homogenate and affect the result. Some nuclei, however, appear to contain lactic dehydrogenase activity.

b. Mitochondria. In 1934 Bensley and Hoerr[60] isolated mitochondria from homogenated tissue for the first time. This technique was subsequently developed by Claude and his co-workers.[61] Bensley and Hoerr originally used saline homogenates, but mitochondria were greatly distorted by the saline and were often difficult to identify. Later workers used sucrose solutions and obtained what was called a "large granule fraction" which in fact was practically pure mitochondria.

The main steps involved in the process of differential centrifugation may prove of interest. The homogenate is first centrifuged at $1000 \times g$ (i.e., about 700 to 800 r.p.m.). If this is done about three times, spinning for 10 minutes each time, all the nuclei, red blood cells, and cell fragments are centrifuged out of the homogenate. The fractions from the last two runs consist almost entirely of nuclei. The large granule fraction is now obtained by centrifuging for about 30 minutes at a force of about $18{,}000 \times g$, i.e., somewhere around 15,000 r.p.m. Some small granules or microsomes also come down, but, if the centrifugate is resuspended in sucrose and recentrifuged, pure mitochondria may be obtained. If the supernatant from this is now centrifuged at $40{,}000 \times g$, particles of 100 mμ in size will come down; these are the microsomes. There are, of course, even smaller particles.

The isolation of mitochondria has given a great impetus to the study of the distribution of enzymes within the cell. Hogeboom and his colleagues,[62-66] Kennedy and Lehninger,[67] Palade and Claude (see Claude[61]) and others have made important contributions to this aspect of histochemistry.

In a later part of this chapter reference will be made to the question of

60 R. R. Bensley and H. H. Hoerr, *Anat. Record* **60,** 251 (1934).
61 A. Claude, *Cold Spring Harbor Symposia Quant. Biol.* **9,** 263 (1941).
62 G. H. Hogeboom, W. C. Schneider, and G. E. Palade, *J. Biol. Chem.* **172,** 223 (1948).
63 G. H. Hogeboom, *J. Biol. Chem.* **177,** 847 (1949).
64 W. C. Schneider, *J. Biol. Chem.* **165,** 585 (1946).
65 W. C. Schneider, A. Claude, and G. H. Hogeboom, *J. Biol. Chem.* **172,** 451 (1948).
66 G. H. Hogeboom, *Federation Proc.* **10,** 640 (1951).
67 E. P. Kennedy and A. L. Lehninger, *J. Biol. Chem.* **179,** 957 (1949).

intracellular localization of vitamins. At this point, however, those vitamins which are concerned with enzymic activity will be mentioned. These constitute in particular certain members of the vitamin B complex—aneurin (thiamine), riboflavin, nicotinamide, pantothenic acid, and pyridoxin. Four of these are particularly relevant at the moment; thiamine is part of the enzyme cocarboxylase; riboflavin is part of cytochrome reductase; nicotinamide is an important constituent of coenzyme I; and pantothenic acid, of coenzyme A. The last two enzymes are all-important links between the Krebs tricarboxylic acid cycle and the cytochrome system. Pyridoxin is a constituent of transaminase, the enzyme which plays an important part in the protein metabolism of the cell.

These observations are significant in connection with the work which has been carried out in the last few years (Hogeboom,[63] Schneider,[64] Schneider *et al.*[65]) which suggests that localized in the mitochondria are a number of enzymes which are specifically concerned with aerobic metabolism in the cell. According to Hogeboom,[66] practically all the cytochrome oxidase and the succinic dehydrogenase activity of the cell are found in the mitochondria, and the residual activity found in other fractions is probably due to contamination with mitochondria. Hogeboom lists the other enzymes which have been found to be concentrated in these organelles as follows: oxaloacetic oxidase, octanoic acid oxidase, glutamic dehydrogenase, diphosphopyridine nucleotide cytochrome reductase, triphosphopyridine nucleotide cytochrome reductase, adenosinetriphosphatase, catalase, and uricase. He also lists cytochrome c, riboflavin, and pyridoxin. Most of the activity of the system catalyzing the synthesis of *p*-aminobenzoic acid and glycine and also myokinase were also found in mitochondria. Kielley and Kielley (according to Hogeboom) have also found that mitochondria in the presence of an oxidizable substrate such as α-ketoglutarate are able to synthesize adenosinetriphosphate from adenosine 5-phosphate and inorganic phosphate. In addition, Mauer and Greco[68] have found that mitochondria contain benzoylarginine amidase. Higgins *et al.*[69] have shown a high concentration of coenzyme A and, at acid pH's, leucine amidase activity in these organelles. Potter *et al.*[70] have pointed out that the use of a Waring blendor for the production of tissue homogenates is undesirable because it may break up some nuclei and mitochondria to a size that results in their contamination of centrifugal fractions in which they would not otherwise be found. It is of interest that freshly isolated mitochondria show very little adenosinetriphosphatase activity, but if they are kept in an icebox

[68] M. E. Mauer and A. E. Greco, *J. Natl. Cancer Insl.* **12,** 37 (1951).
[69] H. Higgins, J. A. Miller, J. M. Price, and F. M. Strong, *Proc. Soc. Exptl. Biol. Med.* **75,** 462 (1950).
[70] V. R. Potter, R. O. Recknagel, and R. B. Hulbert, *Federation Proc.* **10,** 646 (1951).

for a time considerable activity is demonstrated, or if they are broken up in a Waring blendor the relatively low malic dehydrogenase activity of the intact mitochondria is greatly enhanced. Presumably we are seeing in these two experiments the effect of selective permeability of a mitochondrial membrane. It is of interest that Zetterstum[71] has shown that anaerobic oxidation by kidney mitochondria is activated by phosphorylated vitamin D_2, a fact which suggests that phosphatase activated by this vitamin may be present in mitochondria.

It is also interesting that the mitochondria have not been found to have any glycolytic activity, a property which is probably diffused throughout the cytoplasm. It was mentioned earlier that Allen and Bourne[9] had shown that aldolase (zymohexase), which is an enzyme responsible for one of the steps in the breakdown of glucose to pyruvic acid, is diffused through the cell and is not localized in any particular cell organelle.

The third particulate fraction in differential centrifugation is composed of microsomes which contain a high concentration of pentosenucleic acid (six times that of mitochondria), and lipid (40 %). In addition they contain the DPN and TPN cytochrome-reductase enzymes and most of the esterase (lipase?) activity of whole liver. According to Hogeboom,[66] the microsomes also play a role in other enzyme reactions, e.g., anaerobic glycolysis, oxidation of oxalacetate, and reductive cleavage of *p*-dimethylaminobenzene. In the final fraction, the supernatant, are found the enzymes involved in anaerobic glycolysis together with some isocitric dehydrogenase, acid and alkaline phosphatase, catalase, and cytochrome c. There are also present, of course, a number of soluble proteins.

All these results are of great interest, but can we use them to explain any aspects of cell physiology? Some tentative suggestions as to the relations between mitochondria and cytoplasm are given below.

Glycolysis proceeds to the pyruvic acid level in the cytoplasm, and then the pyruvic acid or the 2-carbon substances produced from it must come into contact with, or enter into, the mitochondria for the next stage of respiration to take place. The 2-carbon substances are drawn into the Krebs tricarboxylic acid cycle within or on the surface of the mitochondria, become condensed with oxaloacetic acid to form citric acid, and then pass round the cycle giving off (for each molecule of pyruvic acid entering the system) three molecules of carbon dioxide and five pairs of hydrogen atoms. These pairs of hydrogen atoms are swept away from the Krebs cycle and, through the aid of coenzyme I, coenzyme II, and cytochrome c form water and simultaneously reduce the cytochrome which is reoxidized by cytochrome oxidase. It should be stressed that the whole of this process, from pyruvic acid to the production of CO_2 and water, can be carried out by mitochondria, which appear to have sufficient of all the enzymes necessary for

[71] R. Zetterstum, *Acta Chem. Scand.* **5**, 343 (1951).

this process. It is probable that 80 to 90% of the aerobic metabolism of the cell, from pyruvic acid onwards, takes place in the mitochondria, assuming that the claimed partition of enzymes between them and cytoplasm is correct.

We know that mitochondria break up after injury to the cell or after treatment with respiratory depressants (e.g., chloroform), and in this process we have a possible method of respiratory control of the cell. Since respiration can progress only if the enzyme systems in the mitochondria are fed with pyruvic acid or 2-carbon substances, it is obvious that the speed of respiration will be governed by the area of mitochondrial surface presented to the cytoplasm. It is known that mitochondrial filaments can break up into granules and that granules and batonettes can link up to form filaments, and thus we have a method whereby a cell can control its rate of respiration. The breaking up of the mitochondria in a cell caused by respiratory depressants might be thought of as an "attempt" by the cell to counteract the depressing effect by trying to speed up the rate of respiration. Some mitochondria are also known to be capable of independent movement in the cell, and increase of this movement would provide another method of increasing respiration, since, once the pyruvic acid in the cytoplasm adjacent to the mitochondria is used up, the rate of respiration would depend upon the rate at which pyruvic acid could diffuse from a distance. However, this delay is overcome in some cells (notably in tissue culture cells) by a constant writhing and twisting of the mitochondria. In some living cells, where the mitochondria are not very active, streaming movements of the protoplasm help to bring fresh pyruvic acid into contact with the mitochondria. It is of interest that cyanide inhibits cellular respiration and that it also stops the movement of mitochondria in the living cell. The fact that mitochondria have a fairly constant diameter may be significant. They tend not to increase in girth but usually in length. This may be of significance since, if the mitochondria were too thick, only superficial portions would get an opportunity to react with the pyruvic acid and the more centrally placed enzymes would be wasted, metabolically speaking. In normally fed animals there seems to be a constant number of mitochondria in the cells of particular organs (Thurlow[72]), and their fundamental place in the cell metabolism is illustrated by the fact that they occur even in such a humble member of the animal kingdom as the amoeba.

The orderly functioning of the enzyme systems in mitochondria is dependent upon the structural organization of the mitochondria, for de Robertis *et al.*[73] find that any treatment which causes dissolution of the mitochondria (e.g., freezing) greatly reduces the succinic dehydrogenase activity and oxygen uptake of cells.

[72] W. Thurlow, *Carnegie Inst. Wash. Pub.* **6**, 116, 35 (1917).

[73] E. de Robertis, W. W. Nowinski, and F. A. Saez, General Cytology, W. B. Saunders, Philadelphia, 1948.

In certain insects the wings beat at an extraordinary speed, which suggests that there must be very efficient metabolic organization in the wing muscles to permit this output of energy. Histological examination of these wing muscles shows that lying between the fibrils in the individual fibers are rows of bodies known as sarcosomes, which are mitochondrial in nature. They have been found to contain the same respiratory enzymes as mitochondria, so we can conceive of the amazing efficiency of such an arrangement in which each contractile element is surrounded by bodies crammed with respiratory enzymes. Malaty and Bourne[74] have also shown by histochemical methods that the sarcosomes of mammalian skeletal and heart muscle contain succinic dehydrogenase.

So far we have referred only to the respiratory activity of mitochondria, but the existence in the organelles of the enzymes concerned with the Krebs tricarboxylic acid cycle confers upon them other potentialities, for the Krebs cycle is spoken of as the meeting place of carbohydrate, protein, and fat metabolism, and the enzymes present in mitochondria are those which can catalyze reactions which are able to supply energy for synthesis. So we may conceive of the mitochondria as being able to function as centers in the cell capable of synthesizing protein, fats, and lipids. Bensley[75] has also brought forth evidence that mitochondria are concerned with fat metabolism, and all the fatty acid oxidase of the liver cell has been found in them (Kennedy and Lehninger[67]). There is other evidence of the synthetic ability of mitochondria; Kennedy and Lehninger also found that isolated mitochondria were able to incorporate inorganic phosphate (containing radiophosphorus, P^{32}) into pentosenucleic acid, phospholipid, and "an unidentified acid-insoluble 'phosphoprotein' fraction coupled to the oxidation of substrates of the Krebs tricarboxylic acid cycle."

There is sufficient cytological evidence that the mitochondria are concerned with breakdown or synthesis of protein, fat, and carbohydrate. Horning's[76] work on the amoeba showed an association between the mitochondria and the digestion of food in the creature, and Horning and Petrie[77] showed that in germinating seeds mitochondria are associated with the breakdown of starch. The authors suggested that the mitochondria might liberate diastase, which may indeed be the case, but the former may be aggregated round the starch grains because the breakdown of the grains results in the production of glycogen and/or pyruvic acid or 2-carbon substances and they are able to metabolize it further. Horning also believes that the zymogen granules of the pancreas form in association with the

[74] A. H. Malaty and G. H. Bourne, *Proc. Nutrition Soc. British J. Nutrition* **6,** ix (1952).

[75] R. R. Bensley, *Anat. Record* **98,** 609 (1947).

[76] E. S. Horning, *Australian J. Exptl. Biol. Med. Sci.* **3,** 149 (1926).

[77] E. S. Horning and F. Petrie, *Proc. Roy. Soc.* (*London*) **B 102,** 188 (1927).

mitochondria. Horning and Richardson[78] showed that in *Opalina* mitochondria are associated with the formation of what appear to be protein bodies. A number of authors have presented cytological descriptions of small granules clustered round mitochondria, and these are claimed to be the products of synthesis of the mitochondria. However, two types of microsome (small cytoplasmic elements, about 100 to 200 mμ in diameter) are known, one of which is composed mainly of glycogen or glycogen associated with a protein, and it may be that the particles around the mitochondria are actually collections of this type of microsome undergoing glycolysis and providing pyruvic acid for the mitochondria to metabolize. Nevertheless, it is possible that these small particles represent the products of synthesis by the mitochondria, in which case they would be removed from the mitochondrial surface as they are formed and segregated in the Golgi material. Otherwise the accumulation of the synthetic products at the mitochondrial surface would block further synthesis.

Hirsch[79] has actually seen granules on the mitochondria detach themselves after having been seen at this site for 10 to 70 minutes. Duthie has seen evidence of a similar activity in fixed preparations of salivary glands and in goblet cells of the intestine. There are numerous references in the literature (see Hirsch[79]) to the production of granules by mitochondria and their absorption and possible subsequent elaboration in the Golgi region.

Practically all the evidence available supports the conception that the mitochondria may be regarded as the main synthetic centers of the cell as well as the main respiratory centers.

Claude[80] has calculated that a typical liver mitochondrion could contain about one million protein molecules and that in each mitochondrion there may be twenty-five different enzymatic systems. That means that each system could be duplicated 2000 times in each mitochondrion, which certainly lends support to Claude's description of the mitochondria as the "power plants" of the cell.

The conception of the mitochondria as synthetic centers of the cell is not new; R. J. Ludford[81] wrote more than twenty years ago: "At the mitochondrial-cytoplasmic surface, syntheses by enzymes occur. The resulting products continually diffuse into the cytoplasm, preventing an accumulation at the surface of the mitochondria. At the Golgi apparatus the elaborated products are concentrated into droplets preliminary to their elimination."

[78] E. S. Horning and K. C. Richardson, *Arch. exptl. Zellforsch. Gewebezücht.* **10,** 488 (1929).

[79] G. C. Hirsch, Form und Stoffwechsel der Golgikorper, Protoplasma Monographs, Gebrüder Borntraeger, Berlin, 1939.

[80] A. Claude, *Advances in Protein Chemistry* **5,** 423 (1949).

[81] R. J. Ludford, *J. Roy. Microscop. Soc.* **103,** 288 (1926).

It is of interest that, in addition to their enzymic content, mitochondria possess fats and lipoids, and naturally not all their protein is enzymic in nature. In addition, mitochondria are not permanent; they can be substantially reduced in numbers by starvation. Therefore we might simply consider them, as Bensley does, as a reserve store of metabolic material on which the cell can draw in time of need. Such structures provide suitable surfaces for the adsorption and possible absorption of various enzymes. Once such enzymes are incorporated into the mitochondria the latter are then able to control the metabolic processes of the cell by presenting more or less surface to the cytoplasm.

c. Golgi Apparatus. The establishment of mitochondria as the dynamic centers of the cell appears to reduce the apparent importance of the Golgi apparatus, formerly regarded as being the dominant cytoplasmic organ in this respect, particularly from the point of view of cell secretion. The Golgi apparatus (discovered by Golgi in 1898) is an area in the cytoplasm situated near the nucleus which is rich in lipoidal materials. There has been, as mentioned previously, considerable dissention as to its exact form, but there seems to be little doubt that the first visible products of secretion in the cell always appear in association with the Golgi material. There is no evidence that this material has the enzymic equipment necessary for the wide variety of syntheses that cells must perform. It is true that some cells appear to contain the enzymes alkaline and acid glycerophosphatase in this region, but much more than this is required before an intrinsic synthetic ability can be accredited to the Golgi apparatus. However, it is possible and even probable that small particles of material synthesized by the mitochondria pass to the Golgi region and there become segregated into visible droplets. We must not forget Hirsch's own observation in which he saw in living pancreatic cells fine droplets passing from the mitochondria to the Golgi vacuoles. As will be shown later, there is evidence that vitamin C is often aggregated in the Golgi region of cells, and visible toxic substances, certain dyes, etc., are all segregated in the same region. It seems likely that the Golgi material is, in fact, a segregating region with a primary function of protecting the cell from the effect of diffusible toxic substance and a secondary function of aggregating into droplets substances synthesized by the mitochondria which, if left at the site of synthesis, would by the law of mass action bring the synthetic processes to a standstill.

Other attempts to localize enzymes in tissues have been made by Linderstrøm-Lang and his colleagues, but most of their results demonstrated the *amounts* of enzymes in different regions of a histological specimen and could not be applied either qualitatively or quantitatively to the demonstration of the distribution of enzymes within the mammalian cell. Nevertheless, they have obtained results of cytochemical interest with large cells such as eggs and with amoebae.

Holter and Doyle,[82] for example, centrifuged *Amoeba proteus* so that the mitochondria were stratified in broad zones in the middle region of the cell and the amount of amylase present in sections taken at different levels at right angles to the plane of stratification was estimated. The region where the mitochondria were aggregated was the region where most amylase was found, so it seemed reasonable to assume that the enzyme was contained by the mitochondria. This represents, then, the localization of yet another enzyme in this cell organelle. Further work by Holter and his colleagues[83] has shown that proteinase is distributed in the large amoeba, *Chaos chaos*, on components in the cytoplasm which are subject to centrifugal action and which appear to be mitochondria. They also obtained evidence that succinic dehydrogenase was localized mainly in the same structures. This finding confirms that described earlier in this chapter based on differential centrifugation of tissue homogenates. There is some evidence (Krahl[84]) that some enzymes, including cytochrome oxidase, are found mainly in the ground substance of the cytoplasm or in the microsomes in unfertilized or recently fertilized eggs but that as development continues these enzymes become progressively incorporated into mitochondria.

III. Vitamins

Intracellular localization of vitamins was attempted in a number of cases soon after a particular vitamin was isolated and its chemical identity established. Only a few vitamins have been localized in cells by histochemical procedures, but differential contrifugation of tissue homogenates has given information on the distribution of various vitamins in different cell constituents.

1. Vitamin A

The first intracellular localization of a vitamin was described by Querner in 1932.[85] He found in tissues that certain isotropic fat droplets gave a fluorescence, which quickly faded, with ultraviolet light. He believed that this fluorescence was due to certain fat-soluble pigments, e.g., carotenoids. Later Querner[86] found this fluorescence to be present in liver, retina, adrenal cortex, and pituitary gland of animals and said that it was due to vitamin A.

Hirt[87] claims to have described bright yellow fluorescent droplets in the

[82] H. Holter and W. L. Doyle, *Compt. rend. trav. lab. Carlsberg. Sér. chim.* **22,** 219 (1938).
[83] H. Holter and K. Linderstrøm-Lang, *Physiol. Revs.* **31,** 432 (1950).
[84] M. E. Krahl, *Biol. Bull.* **98,** 176 (1950).
[85] F. Querner, quoted from *Anat. Anz.* **78,** 289 (1932).
[86] F. Querner, *Klin. Wochschr.* **14,** 1213 (1935).
[87] A. Hirt. *Anat. Anz.* **87,** 97 (1940).

walls of the sinusoids of the liver in living animals and to have identified them with vitamin A in 1939.

Bourne[88] in 1935 and Joyet-Lavergne[89] in the same year demonstrated vitamin A in cells histochemically and found the vitamin present in a wide variety of cells in the mitochondria. This is of interest in view of the fact that Gatenby[90] as long ago as 1919 had observed yellow carotenoid pigment in the mitochondria of the cells of the snail, *Limnoea*. Both Bourne and Joyet-Lavergne modified the Carr-Price test for vitamin A. Bourne cut frozen sections of liver of unfixed tissues and mounted them on a slide in a drop of Carr-Price reagent (antimony trichloride in chloroform). By this method vitamin A gives a bright blue color. Carotene also gives a reaction with the Carr-Price reagent, but Anderson and Levine[91] have stated that at room temperature only vitamin A gives the reaction whereas carotene does not do so until the temperature is raised to 60°. Since all preparations were carried out at room temperature, it seems likely that in this instance only vitamin A was being demonstrated, although for various reasons, including the fact that extracted mitochondria have a yellow appearance, it seems likely that mitochondria contain some carotenoid pigment as well. It is of interest, too, that the mitochondria of germ cells also contain carotene (Popper[92]). Carotenoids, in fact, are widely distributed in the organs of reproduction of both animals and plants (Goodwin,[93]), and they appear to be mobilized at sexual maturity and during the spawning season. The author suggests that carotenoids may have some function other than that of their vitamin A activity.

The presence of vitamin A in mitochondria was confirmed by Goerner in 1938[94] and 1939.[95] After extracting mitochondria by Bensley and Hoerr's method, he showed that 27 to 30% of their weight was made up of lipoids which contained vitamin A. He found that 100 mg. of mitochondrial substance contained 249 to 910 U.S.P. units of the vitamin. He also investigated the effect of injection of dibenzanthracene on the vitamin A content of liver mitochondria and found that it was decreased. He believed that under the influence of the dibenzanthracene the liver appeared to lose its ability to split carotene into vitamin A. Goerner also found that the mitochondria of cells of liver tumors formed by injecting azotoluene dyes into animals contained no vitamin A whereas those of the surrounding healthy cells contained the vitamin. Furthermore, he found that the injection of

[88] G. H. Bourne, *Australian J. Exptl. Biol. Med. Sci.* **13,** 239 (1935).

[89] M. P. Joyet-Lavergne, *Protoplasma* **23,** 50 (1935).

[90] J. B. Gatenby, *Quart. J. Microscop. Sci.* **63,** 445 (1919).

[91] G. Anderson and V. E. Levine, *Proc. Soc. Exptl. Biol. Med.* **32,** 737 (1935).

[92] H. Popper, *Physiol. Revs.* **24,** 205 (1944).

[93] W. T. Goodwin, *Biol. Revs. Cambridge Phil. Soc.* **25,** 391 (1950).

[94] A. J. Goerner, *J. Biol. Chem.* **123,** 57 (1938).

[95] A. J. Goerner, *J. Biol. Chem.* **128,** 559 (1939).

vitamin A increased the content of this vitamin in the mitochondria of healthy cells but not of tumor cells.

In 1941 Popper[96] published an extensive series of investigations of the histological distribution of vitamin A based on the localization of the fading green fluorescence produced by the action of ultraviolet light on the vitamin.

In the liver Popper found that the epithelial cells contained numerous fine, fat droplets which in many cases were aggregated near the borders adjacent to the sinusoids and were often arranged like a string of beads. Cells were seen in which all stages of enlargement of these drops were found until they filled the cytoplasm. All stages of development of these drops were found to give strong vitamin A fluorescence. Where a single large, fat droplet filled the cell, Popper found that it had a less intense fluorescence than the smaller drops. He also found that liver cells contained fine bars and droplets which also gave a strong fluorescence, and these he admitted were in form and distribution very like mitochondria. Popper also found fine droplets in the Kupffer cells which gave a positive fluorescence for vitamin A. He found that in vitamin A depletion the vitamin was retained longest by the Kupffer cells, and that in vitamin A feeding these cells were repleted before the epithelial cells. Popper infers from this that the vitamin performs some important function in these cells, and he suggests that they may extract vitamin A from the blood and pass it to the epithelial cells.

Popper has investigated the distribution of vitamin A by this fluorescence method in a number of tissues (see plate IV, *A-F*, and plate V, *C*, *E*, *F*, and *G*). In the adrenal of the rat he found it constantly present in the small lipoid droplets of the zona fasciculata but not in the cells of the glomerulosa. In the human being, however, it was present in small amounts in the glomerulosa as well. In the Leydig cells of the testis fine droplets containing the vitamin were found, a positive reaction developing with sexual maturity and decreasing with involution. In the human ovary he found that vitamin A appeared in the third month of postnatal life. The vitamin was present as fine droplets which gradually increased in size and number in the granulosa cells and in maturity as a corpus luteum formed from them. It is of interest that he found that the corpora lutea of pregnancy showed only carotene and practically no vitamin A fluorescence, whereas the corpora atretica gave a strong reaction. The appearance of vitamin A in the ovarian cells appeared to be correlated with sex hormone activity.

In other female tissue it was found that the lipids of the endometrium occasionally showed the presence of vitamin A. In the non-lactating mammary gland the epithelial cells showed a strong fluorescence.

The epithelial cells of the small intestine gave a reaction only after feeding with vitamin A, and this was most intense in the upper part.

[96] H. Popper, *Arch. Path.* **31,** 766 (1941).

Curiously enough, the epithelial cells of the skin gave no reaction. This may be thought a surprising fact, considering the changes which skin undergoes in vitamin A deficiency, but it may be that those tissues which store the least amount of the vitamin are those which most readily show the effects of its deficiency.

In the eye, vitamin A fluorescence was obtained from the rod and cone layer of the retina, very intense in the light-adapted eye and less intense in the dark-adapted eye. This is in keeping with the view that the protein-vitamin complex, retinene, which is formed from rhodopsin under the influence of light, liberates vitamin A, which subsequently becomes resynthesized into rhodopsin.

Popper has been to some pains to ascertain the reliability of the fluorescence method for demonstrating vitamin A. He found that fluorescence is given in the wavelength range of ultraviolet at which vitamin A absorbs most strongly. The fluorescence was a bright green which faded quickly. Carotene under the same conditions gave a dull green fluorescence which faded only slowly. Anhydro (cyclized) vitamin A, which has similar chemical properties to the vitamin but is biologically inactive, gave a dark brown fluorescence. A variety of fats and oils and fat-soluble hormones and vitamins failed to give a fading green fluorescence. Furthermore, the fading green fluorescence decreased progressively when an animal was on a vitamin A-deficient diet and reappeared when the vitamin was administered. There seems, therefore, to be reasonable ground for accepting the specificity of the reaction.

Very little is known concerning the exact role of vitamin A in cells. Its presence in mitochondria may indicate that it plays a part in some oxidation-reduction system. This is suggested by the presence of five unsaturated bonds in its molecule. Whether it plays a part in the building up and maintenance of cell structure is not certain, but it has been found to quicken the growth and prolong the life of tissue cultures (Bisceglie[97] and Baker[98]). In this connection Batchelder[99] has shown that a plentiful supply of vitamin A delays the onset of senility in the rat.

2. Vitamin C

The second vitamin to be demonstrated histochemically was vitamin C (Bourne;[100, 101, 102] see plate V, *A*). This work was followed by that of

[97] V. Bisceglie, *Arch. Entwicklungsmech. Organ.* **108,** 708 (1926).

[98] L. E. Baker, *Proc. Soc. Exptl. Biol. Med.* **33,** 124 (1935).

[99] E. L. Batchelder, *Am. J. Physiol.* **109,** 430 (1934).

[100] G. H. Bourne, *Nature* **133,** 874 (1933).

[101] G. H. Bourne, *Australian J. Exptl. Biol. Med. Sci.* **11,** 261 (1933).

[102] G. H. Bourne, *Nature* **132,** 859 (1934).

Leblond[103] and by a series of papers by Leblond and Giroud.[104] The principle of the technique is based upon the fact that the vitamin is a powerful reducing agent and that it will reduce silver nitrate not only in alkaline solution, as many other biological substancet do, but also in neutral and acid solution.

The first preparations by Bourne[100, 101] were made with neutral silver nitrate. First attempts to demonstrate the vitamin were made on the adrenals (which are extremely rich in the vitamin) of cat and guinea pig which had been fixed in the vapor of formaldehyde. The results obtained by this method were unreliable, because it is fairly certain that an appreciable proportion of the vitamin was inactivated by the fixative. The next preparations[102] were made on the adrenal glands of the human fetus with a 2% solution of silver nitrate in 70% alcohol which fixed the tissue and was reduced by the vitamin more or less simultaneously. The fetal cortex was found to give a stronger reaction than other parts of the gland. In the cells of the gland the granules were aggregated particularly around the nuclei and around lipoid droplets. Other preparations were made using chloroform vapor to fix the tissue prior to placing the gland in silver nitrate. In the following year Leblond and Giroud, and Leblond and Bourne used acidified silver nitrate to increase the specificity of the reaction. Eventually a precise technique was established not only to reduce the possibility of the reaction's occurring non-specifically but also to prevent light from affecting the result.

In 1928 Szent-Györgyi[105] had shown that the adrenal gland rapidly went black when placed in silver nitrate solution, indicating that it contained a powerful reducing substance. The work of the Dunn Nutritional Laboratory at Cambridge later established that this silver nitrate reaction was due entirely to vitamin C. It was shown, for example, that ability to reduce silver nitrate ran parallel with the antiscorbutic activity of the gland and that it also ran parallel with the ability of the gland extracts to reduce dichlorophenol indophenol, a reaction which was shown to be specific for vitamin C under certain conditions. Moore and Ray[106] from the same laboratory also showed that in scurvy the silver nitrate reaction of the adrenal disappeared, a finding which has since been confirmed by Bourne and by Giroud and Leblond.

It was also established that a wide variety of naturally occurring reducing substances, though able to reduce silver nitrate in alkaline solution,

[103] C. P. Leblond, La vitamine C dans l'organisme, Thesis, University of Paris, 1934.

[104] A. Giroud, L'acide ascorbique dans la cellule et les tissues, Protoplasma Monographs, Gebrüder Borntraeger, Berlin, 1938.

[105] A. Szent-Györgyi, *Biochem. J.* **22,** 1387 (1928).

[106] T. Moore and S. N. Ray, *Nature* **130,** 997 (1932).

were not able to reduce it in the acid solution or would do so only after a long time in the light.

The final technique established for the demonstration of this vitamin was, first, that scrupulous chemical cleanliness was required and, second, that impregnation should be for short periods of time (not more than 20 minutes) and should be carried out in the dark. After impregnation the tissues were treated with photographic "hypo" to remove silver compounds which might subsequently be reduced by the action of light.

There is now a considerable literature on the cytology of vitamin C, and it is of course impossible to refer to it all. Not only has the vitamin been found in practically all cells of most multicellular animals but it has been demonstrated in protozoa, fungi, and algae by Allen and Bourne[107] and Giroud and Leblond and by other workers in the tissues of higher plants.[104] François and Rabasy[108] found by the silver nitrate method that vitamin C was present in all eye tissues except the epithelium covering the ciliary process. They suggest that the ascorbic acid of the eye is in a closed circuit without continuous supplies of ascorbic acid from the blood being necessary. Their view is that the oxidized ascorbic acid of the lens passes into the aqueous humour as required, being reduced again there. It passes back to the lens in the reduced form, and so on.

Probably the greatest interest concerning the cytology of vitamin C is its alleged association with the Golgi apparatus in cells. Bourne[100, 101] first drew attention to the fact that vitamin C preparations showed aggregates of reduced silver in the Golgi region, but doubt was expressed as to whether this region was the Golgi region. Leblond,[103] however, established that this was so, an observation which was subsequently confirmed by Bourne[109, 110] by Giroud and Leblond (in various papers), and by Tonutti.[111] This work led to the assumption that in a large number of cells vitamin C may be associated intimately with the Golgi material. Barnett and Bourne[112] observed that, during the histogenesis of chick embryos, in many cells vitamin C appears to be localized in the Golgi region. This was particularly evident in nerve cells where in addition it may occur along the developing axon. In dividing cartilage cells of chick embryos, the disposition of vitamin C granules around the chromosomes and the spindle figure was strikingly similar to the disposition of the Golgi material. The relationship between this region and vitamin C staining granules has been demonstrated in a

[107] R. J. Allen and G. H. Bourne, *Australian J. Exptl. Biol. Med. Sci.* **13,** 165 (1935).
[108] J. François and M. Rabasy, *Ann. oculist.* **183,** 820 (1950).
[109] G. H. Bourne, *Nature* **166,** 549 (1950).
[110] G. H. Bourne, Cytology and Cell Physiology, Clarendon Press, Oxford, 2nd ed., 1951.
[111] E. Tonutti, *Protoplasma* **31,** 151 (1938).
[112] S. A. Barnett and G. H. Bourne, *Quart. J. Microscop. Sci.* **83,** 457 (1941).

number of cells; the reader is referred to a recent review on this subject.[110] Perhaps one of the most significant experiments is that of Tonutti,[111] who injected the dye trypan blue (which is known to be segregated in the Golgi region of cells) and vitamin C together into a pregnant and a normal animal. The preparations were subsequently stained to demonstrate vitamin C. Sections of kidney and placenta showed that the vitamin C granules were localized in the same region as the trypan blue.

Bourne[110] also showed that vitamin C granules in the cells of an ultracentrifuged adrenal gland were stratified at the same level as the Golgi substance. Sulkin and Kunz[113] found that, in the autonomic ganglia of cat, guinea pig, rabbit, and man, vitamin C granules (silver nitrate reaction) were either distributed uniformly throughout the cytoplasm or limited to the zone of the Golgi apparatus. In ganglia removed from human patients suffering from hypertension the ascorbic acid of the cells appeared to be localized mainly in the zone of the Golgi apparatus. Ganglion cells that had been subjected to a prolonged stimulation showed a reduction in their content of ascorbic acid.

The claim that vitamin C is localized in the Golgi region has not gone uncontested. Barnett and Fisher[114] pointed out that in chemical models impregnated with vitamin C and then treated with acid/silver nitrate, the black deposit of reduced silver appears to be localized at interfaces. Sosa (personal communication) believed that fixing the tissue with "hypo" after treatment with acid/silver nitrate produced blackening in the Golgi apparatus. Deane and Morse[115] have stated that if more than 3 minutes elapsed between the death of the animal and placing of the tissues into silver nitrate there was an apparent concentration of the vitamin in the Golgi region, but there was no concentration if the tissues were placed in silver nitrate before that time. Bacchus[116] agreed with Deane and Morse in this respect. Pfuhl[116a] considered that a substance as diffusible as vitamin C is not likely to be aggregated in life in a granular form similar to that found in the Golgi region of some cells. It is extremely unlikely, however, that vitamin C is in a highly diffusible condition in the cell, and, in any case, segregation is precisely what is likely to happen to a substance with as potent a reducing power as vitamin C, whether it is diffusible or not. In fact, the more diffusible it is, the more likely it is to be segregated in an organ such as the Golgi complex. Furthermore, a recent publication by Bourne[109] has shown that when the penetration of acid silver nitrate through a tissue "squash" is observed under the microscope the granules produced by the reaction

[113] N. M. Sulkin and A. Kuntz, *Anat. Record* **101,** 33 (1948).

[114] S. A. Barnett and R. B. Fisher, *J. Exptl. Biol.* **20,** 14 (1943).

[115] H. W. Deane and A. Morse, *Anat. Record* **100,** 51 (1948).

[116] M. Bacchus, *Anat. Record* **110,** 495 (1951).

[116a] A. Pfuhl, *Z. mikroskop.-anat. Forsch.* **50,** 299 (1941).

between vitamin C and the reagent appeared suddenly in a clear cytoplasmic background as the reagent penetrated into the cell. There was not at first a general diffuse darkening of the cytoplasm followed by a gradual appearance of the granules, as one would expect if the vitamin was diffused in the cytoplasm and the reaction products subsequently aggregated together. A detailed reply to claims of Barnett and Fisher[114] has already been published by Bourne,[117] and a general reply to criticisms concerning the localization of the vitamin can be found in Bourne's chapter on "Mitochondria and Golgi Complex" published in the second edition of "Cytology and Cell Physiology."[110]

Summing up this controversial point, one can say that at present the evidence for localization of the vitamin in the Golgi complex is strong but that the possibility that it may be an artifact cannot be completely ruled out.

It is of interest that the Golgi region of cells tends to accumulate vitamin C when the cell is engaged in synthetic activity. A possible explanation for this is that, since the metabolism of cells is raised when they are carrying out synthetic processes, unless these various products (which may be produced in the first instance by mitochondria) are protected by being absorbed into a specially segregated, highly reducing area of the cytoplasm, they would be oxidized as rapidly as they are formed. Tonutti[111] thinks, on the other hand, that the Golgi apparatus stores vitamin C and liberates it slowly into the cytoplasm in sufficient amounts to prevent oxidation of cell products.

Vitamin C may also function in other ways. It may help to detoxify toxic substances which enter the cell and tend to be segregated in the Golgi region; it may also modify the synthetic product in some way; or it may become attached to the molecule to make it water-soluble. For example, in the adrenal cortex the molecules of corticosteroid may have vitamin C attached to them in the Golgi complex for this purpose. A corticosteroid with vitamin C attached has been described.

The largest amount of work on the intracellular distribution of vitamins appears to have been carried out on vitamins A and C.

3. Riboflavin

Most other vitamins, notably those of the B complex, have been localized by fluorescence microscopy, although Chevremont and Comhaire[118] described a histochemical method for demonstrating riboflavin. By treating tissues with nascent hydrogen to reduce the riboflavin to leucoflavin and then oxidizing it to rhodoflavin, these authors were able to demonstrate

[117] G. H. Bourne, *Nature*, **153**, 254 (1944).
[118] M. Chevremont and S. Comhaire, *Arch. exptl. Zellforsch. Gewebezücht.* **22,** 658 (1939).

bright red granules in the cytoplasm of cells. The greenish fluorescence of riboflavin to indicate the distribution of the vitamin in living tissues was used by Ellinger and Koschara,[119] von Euler and his colleagues,[120] Hirt and Wimmer,[121] and Metcalf and Patton.[122] None of this work, however, can be regarded as providing details of the intracellular localization of the vitamin. Sjöstrand,[123, 124] who analyzed microspectrographically the green fluorescent light given off by riboflavin under ultraviolet stimulation, showed that the vitamin would be localized, for example, in the choroidea of the frog's eye. The demonstration of riboflavin by a fluorescence method in the guinea pig is shown by plate V, *B*. Ellinger[125] has also shown that another form of riboflavin exists which gives an orange-brown fluorescence with ultraviolet light; this is thought to be a bound form or riboflavin, possibly a flavoprotein enzyme. It is of interest that Metcalf[126] has found that the bound riboflavin in the tissues of the cockroach is converted into the free form by injections of pantothenic acid.

4. Thiamine

In an alkaline solution the greenish fluorescence of riboflavin in tissues is replaced by a bluish-white fluorescence characteristic of aneurin or thiamine. As a result of the application of this fact Sjöstrand[123, 124] was able, by his microspectrographic method, to localize vitamin B_1 (aneurin or thiamine) in the myelin sheath of medulated nerves (see plate V, *D*). Smits and Florijn[127] found that in the red cells of chickens 70 % of the vitamin B_1 pyrophosphate was concentrated in the nuclei. Since these nuclei occupied only 10 % by volume of the cell, there was thus a concentration of vitamin phosphate in the nuclei equal to twenty times that of the cytoplasm. The nuclei in this work were isolated by differential centrifugation.

5. Pterines

Jacobson[128] has found that an orange-yellow fluorescence, which he was able to localize in the argentaffin cells of the alimentary tract, was produced by certain pterines. These were related to the pterine substance, folic acid.

[119] P. Ellinger and W. Koschara, *Ber.* **60,** 315, 808 (1933).
[120] H. von Euler, H. Hellstrom, and E. Adler, *Z. vergleich. Physiol.* **21,** 739 (1935).
[121] A. Hirt and M. Wimmer, *Klin. Wochschr.* **18,** 733, 765 (1939).
[122] R. L. Metcalf and R. L. Patton, *J. Cellular Comp. Physiol.* **19,** 1 (1942).
[123] F. Sjöstrand, *Nature* **150,** 698 (1946).
[124] F. Sjöstrand, *Acta Physiol. Scand.* **12,** 42 (1946).
[125] P. Ellinger, *Biochem. J.* **32,** 376 (1938).
[126] R. L. Metcalf, *J. Cellular Comp. Physiol.* **19,** 1 (1943).
[127] J. Smits and E. Florijn, *Biochim. et Biophys. Acta* **5,** 297 (1950).
[128] W. Jacobson, *J. Path. Bact.* **49,** 1 (1939).

6. Nicotinic Acid

Warburg in 1938[129] pointed out that dihydropyridine (dihydrocoenzyme, which contains nicotinic acid) gives marked white fluorescence with ultraviolet light. In sections of living liver tissue small granules with a whitish fluorescence can be seen in the cytoplasm of the cells.

7. Vitamin K

Vitamin K also gives a characteristic fluorescence under ultraviolet light, but little attention appears to have been given to the use of this property for cytological localization.

Addendum. Xanthine oxidase has been shown recently (Bourne G. H. *Nature.* In press) to be localized in granular form in the cytoplasm of cells in kidney, liver, spleen, and small intestine.

[129] O. Warburg, *Ergeb. Enzymforsch.* **7,** 210 (1938).

PLATES I–V

PLATE I

A. Succinic dehydrogenase in the cortex of rat kidney. All tubule cells rich in enzyme activity. Glomeruli negative. *B*. Succinic dehydrogenase in a skeletal muscle fiber of the rabbit, showing localization of enzyme in sarcosomes between the muscle fibrils (sarcosomes are mitochondrial in origin). *C*. Succinic dehydrogenase in the kidney of the rat. Strong positive reaction given by Henle's loops. Distal parts of collecting tubes completely negative.

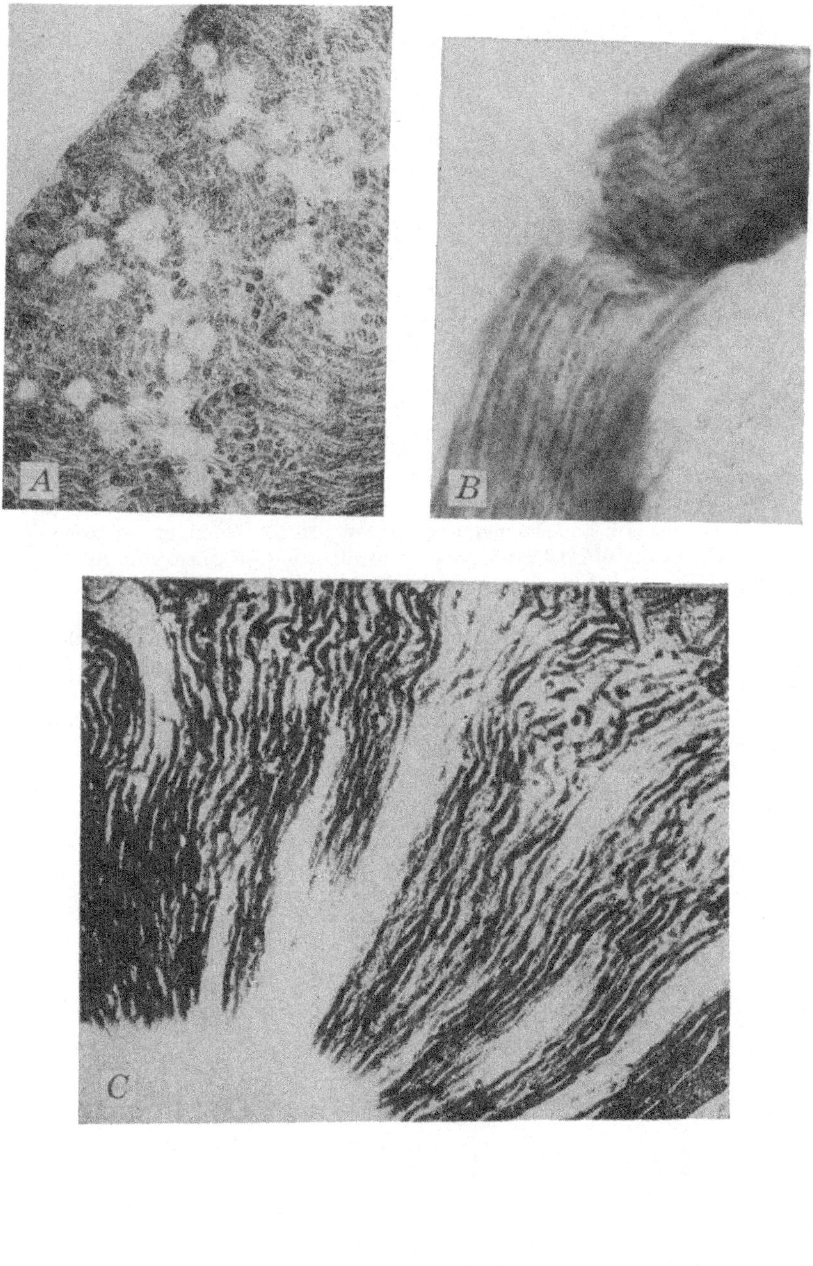
A
B
C

PLATE II

A. Part of the cerebrum of the rat, showing localization of an enzyme splitting vitamin B_1 pyrophosphate in the neurones. pH 6.8–7.0. *B*. Cerebellar cortex of the rat, showing localization of an enzyme splitting vitamin B_1 pyrophosphate. Purkinje cells and glial cells positive. pH 6.5. *C*. Localization of an enzyme splitting adenosine monophosphate in the brain of the rat. Fibers positive. Nerve cells negative except for nucleoli. pH 6.5. *D*. Localization of an enzyme splitting adenosine triphosphate in rat brain, showing positive reaction in glial, neuronal, and vascular wall nuclei. pH 6.5. *E*. Cerebellum of the rat, showing specific localization of an enzyme splitting glycerophosphate (acid phosphatase) in Purkinje cells of rat. pH 5.3. *F*. Single neurones, showing localization of an enzyme splitting glycerophosphate (acid phosphatase). Cytoplasm and nucleolus positive. pH 5.3. *G*. Single neurone from the hypoglossal nucleus of rat brain, showing localization of an enzyme splitting adenosine monophosphate. Except for faint positive nucleolus, cell body and nucleus are unstained. Surrounding fibres are positive pH 6.5. *H*. Simple esterase localized in the Purkinje cells of rabbit cerebellum. pH 9.0. *I*. Single neurone from the hypoglossal nucleus of rat brain, showing localization of an enzyme splitting adenosine triphosphate. Nucleolus strongly positive and cytoplasm lightly positive. pH 6.5.

A, *B*, *C*, *D*, *E*, *F*, *G*, and *I* are by courtesy of D. Naidoo and O. E. Pratt and the *Journal of Neurology, Neurosurgery and Psychiatry*.

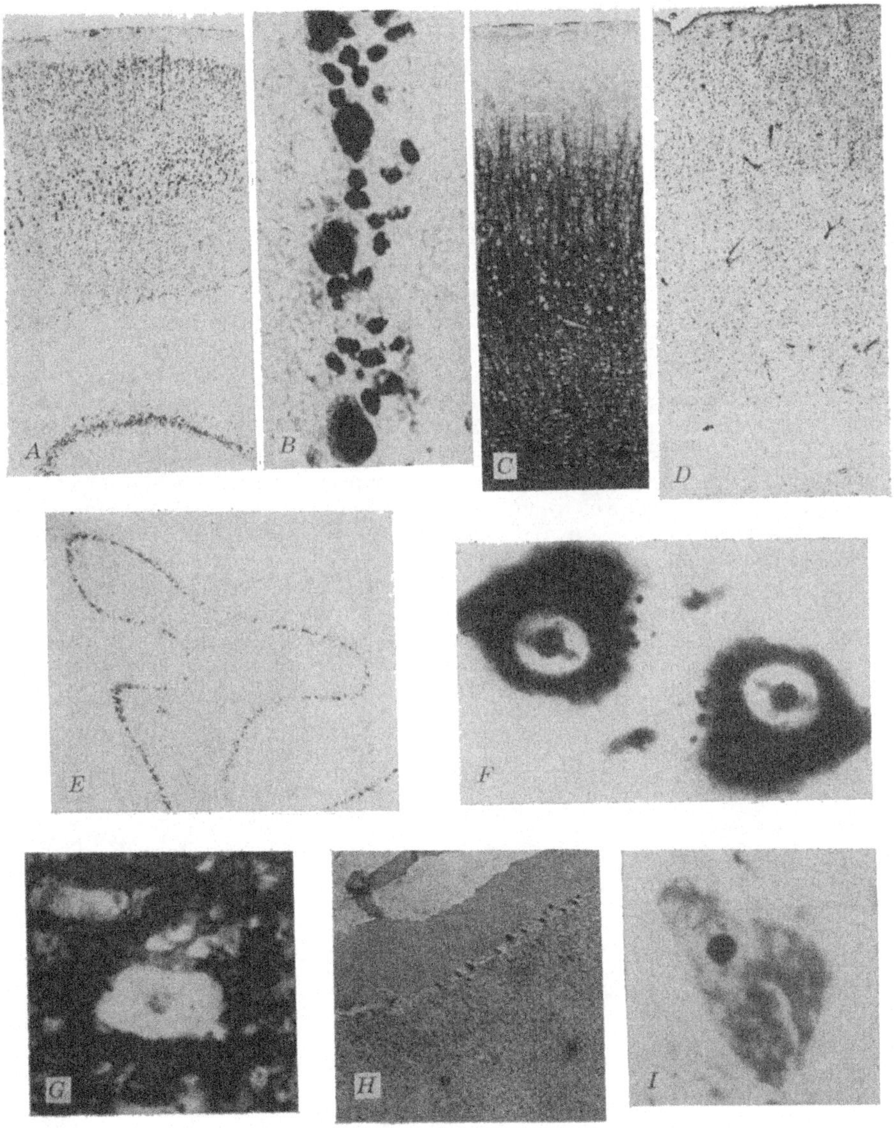
A
B
C
D
E
F
G
H
I

PLATE III

A. Simple esterase in the nasal mucosa of the rabbit. Positive reaction given by epithelium and Bowman's glands. Distal portion of epithelium negative. *B*. Simple esterase in rabbit liver. All cells possess considerable enzyme activity. Cells surrounding the portal vessels have slightly more activity than more central cells. It is possible that some of the reaction in this organ may be due to lipase in the hepatic cell. *C*. Lipase in the nasal mucosa of the rabbit. Only Bowman's glands are positive. Epithelium (at bottom) is negative (cf. *A*). *D*. Simple esterase in the convoluted tubules of kidney. *E*. Simple esterase in the basal part of crypts in rabbit duodenum. Positive reaction is also given by epithelial cells of villi. *F*. Localization of radioactive methionine (containing ^{32}S) in the shafts of developing hair. Walls of follicles also positive. This technique has not yet been applied to vitamins and enzymes, but it is an obvious new development. *G*. Nuclease in cells overlying taste buds in the papilla foliata of the rabbit. Everything else is negative. *H*. Nuclease in the duodenum of the rabbit after accentuation with quinine. Brunner's glands and muscle coat negative. Mucosa, and particular epithelium of villi, positive. *I*. Cholinesterase in small sympathetic ganglion in tongue muscle. Note nuclear-membrane reaction.

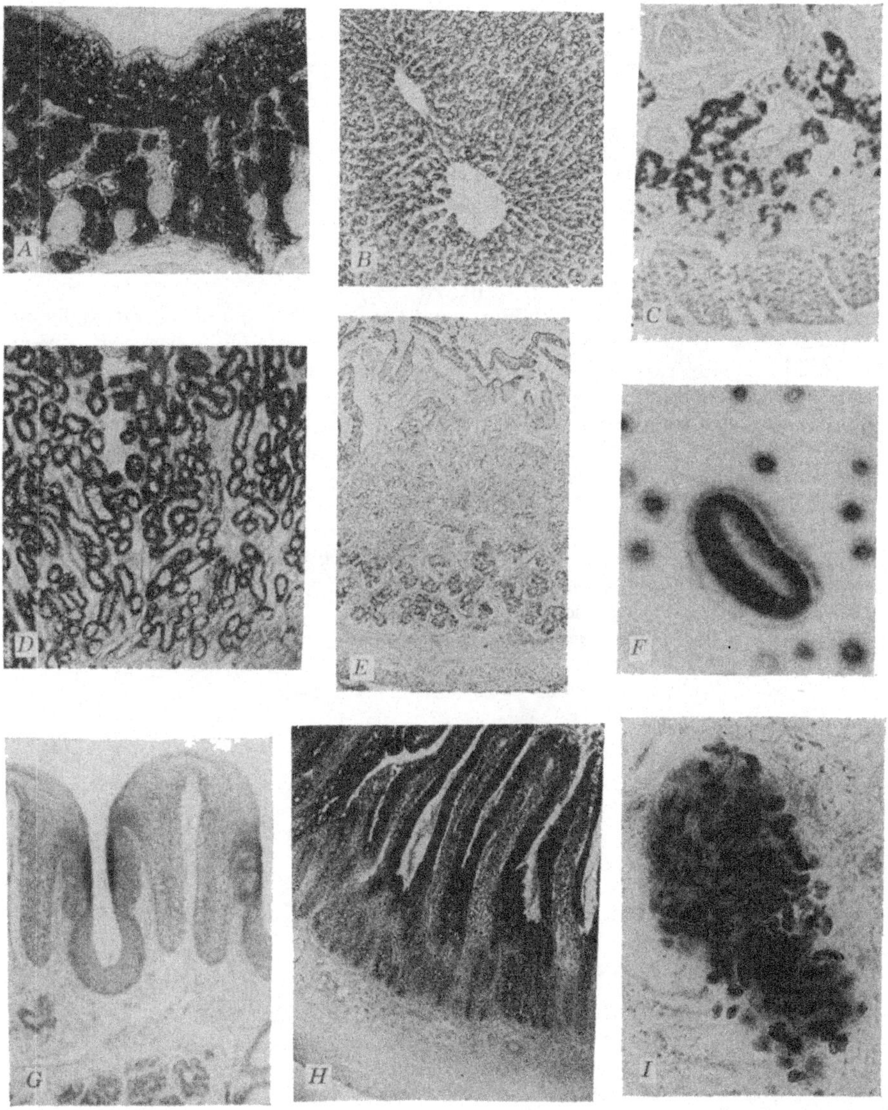
A
B
C
D
E
F
G
H
I

PLATE IV

Vitamin A demonstrated in the liver and the adrenal gland of the rat by fluorescence method. *A*. Liver of a rat deficient in vitamin A. No fluorescence. *B*. Liver of a normal rat. Moderate amounts of vitamin A fluorescence are present in the Kupffer cells and at places in fine droplets at edges of liver cells. *C*. Liver of a rat made hypervitaminotic with vitamin A. Marked vitamin A fluorescence is seen in the Kupffer cells which have apparently proliferated. Moderate amounts of fluorescence in small lipoid droplets within the liver cells. *D*. Adrenal of a vitamin A-deficient rat. No fluorescence. *E*. Adrenal of a well-fed rat. Moderate amounts of vitamin A fluorescence are seen in the lipoid droplets of the zona fasciculata. *F*. Adrenal of a rat made hypervitaminotic with vitamin A. There is striking vitamin A fluorescence of the lipoid droplets of the fascicular layer. The lipoid droplets of the zona glomerulosa are free of vitamin A fluorescence.

Courtesy of Dr. Hans Popper and the *Archives of Pathology*.

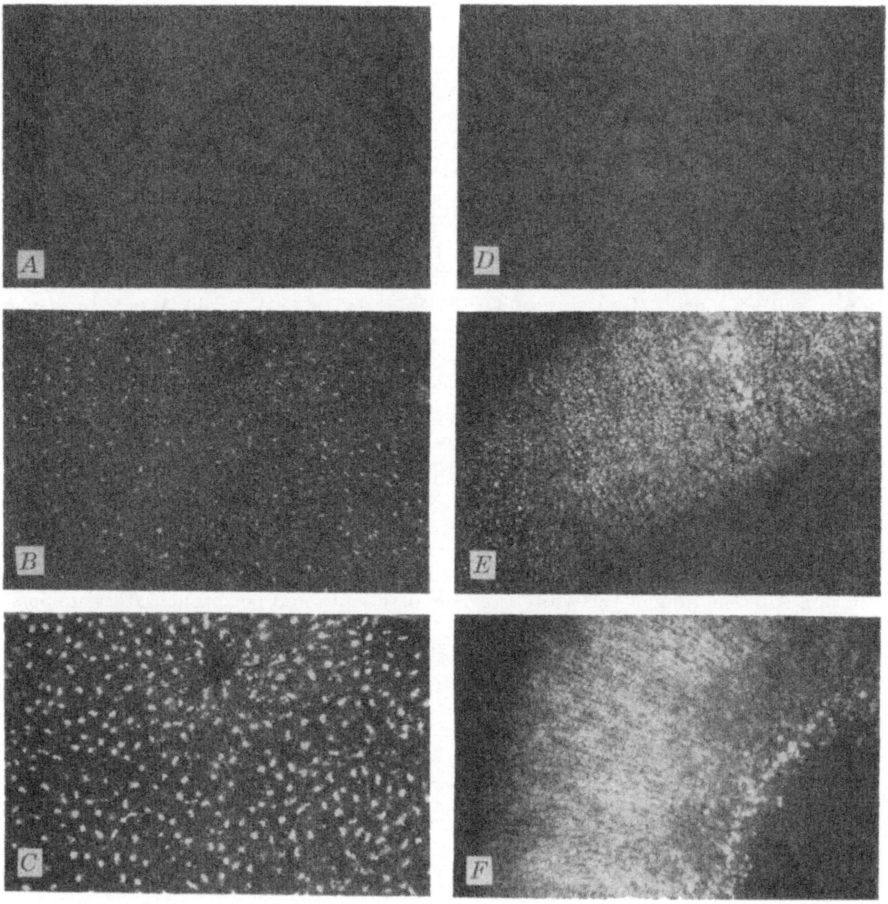
A
B
C
D
E
F

PLATE V

A. Vitamin C in the adrenal cortex. A concentration of granules at the junction of capsule and zona glomerulosa can be seen. Glomerulosa and fasciculata cells show granules associated with the nucleus. *B*. Riboflavin in the convoluted tubule cells of the guinea pig kidney. The granules of riboflavin are similar in form and distribution to mitochondria. (Courtesy of Dr. F. Sjöstrand.) *C*. Vitamin A in villi of the jejunum of the rat after an oral intake of 5000 I.U. of vitamin A. Strong vitamin A fluorescence is shown by droplets within the lumen, and moderate fluorescence by the epithelial lining and by cells in the center of the villi. (Courtesy of Dr. Hans Popper.) *D*. Thiamine (vitamin B_1) in the sciatic nerve of the rabbit. The reaction is probably due to the spontaneous oxidation of thiamine to thiochrome. The reaction is seen to be localized in the myelin sheaths of the nerve fibers. (Courtesy of Dr. F. Sjöstrand.) *E*. Vitamin A in the jejunum of the rat after oral intake of 80,000 I.U. of vitamin A. Particularly strong fluorescence is shown by the lacteals within the villi. (Courtesy of Dr. Hans Popper.) *F*. Vitamin A in the rat ovary. Cords of interstitial cells are rich in vitamin A fluorescence. (Courtesy of Dr. Hans Popper.) *G*. Vitamin A in the lung of the white rat. Alveolar septa and peribronchial tissue positive. (Courtesy of Dr. Hans Popper.)

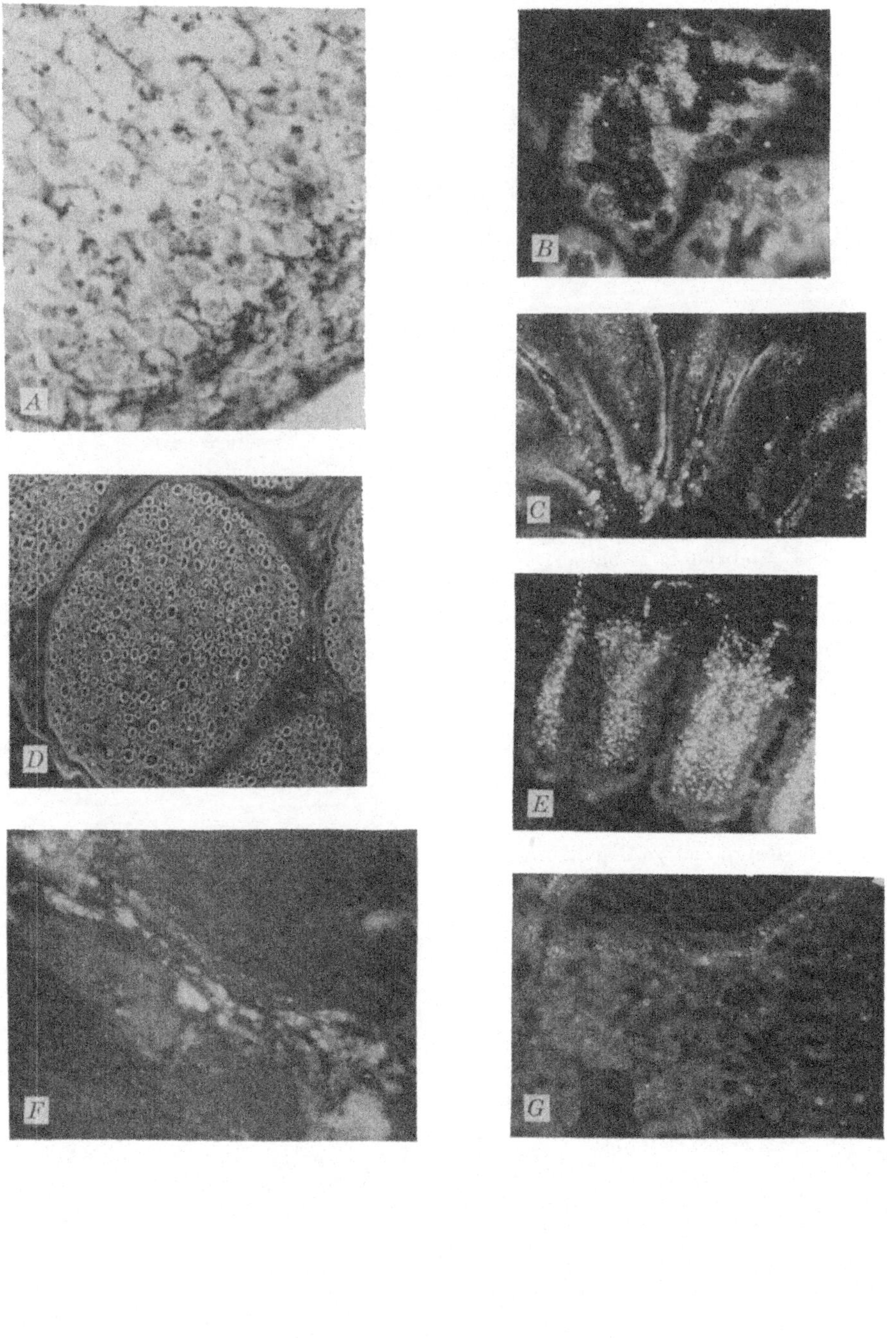
A
B
C
D
E
F
G

CHAPTER 13

Structural Changes In Vitamin Deficiency

GEOFFREY H. BOURNE

Department of Histology, London Hospital Medical College, London, England

I. Introduction

"Nature," said Goethe in *Faust*, "reveals herself best in her abnormalities." The abnormal state of vitamin deficiency therefore provides us with one type of evidence which helps us to obtain some insight into the role the vitamins play in maintaining the normal structure in the tissues and organs of the body. These deficiencies must, of course, first express themselves as biochemical lesions; reference to these will be found elsewhere in this book. Later, visible changes in structure take place, at first microscopic, then macroscopic. In this chapter it is hoped to record some of the microscopic aspects of these changes. Since at least two books have been written on this subject, it is not surprising that in the limited compass of one chapter it is impossible to deal adequately with or indeed even to mention much of the mass of work which has gone before. Some general summaries already exist; the author would like to mention expecially Mason's[1] review (1939), "The Vitamins and the Sex Glands"; Wolbach and Bessey's[2] review (1942); Eddy and Dalldorf's[3] book (1944); and Follis's[4] book (1948); all these have been drawn upon liberally in compiling portions of the present discussion.

II. Vitamin A Deficiency

1. General Effects

Vitamin A deficiency has been characterized (Eddy and Dalldorf[3]) by three cardinal signs: xerophthalmia, dermatosis, and night blindness. These pathological conditions have been recognized ffom the time of Hippocrates, and their occurrence has been known in many parts of the world since then. It is only recently, of course, that knowledge of vitamins has rendered their cause more obvious.

The precise metabolic function of vitamin A is not known, but it may be concerned particularly with some metabolic process peculiar to epithelial cells, since these cells provide one of the earliest manifestations of deficiency of this vitamin. It appears to be concerned with other tissues of ectodermal origin, such as the retina of the eye and the nervous system in general, but there is also some association with the mesodermal structures, such as bone. If vitamin A is added to tissue cultures (Bisceglie[5] and Baker[6]) it increases the rate of growth of fibroblasts, which suggests that it is a growth factor. It is of interest that Fell and Mellanby[7] have shown that, if the long bones of late mouse fetuses are cultivated in a medium to which large amounts of vitamin A are added, the normal growth of the bones in such media is inhibited within 3 days, and the cartilage softens, becomes shrunken, and eventually disappears. In the shaft itself cartilage is replaced by marrow tissue. Although bone growth was retarded by these large amounts of vitamin A, the soft tissues associated with the bone continued to grow normally. Barnicot,[8] too, has shown that if vitamin A acetate is applied to the surface of a young bone it causes local bone resorption. He points out

[1] K. E. Mason, Sex and Internal Secretions, 2nd ed., Baillière, Tindall and Cox, London, 1939.

[2] S. B. Wolbach and O. A. Bessey, *Physiol. Revs.* **22,** 233 (1942).

[3] W. H. Eddy and G. Dalldorf, The Avitaminoses, 3rd. ed., Williams and Wilkins, Baltimore, 1944.

[4] R. H. Follis, The Pathology of Nutritional Diseases, Blackwell, 1948.

[5] V. Bisceglie, *Arch. Entwicklungsmech. Organ.* **108,** 708 (1926).

[6] L. E. Baker, *Proc. Soc. Exptl. Biol. Med.* **33,** 124 (1935).

[7] H. B. Fell and E. Mellanby, *Brit. Med. J.* **ii,** 535 (1950).

[8] N. A. Barnicot, *Nature,* **162,** 848 (1948).

that this result is in accordance with reports that vitamin A deficiency causes overgrowth of bone.

The classical paper on the histological changes in vitamin A deficiency was that of Wolbach and Howe,[9] published as long ago as 1925. Wilson and du Bois[10] had described changes in the epithelial structures in man in vitamin A deficiency in 1922, however, and in the same year Evans discovered (see ref. 1) that female rats on a vitamin A-deficient diet gave vaginal smears which, instead of going through the usual cycle of cornified cells, lymphocytes, nucleated cells, and cornified cells (a change which is coincident with the cyclical changes taking place in the reproductive system, the cornified cells representing the stages of estrus), produced cornified cells all the time. They found that this change appeared earlier than xerophthalmia. Despite the abnormal nature of the vaginal smears, Evans found that estrus and ovulation continued to take place but that if copulation took place during this period both fertilization and implantation were likely to fail. In this way the effect of vitamin A deficiency differs from that of vitamin E deficiency; in the latter fertilization and implantation occur, but resorption of the developing embryo follows.

It is probably better to deal now with the effect of vitamin A deficiency on different organs and tissues rather than continue the story from the historical point of view.

2. Epithelia

Under vitamin A deficiency the various epithelia of the body undergo metaplastic changes. As a general rule, more differentiated epithelia, such as the columnar and ciliated varieties, regress to a simpler type of epithelium of the stratified type (see Fig. 6). Epithelia which are normally stratified become more cornified. Wolbach and Howe[9] showed that the earliest epithelial changes occurred in the respiratory system. Next the ducts of various glands were affected, then the eyes, and about the same time there were changes in the epithelium of the glands associated with the eyes and of the alimentary tract.

Wolfe and Salter[11] showed that in white mice suffering from vitamin A deficiency the stratified keratinized cells which replaced columnar and ciliated epithelia appeared in small clumps below the normal cells. These clumps then grew peripherally and gradually extended beneath the apparently normal epithelium. This epithelium seemed to be able to survive for some time, even though its connection with the underlying connective tissue had been severed by the spreading sheets of keratinized cells.

Wolfe and Salter, like Wolbach and Howe, found that the first signs of

[9] S. B. Wolbach and P. R. Howe, *Proc. Soc. Exptl. Biol. Med.* **22**, 400 (1925).

[10] J. A. Wilson and R. U. du Bois, *Am. J. Diseases Children* **26**, 431 (1922).

[11] D. M. Wolfe and H. P. Salter, *J. Nutrition* **4**, 185 (1931).

epithelial metaplasia occurred in the respiratory mucosa but that these lesions were quickly followed by xerophthalmia. In the nares, the trachea, and the bronchi they found marked desquamation of cells. They found that bronchiectasis was common in the lungs, probably owing to the blocking of the bronchi and bronchioles by masses of desquamated cells. The occurrence of islets of stratified cells among the respiratory epithelium has also been described by de Ruyter[12] and other authors. Metaplastic epithelial changes have been described in the guinea pig and the rat by Wolbach and Howe.[9, 13] Smith[14] has described these same processes in the fox, and Tielder and Miller[15] have described them in the monkey. Epithelial changes have also been described in man, cattle, swine, dog, rabbit (Wolbach[16]), and fowl (Beach[17]). Wolbach has claimed that there is an increased growth activity in epithelia in vitamin A deficiency, but this is not accepted by all workers, notably Friedenwald *et al.*[18] These workers carried out counts of mitotic cells in healing corneal wounds in normal and vitamin A-deficient animals and found that there was 30% less mitosis in the deficient animals.

A striking early metaplastic change is the stratification and keratinization of the ducts of the salivary glands and of the pancreas. Superficial cells are desquamated into the lamina of these ducts and serve to block them. At a later stage the epithelial acini may be involved, although this does not occur in the pancreas where the changes appear to be restricted to the ducts.

The skin of mature animals becomes rougher and more heavily cornified in vitamin A deficiency, and Nicholls[19] has described a hyperkeratinization of the epidermis in human beings which he calls "toad skin" and which he claims is due to vitamin A deficiency. Absence of sweat in human beings suffering from vitamin A deficiency may be due to blocking of the sweat gland ducts by desquamated cells. In animals, blocking of the sebaceous gland ducts leads to degeneration of the hair follicles (Ramalingaswami and Sinclair[20]). These authors have also adduced evidence that "toad skin" is due to deficiency of unsaturated fatty acids.

Wolbach and Bessey[21] have described the sequence of changes which

[12] T. H. de Ruyter, *Acta Brevia Neerland. Physiol. Pharmacol. Microbiol.* **4,** 122 (1934).

[13] S. B. Wolbach and P. R. Howe, *Arch. Path.* **5,** 239 (1928).

[14] S. E. Smith, *J. Nutrition* **24,** 97 (1930).

[15] E. B. Tielder and E. G. Miller, *J. Nutrition* **3,** 121 (1930).

[16] S. B. Wolbach, *J. Am. Med. Assoc.* **108,** 7 (1937).

[17] J. R. Beach, *Science* **58,** 542 (1923).

[18] J. S. Friedenwald, W. Buschke, and M. E. Morris, *J. Nutrition* **29,** 299 (1945).

[19] L. Nicholls, *Indian Med. Gaz.* **68,** 681 (1933).

[20] V. Ramalingaswami and H. M. Sinclair, *Proc. Nutrition Soc., Brit. J. Nutrition* **5,** xi (1951).

[21] S. B. Wolbach and O. A. Bessey, *Physiol. Revs.* **22,** 276 (1942).

take place in a metaplasing epithelium in vitamin A deficiency as "atrophy of the epithelium concerned, reparative proliferation of basal cells, and growth and differentiation of the new products into a stratified keratinising epithelium. Regardless of the original function and structure of the region, this replacement epithelium is identical in all locations, and comparable in all its layers, with epidermis." The authors go on to point out that epithelial cells with what they describe as "chemical" roles, such as the liver cells and the tubules of the kidneys, do not atrophy in vitamin A deficiency nor are they replaced by keratinized epithelium.

3. Alimentary Tract

Both Cramer[22] and de Ruyter[12] have found an atrophy of the mucous glands and disappearance of the goblet cells in the alimentary tract. Generally speaking, however, the changes which take place in the alimentary tract in vitamin A deficiency are not severe. Wolbach and Bessey[21] state that the mucosa of the stomach and intestines does not undergo keratinizing metaplasia, though they think there may be some mucosal atrophy in the intestine. Wolfe and Salter[11] found no changes in the stomach and intestines of the white mice on A-deficient diets, but they found excessive desquamation of stratified squamous cells in the esophagus (see Figs. 8 and 9).

4. Gums

Hyperplastic keratinized gums result from vitamin A deficiency. King[23] and May Mellanby (see E. Mellanby[24]) found that vitamin A was necessary for the proper formation of the epithelium of the gum margin in puppies. In deficient animals the gum epithelium was found to be hyperplastic and to become readily infected whereas the normal gum epithelium was thin and had an intrinsic resistance to infection.

5. Liver and Lymphoid Organs

Changes in the ducts of some of the glands associated with the alimentary tract have been mentioned. Although Wolbach and Howe point out that there is little change in the morphology of the liver cells in vitamin A deficiency, de Ruyter[12] has drawn attention to the fact that both glycogen and fat disappear from the liver cells. De Ruyter has also found changes in the Kupffer cells, i.e., an accumulation of fat, swelling, and some breakdown of the nuclei. He has also found that in the spleen the reticular cells at first swell and then atrophy, and accompanying this is failure on the part of these cells, and also of the Kupffer cells, to phagocytose trypan blue.

[22] W. Cramer, *Lancet* **1,** 1046 (1923).

[23] J. D. King, *Brit. Dental J.* **68,** 349 (1940).

[24] E. Mellanby, *Edinburgh Med. J.* **40,** 197 (1933).

In the Malpighian corpuscles of the spleen there was loss of lymphocytes, and this was found also in the Peyer's lymphoid patches of the ileum and in the small lymphoid nodules underlying the respiratory mucosa. This disappearance of lymphoid cells was reflected in a decreased number of lymphocytes in the blood. Another lymphoid structure that degenerates in vitamin A deficiency is the thymus, which undergoes almost complete atrophy (Wolfe and Salter.[11]) This finding was confrmed by de Ruyter,[12] who found, in addition, that the corpuscles of Hassall had become greatly enlarged and highly keratinized. Jungherr *et al.*[25] have found a focal necrosis and also a cirrhosis of the liver in young A-deficient calves, but they thought that this may have been due to an intercurrent infection which was exacerbated by the deficiency.

6. Eyes

One of the most characteristic and clinically obvious signs of vitamin A deficiency is xerophthalmia. This condition was originally claimed (Mori[26]) to be due to desiccation, as a result of failure of lachrymal secretion. Mori pointed out that in xerophthalmia the lachrymal glands are shrunken and the Harderian glands dilated or empty, and that in the Meibomian glands the epithelium of the ducts become hyperplastic and prevent the passage of the secretion of the gland. Wolbach and Howe,[9] however, stated that keratinization of the cornea and the conjunctiva could occur in the absence of any significant changes in the para-ocular glands, and Findlay[27] suggested that perhaps the keratinization was due to a loss of lysozyme in the tears.

Wolbach and Howe[9] have pointed out that the primary changes effected by vitamin A deficiency in the eye are epithelial in nature. Wolfe and Salter[11] described the eye changes in the deficient mouse and stated that there was first a severe granular exudate which, unlike that found in the rat, was not hemorrhagic. This was followed by the keratinization and desquamation of the epithelium of the palpebral conjunctiva and by a slight keratinization and desquamation of the cornea, together with an infiltration of leucocytes into this site. Although Bowman's membrane in the cornea sometimes persisted, in some animals it completely disappeared (see Figs. 1 and 2). In some cases there was a considerable increase in the thickness of the cornea, and in most cases a considerable increase in the vascularity of the substantia propria of the cornea which became, at the same time, very edematous. In the anterior chamber of the eye a number of polymorphs were found. Mori[26] found that in the rat these changes were followed by a softening of the cornea which became infected and ultimately perforated.

[25] E. L. Jungherr, C. F. Helmboldt, and H. D. Eaton, *J. Dairy Sci.* **33**, 666 (1950).
[26] S. Mori, *Bull. Johns Hopkins Hosp.* **33**, 357 (1922).
[27] G. M. Findlay, *J. Path. Bact.* **24**, 446 (1921).

Mellanby[28] expressed the opinion that these changes in the eye were secondary to a loss of the neurotrophic control which was normally exerted on the cornea by the ophthalmic division of the trigeminal nerve. He has pointed out that where xerophthalmia is present there is always degeneration of the myelin sheath of the trigeminal nerve and that in severe xerophthalmia this nerve shows Wallerian degeneration, which means in fact complete destruction of the nerve as a conducting unit. Ida Mann and A. Pirie[29] have made a detailed study of the changes in the eye of the rabbit in the course of vitamin A deficiency. Observation of the cornea with a slit lamp showed changes to be present before any detectable histological changes could be seen. They found that in deficient rabbits the surface layer of the corneal epithelium was always keratinized but that not much change could be seen in the basal layers of the epithelium or in the substantia propria. They found rather more dividing cells than normal in the basal layers of the cornified epithelium on the outside of the cornea. Keratinization of the lining epithelium of the eyelids was also found. This was likewise present in the conjunctiva, where there was also a great thickening of the conjunctival epithelium and an increase of pigment in the basal cell layers. They found no changes in the Harderian glands, but an atrophy of the tubules and keratinization of the ducts of the lachrymal glands was present.

In addition to changes in the cornea and conjunctiva in vitamin A-deficient rats there are also changes in the retina. These were first described by Tansley[30] and were later also described in rats by Johnson[31] and in the horse by Anderson and Hart.[32] The first histologically detectable change in the retina is the failure of the rod cells to stain differentially with certain dyes. Later the junction between the outer and inner limbs of rods becomes very thin, and the rod may break at this point. Subsequently the rods take on a ragged appearance and may be separated from the pigment epithelium and from themselves by large vacuolar spaces. Changes as extensive as these were not found in the rabbit by Mann and Pirie, but they did find a number of degenerated nuclei both among the rods (outer nuclear layer) and among the bipolar cells (inner nuclear layer). These areas of degeneration were quite small, however, and seemed to be confined to small areas in the central retina. In some cases there was also a breaking up of the rods in these areas. In one animal they found that the retina appeared healthy but one part (the section of the retina distal to the inner nuclear layer) had completely disappeared.

[28] E. Mellanby, *J. Path. Bact.* **37,** 391 (1934).
[29] I. Mann, and A. Pirie, *Am. J. Ophthalmol.* **29,** 801 (1946).
[30] K. Tansley, *Proc. Roy. Soc. (London)* **B114,** 79 (1933).
[31] M. L. Johnson, *J. Exptl. Zoöl.* **81,** 67 (1943).
[32] A. C. Anderson and G. H. Hart, *Am. J. Vet. Research* **4,** 307 (1943).

Warkany and Schaffenberger[33] have shown that a high percentage of young rats born from female rats on a vitamin A-deficient diet showed eye defects. In most of them the eyes were open instead of closed at birth, and they showed a red discoloration between the lids. Histological examination showed a fibrous retrolenticular membrane in place of the vitreous. In all specimens there were also colobomas, eversion, and abnormal structures of the retina with rudimentary development of the iris and of the ocular chambers in many specimens. Defects of the cornea and the conjunctival sac were also found. Johnson[31] found that in mild A deficiency there was a disappearance from the retina of the droplets associated with visual purple. Bergen[34] found that the nuclei of the retina of normal rats contained carotenoids but that these disappeared in A-deficient rats. In more severe deficiency Johnson found that the elements of the retina dissappeared in this order: rod segments, external limiting membrane, outer nuclear layer, epithelial pigment layer, outer molecular layer, and inner nuclear layer.

7. Urogenital System

Wolfe and Salter[11] showed that in the vitamin A-deficient mouse there were extensive changes in the bladder and pelvis of the kidney. They found that the normal epithelium was replaced by one which had become keratinized and that the bladder had become filled with desquamating cells (see Figs. 12 and 13). There was also desquamation of the epithelium in the pelvis, and the kidney itself showed evidence of leucocytic infiltration. There were, however, no morphological changes in the cells of the kidney tubules. It has been pointed out by Bicknell and Prescott[35] that the cells shed by the epithelium of the urinary tract form a nidus around which urinary salts may be deposited, thus building up calculi. A number of authors, e.g., Eddy and Dalldorf,[36] have recorded an increased number of urinary calculi in vitamin A-deficient animals.

Hale[37] in 1935 found ectopic kidneys in the offspring of sows on a vitamin A-deficient diet, and Wilson and Warkany[38, 39, 40] showed that considerable malformations of the urinary tract of the offspring could be obtained by placing the mother on this type of diet. In rats they found fused kidneys (see Fig. 15), stenosis of the uterus, homologous genital ducts, and failure

[33] J. Warkany and E. Schraffenberger, *Proc. Soc. Exptl. Biol. Med.*, **57**, 49 (1944).

[34] P. Bergen, *Proc. Roy. Soc. (London)* **B144,** 606 (1950).

[35] F. Bicknell and F. Prescott, The Vitamins in Medicine, 2nd. ed., Wm. Heinemann, London, 1946.

[36] W. H. Eddy and G. Dalldorf, The Avitaminoses 3rd ed., Williams and Wilkins, Baltimore, 1944.

[37] F. Hale, *Am. J. Ophthalmol.* **18,** 1087 (1935).

[38] J. G. Wilson and J. Warkany, *Proc. Soc. Exptl. Biol. Med.* **64,** 419 (1947).

[39] J. G. Wilson and J. Warkany, *Anat. Record. Suppl.* **97,** 60 (1947).

[40] J. G. Wilson and J. Warkany, *Am. J. Anat.* **83,** 357 (1948).

of male accessory glands to appear (e.g., seminal vesicles and bulbourethral glands), and in some females there was complete lack of vaginal development. Retarded development was exemplified by late partitioning of the cloaca, belated appearance of Müllerian ducts, and slow differentiation of the urogenital sinus. There were also positional abnormalities of the kidneys, ectopic ureteric openings, incomplete caudalward growth of Müllerian ducts, hypospadias, and failure of testicular descent. All the newborn animals and the fetuses older than 18 days which showed malformation of the genitourinary tracts showed a keratinizing metaplasia of the epithelia of these organs.

In male animals placed on an A-deficient diet the most striking changes in the genital system are atrophy of the testes and degenerative changes in the seminiferous tubules (see Mason[1]). These changes were first seen in the mouse by Yamasaki[41] and were described in more detail by Wolfe and Salter[11] in 1931. In the testis they found that there was a desquamation of the epithelium with the formation of large amounts of structureless material containing many clear spaces or vacuoles in the seminiferous tubules. In this material numerous degenerating giant cells were also present. Mason[42, 43] also found giant cells in degenerating vitamin A-deficient testes. Wolfe and Salter found that after a time only the primary germ cells and the sustentacular cells were left in the seminiferous tubules. They noticed no changes in the interstitial cells of Leydig. (Moore and Mark,[44] however, found absence of pigment in these cells in vitamin A deficiency.) The prostate and the seminal vesicles, however, were greatly distended, with desquamated cell masses. In some places they found a downward growth and invasion of the underlying connective tissue by the keratinizing epithelium which had formed in some of the genital ducts. It is of interest that no changes were found in the epithelium of the epididymus. Gross[45] Wolbach and Howe[9] Mason,[46] Evans,[47] Sampson and Korenchevsky,[48] and a number of other authors have described similar changes in the male rat genitalia. Wolbach and Howe[13] have also recorded these changes in the male guinea pig, and Guilbert and Hart[49] have done the same for cattle. Mason[1] has pointed out that in many of these experiments no vitamin E had been added to the experimental diet so that some of the results are suspect. Mason[46] however, was able to establish that testicular degeneration

[41] Y. Yamasaki, *Arch. path. Anat. Physiol.* (*Virchow's*) **245,** 513 (1925).
[42] K. E. Mason, *J. Exptl. Zoöl.* **45,** 159 (1926).
[43] K. E. Mason, *J. Exptl. Zoöl.* **55,** 101 (1930).
[44] R. A. Moore and J. Mark, *J. Exptl. Med.* **64,** 1 (1936).
[45] L. Gross, *J. Path. Bact.* **27,** 27 (1924).
[46] K. E. Mason, *Am. J. Anat.* **52,** 153 (1933).
[47] H. M. Evans, *Am. J. Physiol.* **99,** 477 (1932).
[48] M. M. Sampson and V. Korenchevsky, *J. Path. Bact.* **35,** 875 (1932).
[49] H. P. Guilbert and G. H. Hart, *J. Nutrition* **10,** 409 (1935).

due to vitamin A deficiency could be distinguished from that due to vitamin E deficiency and to inanition. In vitamin E deficiency there is a complete loss of all germ cells in the testes; in vitamin A deficiency there is first a loss of spermatozoa which resembles the state found in inanition. In vitamin A deficiency there is then a progressive loss of layers of germ cells until only a few residual germ cells are left together with the sustentacular or Sertoli cells. Mason[1] pointed out that "nothing is known regarding the role which vitamin A plays in the metabolic activities of the male germ cells." Mason and Wolfe[50] attempted to demonstrate that the gonads may play some part in the utilization of vitamin A by comparing the onset of deficiency symptoms in castrated and non-castrated rats, but they obtained negative results. Mayer and Goddard[51] found that in A-deficient rats gonadotrophic hormone affected the testes in the same way as in normal rats. Apparently, therefore, the testes of deficient rats have not lost their ability to respond to gonadotrophic stimulation, and the cause of testicular failure in A deficiency is therefore localized in the pituitary.

One of the differences between injury to the testes caused by vitamin E deficiency and that caused by vitamin A deficiency is that the former is irreversible whereas the latter is rapidly restored to normal by appropriate treatment with vitamin A.

In the female rat the ovaries were said to suffer pathological changes in vitamin A deficiency (Gross[45] and Guggisberg.[52] This has not been confirmed by later workers (Wolbach and Howe,[9] Evans,[53] and Mason.[46]) Similarly, changes described in the ovaries of mice by Yamasaki[41] have not been confirmed by Wolfe and Salter.[11] Wolbach and Howe,[13] however, have not denied the presence of such changes in the ovary of the guinea pig.[13] Mason[1] expresses surprise at the resistance of the ovary to the effects of vitamin A deficiency. It is of interest that Evans[53] has noticed abnormal ova in the oviducts of mated A deficient rats.

Although the ovary is not particularly affected in A deficiency, the uterus shows changes. As one might expect, these changes are epithelial in nature and take the form of patches of keratinization which extend peripherally. Mason[43] pointed out that this change may make it difficult for spermatozoa to ascend the uterus and so may be a significant factor in the sterility associated with A-deficient animals. Blackfan and Wolbach[54] have recorded keratinization of the uterine epithelium in children suffering from vitamin A deficiency, and keratinization of uterine epithelium has been recorded also by Wolbach and Howe[13] and by others.

[50] K. E. Mason and J. M. Wolfe, *J. Nutrition* **9**, 725 (1935).
[51] J. Mayer and J. M. Goddard, *Proc. Soc. Exptl. Biol. Med.* **76**, 149 (1951).
[52] H. Guggisberg, *Schweiz. med. Wochschr.* **55**, 114 (1925).
[53] H. M. Evans, *J. Biol. Chem.* **77**, 651 (1928).
[54] K. D. Blackfan and S. B. Wolbach, *J. Pediat.* **3**, 679 (1943).

Evans and Bishop[55] in 1922 first showed that vitamin A deficiency led to excessive cornification of the vagina of rats so that the usual estrus smear finally showed cornified cells at all stages of the sexual cycle, which persisted in spite of the deficiency, for such animals should be successfully mated with normal males (Mason and Ellison[56]). Mason and Ellison found that in the normal proestrus vaginal smear the majority of cells were nucleated epithelial cells but that A deficiency resulted in the cornification of these proestral cells as well as loss of mucous from the smear. This cornified condition of the smear lasted for several days. In the normal rat proestrus lasts about 12 hours and is followed by a completely cornified phase which may last about a day. This is followed by a postestrus stage which may last a day and which is signified by the appearance of leucocytes among the cornified cells. Finally the cornified cells become completely replaced by leucocytes and the animal passes into diestrus, in which it may remain for 1 to 1½ days before it passes into proestrus again. The complete cycle takes about 4 or 5 days. The cornified phase, as we have seen, may extend for several days in vitamin A deficiency. There are many fewer leucocytes, and eventually the smear shows continuous cornified cells.

In monkeys on an A-deficient diet the epithelium of the vagina also passes into a state of continuous cornification (Turner and Loew[57]). There is no certain evidence of the effect of vitamin A deficiency on the human vaginal mucosa, but it is of interest that vitamin A has proved of value in the treatment of senile vaginitis in women (Simpson and Mason[58]). The work of Schmidt[59] on the effect of vitamin A deficiency and hormones on repair of the vaginal epithelium of the rat suggests, according to Mason,[1] that the keratinization of the vaginal epithelium represents "a response to localized metabolic disturbances in epithelial cells." Mason also pointed out that the vaginal epithelium loses its ability to undergo mucification in pregnancy and that in fact it loses its responsiveness to the hormonal substances which induce mucification, a finding originally published by Aberle.[60] Mason suggested that this indicated a failure of the mechanism that produces glycoprotein, so presumably vitamin A plays some part in this process. In fact the evidence available suggests that in the absence of vitamin A there is a tendency for epithelial cells to produce scleroproteins instead of their normal glycoproteins.

Sherwood, Brend, and Roper[61] showed that continuous feeding of large

[55] H. M. Evans and K. S. Bishop, *Anat. Record* **23,** 17 (1922).

[56] K. E. Mason and E. T. Ellison, *J. Nutrition* **10,** 1 (1935).

[57] R. G. Turner and E. R. Loew, *J. Nutrition* **5,** 29 (1932).

[58] J. M. Simpson and K. E. Mason, *Am. J. Obstet. Gynecol.* **32,** 125 (1936).

[59] W. Schmidt, *Beitr. path. Anat. u. allgem. Path.* **96,** 129 (1935).

[60] S. D. Aberle, *Am. J. Physiol.* **106,** 267 (1933).

[61] T. C. Sherwood, M. E. Brend, and E. A. Roper, *J. Nutrition,* **11,** 593 (1936).

amounts of carotene to rats caused suppression of estrus and the appearance of a prolonged dioestrus vaginal smear. Such a condition is also produced by castration (i.e., lack of follicular hormone). The reverse effect, excessive cornification of the vagina, can be produced by the administration of excessive estrogenic hormone and by lack of vitamin A. These facts suggest the possibility of some association between the female sex hormone and vitamin A.

Early workers have shown that vitamin A deficiency causes a considerable interference in reproduction. Many of their findings, however, are subject to the criticism (see Mason,[62]) that the diet was not adequately controlled.

Although it is sometimes possible to breed female animals which are suffering, but not too severely, from vitamin A deficiency, once the deficiency passes a certain stage, death of the fetuses occurs, primarily as a result of the effect of the deficiency on the maternal cells. The decidua become necrotic and often infected, and the fetuses die because of their destroyed blood supply. Mason[62] investigated this problem and pointed out that the effect of vitamin A deficiency is fundamentally different from that of vitamin E deficiency, since the effect of the latter is shown directly in the embryo, the death of which is not due to prior degeneration of the placenta. Mason found that, even in embryos which survived, the gestation period was prolonged and that there was a decreased vitality in the fetuses and decreased tone in the uterine and abdominal musculature. These are all conditions which make for difficult and prolonged labor, accentuated by the cornification of the vagina which makes it difficult for the fetus to be expelled.

8. Embryonic Tissues

Some abnormalities in the young of vitamin A-deficient mothers have already been mentioned. Hale[37, 63] found that the progeny of vitamin A-deficient sows suffered from harelips, cleft palates, accessory earlike structures, malformed hind legs, and subcutaneous cysts. In addition the kidneys and the sex glands had remained in their embryonic position.

Subsequently, Warkany and his co-workers described a series of fetal abnormalities in the young of vitamin A-deficient rats. Wilson and Barch,[64] for instance, have pointed out that malformations which involve the eye, the cardiovascular system, and the genito-urinary tract are commonly found in such progeny, and Wilson and Warkany[65] have described keratinizing metaplasia of the genital tract in the 18-day fetal rats from A-deficient

[62] K. E. Mason, *Am. J. Anat.* **57**, 303 (1935).
[63] F. Hale, *J. Heredity* **24**, 105 (1933).
[64] J. G. Wilson and S. Barch, *Proc. Soc. Exptl. Biol. Med.* **72**, 687 (1950).
[65] J. G. Wilson and J. Warkany, *Proc. Soc. Exptl. Biol. Med.* **64**, 419 (1947).

mothers. Wilson and Warkany[66, 67] have also described cardiovascular anomalies in fetal and newborn rats from A-deficient mothers. In one of the commonest of these abnormalities the abnormal pattern of arches was retained but the right subclavian artery "arose distally from the descending aorta and coursed dorsal to the oesophagus to reach the right shoulder." In some animals the 4th and/or 6th aortic arches were retained. Supernumerary arches were also found. In some animals the interventricular septum of the heart failed to close, leaving an interventricular foramen (see Fig. 14). There was also in some cases a general retardation of myocardial development, and the wall of the heart developed a highly trabeculated, spongy appearance. The authors concluded that the anomalies which they observed resulted from interference with the normal processes "occurring on or subsequent to the 12th day of gestation." In their second paper on this subject[68] Wilson and Warkany pointed out that many of the cardiovascular abnormalities they have recorded in A-deficient rats are very similar to those found in infants.

9. Endocrines

There is little evidence that vitamin A deficiency interferes with hormone *production* or with the normal structure of the endocrines. Giedosz[69] claims that there is an increased number of eosinophilic and basophilic cells in the pituitary in such a deficiency and an increase in the zona glomerulosa of the adrenal. On the other hand Mitzkewitsch[70] and Frank[71] have claimed that there are no cytological changes in the pituitaries of A-deficient rats. Jungherr *et al.*[25] found a decrease in the number of chromophilic cells of the pituitary in young A-deficient calves. They found also that the thyroid was hyperplastic.

10. Nervous System

There is ample evidence that vitamin A deficiency results in pathological changes in the nervous system. The first paper on this subject appears to be that of Mellanby,[72] who described degenerative changes in the spinal cords of A-deficient puppies. Elaborating on this subject in 1933, Mellanby[24] said that the spinal cords of such puppies showed demyelination and disappearance of fibers in certain tracts. Somewhat similar changes were caused by ergot and also by neurotoxic substances present in pulses, but the changes induced by these substances could be prevented by vitamin A. He

[66] J. G. Wilson and J. Warkany, *Am. J. Anat.* **85,** 355 (1949).
[67] J. G. Wilson and J. Warkany, *Am. J. Anat.* **85,** 113 (1949).
[68] J. G. Wilson and J. Warkany, *Pediatrics* p. 708 (April, 1950).
[69] B. Giedosz, *Compt. rend. soc. biol.* **120,** 557 (1935).
[70] M. E. Mitskewitsch, *Arch. exptl. Path. Pharmakol.* **174,** 339 (1934).
[71] M. Frank, *Arch. Kinderheilk.* **110,** 91 (1937).
[72] E. Mellanby, *Proc. Physiol. Soc., J. Physiol. (London)* **61,** xxiv (1925).

also found degeneration of peripheral nerves, e.g., the afferent spinal nerves, the optic nerves, and the 8th trigeminal nerve. He thought that loss of the sense of balance in vitamin A-deficient animals may have been due to degeneration of the vestibular division of the 8th nerve. He believed that the xerophthalmia of this deficiency may have been due primarily to degeneration of the afferent fibers of the trigeminal nerve. However, Rao[73] has shown that administration of vitamin A will cure xerophthalmia while the trigeminal nerve is still degenerated. In another paper Mellanby[28] described degenerative changes in the Gasserian ganglion and associated nerve fibers in vitamin A deficiency.

In addition to Mellanby's results with dogs, Aberle[74] and Setterfield and Sutton[75] have described degeneration of medullary sheaths of sensory tracts at the periphery of the cord. In some cases this degeneration was found in the posterior columns, and in others the posterior roots were involved. These changes were found to be associated with marked weakness or paralysis of the extremities. These authors were able to obtain a clinical cure of this condition with cod-liver oil while it was still possible to demonstrate lesions in the nervous system. Aberle suggests that this result may be due to the fact that the regenerated axis cylinder functions before remyelination of the fiber is complete. He points out that stimulation of the motor cortex in embryonic animals may elicit motor responses before the corticospinal tracts are myelinated. Modern research on nerve regeneration supports Aberle's explanation.

Irving and Richards[76] have also investigated A-deficiency in the rat and found neurological lesions after 21 days of deficiency (before, in fact, there is any obvious slackening of the growth rate). They found that these lesions appeared characteristically in certain tracts of the medulla of the brain. The region where they occurred was that which would be occupied by the fillet in higher sections, and below the pyramidal decussation. The fibers from this region become anterolateral columns. They point out that in the rat the pyramidal tracts decuss completely.

Nicholls,[77] who found nerve and cord degeneration in A-deficient humans, and Rao,[73] who found nervous degeneration in A-deficient rabbits, were among other workers who have studied this aspect of vitamin A deficiency.

One of the criticisms of the work on the effect of vitamin A deficiency on the nervous system is that the Marchi method used for demonstrating degenerated myelin is unreliable, but Setterfield and Sutton[75] used polarized

[73] M. V. R. Rao, *Indian J. Med. Research* **27,** 731 (1940).
[74] S. B. D. Aberle, *J. Nutrition* **7,** 445 (1934).
[75] H. E. Setterfield and T. S. Sutton, *J. Nutrition* **9,** 645 (1935).
[76] J. T. Irving and M. B. Richards, *Proc. Physiol. Soc., J. Physiol. (London)* **89,** 112 (1940).
[77] L. Nicholls, *Indian Med. Gaz.* **70,** 1 (1935).

light to examine the nervous tissue from such animals and were able to confirm degeneration in the femoral and sciatic nerves. Extraction of the lipids from the degenerated myelin of nervous tissue of A-deficient animals has shown that there was a decrease in the saturated and conjugated lipids.

11. Bones

A consideration of the effects of vitamin A deficiency upon bone growth may be appropriately considered at this point, since many of the degenerative changes recorded in the nervous system in this condition have been attributed by Wolbach[78, 79] to pressure on the nerves by irregular bone growth.

Vitamin A deficiency has a direct effect on the growth of bone. In the long bones all the "epithelial-cartilage sequences" cease. According to Wolbach, "cells fail to divide and cells which have become almost mature fail to grow any further. Cells which have become vesicular, however, continue the various changes they normally pass through and then they disappear. Vascular penetration can then take place at this point. The trabeculae which were formed and calcified before the deficiency became established become resorbed mainly due to the activity of osteoclasts. The matrix of the cartilage at the epiphyseal line now becomes calcified to 20 or 30 μ back from the line, finally a bony plate forms across the line (see Figs. 10 and 11)." The same result is obtained with simple pyridoxine deficiency, according to Wolbach. Bourne and Harris[80] have seen a similar effect in rats on a nicotinic acid-deficient diet. It is of interest that in vitamin A deficiency endochondral bone growth ceases before soft tissue growth. According to Wolbach, "It is this disproportionate growth of the skeleton and soft tissues which is responsible for the early disastrous consequences in the young of rats, dogs, guinea pigs and chicks, because the cranial cavity and the spinal canal become too small for the central nervous system which continues to grow at approximately the normal rate until general inanition supervenes. The result of the disproportionate growth of bone and central nervous system is over-crowding of spinal canal and cranial cavity, paralysis resulting from mechanical pressure upon the brain, spinal cord and nerve roots. The consequences are exhibited by multiple herniations of the cerebrum and cerebellum into arachnoidal villi, dislocation of the brain as a whole toward the *foramen magnum* and buckling and herniation of nerve roots into intervertebral foramina and into the bodies of vertebrae. The latter is the result of atrophy of bone, caused by the pressure of the buckled nerve root trunks" (see Figs. 4 and 5).

[78] S. B. Wolbach and O. A. Bessey, *Arch. Path.* **32,** 689 (1941).
[79] S. B. Wolbach, *J. Bone Joint Surg.* **29,** 171 (1947).
[80] G. H. Bourne and L. J. Harris, *Proc. Nutrition Soc., Brit. J. Nutrition* **4,** 60 (1951).

Not only does endochondral bone growth cease but remodeling of existing bone comes to a stop so that the long bones become shorter as well as thicker and more imperfectly shaped.

Wolbach and Bessey[78] appear to have been the first to claim that disordered bone growth in vitamin A deficiency was responsible for nervous tissue degeneration, although Mellanby[81] and Loch[82] found an increase in the periostal bone of the labyrinthine capsule and the formation of exostoses in the internal auditory meatus. Mellanby[83] described degeneration of the VIIIth cranial nerve (both divisions, but especially the cochlear division), the sensory fibers of the Vth (all divisions), the optic nerve, and the olfactory nerve. The IIIrd, IVth, third branch of the Vth, VIth, VIIIth, IXth, Xth, and XIIth tend to escape degenerative changes. According to Mellanby, the factors concerned in the production of these changes were as follows: first, direct pressure of the overgrown bones on nerve cells and fibers; second, the restriction of blood supply to nerve cells and fibers by overgrowth of bone; third, increased intracranial pressure due to the same cause; and fourth, a possible direct effect of vitamin A deficiency on the nerve cells themselves. Mellanby has also pointed out that motor nerve axons seem to survive pressures which are capable of destroying the axons of sensory nerves.

Hess[84] and his co-workers, as long ago as 1921, had reported cessation of bone growth in A-deficient rats, a fact which was later confirmed by Wolbach and Howe.[9] Perlman and Willard[85] have also observed bone changes in vitamin A-deficient animals.

Hertz[86] studied the effect of vitamin A deficiency on the healing of experimental fractures in the rat. He found a delay in the absorption of the fracture hematoma. There was no effect on the inflammatory reactions on the part of the cambial layer of the periosteum, the endosteum, and the marrow, all of which appeared to pass normally through their various phases. The only direct effect of the deficiency appeared to be a poor development of cartilage in the fracture callus.

12. Hypervitaminosis A

Excessive administration of vitamin A can cause tissue damage (see Collazo *et al.*[87] Wolbach and Bessey[21] reported fractures of bones in young rats in 6 to 11 days by the daily administration of excessive doses of vitamin

[81] E. Mellanby, *J. Physiol.* (*London*) **94,** 380 (1938).
[82] W. E. Loch, *Monatsschr. Ohrenheilk. Laryngo-Rhinol.* **73,** 542 (1938).
[83] E. Mellanby, *J. Physiol.* (*London*) **101,** 408, (1943).
[84] A. F. Hess, *J. Biol. Chem.* **47,** 395 (1921).
[85] H. B. Perlman and J. Willard, *Ann. Otol. Rhinol. & Laryngol.* **50,** 349 (1941).
[86] J. Hertz, Studies in the Healing of Fractures, Oxford University Press, 1936.
[87] J. A. Collazo, P. Rubino, and B. Varela Fuentes, *Deut. med. Wochschr.* **55,** 1794 (1929).

A. Further details are given by Wolbach.[79] Wolbach's conclusions about the relation of vitamin A to bone growth is that it is determined by epiphyseal-cartilage activities for which vitamin A is essential: ". . . We may deduce that in the maturation of cartilage cells there is produced an inductor factor which is responsible for the maintenance or remodelling of bone during the growth period. Because the accelerated remodelling process caused by excess of vitamin adheres precisely to the normal growth pattern, bone which disappears during growth may be referred to as competent bone" (see Figs. 3 and 7).

Wolbach and Maddock[88] used the changes produced by hypervitaminosis A to assess the vitamin A potency of synthetic vitamin A. Those tested included vitamin A acetate, which was as active as the natural vitamin; vitamin A methyl ether, which was less active; and vitamin A phenyl ether, which did not appear to have any activity. Rodahl[89] has described soft tissue damage in vitamin A excess. He found that there was a general hyperemia and that erythrocytes could be found scattered outside capillaries in various organs and particularly in the space surrounding the glomerulus in Bowman's capsules of the kidney. Fatty material was found to accumulate in the liver, and there was an increase of lipoid in the adrenal cortex.

III. Vitamin B_1 Deficiency

1. General Effects

Despite the biochemical importance of vitamin B_1, which appears to be concerned with most of the oxidative decarboxylations of the body, and the fact that in vitamin B_1 deficiency pyruvic acid accumulates in the tissues, forming what Peters calls a "biochemical lesion," there is surprisingly little histological change in deficient animals. Nevertheless, changes in some tissues have been described.

2. Heart

One of the characteristic clinical symptoms of vitamin B_1 deficiency is a bradycardia, thought by some to be due to overactivity of the vagus, and first described by Drury, Harris, and Maudsley.[90] However, pathological changes occur in the heart itself, and it may be that the slowness of the heart's beat is due partly to this. Not only is the heart rate slowed but changes are also found in the electrocardiographic records. These have been described in five species of mammal, including cats, dogs, monkeys, and pigs. Wintrobe and his colleagues[91] have listed the electrocardiographic

[88] S. B. Wolbach and C. L. Maddock, *Proc. Soc. Exptl. Biol. Med.* **77,** 825 (1951).

[89] K. Rodahl, *J. Nutrition* **41,** 399 (1950).

[90] A. N. Drury, L. J. Harris, and C. Maudsley, *Biochem. J.* **24,** 1632 (1930).

[91] M. M. Wintrobe, R. Alcayaga, S. Humphreys, and R. H. Follis, *Bull Johns Hopkins Hosp.* **73,** 169 (1943).

changes in the swine heart as consisting of prolonged P-R intervals, abnormalities in the P wave, increase of T_4, nodal and ventricular premature beats, A. V. dissociation, complete block, and auricular fibrillation. All these indicate damage to the myocardium.

Such changes have been described in rats (Ashburn and Lowry[92]), in sows (Follis *et al.*[93]), in foxes (Evans *et al.*[94]), and in dogs (Swank *et al.*[95]). Actual destruction of the myocardial fibers occurs. First the fibers lose their cross striations, then there is vacuolation, then hyalinization of the myofibrils, and finally necrosis. The necrotic changes coincide with the migration into the myocardium of polymorphs and mononuclears (probably macrophages (see Fig. 16). The necroses, according to Follis,[4] may occur in foci of varying size or may exist diffusely over large areas of myocardium, and they may occur in either auricular or ventricular muscle. If the animal recovers from the deficiency, these necrotic patches heal by scarring (see Fig. 17). The areas of connective tissue fibers represented by these scars may be quite extensive, more so in animals with chronic vitamin B_1 deficiency. It is surprising that no comparable lesions have been found in skeletal muscle, but Follis[96] was able to demonstrate necrotic patches in this tissue when potassium deficiency was added to the thiamine deficiency.

Follis[4] points out that despite the recorded myocardial damage it does not seem sufficient to account for the bradycardia and the electrocardiographic changes found in deficient animals. He quotes the experiments of Lu[97] and Haynes and Weiss,[98] who administered large quantities of pyruvic acid and related substances to both normal and vitamin B_1-deficient rats and found that they exercised little effect on the action of the heart. From these experiments, therefore, it seemed unlikely that accumulation of these metabolites is responsible for the bradycardia and the electrocardiographic changes. It seems more likely that difficulty in carrying carbohydrate oxidation beyond the pyruvate level may cause failure of the heart muscle to respond properly. Perhaps there is some interference with the conducting system of the heart, or, as mentioned before, there may be physical damage to the muscle.

3. Nervous System

This leads us to the effect of vitamin B_1 deficiency on nervous tissue. Eijkman[99] (1906), in describing the pathology of the beriberi-like disease

[92] L. L. Ashburn and J. V. Lowry, *Arch. Path.* **37,** 169 (1943).
[93] R. H. Follis, M. H. Miller, M. M. Wintrobe, and H. J. Stein, *Am. J. Path.* **19,** 341 (1943).
[94] C. A. Evans, W. E. Carlson, and R. G. Green, *Am. J. Path.* **18,** 79 (1942).
[95] R. L. Swank, R. R. Porter, and A. Yeomans, *Am. Heart J.* **22,** 154 (1941).
[96] R. H. Follis, *Bull. Johns Hopkins Hosp.* **71,** 235 (1942).
[97] G. D. Lu, *Biochem. J.* **33,** 774 (1939).
[98] F. W. Haynes and S. Weiss, *Am. Heart J.* **20,** 24 (1940).
[99] C. Eijkman, *Arch. Hyg.* **58,** 150 (1906).

which he had been able to induce in fowls, stressed that the damage to the peripheral nerves affected both sensory and motor portions and produced a picture similar to the non-inflammatory atrophic degeneration observed after a nerve is cut. In other words, B_1 deficiency produced simple Wallerian degeneration. Eijkman also found degenerative changes in the spinal cord and in the spinal nerve roots. These changes in the nervous system confirmed his diagnosis of the disease as a polyneuritis. However, as Follis[4] and other authors from McCollum[100] on have pointed out, Eijkman's diet was deficient in a number of nutrients apart from vitamin B_1, and one cannot be sure, therefore, that the changes he described are attributable solely to the deficiency of this vitamin.

Wolbach and Bessey[21] have pointed out that degeneration of peripheral nerves occurs more often in chronic vitamin B_1 deficiency than in the acute state. However, Prickett,[101] Prickett and co-workers,[102] Davison and Stone,[103] and Engel and Phillips,[104] found no changes in the peripheral nerves of vitamin B_1 deficiency in rats, and Berry and his co-workers[105] could find no signs of degeneration in the peripheral nerves of cats which had been on a vitamin B_1-deficient diet for as long as 116 days. Nor were any differences to be found in the ability of cut or injured nerves of these animals to regenerate. Follis *et al.*[93] and Wintrobe *et al.*[106] also could find no signs of nerve degeneration in B_1-deficient swine. Coupled with these findings we might record the work of Burt,[107, 108] who found that in tissue cultures of spinal ganglia the addition of vitamin B_1 to B_1-deficient or to normal plasma caused no improvement in axon growth. He also found that nerve growth was not suppressed by the presence of the various intermediary metabolites which accumulate in vitamin B_1-deficient animals.

In the face of this evidence it is difficult to accept the thesis that one of the characteristics of vitamin B_1 deficiency is structural damage to the nervous system. It is this conclusion, no doubt, which led Meiklejohn[109] to suggest that vitamin B_1 did not deserve to be called the antineuritic vitamin.

However, when we consider birds we find that Swank and Prados[110] have produced evidence that in the pigeon neuritic changes inevitably accompany

[100] E. V. McCollum and M. Davis, *J. Biol. Chem.* **23,** 181 (1915).
[101] C. O. Prickett, *Am. J. Physiol.* **107,** 459 (1934).
[102] C. O. Prickett, W. D. Salmon, and G. A. Schrader, *Am. J. Path.* **15,** 251 (1939).
[103] C. Davison and L. Stone, *Arch. Path.* **23,** 207 (1937).
[104] R. W. Engel and P. H. Phillips, *J. Nutrition* **16,** 585 (1938).
[105] C. Berry, C. Neumann, and J. C. Hinsey, *J. Neurophysiol.* **8,** 315 (1945).
[106] M. M. Wintrobe, R. H. Follis, S. Humphreys, H. Stein, and M. Lauritsen, *J. Nutrition* **28,** 283 (1944).
[107] A. S. Burt, *Proc. Soc. Exptl. Biol. Med.* **54,** 191 (1943).
[108] A. S. Burt, *J. Cell. Comp. Physiol.* **21,** 145 (1943).
[109] A. P. Meiklejohn, *New Engl. J. Med.* **223,** 265 (1940).
[110] R. L. Swank and M. Prados, *Arch. Neurol. Psychiat.* **47,** 97 (1942).

vitamin B deficiency. Nevertheless, even their experiments are not completely free from criticism. The earliest changes which they could see in the peripheral nerves were, first, swelling, and then increased basophility of the neurofibrils. These changes they found could occur in the peripheral portion of many axons originating from the vestibular nucleus, although no change could be seen in the proximal parts of the axons and of the cell bodies themselves. Swank[111] believes that changes in the myelin sheath are secondary to those which occur in the axis cylinder, but Zimmerman[112] holds that the reverse is true. It is of interest that Shimizu[113] has found enzymic changes in nervous tissue of B_1-deficient pigeons by histochemical methods. He claimed that first the ventral horn cells of the spinal cord shrunk, and then they showed an increase of alkaline phosphatase activity. As the deficiency reached its maximum there was an increase of the enzyme in the nerve cells of the fourth ventricle, particularly those in the *formatio reticularis*, which were only slightly active in normal birds. There was also a general increase of the activity of the enzyme in the cerebellum and particularly in the shrunken Purkinje cells. Ganglion cells also showed an increase in activity, but in the cells of the ganglionic capsules this activity was reduced. Shimizu also found a decrease of acid phosphatase in axis cylinders in nerves of B_1-deficient animals.

Despite the negative neurological findings in mammals by a number of authors, there is still a considerable evidence from other authors that peripheral nerve degeneration accompanies B_1 deficiency. Leblond and Chaulin-Servinière,[114] found characteristic Wallerian degeneration and eventually damage which was irreparable, in the nerves of monkeys suffering from spontaneous beriberi. They used polarized light and Sudan III staining in frozen sections of nerves to demonstrate degeneration, and their results can therefore be regarded as perhaps being more significant than some others. Street and co-workers[115] found that in chronically deficient dogs there was degeneration of the myelin in peripheral nerves, and also in the posterior columns of the spinal cords, and glial scars were found in the dorso-spino-thalamic tracts.

The evidence for damage to the brain in B_1 deficiency seems less equivocal, for Prados and Swank[116] have described foci of hemorrhage in the vestibular nuclei, Deiter's nuclei, Bechterew's nuclei, and the *nucleus soli-*

[111] R. L. Swank, *J. Exptl. Med.* **71,** 683 (1940).
[112] M. Zimmerman, The Rôle of Nutritional Deficiency in Nervous and Mental Diseases, Williams and Wilkins, Baltimore, 1942.
[113] N. Shimizu, Y. Handa, H. Jiro, and T. Kumamato, *Proc. Soc. Exptl. Biol. Med.* **75,** 696 (1950).
[114] C. P. Leblond and J. Chaulin-Servinière, *Am. J. Med. Sci.* **203** 100 (1942).
[115] H. P. Street, G. R. Cowgill, and H. M. Zimmerman, *Yale J. Biol. and Med.* **13,** 293 (1941).
[116] M. Prados and R. L. Swank, *Arch. Neurol Psychiat.* **47,** 626 (1942).

tarius in B_1-deficient rats. Other changes recorded in these nuclei were chromatolysis, or dissolution of the Nissl substance, of the nerve cells. In kittens suffering from acute B_1-deficiency not only chromatolysis but necrosis of neurones have been observed. Oligodendrocytes were seen to be swollen, and the vestibular nuclei were seen to contain small hemorrhages.

Hemorrhagic lesions in the brain (pons, medulla, and cerebellum) of rats and pigeons have also been described by Prickett[101] and by Church.[117] Zimmerman[118] and Alexander[119] believe that the lesions of Wernicke's disease in the human central nervous system are identical with those of vitamin B_1 deficiency.

Follis[4] has summarized the main pathological symptoms of vitamin B_1 deficiency in man. Mild scarring, hydropic degeneration, and fatty infiltration are the only microscopic changes which have been seen in the beriberi heart. Peripheral nerves have been found to be degenerated, and loss of myelin from the nerve roots and degenerative changes in the tracts of the spinal cord have been recorded. The similarity between changes in the nervous system due to vitamin B_1 deficiency and those due to Wernicke's disease has been mentioned. Also, it has been shown that in Chastek's paralysis of foxes, which is curable with vitamin B_1, the brain lesions are identical with those of Wernicke's disease (Evans *et al.*[94]).

4. Alimentary Tract

Other changes in vitamin B_1 deficiency are found in the gastrointestinal tract. Lack of muscular tone has been often observed, and McCarrison[120] has recorded degeneration of Auerbach's nerve plexus situated between the longitudinal and circular musculature of the gut. McCarrison also observed that ulcers occurred in the alimentary tracts of his deficient animals. Dalldorf and Kellogg[121] found that animals on a partial deficiency of vitamin B_1 for long periods of time also developed ulcers. Thatcher, Sure, and Lee[122] were impressed with the constancy with which gastric ulcers appeared in animals on a vitamin B_1-deficient diet. This has been confirmed by a number of workers, including Schiodt,[123] Drummond and co-workers,[124] and Hawes and Vivier.[125] Drummond and his co-workers found not only ulcers but also erosions in their thousands of experimental animals (also

[117] C. F. Church, *Am. J. Physiol.* **111,** 660 (1935).
[118] H. M. Zimmerman, *Yale J. Biol. and Med.* **12,** 23 (1939).
[119] L. Alexander, *Am. J. Path.* **16,** 61 (1940).
[120] R. McCarrison, *Brit. Med. J.* **1,** 966 (1931).
[121] G. Dalldorf and M. Kellogg, *J. Exptl. Med.* **56,** 391 (1932).
[122] H. S. Thatcher, B. Sure, and J. Lee, *Bull. Agr. Expt. Sta. Univ. Arkansas* **356,** (1938).
[123] E. Schiodt, *Acta Med. Scand.* **84,** 456 (1935).
[124] J. C. Drummond, *J. Hyg.* **38,** 356 (1938).
[125] E L. Hawes and A. J. Vivier, *Am. J. Path.* **12,** 689 (1936).

see Dalldorf and Kellogg[121]). In monkeys with spontaneous beriberi Leblond and Chaulin-Servinière[114] found atony of the gastrointestinal musculature with diffuse redness and echymotic points, which appeared in some cases as well-defined ulcers, especially in the pyloric region of the stomach. There was enlargement of the mesenteric lymph nodes and fatty degeneration of the liver.

5. Endocrines

Structural changes in the thyroid gland have been described in vitamin B_1 deficiency by Hardhausen and Schultze.[126] Associated with it was an increase of colloid. Hyperplasia of the gland has also been recorded (Higgins[127]). If the deficiency is prolonged, the thyroid appears to atrophy. Other workers (Carpenter and Sharpless[128] and Harris and Remington[129]) have not been able to support this work, however.

Ogata[130] found changes in a number of the endocrine glands in vitamin B_1 deficiency. He found atrophy of the thymus (confirmed by Leblond and Chaulin-Servinière[114]), the pituitary, and the spleen, and hypertrophy of the islets of Langerhans (confirmed by Wolbach[131] and Ueno[132]). Marburg[133] found an increase in the basophilic cells of the pituitaries of Chinese coolies dying of beriberi. Ogata also claimed that the adrenal glands become hypertrophic and that lesions were present in the associated sympathetic ganglia. In the early stages of the deficiency there was considerably mitotic activity on the part of the cortical cells of the adrenal.

Deane and Shaw[134] found that 3 weeks vitamin B_1 deficiency in the rat produced no change in the zona glomerulosa, but it caused a loss of lipoid from the zona fasciculata and the fat-free transitional zone disappeared. After 4 to 4½ weeks the glomerulosa and its lipoid content were still unaffected, but nearly all microscopically detectable lipoid had gone from the zona fasciculata. The mitochondria in this zone were swollen and irregular in shape and stained badly. Mitochondria were also swollen, vesiculated, and irregular in the liver cells, which were smaller than normal. Thiamine deficiency thus rapidly exhausts the adrenal cortex. It does this much more rapidly than simple inanition. Goodsell[135, 136] has found changes also in the adrenal gland of the dog in vitamin B_1 deficiency.

126 G. Hardhausen and E. Schultze, *Arch. Exptl. Path. Pharmakol.* **191,** 570 (1939).
127 G. M. Higgins, *J. Nutrition* **16,** 347 (1943).
128 M. D. Carpenter and G. R. Sharpless, *J. Nutrition* **13,** 234 (1937).
129 P. L. Harris and R. E. Remington, *J. Nutrition* **17,** 31 (1939).
130 T. S. Ogata, *Verhandl. Japan. path. Ges.* **10,** 1 (1920).
131 S. B. Wolbach, Growth and Development of the Child, Part 3: Nutrition, The Century Co., New York, 1932.
132 J. Ueno, *Japan. J. Obstet. Gynecol.* **17,** 267 (1934).
133 O. Marburg, *Wien. Arch. inn. Med.* **29,** 1 (1936).
134 H. W. Deane and J. H. Shaw, *J. Nutrition* **34,** 1 (1947).
135 E. Goodsell, *Am. J. Physiol.* **134,** 119 (1941).
136 E. Goodsell, *Am. J. Physiol.* **134,** 125 (1941)

6. Reproductive Organs

In B_1 deficiency or in deficiency of any other vitamins of the B complex there is complete cessation of spermatogenesis in the testis, together with a substantial reduction of the size of the seminiferous tubules. According to Mason[1] the lumena of the tubules become filled with masses of primitive generative and Sertoli (sustentacular) cells. The spermatozoa and the spermatids (both more highly differentiated cells) are lost much earlier. The capsule and the capsular trabeculae are thickened. In later stages of the deficiency the wall of the tubules seems to be composed only of an incomplete layer of cells (spermatocytes and Sertoli cells). In others the lumen is completely obliterated by a more or less solid core of degenerated material. The interstitial (or Leydig) cells may be hypertrophied and show increased lipoid content (McCarrison[137] and Findlay[138]). Other workers claimed a hyperplasia (Korenchevsky[139]), whereas Porter[140] claimed that they are atrophied. Haulbert[141] claimed that they showed increased pigmentation. Marrian and Parkes[142] were not able to find any changes in these cells.

Much of the earlier work on the effect of vitamin B_1 deficiency on the male generative system has been criticized by Mason[1] because he believes that many of the results obtained were due to an associated vitamin E deficiency. He points out that Marburg[133] had noted no significant changes in the testes of Chinese coolies dying of beriberi and that Mattill[143] and Evans[144] found motile sperms in the epididymus and normal testes in rats suffering from a severe deficiency of vitamin B_1.

Moore and Samuels[145] found atrophy of the prostate and of the seminal vesicles in B_1-deficient rats, although the testes in these animals appeared normal. The same changes were found in animals suffering from inanition.

Mason,[1] Marrian and Parkes,[142] Parkes,[146] Ueno,[132] and Shin[147] made studies in vitamin B: deficiency in the female mouse. They found that in young female animals placed on a deficient diet, growth stopped, the opening of the vaginal orifice was retarded, sexual maturity was delayed and the animal passed into a state of anestrus. Various workers (see Mason[1]) have shown that monkeys, rabbits, mice, and rats on a B_1-deficient diet show atrophy of the ovaries and the uterus. Marrian and Parkes[142] found that

137 R. McCarrison, *Indian J. Med. Research*, **6**, 275 (1919).
138 G. M. Findlay, *J. Path. Bact.* **24**, 175 (1921).
139 V. Korenchevsky, *J. Path. Bact.* **24**, 175 (1921).
140 P. Porter, *Compt. rend.* **170**, 755 (1920).
141 G. Haulbert, *Paris méd.* **33**, 473 (1919).
142 G. F. Marrian and A. S. Parkes, *J. Roy. Microscop. Soc.* **48**, 257 (1928).
143 H. A. Mattill, *Am. J. Physiol.* **79**, 305 (1927).
144 H. M. Evans, *J. Nutrition* **1**, 1 (1928).
145 C. R. Moore and L. T. Samuels, *Am. J. Physiol.* **96**, 278 (1931).
146 A. S. Parkes, *Quart. J. Exptl. Physiol.* **18**, 397 (1928).
147 H. Shin, *J. Chosen Med. Assoc.* **23**, 77 (1933).

atresia of follicles of all sizes occurred in the ovary, there were no large follicles at all, there was vacuolar degeneration of the ova, and no recent corpora lutea. In the uterus the mucosal cells were small, the endometrium was reduced in thickness, and the cells of the uterine glands were inactive.

These degenerations may be due to failure in production of gonadotrophic hormones from the pituitary and may perhaps be correlated with the changes in the pituitary. A few such changes are recorded in the literature.

Ueno[132] attributed the genital atrophy of the female in B_1 deficiency to a primary degeneration of the sympathetic nerves passing to these organs, but Mason was sceptical of this interpretation.

In pregnant animals vitamin B_1 deficiency may induce resorption of the fetus. There may be prolongation of pregnancy, and there may be difficulty in bearing the young. Miscarriages are frequent. These changes are similar to those caused by simple inanition.

Vitamin B_1 deficiency appears to interfere with lactation, but there is some doubt as to whether this is specific.

7. Bone

Rats placed on a vitamin B_1-deficient diet cease growing after about 3 weeks (Hertz[86]). In such animals the epiphyses showed a considerable reduction in height in the proliferative cartilage zone (i.e., three rows of cells compared with ten to fourteen in normal animals). There was a great reduction of trabeculae, or they were completely absent. The marrow cavity was lined by cartilage in places. These results of Hertz confirm those previously described by Shipley, McCollum, and Simmonds[148] in 1921.

Hertz found that, if he fractured a long bone, normal growth of the bone was inhibited but the process of repair proceeded normally with respect to both structure and time. Although endochondral ossification had ceased at the epiphysis, it continued in the callus.

These findings suggest again that the B_1 deficiency causes an interference with the normal functioning of the pituitary gland. The failure of epiphyseal growth is presumably due to the failure of production of the pituitary growth hormone, but the intrinsic ability of bone tissues to proliferate and repair is unaffected.

IV. Riboflavin Deficiency

1. Skin

The most obvious change in riboflavin-deficient rats occurs in the skin. According to Follis[4] the fur becomes ragged, and the hairs become uneven in length and encrusted with a red-brown substance. The hair then begins to fall out, areas of alopecia develop and the skin then becomes scaly.

[148] P. G. Shipley, E. V. McCollum, and N. Simmonds, *J. Biol. Chem.* **45**, 343 (1921).

Microscopical examination of the skin of such animals shows that the epidermis is undergoing a process of atrophy. At first there is a hyperkeratosis and the cells of the sebaceous glands become swollen and then atrophy. Atrophy is also the fate of the hair follicles. The epidermis decreases greatly in thickness, and the filiform papillae on the tongue become defective. The changes in the mouse skin are fundamentally the same as those in the rat (Lippincott and Morris[149]), and there may be accumulations of leucocytes in the skin. Hamsters, dogs, pigs, and monkeys all show skin changes in riboflavin deficiency (Routh and Houchin,[150] Potter *et al.*,[151] Wintrobe *et al.*,[152] and Waisman[153]). In riboflavin deficiency in human beings skin changes in the mucocutaneous junctions are reported in the literature (Sebrell *et al.*[154] and others). These changes include desquamation of the epithelium of the lips and the formation of fissures in the mouth angles.

2. Eyes

Ocular changes also occur in riboflavin deficiency. At first there is a growing of capillary blood vessels from the limbic plexus into the cornea (corneal vascularization, first described by Bessey and Wolbach[155]). Wolbach and Bessey[21] summarized their results as follows: "The first invading capillaries lie just beneath the epithelium and extend towards the centre of the cornea. Later, they penetrate the tunica propria. As the deficiency progresses, edema, cellular infiltration and separation of the fibers of the tunica propria, with resulting corneal cloudiness, appear. The corneal epithelium shows no gross changes until late in the deficiency and then undergoes degenerative changes which are regarded as secondary to conditions in the tunica propria. In advanced cases, the collagen fibers of the tunica propria become fused and hyalinised and newly formed fibroblasts appear in the zone of capillary ingrowth and infiltration. The superficial epithelium over such areas becomes markedly changed and separated from the deeper layers. Necrosis and ulceration may follow. Prompt regression of all lesions follows riboflavin administration. The cornea may become clear and a diminished blood flow through the newly formed capillaries becomes evident within a few hours; and unless ulceration has occurred, the eye appears histologically normal within a week, except for the presence of collapsed capillaries, demonstrable only by injection methods."

[149] S. W. Lippincott and H. P. Morris, *J. Natl. Cancer Inst.* **2,** 39 (1941).
[150] J. I. Routh and O. B. Houchin, *Federation Proc.* **1,** 191 (1942).
[151] R. L. Potter, A. E. Axelrod, and C. A. Elvehjem, *J. Nutrition* **24,** 449 (1942).
[152] M. M. Wintrobe, W. Buschke, R. H. Follis, and S. Humphreys, *Bull. Johns Hopkins Hosp.* **75,** 102 (1944).
[153] H. A. Waisman, *Proc. Soc. Exptl. Biol. Med.* **55,** 69 (1944).
[154] W. H. Sebrell and R. E. Butler, *Public Health Repts.* (*U. S.*) **56,** 510 (1941).
[155] O. A. Bessey and S. B. Wolbach, *J. Exptl. Med.* **69,** 1 (1939).

In addition to these changes a number of workers have found that cataracts develop in riboflavin-deficient animals. Day and co-workers,[156] Wintrobe and co-workers,[152] and Bourne and Pyke[157] were able to confirm Day's work, but some other workers were not so successful. In the development of such cataracts the earliest signs of change were found to be epithelial proliferation and degeneration of fibers. Then the fibers broke down completely and the lens became an opaque amorphous mass. Pirie[158] found that histological changes in the eyes in riboflavin deficiency were similar to those produced by tryptophan deficiency.

Most of the organs and tissues show some changes in riboflavin deficiency, but according to Wolbach and Bessey[21] these changes are due simply to inanition and are not specific for deficiency of the vitamin.

3. Nervous System

Some authors have recorded changes in the nervous system in riboflavin-deficient animals. Lippincott and Morris[159] have recorded degeneration of the dorsal columns in the spinal cord, and degeneration of the peripheral nerves has also been described (Street *et al.*[115] and Wintrobe *et al.*[152]).

4. Blood Formation

There are reports in the literature that blood formation is impaired in riboflavin deficiency. Miller and Rhoads[160] produced a blood condition in pigs which resembled a macrocytic anemia, and György[161] claimed that riboflavin increased hemoglobin production in anemic dogs. Spector *et al.*[162] have shown by careful experiments that dogs completely deficient in riboflavin became severely anemic after even slight bleeding. Spector believes that the size of new blood cells is determined at least in part by riboflavin and that this vitamin is probably concerned with the metabolism and possibly the arrangement of the amino acids in the molecules of hemoglobin. It is of interest also that Cottingham and Mills[163] have shown that phagocytosis is impaired in riboflavin deficiency.

5. Embryonic Tissues

Within recent years striking changes have been found in embryos born from riboflavin-deficient mothers. Shortening of the mandible and of all the bones of the fingers and toes have been described. In severe deficiency there

[156] P. L. Day, W. J. Darby, and K. W. Cosgrove, *J. Nutrition* **15,** 83 (1938).
[157] M. C. Bourne and M. A. Pyke, *Biochem J.* **29,** 1865 (1935).
[158] A. Pirie, *Brit. J. Nutrition* **2,** 14 (1948).
[159] S. W. Lippincott and H. P. Morris, *J. Natl. Cancer Inst.* **2,** 39 (1941).
[160] D. K. Miller and C. P. Rhoads, *Proc. Soc. Exptl. Biol. Med.* **30,** 540 (1930).
[161] P. György, *Am. J. Physiol.* **122,** 154 (1938).
[162] H. Spector, A. R. Maass, L. Michaud, C. A. Elvehjem, and E. B. Hart, *J. Biol. Chem.* **150,** 75 (1943).

is actually a disappearance of some of these bones, and fusion of ribs and cleft palate have also been noted (see Warkany *et al.*[164-168]) (see Figs. 23–28). Giroud and co-workers[169-171] found that abortion was frequent in riboflavin deficiency and that monsters often were born (Giroud[169]). Lepkovsky *et al.*[172] had previously showed that riboflavin deficiency in the chick embryo produced a degeneration of the mesonephros and the development of edema and anemia.

6. Endocrines and Liver

Deane and Shaw[134] have shown that in riboflavin deficiency there is a mild stimulation of the adrenal in the early stages which later disappears. Normal lipoid with a normal content of ketosteroid appeared to persist through most of the deficiency. Birefringent lipoid was increased in the outer fasciculata (in which the cells are reduced in size), but the mitochondria were normal. Livers in riboflavin-deficient animals showed a fatty infiltration in the early stages of deficiency, but as the deficiency increased this fat disappeared. Glycogen was present in cells near the central veins of the lobule, and the mitochondria of the liver cells were normal.

V. Nicotinic Acid Deficiency

1. General Effects

Chittenden and Underhill[173] in 1917 described pathological changes in dogs which closely resembled those of human pellagra. There was a hemorrhagic condition of most of the colon and rectum. Later Goldberger and Wheeler[174] produced what they called "black tongue" (canine pellagra) by feeding dogs with a human pellagra-producing diet. The pathology of these dogs was described by Denton.[175] He found degeneration of the mucous

[163] H. Cottingham and C. A. Mills, *J. Immunol.* **47,** 493 (1943).
[164] J. Warkany and R. C. Nelson, *Science* **92,** 383 (1940).
[165] J. Warkany and R. C. Nelson, *Anat. Record* **79,** 83 (1941).
[166] J. Warkany and R. C. Nelson, *J. Nutrition* **23,** 321 (1942).
[167] J. Warkany, R. C. Nelson, and E. Schraffenberger, *Am. J. Diseases Children* **65,** 882 (1943).
[168] J. Warkany and E. Schraffenberger, *Proc. Soc. Exptl. Biol. Med.* **54,** 92 (1943).
[169] A. Giroud, XIIe Congrès international de zoologie, Paris, 1948.
[170] A. Giroud, J. Ettori, G. Lévy, and J. Boisselot, *Rev. intern. vitaminol.* **21,** 261 (1949).
[171] A. Giroud, J. Ettori, G. Lévy, and J. Boisselot, *Rev. intern. vitaminol.* **21,** 255 (1949).
[172] S. Lepkovsky, *Hilgardia* **11,** 559 (1938).
[173] R. H. Chittenden and F. P. Underhill, *Am. J. Physiol.* **44,** 13 (1917).
[174] J. Goldberger and G. A. Wheeler, *Public Health Repts.* (*U.S.*) **42,** 172 (1928).
[175] J. Denton, *Am. J. Path.* **4,** 341 (1928).

membranes and changes in the skin. In man (Denton[176] and Moore *et al.*[177]) there is first of all, in the skin, a rarefaction of the superficial parts of the subepithelial connective tissue (corium) and a dilatation of the blood vessels which produces an erythema. There is also keratinization of the epidermis. The epidermis may separate from the dermis. The sebaceous glands and hair follicles atrophy, but the sudoriparous glands do not seem to be affected. Peripheral nerves in the skin appear to degenerate. Necrotic changes in the skin may also occur.

The skin changes are exacerbated by the sunlight.

2. Alimentary Tract

The small intestine in nicotinic acid deficiency becomes red, its walls become thickened, and it is covered by grey cyst-like structures which are produced by distension of the crypts of Lieberkühn. The fluid content of these cysts is the normal secretion of the crypts (Hertzenberg[178]). The mucosa of the colon may also be infiltrated with various leucocytes, particularly lymphocytes, plasma cells, and eosinophiles (Denton[175]).

3. Nervous System

The nerve plexuses associated with the colon show signs of degeneration, and other nervous changes occur, though they generally appear rather late in the deficiency. They are more usually central, as distinct from those of beriberi, which are more commonly peripheral.

In the cerebral cortex of brain, foci of pyramidal cells show displacement of the nucleus, accumulation of fat, and chromatolysis. These changes tend to occur more commonly in the frontal lobe. The spinal cord changes are most marked in the columns of Goll, although other tracts may be affected. The changes take the form of diminishment in the number of nerve cells and the replacement of the presumably degenerated cells by a glial scar (Orton and Beader[179]).

Pathological changes in pig pellagra and canine blacktongue are in general similar to those found in human pellagra.

4. Kidney

Although rats do not appear to develop characteristic skin lesions on a nicotinic acid-deficient diet, Bourne and Harris[80] found that they suffered from a profound tubular degeneration of the kidneys. Other changes found were atrophy of pituitary adrenals and testes and a complete cessation of

[176] J. Denton, *Am. J. Trop. Med.* **5**, 173 (1925).

[177] R. A. Moore, T. D. Spies, and Z. K. Cooper, *Arch. Dermatol. Syphilol.* **46,** 100 (1942).

[178] H. Hertzenberg, *Beitr. path. Anat. u. allgem. Path.* **96,** 97 (1935).

[179] S. T. Orton and L. Beader, *Bull. Neurol. Inst. N. Y.* **1,** 506 (1931).

endochondral ossification at the costochondral junctions. There was also a loss of HCl-secreting cells from the stomach. The endocrine and reproductive changes resemble those of simple inanition and these at least are probably not due specifically to the deficiency of nicotinic acid.

5. Tissue Culture

The precise function of nicotinic acid in cells is still in some doubt, although it has been shown to be present in cells as a coenzyme and Hall[180] tried the effects of nicotinic acid and its homologues on cultured embryonic chick heart. The first homologue, pyridine-3-sulfonic acid, inhibited the migration of cells, and this effect could not be reversed by nicotinic acid. After 96 hours, however, some flat, dark cells did migrate from the explant. The second homologue, acetyl-3-pyridine, stimulated migration in low concentration but inhibited it in high concentration. After 48 hours of inhibition large yellow cells appeared from the explant which later on became dark and rounded. It was of interest that both compounds inhibited the uptake of P^{32} added to the substrate as a tracer.

VI. Pyridoxine Deficiency

1. Skin

Pyridoxine deficiency produces a dermatitis in rats (Sullivan and Nicholls[181]). The dorsal surfaces of the hind paws and to a lesser extent the front paws become erythematous. The plantar surface then becomes affected in the same way, and hyperkeratosis and scaling of the epidermis sets in (see Figs. 33 and 34). The same changes then develop in the skin in other parts of the body, notably on the ears, nose, chin, and parts of the thorax. The corium is edematous in the dermatotic regions and is infiltrated with leucocytes (chiefly monocytes and lymphocytes). Under the microscope the epidermis shows an increased thickness in all layers. Mitosis is very active in the basal layers. The hair follicles and the sebaceous glands do not appear to be specifically affected, although in the later stages of deficiency they often become infected (see Figs. 35 and 36). Skin lesions have also been found in deficient rats by Agnew.[182] He also found a hematuria and an increase in the weights of the heart and kidneys relative to the body weight.

2. Adrenals and Liver

Antopol and Unna[183] found atrophy of the zona reticularis in the adrenal, and Deane and Shaw[134] found that after 3 weeks of deficiency the fasciculate

[180] W. Hall, *J. Gen. Physiol.* **34,** 75 (1950).
[181] M. Sullivan and J. Nicholls, *J. Investigative Dermatol.* **3,** 317 (1940).
[182] L. R. C. Agnew, *Brit. J. Nutrition* **3,** 217 (1949).
[183] W. Antopol and K. Unna, *Arch. Path.* **33,** 241 (1942).

and reticulate zones of the rat adrenal showed increased lipoid content while the glomerulosa appeared normal. After 6 to 10 weeks of deficiency they found that the lipoid content of all zones was normal. The livers of such rats contained no fat but possessed a moderate amount of glycogen which was more concentrated at the periphery of the liver lobule than at the center. Richards[184] found atrophy of the thymus in pyridoxine-deficient rats.

3. Sites of Lesions

Wolbach and Bessey[21] claim to have seen lesions in their experimentally deficient rats in heart, eye, thyroid, parathyroid, trachea and lungs, para-ocular glands, salivary glands, stomach, intestines, liver, spleen, kidneys, pancreas, adrenals, ovary and testis, bone and bone marrow, lymph nodes, skeletal muscle, and nervous system; but these changes, they claim, are non-specific and are due only to the processes of inanition (see Figs. 29, 30, and 37). There appears to be some evidence that the dermatitis of pyridoxine deficiency can be cured by unsaturated fatty acids (Scheider[185]).

4. Dogs

Dogs on a pyridoxine-deficient diet suffer from microcytic hypochromic anemia (Street *et al.*[186]), although Follis[4] drew attention to the possible relation of folic acid to this condition. Follis has described a similar anemia in swine, pointing out that the number of red cells falls by 50% in a few weeks and that it is accompanied by anisocytosis. Pyridoxine treatment produced an immediate reticulocyte response. As in the dog, the anemia was not completely cured by pyridoxine alone.

5. Nervous System

Follis and Wintrobe[187] have found demyelinization of peripheral nerves in the pig on a pyridoxine-deficient diet, with some possible axone damage (Figs. 31 and 32). The myelin degeneration is progressive, and the dorsal root fibers and dorsal columns of the spinal cord also become affected (see Fig. 38). The nerve cells in these regions show chromatolysis, atrophy, and eventually necrosis (see also Fig. 38a).

Ramalingaswami and Sinclair[188] also found changes in the nervous system together with pathological alteration of the skin and hemopoietic systems in pyridoxine deficiency. They found frequent severe lesions of nose, lips, and angles of the mouth and, less frequently, lesions of the anogenital

[184] M. B. Richards, *Brit. J. Nutrition* **3,** 109, 132 (1949).
[185] H. Scheider, *J. Biol. Chem.* **132,** 539 (1940).
[186] H. R. G. Street, G. R. Cowgill, and H. M. Zimmerman, *J. Nutrition* **21,** 275 (1941).
[187] R. H. Follis and M. M. Wintrobe, *J. Exptl. Med.* **81,** 539 (1945).
[188] V. Ramalingaswami and H. M. Sinclair, *Brit. J. Nutrition* **4,** 13 (1950).

regions. At the mucocutaneous junctions they found that the earliest changes were erythema and edema, followed by hyperkeratosis and fissuring. These changes were described in rats.

VII. Choline Deficiency

Choline and methionine are sources of methyl groups and are therefore important in the body in the processes of transmethylation, particularly so since the body is said not to be able to produce methyl groups itself. Choline is apparently necessary for fat transport. Deficiency of choline therefore results in an accumulation of fat in the liver, beginning as small droplets which eventually run together to form a large droplet of fat which occupies the whole of the hepatic cell, pushing the nucleus to one side (see Fig. 18). This damage to the liver cell eventually leads to the excessive production of fibrous tissue (scarring), and the liver becomes cirrhotic. A waxy pigment also develops in the liver in choline-deficient rats. The kidneys in such animals show dilatation of cortical blood vessels, sometimes with hemorrhage and necrosis of the renal tubules (see Fig. 19), a condition not unlike that found by Bourne and Harris[80] in rats on a nicotinic and deficient diet. Eye changes may also be present in rats in choline deficiency, and there may be hemorrhages into the lens and vitreous and ciliary processes, which also become swollen and appear edematous (Bellows and Chinn[189]).

VIII. Biotin Deficiency

Biotin deficiency has been induced in chickens, rats, mice, monkeys, dogs, and man. In the rat there is a characteristic dermatosis which is erythematous, scaly, greasy, and itchy. Growth ceases, posture is unusual, and there is a high-stepping spastic type of gait. The hair follicles around the eyes degenerate, and the rat takes on a spectacled appearance (Sullivan and co-workers[190, 191]). Microscopical observation of the dermatotic skin shows that hyperkeratosis is present and that acanthosis (increase in the layer of prickle cells) occurs. There is an infiltration of leucocytes into the dermis which becomes edematous. The hair follicles become blocked by masses of keratotic material, the follicle may degenerate, and the epidermis may eventually atrophy.

Okey *et al.*[192] found that biotin-deficient rats became extremely sensitive to being touched. They lost both their subcutaneous and their visceral fat, and they also failed to store fat in the liver. These workers found that in castrated males implantation of diethylstilbestrol delayed the development

[189] J. G. Bellows and H. Chin, *Arch. ophthalmol. Paris.* [N.S.] **30,** 105 (1943).
[190] M. Sullivan and J. Nicholls, *Arch. Dermatol. Syphilol.* **45,** 295 (1942).
[191] M. Sullivan, L. Kolb, and J. Nicholls, *Bull. Johns Hopkins Hosp.* **70,** 17 (1942).
[192] R. Okey, R. Pencharz, and S. Lepkovsky, *Am. J. Physiol.* **161,** 1 (1950).

of the deficiency but that implantation of testosterone in both castrated males and females increased the deficiency.

Despite the appearance of nervous incoordination, there are no signs of damage to the central nervous system, but the muscles show signs of atrophy and in some regions there is necrosis of the fibers. In chicks Couch *et al.*[193] found that biotin deficiency caused shortening of the tibiotarsus, the distal end of which was bent posteriorly. The tarsometatarsus was also shortened, and the distal end was bent medially. In some chicks the wing bones were shortened also.

Wilson *et al.*[194] and Montagna[195] found that in biotin deficiency in the mouse normal differentiation and function of the skin were impaired. As in the rat, the pilosebaceous units were blocked, and in the sebaceous cells the lipoid droplets were much larger tha normal. Some cells fragmented, and the lipoid phagocytosed by dermal histiocytes. In the dermis mesenchyme cells, in mast cells, and in leucocytes were increased.

Burt[196] found that, in 236 tissue cultures of embryonic chick neuroblasts, spindle cells, and macrophages, biotin had no growth-stimulating properties.

IX. Pantothenic Acid Deficiency

Pathological changes in pantothenic acid-deficient animals have been described in the rat (Sullivan and Nicholls[197]), the mouse (Jones *et al.*[198]) and the pig (Wintrobe *et al.*[199]).

1. Skin

The skin of the rat, according to Sullivan and Nicholls, showed loss of hair around the ears and the snout, the remainder of the hair, if pigmented, becoming gray. There was hyperkeratosis and increase of the prickle cell layer in the skin (acanthosis). The whole skin eventually became atrophic.

2. Alimentary Tract, Nervous System, and Bone

There was hyperemia of the intestine, particularly in the large intestine. Ulcers developed in the lymphoid follicles of the walls and sometimes led to perforation (Fig. 40). The cells of the epithelium became atrophic, the process starting as small atrophic foci which eventually spread and covered

[193] J. R. Couch, W. W. Cravens, C. A. Elvehjem, and J. G. Halpin, *Anat. Record* **100,** 29 (1948).

[194] J. W. Wilson, E. H. Leduc, and D. H. Winston, *J. Nutrition* **38,** 73 (1949).

[195] W. Montagna, *Proc. Soc. Exptl. Biol. Med.* **73,** 127 (1950).

[196] A. S. Burt, *Proc. Soc. Exptl. Biol. Med.* **54,** 191 (1943).

[197] M. Sullivan and J. Nicholls, *Arch. Dermatol. Syphilol.* **45,** 917 (1945).

[198] J. H. Jones, C. Foster, F. Dorfman, and G. Hunter, *J. Nutrition* **29,** 127 (1945).

[199] M. M. Wintrobe, R. H. Follis, R. Alcayaga, M. Paulson, and S. Humphreys, *Bull. Johns Hopkins Hosp.* **73,** 313 (1943).

the whole of the lower gut. Atrophic cells accumulated in the crypts which became dilated. Finally the epithelium necrosed with the development of ulcers (see also Fig. 41). There were some degenerative changes in the Harderian glands; in the nervous system there was dissolution of the Nissl bodies in the cells of the dorsal root ganglia (see Fig. 42). In the latter stages demyelination and degeneration of the axis cylinder of peripheral nerves occurred. Degeneration of fibers in the dorsal columns of the spinal cord and of the cells in the anterior horns was also present. Nelson *et al.*[200] have summarized the changes occurring in the pantothenic acid-deficient rat as consisting of graying of the hair, dermatosis, porphyrin deposition, and reduced growth of long bones. In the bones of the rat, after 21 to 23 days of deficiency, there was an absorption of trabecular bone, a decreased proliferation of osteoblasts, and edema of the bone marrow. Between 33 and 63 days retardation of osteogenesis occurred, and between 78 and 109 days a heavy layer of bone was formed which sealed off the epiphyseal cartilage.

3. Adrenals

One of the organs most seriously affected by pantothenic acid deficiency is the adrenal gland. According to Ashburn[201] there is in the rat a deposition of hemosiderin in the gland; there is also hemorrhage, together with atrophy of some of the cells, followed by necrosis. The inner part of the zona fasciculata and the whole of the zona reticularis appears to be affected, but the rest of the cortex does not seem to suffer from these changes although many of the cortical cells lose most of their lipoid (see Fig. 43). Melampy[202] found that in pantothenic acid deficiency there was a progressive loss of ketosteroids from the adrenal cortex which became hypertrophied and that there was an accompanying loss of lipoids from the cells. This suggested that pantothenic acid may be essential for the production of cortisone and other adrenal steroids by the adrenal cortex. Winters *et al.*[203] showed that in pantothenic acid-deficient rats there is an upset of the carbohydrate metabolism. Among the effects were an increased sensitivity to insulin and an inability of the fasting animal to deposit liver glycogen. Winters *et al.*[204] also showed that ACTH had no effect on the eosinophile and lymphocyte counts in the blood of pantothenic acid-deficient rats but that cortisone treatment produced the same effect as in normal animals.

[200] N. M. Nelson, E. Sulon, H. Becks, W. W. Wainwright, and H. M. Evans, *Proc. Soc. Exptl. Biol. Med.* **73,** 31 (1950).

[201] L. L. Ashburn, *Public Health Repts.* (*U.S.*) **55,** 1337 (1940).

[202] R. M. Melampy, D. W. E. Cheng, and L. C. Northrop, *Proc. Soc. Exptl. Biol. Med.* **76,** 24 (1951).

[203] R. W. Winters, R. B. Schultz, and W. A. Krehl, *Endocrinology* **50,** 388 (1952).

[204] R. W. Winters, R. B. Schultz, and W. A. Krehl, *Endocrinology* **50,** 377 (1952).

This suggested that in pantothenic acid deficiency the adrenal had lost its ability to produce steroids in response to anterior pituitary stimulation.

Cowgill *et al.*[205] stated that their own and previous work had established that a characteristic lesion developed in the adrenal cortex in response to pantothenic acid deficiency and that this lesion was more severe following treatment with ACTH but that cortisone prevented development microscopically. They concluded that the above work together with functional studies on carbohydrate metabolism supported the view that in pantothenic acid deficiency there was inadequate cortical function. They believe that pantothenic acid acting as coenzyme A is necessary for the synthesis of adrenal steroids.

4. Embryos

Boisselot[206] has found that in female rats on a pantothenic acid-deficient diet there is interference with the estrus cycle and that the deficient pregnant rat gave birth to abnormal young, some of which were born dead. Many of them were devoid of eyes (see Fig. 39), there was no optic nerve, and abnormalities of the brain and vertebral column also occurred (see Figs. 44 and 45). Administration of choline and nicotinic acid to the pantothenic acid-deficient pregnant females exacerbated these conditions.

X. Folic Acid Deficiency

Hematologic changes characteristic of deficiency of this vitamin are referred to elsewhere in this book, but other aspects of folic acid deficiency might be mentioned here. Sunde[207] found that chick embryos derived from hens deprived of folic acid had a high mortality about the twentieth day of incubation. There was syndactylia, deformity of beaks, and bending of the tibiotarsus. Nelson *et al.*[208] also found congenital abnormalities in fetal rats whose mothers were subjected to folic acid deficiency during pregnancy (see Figs. 20, 21, and 22).

There appears to be considerable evidence that folic acid is essential for growth and development, as well as hemopoiesis. This is presumably because of its role in the synthesis of protein and nucleic acids. In the synthesis of protein folic acid appears to facilitate the ability of the body to use formate for the production of amino acids (Plaut *et al.*[209]). It is thought that both folic acid and vitamin B_{12} are concerned in the synthesis of desoxyribonucleic acid (Woods[210]).

205 G. R. Cowgill, R. W. Winters, R. B. Schultz, and W. A. Krehl, *Internat. Rev. Vitamin Research* **23**, 275 (1952).
206 H. Boisselot, *Arch. franç. pediat.* **6**, 1 (1949).
207 K. Sunde, *Ann. méd.* **52**, 225 (1951).
208 M. M. Nelson, C. W. Asling, and H. M. Evans, *Anat. Record* **106**, 309 (1950).
209 G. W. E. Plaut, J. J. Bertheil, and H. A. Lardy, *J. Biol. Chem.* **184**, 795 (1950).
210 D. D. Woods, *J. Pharm. Pharmacol.* **2**, 537 (1950).

XI. Vitamin C Deficiency

1. General Background

The deficiency disease described as scurvy has been known for a very long time. There are even records which date back to the time of the Crusades (see the accounts by De Joinville, in 1250, of the disease which beset the crusading armies of Louis IX before Cairo, published in Menard's 1617 edition of De Joinville's writings; see also Angelenus,[211] 1604; Kramer,[212] 1720; Walter,[213] 1748; Lind,[214] 1753; Linton[215] 1858; Smart,[216] 1888). All these authors described macroscopic changes in tissues and failure of wounds and fractures to heal in scurvy. The first comprehensive account of tissue changes in scurvy were given by Hess[217] in 1920.

2. Clinical Scurvy

Lind found that the blood of persons who had died of scurvy did not clot, whether it was left in the body or whether it was drawn into a vessel. The lungs were "blackish and putrid" and the muscles "stuffed with black corrupted blood." "Some bodies when moved could be heard to grate, and when they were opened the epiphyses of the long bones were found to be entirely separated away. Some persons when alive had been heard to make a "small low noise" when they breathed; in these the cartilages of the sternum were found to be separated from the bony parts of the ribs." In some the ribs were carious, and ligaments of the joints were corroded and loose in nearly all. Abnormalities in the bones appeared to be a constant feature of scurvy; between 1883 and 1918 there were ten publications on post-mortem changes in scurvy in which only the changes in the bones are described. Lind noted also the presence of what Hess[217] subsequently described as "peculiar boggy tumour-like masses of localised oedema," which he regarded as typical of scurvy. He also found that the "breast, belly and several other parts of the body were filled with water or serum." Barlow[218] wrote also that "the muscular walls of the thorax were pale yellow and watery, as though they had been bathed in serum."

Aschoff and Koch[219] and Hess[217] stressed the fact that the characteristic

[211] Angelenus (1604), quoted by Lind, ref. 214.

[212] J. F. Kramer (1720), quoted by Lind, ref. 214.

[213] R. Walter, A Voyage round the World, John and Paul Knapton, London, 1748.

[214] J. Lind, Treatise on Scurvy, S. Crowder, London, 1753.

[215] C. Linton, Medical and Surgical History of the British Army in the Crimea, Accounts and Papers, British Government, Vol. 3, Part II, 1858.

[216] C. Smart, Medical and Surgical History of the War of the Rebellion, Government Printing Office, Washington, 1888.

[217] A. F. Hess, Scurvy Past and Present, J. B. Lippincott, Philadelphia, 1920.

[218] T. Barlow, *Lancet* **2**, 1075 (1894).

[219] L. Aschoff and W. Koch, Skorbut: Eine pathologische anatomische Studie, Gustav Fischer, Jena, 1919.

lesions of scurvy are those of the bones. The epiphyses in all young persons have separated from the bones, and in young children there has been found a beading of the ribs, wrongly ascribed (according to Hess) to the effects of vitamin D deficiency, i.e., rickets. Scorbutic monkeys have also been shown to develop this beading of the ribs (Hart and Lessing[220]).

The most constant feature of scorbutic bones is a subperiosteal hemorrhage which may extend for some considerable distance along the bone. If a scorbutic bone is split longitudinally, the cortex can be seen to be exceedingly thin and very brittle and the marrow to have lost its red color and become yellowish in color and gelatinous in consistency. This changed marrow is known as the "Gerüstmark." Hemorrhages can also be seen in the marrow. The principal change in the bones (as described by Aschoff and Koch and by Hess) takes place at the ends of the diaphyses where junction is normally affected with the cartilage. Ribs show these characteristic changes very well. The following description of microscopical sections of scorbutic bones is taken from the account given by Hess.[217] Normally the junction of bone and cartilage is a straight line. In scurvy the line of junction is wavy and irregular—generally speaking, it is concave towards the bone. At the junction is what is called the "Trümmerfeld" zone, where all the normal tissues of the region are in a state of disorder. Small and large bony trabeculae are scattered irregularly about. Cells are irregular in shape and not arranged in an orderly fashion, and there are signs of past and recent hemorrhages and unrecognizable detritus. Hess said, "the picture is that of weakened bone having been crushed by the pressure of the more compact cartilage." Very few osteoblasts are present in this zone. There are a greater number of spindle or stellate connective tissue cells. Aschoff and Koch[219] described a fibrin mass (partly organized into connective tissue) which covers the mass of detritus. The cartilage of the rib is also abnormal. The usual proliferating columns of cells near the bone have disappeared, but a few may persist near the periphery.

Next to the Trümmerfeld zone is the Gerüstmark (framework marrow), which appears to be of a fibrogelatinous nature. The trabeculae at the head of each bone are greatly reduced in thickness. The osteoblasts which occur along the edges of the trabeculae are said to become spindle-shaped and shriveled. There is no excess of osteoclasts (which are believed to be associated with bone resorption), so the rarefaction of the trabeculae is obviously not due to increased osteoclastic activity. Hess suggested that the thinning of the trabeculae is due to the combination of normal bone resorption and the failure of normal bone regeneration. Aschoff and Koch believed that the "framework marrow" is normally able to make bone trabeculae

[220] C. Hart and O. Lessing, Der Skorbut der kleinern Kinder, Ferdinard Euke, 1913.

but that in scurvy there is a lack of the material required to build the osteoid.

The weakening of the cortex of long bones sometimes results in a separation of the epiphysis from the diaphysis. The separation does not take place at the junction of the two but is actually due to a fracture in the upper part of the diaphysis—a sort of Colles' fracture. The present author has observed that in dissecting out bones from scorbutic guinea pigs great care must be exercised to avoid pulling off the epiphysis from the diaphysis (in this case the separation is at the line of junction). In normal guinea pigs it is very difficult to pull off the epiphysis during dissection. During maceration of bones in 0.5 % KOH it may be seen also that the epiphyses of scorbutic bones come away from the diaphyses much sooner than those of normal bones.

3. Experimental Scurvy

a. General Effects. Clinical scurvy may of course be due not only to vitamin C deficiency but to other deficiencies as well. It is necessary, therefore, to quote evidence obtained from experimental scurvy in which the diet has been controlled before we can regard the histological findings of human scurvy as being due entirely to vitamin C deficiency. The first workers to produce unequivocal experimental scurvy were Holst and Fröhlich[221] (1903), but some time before this Theobald Smith[222] (1895) had pointed out that guinea pigs fed entirely on cereals died in 1 to 2 months from a disease which was characterized by extravasations of blood in the subcutaneous tissues.

Bolle[223] (1902) reported fragility of bones in experimental guinea pigs fed entirely on cow's milk (cow's milk has very little vitamin C, i.e., 1 mg. per 100 ml.).

For some years ship beriberi had been a serious cause of disablement of Norwegian sailors, and in 1907 Holst and Fröhlich[224] attempted to produce the disease in guinea pigs. They got scurvy instead. Of this matter Hess made the following observation: "A diet of polished rice, or other decorticated grain, will lead to the development of scurvy in the guinea-pig, to polyneuritis in the pigeon or fowl, or, according to Holst and Fröhlich, to a combination of these disorders in the hog." It is rather extraordinary that man, monkeys, and guinea pigs are unique in requiring vitamin C in the diet, and Miwa[225] has added the marmot (*Arctomys marmotta*) to the list.

On the question of whether guinea pig scurvy is a disease which can be

[221] H. Hölst and T. Fröhlich (1903), see Holst and Fröhlich, ref. 224.

[222] T. Smith, *U. S. Dept. Agr. Bur. Animal Industry Ann. Rept.* p. 172 (1895).

[223] C. Bolle, *Z. diatet. physik. Therap.* **6,** 354 (1902).

[224] H. Hölst and T. Fröhlich, *J. Hyg.* **7,** 634 (1907).

[225] A. Miwa, *Oriental J. Diseases Infants* **26,** 3 (1939).

regarded as identical with human scurvy, Hess said: "Viewed from these standpoints [i.e., mode of production, pathology, symptomatology and means of cure] the disease is identical in man and in the guinea-pig. The outstanding distinction is the difference in the length of time elapsing before the development of the symptoms. In the child or in the adult it takes about 6 months of the deficient diet before clinical symptoms are manifest and diagnosis can be established; in the guinea-pig the disorder can be recognised two weeks after restricting the diet." This is to be expected, however, considering the life span of man and of the guinea pig. A guinea pig's life is about 3 years; that of a man, about 60 to 70 years. This gives a ratio of about 1:20. The ratio of the times taken to develop scurvy is only about 1:13, so that one would actually expect man to take longer to develop scurvy than he, in fact, does.

b. Bones. The same general pathological changes that appear in human scurvy can be observed in guinea-pig scurvy (Holst and Fröhlich,[224] McCarrison[226]). There are hemorrhages into the joints, separation of epiphyses, brittleness of bone, intra-abdominal hemorrhages, enlargement of the adrenals, etc. Histologically the bone marrow is seen to become Gerüstmark in nature, the bony trabeculae in the spongy part of the bone are decreased, and there is a thinning of the bony cortex. Cells in the rib cartilage are disordered, and the line of junction with the rib is irregular (Schoedel and Nanwerk[227] and Höjer[228]). Pirani *et al.*[229] said that the pathological changes in the knee joints of guinea pigs on a scorbutic diet were due to hemorrhages, active proliferation, and failure to mature of mesenchymal cells and an insufficient production of ground substance and collagen (see Figs. 56–59). Follis[230] states that in scorbutic cartilage and bone the distribution of cytochrome oxidase and phosphatase activity, and of glycogen, mucopolysaccharides, nucleoproteins, and lipids, suggested that the vitamin is intimately related to phosphatase activity and to the formation of mucopolysaccharides in the Gerüstmark and of ribonucleic acid in the osteoblast.

c. Teeth. One of the characteristic changes in guinea pigs is in the teeth. The incisor teeth of rodents have persistently growing pulps; because these creatures are constantly gnawing at hard matter, the edges of the incisors are being steadily worn away. This is compensated for by growth from the pulp which persists throughout life. It has been shown by Jackson and Moore[231] and by Fish and Harris[232] that guinea pigs placed on a scorbutic

[226] R. McCarrison, Studies in Deficiency Diseases, Oxford University Press, 1921.
[227] J. Schoedel and W. Nanwerk, quoted by Hess, ref. 217.
[228] J. A. Höjer, *Acta Paediatr. Suppl.* **3** (1923).
[229] C. L. Pirani, C. G. Bly, and K. Sutherland, *Arch. Pathol.* **49,** 710 (1950).
[230] R. H. Follis, *Bull. Johns Hopkins Hosp.* **89,** 9 (1951).
[231] L. Jackson and J. J. Moore, *J. Infectious Diseases* **19,** 478 (1916).
[232] E. W. Fish and L. J. Harris, *Phil. Trans.* **B 223,** 489 (1934).

diet suffered from characteristic changes in the incisor teeth because of the failure of normal dentine and enamel formation. There was also fibroid degeneration of, and hemorrhage into, the pulp. The odontoblasts became almost unrecognizable. This degeneration began near the base of the pulp and worked up towards the apex. In some cases there were amorphous deposits of calcium in the pulp. The results obtained from the authors quoted and from a number of other authors established the essential nature of vitamin C for the development of the incisor teeth in guinea pigs, and that, in particular, it is essential for the production of predentine which can be regarded as the equivalent in the tooth of the osteoid or fibrous ground substance of bone. Boyle, Bessey, and Howe,[233] however ,claimed that vitamin C is not essential for the calcification of teeth. Later workers on the effect of experimental scurvy on bones have come to the conclusion that vitamin C is essential for the deposition of the osteoid ground substance of bone (Wolbach and Howe,[9] MacLean *et al.*[234] and Bourne[235]). They also showed that, in the absence of vitamin C, osteoblasts dedifferentiated into fibroblasts. The workers on tooth changes in scurvy have also shown that there is a dedifferentiation of the odontoblast and that resorption of the bone of the jaws of the guinea pig also takes place.

d. Cartilage. It is doubtful that vitamin C is necessary for the maintenance and growth of cartilage. Park *et al.*[236] found that even when the bone ceased to produce osteoid material in scurvy the cartilage of the epiphysis continued to grow. On the other hand, Wolbach[21] quoted unpublished work by himself and Boyle in which they found that there was a loss of firmness in cartilage matrix in guinea pigs fed on small amounts of vitamin C for long periods of time. Höjer[228] in 1924 had previously found what he described as "collagen atrophy" of cartilage in scurvy but had found no change in the chondroitin-sulfuric acid content. Little detailed record has been made of cytological changes in scurvy; Bourne[237] has recorded mitochondrial changes in kidney tubule cells, and Granaglia and Ronzani[238] showed that in the epithelial cells of the duodenum of guinea pigs there was an irregular distribution of mitochondria.

e. Muscle. Höjer[228] in his 1924 monograph on scurvy found other changes, apart from those in the bones and teeth. There was muscular atrophy with necrosis, and sometimes an impregnation of the necroses with

[233] D. E. Boyle, O. A. Bessey, and P. R. Howe, *J. Dental Research* **15,** 331 (1936).
[234] D. L. MacLean, M. Sheppard, and E. W. McHenry, *Brit. J. Exptl. Path.* **20,** 451 (1939).
[235] G. H. Bourne, *J. Physiol.* (*London*) **102,** 319 (1943).
[236] E. A. Park, H. G. Guild, D. Hackson, and M. Bond, *Arch. Disease Childhood* **10,** 265 (1935).
[237] G. H. Bourne, *Australian J. Exptl. Biol. Med. Sci.* **13,** 239 (1935).
[238] G. Granaglia and B. Ronzani, *Arch. sci. biol.* (*Italy*) **35,** 286 (1951).

amorphous deposits of calcium. There were numerous intramuscular hemorrhages. Heart muscle is also affected in scurvy (see Lewin,[239] 1871, and Haymen,[240] 1871, who found fatty degeneration of the heart muscle common in human scurvy; Koch,[241] 1889, who found calcium deposits in the heart muscle of scorbutic monkeys). Boyle and Irving[242] observed some hyaline changes in muscles (skeletal) in chronic scurvy; in acute scurvy there were areas where muscle fibrils and sarcoplasm were seen to end suddenly, leaving intact and empty sarcolemmal sheaths, with the endomysial fibers lying between them. They concluded that ascorbic acid is probably essential for the maintenance of the attachment of myofibrils to the sarcolemma (see Figs. 52 and 53; also see Fig. 51).

f. Liver. The liver is also affected in scurvy. Carl von Linné,[243] as long ago as 1742, found changes in the liver in human cases. Höjer pointed out that the liver in experimental scurvy shows hemorrhage and atrophy of both cells and parenchyma. Animals dying of the disease also showed hemorrhage and siderosis in the spleen, and atrophy, necrosis, and calcification in the kidney (Höjer[228]).

g. Blood. Höjer also found a deficient regeneration of neutrophilic polymorph cells. Salle and Rosenberg[244] in 1921 had previously claimed a deficient production of both red and white cells in scurvy. Brandt[245] had noted a prolonged coagulation time of the blood in scorbutic animals.

h. Connective Tissues and Capillaries. The most important discovery made by Höjer was that, in scurvy, there was a widespread deficiency of collagen in the connective tissues throughout the body. He also showed that in experimental guinea pig tuberculosis there was deficient scar production by the lung tissues because they appeared to be unable to produce the normal amount of collagen.

One general fact that emerges from a consideration of the various results quoted is that hemorrhage is constantly found in the various organs and tissues throughout the body and that scurvy has in fact been regarded as typically a hemorrhagic disease. This suggests that some weakness of the blood vessels is present in scurvy. There is support for this contention in the literature. Jackson and Moore[246] stated that there was "marked thinning of the wall" (of the blood vessels) and that in some of the smaller vessels "the wall as a whole had partially melted away, leaving few traces" (see

[239] M. Lewin, *Gaz. méd.* p. 430 (1871).

[240] M. G. Haymen, *Gaz. méd.* p. 126 (1871).

[241] W. Koch, Die Blutkrankenheit in irhen Variantin, Ferdinand Euke, 1889.

[242] P. E. Boyle and J. T. Irving, *Science* **114,** 572 (1951).

[243] C. von Linné, Collegium dieteticum, 1742. See *Universitets Arsskrift, Upsala* **47,** 3 (1907).

[244] V. Salle and W. Rosenberg, *Ergeb. inn. Med. u. Kinderheilk* **19,** 31 (1921).

[245] H. Brandt, *Deut. med. Wochschr.* **45,** 640 (1919).

[246] L. Jackson and J. J. Moore, *J. Infectious Diseases* **19,** 478 (1916).

Figs. 54 and 55). Höjer also claimed that there was a weakness in the vascular walls in scurvy. It has been suggested that the increased fragility of the capillary blood vessels (note that petechial hemorrhages were said to be characteristic of scurvy by the original describers of the complaint) is due to a defect in the intercellular cement substance which binds the endothelial cells of the capillaries together. Bicknell and Prescott[35] stated that "actually the minute structure of the capillary wall is colloidal and might well respond to vitamin C as does collagen." Other workers have not found any observable difference in the permeability of the capillary walls to diffusible substances in scurvy. On the other hand, Findlay[27] claimed to have observed the diapedesis of red cells between the endothelial cells of the capillaries in scorbutic guinea pigs. It is of interest that Hess[217] in 1920 had previously pointed out that the presence of isolated small groups of extravascular red blood cells in various tissues suggested a diapedesis through the blood vessel walls.

Lantoch and Gagné[247] have found deposits of amyloid in the spleens of vitamin C-deficient guinea pigs, but there was no sign of amyloid in other organs.

4. Effect of Scurvy on Repair of Tissues

a. Older Clinical Evidence. Since the evidence so far produced shows that in scurvy the normal functioning of both soft and hard tissues is disturbed, one would expect that repair of injury to any tissue would be delayed also. This fact can be confirmed by reference to the older literature.

The earliest known record of the effects of scurvy on wounds is given by Richard Walter[248] (1748), Chaplain to Lord Anson's expedition round the world (1740–1744). He made the following comment on this subject.

> At other times the whole body, but more especially the legs, were subject to ulcers of the worst kind, attended with rotten bones, and such a luxuriancy of fungous flesh, as yielded no remedy. But a most extra-ordinary circumstance, and what would be scarcely credible upon any single evidence, is that the scars of wounds which had for many years healed, were forced open again by this violent distemper: of this, there was remarkable instance in one of the invalids upon the *Centurion,* who had been wounded above fifty years before at the battle of the Boyne; for though he was cured soon after, and had continued well for a great number of years past, yet on his being attacked by the scurvy, his wounds, in the progress of his disease, broke out afresh, and appeared as they had never been healed; Nay what is still more astonishing, the callous of a broken bone, which had been completely formed for a long time, was found to be hereby dissolved, and the fracture seemed as if it has never been consolidated.

Lind[214] (1753, p. 151) made the following observation: "When one has

[247] E. Lantoch and F. Gagné, *Rev. can. biol.* **10,** 119 (1951).

[248] R. Walter, A. Voyage round the World, John and Paul Knapton, London, 1748.

been confined from exercise by having a fractured bone, or from a bruise or hurt, these weak and debilitated parts become always first scorbutic." Mr. Ives, writing to Lind[212] (p. 151) of the breaking open of wounds in scurvy, referred to the case of a seaman on *H.M.S.Dragon* who had a shattering of the humerus from a Spanish musket ball. Union of the bone and healing of the skin wounds had been brought about by November, but when scurvy broke out on the ship in December, he fell a victim to the disease, the first symptom being the breaking open of the wounds in his arm. Lind also records that the "slightest bruises and wounds of scorbutic persons degenerate in such ulcers . . . distinguished from all others by being so remarkably putrid, bloody and fungous."

Mead[249] quoted the case of a sailor who had suffered from a fractured clavicle which had apparently healed normally and which broke again 4 months later when the sailor was suffering from scurvy. Six months after this, after the sailor had been on a diet of green vegetables for some time, the fracture reunited.

Marriques,[250] Bell,[251] and Callisen[252] all found softening of the callus of old fractures on bones in scurvy, sometimes with a separation of the ends of the bones. Hammick[253] reported a number of cases of spontaneous refracture of bones and pointed out that it was impossible to secure the uniting of a fracture as long as the patient had scurvy. A further example of refracture of a healed broken bone is given by Budd.[254]

Dr. Linton[215] described the case of a grenadier, aged 23, serving in the Crimean War, with a broken humerus. This man broke his arm while carrying a log of wood across some frozen snow. After he had been put in the hospital, old ulcers on his leg opened up and his gums became spongy. The callus which formed at the site of the fracture in the humerus was unusually small, suggesting a reduced inflammatory reaction to injury. Only when he was put on a "good" diet did his fracture unite and his ulcers heal.

Moore[255] has recorded two cases of fractures of the forearm in which healing was prevented because the patients had scurvy. Moore stated that this disease has a "powerful effect in retarding the consolidation of fractures."

Major Charles Smart[216] pointed out that, as a result of the scurvy which

[249] R. Mead, Medical Works, S. Crowder, London, 1762.
[250] A. Marriques (1783), quoted by Hertz, ref. 86.
[251] J. Bell, A System of Surgery, S. Crowder, London, 1788, Vol. VI.
[252] H. Callisen (1798), quoted by Hertz, ref. 86.
[253] S. L. Hammick, Practical Remarks on Computations, Fractures and Strictures of the Urethra, London, 1830.
[254] A. Budd (1840), quoted by Hertz, ref. 86.
[255] W. G. Moore, *Brit. Med. J.* **1**, 443 (1859).

broke out among the combatant armies of the American Civil War, wounds failed to heal, and slight abrasions, such as the rubbing of a shoe, a bruise, the scratch of a mosquito bite, the prick of a splinter, or vaccination scars tended not only not to heal but to turn into scorbutic ulcers.

Eve[256] stated that in the Confederate Armies during the Civil War the occurrence of scurvy "complicated wounds and seriously interfered with surgical operations." Surgical hemorrhage was much more frequent as the war progressed.

Even during World War I, Lobmayer[257] claimed that the Turkish soldiers suffered severely from scurvy and that in those who developed the disease both skin and flesh wounds and fractures healed poorly. In many cases the fractures showed not the slightest sign of the formation of a callus, even after several months. In confirmation of the fact that it was lack of the antiscorbutic substance which was directly responsible for this failure of fractures to unite, he quoted two cases who suffered from pseudoarthrosis of the humerus and who recovered rapidly and completely as soon as they were put on a diet which was rich in antiscorbutic material.

Since these original clinical observations, various authors have called attention to their significance in the treatment of fractures, notably Bier[258, 259] and Kappis.[260] These observations therefore suggest that not only is the normal functioning of hard and soft tissues interfered with in scurvy but the regenerative powers of these tissues is also reduced.

b. Experimental Evidence. For further confirmation of these facts one must turn again to experimental results. The experimental work which has been carried out on the subject of the effect of vitamin C on the regeneration of tissues has been divided into two main groups, one dealing with animals and the other with human beings. The latter group is restricted to experiments and observations on the regeneration of superficial wounds.

(1) *Refracture Experiments.* It might be of interest at this point to mention various refracture experiments which have been carried out in order to test the oft-repeated clinical observations that an old fracture softens and breaks again when a person develops scurvy. The first of these experiments were carried out by Israel and Frankel[261] and by Israel,[262] who confirmed in experimental animals that this would happen. Hertz,[86] however, believed that their experiments were not critically carried out and that

[256] W. Eve, *in* Medical and Surgical History of the War of the Rebellion, Government Printing Office, Washington, 1866.

[257] G. Lobmayer, *Deut. Z. Chir.* **18,** 208 (1918).

[258] A. Bier, *Arch. klin. Chir.* **127,** 1 (1923).

[259] A. Bier, *Arch. klin. Chir.* **138,** 107 (1925).

[260] H. Kappis, *Deut. med. Wochschr.* **53,** 1734 (1927).

[261] A. Israel and R. Frankel, *Klin. Wochschr.* **5,** 94 (1926).

[262] A. Israel, *Arch. klin. Chir.* **142,** 145 (1926).

what really occurred was a new fracture, resulting from the fragility of the bones which ordinarily accompanies scurvy.

One cannot assume, however, that these latter experiments necessarily invalidate completely the older clinical observations. The diets on which these men were placed in order to secure union of the fractures and the diets they received subsequently, and until they developed scurvy, were almost certainly very low in vitamin C, and it seems likely that an imperfect union of the bones occurred in the first place. In many cases the refracture took place only a few months after the original fracture, and it is probable that these people had been on an almost scorbutic diet all the time and that the fracture had never healed properly. On the other hand, Walter's (1748) observation of the softening of a 50-year-old healed fracture must be regarded with considerable suspicion. It was almost certainly a new fracture, and it seems hardly likely that the person afflicted could remember the exact site of his injury after such a passage of time. Murray and Kodicek[263] have obtained no evidence that healed fractures undergo refracture in scurvy.

Some of the animal experiments on the effect of scurvy on regeneration of tissues were carried out on skin wounds and some on fractured or injured bones. Both series of experiments are included because, as will be seen later, there are some similar processes involved in the healing of both types of tissue.

(2) *Skin Wounds.* It seems well established that there is a delay in the healing of skin wounds in vitamin C deficiency. The first experiment on this subject was carried out by Ishido[264] in 1923. Although he found considerable delay in the healing of wounds, he found also that, even if the animal remained scorbutic, eventually the wounds would heal. Further discussion of this question appears later.

Saitta[265] also found delayed healing of wounds in scorbutic guinea pigs and claimed that vitamin C applied externally to the wounds would accelerate healing. Lauber[266] found that direct application of vitamin C to skin wounds in mice did the same. Against these results he found that injections of vitamin C into normal guinea pigs had no effect on the speed at which wounds heal. This is not surprising because the guinea pigs were already receiving their optimum dose of vitamin C, and it has been shown by Bourne[267] that in animals which are already receiving the optimum amount of vitamin C there is no increase in the amount of bone regenerated when additional vitamin C is given. Likewise it has been found by Proto[268] that

[263] P. D. F. Murray and E. Kodicek, *J. Anat.* **82,** 158 (1949).
[264] B. Ishido, *Arch. path. Anat. Physiol.* (*Virchow's*) **240,** 241 (1923).
[265] S. Saitta, *Scr. biol. Castaldi,* **4,** 301 (1929).
[266] H. J. Lauber, *Beitr. klin. Chir* (*Bruns'*) **158,** 293 (1933).
[267] G. H. Bourne, *Quart. J. Exptl. Physiol.* **31,** 319 (1942).
[268] M. Proto, *Ann. ital. chir.* **15,** 57 (1936).

the injection or local application of vitamin C to skin wounds in pigeons does not affect the rate of wound healing (presumably because pigeons synthesize their own vitamin C). Mice also appear to be able to do the same, and therefore it is surprising that Lauber[266] detected any result from the treatment of wounds in these animals with the vitamin. Rabbits constitute another group of animals which seem to be able to synthesize their own vitamin C, and therefore one could anticipate that Mann and Pullinger[269] would be unable to find that local application of vitamin C had any stimulating effect on the regeneration of a cornea which had previously been burnt with mustard gas.

Lanman and Ingalls[270] carried out direct measurements of the force required to break abdominal wounds in normal and scorbutic animals. Their measurements were made by distending the abdomen with air and noting the pressure on a mercury manometer at which the wound burst. Similar observations were made on wounds in the stomach wall. The pressures in millimeters at which rupturing took place are shown below:

	2 mg. vitamin C	0.5 mg. vitamin C
Abdominal wounds	160	65
Gastric wounds	70	30

Taffel and Harvey[271] confirmed these results. Lauber and Rosenfeld[272] showed that if animals on low vitamin C diet are wounded the organs lose most of their vitamin C, presumably because of the increased demand for, and therefore the mobilization of, the vitamin for the healing process. These authors used the acid silver nitrate method for demonstrating vitamin C in tissues (for discussions of the specficity of this reagent, see Leblond,[273] Bourne,[274] Giroud,[275] and Barnett and Bourne[276]). Lauber and Rosenfeld[272] found that in scorbutic animals the vitamin C content of organs and wounds, as shown by the $AgNO_3$ method, was practically nil. The same tissues in guinea pigs which had been given 100 mg. of ascorbic acid daily stained strongly.

The results of Bartlett *et al.*,[277] derived from quantitative measurements

[269] I. Mann and B. P. Pullinger, *Brit. J. Ophthalmol.* **24,** 444 (1940).

[270] T. H. Lanman and T. H. Ingalls, *Am. J. Surg.* **105,** 616 (1937).

[271] M. Taffel and S. C. Harvey, *Proc. Soc. Exptl. Biol. Med.* **38,** 518 (1938).

[272] H. J. Lauber and W. Rosenfeld, *Klin. Wochschr.* **17,** 1587 (1938).

[273] C. P. Leblond, La vitamine C dans l'organisme, Thesis, University of Paris, 1934.

[274] G. H. Bourne, *Anat. Record* **66,** 370 (1936).

[275] A. Giroud, L'acide ascorbique dans la cellule et les tissues, Protoplasma Monographs, Gebrüder Borntraeger, Berlin, 1938.

[276] S. A. Barnett and G. H. Bourne, *J. Anat.* **75,** 251 (1941).

[277] M. K. Bartlett, C. M. Jones, and A. E. Ryan, *New Engl. J. Med.* **226,** 469 (1942)

of the vitamin C content of wounds, gave some support to the preceding work. They also found that the strength of a scar is proportional to its vitamin C content. For example, the scars of animals receiving 33 mg. of vitamin C twice daily contained on an average 7.64 mg. of vitamin C per 100 g. of tissue and burst at a pressure of 258 mm. of mercury. In guinea pigs on a scorbutic diet the vitamin C content of scars averaged 0.31 mg. per 100 g. and the wounds burst at 127 mm. of mercury. Hunt[278] and Hartzell and Stone[279] also found that vitamin C deficiency lowered the tensile strength of skin wounds.

Experiments on human patients[280] showed that those with low plasma vitamin C (below 0.20 mg. per 100 cc.) had wounds with low tensile strength. It was also shown that when such patients were given large amounts of vitamin C the healing of fascial scars was much faster than the healing of skin scars. Wolfer and Hoebel[280] had previously shown that patients with low values for plasma vitamin C showed the slowest healing of surgical wounds. For further supporting clinical observations, see Archer and Graham,[281] Hunt,[278] and Crandon *et al.*[282]

Crandon, Lund, and Dill studied wound-healing in an experimental human subject. After the subject had been on a vitamin C-deficient diet for 3 months and the plasma vitamin C had been nil for 44 days, a wound made in the back showed good healing within 10 days. There appeared to be abundant intercellular substance and capillary formation. After about 6 months, however, a similar wound made in the back showed no sign of healing. Bourne[283] tested the tensile strength of skin wounds of guinea pigs on graded doses of vitamin C and found the following results.

No vitamin C	25.3 ± 7.40 g.
0.25 mg. vitamin C	79.3 ± 10.40 g.
0.50 mg. vitamin C	118.5 ± 18.00 g.
1.00 mg. vitamin C	137.7 ± 13.90 g.
2.00 mg. vitamin C	338.7 ± 26.80 g.
4.00 mg. vitamin C	296.9 ± 18.50 g.

Comparison of blood plasma vitamin C with tensile strength of wounds in some of these animals showed a progressive decrease of tensile strength as the plasma level fell.

Animals which had wounds of low tensile strength not only showed less fibrous tissue in the scar but a greater proportion of that present was immature precollagen (reticulin) (see Figs. 45–50).

278 A. Hunt, *Brit. J. Surg.* **28,** 436 (1941).
279 J. B. Hartzell and W. E. Stone, *Surg. Gynecol. Obstet.* **75,** 1 (1942).
280 J. A. Wolfer and F. C. Hoebel, *Surg. Gynecol. Obstet.* **69,** 745 (1940).
281 H. E. Archer and G. Graham, *Lancet* **2,** 364 (1936).
282 T. H. Crandon, C. C. Lund, and D. B. Dill, *New Engl. J. Med.* **223,** 353 (1940).
283 G. H. Bourne, *Lancet,* **ii,** 661 (1942).

Campbell *et al.*[284] and Campbell and Ferguson[285] found that in a scorbutic animal healing of a cautery injury to the cornea was due to a delay of epithelialization because a suitable fibrous substrate did not form. Galloway *et al.*[286] found that a scorbutic condition delayed the healing of skin wounds but not those of the gum or cornea. They stated that regeneration of an epithelium can take place normally in scorbutic animals but that where the healing needs new collagenous tissue, as in skin, epithelialization will be retarded. Bunting and White[287] found in skin wounds of scorbutic guinea pigs that connective tissue cells failed to mature, that alkaline phosphatase was often absent from scorbutic wounds, but that more mucopolysaccharides were present. Bradfield and Kodicek[288] found, however, that the mucopolysaccharides and also the precollagen formed in scurvy were abnormal in nature.

It seems fairly well established from the literature quoted that vitamin C is essential for the healing of skin wounds. It can be shown that it is likewise essential for the healing of bone.

(3) *Bone Injuries.* The earliest experimental observation on the importance of vitamin C in the healing of bone was that of Shinya[289] in 1922. He concluded that bones from scorbutic guinea pigs cannot be transplanted into normal animals. He attempted also to transplant sound bones into scorbutic animals, but his animals died of scurvy before results could be obtained.

Ferraris and Lewi[290] found that in scorbutic animals there was not only a delay in the reformation of bone after a fracture but that there appeared to be inhibition of the normal process of removal of cellular debris. The hematoma which formed as a result of the fracture persisted for a very long time.

Watanabe[291] sawed furrows in the skulls of guinea pigs suffering from acute scuvy. As each animal died from scurvy, he examined its skull histologically. He found that the scorbutic animal showed practically no power of regeneration. Watanabe carried out another experiment which is of special interest. He fed animals on insufficient amounts of a complete diet and found that the injured bone of the skull apparently regenerated new bone just as effectively as in animals receiving large amounts of a normal diet. Watanabe also found that rats which were placed on a diet deficient in vitamins A, B, and C were slow in regenerating bone. One can hardly

[284] F. W. Campbell, I. D. Ferguson, and R. C. Garry, *Brit. J. Nutrition* **4,** 32 (1950).
[285] F. W. Campbell and I. D. Ferguson, *Brit. J. Ophthmol.* **34,** 329 (1950).
[286] N. M. Galloway, R. C. Garry, and A. D. Hitchin, *Brit. J. Nutrition* **2,** 228 (1948).
[287] H. Bunting and R. F. White, *Arch. Path.* **49,** 500 (1948).
[288] J. R. G. Bradfield and E. Kodicek, *Biochem. J.* **49,** xvii (1951).
[289] D. H. Shinya (1922), quoted by Hertz, ref. 86.
[290] C. Ferraris and M. Lewi, (1923), quoted by Hertz, ref. 86.
[291] T. Watanabe, *Arch. path. Anat. Physiol.* (*Virchow's*) **251,** 281 (1924).

claim under the conditions of this experiment, however, that it shows a necessity for vitamin C in healing of fractures in the rat, because there was a simultaneous deficiency of two other vitamins, of which vitamin A is known definitely to affect bone growth. In any case rats cannot be given scurvy because they synthesize their own vitamin C. Similar results were obtained by Schilozew[292] and by Roegholt.[293]

Wolbach and Howe[9] investigated the effect of vitamin C deficiency on the regeneration of bones in small saw cuts made in the femora of guinea pigs. In the scorbutic animals there was complete lack of formation of osteoid trabeculae, and although there was some fibrous organization of the clot resulting from the injury there was no penetration into it of capillary blood vessels.

Bier[259] described what he claimed to be the effect of increasing the vitamin C content of the diet of patients suffering from non-union of fractures. One patient, who had had a fractured humerus for 10 weeks which had shown no signs of healing, was given raw carrots (rich in carotene but containing extremely little vitamin C) and lemons (rich in vitamin C) every day. Within another 10 weeks he was cured. Another patient, who showed no sign of callus formation after an osteotomy, obtained complete regeneration of the missing bone within 6 weeks after starting on the above diet. In a third patient consolidation after osteotomy which had not occurred after 6½ months was secured after 4 weeks by giving a diet rich in vitamin C. This improvement might be due to the joint action of vitamins A and C but certainly is not necessarily to be attributed to vitamin C alone.

Jeney and Korpàssy[294] found that wounds made in the heads of various scorbutic animals showed retardation of healing of connective tissue and bone but that there was no disturbance in the formation of new skin. Hunt[278] also showed that there is little interference with epidermal growth in vitamin C deficiency.

Hankey[295, 296] found that on a vitamin C-free diet both rabbits and guinea pigs showed retardation of fracture-healing. In guinea pigs the retardation was severe and often incomplete. In rabbits healing was usually carried to completion but it was considerably delayed. This is a rather surprising result, since rabbits are believed to synthesize enough vitamin C for their normal requirements. Lauber[297] and Lauber and his co-workers[298] stated that, once the normal requirements of a rabbit were satisfied, extra

[292] S. P. Shilozew, *Deut. Z. Chir.* **209,** 320 (1928).
[293] M. M. Roegholt, *Arch. klin. Chir.* **168,** 783 (1932).
[294] A. Jeney and B. Korpassy, *Zentr. Chir.* **61,** 2836 (1934).
[295] H. Hankey, *Deut. Z. Chir.* **245,** 530 (1935).
[296] H. Hankey, *Klin. Wochschr.* **15,** 1121 (1936).
[297] H. J. Lauber, *Beitr. klin. Chir.* (*Bruns'*) **158,** 293 (1936).
[298] H. J. Lauber, H. Nafziger, and T. Bersin, *Klin. Wochschr.* **16,** 1313 (1937).

vitamin C had no further effect on the rate of regeneration of injured bones. It is difficult to appreciate that a rabbit has a normal requirement for vitamin C if it synthesizes the vitamin, but Giangrasso[299, 300] and Giangrasso and Gangitano[301] were also able to secure more rapid regeneration of fractured bones in rabbits by giving vitamin C. The apparent explanation of these facts may be that the healing of a fractured bone calls for more vitamin C than the animal is able to manufacture itself. It has been shown by Bourne[302] that vitamin C does not have any effect on the regeneration of bony tissue in 1-mm. holes bored in the femora of rats (which also synthesize their vitamin C), but there does not appear to be any information as to whether a rat can synthesize enough vitamin C to meet all its needs during repair of a more serious injury such as a fracture.

Hertz[86] has investigated in great detail the effect of vitamin C deficiency on the regeneration of bone in the fractured fibula of a guinea pig. He found delayed absorption of the fracture hematoma, deficient production of osteoid trabeculae, and increased necrosis of the broken ends. Hertz also attempted to repeat Israel and Frankel's refracture experiments, but with no success. Lexer[303] obtained similar results, but he stressed the deficiency of blood supply as a cause of the failure of scorbutic tissues to heal.

Bourne[304] in 1948 also studied the relationship of vitamin C deficiency to certain aspects of regeneration of injured skull and leg bones. In deficient animals there is a delay in healing, and it was found that the migration of polymorphs and macrophages into the injured area was greatly reduced. (See Figs. 60–65). Cottingham and Mills also showed that phagocytosis decreased in scorbutic animals. Murray and Kodicek in a series of papers[263, 305, 306] have investigated in great detail the effect of partial vitamin C deficiency on guinea-pig tissue. They found delay in formation of fracture calluses in bone, failure of consolidation of callus trabeculae (when they did form) into compact bone, and a considerable enlargement of the callus. Stiffening of the knees, which could be bent only painfully and with great difficulty, was also seen. In these partially deficient animals damaged muscle fibers degenerated instead of regenerating, and in some animals this resulted in the disappearance of a large part of the musculature of a limb. Such muscle loss was replaced by hyperplastic connective tissue which was completely or nearly avascular.

[299] G. Giangrasso, *Boll. soc. ital. biol. sper.* **14,** 522 (1939).
[300] G. Giangrasso, *Boll. soc. ital. biol. sper.* **14,** 525 (1939).
[301] G. Giangrasso and A. Gangitano, *Boll. soc. ital. biol. sper.* **14,** 531 (1939).
[302] G. H. Bourne, *J. Physiol. (London)* **101,** 327 (1942).
[303] E. W. Lexer, *Klin. Wochschr.* **18,** 208 (1939).
[304] G. H. Bourne, *J. Anat.* **82,** 208 (1949).
[305] P. D. F. Murray and E. Kodicek, *J. Anat.* **83,** 205 (1949).
[306] P. D. F. Murray and E. Kodicek, *J. Anat.* **83,** 285 (1949).

They also found that guinea pigs completely deprived of vitamin C showed classical changes. The proximal ends of the tibial diaphyses were destroyed by repeated microfractures, and as a result the epiphysis came to overlap (both laterally and medially) the narrow zone of the tibial shaft which was then in contact with it. If animals in which such changes had occurred were given adequate amounts of the vitamin, the long bones became restored more or less to normal. The changes produced were a formation of a subperiosteal thickening in the periosteal cambium between the overhanging epiphysis, the fibrous layer of the periosteum and the old diaphyseal wall, the formation of cartilage (derived from periosteum) around the proximal end of the diaphysis, and its replacement by endochondral bone. Trabecular bone was also found to form in the cartilage (see Figs. 66, 67, and 68).

5. Estrogens and Bone Growth in Scurvy

Estrogenic substances are known to affect bone growth and development. Silberberg and Silberberg[307, 308, 309] have investigated these effects on the bones of scorbutic guinea pigs. Estrogens inhibit the growth of the skeleton in normal animals, and this inhibition is unaffected by the scorbutic condition. In scorbutic animals, however, the calcification and hyalinization of the ground substance is almost completely prevented. The bones are not as fragile as in scorbutic animals without estrogen. The authors believe that this is due to the fact that the scorbutic condition prevent the conversion of intercellular matrix into bone. Another effect of estrogens, the thickening of the walls of the blood vessels, could not be induced in scorbutic animals. One finding of particular interest was that estrogens prolonged the survival of scorbutic guinea pigs and incidentally reduced the severity of the scorbutic symptoms. This latter effect was probably due to the inhibition of skeletal growth by estrogen treatment.

6. Vitamin C and Collagen

Wolbach and Bessey[21] make the following statement about the relationship of vitamin C to the supporting tissues.

> All intercellular substances of supporting tissues, bone, cartilage, fibrous connective tissue and dentine have a common sub-structure of collagen. By collagen we wish to indicate a number of compounds of similar, though not necessarily identical composition, which react similarly to staining technique. It is this protein sub-structure which in scurvy is either not produced or is produced in defective form.

[307] M. Silberberg and R. Silberberg, *Arch. Path.* **36,** 512 (1943).
[308] M. Silberberg and R. Silberberg, *Am. J. Path.* **24,** 1019 (1948).
[309] M. Silberberg and R. Silberberg, *Anat. Record* **102,** 141 (1948).

The work of several authors supports this view. Hunt[278] states also that fibers may be formed in vitamin C deficiency but they are of a precollagenous (reticular) nature and are of low tensile strength.

These results suggest a direct connection between vitamin C and the production of collagen. The reaction of scorbutic tissues to bacterial infection shows lessened production of collagen fibers (see the results of tubercular infection of the lungs of experimental animals by Greene *et al.*[310] and de Savitsch *et al.*[311] and on the intestine by McConkey and Smith[312]). Jeney and Törö[313] investigated the effect of vitamin C upon the production of collagen *in vitro*. They found that, if vitamin C was added to the culture medium in which fibroblasts were being grown, fibers were produced much more rapidly. Mazoué[314] investigated the effects of vitamin C on the organization of a subcutaneous clot produced by severing limb muscles and on the production of aseptic giant cell granulomata produced by injecting kieselguhr into the peritoneal cavity. The most striking result was the lack of production of collagen fibers in vitamin C deficiency in both these conditions. There was also retarded leucocyte infiltration, organization of fibroblasts, and in particular giant cell formation. The administration of vitamin C caused the return of all these processes to normal.

Later Mazoué[315] studied the numerical relationship of the amount of vitamin C (given as lemon juice) to the production of fibroblasts and collagen fibers around intraperitoneal injections of kieselguhr in guinea pigs. In animals on a scorbutic diet 7.5 days elapsed after the injection before the first fibroblasts appeared around the kieselguhr, and 9 days before the first collagen fibers appeared. In animals given 1 cc. of lemon juice per day in addition to the scorbutic diet (equal to 0.5 to 0.7 mg. of vitamin C), the first fibroblasts appeared on the third day and the first collagen fibers on the fifth day. In guinea pigs receiving 3 cc. of lemon juice daily in addition to the scorbutic diet (equal to 1.5 to 2.1 mg. of vitamin C), the first fibroblasts appeared on the second day and the first collagen fibers after about 2½ days. According to Bicknell and Prescott,[35] collagen fibers form in an injury in a scorbutic animal within 24 hours of the administration of vitamin C.

Querido and Gaillard[316] also demonstrated the relation between vitamin C and collagen formation *in vitro*. They cultured osteogenetic cells from the chick in three types of plasma: (1) from scorbutic guinea pigs; (2) from

[310] M. R. Greene, J. Steiner, and R. Kramer, *Am. Rev. Tuberc.* **33,** 585 (1936).

[311] E. de Savitsch, J. D. Stewart, L. Hanson, and E. N. Walsh, *Trans. Natl. Tuberc. Assoc.* **30,** 130 (1934).

[312] M. McConkey and D. T. Smith, *J. Exptl. Med.* **58,** 87 (1934).

[313] A. Jeney and E. Törö, *Arch. path. Anat. Physiol.* (*Virchow's*) **298,** 87 (1934).

[314] H. Mazoué, *Arch. Anat. Microscop.* **33,** 129 (1937).

[315] H. Mazoué, *Arch. Anat. Microscop.* **35,** 91 (1939).

[316] A. Querido and P. J. Gaillard, *Acta Brevia Neerland. Physiol.* **9,** 70 (1939).

scorbutic guinea pigs but with crystalline vitamin C added; (3) from scorbutic guinea pigs which had been treated with large doses of vitamin C for a few days before the experiment. Hardly any fibers were formed by the cells in the first plasma, but in the second and third plasmas, fibers were formed as rapidly as in normal plasma. The cells in the scorbutic plasma showed fatty degeneration.

In the regeneration of bone, fibrous ossein (related to collagen) is laid down on a basis for calcification, and it was shown by Bourne[267] that the production of such fibers (osteoid) in injured areas (drill holes 1 mm. in diameter) in the femurs of scorbutic guinea pigs was directly proportional to the amount of pure synthetic ascorbic acid administered. The optimum level of osteoid formation was found to take place at 2.0 mg. of ascorbic acid per day. The degree of regeneration was measured by cutting sections through the center of the hole, measuring the area of trabeculae (osteoid) present in the central sections, and comparing them with the total area of the hole. In this way an index figure for regeneration was obtained. The results were as follows.

	Trabecular index × 100
No vitamin C	7.73 ± 2.25
0.25 mg. vitamin C	6.70 ± 2.30
0.50 mg. vitamin C	9.74 ± 2.48
1.00 mg. vitamin C	19.41 ± 3.53
2.00 mg. vitamin C	23.73 ± 6.75
4.00 mg. vitamin C	18.09 ± 4.48

Maximum production of bone was therefore secured at a level of 2 mg. of vitamin C per day.

7. Vitamin C and Calcification

The question also arises as to whether ascorbic acid, in addition to its established relation to formation of collagen and bone matrix, is also concerned with calcification. Salter and Aub[317] had originally shown that a scorbutic diet prevented deposition of calcium in bones, but they did not show that vitamin C allowed calcification to take place (because pure vitamin C was not available at the time). Boyle[318] and Boyle, Bessey, and Howe[233] stated that the process of calcification continued in the teeth of guinea pigs when they had been on a scorbutic diet for some time. Wolbach and Bessey[21] stated that there was a generally accepted view that vitamin C played no part in the process of calcification. Bourne[319] investigated this

[317] W. T. Salter and J. C. Aub, *Arch. Path.* **11,** 380 (1931).
[318] P. E. Boyle, *Am. J. Path.* **14,** 843 (1938).
[319] G. H. Bourne, *Quart. J. Exptl. Physiol.* **32,** 1 (1943).

problem in more detail. The conclusions derived from this work were that less bone salt is laid down in normal and regenerating bone in scorbutic animals than in those in which pure vitamin C is given. But the deposition of bone salt is, of course, an orderly and timed process in normal animals. That is to say, bone salt does not appear to be deposited (except in severe scurvy) until there is an adequate fibrous matrix to receive it. Urist and McLean[320] have shown that as osteoid trabeculae are in the process of being formed they already have a deposit of bone salt. The production of fibrous matrix of bone and the deposition of bone salt are therefore probably simultaneous processes. So the deficiency of bone salt in scorbutic animals is due to the fact that there is only a limited amount of matrix produced. It would seem then, provided that there is sufficient vitamin C to produce matrix, that the matrix will be calcified. The apparent failure of long bones to deposit bone salt in scurvy therefore may be due to the fact that no matrix had developed for its reception.

The function of vitamin C in bone formation appears to be to facilitate the production not just of bone matrix but of bone matrix impregnated with phosphatase as well as bone salt. There is no evidence, however, that vitamin C can be regarded as a coenzyme of phosphatase in calcificatory processes or even, for that matter, that phosphatase is concerned with the process of calcification. It may be concerned primarily with the formation of bone matrix. Fell and Danielli,[321] for instance, have shown that first-formed fibres in healing skin wounds are also phosphatase-impregnated.

8. Reproductive Organs

In sex glands it has been found (Lindsay and Medes[322]) that there is a progressive degeneration of the germinal epithelium of the testes in scorbutic animals, but it is a change very similar to that found in inanition. Goettsch[323] has found motile spermatozoa in guinea pigs dying of scurvy, and Mason[1] confirms that the testicular degeneration of scurvy is non-specific. In the female the estrus cycle persists until the later stages of deficiency. Some degeneration of ovarian tissue has been noted in scorbutic animals, but again this is thought not to be a direct or specific effect of vitamin C deficiency.

There are two schools of thought about the effect of vitamin C deficiency on pregnancy (see Mason[1]). According to one group the increased metabolic turnover increases the stress of scurvy in pregnant animals; another group of workers, however, have shown that the histological lesions of scurvy are much less severe in pregnant animals. This protective effect of pregnancy

[320] M. R. Urist and F. C. McLean, *J. Bone Joint Surg.* **23,** 1 (1941).
[321] H. B. Fell and J. F. Danielli, *Brit. J. Exptl. Path.* **24,** 196 (1943).
[322] B. Lindsay and G. Medes, *Am. J. Anat.* **37,** 213 (1926).
[323] M. Goettsch, *Am. J. Physiol.* **95,** 64 (1940).

has been attributed to synthesis of vitamin C by the fetal tissues (Mouriquand and Schoen[324]) and by luteal tissue (Bourne[325]). Giroud[275] has claimed that vitamin C may be synthesized by the fetuses of some species of animals and not by others. Ray[326] and Barnett and Bourne[276] have shown that developing chick embryos synthesize vitamin C.

XII. Vitamin D Deficiency

1. Rickets

Vitamin D deficiency produces changes in bone growth which result in the condition known as rickets, the principal morphological characters of which, in the human, were first described by Pommer[327] in 1885. A full review of the literature of human rickets is given by Hess.[328]

Mellanby[329] in 1918 first described experimental rickets, using the puppy as his experimental animal, and shortly afterwards McCollum *et al.*[330] described the same disease in the rat. It is of interest that Mellanby later found that confinement of his dogs predisposed them toward development of rickets. Also, it is of interest that the rat is normally resistant to vitamin D deficiency and for such a condition to express itself in this animal there must be concomitant unbalance of the calcium-phosphorus ratio or else a deficiency of these elements (Shohl and Wolbach[331]). McCollum and his co-workers[330] showed that cod-liver oil added to the diet of vitamin D-deficient rats induced calcification to take place.

Wolbach and Bessey[21] have pointed out that the pathological changes in the bone-forming tissues in vitamin D deficiency is due to quantitative changes in serum calcium and/or the inorganic phorphorus content of the blood. The process by which this is brought about seems to have been determined (Nicolaysen[332] and Logan[333]). It is apparent that in vitamin D deficiency there is reduced absorption of calcium from the alimentary tract, although in general phosphorus absorption does not seem to be affected. Confirmation of these results has been obtained with radioactive calcium (Cohn and Greenburg[334]).

[324] G. Mouriquand and J. Schoen, *Compt. rend.* **197,** 203 (1933).
[325] G. H. Bourne, *Nature* **135,** 148 (1935).
[326] S. N. Ray, *Biochem. J.* **28,** 189 (1934).
[327] G. Pommer, Untersuchungen uber Osteomalacie und Rachitis, F. C. W. Vogel, Leipzig, 1885.
[328] A. F. Hess, Rickets, Osteomalacea and Tetany, Lea & Febiger, Philadelphia, 1929.
[329] E. Mellanby, *J. Physiol. (London)* **52,** *Proc.* xi, liii (1918).
[330] E. V. McCollum, N. Simmonds, H. T. Parsons, P. G. Shipley, and E. A. Park, *J. Biol. Chem.* **45,** 333 (1921).
[331] M. T. Shohl and S. B. Wolbach, *J. Nutrition* **11,** 275 (1936).
[332] R. Nicolaysen, *Biochem J.* **31,** 122 (1837).
[333] M. A. Logan, *Physiol. Revs.* **20,** 522 (1940).
[334] W. E. Cohn and D. M. Greenberg, *J. Biol. Chem.* **130,** 625 (1939).

Presumably this defective absorption is sufficient to explain the bone changes in rickets, but there has been some question as to whether vitamin D does not also have some local action in bone. Follis[4] says that the balance of evidence is against this, but Cohn and Greenburg and Shimotori and Morgan[335] have suggested that the mechanism in bone which converts inorganic phosphorus into the organic form is affected in rickets. Shipley[336] and Shipley and colleagues[337] found, however, that pieces of bone from rachitic animals would calcify if they were placed in solutions containing an optimum solution of calcium and phosphorus but not if the concentration was not optimal. It is of interest that estrogens stimulate endosteal bone growth in mice but this reaction is not affected by presence or absence of vitamin D (Segaloff and Cahill[338]). Migicovsky[339] showed by the use of radioactive calcium that vitamin D did not exert a direct effect on the mineralization of bone.

Most animals appear to require vitamin D. Rickets has been induced in a large number of different species of animals. Most work on this subject, however, apart from Mellanby's, appears to have been carried out on rats. The characteristic epiphyseal changes which take place in vitamin D deficiency have been described by a variety of authors.

2. Bone Changes

In normal growth the epiphyseal cartilage undergoes proliferation and the diaphyseal edge of it is subject to continual endochondral ossification. As the cartilage cells approach the diaphysis the hypertrophy is accompanied by loss of chondroitin sulfuric acid from the cartilage matrix and accumulation of alkaline phosphatase in its place. Then the cells closest to the diaphysis degenerate and capillary loops pass into the degenerated lines of cartilage cells. There has been some discussion as to whether capillaries penetrate only into spaces in which the cells are dead or whether they are more aggressive and can penetrate into spaces occupied by active cartilage cells as Park[340] and Dodds[341] suggest. Harris[342] and Ham and Lewis[343] believe that the cartilage cells always die before the spaces are invaded by capillary loops. This theory is supported by Wolbach and Bessey,[21] who claim that the pattern of vascular distribution in the invaded cartilage is directly

[335] N. Shimotori and A. F. Morgan, *J. Biol. Chem.* **147,** 201 (1943).
[336] P. G. Shipley, *Bull. Johns Hopkins Hosp.* **35,** 304 (1924).
[337] P. G. Shipley, B. Kramer, and J. Howland, *Biochem J.* **20,** 379 (1926).
[338] A. Segaloff and W. M. Cahill, *Proc. Soc. Exptl. Biol. Med.* **54,** 162 (1943).
[339] B. B. Migicovsky, *Arch. Biochem.* **28,** 324 (1950).
[340] E. A. Park, The Harvey Lectures, Williams and Wilkins, Baltimore, 1938.
[341] G. S. Dodds, *Am. J. Anat.* **50,** 97 (1932).
[342] H. A. Harris, Bone Growth in Health and Disease, Oxford University Press, 1933.
[343] A. W. Ham and M. D. Lewis, *Brit. J. Exptl. Path.* **15,** 228 (1934).

dependent upon the distribution of cartilage cells which have completed their cycle.

Between the invading loops of cartilage, calcification of the matrix takes place and the calcified osteoid is deposited in the lacunae which the cartilage cells originally inhabited and which have now been invaded by capillaries.

In vitamin D deficiency the aligned cartilage cells do not die and capillaries do not appear to penetrate into the lacunae. There is no calcification of the cartilage columns between the cartilage cells, and although some matrix may be laid down it is uncalcified. There is also a resorption of bone formed before the deficiency began. Although endochondral ossification ceases, multiplication of the epiphyseal cartilage continues and so a broad zone is built up which has been described as a "rachitic metaphysis." Not all penetration by capillaries ceases, and some penetrate into the cartilagenous area. Similarly, tongues of cartilage may grow down into the diaphyseal marrow. Trabeculae made up of uncalcified osteoid are also present in this zone. These are due to the deposition of the material on detached groups of cartilage cells. The jumble of tissues in the epiphyseal region shows clinically as a swelling or in the case of the ribs (costochondral junctions) as a beading (see Figs. 69, 70, and 71).

When vitamin D is given to deficient animals there is a calcification in the cartilage nearest the rachitic metaphysis and the width of this calcified area appears to depend upon the amount of vitamin D administered. This fact is applied in the "line test" for assay of vitamin D. Rachitic animals are treated with a known amount of vitamin D to produce a base line. Then substances containing unknown amounts of vitamin D are fed to another group of rachitic animals, usually rats. At the conclusion of the experiment the rats are killed and their long bones dissected out and split in two. The bones are then placed in silver nitrate, and as the line of calcification in both groups becomes black its width is measured. Hence by comparison of the size of the line in the two groups the activity of the vitamin D in the unknown sample can be calculated.

Within 24 hr. of the administration of vitamin D to rachitic animals regenerating cartilage cells can be seen near the diaphysis, and by 2 days there is considerable capillary penetration into the cartilage. Remodeling then goes on, and most signs of the deficiency are removed, but Reed and Reed[344] have pointed out that, in rats at least, X-ray studies show that the rachitic pattern in the epiphyses may never be completely lost. Follis[4] and other authors have also pointed out that bones which have been seriously distorted during the incidence of the disease rarely return to normal.

[344] C. E. Reed and B. P. Reed, *Am. J. Physiol.* **138,** 34 (1942).

3. Teeth

Changes also take place in the teeth in vitamin D deficiency. Weinmann and Schour[345] have described a "calcio-traumatic line" in incisor teeth of deficient rats which is due to interference with the calcification of the dentine at these sites. There is also retardation of formation of bone predentine and cementum. The enamel organ atrophies and frequently undergoes cystic changes in the process.

The primary change in vitamin D deficiency, however, is that of the bones. Other changes, when they do occur, are secondary to those of the bone changes and sometimes are caused by pressure change.

4. Bone Repair

Hertz[86] investigated the effect of vitamin D deficiency in the healing of fractures and was not able to find any evidence that the chondrocytes in his fracture calluses were able to undergo metaplasia directly into bone cells in the way that has been described by various authors in the past (Muller[346]). On the other hand his specimens showed some evidence of direct metaplasia of cartilage into osteoid (chondro-osteoid of von Recklinghausen), although the metaplasia was never complete. He found little or no calcification in the cartilage of the vitamin D-deficient callus. Another finding was that there was a hyperproduction of both osteoid and cartilage.

5. Soft Tissues

The parathyroid undergoes hyperplasia in D deficiency, and according to de Robertis[347] this is accompanied by an increased complexity of the Golgi apparatus. There are few records of changes in the soft tissues or in the reproductive glands, although Richter and Rice[348] have reported that vitamin D deficiency induced dioestrus in the rat and Arvey and Gabe[349] have found that rachitic animals have small, pale, inactive thyroids with the follicles loaded with colloid and the follicular cells flattened. Theopold[350] found an increase of glycogen in the liver, a deposition of iron in the spleen and in the lymphatic ganglia, and atrophy of the ducts of the submaxillary glands. There was a reduction in the number of red cells and of reticulocytes but an increase in the alkaline phosphatase content of intestine, kidneys, and costochondral junctions in rachitic rats. It is of interest that Hensborough and Nicholas[351] found that vitamin D injected together with P^{32}

[345] J. O. Weinmann and I. Schour, *Am. J. Pathol.* **21,** 833 (1945).

[346] H. Muller, *Z. wiss Zoöl.* **9,** 147 (1858).

[347] E. de Robertis, *Anat. Record* **79,** 417 (1941).

[348] C. P. Richter and K. K. Rice, *Am. J. Physiol.* **139,** 693 (1943).

[349] L. Arvey and M. Gabe, *J. Physiol. (London)* **42,** 835 (1950).

[350] W. Theopold, *Monatsschr. Kinderheilk.* **98,** 227 (1951).

[351] L. A. Hensborough and P. A. Nicholas, *J. Expt. Zoöl.* **112,** 185 (1949).

(radioactive phosphorus) into the air chamber of a developing hen's egg facilitated the incorporation of the P^{32} into the developing embryo. Autoradiographic studies showed P^{32} to be present in all parts of the embryo and to be greatest in the regions of greatest concentration of cells.

XIII. Vitamin E Deficiency

1. General Description

The existence of a "hitherto unknown dietary factor essential for reproduction" was described by Evans and Bishop[352] in 1922 and in subsequent papers. It was discovered first in the rat, and detailed studies of the effect of the deficiency in this animal were carried out by Mason.[1] Later Goettsch and Pappenheimer[353] described vitamin E deficiency in guinea pigs and rabbits, and in 1931 they attributed "crazy chick disease" to a deficiency of this vitamin. Anderson *et al.*[354] found that dogs appeared to need vitamin E in their diet.

2. Embryo

Interference with reproduction is the outstanding effect of vitamin E deficiency. In the female there is no effect on the various processes of estrus and ovulation or even implantation of the fertilized embryo. However, eventually the fetus is resorbed. At about the seventh day of gestation the mesodermal tissues of the embryo show signs of retarded development. In the developing embryo the first sign of trouble is usually indicated when the ectodermal cavity does not appear. Failure of ectodermal activity is expressed too by failure of the ectoplacentral and amniotic cavities to develop, while mesodermal failure is expressed by lack of formation of blood islands and hemopoiesis in general. The liver develops irregularly, and blood cells are absent from heart and blood vessels. There is failure (or decreased activity) of the fetal vessels to invade the maternal tissues, and it is this inadequate connection between mother and embryo that is probably responsible for asphyxia and starvation of the latter. Soumalainen[355] found that the embryos of mouse and hedgehog are partly nourished in the uterus by the histiotrophe (the part of the endometrium not occupied by the decidua) and that in vitamin E deficiency the histiotrophe does not form. Histochemical studies of the localization of iron have shown that this metal decreases in deficient animals just before resorption of the embryo begins. During resorption large amounts of iron appear in the yolk sac. The fetus eventually becomes necrotic and is resorbed. Vitamin E has

[352] H. M. Evans and K. S. Bishop, *Science* **56,** 650 (1922).

[353] M. Goettsch and A. M. Pappenheimer, *J. Exptl. Med.* **54,** 145 (1930).

[354] H. D. Anderson, C. A. Elvehjem, and J. E. Gonce, *Proc. Soc. Exptl. Biol. Med.* **42,** 750 (1939).

[355] P. Suomalainen, *Nature* **165,** 364 (1950).

also been found necessary for hatchability of the eggs of the domestic fowl. The developing E-deficient chick embryo shows a retardation of growth and does not differentiate properly. At the fourth day of incubation there is a distintegration of the blood vessels of the blastoderm, hemorrhage into the coelome, and the formation of a dense ridge of tissue, described as a lethal ring, in the blastoderm around the embryo. This prevents the expansion of the embryonic membranes and interferes with the vitelline circulation. Even if embryos survive the critical lethal ring period, they die within the next day or two from internal hemorrhage. This account of vitamin E deficiency is taken from Mason[1] (see Figs. 73 and 74).

3. Male Reproductive Organs

Mason also listed the workers who have described changes in the testes of male rats on a vitamin E-deficient diet. He quoted the list of progressive changes in the sperm in this deficiency, first described by Evans and Burr.[356] These are:

1. Sperm normal in number, morphology, and motility, but fertilizing power lost.
2. Sperm morphology normal but motility lost.
3. Sperm fused in groups, invested by Sertoli cell cuffs, and showing abnormal staining reactions.
4. Sperm absent from ejaculate.
5. Loss of ability of animal to form a copulation plug.
6. Loss of all sex interest by the animal.

Accompanying these changes we find chromatolysis and fusion of sperms. The spermatids then become vesicular and finally degenerate. The chromatin of the spermatocytes shows liquefaction, and it aggregates in the form of crescents to one side of the structure. Sometimes coalescence of these cells builds up large multinucleate masses. In general the Sertoli cells are not affected, although some show signs of damage. Eventually, the Sertoli cells become the sole occupants of the seminiferous tubules. There seems to be no particular change in the interstitial cells, although Marchesi[357] believed that they may proliferate and Ringsted[358] that in advanced vitamin E deficiency they atrophy. One characteristic of testicular degeneration in avitaminosis is that it cannot be stopped from running to completion (see Fig. 72).

The changes so far recorded, particularly those concerned with the testes, suggest that there is some intimate connection between vitamin E and the functioning of the nucleus. Since in both the embryo and the testis there is rapid cellular multiplication, it is legitimate to speculate that this might

[356] H. M. Evans and G. O. Burr, *Mem. Univ. Calif.* **8,** 1 (1927).
[357] F. Marchesi, *Sperimentale* **89,** 119 (1935).
[358] A. Ringsted (1936), quoted by Mason, ref. 1.

be concerned with the process of synthesis of desoxyribosenucleic acid, a process which takes place every time a cell divides and which incidentally is known to be affected by X-radiation (Howard and Pelc[359]).

In view of the fact that the intersitial tissue is undamaged, it is not surprising that accessory sex glands are unaffected by the deficiency.

4. Pituitary Gland

The anterior pituitary gland shows changes in rats on a vitamin E-free diet. The basophilic cells become modified in a fashion not unlike that produced by undescended testes or by castration (see Nelson,[360] Geller,[361] and Gierhake[362]). Mason discussed this work and pointed out that pituitaries from such animals when transplanted into immature rats and mice caused precocious sexual maturity. This led to the view that vitamin E controls the endocrine function of the anterior pituitary gland. For further discussion of this problem see Mason's work.

5. Muscle

In 1930 Goetsch and Pappenheimer[353] described a "nutritional muscular dystrophy" in rabbits and in guinea pigs which has now been referred to vitamin E deficiency (Olcott[363]). Paralysis in vitamin E-deficient animals had, however, first been described by Evans and Burr[364] (see also Demole and Pfaltz[365]). Pappenheimer[366, 367] subsequently described in more detail the pathology of nutritional muscular dystrophy due to vitamin E deficiency in young rats and mice, and Mackenzie and McCollum[368] have shown that the same disease in young rabbits can be cured with vitamin E. They showed that there is some interference with the creatine metabolism of the muscles of such animals. Kaunitz and Pappenheimer[369] later investigated oxygen uptake in vitamin E deficiency, and Fenn and Goettsch[370] studied electrolyte changes in this condition. The conclusion drawn from these observations is that there is an increased uptake of oxygen by vitamin E deficient muscle which occurs some time before any histological change can be observed. Simultaneously there is an increase in muscle sodium and

359 A. Howard and S. Pelc *in* The Chemistry and Physiology of the Nucleus, Suppl. 2 of *Exptl. Cell Research*, Academic Press, New York. 1952
360 W. O. Nelson, *Anat. Record* **56**, 241 (1933).
361 F. C. Geller, *Arch. Gynäkol.* **156**, 345 (1934).
362 E. Gierhake, *Deut. med. Wochschr.* **61**, 1674 (1935).
363 H. S Olcott, *J. Nutrition* **15**, 221 (1938).
364 H. M. Evans and G. Burr, *J. Biol. Chem.* **76**, 263 (1928).
365 V. Demole and H. Pfaltz, *Rev. méd. Suisse romandes* **60**, 464 (1940).
366 A. M. Pappenheimer, *Am. J. Path.* **15**, 179 (1939).
367 A. M. Pappenheimer, *Am. J. Path.* **18**, 169 (1942).
368 C. G. Mackenzie and E. V. McCollum, *J. Nutrition* **19**, 345 (1940).
369 H. Kaunitz and A. M. Pappenheimer, *Am. J. Physiol.* **138**, 328 (1943).
370 W. O. Fenn and M. Goettsch, *J. Biol. Chem.* **124**, 41 (1944).

chloride and a reduction in muscle creatinine. Lu, Emerson, and Evans[371] have also shown that the ability of muscle to carry out phosphorylation of glycogen is reduced. Other chemical changes in muscle are an increase in fat, lipid, and cholesterol (Morgulis *et al.*[372]), and an increase of calcium and phosphorus. The calcium can be demonstrated histologically in the muscle fibers when they become necrotic.

Follis[4] has described the histological changes in muscle in E deficient rats as follows. There is swelling and hyalinization of muscle fibers, followed by necrosis. The fluid of the intercellular spaces may be viscous and it contains extra protein. In this process there is a rapid and striking loss of cross striations of the muscle fibers which become amorphous masses. In the final stages the muscle fibers contain only granular debris and newly formed nuclei (Telford *et al.*[373]) (see Figs. 75, 76, 78, 79, and 80). It is of interest that considerable histological degeneration of muscles may occur in vitamin E deficient animals without clinical symptoms. The sarcolemmal nuclei proliferate and there is leucocytic infiltration. There is also an increase of fat. Regeneration is prompt with administration of vitamin E. Telford has described a loss of motor end plates from the muscles of deficient animals and their replacement on administration of the vitamin. Martin and Moore[374] have described necrosis of the tubular epithelium of the kidney of rats on a vitamin E-deficient diet.

6. Pigment and Nervous System

There is a widespread deposit of a golden-colored fluorescent pigment (Moore and Wang[375]) in vitamin E deficiency. It was described first in the uterus and may appear also in skeletal muscle. It appears to be associated with muscular dystrophy. Lymph nodes also contain the pigment. Moore and Wang believe that this pigment arises *in vivo* from the abnormal oxidation of protein. They state that the presence of oxidized and deaminated groups derived from tryptophan is to be suspected. The pigment has some similarities with hemofuscin. It has some properties which differentiate it from ceroid, a brown fluorescent pigment present in rats suffering from experimental cirrhosis of the liver, although Mason and Emmel[376] have drawn attention to certain similarities between the two pigments.

Interference with pigment metabolism is shown by the fact that the iron-containing pigment which discolors the incisor teeth of rats is normal in rats which are maintained on a diet deficient in both fat and vitamin E,

[371] G. D. Lu, G. A. Emerson and H. M. Evans, *Am. J. Physiol.* **133,** 367 (1941).

[372] S. Morgulis, V. M. Wilder, H. C. Spencer, and S. H. Eppstein, *J. Biol. Chem.* **124,** 755 (1938).

[373] I. R. Telford, *Anat. Record* **81,** 171 (1941).

[374] A. J. R. Martin and T. Moore, *J. Hyg.* **39,** 643 (1939).

[375] T. Moore and Y. Wang, *Brit. J. Nutrition* **1,** 53 (1947).

[376] K. E. Mason and A. F. Emmel, *Anat. Record* **92,** 33 (1945)

but if fat is added to the diet the pigment is not deposited upon the teeth (Moore[377] and Dam and Granados[378]).

Mason *et al.*[379] showed that if rats were raised on a vitamin E-deficient diet which contained 20% of cod-liver oil there was an accumulation of acid-fast pigment in the adipose tissue where it was localized particularly in developing fat cells. There were also localized proliferations of cells which resulted in the formation of pigmented cell clusters, the cells of which merged to form multinucleated giant cells. Accumulations of pigmented macrophages also occurred.

Granados *et al.*[380] found that it was the unsaturated fraction of cod-liver oil that was most active in this respect when the diet was simultaneously deficient in vitamin E.

Lipschutz[381] in 1936 reported lesions of the vestibular nuclei and their connections in vitamin E-deficient rats and also in the extrapyramidal tracts, the proprioceptive tracts, and the ventral horn cells of the spinal cord. There has been some disagreement about the incidence and extent of neurological changes, but the variations in results obtained may be explained by the work of Luttrell and Mason.[382] They found that after some months on a vitamin E-deficient diet young rats showed evidence of hypalgesia and progressive paresis. Autopsy showed gliosis, distortion of axon pattern, and demyelinization in the posterior columns of the spinal cord (fasciculi cuneatus and gracilis) and of the proximal parts of the posterior roots of the cervical, thoracic, and lumbar segments of the spinal cord. When oxidized lard was used instead of fresh lard in the diet, however, the onset of symptoms was delayed and the severity of the neurological changes diminished (see Figs. 81 and 82).

In mice, however, although raising the amount of cod-liver oil (and thus the unsaturated fat) in the diet from 2% to 10% or 20% caused an increased incidence of myopathy and paralysis, no changes were observed in the spinal cord (Tobin[383]). For a review of the effect of unsaturated fat in vitamin E-deficient diets see Pappenheimer.[384]

Tobin also showed that raising the cod-liver oil percentage in the vitamin E deficient diet caused an increase in acid-fast pigment in various tissues. Tobin and Birnbaum[385] had previously shown that a brown degeneration

[377] T. Moore, *Biochem. J.* **37,** 112 (1943).
[378] H. Dam and H. Granados, *Science* **102,** 327 (1945).
[379] K. E. Mason, H. Dam, and H. Granados, *Anat. Record* **94,** 265 (1946).
[380] H. Granados, K. E. Mason, and H. Dam, *Acta Path. Microbiol. Scand.* **24,** 86 (1947).
[381] M. D. Lipschutz, *Rev. neurol.* **65,** 221 (1936).
[382] C. N. Luttrell and K. E. Mason, *Ann. N. Y. Acad. Sci.* **52,** 113 (1949).
[383] C. E. Tobin, *Arch. Path.* **50,** 1 (1950).
[384] A. M. Pappenheimer, *Physiol. Revs.* **23,** 37 (1943).
[385] C. E. Tobin and J. F. Birnbaum, *Arch. Path.* **44,** 269 (1947).

was induced in the adrenal glands of mice on a similar diet. The pigment was found in all zones of the adrenal cortex as well as in the medulla in advanced stages. This pigment, from its staining reactions, was similar to that found in rats after treatment with a similar diet and to the ceroid pigment of rats suffering from dietary cirrhosis (see Fig. 77). Mason and Hartsough[386] found that in mink suffering from steatitis there were deposits of acid-fast yellow-brown pigment in lymph nodes, adrenals, and adipose tissue. The tissues of these animals were found to be low in tocopherol, and investigation of the diet showed a tocopherol deficiency also. The authors consider that a diet high in unsaturated fat and low in vitamin E plays an important role in the genesis of steatitis in the mink. It is of interest that it has been shown (Bourne[387, 388]) that the Australian opossum in its natural state may show accumulations of yellow and brown acid-fast pigment in adipose tissue and adrenals. Whether this could be attributed to a dietary aberration it is difficult to say.

7. Heart

Houchin and Smith[389] have described death in vitamin E deficient rabbits as being due to myocardial failure. They have also described an increased uptake of oxygen by the cardiac muscle of vitamin E deficient hamsters. It had previously been reported by Freire[390] that vitamin E deficient rats showed foci of hyaline necrosis in the myocardium. Freire and Magahas[391] confirmed this result and found accompanying changes in the electrocardiogram. Subsequently Mason and Emmel[376] described necrosis of the myocardium and fibrous replacement of the damaged tissue.

Gatz and Houchin[392] found that the degenerative processes in the heart muscle of E deficient rabbits resembled those in skeletal muscle. In the cardiac tissue which showed considerable increase of oxygen consumption the capillaries and other vessels were prominent and distended with blood. Constriction bands developed around some fibers. The view of the authors is that muscle metabolism in vitamin E deficiency becomes increased prior to the appearance of necrosis.

One of the characteristics of degenerating cardiac fibers in this deficiency is the appearance of vacuoles in the sarcoplasm. The nature of these vacuoles is not known, but they do not contain fat or acid-fast pigments. As the droplets enlarge they separate the myofibrillae, and as a result the two bands became more granular in appearance. There is then a progressive

[386] K. E. Mason and G. R. Hartsough, *J. Am. Vet. Med. Assoc.* **119,** 72 (1951).
[387] G. H. Bourne, *Nature* **134,** 664 (1934).
[388] G. H. Bourne, The Mammalian Adrenal Gland, Oxford University Press, (1949).
[389] O. P. Houchin and P. M. Smith, *Am. J. Physiol.* **141,** 242 (1944).
[390] S. A. Freire, *Brasil-med.* **55,** 308 (1941).
[391] S. A. Freire and A. Magahas, *Rev. brasil. biol.* **3,** 91 (1943).
[392] A. J. Gatz and O. B. Houchin, *Anat. Record* **110,** 249 (1951).

loss of myofibrillae as the vacuoles coalesce, and finally only a sarcolemmal sac full of fluid and containing little else beyond a nucleus is found. There is little evidence of degeneration of nerve endings in cardiac muscle. Filer *et al.*[393] found that vitamin E deficiency combined with a low fat diet led to slight but consistent changes in the electrocardiogram and pneumocardiogram. The changes were a reduction in amplitude of the R and T waves, inversion of the T wave and shortening of the time for initiation of ventricular ejection from the heart.

Gatz and Houchin also examined the thyroid gland histologically in view of the increased oxygen uptake by muscle. They found that the thyroids appeared relatively inactive as judged by the accumulation of colloid in the follicles and the flattened nature of the follicular epithelium.

8. Liver

Chevrels[394] found that in vitamin E deficiency in rabbits there was fatty infiltration in the center of the liver lobules. The glycogen stores were reduced, and the infiltrated regions were poor in ribonuclease. Alkaline phosphatase was also found to be present in the central zones of the lobules and in the biliary canals. Linden and Himsworth[395] found that a double deficiency of vitamin E and sulfur-containing amino acids caused massive necrosis of the liver.

9. Embryonic Eyes

Allison and Orent Keiles[396] found that the eyes of rats born of vitamin E deficient mothers showed retrolental fibroplasia such as is sometimes seen in premature human infants. The eyes were unusually small and the eyelids failed to open. There was an opaque white membrane behind the pupil in animals so affected, and in one animal the pupil was filled by a clot of blood.

XIV. Deficiency of Other Factors

Deficiency in other nutrients, e.g., minerals, protein, etc., also causes structural changes in tissues. Deficiency of individual amino acids (e.g., tryptophan) produces well-defined histological alteration, and deficiency of certain essential fatty acids, the importance of which was first described by Burr and Burr[397, 398] also results in pathological changes in tissues. In female rats lack of these fatty acids causes cessation of estrus, fetal resorp-

[393] L. J. Filer, R. E. Rumery, P. N. G. Yu, and K. E. Mason, *Ann. N. Y. Acad. Sci.* **52,** 284 (1949).

[394] M. L. Chevrels, *Compt. rend.* **232,** 1024 (1951).

[395] O. Linden and H. P. Himsworth, *Brit. J. Exptl. Path.* **31,** 651 (1950).

[396] E. C. Allison and E. Orent Keiles, *Proc. Soc. Exptl. Biol. Med.* **76,** 295 (1951).

[393] G. O. Burr and M. M. Burr, *J. Biol. Chem.* **82,** 345 (1929).

[398] G. O. Burr and M. M. Burr, *J. Biol. Chem.* **86,** 587 (1930).

tion, extended gestation, impaired lactation, and production of weak and defective young.

Male rats on this deficiency show degenerative changes in the testes, with poor growth and development of the animals as a whole. It is thought that fatty acid deficiency may result in inadequate production of sex hormones. The fatty acids in question are linoleic, linolenic, and arachidonic acids which, according to Shoenheimer and Rittenberg,[399] are not synthesized in the body. Other tissue changes induced by deficiency of these elements include hyperkeratosis of the skin and damage to the tubular epithelium of the kidney.

Ramalingaswami and Sinclair[400] found that deficiency of essential fatty acids in rats was characterized by failure of growth and by lesions of the skin and muco-cutaneous junctions. The changes in the lips, the angles of the mouth, and in the paws were similar to those found in pyridoxine deficiency. They also found a similarity in the two deficiencies in the elevation of the erythrocyte count and the reduction of the mean corpuscular volume, though these changes were less in essential fatty acid deficiency. In the lungs in the fatty acid deficiency there was accumulation of large foamy phagocytic cells which resemble "heart failure" cells in certain respects. As mentioned before, these authors feel[20] that the condition of "phrynoderma" described by Nicholls resembles fatty acid deficiency more than it does vitamin A deficiency (see Figs. 84 and 85).

Adendum: Vitamin C. There is an absence of extra-cellular metachromatic substances in scorbutic wounds but these appear 6 hours after injection of vitamin C[401]. However, increased glycoproteins have been recorded in scorbutic ground substance (which is also said to be depolymerized.[402] Recent studies[403] have shown a general increase in mucopolysaccharides in scorbutic ground substance, the latter being in an abnormal physical condition. The water-binding capacity of the skin was increased. In wounds a depolymerized, water-soluble, periodic-Schiff-positive, non-metachromatic ground substance was present. A similar substance was present in the fibroblasts. Adrenalectomy aggravated the scorbutic condition but both cortisone injections and local application of vitamin C caused gelation of ground substance. Many authors (see ref. 404) showed that cortisone injections induced connective tissue changes similar to scurvy but these are not prevented by treatment with vitamin C (Bourne[404]).

[399] R. Schoenheimer and D. Rittenberg, *Physiol. Revs.* **20,** 218 (1940).
[400] V. Ramalingaswami and H. M. Sinclair, *Brit. J. Nutrition* **5,** x (1951).
[401] J. R. Penney and B. M. Balfour, *J. Path. Bact.* **61,** 171 (1949).
[402] I. Gersh and H. R. Catchpole, *Amer. J. Anat.* **85,** 457 (1949).
[403] B. H. Persson, *Supplement 2, Acta Societ. Medic. Upsaliensis* (1953).
[404] G. H. Bourne, *Internat. Rev. Vit. Res.* **24,** 318 (1952).

PLATE I

FIG. 1. Anteroposterior section of the retina of a control rat.

FIG. 2. Anteroposterior section of the retina in a vitamin A-deficient rat, showing marked reduction in thickness and poor staining of rod and cone layer.

FIG. 3. Transverse section of the upper third of the tibia of a rat receiving excess vitamin A (1250 I.U. of vitamin A daily from 21st to 28th day of age). Resorption of cortex externally, and newly formed, partly calcified bone deposited internally.

FIG. 4. Undersurface of the brain of a normal young puppy to show optic nerves.

FIG. 5. Undersurface of the brain of a vitamin A-deficient puppy (litter mate). Note the tortuosity of optic nerves in the orbit, also compression deformities of cerebrum and cerebellum. The tortuous optic tract is evidence of failure of dimensional increase of the long bony orbit in a region which excludes deformation by pressure of bone.

FIG. 6. Lung of a vitamin A-deficient rat, showing focal replacement of the columnar ciliated epithelium of bronchus by stratified keratinizing epithelium, and infiltration of submucosa.

FIG. 7. Fractured femur near the distal epiphysis of a rat which received 1250 I.U. per gram of pure vitamin A for 7 days, following weaning at 21 days of age.

FIG. 8. Fore-stomach of a normal rat. Note normal squamous keratinizing epithelium, thin and relatively avascular *lamina propria*, well-developed *muscularis mucosae*, submucosa with less connective tissue and numerous blood vessels, and circular and longitudinal layers of the *muscularis externor*.

FIG. 9. Fore-stomach of a vitamin A-deficient rat, showing severe epithelial hyperplasia and hyperkeratosis, thickening and papillary prolongation of *lamina propria*, atrophy and penetration by blood vessels of *muscularis mucosae*. Continuity of *muscularis mucosae* is unbroken.

Figures 3, 4, 5, and 7 are by courtesy of S. B. Wolbach and the *Journal of Bone and Joint Surgery*. Figures 1, 2, 6, 8, and 9 are by courtesy of H. M. Sinclair and V. Ramalingaswami.

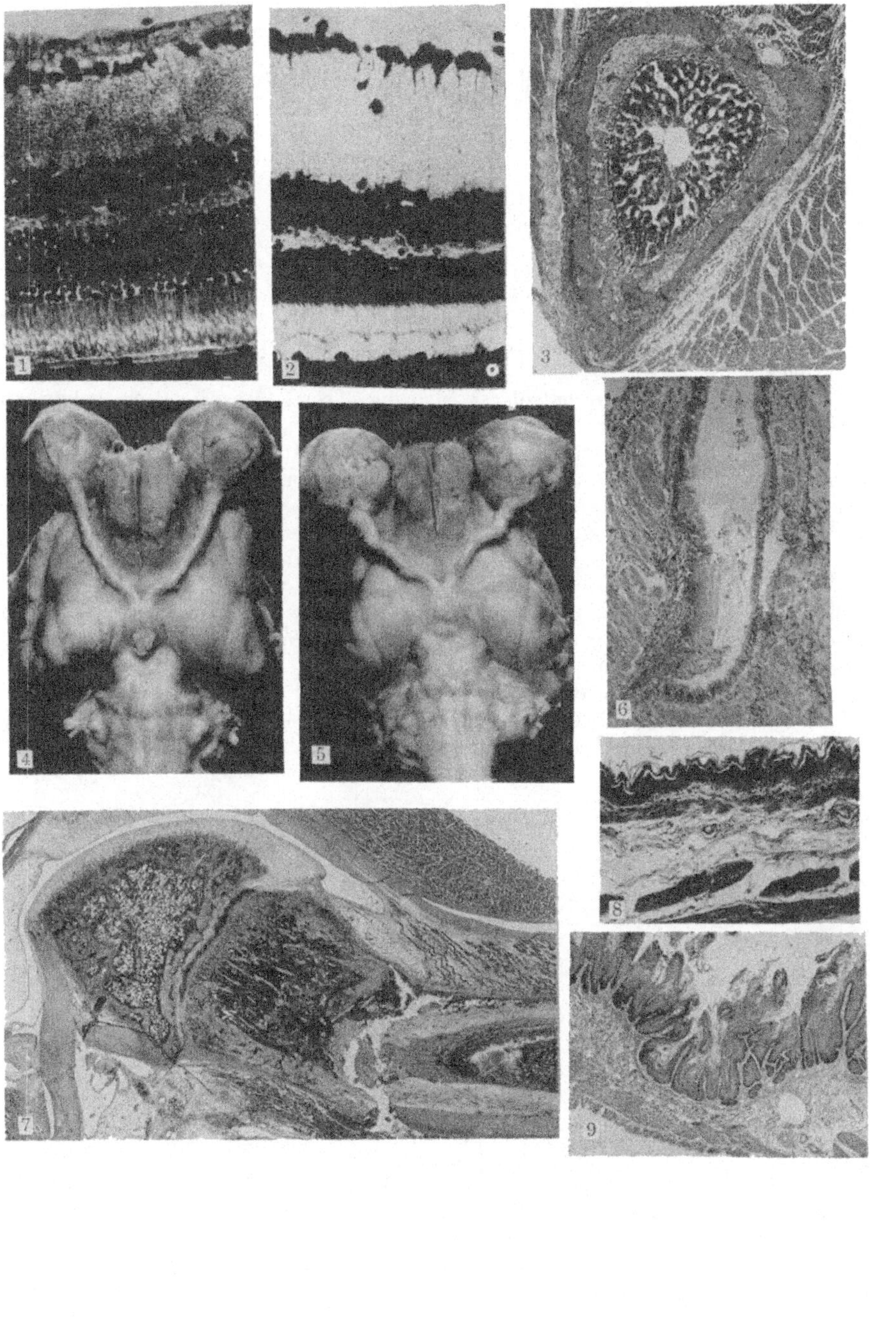
1
2
3
4
5
6
7
8
9

PLATE II

FIG. 10. Proximal epiphysis of the tibia of a normal young puppy.

FIG. 11. Proximal epiphysis of the tibia of a vitamin A-deficient puppy (litter mate). Note reduction of bony trabeculae due to resorption and disorganization of cartilage plate.

FIG. 12. Bladder of a control rat, showing normal transitional epithelium and well-developed muscularis coat.

FIG. 13. Bladder of a vitamin A-deficient rat, showing marked keratinization and increase in thickness of epithelium, with proliferation of subepithelial connective tissue and degeneration of muscle layers.

FIG. 14. Section through the heart of a young rat whose mother was deficient in vitamin A. Note failure of separation of the two ventricles.

FIG. 15. a, section through the body of a normal newborn rat to show normal kidneys. b and c, section through the bodies of newborn rats from a vitamin A-deficient mother. Note failure of pelvis to open in both kidneys in b, and joining of both to form a single horseshoe-shaped kidney in c.

Figures 10 and 11 are by courtesy of S. B. Wolbach and the *Journal of Bone and Joint Surgery*. Figures 12 and 13 are by courtesy of H. M. Sinclair and V. Ramalingaswami. Figures 14 and 15 are by courtesy of J. Warkany.

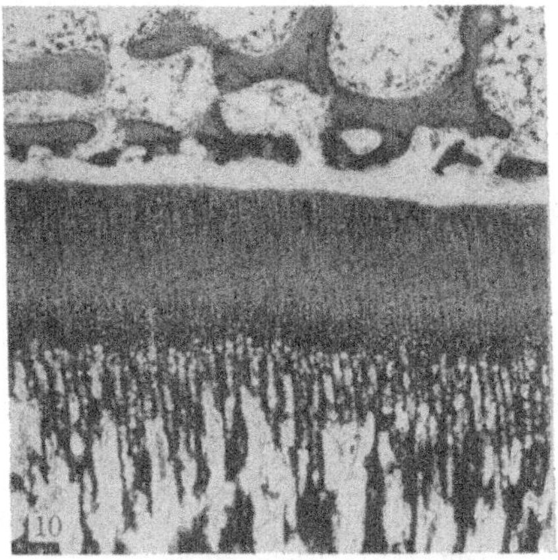
10

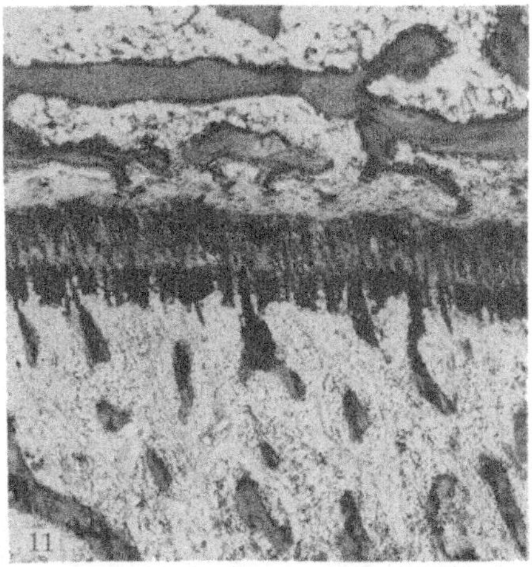
11

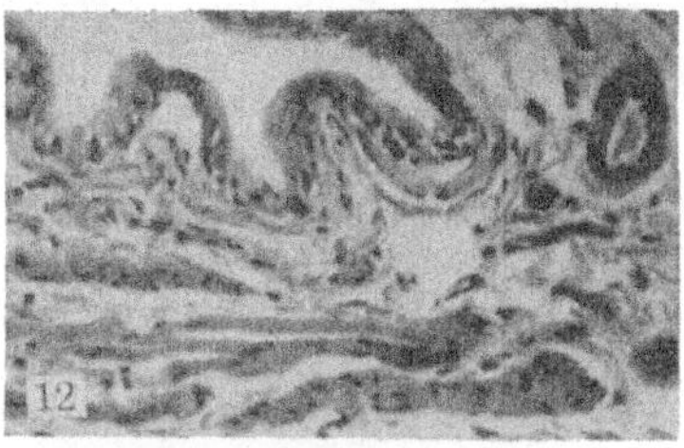
12

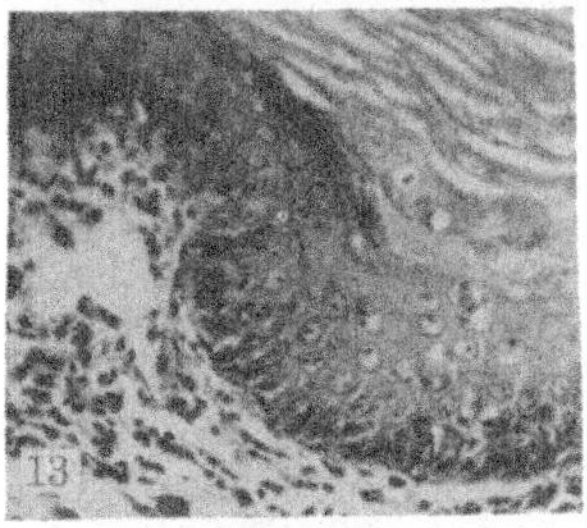
13

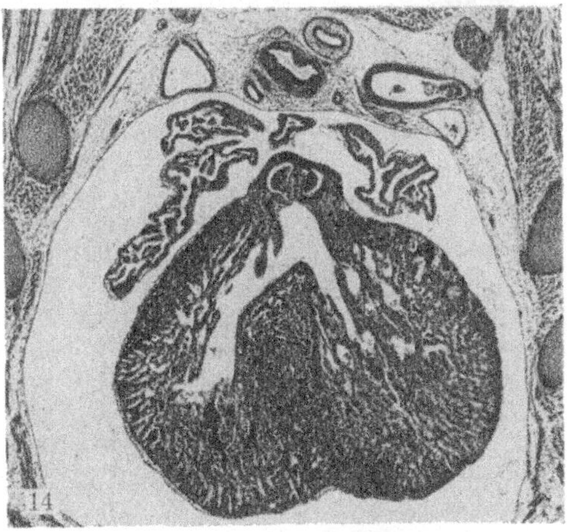
14

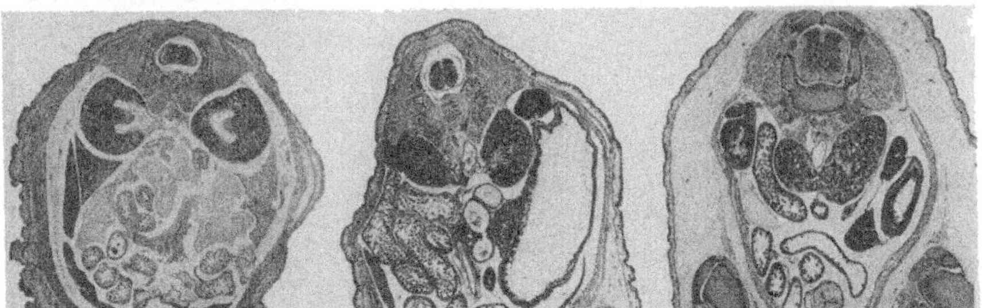
15a
15b
15c

PLATE III

FIG. 16. Portion of the auricular myocardium of a vitamin B_1-deficient pig. Note diffuse infiltration with leucocytes, mostly mononuclears. Some fibers show necrosis.

FIG. 17. Ventricular myocardium of a pig on a vitamin B_1-deficient diet for 246 days. Fibrous scar tissue (stained black) is apparent.

FIG. 18. Liver of a rat suffering from choline deficiency (low-protein, high-fat diet). Note extensive fatty infiltration of cells. Fibrous tissue has developed which divides organ up into lobules which bear little relation to the normal anatomical lobular arrangement.

FIG. 19. Cortex of kidney from a weanling rat on a choline-deficient diet for 8 days. Increased thickening of capsule and a focus of necrotic tubules beneath capsule.

FIGS. 20, 21, and 22. Sections through the head of a fetal rat of a folic acid-deficient mother. Distortion and deorientation of optic cups may be seen.

Figures 16, 17, 18, and 19 are from "The Pathology of Nutritional Disases" by R. H. Follis, by courtesy of R. H. Follis and Blackwell Scientific Publications. Figures 20, 21, and 22 are by courtesy of A. Giroud.

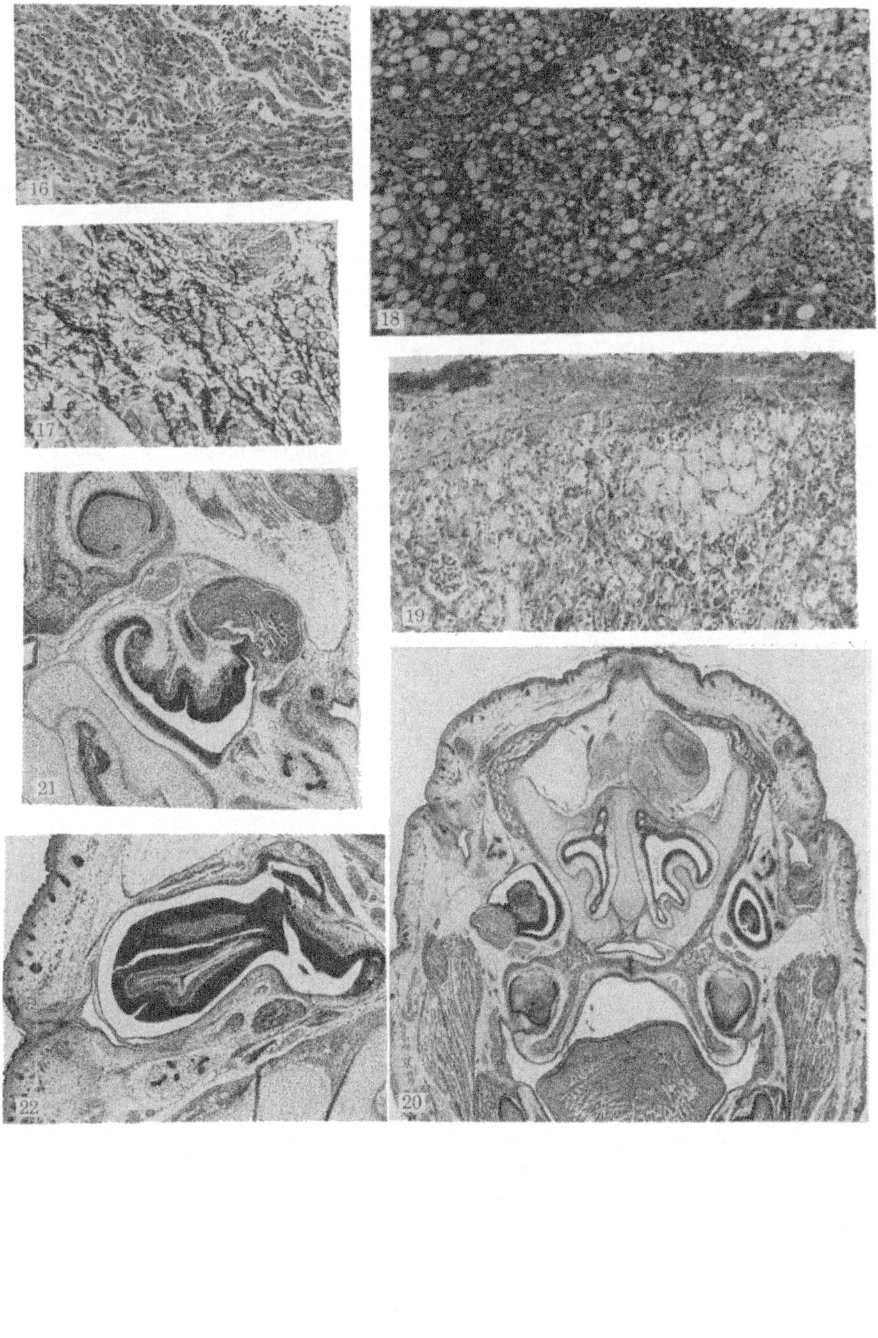
16
17
18
19
21
22
20

Plate IV

Fig. 23. Section of the leg of a newborn rat from a riboflavin-deficient mother. The tibia is shortened as a result of reduction of the osseus part; cartilaginous ends are of normal length. Lines of ossification are irregular, and large cartilaginous islands are present in the metaphysis.

Fig. 24. A small, completely cartilaginous tibia in the leg of a newborn rat from a riboflavin-deficient mother.

Fig. 25. Focus of ossification in the cartilage of the tibia of a newborn rat from a riboflavin-deficient mother. The focus is quite abnormal in position. It is situated in a lateral bulge of the tibia and is covered by a thin shell or bone.

Fig. 26. Section through the hand of a newborn rat from a riboflavin-deficient mother. The skeletal elements of the central digits are seen to be fused into a broad plate of cartilage, and normal articulations are missing.

Fig. 26a. Section through the lower arm and forefoot of a newborn rat from a riboflavin-deficient mother. The skeleton of the hand is abnormal, and a cartilaginous bridge may be seen linking radius and ulna.

Fig. 27. Normal foot and abnormal foot from a newborn rat of a riboflavin-deficient mother.

Fig. 28. Notice the varying degrees of cleft palate in newborn rats from riboflavin-deficient mothers.

Figures 23, 24, 25, and 26a are by courtesy of J. W. Warkany. Figures 26, 27, and 28 are by courtesy of A. Giroud.

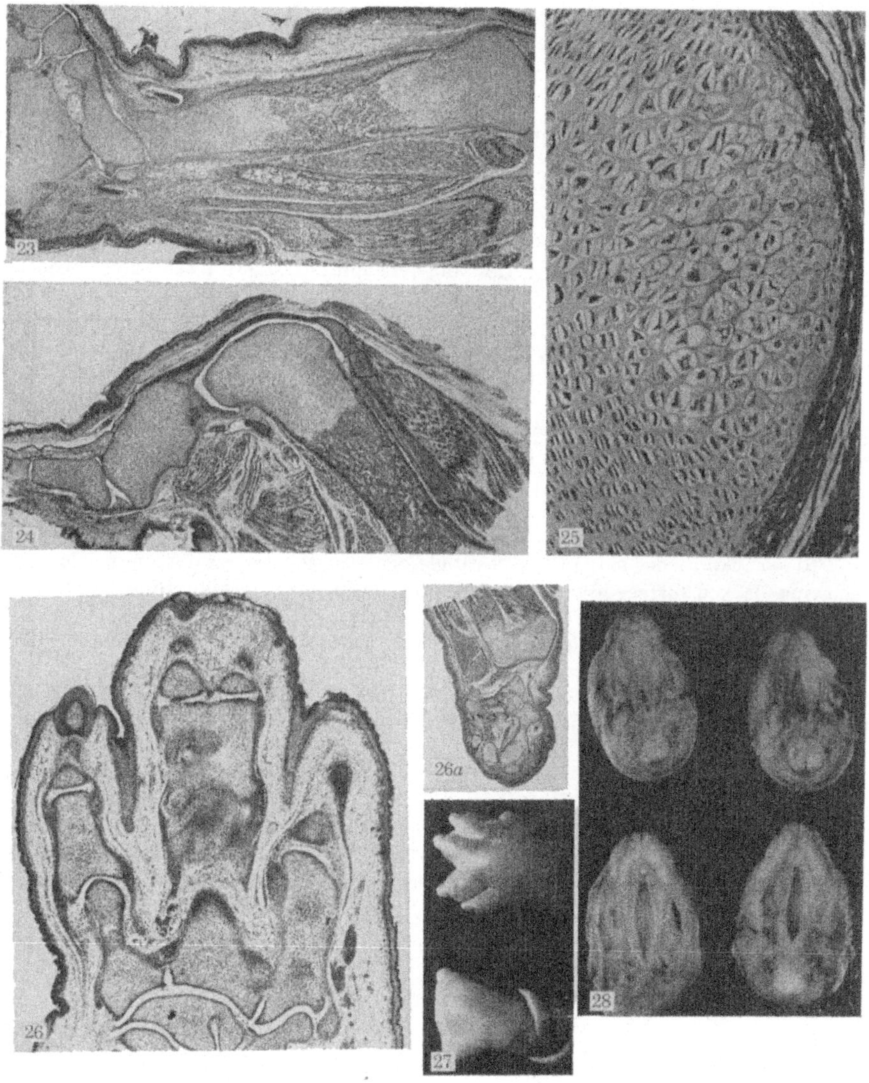
23
24
25
26
26a
27
28

PLATE V

FIG. 29. Spleen of a pyridoxine-deficient pig treated with pyridoxine. Animal had been anemic.

FIG. 30. Spleen of a pyridoxine-deficient pig. Reduction of red pulp and deposition of hemosiderin in pulp capsule and trabeculae. Animal was suffering from severe microcytic anemia.

FIG. 31. Sciatic nerve from a pyridoxine-deficient pig, showing extensive degeneration of myelin sheaths.

FIG. 32. Sciatic nerve from a pyridoxine-deficient pig, showing degeneration and fragmentation of axis cylinder.

FIG. 33. Skin from the dorsum of the hind-paw of a normal rat.

FIG. 34. Skin from the dorsum of the hind-paw of a pyridoxine-deficient rat, showing severe hyperkeratosis and acanthosis of epidermis with vascular proliferation and midcellular infiltration of the corium. Note waviness of the epidermis and the nuclear remnant in the *stratum corneum*.

FIG. 35. Cutaneous side of the vestibule of the nose of a control rat, showing large, tactile hair follicles.

FIG. 36. Cutaneous side of the vestibule of the nose of a pyridoxine-deficient rat, showing marked surface hyperkeratosis and acanthosis of the epidermis. Vacuolation of prickle cell layer can be seen in some places.

FIG. 37. Liver of a pyridoxine-deficient pig. Extensive fatty infiltration of central portions of liver lobules and deposition of hemosiderin in the periphery.

FIG. 38. Transverse section through the lumbar region of the spinal cord of a pyridoxine-deficient pig. There is obvious degeneration of the dorsal columns. Note also myelin degeneration of the dorsal roots.

FIG. 38a. Cells of dorsal root ganglion from a pyridoxine-deficient pig. Cells are shrunken, and some are necrotic.

Figures 29, 30, 31, 32, 37, 38, and 38a are from "The Pathology of Nutritional Diseases" by R. H. Follis, by courtesy of R. H. Follis and Blackwell Scientific Publications. Figures 33, 34, 35, and 36 are by courtesy of V. Ramalingaswami and H. M. Sinclair.

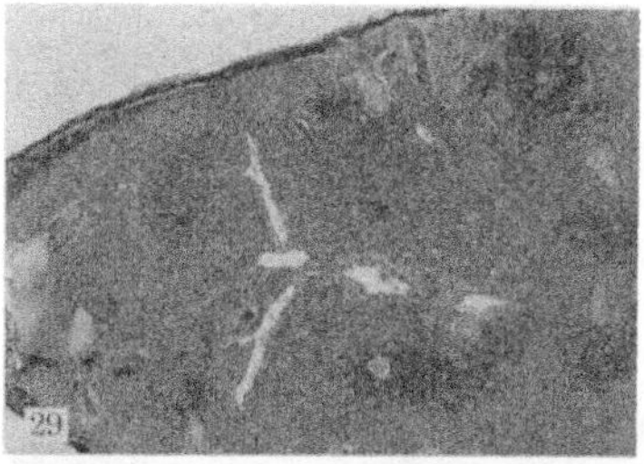
29

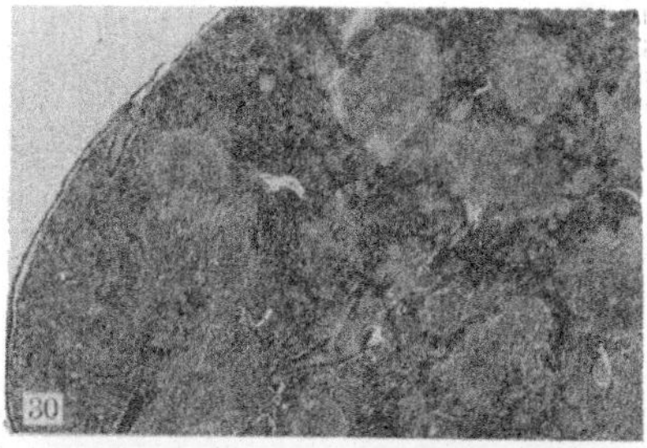
30

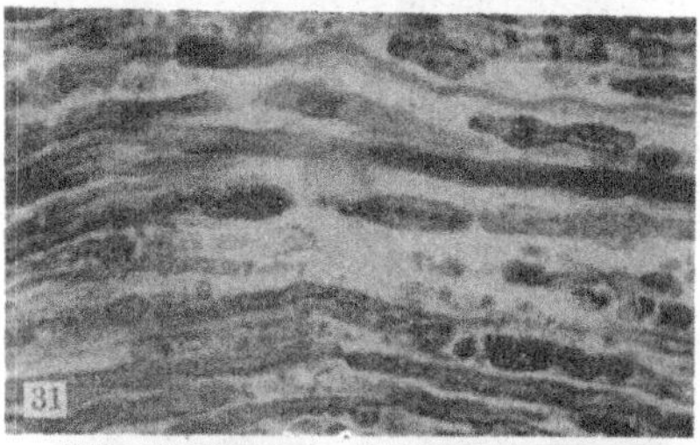
31

32

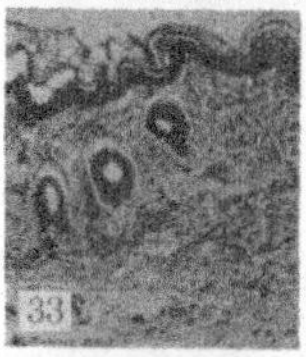
33

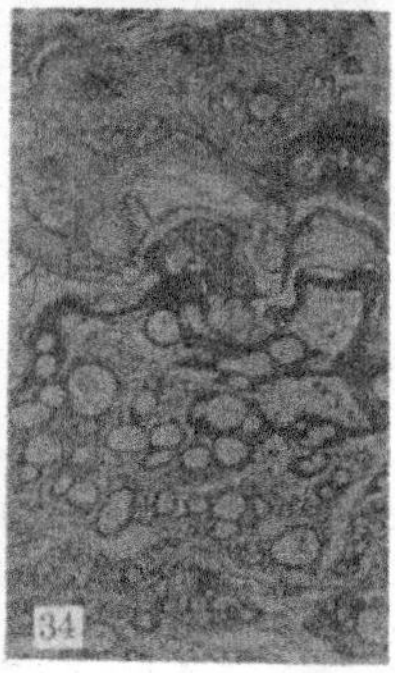
34

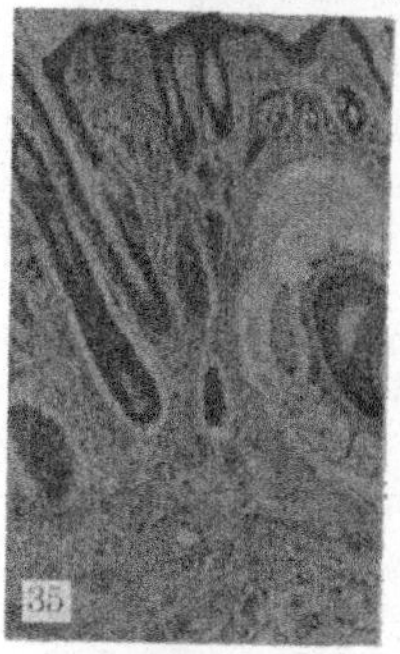
35

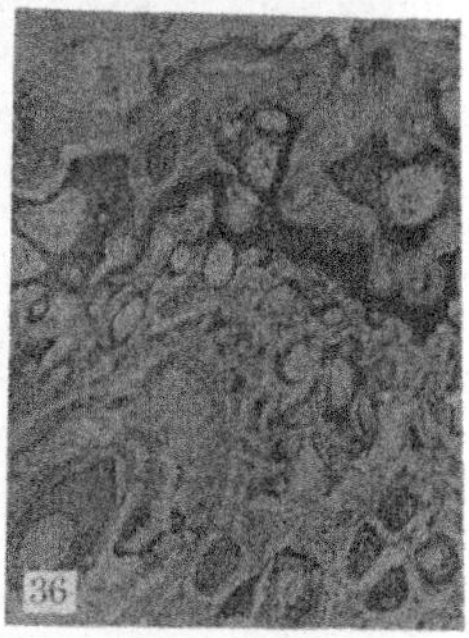
36

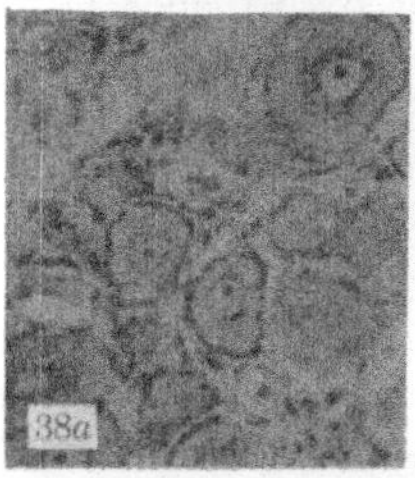
38a

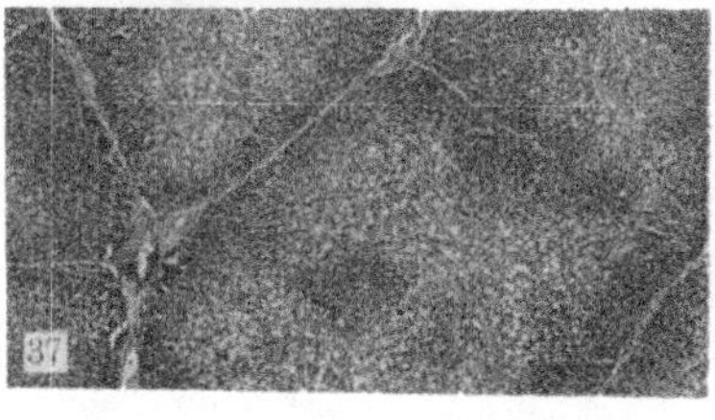
37

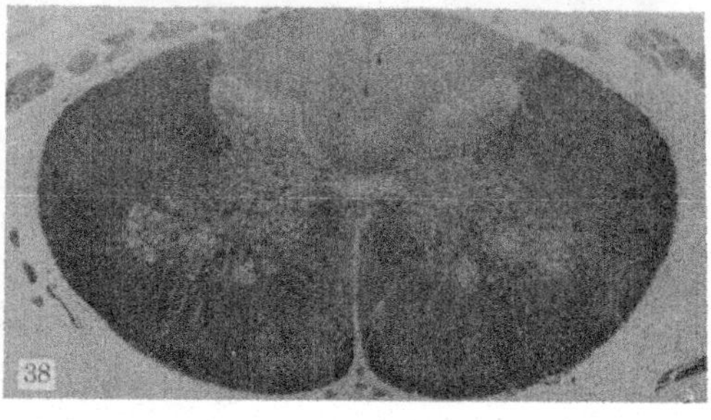
38

PLATE VI

FIG. 39. Failure of the development of the eye in the fetal rat on the left (from a pantothenic acid-deficient mother). Compare with the normal eye in the fetus on the right.

FIG. 40. Two abcesses in the lymphoid follicle in the colon of a pantothenic acid-deficient pig.

FIG. 41. Beginning of ulceration in the colon of a pantothenic acid-deficient pig. Cystic glands may be seen in the mucosa.

FIG. 42. Dorsal root ganglion of a pantothenic acid-deficient pig. Note beginnings of chromatolysis in bottom right cell.

FIG. 43. Adrenal gland of a pantothenic acid-deficient rat. Zona glomerulosa and outer zona fasciculata are normal. Inner fasciculata and outer reticularis show a focus of necrosis. Inner part of zona reticularis appears normal.

FIG. 44. Section through the head of the embryo of a rat on a pantothenic acid-deficient diet. There is a failure of the bones of the vault of the cranium to develop, and a condition known as "exencephaly" has resulted. The brain is surrounded by skin and brain membranes only.

FIG. 45. Fetuses from a pantothenic acid-deficient rat. They show edema, anophthalmia, and exencephaly.

Figures 39, 44, and 45 are by courtesy of A. Giroud. Figures 40–43 are from 'The Pathology of Nutritional Diseases" by courtesy of R. H. Follis and Blackwell Scientific Publications.

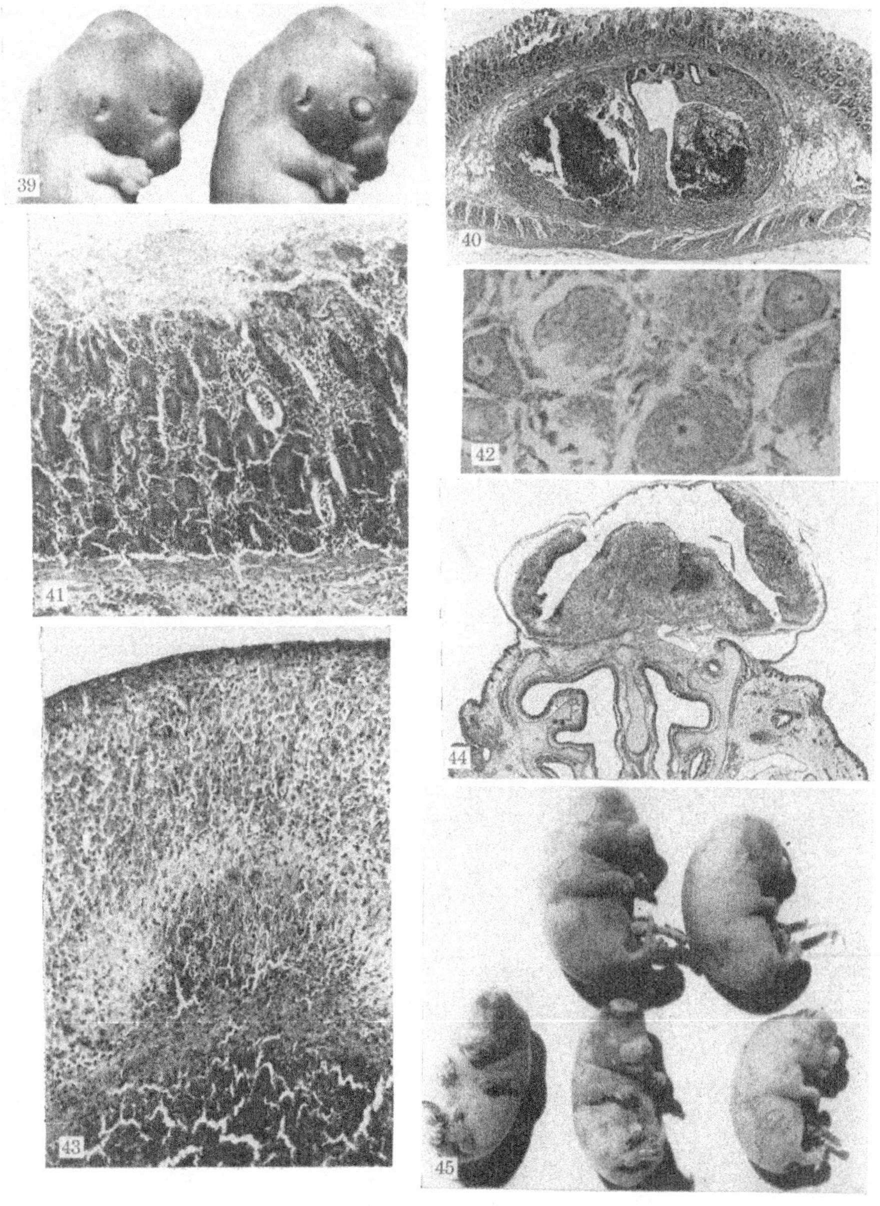
39
40
41
42
43
44
45

PLATE VII

FIG. 46. Skin wound in a normal guinea pig, after 1 week's healing.

FIG. 47. Skin wound in a scorbutic guinea pig, after 1 week's healing. Note paler staining, particularly of connective tissue, which is edematous. Note also absence of scar tissue. Although the epithelium has grown across the wound, beneath it the edges are separated by a space.

FIG. 48. Fibrous tissue of a scar of an animal receiving 1 mg. of vitamin C daily, after healing for 1 week, stained to demonstrate reticulin. Note that all fibers of the scar are in the reticular condition.

FIG. 49. Reticulin preparation of a skin wound in a guinea pig receiving 2 mg. of vitamin C daily, after healing for 1 week. Note equal intensity of reaction by scar tissue and surrounding mature connective tissue.

FIG. 50. Scar of a guinea pig receiving 2 mg. of vitamin C daily, after healing for 1 week. Note great reduction in number of reticulin fibers. (Compare fig. 48)

FIG. 51. Degenerated muscle fibers of the leg in a guinea pig suffering from partial vitamin C deficiency. Empty sarcolemmal envelopes can be seen containing wandering cells (near site of bone injury).

FIG. 52. Note the loss of cross striations and degeneration of muscle fibers in a guinea pig suffering from vitamin C deficiency.

FIG. 53. Tendino-muscular attachment in a vitamin C-deficient guinea pig. Note retraction of sarcoplasm from region of attachment, leaving muscle fibers attached by empty sarcolemmal tubes.

FIG. 54. An artery in a vitamin C-deficient guinea pig near the site of a bone injury. Note vaccuolation of endothelium.

FIG. 55. An artery in a vitamin C-deficient guinea pig, showing vacuolation of both endothelium and connective tissue of wall (near site of bone injury).

Figures 51, 54, and 55 are by courtesy of P. D. F. Murray, E. Kodicek, and the *Journal of Anatomy*. Figures 52 and 53 are by courtesy of P. E. Boyle, J. T. Irving, and the *Proceedings of the Society for Experimental Biology and Medicine*.

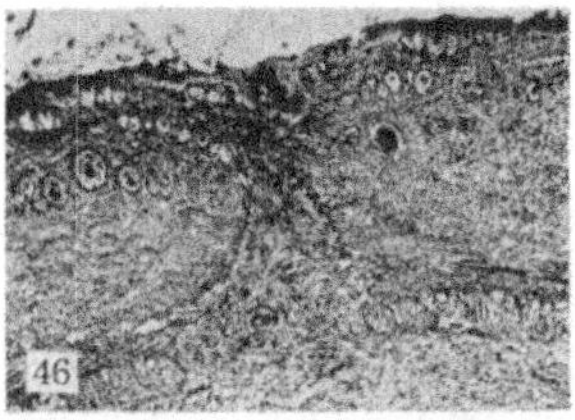
46

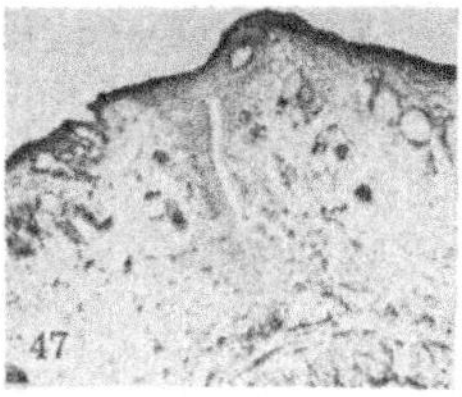
47

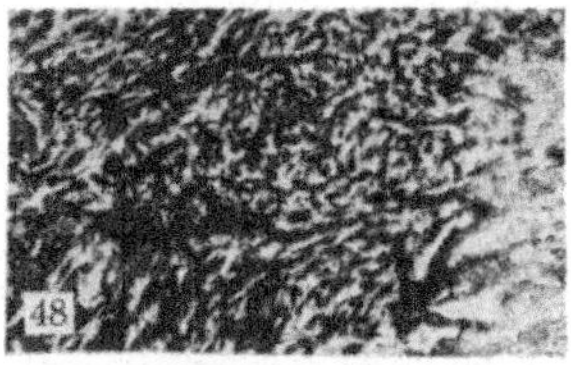
48

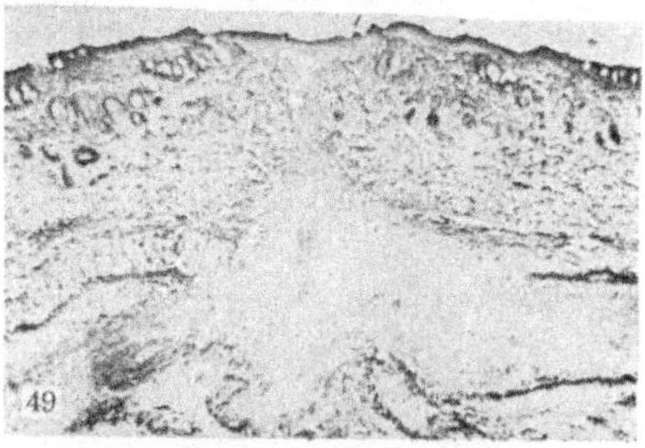
49

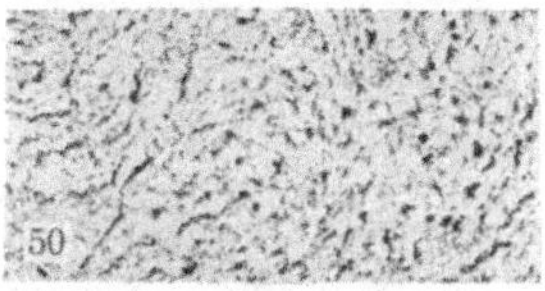
50

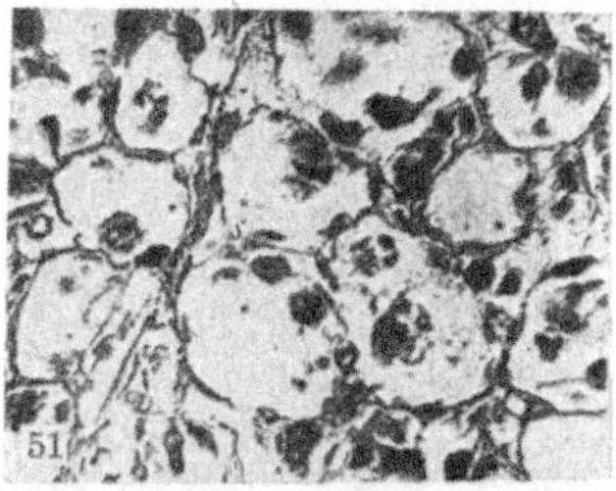
51

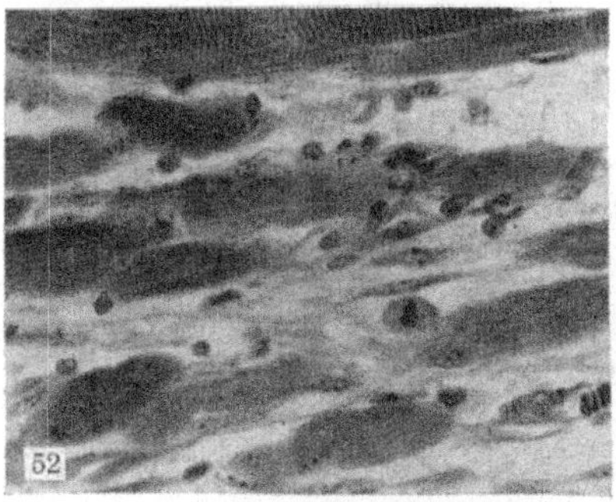
52

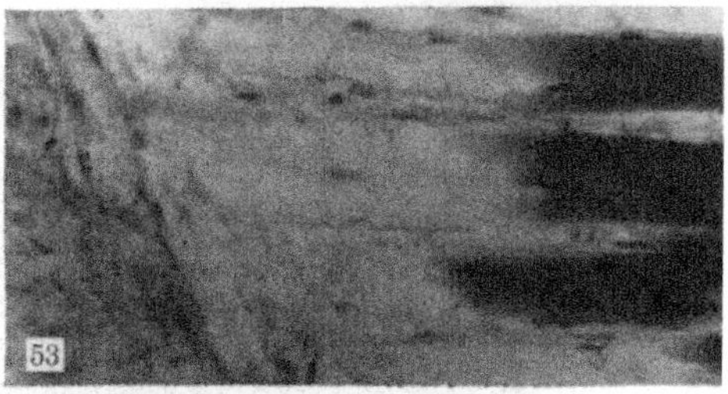
53

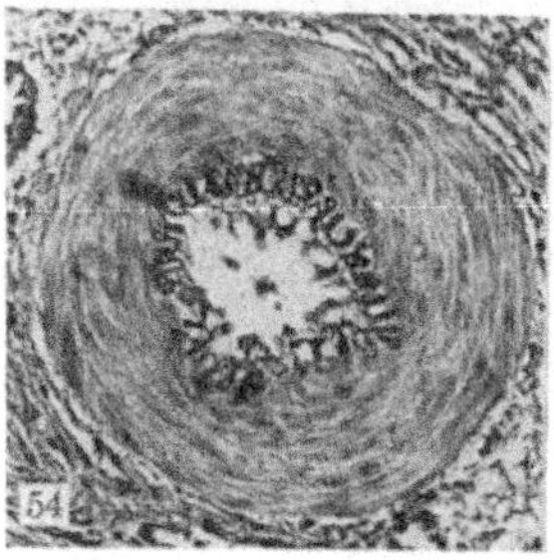
54

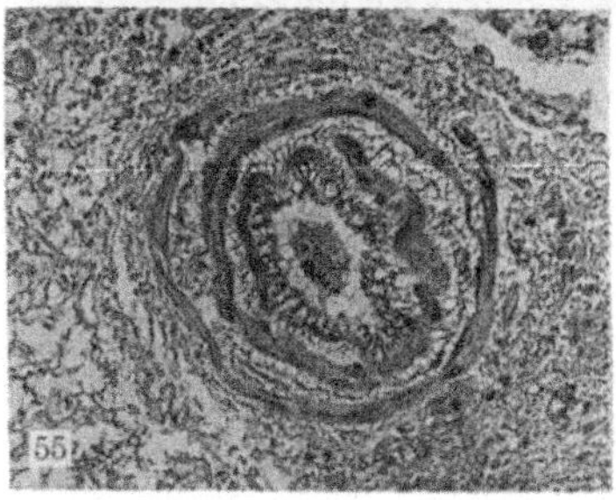
55

PLATE VIII

FIG. 56. Costochondral junction of a normal guinea pig, stained to demonstrate bone. Note numerous bony trabeculae. *A*, cartilaginous part of rib. *B*, line of bone formation. *C*, trabeculae. *D*, cortex of bony part of rib.

FIG. 57. Costochondral junction from a scorbutic guinea pig. Note great reduction of trabeculae. *A*, cartilaginous part of rib. *B*, line of abortive calcification. *C*, continuous line of bone at junction. *D*, trabeculae. *E*, cortex of bony part of rib.

FIG. 58. Phosphatase preparation of the costochondral junction of a normal guinea pig. Note broad band of phosphatase activity at junction continuous with that of periosteum. *A*, phosphatase band. *B*, phosphatase-active periosteum. *C*, phosphatase-active bone marrow. A slight spurious reaction is given by the cartilaginous part of the rib.

FIG. 59. Costochondral junction of a scorbutic guinea pig. Note great reduction of thickness of phosphatase band at junction. Reaction in bone marrow appears undiminished. This is in keeping with other experiments which showed, judged by histochemical reactions, that scurvy did not affect phosphatase activity of soft tissues. *A*, cartilaginous part of rib. *B*, line of phosphatase at junction. *C*, muscle. *D*, periosteum.

FIGS. 60–63. Effects of graded doses of vitamin C on the response of the periosteum of guinea pig femur to an injury inflicted 1 week previously.

FIG. 60. Scorbutic animal. Scarcely any reaction by cellular layer of periosteum and tearing of fibrous layer. *A*, cellular layer. *B*, fibrous layer.

FIG. 61. Animal receiving 0.25 mg. of vitamin C daily. Note greater thickness of cellular layer of periosteum, indicating greater cell multiplication. *A*, cellular layer of periosteum. *B*, fibrous layer of periosteum.

FIG. 62. Animal receiving 1.0 mg. of vitamin C daily. Note great thickness of cellular layer of periosteum and beginning of formation of bony trabeculae. *A*, fibrous layer of periosteum. *B*, bony trabeculae forming.

FIG. 63. Animal receiving 2 mg. of vitamin C daily. Note solid bony trabeculae (*A*) formed by cellular layer of periosteum.

FIG. 64. A 1-mm. hole in the femur of a scorbutic guinea pig, 1 week after injury. Note complete absence of any repair activity.

FIG. 65. A 1-mm. hole in the femur of a guinea pig receiving 2 mg. of vitamin C daily. Undecalcified section stained by the Gomori method for alkaline phosphatase. Note that the hole is filled by phosphatase-positive fibers and that in several places intensely phosphatase-positive bony trabeculae have formed.

FIG. 66. Transverse section of the tibia of a normal guinea pig; cf. Fig. 68.

FIG. 67. From a partially vitamin C-deficient guinea pig. Frontal section of the proximal end of the fibular diaphysis, showing a microfracture and new subperiosteal bone developing in a widened cambial (cellular) layer of the periosteum.

FIG. 68. From a partially vitamin C-deficient guinea pig. Transverse section of tibia showing great subperiosteal thickening and porosis of the old bone.

Figures 66, 67, and 68 are by courtesy of P. D. F. Murray, E. Kodicek, and the *Journal of Anatomy*.

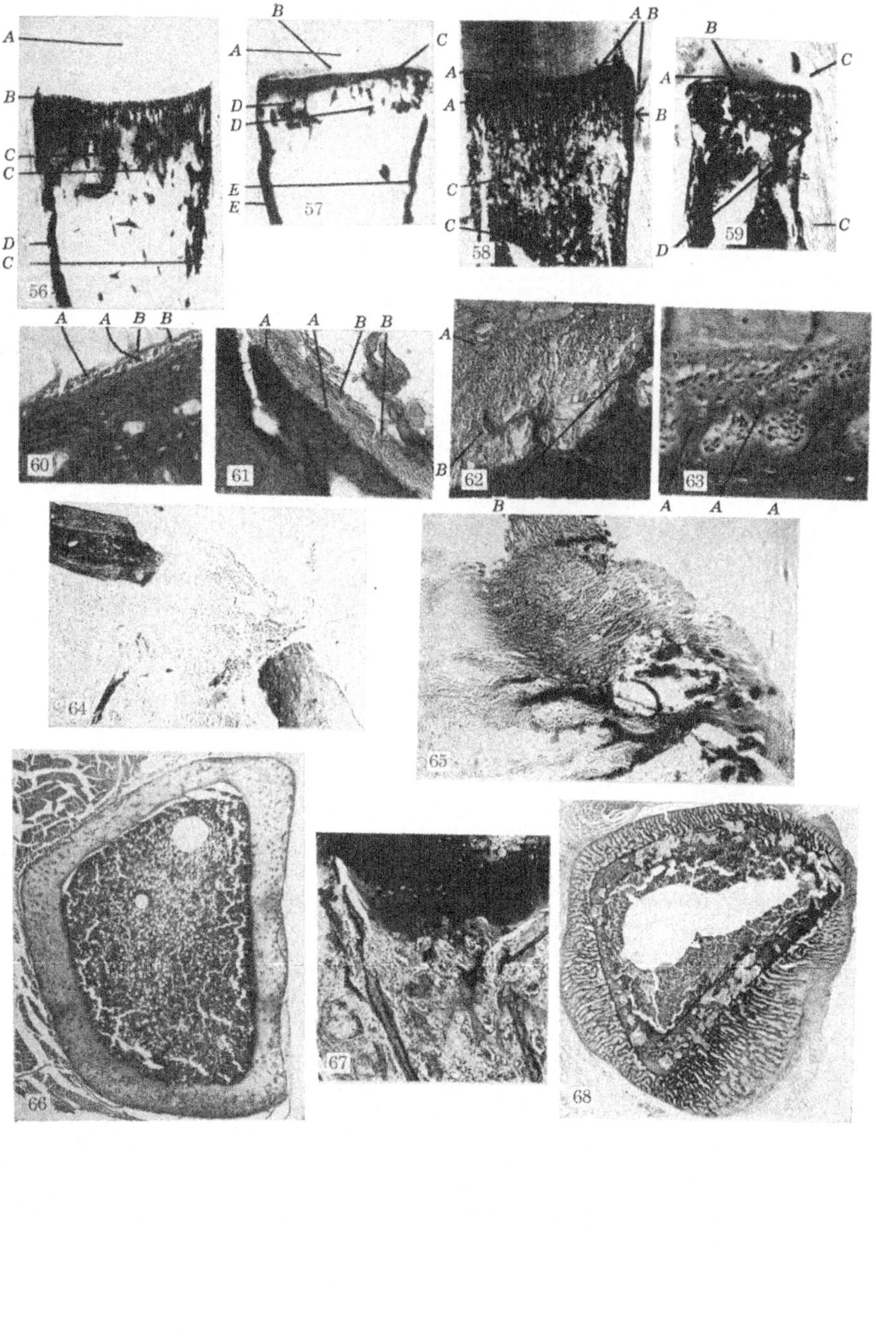
56
57
58
59
60
61
62
63
64
65
66
67
68

PLATE IX

FIG. 69. Costochondral junction from a 19-months-old white baby, showing healed rickets. The junction is normal except that the zone of hypertrophic proliferative cells is increased in width and in it is a line of dark-staining calcified cartilaginous material. This demonstrates the resumption of the deposition of lime salts at the place where they should have been laid down, had rickets not been present.

FIG. 70. Costochondral junction from a 7-months-old colored child, showing moderate rickets. Note increase in width of junction, especially in zone of proliferative cartilage cells, irregularity of calcification, tongues of cartilage projecting down toward the shaft.

FIG. 71. Costochondral junction of an 11-months-old colored child, showing severe rickets. Note extreme swelling of junction and complete disorganization due to collapse of cartilage and trabeculae.

Figures 69, 70, and 71 are from "The Pathology of Nutritional Diseases" by courtesy of R. H. Follis and Blackwell Scientific Publications.

69

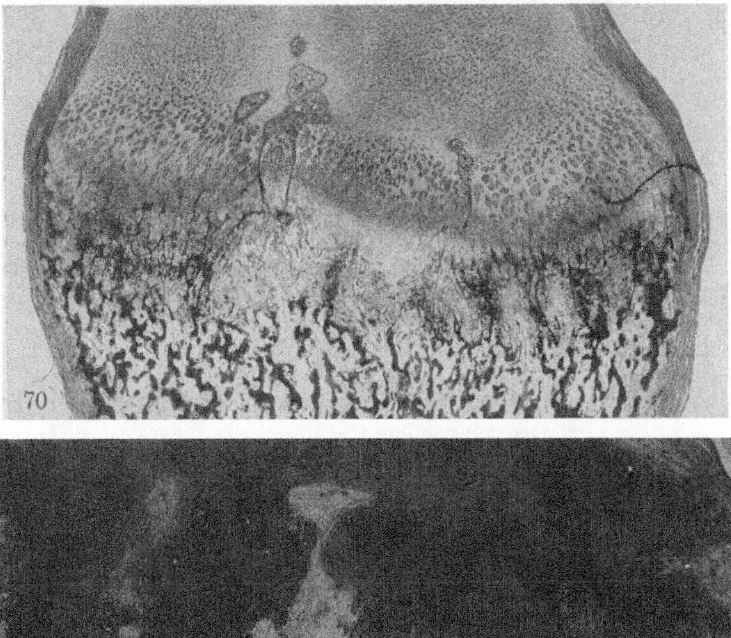
70

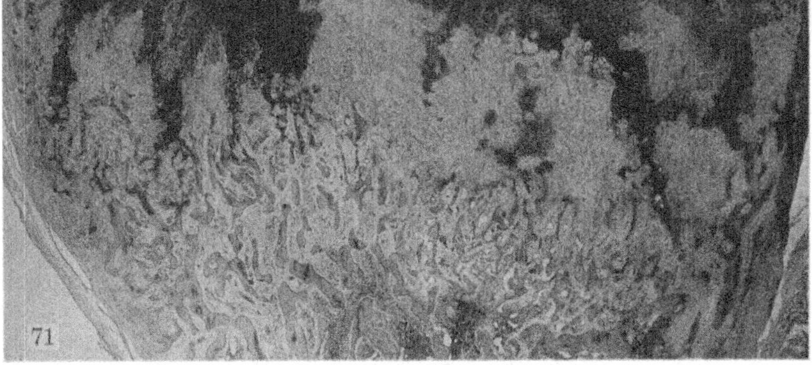
71

PLATE X

FIG. 72. Testis of a vitamin E-deficient rat, showing almost complete loss of spermatogenic elements. Note persistence of interstitial cells of Leydig.

FIG. 73. Section through the placenta and fetus from a normal female rat at 16th day.

FIG. 74. Section through the placenta and foetus from a vitamin E-deficient female at 16th day. Note reduction in size of placenta. Foetus is dead and undergoing extensive autolytic changes.

FIG. 75. Portion of the myocardium of a rat on a vitamin E-deficient diet for more than a year. Note loss of myocardial fibers, round-cell infiltration, and fibrosis.

FIG. 76. Striated muscle from a vitamin E-deficient hamster. Note destruction of muscle fibers, cellular infiltration, and sarcolemmal proliferation.

FIG. 77. Para-ovarial fat from a vitamin E-deficient rat fed a high cod-liver oil diet, showing accumulation of pigment-containing cells.

FIG. 78. Section through striated muscle from a vitamin E-deficient mouse receiving a high proportion of cod-liver oil in its diet. Note calcified area surrounded by proliferation of sarcolemmal nuclei within a necrotic muscle fiber.

FIG. 79. Two necrotic striated muscle fibers from a mouse receiving the same treatment as that in Fig. 78. Note loss of striations, clumping of sarcoplasm, and proliferation of sarcolemmal nuclei.

FIG. 80. Cross section of striated muscle fibers from a young mouse receiving the same treatment as that in Fig. 78. Note enlarged sarcolemmal nuclei and degenerating muscle fibers with centrally placed nuclei among normal muscle fibers.

FIG. 81. Portion of a cervical section of the spinal cord of a normal rat, showing fasciculus gracilis.

FIG. 82. Same portion of the spinal cord of a rat on a vitamin E-deficient, high-fat diet. Note extensive gliosis of fasciculus gracilis.

FIG. 83. Skin from the abdomen of a fatty acid-deficient rat after 23 weeks of deficiency. Note plugging of hair follicle with dense, concentrically arranged layers of keratin, greatly enlarged and plump sebaceous glands.

FIG. 84. Skin of the dorsum of the hind-paw of a normal rat.

FIG. 85. Skin of the dorsum of the hind-paw of a fatty acid-deficient rat after 23 weeks of deficiency, showing acanthosis and hyperkeratosis of epidermis.

Figures 72–76 are by courtesy of K. E. Mason, R. H. Follis, and Blackwell Scientific Publications. Figure 77 by courtesy of K. E. Mason. Figures 78–80 are by courtesy of K. E. Mason, C. E. Tobin, and *Archives of Pathology*. Figures 81 and 82 are by courtesy of K. E. Mason and the *Ann. New York Acad. Sci.* Figures 83–85 are by courtesy of V. Ramalingaswami, and H. M. Sinclair.

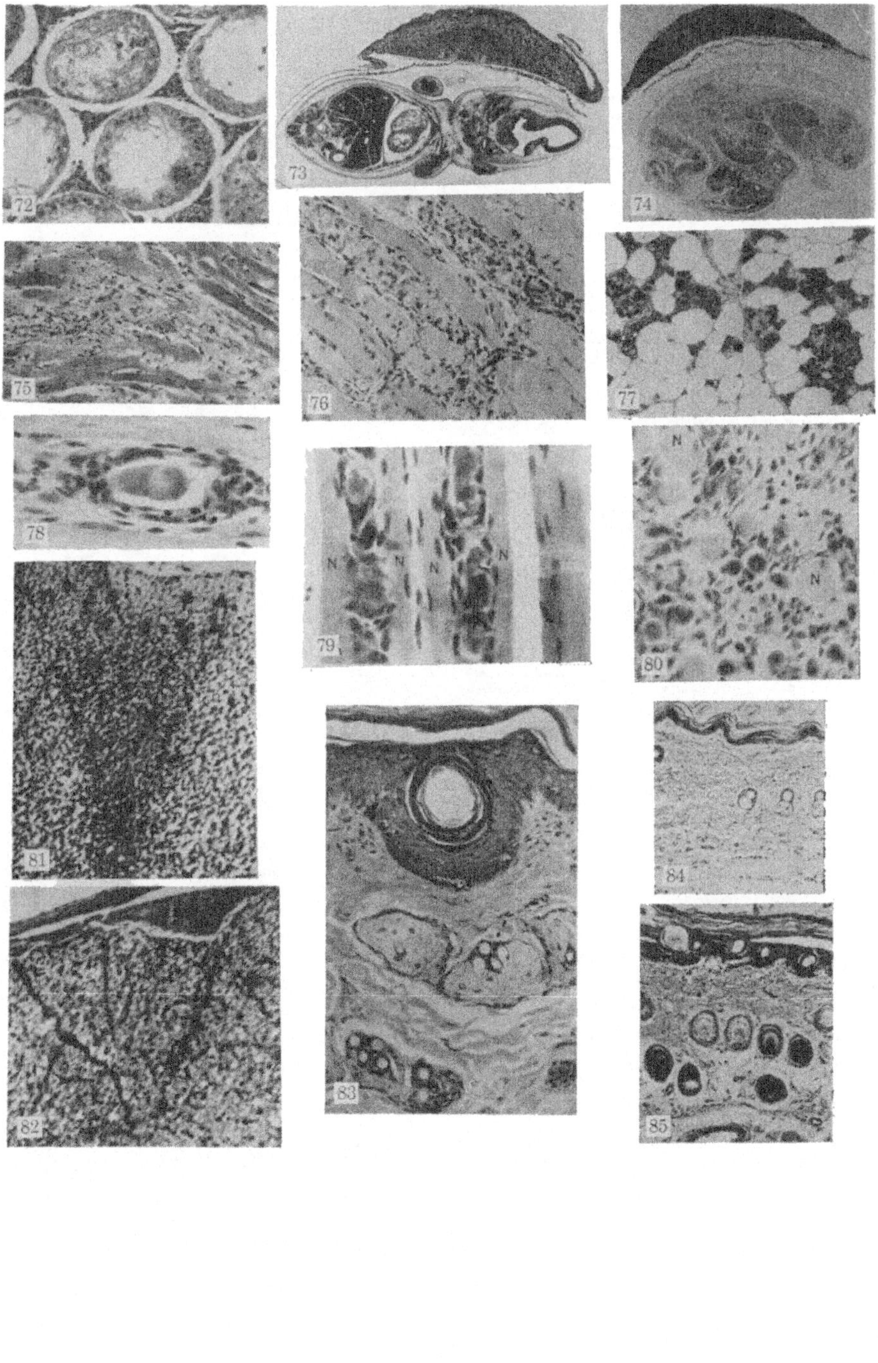
72
73
74
75
76
77
78
79
80
81
82
83
84
85

CHAPTER 14

Microbiology of Digestion

D. P. CUTHBERTSON AND A. T. PHILLIPSON

Rowett Research Institute, Aberdeen, Scotland

I. Introduction

All animals probably harbor bacteria and sometimes protozoans and other primitive organisms in their alimentary tracts; this state may be considered normal. The balance maintained by the host and these microorganisms is undoubtedly influenced by the character of the alimentary tract and the food eaten by the host. Therefore, in considering the role of microorganisms in a particular host, reference must repeatedly be made to the physical and chemical state of the food. This being so, it is understandable that different species of animal with different feeding habits should possess fauna and flora that may be characteristic of the species. The feeding habits, however, may not be the only factor concerned, for the environment provided by the host may in itself also influence the character of the alimentary microorganisms.

In the higher organisms the caecum and the colon are the common sites of microbial digestion, but in ruminants in particular, and to a less extent in certain other animals such as the hippopotamus and kangaroo, the stomach is so enlarged that additional provision is made for the temporary retention of large quantities of coarse fibrous foodstuffs.

In the ruminant, food passes slowly from the rumen and reticulum through the omasum to the abomasum, and it is not until it reaches this organ that it comes into contact with gastric juice. The rumen and the reticulum provide an ideal environment for microorganisms, and saliva from the parotid glands is constantly secreted, thus forming a slightly alkaline bicarbonate phosphate buffer in the forepart of the stomach system. As the food enters the rumen first, the substrates available to the organisms living in this organ are plentiful. The characters of the population of the caecum and colon differ from those of the rumen because only those organisms that can survive the environment of the abomasum and small intestines can establish themselves in the large gut. Similarly, since easily available substrates may wholly or partially disappear as the food residues pass to the large intestine, the density of the population is likely to be less there than in the rumen.

The relative importance of the capacity of the caecum and colon varies with the species; thus the greatest relative capacity of the caecum and colon is found in true non-ruminating herbivores such as the horse, the rabbit, and the koala bear. These animals possess the necessary capacity in their large intestine to allow microbial decomposition of bulky vegetable residues to take place.

The large gut is relatively smaller in omnivores, but even so, as in the pig, the rat, and man and other primates, the capacity is sufficient to make microbial digestion on a significant scale possible. Only in the true carnivores is the relative size of the caecum and colon so small that any microbial activity that occurs is quantitatively unimportant compared to the action of the digestive juices of the stomach and the small intestine.

No systematic account of the biosynthesis of vitamins or their utilization by microorganisms of the alimentary tract will be given here, as this subject is dealt with in Chapter 9.

The large heat increment of the ruminant which occurs after eating may be wasteful, for the animal has little or no control over this portion of the calorific value of its food. Thus the SDA in a carnivore is chiefly due to the heat liberated in deaminating the surplus amino acids resulting from digestion; in the ruminant it is due to the exothermic heat of fermentation liberated in the rumen but more especially to the rapid metabolism of short-chain fatty acids absorbed from that organ. Although undoubtedly of value in cold weather or to an animal on a maintenance ration, this heat may be a burden in a heavy yielder in a warmer climate.

II. The "Germ-Free" State

Pasteur's remark in 1885[1] that the presence of bacteria in the alimentary tract might be an essential condition for normal life prompted Nuttall and Thierfelder[2] to attempt to rear guinea pigs delivered at full term by Caesarian section under sterile or "germ-free" conditions. Some of these guinea pigs lived for 13 days with an alimentary tract from which no bacteria could be isolated by cultural means at post-mortem examination. Since then several other investigations on similar lines have been made and greater success has been achieved as knowledge on the nutritional requirements, particularly of B vitamins, has increased. Küster[3] experimented with goats; Cohendy and Wollman[4] and Glimstedt[5] with guinea pigs; Balzam[6] with chickens: and all these experiments suggest that, provided the diet is adequate, the presence of bacteria is not essential for life.

Recently a study of the germ-free state has been undertaken on a grand scale by Reyniers and his colleagues.[7, 8] A technique for rearing germ-free animals has been developed and used to raise guinea pigs, rats, and chickens to maturity.[9] Chickens present fewer difficulties than mammals as they have no suckling period. Most of the difficulties encountered in rearing germ-free rats are in feeding the newly delivered full-term ratlings by hand. Under these conditions germ-free ratlings do not grow so well as normally suckled rats and there is quite a substantial death rate. A comparison of the anatomical development of germ-free and normal rats shows that the lymphatic system is less developed in the germ-free animal and the caecum is often enlarged. In germ-free chickens the proventiculus, the small intestines, and the caecum are smaller than in control birds. Histologically in the intestine this seems to be due to a decrease of the lymphoid tissue, particularly in the caudal parts. The thymus is enlarged.

Analysis of some B vitamins in the caeca of germ-free and control birds showed that there was as much riboflavin, niacin, biotin, and pantothenate in the germ-free birds as in the controls, and appreciable quantities of folic

[1] L. Pasteur, *Compt. rend.* **100,** 68 (1885).

[2] G. H. F. Nuttall and H. Thierfelder, *Z. physiol. Chem.* **21,** 109 (1895–1896).

[3] E. Küster, *Arb. Reichsgesundh.* **48,** 1 (1914).

[4] M. Cohendy and E. Wollman, *Compt. rend.* **158,** 1283 (1914).

[5] G. Glimstedt, *Skand. Arch. Physiol.* **73,** 48 (1936).

[6] N. Balzam, *Ann. physiol. physicochim. biol.* **13,** 370 (1937).

[7] J. A. Reyniers, *Lobund Repts.* **1,** University of Notre Dame, Indiana (November, 1946).

[8] J. A. Reyniers, *Lobund Repts.* **2,** University of Notre Dame, Indiana (February, 1949).

[9] J. A. Reyniers, *Proc. N. Y. State Assoc. Pub. Health Lab.* **28,** 60 (1949).

acid, vitamin B_{12}, and vitamin C were present.[10] This shows that a concentration of vitamins in the alimentary tract may not be due solely to microorganisms. In addition, Luckey[11] has found that germ-free chicks maintained on a ration deficient in niacin, biotin, pantothenic acid, and folic acid excrete more of these factors in their feces than do normally infected controls.

These studies show that germ-free existence is possible. In addition they throw fresh light on the question of intestinal synthesis of B vitamins, for the concentration of B vitamins in the gut does not necessarily imply bacterial synthesis.

III. The Establishment of the Microbial Population in the Alimentary Tract

1. In Human Infants

At birth there are no organisms in the intestinal tract unless there has been a small degree of prenatal contamination, possibly from the vagina. The meconium is sterile or virtually so. Organisms appear in the feces within 24 to 28 hr., the type of microflora depending on whether the child is breast-fed or artificially fed. In the breast-fed child the predominant organism in the feces is the Gram-positive *Lactobacillus bifidus*, which is present in almost pure culture (Tissier[12] and Cruickshank[13]). There are a few enterococci and Gram-negative coliform bacilli. The pH of the feces is on the acid side of neutrality. In the artificially fed infant—infants fed on cow's milk preparations—the fecal flora is heterogeneous and consists of a mixture of Gram-positive and Gram-negative organisms. Under such circumstances the pH of the feces is invariably on the alkaline side.

The establishment of a lactobacillary flora in the feces of the breast-fed child is generally ascribed to the greater lactose content of human milk as compared with cow's milk. As the administration of lactose and the establishment of a lactobacillary flora are so closely related, it is obviously difficult to disentangle the biological value of the former from that of the latter. Thus, although no positive evidence of the nutritional value of a lactobacillary intestinal flora to the human infant is available, the biological value of lactose has been stressed by various workers, e.g., Jarvis.[14] More convincing evidence of the value of lactose is afforded by animal experi-

[10] J. A. Reyniers, P. C. Trexler, R. F. Ervin, M. Wagner, H. A. Gordon, T. D. Luckey, R. A. Brown, G. J. Mannering, and C. J. Campbell, *J. Nutrition* **41,** 31 (1950).

[11] T. D. Luckey, personal communication, 1951.

[12] H. Tissier, La Flore Intestinale des Nourrissons, Paris, 1900, quoted by R. Cruickshank, ref. 13.

[13] R. Cruickshank, *J. Hyg.* **24,** 241 (1925).

[14] B. W. Jarvis, *Am. J. Diseases Children* **40,** 993 (1930).

ments.[15, 16] Johansson and Sarles[17] claim that on a diet with lactose there is a greater synthesis of vitamins. They quote Rettger and Horton as having shown that lactobacilli appear in the stools of rats which are fed on a vitamin B-deficient diet. The health of the rats was maintained by refection, and it was suggested that the lactobacilli synthesized the deficient vitamin B complex.

These findings, taken in conjunction with the modern concept of the importance of the pattern of the normal bacterial flora in the body economy, might suggest that the lactobacillary flora of the breast-fed infant is of intrinsic value. It might be argued that it would synthesize very different nutrients from those synthesized by the heterogeneous intestinal flora of the artificially fed baby.

Although there is no really satisfying investigation concerning the localization of this flora in the intestinal tract of breast-fed infants, the investigations of Barbero *et al.*,[18] using double-lumen Miller-Abbott tubes, have permitted specimens to be withdrawn from various levels of the intestine. In general the duodenum was relatively sterile, but bacteria, mainly enterococci, were found in the jejunum and in increasing numbers toward the ileum. There was marked bacterial proliferation in the caecum and distal to it. The pH reached a peak of 7.0 to 8.0 in the caecum and tended to fall in the distal portion of the colon. This change toward the distinctly acid level was more pronounced in breast-fed infants than in infants fed cow's milk. *Lactobacillus bifidus* was never cultured above the terminal ileum but was usually cultured in the large intestine of both breast-fed infants and infants fed cow's milk. The flora of the breast-fed infant consisted of over 90 % of Gram-positive bacilli, whereas that of the infant on cow's milk was variable in its bacterial flora. The feeding of brands of "adapted" milk failed to produce stools with an acidity and bacterial flora similar to those of breast-fed infants.

2. In Ruminants

The microorganisms of the rumen in the newborn calf differ from those of the adult (Pounden and Hibbs;[19] Huhtanen, Saunders, and Gall[20]). Calves that are reared in isolation may fail to acquire protozoans in the rumen and other microorganisms which are normally present. Such animals have been described as "pot-bellied" and as having rough coats, although

[15] L. R. Dragsted and S. C. Peacock, *Am. J. Physiol.* **64,** 424 (1923).

[16] O. L. Kline, J. A. Keenan, C. A. Elvehjem, and E. B. Hart, *J. Biol. Chem.* **98,** 121 (1932).

[17] K. R. Johansson and W. B. Sarles, *Bact. Revs.* **13,** 25 (1949).

[18] G. J. Barbero, G. Runge, D. Fischer, M. N. Crawford, F. E. Torres, and P. György, *J. Pediat.* **40,** 152 (1952).

[19] W. D. Pounden and J. W. Hibbs, *J. Dairy Sci.* **31,** 1041 (1948).

[20] C. N. Huhtanen, R. K. Saunders, and L. S. Gall, *J. Animal Sci.* **10,** 1049 (1951).

no difference in weight occurs.[21] Apparently it is an old practice in Sweden to use "cud" inoculations for the treatment of "unthrifty" beasts, the cuds being obtained from different areas where the animals are normal. Bacteriological investigations[20] on 200 calves and 200 adult cattle have demonstrated nine organisms that are characteristic of the rumen of calves. These seldom occur in the ruman of adult cows, although two types are recognized as occurring in the rumen at all ages. Adult forms, of which eleven have been investigated, appear in the rumen as early as 2 months after birth and increase in predominance with age. The common bacteria of the rumen of calves are those that produce lactic acid and are fast-growing compared to adult forms. The latter produce short-chain fatty acids. Hay or roughage in the ration appears important for the early establishment of the normal rumen population in calves. The same rumen microorganisms were established in calves on pasture as have been observed in calves that were fed hay. There is little direct evidence that rumen inoculations have any influence on the health of calves over and above the beneficial effects that result from early feeding of good roughage. Protozoans and organisms associated with hay ingestion are generally absent from the rumens of calves on strictly grain rations. Digestive upsets and diarrhea frequently noted in young calves are apparently prevented as effectively by hay feeding as by rumen inoculation.[22]

McCandless and Dye[23] have studied the decrease in the blood sugar that occurs as the young ruminant develops, and they compared this to the known facts concerning the development of the rumen and the fermentations that proceed within this organ. They suggest that the decrease in blood sugar is due to the relative increase in the quantity of short-chain fatty acids absorbed.

3. In Chickens

Newly hatched chicks harbor very few bacteria, but within 21 hr. Shapiro and Sarles[24] and their colleague Rhodes[25] found gas-producing organisms, although feeding did not begin until they were 24 hr. old. After feeding, the numbers of bacteria rapidly increased and reached a peak 16 hr. after feeding. The caeca contained the greatest numbers and on either side of them the counts declined progressively.

The greater part of the population of the caecum is made up of coliform organisms with lactobacilli as the most important groups of bacteria, enterococci being present in smaller numbers. *B. coli* was the dominant coli-

[21] W. D. Pounden and J. W. Hibbs, *J. Dairy Sci.* **33,** 639 (1950).
[22] W. D. Pounden and J. W. Hibbs, *J. Dairy Sci.* **31,** 1051 (1948).
[23] E. L. McCandless and J. A. Dye, *Am. J. Physiol.* **162,** 434 (1950).
[24] S. K. Shapiro and W. B. Sarles, *J. Bact.* **58,** 531 (1949).
[25] S. K. Shapiro, R. A. Rhodes, and W. B. Sarles, *J. Bact.* **58,** 689 (1949).

form there. *Streptococcus faecalis* was also present in relatively large numbers in the large gut.

IV. Protozoans and the Host

Two main associations will be discussed: (1) flagellates in termites and (2) ciliates in mammals.

1. Flagellates in Termites

Interesting observations on the flagellate populations of termites provide evidence of the relatively enormous numbers which may be found. For example, in the hind gut of *Zootermopsis* they may constitute as much as a third of the weight of the termite. Despite these teeming numbers the host remains healthy.[26] A wood-eating cockroach related to the termite has been found which also contains numerous protozoans.

So far no significant evidence of cellulose digestion by bacteria has been found in termites which contain protozoans, and Baldacci[27] could not demonstrate cellulolytic bacteria in defaunated termites.

In the light of the available evidence Hungate[28] considers that the protozoans and the termite are mutually linked in the chain of biochemical events leading to the complete oxidation of the cellulose and hemicellulose in the wood. The termite comminutes the wood and transports it to the protozoans as small particles, provides a fermentative chamber, and through its own oxygen consumption reduces the oxygen tension around the protozoans, thus aiding in the maintenance of anaerobic conditions. The insoluble carbohydrates are digested and fermented by the protozoans, thus providing them with energy, and the fermentation products, notably acetic acid, are absorbed and oxidized by the termite, which not only is enabled to satisfy its own energy requirements but also to remove the metabolic products of the protozoans from the gut which would become toxic to the protozon if allowed to accumulate.

In their nitrogen metabolism the termites show a remarkable economy. It is possible that the protozoans rather than their nitrogenous wastes may be a source of nitrogen for the termite. Proctodeal feeding provides for the transmission of the flagellates to the termite and permits their digestion, but quantitative experiments are needed to evaluate the importance of proctodeal feeding in terms of nitrogen to the termite.

2. Ciliates in Mammals

Ciliates are found in the rumen of domestic ruminants including the camel and reindeer and have been reported in a large number of wild

[26] L. I. Katzin and H. Kirby, *J. Parasitol.* **25,** 444 (1939).

[27] E. Baldacci, quoted by R. E. Hungate, ref. 28.

[28] R. E. Hungate, *Ann. Rev. Microbiol.* **4,** 53 (1950).

ruminants. Cattle and sheep ciliates appear to be endogenous to the rumen, for no one has found them in the feces or on pasture or other fodders.

Six principle genera are found in the rumen of both domestic and wild ruminants, but here they will simply be referred to as those belonging to the order Holotrichida and those belonging to the order Oligotrichida. Equally specialized forms are found in the caecum and colon of horses. According to Buisson[29] the African elephant, the rhinoceros, the American tapir, and the guinea pig, and among the primates the chimpanzee and the gorilla, harbor various species of ciliate in the large intestine. *Balantidium coli*, a ciliate usually considered to be pathogenic, has been encountered in the macaque and the orangutan. Buisson states that the ciliates soon disappear from the large intestine of the great apes when kept in captivity but that they are found in naturally living specimens.

Gruby and Delafond,[30] who discovered the rumen ciliates, observed that in the abomasum such organisms were found dead and disintegrating, an observation which has been repeatedly confirmed. Ferber[31] also observed that even in the omasum the ciliates appeared to be dead and that no ciliates could be detected in the small or large intestines; these observations have been confirmed by Masson.[32] It is common knowledge that the oligotrich ciliates soon die when they are taken from the rumen and placed on a slide, as they are sensitive to temperature. Holotrich organisms are more resistant. These organisms are susceptible to changes of pH as was demonstrated by Knoth[33] by *in vitro* studies. The disappearance of ciliates from the rumen of sheep fed on a ration rich in starch with only small quantities of hay was observed by Masson;[34] these animals developed a pH between 4 and 5 in the rumen, owing to the high concentrations of lactic acid. (Phillipson[35]). Their subsequent reappearance was associated with an increased pH of the rumen contents.

There is no satisfying evidence that ciliates or encysted forms are excreted in the feces or that there is any development of resistant forms. Fantham[36] is the only one of many observers who claims to have found such a stage in their life cycle. Oral transmission appears to be the route of contact.[37, 38]

Both young sheep and goats have been maintained in a defaunated

[29] J. Buisson, Les Infusoires Ciliés de Tube Digestif du l'Homme et des Mammifères, Amadée Legrand, Paris, 1923.

[30] D. Gruby and O. Delafond, *Compt. rend.* **17,** 1305 (1843).

[31] K. E. Ferber, *Z. Tierzücht. Züchtungsbiol.* **12,** 31 (1928).

[32] M. J. Masson, *Brit. J. Nutrion* **4,** viii, Proc. (1950).

[33] M. Knoth, *Z. Parasitenk.* **1,** 262 (1928).

[34] M. J. Masson, *Research (London)* **4,** 73 (1951).

[35] A. T. Phillipson, *Brit. J. Nutrition* **6,** 190 (1952).

[36] H. B. Fantham, *S. African J. Sci.* **19,** 332 (1922).

[37] E. R. Becker, *Quart. Rev. Biol.* **7,** 282 (1932).

[38] E. Mangold, *Biedermanns Zentr.* **A3,** 161 (1933).

condition over a sufficiently long period to observe that their growth rate is not seriously impaired by the elimination of the protozoans and that they digest as much cellulose as normal animals. This has tended to have a damping effect upon investigations into their metabolism, and until recently emphasis has been placed upon the bacteria rather than upon the protozoans. Hungate[28] postulated that the role of the ciliates is taken over by the bacteria in their absence so that the host does not suffer. The numbers present are so large that it is impossible to believe that they do not contribute to the chemical changes that go on in the rumen, but for the present it is only possible to discuss what is known of their metabolism and to guess at the significance of this to the host.

There appears to be selective ingestion of particulate matter by ciliates. Large ciliates and the smaller Entodinia ingest starch, and the larger forms ingest plant fragments, particularly those containing chlorophyll. Holotrichs, however, do not appear to ingest much particulate matter.

Much attention has been given to the question of whether ciliates digest cellulose, but until Hungate[39, 40] was able to grow cultures of several species of Diplodinia no satisfactory evidence was produced that disintegration of cellulose was due to a cellulase produced by the species in question rather than by bacteria ingested with the plant particles. This is based on the evidence that suspensions of Diplodinia grown in clone cultures, when washed practically free of bacteria and debris, produced reducing substances from cellulose whereas the washings did not. Similarly he observed that a polysaccharide was deposited in the bodies of *Diplodinium maggii* within 2 hr. of feeding pure cellulose, which was considered too rapid a change to be due to bacterial action.

The disintegration of starch by ciliates is open to the same controversy. Trier[41] described the ingestion and disintegration of starch granules, which resulted in the laying down of granular glycogen in the body of the ciliates. Usuelli[42] estimated that as much as 84 % of 10 g. of starch introduced into the rumen of a sheep was ingested by the ciliates within 20 min. This estimate no doubt has a large margin of error: it does indicate, however, that as far as starch is concerned ciliates may play an important part in its fate in the rumen. Van der Wath and Myburgh[43] followed the disintegration of starch granules inside Diplodinia and came to the conclusion that it was due to iodophilic bacteria ingested with the grains and that the bacteria themselves were subsequently digested by the ciliate. Schlottke[44] was able

[39] R. E. Hungate, *Biol. Bull.* **83,** 303 (1942).
[40] R. E. Hungate, *Biol. Bull.* **84,** 157 (1943).
[41] H. J. Trier, *Z. vergleich. Physiol.* **4,** 305 (1926).
[42] F. Usuelli, *Clin. vet.* **53,** 545, 625, 787 (1930).
[43] J. G. van der Wath and S. J. Myburgh, *Onderstepoort J. Vet. Sci. Animal Ind.* **17,** 61 (1941).
[44] E. Schlottke, quoted by R. E. Hungate, ref. 28

to demonstrate more amylase activity in extracts of rumen ciliates than in extracts of rumen bacteria. Oxford[45] has shown that the large quantities of polysaccharide stored by isotrichs of the rumen are in fact nearer to starch than to glycogen as regards chemical structure. Holotrichs utilize a variety of sugars and rapidly convert them to this protozoan polysaccharide (Masson and Oxford[46]), but oligotrichs utilize them more slowly. It has been taken for granted that the rumen ciliates ingest and digest bacteria, but the granular material that can be seen within the cytoplasm of these organisms appears to be largely polysaccharide granules, and recent investigations of Sugden[47] throw doubt on the ingestion of bacteria by rumen holotrich ciliates, for after considerable observation she has not been able to observe this to occur. Indeed the ciliates appear to be able to select or reject particulate material with a great deal of precision.

Schulze[48] suggested that the large quantities of stored polysaccharide, rather than representing the major portion of the food used, may be an adaptation which permits the protozoans to continue fermentation when carbohydrate food is not readily available. Heald and Oxford[48a] have shown that this is so as far as the Holotrich ciliates of the rumen are concerned. Several species, maintained in virtually a bacteria-free condition, were able to ferment the sugars normally present in grass and other fodders, and acetic, butyric, and lactic acids together with carbon dioxide and hydrogen were produced. More glucose was used to form glucosan than was fermented, but even so isotope experiments showed that endogenous fermentation of glucosan was not suppressed by the presence of glucose and it persisted when no glucose was present.

It has been suggested that the ruminant is like a plankton feeder with the added adaptation that it rears the planktonic forms and utilizes not only their cells but also the fermentation products of their growth.

The relation of the numbers and kinds of ciliates present in the rumen or large gut has received considerable attention, but owing to the difficulties of sampling, the results are questionable; even so the differences recorded are so great that it seems unlikely that they can be due merely to experimental error. Thus Mowry and Becker[49] record figures for ciliates present in the rumen of the order of 200,000 per milliliter for sheep fed on hay alone, whereas in sheep receiving hay with additional starch and protein the numbers rose to 2,000,000 per milliliter. Ferber[50] found that the numbers of ciliates in the rumen of sheep and goats increased during the later stages

[45] A. E. Oxford, *J. Gen. Microbiol.* **5,** 83 (1951).
[46] F. M. Masson and A. E. Oxford, *J. Gen. Microbiol.* **5,** 664 (1951).
[47] B. Sugden, personal communication, 1952.
[48] P. Schulze, quoted by R. E. Hungate, ref. 28.
[48a] P. J. Heald and A. E. Oxford, *Biochem. J.* **53,** 506 (1953).
[49] H. A. Mowry and E. R. Becker, *Iowa State Coll. J. Sci.* **5,** 35 (1930).
[50] K. E. Ferber, *Z. Tierzücht. Züchtungsbiol.* **15,** 375 (1929).

of pregnancy and during lactation, but it is not clear whether this is an appetite effect or not. Van der Wath and Myburgh,[43] working in South Africa, found that the numbers of ciliates in the rumen of sheep were related to the ration, for when the ration consisted of wheat straw the numbers present were 100,000 per milliliter or less; when lucerne hay or green lucerne was given, the numbers varied from 200,000 to 500,000 per milliliter; and crushed maize added to either straw or lucerne hay produced counts of 700,000 to 800,000 per milliliter in the former, or over 2,000,000 per milliliter in the latter instance. Similarly, sheep grazing on the veldt harbored the most ciliates in January (summer) and the least in July (winter). Analysis of the grass showed that the digestibility of the dry matter was greatest in July, although the percentage of nitrogen was 0.3 % lower than in October. The quantity consumed, however, is stated to be more in July.

Van der Wath and Myburgh,[43] by using differential counts, were able to show that the proportion of ciliates classified as Entodinia was greater in browsing than in grazing antelopes under natural conditions and related this to the fact that, since the smaller Entodinia cannot ingest such large particles as the larger Diplodinia, they are more reliant upon diets rich in protein and carbohydrate. An exception was the steenbock—a grazing antelope—but these animals are highly selective feeders and eat only the greenest and most delicate parts of the plant, and it may be that the requirements for a high Entodinia population are satisfied. It is interesting that the Diplodinia—some of which, in the bovine, appear to produce a cellulase—are less dependent upon fresh green material in the diet which is presumably rich in soluble carbohydrate.

Adams[51] has made accurate estimates of the ciliates present in the large gut of the horse and finds that their concentration is the highest in the third part of the colon. Her work was done on slaughter-house specimens and so represents the numbers present in the fasted horse. The interesting observation was made that the species of ciliate present in the caecum and the ventral part of the colon are considerably different from those present in the dorsal part.

A detailed study of the protozoan populations of the horse by Oxford[52] has shown that storage of iodophilic polysaccharide occurs in the large ciliate *Cycloposthium* which is abundantly present in the ventral colon.

The ciliates are thus fascinating and highly organized entities, possessing a highly active metabolic cycle. They are present in the rumen in large numbers, so much so that 10 to 20 % of the total rumen nitrogen has been estimated to be present as ciliate nitrogen in the rumen of sheep.

The fact that specialized forms are present in large numbers in the colonic

[51] K. M. G. Adam, *Parasitology* **41,** 301 (1951).

[52] A. E. Oxford, personal communication, 1952.

contents of the horse and that they are present in a wide variety of wild herbivores makes their biological position unique, and it is disappointing that so far no one has been able to fit them securely into the cycle of events that constitute microbial digestion, for although in the sheep they can be dispensed with without any apparent loss to the host, there can be no doubt that they have a place. The fact that the holotrich ciliates rapidly synthesize polysaccharide from soluble sugars and that the oligotrich ciliates readily digest starch is some indication of their activity, but until more is known of their metabolism it is impossible to assess these properties in terms of the total digestion of the animal.

V. Bacteria and the Host

General Discussion

Alimentary bacteria can be beneficial to the host if they decompose otherwise indigestible materials such as cellulose, or if they synthesize molecules or specific groups which the host requires and which are absent in the diet or present in a quantity insufficient to meet requirements. Both attributes are of benefit to the host only if the products of disintegration or of synthesis can be absorbed from the alimentary tract. On this definition bacteria may be useful if they ferment cellulose, pentosans, or even raw tuber starch, which are otherwise indigestible, or if they synthesize members of the B complex when these are not present in the food in adequate quantities. The changes wrought in proteins and carbohydrates of the food are beneficial to the host if they are essential to support the life of the bacteria which perform the aforementioned feats. It is, perhaps, arrogant to set oneself up as a judge of the benefits that accrue from a relationship between two partners—in this instance the host and its alimentary microflora—and it is wiser to examine such a partnership in a more objective spirit in order to attempt to understand the physicochemical basis of its equilibrium. Any other attitude reflects a tendentious approach. This being so, this section will be devoted to a discussion of the occurrence and metabolism of bacteria present in the alimentary tract so far as is known. Only organisms which are presumed to be present in numerically adequate numbers to be significant will be discussed. Some recent workers (Moir and Masson[53]) have considered it best to describe the morphological characters rather than to name them at this stage of development.

In this field it has been of the greatest advantage to use all the resources of direct microscopy while trying to dispel some of the limitations of the cultural approach. The direct microscopic method permitted Baker and colleagues[54, 55] to make observations on the structural integration of the

[53] R. J. Moir and M. J. Masson, *J. Path. Bact.* **64,** 343 (1952).

[54] F. Baker and H. Nasr, *J. Roy. Microscop. Soc.* **67,** 27 (1947).

[55] F. Baker, H. Nasr, F. Morrice, and J. Bruce, *J. Path. Bact.* **62,** 617 (1950).

substrate materials and the visible sites and agents of attack. This has been a great service even though it deals only with structural changes. The microbial breakdown of structural cellulose is accompanied by the following features: (1) simultaneous loss of histochemical reactions and optical birefringence occurs in the affected structures; (2) the microbial origin of the process of breakdown is demonstrated by the formation of clear-cut enzymic cavities around the responsible microorganisms; (3) the histological components of fodder-plant materials, such as grasses, are digested in a certain order and at widely different rates; (4) lignin, cutin, and other encrusting substances are especially prominent in undigested residues; (5) there is present a constant association in the same natural habitat of several morphologically distinguishable types of microorganisms.

The number of bacteria in the rumen has been estimated by slide counts and by a dilution method using media developed by Gall *et al.*[56] Counts as high as 50 billion per gram of fresh rumen contents have been recorded from cattle and sheep on winter rations and from 85 to 96 billion from animals at pasture. Gall and Huhtanen[57] have described in detail five strains of lactobacilli which have been isolated repeatedly from the rumen of many animals. These are quoted as only five out of twenty types of organism that can be regularly isolated and which are present in numbers greater than 1 million per gram of fresh rumen contents. These observations, although as yet incomplete, indicate that it will soon be possible to refer any organism of the rumen against standard cultures which have been selected on the following criteria: (1) they are anaerobic bacteria; (2) they are present in quantity greater than 1 million per gram of fresh tissue; (3) they have been isolated on at least ten occasions from two or more animals; (4) they have been isolated from animals in at least two geographic areas; and (5) they produce end products of fermentation known to occur in the rumen from substrates known to be present in the rumen; intermediary products of fermentation are presumably included.

2. Fermentation of Cellulose

The isolation of cellulose-fermenting bacteria from the alimentary tract has received a considerable amount of attention, but although several organisms have been studied in the past there is considerable doubt whether they are effective inhabitants.

The cellulose-fermenting bacteria are notoriously difficult to isolate in pure culture, but recently Hungate,[58] Gall and her colleagues,[57] and Sijpesteijn[59] have been able to study the fermentation of cellulose by bacteria

[56] L. S. Gall, C. N. Stark, and J. K. Loosli, *J. Dairy Sci.* **30,** 891 (1947).
[57] L. S. Gall and C. N. Huhtanen, *J. Dairy Sci.* **34,** 353 (1951).
[58] R. E. Hungate, *J. Bact.* **53,** 631 (1947).
[59] A. K. Sijpesteijn, *J. Gen. Microbiol.* **5,** 869 (1951).

in vitro under conditions that simulate the rumen. All are anaerobes or facultative anaerobes, and their fermentation products include short-chain fatty acids, carbon dioxide, and hydrogen. From Baker and Harriss'[60] observations on the disintegration of plant tissues containing cellulose it would appear that coccoid forms are principally responsible, yet Baker and Harriss draw attention to the pleomorphic nature of the organisms concerned. Under these circumstances it would be unwise to dogmatize as to the morphological nature of the organisms responsible for cellulose fermentation in the rumen, as those using the cultural approach find rods more active than coccal forms. It is necessary to recognize the fact that all those that have so far received sufficient study produce short-chain fatty acids as products of activity. Marston's[61] work suggests that the proportion of the energy of the substrate that is incorporated in the bacteria themselves is approximately 6% of the total used. No comparable work has been done on the cellulose-fermenting organisms present in the large intestines of other species, although Baker and Martin[62] from their description of the organisms present in the large intestines of the horse and guinea pig found that similar morphological types could be observed attached to plant fragments in these organs.

Hummel, Shepherd, and Macy[63] found that older children were able to decompose more cellulose and hemicellulose than younger children, and this they attributed to the greater intestinal length and greater opportunity for bacterial decomposition of fiber. For children of 4 to 12 years 5 to 7 g. dietary fiber daily caused no undesirable effects on the absorption of nitrogen or minerals.

The work of Trautmann and Asher[64] has shown quite clearly that pure reprecipitated cellulose fibers can easily be digested by cellulose-destroying bacteria in the caecum of the pig but that cellulose acetate and methyl cellulose are completely undigested. Plant cellulose can also be brought slowly into solution in the caecum. The optimal conditions for digestion are freedom from encrusting substances and maximum surface area in relation to weight.

Vartiovaara, Roine, and Poijärvi[65] have investigated the possibility of accelerating microbial digestion of cellulose in pigs. The amounts normally digested appear to vary quite considerably, the quantities digested in a given instance being dependent on the species of animal, the character of the cellulosic material, and the stability of the microbial association established.

[60] F. Baker and S. T. Harriss, *Nutrition Abstract Revs.* **17,** 3 (1947).
[61] H. R. Marston, *Biochem. J.* **42,** 564 (1948).
[62] F. Baker and R. Martin, *Zentr. Bakt. Parasitenk.* Abt. II, **99,** 400 (1939).
[63] F. C. Hummel, M. L. Shepherd, and I. G. Macy, *J. Nutrition* **25,** 59 (1943).
[64] A. Trautmann and T. Asher, *Biedermanns Zentr.* **B14,** 353 (1942).
[65] A. Vartiovaara, P. Rione, and I. Poijärvi, *Maataloustieteellinen Aikakauskirja* **16,** 75 (1944), quoted from *Nutrition Abstracts & Revs.* **17,** 410 (1947).

Enrichment cultures of cellulose-decomposing bacteria from the digestive tract were fed to pigs, and these appeared after a time to improve digestion of the cellulose.

If the products of bacterial decomposition of cellulose in the large intestine of omnivores, which are most probably acetic and propionic, acids, can be utilized as productively by pigs as by ruminants, then fodder cellulose should have a value for fattening pigs not much inferior to that of oats.

A survey of the feces of human beings by Hirschberg[66] revealed that over one-third of the specimens showed fungi of the *Aspergillus* group which could decompose cellulose. There was no difference in this respect between normal and constipated persons.

3. Fermentation of Starch

In the alimentary tract of mammals, starch is digested either by the amylolytic enzymes secreted by the digestive glands or by bacterial action; and the extent to which one or the other of these preponderates may depend on several factors, particularly on the morphology of the gut and the structure of the raw starch granule. Thus in non-ruminants, except the hamster, untreated cereal starches are broken down in the small intestine by the secretions of the digestive glands so that little starch accumulates in the caecum and bacterial action plays only a minor role in breakdown. Raw tuber starches, except tapioca starch, may completely escape the action of the digestive juices and be attacked by bacteria in the caecum. Ground raw potato starch given to rats is digested in the small gut and the animals fail to refect; i.e., they fail to grow and survive in the absence of the vitamin B complex. It is probable that the hydrolytic breakdown of starch by digestive enzymes results in a more rapid and efficient utilization of starch as a source of energy. In the hamster, raw tuber starches are also broken down by bacteria in the gastric diverticulum and in the caecum. In ruminants, raw starches and starch products are broken down in the rumen by bacterial action.[55]

Characteristic bacteria are associated with the breakdown of starch. In the pig's caecum *Clostridium butyricum* was found to be the primary agent of breakdown, and a strain isolated from that source synthesized members of the B complex, other than biotin and *p*-aminobenzoic acid, required for its own growth.[55] Pigs given vitamin B-free diets were able to grow and survive if untreated potato starch was added to the diet, but not if the potato starch was rendered soluble by steaming or if untreated maize starch was substituted.[67] Coccoid forms were seen clustered round starch granules in the rumen of sheep by Van der Wath and Myburgh,[43] and several types

[66] N. Hirschberg, *Am. J. Digestive Diseases* **9**, 200 (1942).
[67] H. Nasr, *Brit. J. Nutrition* **4**, v, Proc. (1950).

have been isolated by MacPherson[68] from the rumen of sheep feeding on starch-rich rations. These appear to be similar to *S'reptococcus bovis* but can be distinguished by their antigenic reactions. Similar organisms have also been isolated by her from the caecum of the horse and the chick. Thus, although *Clostridium butyricum* is present in large numbers in the caecum of the pig fed on potato starch it does not follow that it is the only organism concerned with its disintegration. Masson[32, 34] observed large numbers of *Cl. butyricum* in the rumen of sheep fed on a ration rich in flaked maize, yet coccoid forms, whose chemical properties were similar to those of *S. bovis*, were also present and, as far as can be judged by the high concentrations of lactic acid formed, were numerically highly significant.

The property of storing a polysaccharide within its cytoplasm is characteristic of the bacteria associated with the hydrolysis of starch in both the rumen and the caecum. *Cl. butyricum* is known to produce an amylase *in vitro*,[69] and a free amylase that occurs in the rumen of starch-fed sheep is presumably of bacterial origin, as no amylase is present in sheep's saliva.[70] Similarly an amylase is produced *in vitro* by starch-splitting cocci,[68] and although this enzyme hydrolyzes starch as far as the dextrin stage the end products of fermentation include acetic, propionic, and butyric acids.

In an investigation of the fate of starch in the bird, Baker, Carpenter, and Duckworth[71] have shown that potato starch reaches the caecum in large amounts, maize starch only in small amounts. In the caecum starch granules are attacked by large and small iodophile sporing rods, morphologically resembling *Cl. butyricum*. The number of sporing rods was greatest in birds fed on potato starch and least in birds fed on diets with lactose as the chief source of carbohydrate. Birds fed maize starch presented an intermediate picture.

4. Fermentation of Pentosans

Pentosans, of which the chief component is xylan, form some 16 to 20% of the dry matter of grass and hay (Fraps,[73] Hallsworth,[74] and Ekelund[75]) and are digested to the extent of 50% in the alimentary tract of the ruminant.

Although pentosan-decomposing microorganisms are widespread in na-

[68] M. J. MacPherson, personal communication, 1952.
[69] W. J. Whelan and H. Nasr, *Biochem. J.* **48,** 416 (1951).
[70] H. Nasr, *J. Agr. Sci.* **40,** 308 (1950).
[71] F. Baker, K. J. Carpenter, and J. Duckworth, personal communication, 1952.
[71a] S. R. Elsden, *J. Exptl. Biol.* **22,** 51 (1945).
[72] A. T. Johns, *J. Gen. Microbiol.* **5,** 317 (1951).
[72a] A. K. Sijpesteijn and S. R. Elsden, *Biochem. J.* **52,** 41 (1952).
[73] G. S. Fraps, *Bull. Texas Agr. Expt. Sta.* **418,** (1930).
[74] E. G. Hallsworth, *J. Soc. Chem. Ind. (London)* **58,** 357 (1939).
[75] S. Ekelund, *Ann. Roy. Agr. Coll. Sweden* **16,** 179 (1949).

ture (Patrick and Werkman[76]), there is little or no information concerning the types and numbers of similar microorganisms in the ruminant, nor indeed of the quantitative importance of the fermentation of these substances.

The simple components xylose, glucuronic acid, and arabinose are rapidly fermented by rumen microorganisms with the formation of the volatile fatty acids (Heald[77] and McNaught[78]), acetic, propionic, and butyric acids, with acetic acid predominating.

The report (D'yakov and Ivankin[79]) that xylose sirups (wood sugar) fed to sheep on a poor hay diet can convert a negative nitrogen balance to a positive one, while needing confirmation, may find some explanation in the findings of McNaught [8] that xylose serves as a suitable energy source for the synthesis of protein from urea by rumen microorganisms. With regard to the quantities of pentosans, and xylan in particular, fermented in the rumen, Marshall[80] has shown that some 30 to 40 % of the furfural-yielding components of meadow hay disappear in the rumen and the omasum. Heald[81] used a more precise estimation of xylose present in hydrolyzates of food leaving the abomasum. He finds a similar percentage digested and calculates that some 60 to 80 g. of xylan may be fermented every 24 hr. in the rumen of a sheep at pasture. If the products are mainly acetic acid (Sellière[82]), this quantity is of nutritional significance.

5. Fermentation of Pectins

Investigations by Werch and Ivy[83] have shown that pectin is broken down in the alimentary tract of both dog and man. Ileostomy experiments have shown that the breakdown of the pectin occurred chiefly in the colon and not in the small intestine and that bacterial enzymes were largely responsible for the decomposition.

Leroy and Michaux[84] found that sheep digested a high proportion of the "pectic substances" included in such foodstuffs as apple pulp, sugar beet pulp, hay, and straw. It was found that a sheep feeding on hay consumed from 75 to 102 g. daily of pectic substances (Michaux).[85] The proportion of pectic acid to pectin is high in apple and sugar beet pulp, the ratio being

[76] R. Patrick and C. H. Werkman, *Iowa State Coll. J. Sci.* **7,** 407 (1933).
[77] P. J. Heald, *Biochem. J.* **50,** 503 (1952).
[78] M. L. McNaught, *Biochem. J.* **49,** 325 (1951).
[79] M. I. D'yakov and V. K. Ivankin, *Probleme Tierzucht* (*U.S.S.R.*) **4,** 76 (1933) [*C. A.* **30,** 4231 (1936)].
[80] R. A. Marshall, *Brit. J. Nutrition* **3,** 1 (1949).
[81] P. J. Heald, Thesis, Aberdeen University, 1952.
[82] G. Sellière, *Compt. rend. soc. biol.* **68,** 989 (1910).
[83] S. C. Werch and A. C. Ivy, *Am. J. Digestive Diseases* **8,** 101 (1941).
[84] A. M. Leroy and A. Michaux, *Compt. rend.* **229,** 1034 (1949).
[85] A. Michaux, *Compt. rend.* **230,** 2051 (1950).

approximately 1:10; the ratio of these substances in the feces of sheep fed on these foods was 1:0.5. In hay, however, pectin predominates. The digestibility of pectic substances in sheep was approximately 75% or over for hay and 90% for apple and sugar beet pulp, but in a lamb the digestibility was considerably less.[86]

VI. The Results of Microbial Digestion

There are several ways in which comparisons between the results of microbial digestion have been made based on the quantities of the products of fermentation formed in relation to bodyweight or more simply on the amounts of cellulosic fiber disappearing from the food as it passes through the gut. These comparisons are not valid on their own account, for strictly controlled tests have not yet been made. However, the evidence available sustains the widely held view that microbial digestion plays a more important role in the ruminant than in animals where microbial digestion is confined primarily to the large intestine.

The studies of Crampton *et al.*[87] on the comparative value of foodstuffs of low fiber to human subjects, rats, guinea pigs, swine, and sheep so far suggest that, of themselves, the anatomical differences in the digestive systems of the five species studied do not result in large differences in digestive capacity with respect to diets of negligible crude fiber content.

Comparisons may also be made of the quantities of methane produced and the quantities of fatty acid present in the digesta. In comparing the capabilities of different species of digesting cellulose, attention must be paid to the form in which the cellulose is offered, with particular reference to encrusting substances. It is well known that the age of the plant, which determines the degree of lignification of the structural components of the plant, influences markedly the digestibility of the cellulose.

Pure cellulose or the cellulose of straw after alkali or sulfite treatment is almost completely digested (80% or more) by the ruminant and also by the pig.[88] When the digestibility of the fibers of straw is compared, however, the ruminant digests more than the horse, and the digestibility in the pig is very low and variable.[89]

The quantity of cellulose digested in man depends upon the form in which it is given, with particular reference to the extent of encrusting substances such as lignin. Little, if any, is digested from heavily lignified plant tissues, as the experiments of Rubner and his colleagues showed in Germany in World War I, yet the cellulose contained in such foods as lettuce, cabbage,

[86] A. Michaux, *Compt. rend.* **232,** 121 (1951).

[87] E. W. Crampton, M. I. Irwin, L. E. Lloyd, and H. R. Nielson, *J. Nutrition* **43,** 541 (1951).

[88] H. E. Woodman and R. E. Evans, *J. Agr. Sci.* **37,** 202, 211 (1947).

[89] H. E. Woodman, *Biol. Revs. Biol. Proc. Cambridge Phil. Soc.* **5,** 273 (1930).

carrots, and low extraction bread apparently disappears to a large extent as it passes through the gut, although the quantity in the total food is small.[90]

The quantity of methane evolved by cattle bears a close relation to the quantity of digestible carbohydrate in the ration.[91] Ritzman and Benedict[92] compared the quantities of methane evolved from the ruminant and other herbivores and found that when hay was fed to the horse and the Indian elephant more methane appeared per unit of body weight from the ruminant than was produced by either of the others.

A comparison of the fatty acids in the digesta of a variety of animals was made by Elsden *et al.*,[93] and it appeared that the greatest proportion per unit of body weight was in ruminants and that the horse, the pig, the rabbit, the rat, and the dog followed in that order. These observations do not take account of what is absorbed.

Whenever microbial digestion occurs, there appears a mixture of short-chain fatty acids in the contents of the stomach or gut. The acids known to be present are acetic, propionic, and butyric acids, and of these acetic predominates. This mixture has been found in the large intestines of the horse, the pig, the sheep, the rabbit, the rat, and the dog, and short-chain fatty acids are known to be present in man. They also occur in the rumen and the reticulum of the sheep, the goat, the ox, and the red deer. The concentration and the relative proportions of these acids in the rumen vary with the interval after feeding, and the nature of the food also influences these proportions. Diets rich in starch or soluble carbohydrate support a lactic acid-producing flora, and this acid may accumulate temporarily in the rumen. Under most circumstances lactic acid is unstable and is fermented, presumably by members of the group of propionic bacteria, with the production of acetic and propionic acids.[94] Under conditions of heavy starch feeding, however, lactate may remain in the rumen in considerable concentrations and the pH may then become very acid. The fate of lactate is not fully explained in the rumen, and much work still remains to be done. Death may occur in sheep allowed unlimited quantities of wheat; the exciting cause of death, however, still needs verification.

The occurrence of propionic acid in the rumen has led to a search for members of the propionic acid group. Elsden[71a] in 1945 first isolated an organism which was capable of fermenting glucose and lactic acid with the production of acetic and propionic acids. Johns[72] has recently made a detailed study of organisms from the rumen of sheep similar to *Vellionella*

[90] E. Mangold, *Nutrition Abstracts & Revs.* **3**, 647 (1934).

[91] J. W. Bratzler and E. B. Forbes, *J. Nutrition* **19**, 611 (1940).

[92] E. G. Ritzman and F. G. Benedict, *Carnegie Inst. Wash. Pub.* **494**, (1938).

[93] S. R. Elsden, M. W. S. Hitchcock, R. A. Marshall, and A. T. Phillipson, *J. Exptl. Biol.* **22**, 191 (1945–1946).

[94] S. R. Elsden and A. T. Phillipson, *Ann. Rev. Biochem.* **17**, 705, (1948).

gazogens which ferment glucose and lactate. The mechanism of propionic acid formation depends upon the fixation of carbon dioxide with pyruvic acid; oxaloacetate so formed passes via malate and fumarate to succinate, subsequent decarboxylation producing propionate. Varying proportions of propionate and acetate formed appear to depend upon the carbon dioxide tension and pH values. The occurrence of succinate in the rumen has been reported by Sijpesteijn and Elsden[72a] and the disappearance of added succinate is accompanied by an increase in concentration of propionic acid.

Recently the occurrence of branched isomers of butyric and valeric acids have been found in the rumen by El Shazly.[95] The quantity present is related to the nature and quantity of the protein fed in the ration; thus, soluble casein causes an increase in these compounds.

Absorption of acetic, propionic, and butyric acids is known to occur across the stratified squamous epithelium lining the rumen, and this absorption of short-chain fatty acids—or, to be more precise, steam-volatile fatty acids, according to the actual method of determination—also takes place from the reticulum, the omasum, the caecum, and the colon of the sheep, and from the large intestine of the horse, the pig, the rabbit, and the dog.[94] In this way some of the products of the metabolism of the microbial population are continually removed from the milieu, thus satisfying one of the essential requirements for the continuing activity of a mixed culture of organisms.

A further mechanism which helps to maintain stability of the milieu of the rumen, and possibly of the large intestine, is that as the pH falls owing to the production of acidic products of fermentation the proportion of free fatty acid to anion increases, and as the free fatty acid penetrates the epithelial cells lining the rumen more rapidly than the anion, so the rate of absorption increases and the pH is prevented from becoming too acid. The fact that large quantities of lactic acid may be associated with a pH of the rumen that is more acid than normal, i.e., a pH below 5, suggests that lactic acid as such penetrates the epithelial membrane at a slower rate than the other fatty acids.[94]

The fatty acids originate from the decomposition of both carbohydrate and protein. McDonald[96, 97] found that ammonia is normally present in small concentrations in the rumen liquor and the introduction of protein causes an increase. This increase is associated with the production of fatty acids, which El Shazly[95] has shown includes isomers of butyric and valeric acids and carbon dioxide.

Cuthbertson and Chalmers[98] found that when a first-class soluble protein

[95] K. El Shazly, *Biochem. J.* **51,** 647 (1952).

[96] I. W. McDonald, *Biochem. J.* **42,** 584 (1948).

[97] I. W. McDonald, Thesis, Cambridge University, 1948.

[98] D. P. Cuthbertson and M. I. Chalmers, *Biochem. J.* **46,** xvii (1950).

was given to sheep in need of protein they did not make satisfactory use of it, but if the protein was given straight into the duodenum rather than into the rumen this defective utilization no longer took place. Chalmers and Synge[99] followed up these observations and found that the physical character of the protein in terms of its solubility seemed to be a factor, thus confirming the work of McDonald on zein, an insoluble protein. Chalmers and Synge also found that compared with casein there was a smaller rise in rumen ammonia, and therefore loss of nitrogen, when herring meal was fed.

Most animal fodders include a proportion of nitrogen in a non-protein form. In pasture grass the bulk of this is in the form of basic substances and well-known amino acids in the free state (Synge).[100] Amino acids and added urea undergo rapid deamination in the rumen. How much of the ammonia nitrogen so formed is incorporated into bacterial protein probably depends upon the food which, in turn, influences the numbers and kinds of bacteria present. Pearson and Smith[101,102] found that an increase in protein accompanied the addition of urea and starch to rumen contents, while rations in which urea forms practically the sole source of nitrogen have been shown to sustain weight increases in sheep.[112] Synge and his colleagues (*J. Agric. Sci.* in press) find that when high protein rations are fed to sheep the concentrations of ammonia produced in the rumen are considerable; much of this nitrogen is excreted in the urine. Starch, added to the ration, reduces the ammonia formed in the rumen. The difference in these rations clearly leads to different utilisation of the available nitrogen. McDonald[103] calculated that in a ratio of which zein formed the main source of protein approximately 40% of the nitrogen was converted to bacterial nitrogen.

A glucuronide-decomposing enzyme has been found by Levvy and his colleagues[104] as a normal constituent of the medium-sized microorganisms in the sheep rumen. This enzyme has been purified and the nature of its action is being studied. It is believed that its function may be to assist in the digestion of plant polyuronides, a view which is strengthened by the finding of similar enzymes in large intestine of the horse and in the alimentary tract of other species in those zones where bacteria are found.[105]

The gas mixture in the rumen is composed principally of carbon dioxide and methane, although traces of oxygen, nitrogen, and hydrogen sulfides are reported under normal feeding conditions. The proportions of methane and carbon dioxide fluctuate considerably. Rumen microorganisms appear

[99] M. I. Chalmers and R. L. M. Synge, *Brit. J. Nutrition* **4,** ix (1950).
[100] R. L. M. Synge, *Biochem. J.* **49,** 642 (1951).
[101] R. M. Pearson and J. A. B. Smith, *Biochem. J.* **37,** 142, 148, 153 (1943).
[102] M. L. McNaught and J. A. B. Smith, *Nutrition Abstracts & Revs.* **17,** 18 (1947).
[103] I. W. McDonald, *J. Physiol.* (*London*) **107,** 21P (1948).
[104] M. C. Karunairatnam and G. A. Levvy, *Biochem. J.* **49,** 210 (1949).
[105] C. A. Marsh, F. Alexander, and G. A. Levvy, *Nature* **170,** 163 (1952).

to require a high concentration of carbon dioxide for their growth. The most significant relationship, so far as methane is concerned, seems to be with the quantity of carbohydrate decomposed. Methane is evolved most rapidly during the first 4 hr. after feeding (Washburn and Brody[106]); thereafter the rate steadily declines, but it takes 3 to 5 days before all the methane disappears from the rumen. When feeding is resumed after prolonged fasting hydrogen appears to be evolved in varying quantities. After 3 to 4 days this hydrogen disappears from the rumen gas mixture and is replaced by methane. Similar results may be obtained by emptying the rumen after 3 days' fast and replacing the contents with a solution of sodium bicarbonate before feeding is resumed (Pilgrim[107]).

The probable role of hydrogen in the formation of methane in the rumen is discussed by Beijer,[108] who showed that the addition of formate and succinate to the rumen liquor of goats produced a marked evolution of methane as well as some increase in carbon dioxide and hydrogen. Beijer[108] suggests that both acids act as hydrogen donators for the bacterial reduction of carbon dioxide. Lewis[108a, b] has shown that, in addition to formate and succinate, lactate, citrate, glucose, malate, and mannitol are hydrogen donators for the reduction of nitrate by washed suspensions of mixed rumen bacteria. Acetate, propionate, and *n*-butyrate are inactive both as a source of hydrogen for methane formation, according to Beijer,[108] and for nitrate production. Nitrate reduction in the rumen can lead to fatal methemoglobinemia if sufficient is present in the fodder.

VII. The Fate of Alimentary Microorganisms

The ciliates and bacteria of the rumen pass with the food residues through the alimentary tract. Consequently they are exposed to the action of proteolytic enzymes in the abomasum and small intestines. The pH of the abomasum is in the region of 3, and here the ciliate protozoans disintegrate. The effect of the acidity and peptic activity of the abomasal contents is less marked on the bacteria. Masson's[32] observations on the abomasal organisms in comparison to those of the rumen in the same sheep show that the bacteria lose definition and also fail to stain satisfactorily so that their appearance has been described as that of a dying culture. It is exceedingly unlikely that all bacteria are killed by the change in pH, by the action of pepsin, or by the subsequent action of the duodenal and intestinal juices; no cultural work has been done either on the abomasal contents or the contents of the small intestine, and until estimations of the numbers of

[106] L. E. Washburn and S. Brody, *Bull. Missouri Agr. Expt. Sta.* **263** (1937)
[107] A. F. Pilgrim, *Australian J. Sci. Research* **B1**, 130 (1948).
[108] N. H. Beijer, *Nature* **170,** 576 (1952).
[108a] D. Lewis, *Biochem. J.* **48,** 175 (1951).
[108b] D. Lewis, *Biochem. J.* **49,** 148 (1951).

viable bacteria present in these organs have been made, there is no sure evidence to go on.

The fact that urea can be used to replace part of the protein of the ration without loss in production has been interpreted as meaning that the nitrogen so administered is largely converted to bacterial protein and that the bacteria are subsequently digested. There is no doubt that the addition of urea together with starch to a ration otherwise low in nitrogen causes a marked increase in the numbers of organisms present in the rumen. It is also known that the inclusion of methionine with the urea[109, 110] causes an even greater increased initial count and that the bacterial nitrogen so supplied has a high feeding value to rats, but although this offers a strong suggestion that the bacterial protein synthesized from these materials has a nutritive value to the host the direct proof is lacking. Similarly it has been shown that the biological value of dried rumen bacteria as a source of protein to rats is of the order of 80 or more, which indicates its potential value to ruminants after digestion in the intestines.[102] The conversion of non-protein nitrogen into amino acids is proved by the fact that sheep on a purified diet with urea as virtually the sole source of nitrogen in the rumen synthesized the ten amino acids, including methionine, which are known to be essential to the rat.[111] In another experiment sheep maintained on such a ration increased in weight by approximately ½ to 1 lb. per week, but removal of inorganic sulfur from the ration resulted in loss of weight and the animals were then in a state of negative nitrogen balance.[112] The introduction of labeled inorganic sulfur into the rumen of the cow, the sheep, and the goat resulted in its appearance in sulfur-containing amino acids in the proteins of the rumen, plasma and milk.[113, 114] Experiments such as these leave no doubt that some of the rumen microorganisms disintegrate as they pass through the alimentary tract and that the animal derives amino acids from this process. It must not be assumed, however, that all rumen bacteria are digested, for the observations of Pounden, Ferguson, and Hibbs[115] suggest that different organisms are not equally susceptible to the action of the digestive juices. It has often been suggested that one of the roles of the protozoans is to ingest bacteria and digest them so that the ultimate product of rumen synthesis of protein is protozoan nitrogen. There is no doubt that the protozoans disintegrate in the abomasum, and this hy-

[109] J. K. Loosli and L. E. Harris, *J. Animal Sci.* **4,** 435 (1945).

[110] V. J. Williams and R. J. Moir, *Australian J. Sci. Research* **4,** 377 (1951).

[111] J. K. Loosli, H. H. Williams, W. E. Thomas, F. H. Ferris, and L. A. Maynard, *Science* **110,** 144 (1949).

[112] W. E. Thomas, J. K. Loosli, H. H. Williams, and L. A. Maynard, *J. Nutrition* **43,** 515 (1951).

[113] R. J. Bloch and J. A. Stekol, *Proc. Soc. Exptl. Biol. Med.* **73,** 391 (1950).

[114] R. J. Bloch, J. A. Stekol, and J. K. Loosli, *Arch. Biochem. Biophys.* **33** (1950).

[115] W. D. Pounden, L. C. Ferguson, and J. W. Hibbs, *J. Dairy Sci.* **33,** 565 (1950).

pothesis would be attractive if it were not for the recent observations of Sugden,[47] who failed to find any evidence for the protozoan ingestion of rumen bacteria.

The question of the disintegration of bacteria in the alimentary tract is important, for the liberation of their intracellular products, i.e., polysaccharides, as well as the protoplasmic nitrogen presumably depends on this disintegration. Consequently it is disappointing that so little direct work has been done upon it, and the hypothesis that rumen bacteria are digested as they pass through the abomasum and small intestine rests on circumstantial evidence. The quantity of protozoan and bacterial polysaccharide liberated in the small intestine by the disintegration of microorganisms is of considerable interest from the nutritional point of view. Heald[116] has found that from 5 to 6 g. of polysaccharide passes to the duodenum per 24 hr., or 12 in the hay-fed sheep. There is no estimate yet of ciliate polysaccharide. It should be noted that Heald's figure relates only to hay-fed sheep, for as polysaccharide synthesis appears to be greater when starch is fed its importance may be correspondingly enhanced.

Polysaccharides are stored by microorganisms for their own needs, and the fact that some of these may become available to the host is purely fortuitous. Elsden[117] observed that after the introduction of glucose into the rumen the strong iodophilic reaction which could be observed by spinning down the iodine-treated organisms became weaker with the time after dosage. Similarly the glycogen-like polysaccharides stored by the organism observed by Quin[118]—now thought to belong to the group of Selenomonas—become depleted of polysaccharide if no glucose is provided.

The same problem confronts us in discussing the bacteria as products of digestion in the large intestine. Here the difficulty is greater, as there is no obvious reason why the caecal and colonic bacteria should die and disintegrate at all. It is well known that most of the feces in man are composed of bacterial cells, both viable and dead. In the herbivora there is little information on the fate of caecal bacteria; do they die off as they pass into the lower parts of the colon? If they do, has this any relation to the loss of water from the food residues? Alternatively, do they suffer from lack of substrate? Considering the robust nature of bacteria compared to the more highly developed ciliates it seems improbable that a comparatively small change in the moisture of their environment would seriously prejudice their existence. There is no answer to the question, but it is perhaps relevant to consider the coprophagic habits of the rabbit. This animal passes two sorts of feces; soft, moist feces containing approximately 28 to 30% of the dry

[116] P. J. Heald, *Brit. J. Nutrition* **5**, 84 (1951).
[117] See ref. 71a.
[118] J. I. Quin, *Onderstepoort J. Vet. Sci. Animal Ind.* **18**, 91 (1943).

matter as protein, and hard, dry feces containing much less protein.[119] Bacteriological studies[120] have indicated that the significance of coprophagy for the rabbit may lie not in the replenishment of the flora of the digestive tract *per os*, but in bacterial synthesis of some substance important for the nutrition of the rabbit. The growth rate of rabbits kept in cages that prevented coprography, was reduced when the diet was hat and roots; inclusion of fresh grass in this ration allowed normal growth although it was inadequate for guinea-pigs and mice which died. How these two types of feces are formed is not known, but it has been observed that the rabbit re-ingests the soft type of feces, or most of them, and excretes the hard type. The interpretation offered for this remarkable habit is that the rabbit gains bacterial protein and B vitamins by eating the soft feces. The physiology of this process is most obscure. There seems to be no distinct separation of food in the alimentary tract during its first and second passage through the gut, as was at first thought, for barium fed to rabbits has been found to pass through the gut once, twice, or even three times. This being so, it may be that there is a rhythm of activity in the large intestine whereby digesta in the caecum can pass to the anus rapidly or slowly. If it passes rapidly, the soft type of feces appear which are reingested; if it passes slowly, then hard, dry feces appear. The rate of passage therefore may be the important factor in determining how much of the nitrogen in the digesta appears in the feces, and if this is so then it seems that some of the bacterial nitrogen may be made available to the host, provided that there is sufficient time and opportunity for desiccation even without reingestion.

The occurrence of refection in the rat and in the pig is not necessarily associated with the reingestion of feces. The essential point seems to be the appearance of a certain type of flora in the caecum which will synthesize B vitamins; this is associated with the feeding of raw potato starch to non-ruminants. The fact that an otherwise B-deficient diet can be supplemented by this means implies that extracellular B vitamins are formed which can be absorbed by the host.

The large intestine as an organ of digestion has been largely neglected. Considering its importance in herbivores and even possibly in man, it remains an interesting zone for research.

VIII. The Intestinal Flora of Man

1. General Discussion

In 1907 Schmidt,[121] reviewing the existing knowledge at that time, stated that the chemical processes which occur in the decomposition of the chyme

[119] A. Eden, *Nature* **145**, 628 (1940).

[120] I. Frank, U. Hadeler, and W. Harder, *Arch. ges. Physiol.* (*Pflügers*) **253**, 173 (1951).

[121] A. Schmidt *in* Van Noorden, Metabolism and Practical Medicine, Wm. Heinemann, London, 1907, Vol. 2, p. 169.

in the intestines are: fermentation of carbohydrates, "putrefaction" of protein, and conversion of the fats into the lower fatty acids. It was also stated that fermentation of carbohydrates takes place normally, both in the lower part of the small intestine and in the colon, but that "putrefaction" of protein occurs exclusively in the large intestine. The ileocaecal valve was considered to form a sharp line of demarcation, above which "putrefaction" of protein never occurred, except under abnormal conditions. In the caecum and in the ascending colon, which were the sites of most active decomposition, fermentation and "putrefaction" were regarded as taking place together. The latter was believed to exceed the former, then to decrease again in the last portion of the colon, where the feces become inspissated. In correspondence with this the bacteria which flourished abundantly in the caecum were stated to diminish gradually in numbers further down the tract.

The term "putrefactive" later gave way to "proteolytic," but this did not clarify matters, for no sharp distinction has apparently be made between proteolytic and saccharolytic bacteria, at least in the feces.[122]

Although vitamin synthesis will be discussed elsewhere, it should be noted that just as the bacteria of the ruminant and other herbivores synthesize vitamins in substantial amounts, so it is highly probable that in man vitamins K, B complex, and E are synthesized in amounts which help to meet requirement. Indeed some evidence for this already exists. On the other hand, certain intestinal bacteria destroy vitamins, e.g., vitamin C and nicotinic acid.

In rats it has been found that the giving of large quantities of lactose or dextrin in the diet can cause the intestinal flora so to change that it contains 90% or more of aciduric bacilli.[123-125] These two carbohydrates are acted upon very slowly and consequently probably pass to the large intestine where they are acted upon by the aciduric bacilli—notably *Lactobacillus acidophilus*—with the formation of a large amount of lactic acid; the presence of this acid is said to be unfavorable to the persistence of the proteolytic bacteria.

Schmidt[121] described the products of fermentation of carbohydrate as consisting of the gases carbon dioxide, hydrogen and methane and lower volatile fatty acids which were for the most part absorbed by the intestinal wall. The fermentation products that were not absorbed were excreted, either mixed with the feces or as flatus. "Putrefaction" of protein produced

[122] G. S. Wilson and A. A. Miles, *in* Topley and Wilson, Principles of Bacteriology and Immunity, Arnold, London, 1947, Vol. 2.

[123] P. L. F. Rettger and H. A. Cheptin, A Treatise on the Transformation of the Intestinal Flora with Special Reference to the Implantation of Acidophilus, New Haven, 1921, quoted by G. S. Wilson and A. A. Miles, ref. 122.

[124] P. R. Cannon and B. W. McNease, *J. Infectious Diseases* **32,** 175 (1923).

[125] R. Cruickshank, *Brit. J. Exptl. Path.* **9,** 318 (1928).

ammonia, hydrogen sulfide, carbon dioxide, and hydrogen and also a number of characteristic bodies such as the aromatic oxy acids phenol, indol, and skatol. These last were also absorbed by the intestinal wall and part of the gases were said to be lost in the expired air.

Between this older literature and recent observations there is little of importance to describe, but in 1946 Baker and Palmer[126] showed that the human caecum and colon have an abundant iodophile population (giant clostridia, giant diplococci, slender vibrios, small rods and cocci, etc.). These microorganisms were seen on the surfaces and in the interior of ingested plant structures but not on intact or partially digested muscle fibers, pointing to a specific relationship to decomposition of vegetable material, and further observations suggested their capability of attacking true cellulose. From comparative observations on other omnivora it would be reasonable to assume that lower volatile fatty acids are produced, B vitamins are synthesized, and extracellular enzymes are liberated. Although there is no direct evidence of absorption of these fermentation products, it is presumed that they are of value to the organism and in the case of the B vitamins may reduce requirements.

Adult human beings are not so constructed that they can exist on diets leaving no residue, or even so little a residue as milk provides. Habituation to a vegetarian diet may be thought to bring with it an improvement in the so-called utilization of such food, but it will be remembered that Voit's[127] laboratory man made just as good use of such food as a vegetarian of many years' standing. Owing to its "wrapping" of cellulose, vegetable protein is not generally so fully used as animal protein. Rubner[128] found that in the human subject on a diet of 3077 g. peeled potatoes, taken with butter, vinegar, and oil, 32.2% of the nitrogen was lost in the feces and 7.6% of the carbohydrate.

Where there is much cellulose or starch which has escaped digestion higher up, there is considerable microbial activity in the large intestine.

The effect of low fat intakes and of crude fiber has been studied by Walker[129] by means of balance studies on a healthy European male. When a large increase was made in the crude fiber content of equal fat diets, the amount of fecal fat excreted detectably increased. This was considered not to have been derived directly from the food. The possibility was raised by Walker that it might be of microbial origin.

It is well known that in chronic diarrhea there is an increase in fecal nitrogen per 24-hr. period, and this also occurs in shorter periods of gastro-

[126] F. Baker and B. Palmer, *Nature* **157**, 412 (1946).
[127] C. Voit, *Z. Biol.* **25**, 232 (1889).
[128] R. Rubner, *Z. Biol.* **19**, 45 (1883).
[129] A. R. P. Walker, *Nature* **164**, 825 (1949).

intestinal upset characterized by loose stools.[130] In chronic constipation the fecal losses are said to drop to 75 to 50 % of that observed in normal digestion. It is presumed that there is more time for microbial activity and absorption of the products of the activity. Toscani and Whedon[131] found that in normal subjects even on a constant dietary regime and with regularity in bowel movements, there was a considerable variation in fecal nitrogen. However, it is suggested by these authors that variation in fecal nitrogen is less marked in the same individuals than among the several individuals on the same dietary intake.

2. Intestinal Flora in Infancy and Resistance to Infection

It has been shown that breast-fed infants have a decided advantage over bottle-fed infants in their resistance to infections of various kinds.[132, 133, 134] This resistance is most striking in gastroenteritis of infancy which rarely affects a breast-fed child. Recent work has demonstrated that infantile gastroenteritis is intimately associated with certain types of *B. coli*.[135, 136] In 1925, Cruickshank[137] showed that *B. coli* fails to grow in a medium the pH of which is more acid than 5.0, and it does not survive if grown in association with *Lactobacillus bifidus* in a carbohydrate-containing medium. Such a finding supports the view that the acid pH and/or lactobacillary flora of the intestinal contents of the breast-fed child may be inhibitory to the growth of the types of *B. coli* which are associated with diarrhea and vomiting of infancy. Such a mechanism, mediated through intestinal environmental conditions, would result in an "immunity" localized to the intestinal tract. On the other hand, it is possible that the biological value of lactose and/or a lactobacillary flora discussed above may, in the breast-fed child, play a part in maintaining a more general resistance to infection.

From a consideration of the predominance of *Lactobacillus bifidus* in the feces of breast-fed but not of artificially fed infants, a chromatographic study was made by Ross[138] of the amino acids in the feces of breast-fed and artificially fed (dried milk) infants. In the breast-fed infant's feces,

[130] S. H. Bassett, E. H. Kentmann, H. van Z. Hyde, H. E. van Alstine, and E. Russ, *J. Clin. Invest.* **18,** 101 (1939).

[131] V. Toscani and G. D. Whedon, *J. Nutrition* **45,** 119 (1951).

[132] C. G. Grulee, H. N. Stanford, and P. H. Herron, *J. Am. Med. Assoc.* **103,** 735 (1934).

[133] O. B. Cordua, *Arch. Pediat.* **52,** 845 (1935).

[134] D. Levi, *Brit. Med. J.* **1,** 963 (1941).

[135] J. Bray, *J. Path. Bact.* **57,** 239 (1945).

[136] J. Taylor, B. W. Powell, and J. Wright, *Brit. Med. J.* **2,** 117 (1949).

[137] R. Cruickshank, *J. Hyg. Camb.*, **24,** 241 (1925).

[138] C. A. Ross, *Lancet* **258,** 716 (1950).

alanine, glycine, glutamic acid, valine, and leucine were always present; serine, threonine, aspartic acid, and one or more of the lysine, ornithine, and arginine group were usually present. In the feces of the artificially fed infant only valine was present in all samples; glutamic acid, leucine, alanine, and glycine were frequently present, and occasionally traces of lysine and aspartic acid.

Of 75 infants admitted to hospital with a diagnosis of gastroenteritis, Ross[139] found that those showing a large proportion of Gram-negative organisms in their fecal smears were very ill.

3. Sprue, Idiopathic Steatorrhea, and Coeliac Disease

In these conditions of unknown origin the feces contain excessive amounts of fatty acids and calcium soaps, owing to inability to absorb the normally hydrolyzed fat. Because of this there is a secondary loss of minerals such as calcium, phosphorus, and iron, and a deficiency develops, particularly of the B group of vitamins.

From an examination of a large number of cases of persistent steatorrhea (coeliac disease) in children, Anderson[140] found two additional objective findings of importance. There were (1) a deficiency of amylase in the duodenal juice and (2) an excess of starch in the feces when starch was given. Difficulty in digesting starch was a constant finding in this condition, and these observations have been used to suggest the rationale for a diet high in protein, low in starch, and with "abundant vitamins." Such a diet appears to be well tolerated.

From our knowledge of starch digestion in other environments, it is presumed that if these findings be general then the failure to digest starch through the normal digestive channels may well be accompanied by considerable microbial activity in the large intestine, extending to higher levels in the small intestine and possibly with biosynthesis of vitamins.

The situation has recently been clarified considerably by the finding that the removal of wheat and rye flour from the diet of children with coeliac disease has resulted in rapid improvement, both clinically and biochemically,[141] and that deterioration followed the reintroduction into the diet of wheat flour or wheat gluten, but wheat starch had no effect.[142] As reported by Weyers, van de Kamer, and Dicke at the First International Congress on Dietetics at Amsterdam, the decisive factor is the protein gliadin.

[139] C. A. Ross, *Brit. Med. J.* **1,** 465 (1950).
[140] D. H. Anderson, *J. Pediat.* **30,** 564 (1947).
[141] W. K. Dicke, Coeliake, Thesis, Utrecht, 1950.
[142] C. M. Anderson, A. C. Frazer, J. M. French, J. W. Gerrard, H. G. Sammons, and J. M. Smellie, *Lancet* **1,** 836 (1952).

IX. Antibiotics

It has been established that the value of orally administered aureomycin or other broad "spectrum" antibiotics for infective states in the field of human medicine is countered by their seemingly greatest potential weakness as therapeutic agents, namely their ability to induce a measure of "sterilization" of the alimentary tract. It is reported that patients subjected to continuous treatment by the oral route may develop diarrhea with a heavy growth of *Candida albicans* in their feces. The end situation of "cure" may be coupled with side effects which are sometimes as disagreeable as the initial disease, and occasionally deaths due to *Candida* infections (moniliasis) may result.[143]

It is to be noted that aureomycin and penicillin have been considered to have some therapeutic value in the treatment of Addisonian pernicious anemia.[144, 145] The suggestion is that the change in intestinal flora produced by these antibiotics may involve the appearance of organisms which synthesize hemopoetic material, or the suppression of organisms which either use up or interfere with the absorption or utilization of hemopoetic substances. In this connection particular interest is centered on vitamin B_{12} and folic acid. Thus Dyke and Hind[146] have shown that certain types of lactobacilli present in the intestinal tract can produce vitamin B_{12}. There is as yet, however, no evidence that vitamin B_{12} can be absorbed from the large intestine, and the precise role played by vitamin B_{12} in the etiology of pernicious anemia is still uncertain. It should also be recorded here that Frazer[147] has pointed out that in sprue there is a heavy invasion of the small intestine with organisms from the colon. He suggests that the vitamin B deficiencies and hemotological changes found in this disease may arise from competition between the host and the abnormal flora for these vitamins.

In the treatment of patients with the antibiotics aureomycin, chloramphenicol, and terramycin certain complications have been reported which have been ascribed to vitamin B deficiency—the result of bacterial destruction that interferes with the normal production of the vitamin B complex. Harris[148] and Tomaszewski[149] found that these complications were less severe when patients were receiving vitamin B complex. It has also been argued

[143] K. B. Raper, A Decade of Antibiotics in America. Presidential address to the Mycological Society of America, 1951.

[144] H. Lichtman, J. Watson, V. Ginsberg, J. V. Pierce, E. L. R. Stokstad, and T. H. Jukes, *Proc. Soc. Exptl. Biol. Med.*, **72,** 643 (1949).

[145] H. Foy, A. Kondi, and A. Hargreaves, *Brit. Med. J.* **1,** 380 (1951).

[146] W. J. C. Dyke and H. G. Hind, *Lancet* **1,** 486 (1950).

[147] A. C. Frazer, *Brit. Med. J.* **2,** 731 (1949).

[148] H. J. Harris, *J. Am. Med. Assoc.* **142,** 161 (1950).

[149] T. Tomaszewski, *Brit. Med. J.* **1,** 388 (1951).

that the emergence of a fungal flora in the bowel might explain the changes which occur. But milder effects have shown very considerable promise, particularly in the field of lower animal nutrition. This interesting field of development stems back to the observations of Stokstad *et al.* (1949),[150] and Stokstad and Jukes (1950),[151] which demonstrated the profound growth stimulation in birds and other animals of oral aureomycin over and above that due to associated vitamin B_{12} where animal protein factor supplements of the fermentation industry were used in the test. To aureomycin there has now been added terramycin, penicillin, and bacitracin. Polymyxin D has been shown to exert some growth stimulation in chicks. Streptomycin may be used, but optimal levels for this and the other antibiotics are still uncertain, particularly in relation to stage of growth.

In the presence of the antibiotic the amount of vitamin B_{12} required is often reduced, the antibiotic appearing to exert a "sparing action." This does not obviate the need of a dietary source of this factor for normal growth in young animals or for reproduction. Nor does the feeding of vitamin B_{12} and antibiotics at optimal levels appear to alter the need for other members of the B complex.

In addition to improved growth responses in young rats, birds, and pigs, it is claimed that supplementing the diet with an antibiotic benefits young calves. As the commensal microorganisms of the rumen play an important role in older animals, it is not unexpected to find evidence that an antibiotic such as aureomycin interferes with the pattern of the normal flora sufficiently to alter the course of microbial digestion in these older animals.[152]

The general trend of experiments with pigs and poultry in the United States has indicated that faster and more economic growth is obtained and that earlier marketability is attained without loss in meat quality. The animals have also been found to be less subject to enteric disease. Jukes[153] has stated that the practical level for aureomycin for feeding is about 10 g. per ton of feed. The levels for procaine penicillin, bacitracin, and terramycin are about the same. Investigators in the United Kingdom at this early stage in their experience of antibiotics in relation to their feeding practices and management have found that the addition of penicillin or aureomycin to the diet of fattening pigs results in an increase in live weight gain of about 8 % on a diet containing fish meal. With this goes an increase in efficiency of feed conversion of about 6 %. When the diet is poorer qualita-

[150] E. L. R. Stokstad, T. H. Jukes, J. Pierce, A. C. Page, Jr., and A. L. Franklin, *J. Biol. Chem.* **180,** 647 (1949).

[151] E. L. R. Stokstad and T. H. Jukes, *Proc. Soc. Exptl. Biol. Med.*, **73,** 523 (1950).

[152] M. C. Bell, C. K. Whitehair, and W. D. Gallup, *Proc. Soc. Exptl. Biol. Med.* **76,** 284 (1951).

[153] T. H. Jukes, Nutrition Symposium, Second Presentation, Proceedings of the Semi-Annual Meeting of the Nutrition Council of the American Feed Manufacturers Association, Nov. 26–28, 1950, pp. 13–15.

tively but not quantitatively, e.g., by exclusion of fish meal, the degree of improvement is greater. No evidence of deterioration in carcass quality has resulted from the inclusion of either of these two antibiotics, nor is it considered that the meat obtained from these animals will have any adverse effect on human beings.

The full potentiality of these antibiotics is appreciably reduced by a restrictive scale of feeding. Unthrifty weanling pigs often respond remarkably to antibiotic supplementation, and scouring is frequently controlled. but an antibiotic will not make a bad ration good.

Although not known to be harmful, antibiotic feed supplements are not recommended for use in rations of breeding stock on the basis of lack of response and cost. According to some Canadian workers, antibiotic supplements appear to spare protein, as pigs on a 13 % protein ration with antibiotic grew as rapidly as those on a 17 % protein ration without antibiotic. The largest profit on feed costs in the Canadian experiment resulted when a ration containing 15 % protein with antibiotic was fed during the growing period to 110 lb.[154] According to Robinson of the Department of Agriculture of Queen's University, Belfast, antibiotics increase the digestibility of protein by pigs. There is accumulating evidence from the United States that these antibiotic supplements are also useful in rations for artificial feeding of suckling pigs.

Carpenter and Duckworth[155] found that, although growth in chicks could be accelerated considerably by aureomycin, by 21 weeks of age they were only 8 % heavier, and they did not come into lay any earlier nor was their level of egg production or size advanced.

The recent findings of the research groups at the National Institute for Research in Dairying and Glaxo Laboratories are most interesting. They found that birds housed in accommodation where poultry had not previously been kept showed good growth and no response to penicillin, whereas birds of the same origin and on the same diet living in an established poultry building showed as good growth only when procaine penicillin was added to their diet.[156] Davis and Briggs also failed to obtain a response with chicks housed in new quarters.[157] This suggests that the effect of an antibiotic may well depend on the release of an inhibition caused by a low-grade endemic "infection." Indeed the very recent report by the United Kingdom team (Coates[158]) has provided evidence that this infection is transmissible

[154] J. P. Bowland and L. W. McElroy, *The Press Bulletin* **37**, Nos. 2, 7, and 9, Department of Extension University of Alberta, 1952.

[155] K. J. Carpenter and J. Duckworth, *Nature* **169**, 669 (1952).

[156] M. E. Coates, C. D. Dickinson, G. F. Harrison, S. K. Kon, J. W. G. Porter, S. H. Cummins, and W. F. J. Cuthbertson, *J. Sci. Food Agr.* **3**, 43 (1952).

[157] R. L. Davis and G. M. Briggs, *Poultry Sci.* **30**, 767 (1951).

[158] M. E. Coates, Report on Discussion on Antibiotics in Animal Nutrition at Belfast Meeting of British Association, Sept. 9, 1952, *Nature* **170**, 868 (1952).

and is counteracted by an antibiotic such as penicillin. It must be recorded that no sign of infection other than a slight growth depression has ever been observed. It was also reported by these British workers that Reyniers and his group have found no response to an antibiotic in germ-free animals. This observation fits the findings of Coates *et al.*[156] and raises the question of general hygiene and the effect of a low-grade endemic infection adversely affecting growth and production in certain situations. Thus antibiotics would appear to be most generally effective where there is suboptimal nutrition due to a low-grade infection, or because they directly or indirectly supply some as yet unidentified factor. The nature of the responsible organism, if that is the agent, has not been determined. It should be noted in passing that penicillin fed at the rate of 6.8 mg. per pound of feed inhibited the development of *Ascaridia galli* in the intestine of chickens.[159]

An unexpected finding occurred in an *in vitro* study of the effect of penicillin, streptomycin, neomycin, and chloromycetin on cellulose digestion in rumen contents. In the concentrations used, penicillin stimulated the cellulolytic rumen microorganisms. At the lower concentrations, neomycin was stimulatory in all concentrations; streptomycin was slightly stimulatory in the lowest concentrations; and chloromycetin adversely affected the microorganisms.[160]

Antibiotics decrease dietary requirements for certain vitamins; e.g., chicks need less vitamin B_{12} and rats less vitamin B_1 if they receive penicillin. But, as the effect of the antibiotic can apparently occur in the presence of a diet containing all known requirements of vitamins, it seems unlikely that the action of the antibiotic is through limitation of the competition of microorganisms for these nutrients. W. J. F. Cuthbertson has recently pointed out that, should normal practical rations contain insufficient amounts of some unidentified growth factors, the effects of antibiotics could be explained as involving a decrease in the need for these factors.[161] The addition of a rich source of unidentified factor to the diet should in that event improve growth; further, antibiotics should no longer increase the growth of animals given this fortified ration. He then reported that a freeze-dried liver preparation will improve chick growth, and in its presence the growth response to penicillin is less than usual. This would appear to support his view that antibiotics may act by decreasing the need for certain growth factors as yet unidentified.

It is difficult to appreciate fully the growth response which Stokstad and Jukes[151] found in chicks with cultures in which the aureomycin as measured

[159] A. C. Todd, *Poultry Sci.* **30,** 763 (1951).

[160] R. H. Wassermann, C. W. Duncan, E. S. Churchill, and C. F. Huffman, *J. Dairy Sci.* **35,** 571 (1952).

[161] W. J. F. Cuthbertson, Report on Discussion on Antibiotics in Animal Nutrition at Belfast Meeting of British Association, Sept. 9, 1952, *Nature* **170,** 869 (1952).

by antibiotic potency was destroyed by alkaline hydrolysis. Recent information on somewhat comparable effects through the administration of surface acting agents makes for caution in interpretation.

The experiments of Dixon and Thayer[162] indicate that intramuscular injections of aureomycin and procaine penicillin G are effective in enhancing growth, possibly by gaining access to the bowel, and that functional caeca are not essential for the growth-promoting action in the bird.

In all this work it is very necessary to appreciate the influence of (1) the general system of hygiene and possibility of low-grade infection obtaining in the unit investigated, (2) the possibility of hitherto unidentified factors, and (3) the character of the basal diet in respect to its quality and quantity. For example, the nature of the starch in the diet can influence profoundly the amount of carbohydrate reaching the caecum and the activities of the microflora there, and the basal diet may itself contain some antibiotic material. In a personal communication E. W. Crampton advised the authors that "our observation has been that where fishmeal has been included in the ration in reasonable quantities, there has been very little effect from antibiotic added in addition." This is possibly due to some fermentation during storage. Crampton points out that there should be both a good ration and a poor ration in each experiment with an antibiotic, for where the rations have been good the results of feeding have usually been of the order of 10 to 20 % whereas if the ration is obviously deficient in many things the effects may be as high as 40 to 45 %. But it is unwise practice to rely on antibiotics to make a poor diet good, though they do exert part of their beneficial effect by economizing on certain vitamins. Their best role probably lies in the rapidly growing period of life.

What seemed an interesting development appeared to follow the surmise that since the non-ionic detergents are akin in chemical structure to certain antibiotics they might possibly have a comparable action on the growth of farm animals. The earlier information reported supported this.[163, 164] But on wider trial the beneficial effects have not been substantiated. The writers cannot but be dubious of the ultimate benefit to be derived by agriculture from such agents, for experience indicates that there is generally a natural reaction on the part of the organism to establish the *status quo ante* when disturbed.

The authors acknowledge with gratitude the assistance they have received from Dr. P. J. Heald, Dr. C. A. Ross, Miss R. M. Campbell, and Miss R. C. Jamieson in the preparation of this chapter.

[162] J. Dixon and R. H. Thayer, *Poultry Sci.*, **30**, 910, *Proc.* (1951).
[163] C. M. Ely, *Science* **114**, 523 (1951).
[164] *Fertiliser, Feeding Stuffs, Farm Supplies J.* **38**, No. 9, 271 (1952).

CHAPTER 15

The Nutrition of Invertebrate Animals

GEORGE W. KIDDER

Biological Laboratory, Amherst College, Amherst, Massachusetts

I. Introduction

Our knowledge of invertebrate nutrition is very fragmentary indeed. Of the many attempts to culture members of the phyla of invertebrates, relatively few have yielded results which contribute useful biochemical information. These few cases are drawn mainly from the Protozoa and the Insecta. The very fact that aseptic culture is often difficult or impossible to obtain indicates our ignorance of many conditions and biochemical factors of importance in the nutrition of organisms.

Nutritional studies on aquatic forms are usually of little value unless the forms are cultured in the absence of associated microorganisms. Only then can one arrive at conclusions regarding the limitations of synthetic enzymes, which is the real core of the science of nutrition. Pure cultures (bacteria-free) have been obtained with a small number of animal microorganisms (true Protozoa), but, with one exception, additions of crude extracts of natural materials are necessary to maintain growth. This is the situation in all other invertebrates so far reported, with the single exception of *Drosophila*,[1] and even in this case certain unknowns of nutritional importance are possible contaminants of agar, cholesterol, lecithin, and commercial nucleic acid.

Valuable information along nutritional lines has been obtained in certain insect species where food materials are taken in the dry state by the organisms. Under these conditions bacterial and fungal growth is greatly reduced and the synthetic capacities of the microorganisms cannot be fully exploited by the insect under study. Even in these cases, no really definitive conclusions can be made regarding the quantitative aspects of nutritional requirements, owing to possible contributions from intestinal microorganisms.

Up to the present time discussions of comparative nutrition have dealt almost exclusively with different species of vertebrates, most frequently birds and mammals. Many of the observations upon which these discussions were based appear to be open to qualifications, owing to the unknown contributions of intestinal flora and fauna. These qualifications will assume most importance in quantitative evaluations of nutritionally active components of the diet required in minute amounts for metabolism. This problem has been recognized for some time, and one method in common use to minimize the contributions of intestinal microorganism is to diminish their numbers by administration of sulfonamides and/or antibiotics. One may question this type of procedure as possibly producing abnormal responses in the animal under study, but on the whole it may be justifiable until careful and critical work is done on bacteria- and fungus-free experimental vertebrates.

The present review will deal with reports on the nutrition of those invertebrates in which it has been found possible to control or to minimize the chemical contributions of bacteria, yeasts, and molds. Various reports where such control was lacking have been included in previous reviews.[2, 3]

Exclusion of the "plant flagellates" seems justified. On morphological grounds Calkins[4] rejected all chlorophyll-bearing flagellates and their close relatives from the phylum Protozoa, retaining only the animal-like microorganisms. This stand has rarely been followed by systematists. Most persist in classifying the "plant flagellates" among the Protozoa in zoological works and as lower plants in botanical works. Although one certainly recognizes the structural relationships between these organisms and typical "animal flagellates," and also their evolutionary relationships, nevertheless

[1] T. Hinton, D. T. Noyes, and J. Ellis, *Physiol. Zoöl.* **24,** 335 (1951).

[2] W. Trager, *Biol. Rev.* **47,** 148 (1947).

[3] C. L. Prosser, *in* Prosser, Comparative Animal Physiology, W. B. Saunders Co., Philadelphia, 1950.

[4] G. N. Calkins, Biology of the Protozoa, Lea and Febiger, Philadelphia, 1933.

the nutritional patterns which have been discovered make a discussion of the "phytoflagellates" as invertebrates inappropriate. For rather recent reviews on the biochemistry and nutrition of these interesting organisms, reference may be made to Lwoff[5] and to Hutner and Provasoli.[6] The types which it seems justifiable to include, therefore, will be members of the phylum Protozoa (in terms of Calkins) and all the other phyla of the animal kingdom save the Chordata. Although the number of groups here encompassed is large indeed, the number of investigations of a nutritional nature is limited to very few types.

II. Protozoa (Animal Microorganisms)

A number of extensive reviews on the biochemistry and nutrition of the various groups of Protozoa have been written, some rather recently.[7, 8, 9] Although very few species of this large phylum of animals have been studied biochemically, some interesting information is available for each major subdivision. For only one genus, *Tetrahymena*, do we have complete information on nutritional requirements, however.

1. Zoomastigophora (Animal Flagellates)

In the animal flagellates, we find our knowledge of a biochemical nature limited largely to two families, the Trypanosomidae (class Protomastigota) and the Trichomonadidae (class Metamastigota). There are two reasons why representatives of these families have yielded some biochemical information. The members are all parasites and it has been found possible to achieve asepsis fairly readily; and there are important pathogenic parasites of mammals, including man, in each group. The wealth of free living forms that exist offer important material for future investigations.

a. Trypanosomidae. (1) *Hematin.* Probably the oldest and best-known nutritional requirement of the parasitic genera *Trypanosoma* and *Leishmania* is blood. Novy and MacNeal[10] succeeded in culturing *T. lewisi* from the rat and *T. brucei* from cattle on blood agar, and Zotta[11] demonstrated the stimulatory action of the blood fraction. In 1933, M. Lwoff[12] reported the growth of *Strigomonas fasciculata* from the digestive tract of the mos-

[5] A. Lwoff, L'Evolution Physiologique, Hermann et Cie., Paris, 1944.

[6] S. H. Hutner and L. Provasoli, *in* Lwoff, Biochemistry and Physiology of Protozoa, Academic Press, New York, 1951.

[7] M. Lwoff, *in* Lwoff, Biochemistry and Physiology of Protozoa, Academic Press, New York, 1951.

[8] R. W. McKee, *in* Lwoff, Biochemistry and Physiology of Protozoa, Academic Press, New York, 1951.

[9] G. W. Kidder and V. C. Dewey, *in* Lwoff, Biochemistry and Physiology of Protozoa, Academic Press, New York, 1951.

[10] F. G. Novy and W. J. MacNeal, *J. Infectious Diseases* **2**, 256 (1950).

[11] G. Zotta, *Compt. rend. soc. biol.* **88**, 913 (1923).

[12] M. Lwoff, *Ann. inst. Pasteur* **51**, 55 (1933).

quito on peptone to which either small amounts of blood or the active principle of the blood, protohematin, had been added.

Relatively few organisms lack the ability to synthesize hematin (certain Trypanosomidae, the blood-sucking redivid *Triatoma*, and certain bacteria). For these organisms hematin is a true growth factor.[7]

Not all members of the Trypanosomidae are incapable of synthesizing hematin (Table 1). *Strigomonas oncopelti*, *S. parva*, and *S. media* can be grown in peptone without blood or hematin. In addition to hematin the trypanosomes and the leishmanias require additional factors present in blood.[7]

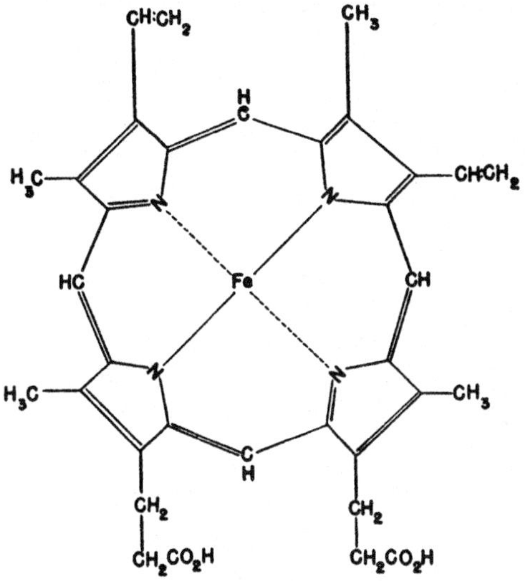

PROTOHEME

FIG. 1

The Lwoffs have shown that the activity of hematin is due to its stimulation of respiration (see M. Lwoff[7]) and not to its peroxidase activity. *Strigomonas fasciculata*, for instance, can insert iron into protoporphyrin and can utilize protohematin. Protoporphyrin appears to be necessary for the synthesis of the respiratory system of aerobic forms, and when organisms lose their ability to carry out its synthesis it must be supplied exogenously if they are to survive.

(2) *Ascorbic Acid.* Certain trypanosomids require the presence of ascorbic acid for multiplication. This factor is ordinarily supplied in the erythrocytes when whole blood is used, and growth fails when only serum, hematin, and peptone are present. The addition of ascorbic acid to such a medium makes growth possible. The organisms of this type are *Leishmania donovani*, certain strains of *L. tropica*, *L. brasiliensis*, and *Trypanosoma cruzi*.

TABLE 1

Growth Factor Requirements of Several Invertebrates and the Rat

	Thiamine	Riboflavin	Niacin	Pantothen	Pyridoxine*	Biotin	PGA	PABA	Choline	Thioctic acid	Carnitine	Hematin	Ascorbic acid	Sterol	Fatty acid	Vitamin B_{12}	Purine	Pyrimidine
Strigomonas oncopelti	+											−	−	−	−			
Strigomonas fasciculata	+											+	−	−	−			
Leishmania tropica (Dim)												+	−					
Leishmania tropica (*Tunis*)												+	+					
Leishmania donovani												+	+					
Trypanosoma cruzi												+	+					
Trichomonas columbae													+	+				
Trichomonas foetus													+†	+†				
Trichomonas vaginalis				+									−	−‡	+			
Endamoeba histolytica	+§	+§	+§	+§	+§			+§	+§					+				
Acanthamoeba castellani	+¶																	
Plasmodium knowlesi								+										
Tetrahymena geleii	+‖	+	+	+	+	−	+	−	−	+	−	−	−	−	−	−	+	+
Paramecium multimicronucleatum																	+**	+**
Colpoda steinii	+	+	+	+	+													
Aedes aegypti	+	+	+	+	+	+	+		+					+				
Tribolium confusum	+	+	+	+	+	+	+		−	−	−			+				
Tenebrio molitor	+	+	+	+	+	+	+		+	−	+			+				
Ephestia kuehniella	+	+	+		+	+	+		+††	−				+	+			
Rat	+	+	+‡‡	+	+	+	+	−	−§§			−	−	−	+	+	−	−

* Or derivatives (pyridoxal, pyridoxamine).
† According to Weiss and Ball[20], their strain did not require either sterol or ascorbic acid.
‡ Stimulatory.
§ Presumed requirement, but individuals not tested singly (by omission).
¶ Requirement for pyrimidine portion of the molecule only.
‖ Not required under some conditions.
** Hydrolyzed nucleic acid used, or guanylic and cytidylic acids.
†† Stimulatory only.
‡‡ Niacin can be synthesized provided that adequate tryptophan is present in the diet.
§§ Choline not required when adequate vitamin B_{12} and methionine present.

The ascorbic acid requirement in the trypanosomids appears to differ from that of many other organisms where a number of reducing substances can take its place (cysteine, glutathione, etc.). No reducing substance which has been used will substitute for ascorbic acid for these requiring flagellates, nor will the ascorbic acid analogues D-isoascorbic acid, D-glucoascorbic acid, and 3-methylascorbic acid.[7] Thus, ascorbic acid appears to be a true vitamin for these flagellates.

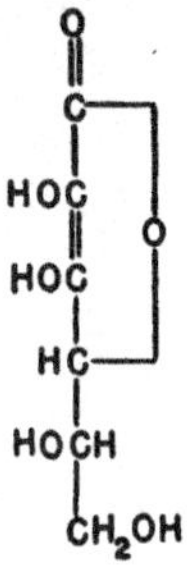

ASCORBIC ACID

FIG. 2

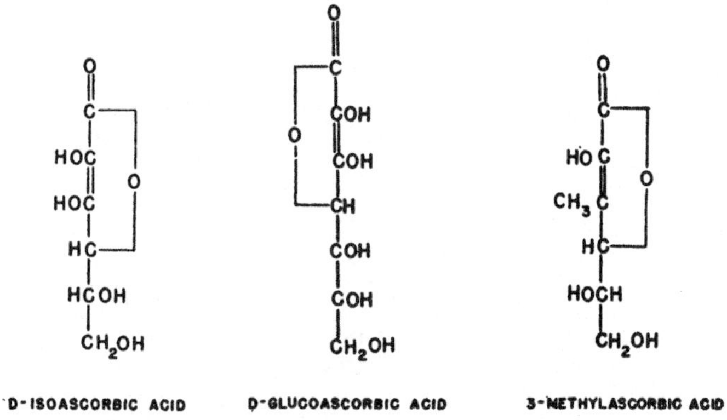

FIG. 3. Ascorbic acid analogues

(3) *Serum Factors*. The various strains of *Leptomonas*, *Leishmania*, and *Trypanosoma* which have been investigated carefully fail to multiply in media devoid of serum. This only means that as yet our knowledge of the biologically active compounds or complexes which these flagellates fail to synthesize is so incomplete that serum must be added as their source. This is the present situation when mammalian and avian tissues are cultivated *in vitro* and, considering the importance of the latter problem, the whole field of tissue culture techniques stands to profit greatly by further intensive investigation of the serum factors required by these flagellates.

Some advance is being made on the serum fractions required by *Trichomonas foetus* (see below).

(4) *Other Growth Factors.* In a few of those trypanosomids where serum factors are synthesized by the organisms and therefore need not be added to the medium, a thiamine requirement has been shown. M. Lwoff[13] was able to show that *Strigomonas oncopelti* would multiply in a medium of hydrolyzed silk peptone and glucose only if thiamine was added. Ordinary peptones could be rendered thiamine-free by heating at pH 9.8. This type of medium supported growth when thiamine was restored. By testing a number of thiamine analogues, M. Lwoff[14] was able to show that both the β-hydroxypropyl and the γ-hydroxypropyl 5-substituted thiazole moiety of the whole molecule was active in place of thiamine. Likewise when the methyl group at position 2 in the pyrimidine portion of the molecule is

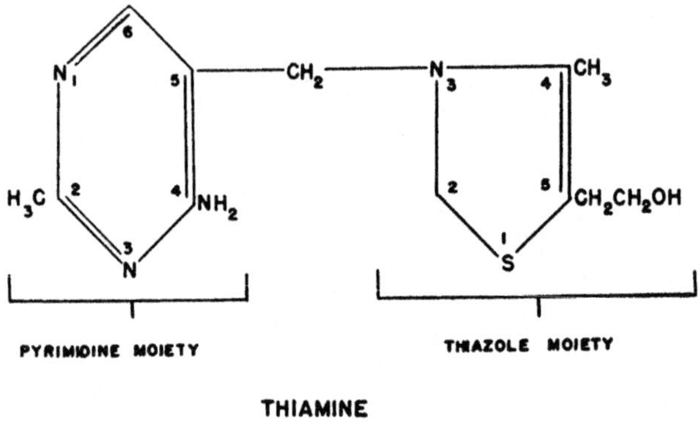

FIG. 4

substituted by an ethyl group, the resulting compound has thiamine activity. On the other hand, unsubstituted position 5 of the thiazole portion of the molecule, the shift of the methyl group from position 2 to position 6 in the pyrimidine moiety, or additional linkage of the pyrimidine to the thiazole through the 4-position N substitution to the 2-position carbon of the thiazole (thiochrome) destroyed all thiamine activity for *S. oncopelti.*

Although a few of the trypanosomids have been investigated directly for a thiamine requirement, M. Lwoff believes it highly probable that they all lack synthetic ability in this respect.

We have no exact information regarding any of the other growth factors of trypanosomids, owing to the fact that all media so far used have been far from chemically defined (peptones, serums, etc.). It is to be hoped that this gap in our knowledge will be filled in the not too distant future.

[13] M. Lwoff, *Compt. rend. soc biol.* **126,** 771 (1937).

[14] M. Lwoff, Recherches sur le pouvoir de synthèse des flagellés trypanosomides, Masson et Cie., Paris, 1940.

b. Trichomonadidae. The genus *Trichomonas* is one of the most ubiquitous of parasitic protozoans. Representatives are found in invertebrates (mollusks, annelids, insects), cold-blooded vertebrates (fish, amphibians, reptiles), and warm-blooded vertebrates (birds and mammals, including man). Most appear to be well adjusted to their host and cause little harm, but a few are truly pathogenic (*T. columbae* in the pigeon, *T. foetus* in female cattle, and *T. vaginalis* in the female human). Two species of free-living trichomonads have been described once each, *Trichomonas keilini*[15] and *Coelotrichomastix convergus*,[16] but no nutritional information regarding them is available.

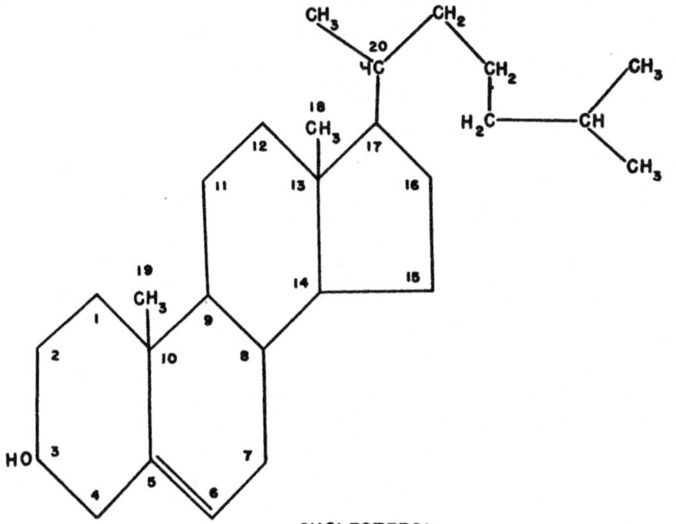

CHOLESTEROL

FIG. 5

(1) *Cholesterol. Trichomonas columbae* was early shown to require fat-soluble materials.[17] Successful cultures may be maintained in nutrient broth plus liver. If the liver is acetone-extracted, however, it no longer supplies the substance necessary for growth. When the acetone extract is added to nutrient broth no growth occurs unless the liver residue is also present. Nutrient broth and acetone-extracted liver was used as the basal medium for the testing of a large number of sterols.[17] It was found that the growth-promoting effect of the acetone extract of liver could be duplicated with highly purified cholesterol. Certain modifications of the cholesterol molecule could be made without affecting the biological activity for *T. columbae;* other modifications rendered it useless to the organism.

[15] A. Bishop, *Parasitology* **27,** 246 (1945).

[16] A. Hollande, *Bull. soc. zool. France* **64,** 114 (1939).

[17] R. Cailleau, *Ann. inst. Pasteur* **59,** 1 (1937).

In the first category of compounds (active) are cholestanol, erganostanol, sitostanol, *cis*-cholestane-3,4-diol, ergosterol, 7-dehydrocholesterol, 22-dihydrocholesterol, γ-dihydroergosterol, ergostenol, α-sistosterol, and cinchol. This indicates that the systems of the organism attach little importance to the unsaturated or saturated conditions of bonds 5–6, 7–8, or 22–23, to the presence or absence of a hydroxyl group at position 4, or to the position of branches on the side chain.

In the second category (inactive) are found such sterols as Δ-5,6-cholestenone, cholestane-3-one, α-3,5,6-cholestanetriol, epicholestanel, dehydroandrosterone, androstene-3,17-diol, testosterone, estrone, equilin, equilenin, and irradiated ergosterol.

It appears likely (although quantitative data on activities were not stressed) from the work of Cailleau that cholesterol represents the true requirement for *T. columbae*, and the other sterols may be substituted only in so far as the organism can change them into the active compound.

Cholesterol is also required by *Trichomonas foetus*, *T. batrachorum*[18] and *Trichomastix colubrorum*.[19] Nothing is known regarding the biochemical and physiological function of this compound.

(2) *Ascorbic Acid.* Cailleau[18, 19] found that ascorbic acid, when added to otherwise inadequate media for *Trichomonas foetus*, *T. columbae*, and *Trichomastix colubrorum*, rendered the media suitable for growth. The serum fraction of these media rapidly lost activity upon aging, and it was this activity which was restored by the addition of ascorbic acid.

The apparent ascorbic acid requirement of the trichomonads appears to be non-specific, however, and the compound functions not as a vitamin but as a reducing agent. This is in contrast to the situation in the trypanosomids (see above). Thus D-isoascorbic acid, D-glucoascorbic acid, and other reducing agents were as active in promoting the growth of the trichomonads as was ascorbic acid itself.

From the more recent report of Weiss and Ball[20] on *Trichomonas foetus*, it is apparent that ascorbic acid, or other strong reducing agents, are not necessary for growth in more synthetic media. They report growth of the flagellates in the medium of Kidder and Dewey[21] in which no ascorbic acid appears. The importance of this paper is in its relationship to amino acid requirements. By the omission technique they report that *T. foetus* requires exogenous arginine, glycine, phenylalanine, tryptophan, histidine, isoleucine, leucine, lysine, threonine, methionine, proline, serine, and valine (Table 2). This represents a fairly typical animal pattern and is similar, on the whole, to the requirements which have been determined for certain of

[18] R. Cailleau, *Compt. rend. soc. biol.* **130,** 1089 (1939); **131,** 964.

[19] R. Cailleau, *Compt. rend. soc. biol.* **127,** 861 (1938); **127,** 1421.

[20] E. D. Weiss and G. H. Ball, *Proc. Soc. Exptl. Biol. Med.* **65,** 278 (1947).

[21] G. W. Kidder and V. C. Dewey, *Physiol. Zoöl.* **18,** 136 (1945).

the ciliated protozoans (see below). The medium based on amino acids was suboptimal for *T. foetus*, however, and could be improved by the addition of serum or enzymatically treated proteins. Weiss and Ball suggest the possibility of a partial strepogenin[22, 23] requirement.

(3) *Linoleic Acid. Trichomonas vaginalis* can be cultured satisfactorily in a number of serum-containing media. Sprince and Kupferberg[24] have

TABLE 2

Amino Acid Requirements of Several Invertebrates as Compared to the Rat and to Man

	Trichomonas foetus	*Tetrahymena geleii*	*Glaucoma scintillans*	*Tribolium confusum*	*Attagenus sp.*	*Aedes aegypti*	Rat	Man
Glycine	+	−	−*	−		+	−	−
Alanine	−	−	−	−		−	−	−
Valine	+	+	+	+	+	−	+	+
Leucine	+	+	+	+	+	+	+	+
Serine	+	+†	−	−		−	−	−
Phenylalanine	+	+	+	+	+	+‡	+	+
Tyrosine	−	−	−	−	−	+‡	−	−
Tryptophan	+	+	+	+	+	+	+	+
Histidine	+	+	+	+	+	+	+	−
Arginine	+	+	+	+	+	+	+§	−
Lysine	+	+	+	+	+	+	+	+
Proline	+	−	+	−		−	−	−
Glutamic acid	−	−	−	−		−	−	−
Aspartic acid	−	−	−	−		−	−	−
Hydroxyproline	−	−	−	−		−	−	−
Isoleucine	+	+	+	+	+	+	+	+
Methionine	+	+	+	+	+	+	+	+
Threonine	+	+	+	+	+	+	+	+
Cystine	−	−	−	−		−	−	−

* Detoxifies alanine, tyrosine, and glutamic acid.
† Serine synthesized (strain W), required (strain E).
‡ Either phenylalanine or tyrosine, but not both, required.
§ Required for growth (young) but not maintenance (adult).

obtained good growth in a medium composed of a pancreatic digest of casein, acetate, cysteine, ascorbic acid, asparagine, ribose, purines and pyrimidines, the B vitamins, and filtered human serum. No attempt was made to determine the necessity of all the ingredients, as the workers were interested in the serum factors. They were able[25] to replace the serum with linoleic acid (ether-soluble fraction) and serum albumin (ether-insoluble

[22] D. W. Woolley, *J. Biol. Chem.* **159,** 753 (1945).
[23] D. W. Woolley, *J. Biol. Chem.* **162,** 383 (1946).
[24] H. Sprince and A. B. Kupferberg, *J. Bact.* **53,** 435 (1947).
[25] H. Sprince and A. B. Kupferberg, *J. Bact.* **53,** 441 (1947).

fraction). Sprince[26] reported that the serum albumin fraction could be precipitated with 50% ethanol.

The importance of the above studies is in showing the requirement for an unsaturated fatty acid (linoleic) and the development of a basal medium for future work of a more precise nature.

2. SARCODINA

Very little of an exact nature is known about the nutrition of the amoebae. By far the most work has been done on the parasitic Endamoebae, and these organisms have exhibited synthetic deficiencies which make the inclusion of very complex (even living) materials obligatory.

Perhaps the most interesting and promising organism in this large group is *Acanthamoeba castellani*, which will grow in heat-sterilized peptone medium.[27, 28] An excellent start was made on the study of its nutritional requirements in 1938, but since the paper of A. Lwoff[29] it has been largely neglected.

Lwoff[29] reported that *A. castellani* could not be grown in peptonized silk unless thiamine was added. He showed, moreover, that this organism was capable of joining the two portions of the molecule (pyrimidine and thiazole). The specific requirement, therefore, is not thiamine itself, which can be synthesized, but certainly the pyrimidine moiety and possibly the thiazole moiety as well.

Aside from its inability to synthesize the constituent parts of the thiamine molecule, nothing further is known regarding the synthetic limitations of *A. castellani*. Students of amoeba nutrition should consider this organism as exceptional material for future investigations, now that our knowledge of biologically active compounds has progressed well beyond that of the period of 1938.

The nutritional requirements of the parasitic amoebae, most frequently *Endamoeba histolytica*, have been of interest to a great many investigators ever since Boeck and Drbohlav[30] first cultured the dysentery amoeba. Aside from *E. invadens,* a reptilian parasite, and in spite of enormous effort, no parasitic form has been successfully cultured *in vitro* in the absence of other living organisms. Even for *E. invadens* fresh snake tissue or minced chick embryo[31] or fresh liver slices[32] must be added. We know, however, that *E. histolytica* multiplies abundantly in sterile abcesses, so it seems

[26] H. Sprince, *J. Bact.* **55,** 169 (1948).
[27] R. Cailleau, *Compt. rend. soc. biol.* **114,** 474 (1933).
[28] R. Cailleau, *Compt. rend. soc. biol.* **116,** 721 (1934).
[29] A. Lwoff, *Compt. rend. soc. biol.* **128,** 455 (1938).
[30] W. C. Boeck and J. Drbohlav, *Am. J. Hyg.* **5,** 371 (1925); *Trans. Roy. Soc. Trop. Med. Hyg.* **18,** 238.
[31] L. Lamy, *Compt. rend.* **226,** 2021 (1948).
[32] M. J. Miller, *Can. J. Comp. Med.* **15,** 268 (1951).

clear that the need for other living organisms in cultures reflects a lack on the part of the entozoic amoeba of synthetic abilities for certain complexes capable of being supplied by the host tissue or associated microorganisms.

Even in the presence of suitable microorganisms, *Endamoeba histolytica* and *E. terrapainae* will not grow unless certain additions are made.[33, 34, 35] Rice starch of certain particle size and serum allowed growth of the amoebae (bacteria present). The serum could be replaced by cholesterol, confirming the earlier findings of Snyder and Meleney[36] and Rees *et al.*,[37] and more efficiently by cholesterol and oleic acid. Hansen and Anderson[38] have published a synthetic meudim for *E. histolytica* which is based on crystalline amino acids, B vitamins, nucleid acid, cholesterol, and rice starch. This medium will not support transplantable growth of the amoebae unless bacteria (organism *t* of Rees was used) are present. Embryonic fluids containing bacteria have also been used.[39]

At the present time we can only say that, for the continued growth of *E. histolytica in vitro*, cholesterol is needed. Cholesterol is inactive in promoting growth in the absence of the B vitamins (thiamine, robiflavin, nicotinic acid, pyridoxine, pantothenic acid, *p*-aminobenzoic acid, inositol, and choline were used[37]), and living organisms must be present (bacteria of various kinds or *Trypanosoma cruzi*[40]). In addition a rather low oxidation-reduction potential is essential for continued growth.[41-44]

3. Sporozoa

Nothing of a nutritional nature is known regarding these parasitic protozoans with the single exception of the malarial organism, *Plasmodium*. The very nature of their habitat and their close association with their hosts preclude precise observations. Even in the malarial organisms our biochemical knowledge is meager indeed, as compared to other more amenable types.

Plasmodium is an intracellular parasite and its association with blood is

[33] A. M. Griffin and W. G. McCarten, *Proc. Soc. Exptl. Biol. Med.* **72,** 645 (1949).

[34] A. M. Griffin and W. G. McCarten, *J. Parasitol.* **36,** 238 (1950); **36,** 253.

[35] A. M. Griffin and L. J. Michini, *J. Parasitol.* **36,** 247 (1950).

[36] T. L. Snyder and H. E. Meleney, *J. Parasitol.* **29,** 278 (1943).

[37] C. W. Rees, J. Bozichevich, L. V. Reardon, and F. S. Daft, *Am. J. Trop. Med.* **24,** 189 (1944).

[38] E. L. Hansen and H. H. Anderson, *Parasitology* **39,** 69 (1948).

[39] E. H. Sadun, I. M. Krupp, and M. G. Everritt, *Proc. Soc. Exptl. Biol. Med.* **80,** 272 (1952).

[40] B. P. Phillips, *Science* **111,** 8 (1950).

[41] C. W. Rees and L. V. Reardon, *Am. J. Trop. Med.* **25,** 109 (1945).

[42] S. L. Chang, *Parasitology* **37,** 101 (1946).

[43] J. C. Schaffer and W. W. Frye, *Am. J. Hyg.* **47,** 214 (1948).

[44] J. C. Schaffer, F. W. Ryden and W. W. Frye, *Am. J. Hyg.* **47,** 345 (1948); *Am. J. Hyg.* **49,** 127 (1949).

firmly established, and one should not hope for information regarding nutritional requirements of the parasite so long as blood or blood fractions are used in the medium. A very few observations of an entirely qualitative nature have emerged, however, during the past few years. Most of the evidence is circumstantial, however, because blood and blood products would be expected to contain vast arrays of biologically active compounds and complexes.

The group at Harvard (see McKee[8] for an excellent review of their work) were able to show an apparent requirement of *P. knowlesi* for glucose and *p*-aminobenzoic acid. These observations culminated intensive study of the growth conditions of the parasite outside the blood stream of the host. At first whole blood and later washed red cells were used. Stimulatory effects were noted by the addition of a number of amino acids, purines, and pyrimidines and certain vitamins, provided that all were added together.[45] Later McKee and Geiman (see ref. 8) demonstrated a beneficial effect of methionine on the multiplication of *P. knowlesi in vitro*.

A distinct advance in this field was made by Trager,[46] who was able to obtain extracellular multiplication of the avian parasite, *Plasmodium lophurae*. By hemolysis of infected duck erythrocytes into a medium containing extracts of duck erythrocytes, he demonstrated favorable effects, on survival and multiplication, of gelatin, adenylic acid, and cozymase. More pronounced beneficial effects were noted upon the addition of adenosinetriphosphate and pyruvic acid.

Although the final elucidation of the nutritional requirements of malarial parasites is not yet in sight, it appears that Trager's results on extracellular cultivation may be the most important development toward this end.

4. Infusoria

This subphylum is usually considered to be the most highly specialized group of the Protozoa. The two classes it contains are the Ciliata (containing by far the greatest number of species) and the Suctoria. The majority of Ciliata and Suctoria are free-living and holozoic. Food habits vary. Some are osmotrophic (astomatous ciliates), many are bacteria feeders, many are carnivorous (feeding on other protozoans) or herbivorous (feeding on filamentous and non-filamentous algae), and many are more or less omnivorous. Feeding in the Suctoria is confined to the protoplasm of living prey (usually small ciliates), and the contents of the prey is removed by means of suctorial tentacles.

The two classes of Infusoria possess cilia at some stage of their development. These locomotor organelles are retained in the Ciliata, but they are

[45] E. G. Ball, C. B. Anfinsen, Q. M. Geiman, R. W. McKee, and R. A. Ormsbee, *Science* **101,** 542 (1945).

[46] W. Trager, *J. Exptl. Med.* **92,** 349 (1950).

resorbed and replaced by the suctorial tentacles in the adult, non-motile Suctoria.

Although ciliates and suctorians have been cultured for many decades, earlier work yielded no useful biochemical information, owing to the inclusion of myriads of bacteria, yeasts, and molds from the watery environment.

The first ciliate to be successfully cultured in the absence of all other microorganisms was *Tetrahymena geleii** (*Glaucoma piriformis*). This task was accomplished by A. Lwoff in 1923.[47] The medium as finally used was heat-sterilized peptone. From this beginning to the present we have gained so much information regarding this organism that today it stands as the only animal which can be grown under rigidly controlled aseptic conditions in a completely chemically defined medium. In other words, we have a clear picture of its synthetic capacities. *Tetrahymena* will be discussed, therefore, in relation to its nutrition and metabolism, as all other ciliates can then be compared to it.

a. Inorganic Requirements. Less is known about the inorganic metabolism of *Tetrahymena* than about its organic metabolism. The need for certain ions can be easily demonstrated by simply omitting them from the medium. By this technique it was shown[48] that 0.1 μ mole of Mg^{++}, 0.3 μ mole of K^{+}, and 0.5 μ mole of $PO_4^{\equiv}$ ions are required for maximum growth in a synthetic medium. The need for iron and copper have been shown,[101] but Ca^{++}, $SO_4^{=}$, and Cl^{-} ions could not be demonstrated as requirements nor could the trace elements Mn^{++}, Zn^{++}, Co^{++}, Fl^{-}, $BL_3^{=}$, or MoO_3. The probable explanation of these results is the inclusion of these elements as contaminants in other portions of the basal medium, especially the amino acid portion.

Recently the chelating agent ethylenediaminetetraacetic acid (EDTA) has been used in connection with this problem,[49] and although the relationships are complex, it is possible to show biological activity for Ca^{++} and Co^{++}. Slater[49a] has used 8-hydroxyquinaline to remove trace elements from the basal medium and has demonstrated a cobalt requirement in glucose-free medium. No such requirement was found, however, when glucose was present.

b. Nitrogen Metabolism. *Tetrahymena* possesses a fairly complete and active battery of proteolytic enzymes. In nature it is probable that most, if not all, of its nitrogen metabolism is satisfied by amino acids derived from bacterial protein by digestion. Such proteins as casein and gelatin are readily attacked.

[47] A. Lwoff, *Comp. rend.* **176**, 928 (1923).
[48] G. W. Kidder, V. C. Dewey, and R. E. Parks, *Physiol. Zoöl.* **24**, 69 (1951).
[49] G. W. Kidder and V. C. Dewey, unpublished observations.
[49a] J. V. Slater, *Physiol. Zool.* **25**, 323 (1952).
* According to the work of Corliss, (*Trans. Amer. Micros. Soc.*, **71:** 159 (1952)) the correct name of this organism is *Tetrahymena pyriformis*.

In a medium containing all the growth factors required by *Tetrahymena geleii* W. and the nitrogen supplied by crystalline amino acids, it was found that growth failed or was negligible if any one of the following amino acids were omitted (Table 2): arginine, histidine, isoleucine, leucine, lysine, methionine, phenylalanine, threonine, tryptophan, and valine.[21, 50] All were found to be absolute requirements for the organism, except arginine. In the absence of arginine, slow and suboptimal growth results, indicating, as in the rat, slow synthesis of this compound. Recent experiments[49b] on arginine synthesis by *Tetrahymena* indicate complete lack of synthetic ability for this amino acid when other components of the medium are chemically defined.

When a number of non-essential amino acids are also included, toxicities become apparent which can be overcome to some extent by glycine and very effectively by serine.[50] Elliott[51] reports that serine is an absolute requirement for *T. geleii* E. This observation has been verified for strain E.[49b] Serine

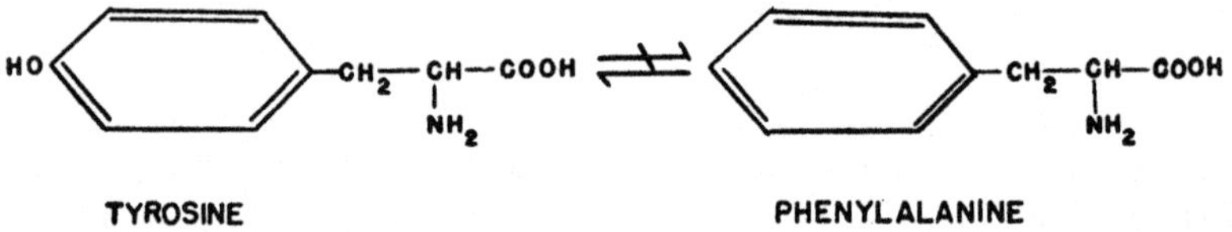

FIG. 6. In this reaction the reduction of tyrosine to phenylalanine is blocked, as indicated by the crossed arrow.

is synthesized from glycine in strain W but the source of the one carbon fragment is still in doubt. Formate as such is inactive in this reaction.

Many amino acid interrelationships have been brought to light by studies on *Tetrahymena*. It was early[52] shown that although tryosine cannot replace phenylalanine it exerts considerable sparing action. This is interpreted as meaning that this organism, like the vertebrates, can oxidize the *p* position of phenylalanine but the reaction is not reversible. Cystine or cysteine and homocystine spare but do not replace methionine,[53] and this organism is incapable of methylating homocysteine, even in the presence of added vitamin B_{12}.[54] Both citrulline and ornithine spare arginine,[9] being precursors

[49b] V. C. Dewey and G. W. Kidder, unpublished observation.

[50] G. W. Kidder and V. C. Dewey, *Arch. Biochem.* **6,** 425 (1945); *Proc. Natl. Acad. Sci. U. S.* **33,** 347 (1947).

[51] A. M. Elliott, *Physiol. Zoöl.* **22,** 337 (1949).

[52] G. W. Kidder, *Ann. N. Y. Acad. Sci.* **49,** 99 (1947).

[53] D. S. Genghof, *Arch. Biochem.* **23,** 85 (1949); *Arch. Biochem. Biophys.* **34,** 112 (1951).

[54] G. W. Kidder, *in* Bourne and Danielli, International Review of Cytology, Vol. 1, Academic Press, New York, 1952.

of proline.[49b] It is known that *Tetrahymena* does not produce urea[55] and the Krebs-Henseleit urea cycle does not function.

Investigations regarding the availability of the optical isomers of most of the essential amino acids have been carried out.[9] When great care is taken to insure the purity of the various isomers, it is found that the unnatural configuration of arginine alone could be utilized by *T. geleii* W. The D isomers of leucine and serine were found to be inhibitory. Elliott *et al.*[56] report that *T. geleii* E. can utilize the D isomers of arginine, methionine, and lysine, but as their report is only in abstract form no proof of purity is given. In the case of methionine it has been found[9] that the D isomer spares, but does not replace, L-methionine. The D isomer can be used for cysteine synthesis but cannot be changed to the natural form for direct incorporation.

Although minor differences between the synthetic capacities for amino acids in *Tetrahymena* and a mammal, such as the rat, do exist, the many points of similarity are striking. One cannot help thinking that the animal enzymatic pattern for amino acid synthesis must have become established very early in the evolution of the whole kingdom. Those organisms with no animal affinities (phytoflagellates, bacteria, molds) may be quite low in synthetic capacities (e.g., the lactic acid bacteria), but there is no consistent pattern to be observed, Where a true animal has been carefully studied, the pattern is always largely the same.

c. Carbon Metabolism. *Tetrahymena geleii* W. can multiply indefinitely in a medium devoid of carbohydrates. Amino acids are deaminated with the production of their principal waste product, ammonia. This organism is capable of utilizing a number of carbohydrates, however, in which case the amino acids are spared. Starch and glycogen are easily hydrolyzed, as is maltose, and glucose and fructose are fermented.[21] *T. geleii* E. and *T. vorax* can ferment galactose, and the latter can hydrolyze lactose. One rather strange thing about the ciliates for which we have information is the entire lack of ability to utilize sucrose. This most common of sugars is inert nutritionally.

Of the organic acids tested on *T. geleii* W. only acetic and pyruvic were stimulatory to growth (see Kidder and Dewey[9]). Fatty acids have very little effect, except under certain conditions where they are complexed, as in the Tweens. When Tween 80 (polyoxyethylene derivative of sorbitan esterified with 1 mole of oleic acid) is added to the synthetic medium, enormous growth stimulation, as judged by optical density, occurs. The cell count and nitrogen content of the protoplasm of such cultures, however, is not greater than non-Tween cultures. Microscopic examination reveals that the ciliates are distended by droplets of fatty acid, a condition which

[55] A. Lwoff and N. Roukhelman, *Compt. rend.* **183,** 156 (1926).

[56] A. M. Elliott, J. F. Hogg, and W. Chung, *Federation Proc.* **11,** 207 (1952).

renders them quite opaque.[57] There is no evidence to show that the fatty acid is utilized, however, and the advisability of the use of Tween is in question. Tween does increase the sensitivity of dose responses when turbidimetric measurements are used, but it renders the organisms inferior for work that involves handling the cells for fractionation, etc.

d. Growth Factors. (1) *B Vitamins.* The first observation of a requirement by ciliates for any of the water-soluble vitamins was made by Lwoff and Lwoff in 1937,[58] when they found that *Tetrahymena geleii* (*Glaucoma piriformis*) could not be cultured in silk peptone unless thiamine was added. The two portions of the molecule (pyrimidine and thiazole) added either singly or together would not permit growth. They concluded, therefore, that this ciliate lacked enzymes for linking the two portions to form the thiamine molecule.

In 1942 Kidder and Dewey[59] reported the growth of *T. geleii* and *T. vorax* in a vitamin-free casein medium supplemented with heat- and alkali-treated alfalfa extract. They recognized the fact that the extract contained many factors but was devoid of thiamine. They postulated a heat- and alkali-stabile "factor S" which catalyzed the synthesis of the thiamine molecule from the pyrimidine and thiazole. Later[60] it was found that *T. geleii* W. could be grown in tryptophan-supplemented casein or gelatin hydrolyzate and heat- and alkali-treated alfalfa extract, without the addition of thiamine. Added thiamine was observed to increase the total yield and individual cell size. When butanol-extracted and heat-treated liver extract was used in place of the alfalfa extract, no growth occurred without the addition of thiamine,[61] provided that the nitrogen source was gelatin hydrolyzate. When crystalline amino acids were used as the nitrogen source, growth occurred without the addition of thiamine or factor S from alfalfa. It was discovered that certain unnatural (D) isomers of amino acids (added as the racemic mixture) made growth possible under these conditions. It was postulated that *Tetrahymena* required some D configuration if an exogenous source of thiamine was absent.

More recently (see Kidder and Dewey[9]) it has been found that in the completely synthetic medium of Dewey *et al.*[62] thiamine is required for growth of *T. geleii*, even though the unnatural isomers of amino acids are present. Heat- and alkali-treated extracts of natural materials, added to the synthetic medium, allow growth to take place. It appears that there is an unknown factor (or factors) necessary for thiamine biosynthesis even in

[57] G. W. Kidder, V. C. Dewey, and M. R. Heinrich, unpublished observation.
[58] A. Lwoff and M. Lwoff, *Compt. rend. soc. biol.* **126,** 644 (1937).
[59] G. W. Kidder and V. C. Dewey, *Growth* **6,** 405 (1942).
[60] G. W. Kidder and V. C. Dewey, *Biol. Bull.* **87,** 121 (1944).
[61] G. W. Kidder and V. C. Dewey, *Biol. Bull.* **89,** 131 (1945).
[62] V. C. Dewey, R. E. Parks, and G. W. Kidder, *Arch. Biochem.* **29,** 281 (1950).

the presence of D-amino acids. This is an interesting problem which should receive more attention.

The function of thiamine (as the pyrophosphate) in *Tetrahymena* has been suggested as the cocarboxylase of pyruvate, as pyruvic acid accumulates in thiamine-deficient cultures.[9, 63] Other organic acids also accumulate in such cultures, one of which is α-ketoglutaric acid.[64] It appears likely that thiamine (along with thioctic acid, see below) functions in the decarboxylation of α-ketoglutaric acid in the same manner as it does with pyruvic. The mediation of pyruvic decarboxylation by thiamine has long been known in birds and mammals, so it is altogether expected that the same function would be found in the ciliates.

From what we know of animals in general it is safe to predict that all ciliates will be found to require thiamine in their diet, under ordinary conditions, but so far only two other genera have been grown under conditions which allow for direct evidence of such a requirement. *Glaucoma scintillans* was found to require thiamine[59] and failed to grow without it, even in the presence of factor S. Using a bacterial "plasmoptyzate" for the growth of *Colpoda steinii* (*duodenaria*),[65, 66] this organism was shown to be dependent upon exogenous thiamine, as well as rather high concentrations of a number of other B vitamins (riboflavin, pantothen, niacin, and pyridoxine).

Tetrahymena geleii and *T. vorax* are also incapable of synthesizing riboflavin, pantothen, pyridoxal or pyridoxamine, pteroylglutamic acid, and thioctic acid.[67] *T. geleii* synthesizes choline, biotin, and vitamin B_{12}.[9]

It is interesting to note that the synthesis of nicotinic acid from tryptophan, which has been found in mammals, is impossible for *T. geleii*.[68] Both tryptophan and niacin are absolute requirements. *Tetrahymena* resembles *Drosophila*[69] in this respect.

A comparison of the activities of pyridoxine, pyridoxal, and pyridoxamine[67] makes it appear likely that the growth-promoting action of pyridoxine is due solely to its spontaneous animation in the medium. Pyridoxal phosphate was found to be less active than pyridoxal, indicating either a function for the aldehyde (or the amine), which had to be obtained by de-

[63] I. A. Tittler, *Anat. Record* **105**, 501 (1949).

[64] V. C. Dewey and G. W. Kidder, *Proc. Soc. Exptl. Biol. Med.* **80**, 302 (1952).

[65] E. L. Tatum, L. Garnjobst, and C. V. Taylor, *J. Cellular Comp. Physiol.* **20**, 211 (1942).

[66] L. Garnjobst, E. L. Tatum, and C. V. Taylor, *J. Cellular Comp. Physiol.* **21**, 199 (1943).

[67] G. W. Kidder and V. C. Dewey, *Arch. Biochem.* **21**, 58 (1949); **21**, 66.

[68] G. W. Kidder, V. C. Dewey, M. B. Andrews, and R. R. Kidder, *J. Nutrition* **37**, 521 (1949).

[69] J. Schultz and G. T. Rudkin, *Federation Proc.* **7**, 185 (1948).

phosphorylation,[9] or permeability difficulties for the phosphorylated compound.

Tetrahymena is dependent upon the whole pteroylglutamic acid molecule,[67, 70] but, like birds and mammals, and unlike many bacteria, it possesses enzymes for removing glutamic acid residues from triglutamates and heptaglutamates. The reduced formyl pteroylglutamic acid (citrovorum factor, leucovorin) is no more active for this organism than is PGA itself.[71]

Tetrahymena can utilize Aminopterin (4-amino-pteroylglutamic acid) and Methopterin (9-methyl pteroylglutamic acid) for growth, the former with about 17% efficiency and the latter with about 2%.[71] This was interpreted as showing the presence of enzymes for the deamination of position 4 of

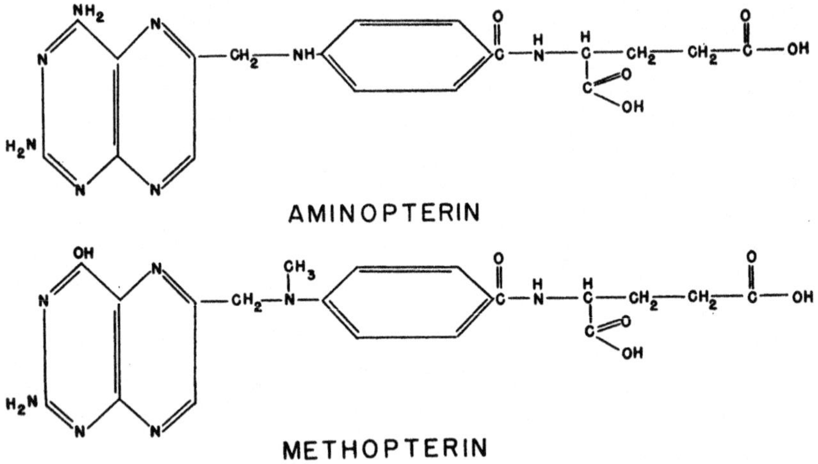

FIG. 7. Pteroylglutamic acid analogues

PGA and the demethylation of position 9.* When both substitutions occur simultaneously (Amethopterin), however, an inhibition results.[72]

Very early in the culture of *T. geleii*[73] the need for a factor or factors from natural sources was recognized. When other requirements had been satisfied, it was possibe to show that the unknown factor was single and could

[70] G. W. Kidder and R. C. Fuller, *Science* **104,** 160 (1946).
[71] G. W. Kidder, V. C. Dewey, and R. E. Parks, *Proc. Soc. Exptl. Biol. Med.* **78,** 88 (1951).
[72] V. C. Dewey, G. W. Kidder, and R. E. Parks, *Proc. Soc. Exptl. Biol. Med.* **78,** 91 (1951).
[73] V. C. Dewey, *Proc. Soc. Exptl. Biol. Med.* **46,** 482 (1941); *Biol. Bull.,* **87,** 107 (1944).

* Recent work (M. R. Heinrich, V. C. Dewey, and G. W. Kidder, *J. Am. Chem. Soc.* in press) has shown that the aminopterin used was contaminated with pteroylglutamic acid. The growth obtained was due to the contaminating growth factor. Therefore, the organism is simply naturally resistant to Aminopterin.

be concentrated. The name protogen was given to this factor,[74] and it was shown[75] to be the same as the acetate factor and the pyruvate oxidation factor for lactic acid bacteria. Recently the Lederle group[76] have crystallized and synthesized protogen and identified it as dimercaptooctanoic acid.

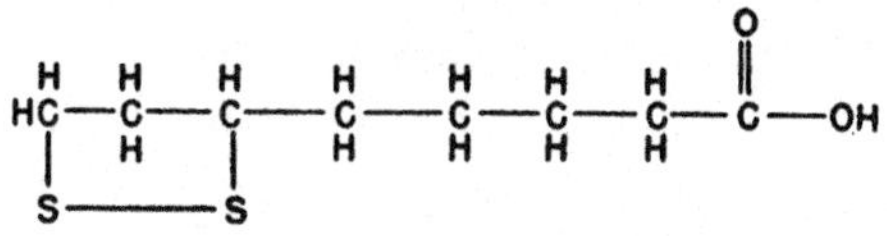

THIOCTIC ACID

FIG. 8

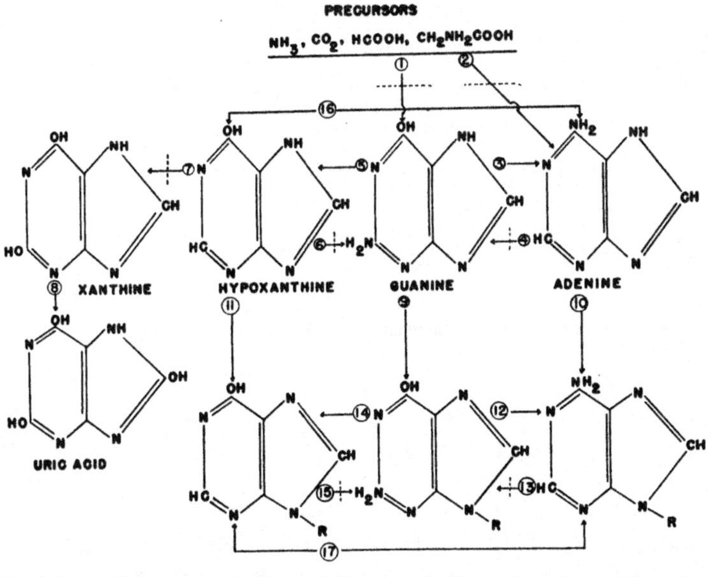

FIG. 9. Schematic representation of the metabolism of purines. The dotted lines represent blocks in specific pathways which occur in *Tetrahymena*. (Modified from Kidder *et al.*[82])

They have named it thioctic acid. Reed *et al.*[77] carried out similar identifications on what they called[77a] α-lipoid acid (= protogen). Publication of

[74] E. L. R. Stokstad, C. E. Hoffmann, M. A. Regan, D. Fordham, and T. H. Jukes, *Arch. Biochem.* **20,** 75 (1949).

[75] E. E. Snell and H. P. Broquist, *Arch. Biochem.* **23,** 326 (1949).

[76] J. A. Brockman, E. L. R. Stokstad, E. L. Patterson, J. V. Pierce, M. E. Macchi, and F. P. Day. *J. Am. Chem. Soc.* **74,** 1868 (1952).

[77] L. J. Reed, B. G. DeBusk, I. C. Gunsalus, and G. H. F. Schnakenberg, *J. Am. Chem. Soc.* **73,** 5920 (1951).

[77a] L. J. Reed, B. G. DeBusk, I. C. Gunsalus, and C. S. Hornberger, *Science* **114,** 93 (1951).

activities of three isomeric configurations[78] make it rather certain that the active vitamin form of thioctic acid is 6,8-dimercaptooctanoic acid. In natural materials there are probably many mixed disulfides with protogen activity.[79] The possible coenzyme form and its functions are discussed in Chapter 19.

(2) *Nucleic Acid Derivatives. Tetrahymena* has enzymatic limitations which make obligatory the inclusion of nucleic acid or certain derivatives in the diet. Yeast or thymus nucleic acid or their hydrolyzates were earlier used to supply missing factors.[80] The enzymatic limitations of the organism were shown to be concerned specifically with the purine guanine and the pyrimidine uracil.[81] No natural purine could replace guanine although the nucleoside and the neuleotide were fully active. Adenine and hypoxanthine and their nucleosides were found to spare guanine but not replace it.[82] Xanthine and uric acid are inert. *Tetrahymena* is incapable of making the

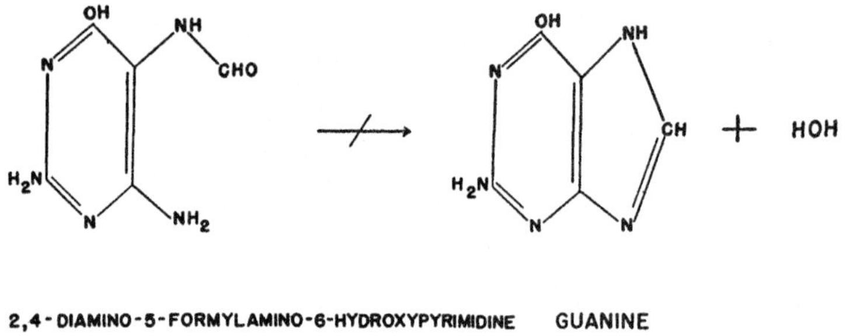

FIG. 10. Tetrahymena does not possess the ability to form the imidazole ring

specific substitutions resulting in guanine and is also incapable of forming the purine ring from precursors. When 2,4-diamino-5-formylamino-6-hydroxypyrimidine is given, it is entirely inert,[82] showing that the organism cannot close the imidazole ring. Likewise both 4-amino-imidazole-5-carboxamide and its N-formyl derivative, 4-formylamino-imidazole-5-carboxamide are inert, showing that closure of the pyrimidine ring to hypoxanthine is impossible.[83] These latter compounds have been found active in supplying the purine need of certain mutant *E. coli*.[84, 85]

[78] M. W. Bullock, J. A. Brockman, E. L. Patterson, J. V. Pierce, and E. L. R. Stokstad, *J. Am. Chem. Soc.* **74,** 3455 (1952).

[79] L. J. Reed, B. G. DeBusk, C. S. Hornberger, and I. C. Gunsalus, *J. Am. Chem Soc.* in press.

[80] G. W. Kidder and V. C. Dewey, *Arch. Biochem.* **8,** 293 (1945).

[81] G. W. Kidder and V. C. Dewey, *Proc. Natl. Acad. Sci. U. S.* **34,** 566 (1948).

[82] G. W. Kidder, V. C. Dewey, R. E. Parks, and M. R. Heinrich, *Proc. Natl. Acad. Sci. U. S.* **36,** 431 (1950).

[83] G. W. Kidder and V. C. Dewey, unpublished observations.

[84] R. Ben-Ishai, B. Volcani, and E. D. Bergmann, *Arch. Biochem.* **32,** 229 (1951).

[85] E. D. Bergmann, R. Ben-Ishai, and B. E. Volcani, *J. Biol. Chem.* **194,** 531 (1952).

Flavin and Graff[86] showed that when C^{14}-labeled guanine formed the sole purine in the diet of *T. galeii* W., both labeled guanine and adenine were found in the isolated nucleic acid, but only the nucleic acid adenine was radioactive when C^{14}-labeled adenine[87] was administered (along with non-labeled guanine, for growth). This simply confirms the results obtained by growth studies. They failed to find any activity in the soluble fractions (containing ATP), however. Heinrich *et al.*[88] reinvestigated the possibility of purine biosynthesis in ATP. They showed that one of the usual purine precursors in birds and mammals, formic acid, was oxidized and not incorporated in the purines in *Tetrahymena*. Formate-C^{14} was given and CO_2 collected from cultures. When C^{14}-labeled guanine was administered, however, all purines isolated, including ATP, had approximately equal activities. Thus it seems certain that *Tetrahymena* has lost all ability to

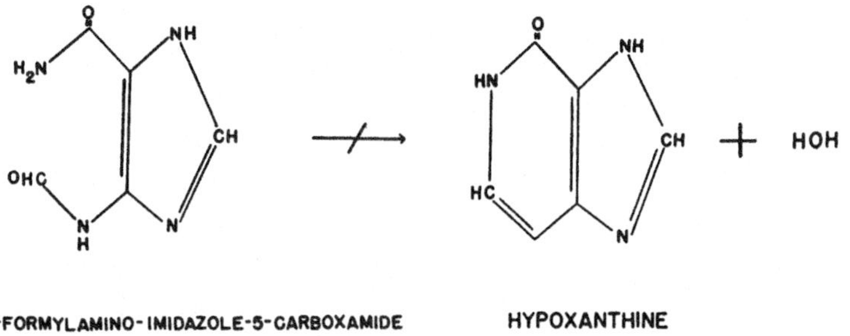

FIG. 11. Tetrahymena does not possess the ability to close the ring for the production of a purine.

form the purine ring for nucleic acid and free nucleotide synthesis. This is in contrast to the condition in birds and mammals where tracer studies have shown that rapid purine synthesis is possible from carbon dioxide, formic acid, and glycine.

One of the most interesting problems, which remains unsolved, is the intermediate or intermediates that form the link between guanine and adenine. Recent experiments[89] have shown that 2,6-diaminopurine is not a logical candidate. Diaminopurine can be oxidized by *Tetrahymena* at position 6 to yield guanine, but it cannot be deaminated at position 2 to yield adenine. Isoguanine is inert. Inasmuch as few organisms, and no other animal, have been shown to possess an absolute guanine requirement, *Tetrahymena* offers exceptional material for studying this problem.

The block in pyrimidine synthesis in *Tetrahymena* involves enzymes con-

[86] M. Flavin and S. Graff, *J. Biol. Chem.* **191**, 55 (1951).
[87] M. Flavin and S. Graff, *J. Biol. Chem.* **192**, 485 (1951).
[88] M. R. Heinrich, G. W. Kidder, and V. C. Dewey, unpublished observations.
[89] V. C. Dewey and G. W. Kidder, unpublished observations.

cerned with shifting substituent groups. Of the natural pyrimidine bases only uracil will satisfy the pyrimidine requirement for nucleic acid synthesis.[81, 82] Cytosine, thymine, and orotic acid will neither spare nor replace uracil. Cytidine or cytidylic acid, however, replace uracil completely. This indicates that cytidine deaminase is present but there is no cytosine deaminase. The inert condition of cytosine indicates that the sugar-coupling enzymes cannot use this compound. Neither thymine nor thymidine will spare uracil, but both are active in another way. They will spare PGA.

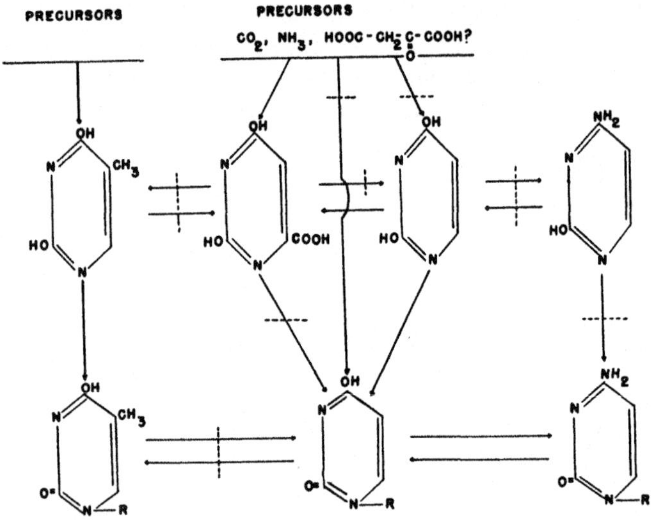

FIG. 12. Schematic representation of the metabolism of pyrimidines. The dotted lines represent blocks in specific pathways which occur in *Tetrahymena*. (Modified from Kidder *et al.*[82])

This was interpreted to mean that thymine is synthesized from non-pyrimidine precursors and that PGA takes part in the reaction.[82]

An interesting phase of pyrimidine metabolism, not completely solved for any organism, is the question of the precursors. *Tetrahymena* offers favorable material for the many problems involved in such a study.

A number of interesting studies on *Tetrahymena* have been made using both purine and pyrimidine analogues. Although these studies have yielded much information regarding enzyme capacity and limitations, their value in a discussion of nutrition is questionable. Readers interested in the details of these studies are referred to the earlier review.[9]

Two other Infusoria have yielded interesting data relating to nucleic acid, or derivatives, as nutritional factors.

In 1942 Johnson and Baker[90] first reported the sterile culture of *Para-*

[90] W. H. Johnson and E. G. S. Baker, *Science* **95**, 333 (1942).

mecium (*P. multimicronucleatum*). The first medium was made up of yeast juice and proteose-peptone. Although the proteose-peptone could be heat-sterilized, the yeast juice could not be heated and retain activity. The yeast juice was later fractionated[91] into heat-labile and heat-stabile portions, both necessary for growth. Recently Johnson[92] has reported being able to substitute either hydrolyzed nucleic acid (RNA or DNA) or a mixture of cytidylic and guanylic acid (all heat-stabile) for the heat-labile fraction of the yeast press juice. No explanation is immediately at hand to account for this observation, but it is reported that the medium (all heat-sterilized) composed of peptone, heat-stabile fraction of yeast press juice, and the nucleic acid derivatives is just as active in promoting growth in *P. aurelia* as in *P. multimicronuleatum*.

The other report of interest deals with the suctorian *Tokophyra infusionum*. This organism is entirely carnivorous and can be cultured aseptically on a diet of living *Tetrahymena* in water. Rudzinska[93] has reported that longevity and transplantability are dependent upon the correct number of food organisms. Recently Lilly and Gilmore[94] have reported that abnormalities and declining embryo production resulting from overfeeding can be overcome by the addition of 0.05% yeast extract, or 100 γ per milliliter of nucleic acid. Even more effective is a mixture of guanylic acid, adenylic acid, cytidylic acid, and uracil. Just a mixture of guanylic acid and uradil is effective, but the purine bases are without effect. It would appear that *Tokophyra* either has an abnormally high requirement of nucleic acid derivative or fails to utilize all the nucleoprotein portion of the food organism under conditions of "luxury living."

(3) *Unknown Factors.* The fact that *Tetrahymena* is the only animal microorganism which can be grown in chemically defined media indicates clearly that a number of biologically active unknown factors, required by organisms which have not been successfully grown, remain to be discovered. As it has been the experience in the past that growth factors discovered through the use of "lower" organisms have usually assumed considerable importance in studies of "higher" organism, attention should be focused in this direction. If and when all the biologically active compounds and complexes are known, the culturing of mammalian tissues in chemically defined media will become an easy actuality. The importance of such an advance is obvious, if for nothing more than its importance to cancer research. The brief summary which follows is presented to call attention to some isolated cases (among a wealth of untouched material) where fruitful

[91] W. H. Johnson and E. L. Tatum, *Arch. Biochem.* **8,** 163 (1945).
[92] W. H. Johnson, *Physiol. Zoöl.* **25,** 10 (1952).
[93] M. A. Rudzinska, *Science* **113,** 10 (1951).
[94] D. M. Lilly and J. P. Gilmore, *Anat. Record* in press.

investigations may be made. The organisms have all been mentioned in this section, but attention was previously directed to the known factors.

Strigomonas: At present these trypanosomids have only been cultured in media containing peptones.

Leishmania and Trypanosoma: All members of these genera require serum factors. Some of these factors appear to be heat-labile, and the serum undoubtedly contains many factors in addition which are required by the organisms.

Trichomonas and Trichomastix: With one exception, all the species of these genera have been cultured only on complex media (beef broth, blood serum, egg, etc.). Even in the case of *Trichomonas foetus* (Weiss and Ball[20]), which certainly needs confirmation, the medium contained extracts of plant materials. For optimum growth unknown substances had to be added in enzymatically digested proteins or serums.

Acanthamoeba: Peptones have been used in the few cases where this organism has been grown. We have no information about its nutritional requirements other than its need for at least one portion of the thiamine molecule.

Endamoeba: So far a few species of this genus have been cultured, but in every case living cells of some sort have been found to be essential constituents of the otherwise complex media.

Tetrahymena: In spite of the fact that rapid, indefinite, transplantable and reproduceable growth can be obtained with *T. geleii* and *T. vorax* on completely synthetic media, good evidence exists that stimulatory (growth-rate) factors exist in natural materials. This must mean that these organisms can synthesize everything they require from the chemicals of the medium, but they take advantage of certain preformed compounds or complexes if they are present. This ability to utilize unknown stimulatory factors in crude materials make the use of these ciliates for the assays of vitamins, amino acids, etc., in natural materials hazardous.

Glaucoma: G. scintillans requires, in addition to its essential amino acids, which are known,[94a] certain unknown compounds found in liver, plant extracts, etc. These unknown substances are heat-stabile and should lend themselves readily to investigation. Moreover, *G. scintillans* produces sufficiently heavy cultures that assay quantifications can be carried out turbidimetrically, a tremendous advantage over many slower techniques.

Paramecium: The nearest approach to controlled cultures of *Paramecium* involves the use of peptone, yeast fractions, and nucleic acid derivatives, none of which can be omitted, and therefore must bring in many unknown factors. All components are heat-stabile, however, and should be amenable to ultimate solution.

[94a] Fuller, R. C., Studies on the Biochemistry of Glaucoma scintillans (Masters' thesis, Amherst College, Mass., 1948).

Colpoda: The problem with *Colpoda* (especially *C. steinii*) appears similar to that with *Paramecium*. No attempts have been made, however, to substitute for the plazmoptyzate medium the medium of Johnson[92] which was successful on two species of *Paramecium*. An obvious advantage of *Colpoda* is its high reproduction rate and its high maximum yield.[95]

Stylonychia and Pleurotricha: These highly specialized hypotrichous ciliates have been grown in fluid media with the addition of *Tetrahymena* as food.[96] In order that growth continue, however, a certain unknown factor from yeast, etc., must be present, indicating that *Tetrahymena* is lacking or deficient in this factor required by the hypotrichs. The active fraction has now been concentrated by chromatographic methods so that a fraction of a microgram per milliliter of fluid (along with the food organism) permits growth.[97]

Euplotes: Lilly *et al.*[98] have reported that *E. patella* requires a factor similar to or identical with the *Stylonychia* factor (from yeast) and in addition two other unknown factors, one from liver and one from bacteria (heat-killed).

The above list of animal microorganisms represents a small fraction only of those whose cultural characteristics have been investigated, even cursorily. The vast numbers which occur as free-living of parasitic organisms offer a wealth of material for future workers. Exploitations of this group, with emphasis on biochemistry and nutrition, should prove of great value to the general field of nutrition.

III. Worms

Of particular interest to students of comparative nutrition are the reports of Dougherty and associates on their experiments with rhabditoid nematodes. These organisms are easily obtained from soil enrichment cultures and appear to offer excellent material with which to work. The forms used by Dougherty and co-workers belong to the genus *Rhabditis*, and although many papers have appeared regarding the culture of various species of this genus, little of a biochemical nature appeared until recently.

The first report from the California group[99] deals with their techniques for establishing pure cultures of *Rhabditis pellio*, a dioecious symbiont of earthworms, and *R. briggsae* (*elegans*), a monoecious soil dweller. These worms were maintained as two-membered cultures using *Escherichia coli* as the food organism. Most difficulty was encountered in rendering worms, with mixed bacterial associates, aseptic. This was finally accomplished by a

[95] G. W. Kidder and C. A. Stuart, *Physiol. Zoöl.* **12,** 329, 341 (1939).
[96] D. M. Lilly, *Physiol. Zoöl.* **15,** 146 (1942).
[97] D. M. Lilly, unpublished observations.
[98] D. M. Lilly, A. Pignataro, and K. Heilshorn, *Anat. Record* **105,** 32 (1949).
[99] E. C. Dougherty and H. G. Calhoun, *Proc. Helminthol. Soc. Wash. D. C.* **15,** 55 (1948).

24-hr. treatment of gravid females with a mixture of penicillin and streptomycin (5000 units of each per milliliter) followed by sterile water rinse and a 1-hr. exposure to 1:1000 aqueous solution of merthiolate. From the merthiolate solution viable larvae were transferred through one or two changes of sterile water (usually containing penicillin) and placed onto slants containing *E. coli*. Pure cultures were uniformly obtained from two-membered cultures by 3- to 4-hr. exposure to streptomycin (5000 units per milliliter in broth).

Pure cultures of *Rhabditis pellio* and *R. briggsae* (*elegans*) have been maintained for many transplants on a complex medium (Difco nutrient agar) supplemented with Seitz-filtered liver extract. In the case of the dioecious *R. pellio* decreased fertility manifested itself eventually, leading to the hypothesis by the authors that the medium was deficient, although

TABLE 3

Growth of Rhabditis briggsae

Chick embryo juice	Amino acid-vitamin medium	Growth	Maturity*	Second-generation larvae
0	Undiluted	0	0	0
Undiluted	0	+	+	+†
1 part	1 part	+	+	+‡
0.8 part	1 part	+	+	+
0.1 part	1 part	+	0	0

* Maturation and eggs at 4 days.
† Very few larvae at 8 days.
‡ Massive cultures of larvae at 8 days.

later[100] they state that the infertility probably was of genetic origin, owing to the imbreeding. The factor (or factors) brought in by the liver was found to be heat-labile and not to correspond to any of the known B vitamins.

Using the hermaphroditic form *Rhabditis briggsae* Dougherty *et al.*[100] have been able to overcome some of the difficulties encountered in the liver extract-containing medium and have obtained as rapid and prolific growth as controls containing living bacteria. This was accomplished in the amino acid medium of Kidder and Dewey[101] to which was added chick embryo juice. Neither the *Tetrahymena* medium alone nor the embryo juice alone gave these results (Table 3). The embryo juice contained the heat-labile factor or factors but was deficient in some components supplied by the *Tetrahymena* medium. Inasmuch as the *Tetrahymena* medium showed no inhibitory properties, the authors felt that it offered an excellent basal medium for further work on the concentration and isolation of the heat-labile factor or factors. They suggest that the real difference between the

[100] E. C. Dougherty, J. C. Raphael, and C. H. Alton, *Proc. Helminthol. Soc. Wash. D. C.* **17**, 1 (1950).

[101] G. W. Kidder and V. C. Dewey, *Arch. Biochem.* **20**, 433 (1949).

dietary requirements of *Rhabditis* and *Tetrahymena*, and indeed birds and mammals, will probably be only one or few complexes which these worms have lost the ability to synthesize, and that they will ultimately be found to be dependent on the "essential" amino acids and the usual B vitamins. Fractionation of the chick embryo juice and of liver extract has been carried out.[101a] An active fraction from both sources, called factor Rb, appears to be protein in nature and is heat-labile. We can look forward with confidence to further valuable information on these nematode worms, especially since the announcement in abstract form of partial success in the culture of the Steinernematid *Neoaplectana glaseri*[102] in defined media. No details were given, however.

IV. Insects

The practical aspects of insect nutrition have been recognized for years, and a host of reports have appeared dealing with the natural food habits, food selectivity, etc. Extensive reviews of this literature are available.[2, 103, 104, 105] From the biochemical standpoint, however, no insect species has been thoroughly worked out as to its dietary requirements. In those species where asepsis has been accomplished (mosquito and fruit fly) in a convincing manner there appears many gaps in our knowledge. In the dry-food insects (certain beetles and moths), although bacterial contamination of the food is gratifyingly low, there always remains the possibility of important contributions by intestinal microorganisms obscuring a true synthetic lack.

Nevertheless the valuable and interesting reports which have been made on this group contribute much to our knowledge of comparative biochemistry and nutrition. In most of the insects which have been critically examined, quantifications are based on larval or growing forms, for it is often the case (certain moths) that the adults feed little or none or if they do will maintain themselves on carbohydrate and water (certain flies). To determine growth requirements, therefore, the actively growing stages are used. So far only a few of the factors (dietary) influencing the complicated process of eclosion are known, but this stage is often used in relation to time, for quantification of results.

1. Diptera

The most exhaustive studies on this order concern themselves with larval and pupal stages of the yellow fever mosquito *Aedes* and the fruit fly *Drosophila*.

[101a] E. C. Dougherty, *Exptl. Parasitol.* **1,** 34 (1951).
[102] N. R. Stoll, *J. Parasitol.* **34,** 12 (1948).
[103] V. B. Wigglesworth, The Principles of Insect Physiology, E. P. Dutton and Co., New York, 1939.
[104] R. Craig and W. M. Hoskins, *Ann. Rev. Biochem.* **9,** 617 (1940).
[105] W. Trager, *Physiol. Revs.* **21,** 1 (1941).

The nitrogen nutrition of *Drosophila* has received some attention although all the amino acid requirements are not as yet satisfactorily demonstrated. It was early shown that the fruit fly could be grown aseptically,[106, 107, 108] but the media used (casein, yeast, etc.) did not produce information regarding amino acid requirements. Schultz *et al.*[109] have reported in abstract form that sterile eggs of *Drosophila* will hatch and the larvae will pupate on a medium based on crystalline amino acids, and Rudkin and Schultz,[110] again in abstract form, report that the amino acid requirements of *Drosophila* are essentially similar to the mammal. Hinton *et al.*[1] carried out an extensive study of *Drosophila* in a chemically defined medium. They found a requirement for arginine, isoleucine, and tryptophan and found that glycine in high concentrations was beneficial. An important finding in the study by the Amherst workers was the serious results of amino acid imbalance and the inhibitory effects of even the required amino acids. This recalls a like finding in the earlier studies of *Tetrahymena*.[6, 50]

Hinton *et al.*[1] report that only the natural isomer of tryptophan is utilized, that D-serine is highly inhibitory and L-serine mildly so, and that citrulline (but not ornithine) shows a sparing effect on arginine. Citrulline also spares biotin. From the data given one can say that certain amino acids normally found in casein (alanine, aspartic acid, hydroxyproline, proline, serine, and tryosine) are not dietary requirements, as growth resulted when these were all omitted from the medium. One suspects also no dietary requirement for glutamic acid and cystine, in the sense that the organisms are incapable of their specific synthesis. It was found, however, that the addition of the non-essential amino acids (except serine) had a favorable effect on growth.

Trager[111, 112] has studied the growth and nutrition of the larvae of the yellow fever mosquito *Aedes aegypti* for many years. His media contained casein, however, so no information regarding the amino acid requirements was obtained. This problem has been taken up by Goldberg and de Meillon.[113] Using crystalline amino acids, they report growth failure for this organism upon the omission from the medium of any one of the nine "essential" amino acids, except phenylalanine. Tyrosine or phenylalanine, but not both, were necessary, especially for melanine formation. The ability to reduce tyrosine to phenylalanine is rare among animals, and this report needs confirmation before acceptance.

106 A. Delcourt and E. Guyénot, *Bull. sci. France Belg.* **45,** 249 (1911).
107 J. Loeb and J. H. Northrop, *J. Biol. Chem.* **23,** 431 (1915).
108 P. Baumberger, *J. Exptl. Zoöl.* **28,** 1 (1919).
109 J. Schultz, P. St. Lawrence, and D. Newmeyer, *Anat. Record* **96,** 540 (1946).
110 G. T. Rudkin and J. Schultz, *Anat. Record* **99,** 613 (1947).
111 W. Trager, *Am. J. Hyg.* **22,** 475 (1935).
112 W. Trager, *Biol. Bull.* **71,** 343 (1936).
113 L. Goldberg and B. de Meillon, *Biochem. J.* **43,** 379 (1948).

The growth factor requirements of the Diptera have received relatively more attention than have the nitrogen requirements. For most of the Diptera studied this information is still fragmentary and often uncertain. Positive findings of requirements can be qualitatively accepted as indicating a need for compounds, over and above that which may be contributed by other organisms or as contaminants of certain components of the diet. The common inclusion of commercial yeast, nucleic acid, lecithin, cholesterol, casein, and even agar with the hope that they represent the pure compounds corresponding to the label on the bottle is precarious for definitive information. Before one can say that any organism *does not* require a certain chemical compound in its diet, all components must be rigorously examined for traces of that chemical.

Of the fat-soluble factors only cholesterol appears to be dietarily important to Diptera. The fleshfly, *Lucilia sericata*,[114] will not grow in ether-extracted meat peptone plus yeast extract but will grow normally in this medium upon the addition of cholesterol. Van't Hoog[115] has shown a cholesterol requirement in *Drosophila*, and Goldberg and de Meillon[116] in *Aedes aegypti*.

The usual water-soluble factors have been found important in Diptera nutrition. *Drosophila* requires thiamine, riboflavin, pyridoxine, pantothen, niacin, biotin, and folic acid,[117, 118, 109] but does not require inositol, thioctic acid, carnitine, or PABA.[1] *Drosophila* is like *Tetrahymena*[68] in its inability to synthesize niacin from tryptophan.[119] Some indication of a possible function of biotin in amino acid metabolism comes from the observation that citrulline exerts a sparing action on the biotin requirement.[1] This was not found to be the case with lecithin, however, although the nature of the experiments does not altogether rule out the possibility. Although whole nucleic acid is usually added to the "synthetic" media used in *Drosophila* culture, its necessity or the necessity of its components has not been clearly demonstrated. Hinton *et al.*[120] have shown, however, that an inversion stock [Inversion (2LR)40d] failed to pupate unless adenine was supplied. No other purine or pyrimidine tried would substitute for adenine. Although it was not used, one would expect hypoxanthine to meet this requirement. This is a very interesting problem, and the report is obviously of a preliminary nature. Further studies on this and other purine- and pyrimidine-requiring stocks are in progress (Ellis, personal communication).

Larvae of *Aedes aegypti* (Table 1) require thiamine, riboflavin, pantothen,

[114] R. P. Hobson, *Biochem. J.* **29,** 2023 (1935).
[115] E. V. van't Hoog, *Z. Vitaminforsch.* **5,** 118 (1936).
[116] L. Goldberg and B. de Meillon, *Biochem. J.* **43,** 372 (1948).
[117] E. G. van't Hoog, *Z. Vitaminforsch.* **4,** 300 (1935).
[118] E. L. Tatum, *Proc. Natl. Acad. Sci. U. S.* **27,** 193 (1941).
[119] J. Schultz and G. T. Rudkin, *Federation Proc.* **7,** 185 (1948).
[120] T. Hinton, J. Ellis, and D. T. Noyes, *Proc. Natl. Acad. Sci. U. S.* **37,** 293 (1951).

pyridoxine, biotin,[121, 122, 123] and, for pupation, folic acid.[124] Trager[123] has shown that at least part of the biotin requirement can be replaced by oleic acid, an oil from plasma or lecithin. Pimelic acid was inactive as a substitute for biotin.

2. Coleoptera

The most extensive work on this group has been done by Fraenkel and his associates. Although Fraenkel does not work with axenic cultures, much valuable information has been accumulated.

The nitrogen nutrition of the beetles remains largely unknown. Casein has been the usual nitrogen source in the dry food (largely starch) of these forms. Recently, however, Lemonde and Bernard[125] have grown the flour beetle *Tribolium confusum* on a diet, the principal source of nitrogen being crystalline amino acids. On the basis of decreased pupation time or no pupation they state that the usual ten "essential" amino acids (arginine, histidine, lysine, tryptophan, phenylalanine, methionine, threonine, leucine, isoleucine, and valine) are dietary requirements for this species (Table 2). These conclusions have been confirmed (Printy and Fraenkel, personal communication). Moore[126] has reported identical requirements for the carpet beetle *Attangenus* sp.

Some beetles have been found to have a carbohydrate requirement. *Tenebrio* failed to grow unless carbohydrates were present in an otherwise complete medium.[127, 128] Fraenkel and co-workers have found carbohydrate requirements in the following beetles: *Tenebrio molitor*,[128] *Silvanus surinamensis*, *Stegobium* (*Sitodrepa*) *paniceum*,[129] and *Palorus ratzeburgi*.[130] No carbohydrate is necessary in the diet of *Tribolium confusum*, *Lasiderma serrecorne*, or *Ptinus tectus*.[129] These requirements or lack of them play a role in determining the distribution of these beetles in nature.

Studies on the fat-soluble substances required by the beetles have been most interesting. Fraenkel[131] and Fraenkel and Blewett[132] made an extensive investigation of the sterol requirement of several beetles. They found that all five beetles studied required a sterol, although some appeared to be less adversely affected on sterol low diets than others. It is of interest that,

[121] W. Trager and Y. SubbaRow, *Biol. Bull.* **75,** 75 (1938).
[122] Y. SubbaRow and W. Trager, *J. Gen. Physiol.* **23,** 561 (1940).
[123] W. Trager, *J. Biol. Chem.* **176,** 1211 (1948).
[124] L. Goldberg, B. de Meillon, and M. Lavoipierre, *J. Exptl. Biol.* **21,** 90 (1945).
[125] A. Lemonde and B. Bernard, *Can. J. Zoöl.* **29,** 80 (1951).
[126] W. Moore, *Ann. Entomol. Soc. Am.* **39,** 513 (1946).
[127] M. Lafon and G. Teissier, *Compt. rend. soc. biol.* **131,** 75 (1939).
[128] G. Fraenkel, M. Blewett, and M. Coles, *Physiol. Zoöl.* **23,** 92 (1950).
[129] G. Fraenkel and M. Blewett, *J. Exptl. Biol.* **20,** 28 (1943).
[130] M. I. Cooper and G. Fraenkel, *Physiol. Zoöl.* **25,** 20 (1952).
[131] G. Fraenkel, *J. Roy. Coll. Sci.* **13,** 59 (1943).
[132] G. Fraenkel and M. Blewett, *Biochem. J.* **37,** 692 (1943).

although all beetles will utilize cholesterol and 7-dehydrochotesterol, most failed to grow on calciferol and others grew but poorly. *Dermestes vulpinus*, the hide beetle, failed to grow in the presence of the plant sterol, sitosterol, or the microorganism sterol, ergosterol. *Tribolium*, *Lasioderma*, *Silvanus*, and *Ptinus* grew well on these sterols as well as the sterol of animal origin (cholesterol).

It may be concluded that probably all insects require dietary sterol and, unlike the vertebrate, cannot use vitamin D. Nothing is yet known of the function of sterols in insects.

Water-soluble vitamins are required by all beetles so far studied.[133-139] As might be expected thiamine, riboflavin, niacin, pyridoxine, and pantothen were first detected as dietary requirements, and it may be concluded that these five are as important for the Coleoptera as for the Diptera and Ciliata.

When it comes to the more active vitamins, folic acid and biotin, there appears to be some divergence, although Cooper and Fraenkel[130] state that this may be due to inclusion of the traces of these vitamins in certain constituents of the diet. It is probably safe also to assign the role of biotin and folic acid supplier to intestinal microorganisms, at least in part. Wherever the tests have been critical, all insects show requirement for all seven B vitamins.

The requirement for choline varies. Thus *Ptinus*, *Lasioderma*, *Tenebrio*, and *Palorus* respond to added choline in the diet, whereas *Tribolium*, *Silvanus* and *Stegobium* do not. In *Stegobium*, growth is good without choline even under aseptic conditions.[140]

Two beetles have been found which require vitamin B_T or carnitine[141, 141a] in the diet. These are *Palorus ratzeburgi* and *Tenebrio molitor*.[128, 130, 139, 142] This compound is widespread in natural materials, being highest in materials of animal origin (except the hen's egg and *Tetrahymena* extracts).[142]

One is tempted to speculate that carnitine might be active in *Tenebrio* and *Palorus* by virtue of labile CH_3 groups. This speculation becomes less

[133] G. Fraenkel and M. Blewett, *Nature* **149,** 301 (1942).
[134] G. Fraenkel and M. Blewett, *Nature* **150,** 177 (1942).
[135] G. Fraenkel and M. Blewett, *Nature* **151,** 703 (1943).
[136] G. Fraenkel and M. Blewett, *Biochem. J.* **37,** 686 (1943).
[137] G. Fraenkel and M. Blewett, *Biochem. J.* **41,** 469 (1947).
[138] P. Ellinger, G. Fraenkel, and M. M. Abdel Kader, *Biochem. J.* **41,** 556 (1947).
[139] G. Fraenkel, M. Blewett, and M. Coles, *Nature* **161,** 981 (1948).
[140] N. C. Pant and G. Fraenkel, *Science* **112,** 498 (1950).
[141] H. E. Carter, P. K. Bhattacharyya, K. R. Weidman, and G. Fraenkel, *Arch. Biochem. Biophys.* **35,** 241 (1952).
[141a] H. E. Carter, P. K. Bhattacharyya, K. R. Weidman, and G. Fraenkel, *Arch. Biochem. Biophys.* **38,** 405 (1952).
[142] G. Fraenkel, *Arch. Biochem. Biophys.* **34,** 457 (1951).

attractive, however, when it is remembered that choline is an absolute requirement for *Palorus* and a partial requirement for *Tenebrio*. For the present, therefore, it must be said that no function is known for this compound, either in these two insects for which it is a dietary requirement or in higher animals where it is invariably present.

3. Lepidoptera

The caterpillars of a few moths have been investigated under conditions which have yielded some valuable information. In some cases only isolated factors of their nutrition is known, whereas somewhat complete information is at hand for others.

In their extensive work on insect nutrition, Fraenkel and his associates have investigated three species of *Ephestia*, one species of *Plodia*, and the clothes moth *Tineola*. For *E. kuehniella* and *E. elutella* these workers[143] found that growth failed in the absence of the usual B vitamins (Table 4), and choline was found to be somewhat beneficial for *E. kuehniella* and necessary for *E. elutelta*. A sterol (cholesterol) is indispensable for both, and growth failed in the absence of wheat germ oil. The saponifiable fraction of the wheat germ oil could be replaced by linolenic or linoleic (but not oleic) acids[144, 145] and is necessary for emergence of the larva and for normal scale development of the adult. The unsaponifiable fraction of wheat germ oil is necessary for larval growth and could be replaced by α-tocopherol[144] or arachidonic acid.[145] It is believed that α-tocopherol acts as an antioxidant (ethyl and propyl gallates and ascorbic acid also produce growth effects). Both species require carbohydrate; *E. kuehniella* is able to utilize both starch and glucose, but *E. elutella* cannot utilize starch.

Both *Ephestia cautella* and the closely related *Plodia interpunctella* appear similar in their nutritional requirements. These two species have not been as thoroughly investigated as *E. kuehniella* and *E. elutella*, but there appears to be only one major nutritional difference between them. In *Plodia* emergence and wing-scale development are normal in the absence of dietary fatty acids, although growth is somewhat retarded. *E. cautella* requires fatty acid. It is impossible to say at this time whether this difference is significant or whether the difference reflects the character of the intestinal microorganisms of *Plodia*.

The clothes moth, *Tineola*, requires the usual B vitamins[146] and a sterol but is able to grow normally in the absence of dietary fat and carbohydrate (Table 4), whereas *Malocosoma neustria* requires carbohydrate.[147] *Aglaia*

[143] G. Fraenkel and M. Blewett, *J. Exptl. Biol.* **22,** 162 (1946).
[144] G. Fraenkel and M. Blewett, *J. Exptl. Biol.* **22,** 172 (1946).
[145] G. Fraenkel and M. Blewett, *Biochem. J.* **41,** 475 (1947).
[146] G. Fraenkel and M. Blewett, *J. Exptl. Biol.* **22,** 156 (1946).
[147] A. C. Evans, *Trans. Roy. Ent. Soc.* (*London*) **89,** 13 (1939).

TABLE 4

Nutritional Requirements of Certain Lepidoptera

	Vitamins										Lipids			Carbohydrate	
	Thiamine	Riboflavin	Niacin	Pyridoxine	Pantothen	Biotin	PGA	Choline	*p*-Amino-benzoic acid	Inositol	Cholesterol	α-Tocoph-erol	Fatty acid	Starch	Glucose
Ephestia kuehniella	+	+	+	+	+	+	+	+(?)	−	−(?)	+	+	+	+*	+*
Ephestia elutella	+	+	+	+	+	+	+	+	−	−	+	+	+	−†	+
Ephestia cautella													+		
Plodia interpunctella						+	+				+		−		+
Tineola bisselliella	+	+	+	+	+	+(?)	+		−	−	+		−		−
Galleria mellonella	+		+										+		
Corcyra cephalonica	+	+		+											
Malacosoma neustria													−	+	
Aglaia urticae													+	−	
Pieris brassicae													+	−	

* Can utilize either starch or glucose.

† Must have glucose and cannot utilize starch.

urticae and *Pieris brassicae*, on the other hand, require fat in the diet but do not require a dietary carbohydrate.[147]

The wax moth, *Galleria mellonella*, in addition to requiring fat in the diet, requires at least thiamine and nicotinic acid,[148-150] and the rice moth, *Gorcyra cephalonica*, requires thiamine, riboflavin, and pyridoxine[151, 152] in addition to unidentified factors.

V. Conclusions

One of the first things which strikes one in reviewing what is known about the biochemistry of nutrition in the invertebrate animals is the meager nature of the data. Of the tens of thousands of species only a few have been investigated in a manner which yields precise information. These few belong to but three phyla, the Protozoa, the Nemathelminthes, and the Arthropoda.

In spite of the small number of types studied, it is impressive that their nutritional patterns should so closely correspond with each other and to those of birds and mammals. One sees a striking similarity between the amino acids needed by all forms. Thus we can say that the "animal's essential amino acid pattern" consists of arginine, isoleucine, leucine, lysine, phenylalanine, methionine, tryptophan, threonine, and valine, and to all but that of man must be added histidine. Additional enzymatic deficiencies in some types have occurred so that serine, glycine, and proline must be supplied exogenously.

The basic B vitamins are regularly required by all animals. Exceptions are rare; for instance, niacin is not required by mammals (in the presence of adequate tryptophan) nor biotin by *Tetrahymena*. The need for thiamine, riboflavin, pyridoxine (or derivatives), pantothen, and folic acid appears to be universal throughout the animal kingdom. Rather rare and specialized requirements are encountered, such as cholesterol for certain trichomonad flagellates and insects; hemin for trypanosomids and at least one blood-sucking insect; carnitine for two species of beetles; thioctic acid for ciliates; and purines and pyrimidines for ciliates.

It is of course quite possible that the above comparisons will have to be modified or abandoned when we have examined the nutritional patterns of more animal types. The current conclusions will have to suffice until more data are gathered. They do have the obvious advantage of lending weight to data gathered from "lower" animals for attention and use of those investigators interested in "higher" animals, including man.

[148] D. Rubenstein and L. Shekun, *Nature* **143**, 1064 (1939).

[149] D. Rubenstein and L. Shekun, *Bull. biol. méd. exptl. U.R.S.S.* **9**, 190 (1940).

[150] M. H. Haydak, *Proc. Minn. Acad. Sci.* **9**, 27 (1941).

[151] P. S. Sarma, B. G. Swami, and M. Sreenivasaya, *Current Sci.* (*India*) **11**, 332 (1942).

[152] P. S. Sarma, *Indian J. Med. Research* **31**, 165 (1943).

CHAPTER 16

Energetics and Metabolic Function

EUGENE P. KENNEDY

Ben May Laboratory for Cancer Research and Department of Biochemistry, University of Chicago

I. Introduction

1. FREE ENERGY AND CELL DYNAMICS

The concept of the dynamic state of constituents of living cells has brought about a revolution in biochemical thought which is in its way comparable to the overthrow of the "classical" physics of the nineteenth century. The approach of present-day investigators to the problems of energy supply and utilization by living organisms is very different from that of even a few decades ago, when the animal body was apt to be considered as merely a kind of heat engine into which certain amounts of foodstuffs were ingested, burned with the release of very precisely ascertainable amounts of heat, and the waste products excreted. It is true that the insight of gifted workers such as F. G. Hopkins made them aware of the limitations of such a mechanistic point of view long before full experimental evidence had been obtained which was to make such a narrow interpretation of biochemical and metabolic phenomena quite untenable. The full realization, however, by the rank-and-file of biological workers that the living body is in a state of continual flux, in which the foodstuffs ingested are coming continually into equilibrium with the tissues themselves, which in turn are continually breaking down and being renewed, must be considered a product of the "era of isotopes."

The explicit and brilliant formulation by Schoenheimer[1] of this dynamic interpretation of life processes has now been thoroughly documented by an overwhelming body of data resulting from isotope tracer studies bearing on nearly every important aspect of intermediary metabolism.

It is a fact which is perfectly obvious to workers in all disciplines of biology that the characteristic activities of living organisms, such as growth and movement, require the continuous expenditure of energy. The continual breakdown and resynthesis of body constituents requires that even adult animals, in which no net growth is taking place, must expend large amounts of free energy for synthetic purposes, simply to achieve a stable, dynamic condition. The synthesis of cell constituents involves not only a diminution of entropy implicit in building up a highly ordered architectural structure from relatively simple building blocks, but it will be seen that the actual chemical steps are in many cases highly endergonic. Metabolic energy must be furnished by coupled reactions, which make available the free energy of catabolic processes, in order to drive these synthetic steps to completion. Although such requirements have of course been understood in general terms for many years, it has been only recently that detailed experimental evidence has been accumulated concerning specific enzymatic reactions.

The present treatment of energy transformations by the organism will be oriented toward enzymatic mechanisms by which body metabolites are catabolized, with the release of free energy in a form which is available for the performance of work or for biosyntheses. Emphasis will be placed on topics which are at present in a rapid state of development, such as oxidative phosphorylation and the relationships between synthetic processes and high-energy bonds, but such subjects as anaerobic glycolysis will not be considered at length, since this pattern of reactions has now been worked

[1] R. Schoenheimer, The Dynamic State of Body Constituents, Harvard University Press, Cambridge, 1942.

out in minute detail. Space will not permit the development of subjects such as the utilization of energy in the performance of muscular work, which would indeed require volumes in itself.

Although most of the experimental data to be considered in the following discussion have been obtained in studies of mammalian tissues, the growing realization that certain master schemes of metabolic pathways are fundamental to the pattern of living of very diverse organisms makes it likely that the conclusions drawn from these data may have quite a general significance.

2. Thermodynamic Considerations

Before proceeding to a discussion of energy requirements in metabolic functions, it may be profitable to review very briefly certain fundamental thermodynamic equations which are involved. This treatment of the fundamental thermodynamic basis of energy transfer is necessarily perfunctory, and the reader is referred to the standard text of Lewis and Randall[2] for a more rigorous presentation.

The total energy content of any system may be considered as consisting of two components, according to equation 1.

$$H = F + TS \tag{1}$$

Here F represents the free energy of the system, T the absolute temperature, and S the entropy function. The free energy component, F, may be described as that portion of the total energy which is available for doing useful isothermal work. *The free energy change is the driving force of chemical reactions.* Every spontaneous process which is observed in nature is accompanied by a net release of free energy. Reactions which involve the release of free energy may be described as exergonic, and those which involve the input of free energy may be called endergonic.

In any reversible reaction such as equation 2,

$$A + B \rightleftharpoons C + D \tag{2}$$

the relative concentrations of reactants and products at equilibrium is given by the equation:

$$K = \frac{(C)(D)}{(A)(B)} \tag{3}$$

In equation 3 K represents the familiar equilibrium constant, and the values in parentheses represent the *activities* of the reaction components. For work of less precision, concentrations may be substituted for the activities.

[2] G. N. Lewis and M. Randall, Thermodynamics and the Free Energy of Chemical Substances, McGraw-Hill Book Co., New York, 1923.

The standard free energy change may be defined as the free energy change which occurs when all reaction components are in their standard states, usually taken to be unit value, e.g., one atmosphere pressure for gases and one molal for solutes. The equilibrium constant is related to the standard free energy change by equation 4,

$$-\Delta F^{\circ} = RT \ln K \tag{4}$$

where R is the gas constant. From this equation, it can be seen that, when ΔF° has a large negative value, K will have a large positive value, so that at equilibrium $A + B$ will be very largely converted into $C + D$.

The concept of *coupled* reactions is of great importance for biological reaction mechanisms. If the transformation of one substance to another by a reaction which is endergonic can proceed by a pathway which involves a highly exergonic reaction, then the net free energy change is simply additive and the reaction may proceed essentially to completion.

$$A \rightarrow C \qquad \Delta F = +3{,}000 \text{ cal.} \tag{5}$$

$$X \rightarrow Y \qquad \Delta F = -10{,}000 \text{ cal.} \tag{6}$$

$$\text{Net: } A + X \rightarrow C + Y \qquad \Delta F = -7{,}000 \text{ cal.} \tag{7}$$

In order for the coupled reaction summarized by equation 7 actually to take place, equations 5 and 6 must be linked by a common reactant.[3] Such a series of equations may be represented by the following:

$$A + X \rightarrow B + Y \qquad \Delta F = -4{,}000 \text{ cal.} \tag{8}$$

$$B \rightarrow C \qquad \Delta F = -3{,}000 \text{ cal.} \tag{9}$$

$$A + X \rightarrow C + Y \qquad \Delta F = -7{,}000 \text{ cal.} \tag{10}$$

An extremely important general source for energy capture by living organisms consists in the oxidation of substrates. The free energy change of such a process is given by equation 11,

$$-\Delta F = N\,\mathrm{F}\,E \tag{11}$$

where F is the faraday, N is the number of electrons involved in the reaction, and E is the potential difference between oxidant and reductant. In many calculations of the free energy changes in biological oxidations, the potential difference is taken to be identical with the difference between the normal potentials (E_0) of oxidant and reductant. The actual potential of a half-cell is related to the normal potential by the familiar equation:

$$E = E_0 + \frac{RT}{N\mathrm{F}} \ln \frac{\text{(oxidized form)}}{\text{(reduced form)}} \tag{12}$$

The determination of the ratio of oxidized to reduced forms of a given redox system in the living cell, or in a complicated enzyme system, is often impossible for technical reasons. It is not likely that the use of the normal potentials, as a first approximation, introduces errors of such magnitude as to invalidate these calculations, which often include other approximations and assumptions which make a high degree of accuracy unattainable.

II. Concept of "Energy-Rich" Phosphate Esters

1. Phosphate Esters and Muscular Contraction

The discovery by Lundsgaard[4] that iodoacetate-poisoned muscle was capable of performing work under conditions in which the breakdown of glucose to lactic acid was completely inhibited furnished one of the fundamental facts which has led to present concepts of energy transfer and utilization. These experiments showed that the performance of muscular work could be correlated with the disappearance of labile phosphate esters. Further investigation brought out the important role of creatine phosphate in providing a reservoir of energy for working muscle. The conversion of creatine phosphate to adenosine triphosphate (ATP)[5] was discovered by Lohmann.[6] The work of Engelhardt[7] indicated that it was the breakdown of ATP itself which was the immediate source of the free energy required for muscular work, giving further evidence of the enormously important biological functions of this nucleotide.

Concomitantly with these discoveries, a closer understanding of the pattern of reactions of anaerobic glycolysis elucidated the exact enzymatic steps which are interlocked with the formation of ATP. Taken together, this nexus of discoveries concerning the generation of ATP in the fermentation of glucose and its utilization in the performance of muscular work must be regarded as one of the fundamental achievements of biochemistry to date.

2. Classification of Phosphate Esters

In a paper which has already begun to assume the status of a classic, Lipmann[8] in 1941 did much to make clear the chemical relationships under-

[3] M. J. Johnson, *in* Lardy, Respiratory Enzymes, Burgess Publishing Co., Minneapolis, 1949, p. 255.

[4] E. Lundsgaard, *Biochem. Z.* **217,** 162 (1930).

[5] The following abbreviations will be used in this article. ATP = adenosine triphosphate; ADP = adenosine diphosphate; DPN_{ox} = oxidized diphosphopyridine nucleotide; DPN_{red} = reduced diphosphopyridine nucleotide; CoA or $\bar{CoA}$—SH = coenzyme A; $\bar{CoA}$—S—$COCH_3$ = acetyl-coenzyme A; FAD = flavin adenine dinucleotide; R—P = low-energy phosphate bond; R$\sim$P = high-energy bond; P_i = inorganic orthophosphate.

[6] K. Lohmann, *Biochem. Z.* **271,** 264 (1934).

[7] W. A. Engelhardt, *Advances in Enzymol.* **6,** 147 (1946).

[8] F. Lipmann, *Advances in Enzymol.* **1,** 99 (1941).

lying the biological functions of compounds such as creatine phosphate and ATP. Lipmann noted that phosphate esters of biological interest could be divided into two classes on the basis of the free energy released on hydrolytic cleavage of the phosphate bond. One group of these substances, which includes acyl phosphates, guanidino phosphates, enol phosphates, and pyrophosphates, is characterized by the release of a large amount of free energy—between 10,000 and 15,000 cal. per mole—upon hydrolysis. In Lipmann's terminology, these compounds are called "energy-rich" esters. Another group of phosphate esters, which includes simple esters of alcohols and sugars, yields much smaller amounts of free energy, in the region of 3000

TABLE 1

THE ENERGY CONTENT OF PHOSPHATE ESTERS

- I. High-Energy Phosphate Esters
  - 1. Carboxyl phosphates
    - Acetyl phosphate
    - 1,3-Diphosphoglyceric acid
  - 2. Enol phosphates
    - Enol phosphopyruvic acid
  - 3. Guanadino phosphates
    - Creatine phosphate
    - Arginine phosphate
  - 4. Compounds containing pyrophosphate linkages
    - ADP, ATP
    - FAD, DPN, TPN
    - Thiamine pyrophosphate
    - Inorganic pyrophosphate
- II. Low-Energy Phosphate Esters
  - 1. Phosphate esters of alcohols
    - Glycerophosphate
    - Hexose phosphates
    - Phosphorylcholine

cal. per mole. These are "energy-poor" phosphate esters. In Table 1 are shown a number of phosphate esters of importance and their classification according to this scheme.

The striking difference between the free energy of hydrolysis of these two groups of compounds has naturally been the object of much speculation from the purely physicochemical point of view. Attempts have been made by Kalckar,[9] Oesper,[10] and Hill and Morales[11] to account for the magnitude of the free energy change noted upon the hydrolysis of energy-rich phosphate esters.

[9] H. M. Kalckar, *Chem. Revs.* **28,** 71 (1941).

[10] P. Oesper, *Arch. Biochem.* **27,** 255 (1950).

[11] T. L. Hill and M. F. Morales, *J. Am. Chem. Soc.* **73,** 1656 (1951).

Kalckar has pointed out that many of these high-energy phosphate esters may be expected to exhibit the phenomenon of opposing resonance. According to the theory of resonance, when a number of possible structures which differ only in the position of electrons, may be written for a compound, then the actual state of the compound is not represented by any single one of these possibilities but rather is a "hybrid" to which all contribute in some degree. The larger the number of structures having approximately the same energy content which contribute to the final state of the substance, the greater is the stabilization brought about resonance. Acetyl phosphate may be taken as an example of a high-energy ester, the component parts of which (acetate + inorganic phosphate) are considerably stabilized by resonance. However, in acetyl phosphate itself, the oxygen bridge cannot participate fully in all the resonance forms of both the phosphate and the acetate moieties of the molecule; the phosphate and the acetate may be said to compete for the oxygen atom, and a number of resonating structures are thus eliminated, with corresponding loss in stability with respect to the breakdown products, acetate and free phosphate. This must mean that a considerable release of free energy occurs on hydrolysis.

Although this explanation of the high-energy content of many of these compounds appears to have quite a general application, Oesper[10] has emphasized the importance of other factors which must be considered. When acetyl phosphate, to continue the example, is hydrolyzed at neutral pH, there is a considerable amount of free energy derived from the neutralization of the carboxylic acid liberated. This amounts to about 3000 cal. per mole. Again in the case of the enol phosphopyruvate, another factor to be considered is the transformation of the free pyruvate, after hydrolysis, to the keto form which is more stable, and this transition may account for a large amount of the total free energy change noted.

III. Central Role of ATP in Energy Transfer

1. Interconversion Reactions of ATP

Adenine polyphosphates appear to be universally distributed in living organisms. The importance of ATP in muscular function has been mentioned above. There is now at hand abundant evidence which points to the idea that ATP plays a predominant, if not indeed a unique, part in the process by which energy is generated, conserved and transferred in forms available for vital processes. One such line of evidence consists in the known interconvertibility of other high-energy compounds with ATP. These interconversions may be represented by the following over-all equations, which summarize reactions occurring in a very wide range of living tissues.

$$\text{ADP} + \text{creatine phosphate} \rightleftharpoons \text{ATP} + \text{creatine} \tag{13}$$

$$\text{ADP} + \text{1,3-diphosphoglycerate} \rightleftharpoons \text{ATP} + \text{3-phosphoglycerate} \tag{14}$$

$$\text{ADP} + \text{enol phosphopyruvate} \rightleftharpoons \text{ATP} + \text{pyruvate} \quad (15)$$

$$\text{ADP} + \text{acetyl phosphate} \rightleftharpoons \text{ATP} + \text{acetate} \quad (16)$$

The existence of these reactions lends credence to the idea that ATP is the currency, so to speak, of the living cell and represents the common denominator of energy transfer reactions. This concept fits in with the finding that ATP is involved in a large number of synthetic reactions, which are of themselves endergonic, but are found to be coupled to the breakdown of ATP in such a fashion that the large amount of free energy released by the cleavage of the ATP pyrophosphate bond forces the synthetic reaction to virtual completion. A number of such reactions will be dealt with later in this chapter.

It is a corollary of this principle of the central position of ATP in the energy metabolism of cells in general that the very diverse metabolic patterns by which living cells obtain their energy must have one feature in common: interconvertibility of a large part of this energy with the high-energy phosphate linkages of ATP. This statement has been found to apply not only to mammals but to yeast cells and a wide variety of bacteria as well.

Although undoubtedly an oversimplification, this interpretation of energy metabolism in terms of the generation of high-energy bonds lends a certain unity and continuity to the study of living cells which might seem superficially completely dissimilar. The microorganism *Thiobacillus thioxidans*, for example, carries out as its principal energy-gaining reaction the oxidation of elementary sulfur to sulfuric acid. During the course of this reaction, the pH falls to very low values, and the organism may live and grow in sulfuric acid solutions of pH well below 1.0. It would be perhaps difficult to find a type of metabolism apparently more different from that carried out by mammalian tissues. Yet investigations of the metabolism of *Thiobacillus* carried out by Vogler and Umbreit[12] led to the discovery that the oxidation of sulfur by this organism was also linked to the generation of ATP, which presumably could be utilized in much the same fashion as in mammalian tissues, for function and for biosynthetic purposes. Other examples of the very wide range of uses to which ATP may be put for biological function are shown in studies by McElroy indicating its participation in photoluminescence, and by Nachmansohn on the production of current by the electric organs of the eel.

So much evidence has been accumulated on the part played by ATP in mechanisms requiring the input of free energy, and so much awareness has arisen on the part of enzymologists of the importance of this substance, that it has become one of the common reagents routinely added to enzyme

[12] K. G. Vogler and W. W. Umbreit, *J. Gen. Physiol.* **26**, 157 (1942).

reaction mixtures. Although this emphasis is undoubtedly justified in view of the facts just examined, the danger now appears that ATP may come to be regarded as the *only* form in which energy is made available for function and synthetic reactions. Overemphasis on adenine nucleotide and phosphate bond mechanisms may lead to the neglect of other reaction pathways which may be of importance in living cells. Very recent work has shown that at least one other major coenzyme may participate in energy transfer and synthetic reactions, without requiring the direct activation of ATP. This coenzyme is Lipmann's coenzyme A. The functions of this substance in metabolic energy transfer will be considered in a later section of this chapter.

Even with these reservations, however, an adequate understanding of how the breakdown of foodstuffs is related to the generation of ATP is essential to an appreciation of how metabolic energy is harnessed in a form useful to the living body. The bulk of catabolic, energy-yielding reactions to be considered here can be arbitrarily divided into two general areas. The first of these comprises the set of fermentative reactions by which carbohydrate is broken down to lactic acid, and the second may be considered to be the cyclic series of reactions by which not only carbohydrate but also fat and protein undergo terminal respiration to carbon dioxide and water. It is now known that both of these reaction sequences lead to the generation of ATP.

IV. Generation of ATP in Anaerobic Glycolysis

1. Oxidation of 3-Phosphoglyceraldehyde

The sequence of reactions by which glucose is broken down to lactic acid in the animal body has been thoroughly worked out, and most of the enzymes involved have been isolated in crystalline form and subjected to study in detail. Several excellent reviews[13, 14, 15] have presented a complete picture of the glycolytic pathway, and details need not be repeated here. Instead, our attention will be directed to two stages only in this process, the two enzymatic steps which involve the production of ATP.

The first reaction is the oxidation of 3-phosphoglyceraldehyde. This reaction has been studied by Warburg and his school,[16, 17, 18] who discovered

[13] F. Dickens *in* Sumner and Myrbäck, The Enzymes, Academic Press, New York, 1951, Vol. II, Part 1, p. 624.

[14] H. A. Lardy *in* Lardy, Respiratory Enzymes, Burgess Publishing Co., Minneapolis, 1949, p. 179.

[15] O. Meyerhof, *Ann. N. Y. Acad. Sci.* **45,** 377 (1944).

[16] O. Warburg and W. Christian, *Biochem. Z.* **301,** 221 (1939).

[17] O. Warburg and W. Christian, *Biochem. Z.* **303,** 40 (1939).

[18] E. Negelein and H. Bromel, *Biochem. Z.*, **303,** 132 (1939).

that inorganic phosphate and diphosphopyridine nucleotide (DPN) were required for the reaction summarized in equation 17.

$$\text{3-Phosphoglyceraldehyde} + DPN_{ox} + P_i \rightleftharpoons \text{1,3-diphosphoglycerate} + DPN_{red} \quad (17)$$

The 1,3-diphosphoglyceric acid then transfers its high-energy phosphate to ATP by the next reaction in the sequence:

$$\text{1,3-Diphosphoglyceric acid} + ADP \rightleftharpoons ATP + \text{3-phosphoglyceric acid} \quad (18)$$

In the presence of arsenate, ATP is no longer formed, and the reactions given are diverted so that 3-phosphoglyceric acid is the only product. Warburg and Christian have postulated that this effect of arsenate may be due to replacement by arsenate of inorganic phosphate in equation 17 with the consequent formation of the acyl arsenate of 3-phosphoglyceric acid, which undergoes spontaneous breakdown, yielding 3-phosphoglyceric acid and arsenate.

The actual reaction mechanism by which inorganic phosphate enters into reaction 17 is of great interest, since it may be considered a model for the formation of a high-energy phosphate bond from inorganic phosphate, coupled to an oxidative reaction. It was at first believed by Warburg that 1,3-diphosphoglyceraldehyde was the intermediate:

$$\text{3-Phosphoglyceraldehyde} + P_i \rightleftharpoons \text{1,3-Diphosphoglyceraldehyde} + H_2O \quad (19)$$

$$\text{1,3-Diphosphoglyceraldehyde} + DPN_{ox} \rightleftharpoons \text{1,3-Diphosphoglycerate} + DPN_{red} \quad (20)$$

Meyerhof and his collaborators[19] have made extensive efforts to determine whether 1,3-diphosphoglyceraldehyde is actually formed during this reaction, and they were able to obtain no evidence to support this hypothesis. Nevertheless, the intermediate formation of 1,3-diphosphoglyceraldehyde has been assumed by many authors.

In view of the negative results of Meyerhof's experiments, and with other enzyme mechanisms which have since been studied as models, it may be possible to formulate another tentative reaction scheme.

$$\text{3-Phosphoglyceraldehyde} + \text{enz.} \rightleftharpoons \text{3-Phosphoglyceraldehyde-enz.} \quad (21)$$

$$\text{3-Phosphoglyceraldehyde-enz.} + DPN_{ox} \rightleftharpoons \text{3-Phosphoglyceric-enz.} + DPN_{red} \quad (22)$$

$$\text{3-Phosphoglyceric-enz.} + P_i \rightleftharpoons \text{1,3-Diphosphoglyceric} + \text{enz.} \quad (23)$$

Although it must be emphasized that definite evidence to support such a mechanism is lacking, interactions of the type suggested are now known to occur between enzyme and substrate in the case of the sucrose phosphorylase (transglucosidase) described by Doudoroff, Hassid, and Barker,[20] and

[19] O. Meyerhof and P. Oesper, *J. Biol. Chem.* **170,** 1 (1947).

[20] M. Doudoroff, H. A. Barker, and W. Z. Hassid, *J. Biol. Chem.* **168,** 725 (1947).

in the action of the enzyme phosphotransacetylase discovered by Stadtman and Barker.[21] All three enzymes may be considered as acting by phosphorylysis of a relatively stable chemical bond between substrate and enzyme (or coenzyme) with the production of the phosphorylated form of the substrate (glucose-1-phosphate, acetyl phosphate, or 1,3-diphosphoglycerate). The sensitivity of all three enzymes to arsenate would also fit in with the analogy. The 3-phosphoglyceraldehyde dehydrogenase is known to be sensitive to sulfhydryl reagents, which suggests that the bond between substrate and enzyme may involve sulfhydryl linkages.

2. Enol Phosphopyruvate

The second reaction stage in anaerobic glycolysis which brings about the phosphorylation of ADP to form ATP is the transformation of enol phosphopyruvate to pyruvate, which occurs according to reaction 24:

$$\text{Enol phosphopyruvate} + \text{ADP} \underset{\text{Mg}^{++}}{\overset{\text{K}^+}{\rightleftharpoons}} \text{pyruvate} + \text{ATP} \qquad (24)$$

Reaction 24 was regarded for some time as being irreversible, but in 1945 Lardy and Ziegler[22] showed that it was reversible and emphasized the requirement of both magnesium and potassium ions for full activity.

3. Efficiency

The efficiency of anaerobic glycolysis, viewed simply as a technical operation employed by the cell to generate high-energy phosphate bonds, may readily be determined. Each mole of glucose which is split to lactate brings about the phosphorylation of 4 moles of ADP to form 4 moles of ATP, since the phosphorylation process takes place at two stages of the reaction pattern, and 2 moles of triose are formed from each mole of glucose. However, 2 moles of ATP are fed back into the operation in order to phosphorylate the hexose prior to splitting. There is thus a net gain of 2 moles of ATP per mole of glucose utilized, which represents a gain of about $2 \times 12{,}000 = 24{,}000$ cal. of free energy. The over-all free energy change for the breakdown of glucose to 2 moles of lactate is about 56,000 cal. The efficiency is about 40%.

As technical operations go, this is rather a high efficiency. But it will be seen that this efficiency is extremely low per mole of glucose utilized, when this yield is compared to the amount of free energy released by the complete oxidation of glucose to carbon dioxide and water, which amounts to almost 700,000 cal.

[21] E. R. Stadtman and H. A. Barker, *J. Biol. Chem.* **184,** 769 (1950).

[22] H. A. Lardy and J. Ziegler, *J. Biol. Chem.* **159,** 343 (1945).

V. Generation of ATP in Terminal Respiration

1. Pathways of Terminal Respiration

The importance of aerobiosis and the necessity of this way of life for higher animals is a direct consequence of the much greater efficiency with which aerobic organisms can utilize available metabolites. This truth was explicitly demonstrated by the experiments of Pasteur,[23] who grew yeast cells in vessels which afforded varying degrees of surface exposure to air per unit volume. In this fashion, varying degrees of aerobiosis were permitted, and Pasteur learned that the amount of cell substance synthesized per unit amount of sugar consumed was much higher under aerobic conditions than under anaerobic conditions. Since synthesis of cell substance requires energy, it was thus shown that the breakdown of sugar was very much more efficient from the point of view of energy capture in aerobiosis than in anerobiosis. It was upon this simple and almost austere experimental framework that the "Pasteur effect" was based, which has since given rise to so many almost fantastically complex theories.[24]

The understanding of how substrates such as simple sugars are burned in animal tissues, under aerobic conditions, to carbon dioxide and water has lagged far behind the study of the anaerobic glycolytic transformations, because of technical difficulties in isolating and studying the enzymes concerned. In the light of present information, these difficulties are seen to stem from the high degree of structural organization of these enzymes, which are bound to cytoplasmic granules known as mitochondria.

On the basis of experimental evidence representing the contributions of a number of workers, Krebs[25] suggested a pathway for the degradation of pyruvate in pigeon breast muscle, which involved the catalytic intermediation of dicarboxylic and tricarboxylic acids in cyclic fashion. This formulation, which has come to be called the Krebs citric acid cycle, has been the subject of intense investigation and exploitation during the last decade. Although at first subject to considerable criticism, a large body of data based principally upon isotope tracer studies has made it certain that the essential reactions of the Krebs cycle are involved in the aerobic degradation not only of pyruvate but also of amino acids and fats in mammalian tissues. Figure 1 illustrates an interpretation of interrelationships of fat, carbohydrate, and protein oxidative metabolism, based on the Krebs citric acid cycle.

Most of the reactions shown in Fig. 1 require no further comment, but attention should be directed to two features of the scheme as shown, which

[23] L. Pasteur, Études sur la Bière, translated by F. Faulkner and D. Robb, Macmillan & Co., London, 1879.

[24] D. Burk, *Cold Spring Harbor Symposia Quant. Biol.* **7**, 420 (1939).

[25] H. A. Krebs, *Advances in Enzymol.* **3**, 191 (1943).

are based on rather recent and important developments. First, it will be seen that the tricarboxylic acid first formed from oxalacetate and "activated acetate" is citric acid. Evans and Slotin[26] studied cell-free enzyme suspensions capable of fixing $C^{11}O_2$ into oxalacetic acid. When α-ketoglutaric acid presumably arising from radioactive oxalacetic acid was isolated from this system, all the radioactivity was found in the carboxyl group adjacent to the carbonyl group. Such an unsymmetrically labeled α-ketoglutarate could presumably not have passed through a "symmetrical" citric

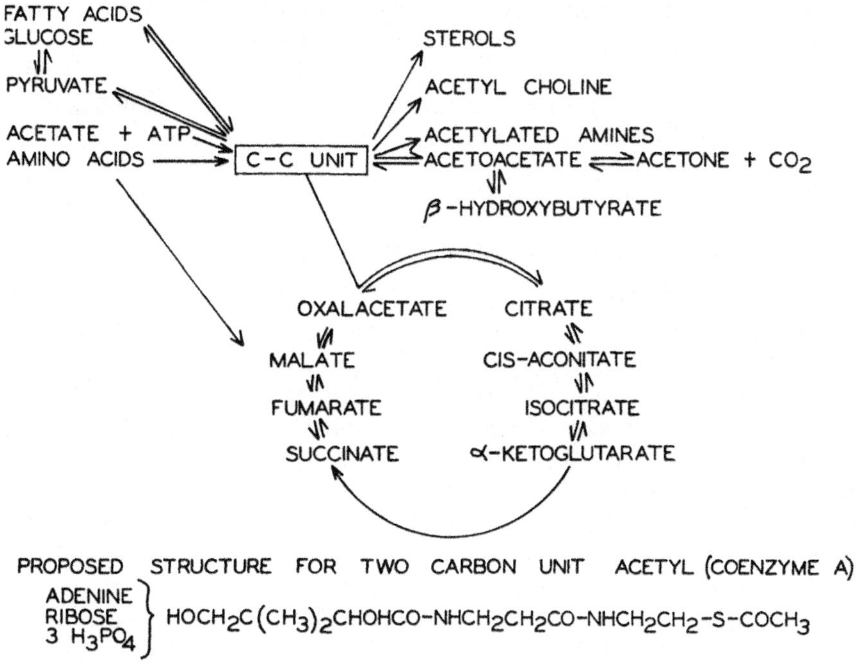

FIG. 1

acid stage. However, Ogston[27] suggested the possibility that a compound like citric acid, when attached to an enzyme surface by a three-point attachment mechanism, could preserve the discrete identity of its carboxyl groups. This hypothesis was tested by Potter and Heidelberger[28] who furnished convincing evidence that this was indeed the case.

The second point of interest is the identity of the "activated acetate" or "2-carbon unit" which has long been the object of speculation and in-

[26] E. A. Evans, Jr. and L. Slotin, *J. Biol. Chem.* **141**, 439 (1941).

[27] A. G. Ogston, *Nature* **162**, 963 (1948).

[28] V. R. Potter and C. Heidelberger, *Nature* **164**, 180 (1949).

quiry. This compound has now been shown to be the acetylated form of Lipmann's coenzyme A, the pantothenic acid-containing coenzyme of acetylation of aromatic amines. Work in the laboratories of Lipmann,[29, 30] Ochoa,[31, 32, 33] and Lynen[34] has led to the discovery of the function of this compound in citrate synthesis. Its implication in many other enzyme systems has already focused attention on this coenzyme, which occupies a central position in so many converging metabolic pathways.

In the Krebs cycle, the oxidation of pyruvate proceeds in a stepwise fashion, with the oxidation of the various tricarboxylic and dicarboxylic acids which are intermediates. The electrons which are removed from the substrate molecules at these oxidative steps are eventually transferred to molecular oxygen. This process of electron transfer is very poorly understood, but it is known to require the participation of a number of oxidation-reduction enzymes in series, with the cytochrome system acting as the ultimate link to molecular oxygen. A model reaction scheme, based on that of Lipmann[35] is described in Fig. 2.

2. Oxidative Phosphorylation in the Krebs Cycle

Since the key reactions of anaerobic glycolysis were found to lead to the formation of ATP, the possibility was immediately suggested that aerobic metabolism is also somehow geared to the production of ATP or other high-energy phosphate bonds. Since by far the largest portion of the free energy derived from the utilization of a substrate such as glucose is released during the aerobic reactions by which lactate (via pyruvate) is oxidized to carbon dioxide and water, this question is of the greatest importance. The early investigations of Belitzer,[36] Kalckar,[37] Ochoa,[38] and others, working for the most part with crude insoluble preparations of heart, kidney, or liver, showed conclusively that inorganic phosphate could be incorporated into ester linkages in phosphorylations coupled to oxidations.

The crude enzyme preparations capable of carrying out this type of

[29] G. D. Novelli and F. Lipmann, *Arch. Biochem.* **14,** 23 (1947).

[30] G. D. Novelli and F. Lipmann, *J. Biol. Chem.* **182,** 213 (1950).

[31] S. Korkes, J. R. Stern, I. C. Gunsalus, and S. Ochoa, *Nature* **166,** 403 (1950).

[32] J. R. Stern, B. Shapiro, E. R. Stoadtman, and S. Ochoa, *J. Biol. Chem.* **193,** 703 (1951).

[33] S. Korkes, A. del Campillo, I. C. Gounsalus, and S. Ochoa, *J. Biol. Chem.* **193,** 721 (1951).

[34] F. Lynen, *Ann.* **574,** 1 (1951).

[35] F. Lipmann, *in* Green, Currents in Biochemical Research, Interscience Publishers, New York, 1946, p. 137.

[36] V. A. Belitzer and K. S. Golovskaga, *Science* **92,** 536 (1940).

[37] H. Kalckar, *Enzymologia* **2,** 47 (1937).

[38] S. Ochoa, *J. Biol. Chem.* **151,** 493 (1943).

oxidation-linked phosphorylation were found to be extremely labile. Treatment with organic solvents even at low temperature, freezing and thawing, drying, or simple storage for a few days in the icebox, brought about complete inactivation of the phosphorylating enzymes, even though oxidative reactions with such substrates as succinate proved to be much more stable. Washed, particulate residues of tissue homogenates, such as those described by Lehninger[39] and Potter[40] have been widely used in the study of oxidative phosphorylation. Such enzyme preparations must be supple-

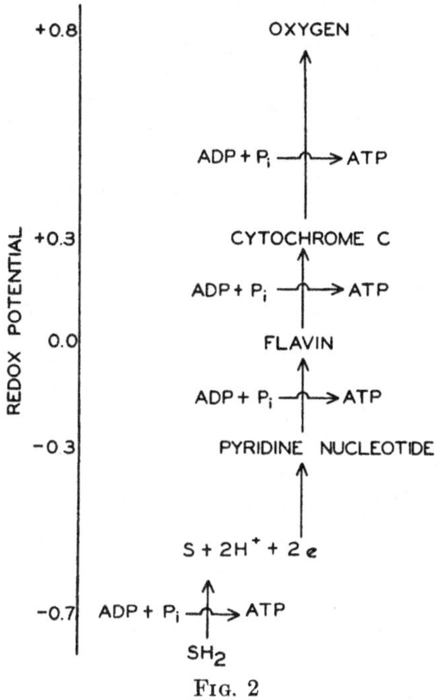

FIG. 2

mented with inorganic phosphate, magnesium ions, adenine nucleotide, and added cytochrome c for optimum activity.

3. Efficiency of Oxidative Phosphorylation

One line of attack on the problem of aerobic phosphorylation relates to the efficiency with which the energy released during the oxidative reactions can bring about the formation of high-energy phosphate linkages. It is of course only the maximum attainable values which are of any real theoretical interest. The direct measurement of ATP formed in the enzyme

[39] A. L. Lehninger and E. P. Kennedy, *J. Biol. Chem.* **173,** 753 (1948).
[40] V. R. Potter, *J. Biol. Chem.* **169,** 17 (1947).

systems studied is usually not feasible, because of the presence of phosphatases which rapidly break down ATP with the reliberation of inorganic phosphate. It has been only recently that Kielley[41] has reported data obtained from systems largely devoid of ATP-ase activity. For the most part, workers have minimized the effect of ATP-ase by the addition to the test system of fluoride, which inhibits ATP-ase activity, and by the use of "trapping systems" such as added glucose and purified hexokinase. By means of such an enzymatic "trap" ATP which is formed by oxidative phosphorylation is immediately converted to glucose-6-phosphate by the action of hexokinase.

$$\text{ADP} + \text{P}_i + 12{,}000 \text{ cal.} \rightarrow \text{ATP} \tag{25}$$

$$\text{ATP} + \text{glucose} \xrightarrow{\text{hexokinase}} \text{Glucose-6-phosphate} \tag{26}$$

For each mole of ATP which is formed, 1 mole of glucose-6-phosphate is produced, and 1 mole of inorganic phosphate disappears. It is thus possible to measure the uptake of inorganic phosphate in isolated enzyme systems which are carrying out oxidative phosphorylations, and to correlate this phosphate uptake with oxygen consumption measured simultaneously in the Warburg manometer. The ratio $\left(\frac{\text{moles of inorganic phosphate esterified}}{\text{gram-atoms of oxygen consumed}}\right)$ is known as the P/O ratio and is an index of the efficiency of the phosphorylating process.

Ochoa, in one of the early experiments of this type, measured the P/O ratio for the oxidation of pyruvate in cell-free extracts of cat heart.[38] He found that more than 2 moles of phosphate were esterified for each gram-atom of oxygen consumed, under conditions such that the breakdown of ATP was not completely inhibited. When suitable corrections were made for the estimated loss due to hydrolysis of the ATP by ATP-ase activity, a value of 3 for the average P/O ratio was obtained. This result is of considerable theoretical significance, since it demonstrates that there must exist at least three successive stages of oxidation-reduction enzymes, through which each pair of electrons must pass, and that the free energy released at each such stage by the oxidative process must be available for ATP synthesis. This picture is fundamentally different from previously known phorphorylations, such as that occurring in the oxidation of 3-phosphoglyceraldehyde, where only 1 mole of phosphate is esterified for each pair of electrons removed. In general, it will be seen that in any case where the phosphorylating mechanism involves the addition of inorganic phosphate to a substrate molecule, with the formation of a low-energy phosphate addition product, and the subsequent oxidation of this addition

[41] W. W. Kielley and R. K. Kielley, *J. Biol. Chem.* **191**, 485 (1951).

product to the level of a high-energy phosphate bond, there can be only 1 mole of phosphate esterified for each pair of hydrogen atoms removed.

This statement of the efficiency of oxidative phosphorylation has not gone unchallenged. Ogston and Smithies[42] in a detailed review of this topic have particularly questioned the validity of corrections of the data for loss of ATP through ATP-ase activity and have presented theoretical arguments which lead them to question the possibility of P/O ratios as high as 3. The evidence at present, however, lends support to the original contention of Ochoa. Phosphate uptake associated with each of the oxidative reactions of the Krebs cycle has been studied separately in a number of laboratories. Table 2 summarizes the data which have been obtained. It will be seen that the oxidation of most of the Krebs cycle intermediates brings about the esterification of 3 atoms of P_i, while for at least one step (the oxidation of α-ketoglutarate) there is evidence that the actual P/O ratio is 4.0. On the other hand, the P/O ratio for the oxidation of succinate may be only 2.0.

The efficiency of the aerobic metabolism of glucose may be calculated from Table 2. The anaerobic splitting of glucose yields 2 moles of lactate. Under aerobic conditions, the oxidation of lactate to pyruvate generates 1 mole of reduced DPN, which is then oxidized, yielding 3 atoms of esterified phosphate. The subsequent oxidation of pyruvate yields another 15 moles of ATP, or a total of 18 moles of ATP per half mole of glucose. Ogston and Smithies[42] have calculated that the free energy change for the complete oxidation of 1 mole of lactate is about 324,000 cal. Of this free energy change, $18 \times 12{,}000$ or 216,000 cal. are recovered in the form of ATP, reckoning each ATP molecule at 12,000 cal. The efficiency, then, is about 67%, which must be considered a very high recovery when the process of aerobic phosphorylations is compared with other "technical" operations designed for energy capture.

4. Phosphorylation on the Substrate Level

Because of the high value of 4.0 which has been observed for the P/O ratio of the oxidation of α-ketoglutarate to succinate, this reaction has attracted considerable attention. In interpreting the results of studies on the oxidation of α-ketoglutarate and the accompanying phosphorylations, the scheme shown in Fig. 2 appears to fit in well with the experimental findings. According to this scheme, esterification of 1 mole of inorganic phosphate is associated with the oxidation of the substrate molecule itself in a fashion fundamentally similar to the phosphorylations associated with the oxidation of 3-phosphoglyceraldehyde in muscle extracts, or of acetalde-

[42] A. G. Ogston and O. Smithies, *Physiol. Revs.* **28,** 283 (1948).

hyde in *Cl. kluyveri*.[47] This may be termed phosphorylation on the substrate level. Hunter[48] discovered that the dismutation of α-ketoglutarate in the presence of ammonia may lead to the esterification of ATP under *anaerobic* conditions. The reaction involved is the following:

$$2\ \alpha\text{-Ketoglutarate} + NH_3 \rightarrow \text{Succinate} + \text{glutamate} + CO_2 \qquad (27)$$

For each mole of α-ketoglutarate oxidized to succinate and CO_2, 1 mole of α-ketoglutarate is reduced to glutamate, and 1 mole of phosphate is incorporated into ATP.

There is reason to believe that the phosphorylation coupled to this anaerobic dismutation reaction is similar in mechanism to that observed in the first stage of the aerobic phosphorylation scheme. Loomis and Lipmann[49] found that the metabolic poison 2,4-dinitrophenol has an "uncoupling"

TABLE 2

P/O RATIOS FOR OXIDATIONS OF THE KREBS CYCLE

Reaction	ΔF calories per gram-atom O_2	P/O ratio
Pyruvate + Oxalacetate → Citrate	−72,500	3
Isocitrate → α-Ketoglutarate	−51,000	3
α-Ketoglutarate → Succinate	−72,500	4
Succinate → Fumarate	−36,500	2
Malate → Oxalacetate	−45,000	3
$DPN_{red} \rightarrow DPN_{ox}$	−49,300	3

Free energy data are those calculated by Kaplan[43] on the basis of data given by Lipmann.[8] P/O data are based on determinations by Cross *et al.*,[44] Hunter and Hixon,[45] and Lehninger.[46]

effect on the oxidative phosphorylation system. In the presence of the inhibitor, oxygen uptake is unimpaired, but phosphate uptake is abolished. Hunter[50] has found that the phosphorylation linked to the anaerobic dismutation is resistant to the effects of dinitrophenol, and that similarly one phosphorylation coupled to the aerobic oxidation of α-ketoglutarate is also dinitrophenol resistant, while the other three phosphorylation steps are inhibited. It is reasonable to suppose that the single dinitrophenol-resistant

[43] N. O. Kaplan *in* Sumner and Myrbäck, The Enzymes, Academic Press, New York, 1951, Vol. II, Part 1, p. 55.

[44] R. J. Cross, G.A. Covo, J. V. Taggart, and D. E. Green, *J. Biol. Chem.* **177,** 655 (1949).

[45] F. E. Hunter, Jr. and W. S. Hixon, *J. Biol. Chem.* **181,** 73 (1949).

[46] A. L. Lehninger and S. W. Smith, *J. Biol. Chem.* **181,** 415 (1949).

[47] E. R. Stadtman and H. A. Barker, *J. Biol. Chem.* **180,** 1095 (1949).

[48] F. E. Hunter, Jr. and W. S. Hixon, *J. Biol. Chem.* **181,** 67 (1949).

[49] W. F. Loomis and F. Lipmann, *J. Biol. Chem.* **173,** 807 (1948).

[50] F. E. Hunter, Jr. and S. Spector, *Federation Proc.* **10,** 201 (1951).

step in the aerobic sequence represents phosphorylation on the substrate level, particularly since other substrate-level phosphorylations are also resistant to dinitrophenol. Phosphorylations which involve the oxidation and reduction of enzymes of the electron transfer system, however, which may be termed phosphorylations on the carrier level, are completely inhibited by low concentrations of dinitrophenol. Thus phosphorylation coupled to the oxidation of succinate or the DPN-linked oxidation of β-hydroxybutyrate is completely abolished by dinitrophenol.

The α-ketoglutarate substrate-level phosphorylation has been reported by Kaufman[51] to occur in soluble extracts, and he has suggested the following reaction scheme to account for the experimental findings.

$$\alpha\text{-Ketoglutarate} + DPN_{ox} + CoA \rightarrow \text{Succinyl-CoA} + CO_2 + DPN_{red} \quad (28)$$

$$\alpha\text{-Ketoglutarate} + NH_3 + DPN_{red} \rightarrow \text{Glutamate} + DPN_{ox} \quad (29)$$

$$\text{Succinyl-CoA} + H_2O \rightarrow \text{Succinate} + CoA \quad (30)$$

$$\text{Succinyl-CoA} + ADP + P_i \rightleftharpoons \text{Succinate} + CoA + ATP \quad (31)$$

Evidence for the formation of an activated succinate arising from α-ketoglutarate oxidation, which may be represented by a succinyl-coenzyme A complex, has also been presented by Sanadi and Littlefield[52] working in Green's laboratory. These workers have found that sulfanilamide may be succinylated by a CoA-requiring mechanism, which appears to be essentially analogous to the acetylation of sulfanilamide by "activated acetate."

5. Phosphorylation on the Carrier Level

Since phosphorylation on the substrate level can account for only 1 mole of phosphate uptake per gram-atom of oxygen consumed, and since the oxidation of substrates such as succinate does not appear to involve phosphorylation on the substrate level at all, the central problem of oxidative phosphorylation in animal tissues hinges on oxidations of carriers in the electron transfer system. Early attempts by Ochoa to demonstrate oxidative phosphorylation coupled to oxidation of DPN_{red} were unsuccessful. However, Friedkin and Lehninger,[53] using cell-free washed residues of rat liver, found that the oxidation of chemically reduced DPN brought about the incorporation of radioactive inorganic phosphate into ATP pyrophosphate linkages. Addition of oxidized DPN was without effect, and the system was sensitive to the usual uncoupling techniques. This must be

[51] S. Kaufman, Symposium on Phosphorus Metabolism, McCollum-Pratt Institute, Baltimore, June 1951.

[52] D. R. Sanadi and J. W. Littlefield, *J. Biol. Chem.* **193,** 683 (1951).

[53] M. E. Friedkin and A. L. Lehninger, *J. Biol. Chem.* **174,** 757 (1948); *ibid.* **178,** 611 (1949).

regarded as the first definite evidence for the participation of the electron transfer system in oxidative phosphorylation.

Lehninger[46] has extended these studies by the use of the β-hydroxybutyrate dehydrogenase system, which is DPN-linked, and in isolated mitochondria oxidizes β-hydroxybutrate in a one-step reaction to acetoacetate, a product which is quite inert in the system studied. The dehydrogenase appears to act merely as a source of enzymatically generated reduced DPN. P/O ratios of 3.0 have been reported by Lehninger. The value of 3.0 would appear to be likely for every DPN or TPN-linked oxidation in

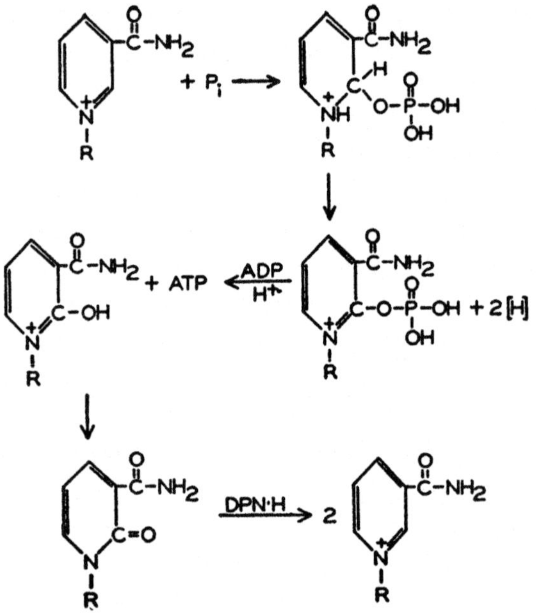

FIG. 3

these preparations; if in addition to the carrier phosphorylations there is also a substrate level phosphorylation, as in the case of α-ketoglutarate oxidation, then the P/O ratio becomes 4.0.

The problem of how oxidation of reduced DPN in these enzyme systems can involve phosphate uptake is at present completely mysterious. No real experimental data bearing on this problem, or indeed upon any carrier-level phosphorylation, is at hand. A suggestion of the kind of mechanism which may be involved has been made by Kaplan[54] and is depicted in Fig. 3. The essential features of this scheme involve the addition of inorganic phos-

[54] N. Kaplan, Symposium on Phosphorus Metabolism, McCollum-Pratt Institute, Baltimore, June 1951.

phate to the pyridine ring of DPN_{ox}, and subsequent conversion of this addition product to a high-energy enol phosphate, which then phosphorylates ADP and is converted to the keto form. The pyridone would then react with a mole of reduced DPN to form 2 moles of DPN, bringing about a cyclic reaction.

The enzymatic oxidation of reduced DPN is widely assumed to involve the participation of yellow enzymes, with riboflavin phosphate or flavin adenine dinucleotide as coenzymes. Since the oxidation of reduced DPN is now known to bring about coupled phosphorylation, it seems reasonable to assume that flavin intermediates are involved in the generation of high-energy phosphate bonds. Hummel and Lindberg[55] have claimed that, in rabbit liver homogenates carrying on oxidative phosphorylation in the presence of inorganic phosphate labeled with P^{32}, the FAD fraction becomes very radioactive, and that the specific activity of this fraction approaches that of the ATP fraction. These investigators have therefore suggested that the phosphate linkages of FAD might in fact be precursors of ATP. Experiments by Kennedy and Lipmann[56] have shown that this possibility does not hold in oxidative phosphorylation in isolated rat liver mitochondria, where the radioactivity of the FAD fraction was found to be negligible under conditions where the ATP became extremely radioactive.

The participation of the flavin enzymes in oxidative phosphorylation seems a certainty, if our ideas concerning the electron transfer mechanisms involved in oxidation of reduced DPN have any validity, but any substantial direct information on this topic is still lacking.

Many attempts have been made to demonstrate phosphorylation coupled to oxidation of reduced cytochrome c. Judah[57] has presented evidence which suggests that phosphorylation occurs when cytochrome c is reduced *in situ* by a non-enzymatic reaction with ascorbic acid, and then subsequently oxidized enzymatically. Slater,[58] on the other hand, has reported experiments which he interprets as meaning that phosphorylation does not occur when cytochrome c is oxidized. Slater's experiments seem to be open to some criticism on technical grounds. In any event, it is difficult to explain the P/O ratio of 2.0 which has been observed in the oxidation of succinate unless one assumes that phosphorylation occurs coupled to oxidations above the level of cytochrome c. The potential span between succinate and cytochrome c is only about 0.3 v., which is not sufficient to generate the 24,000 cal. needed to bring about the formation of 2 moles of ADP to ATP.

The entire field of carrier-level oxidative phosphorylation is at present

[55] J. P. Hummel and O. Lindberg, *J. Biol. Chem.* **180,** 1 1949).
[56] E. P. Kennedy and F. Lipmann, unpublished experiments.
[57] J. D. Judah, *Biochem. J.* **49,** 271 (1951).
[58] E. C. Slater, *Nature* **166,** 982 (1950).

far from satisfactorily explored. The importance of further information on this topic is obvious. It seems likely that the next few years will see intensive efforts made to penetrate this challenging frontier of biochemistry.

6. Phosphorylation and Fatty Acid Oxidation

Most of the work on oxidative phosphorylation has been concerned with the oxidation of intermediates of the Krebs cycle. Although this reaction scheme was originally presented to account for the oxidation of pyruvate, and although the dicarboxylic and tricarboxylic acid intermediates are often considered "carbohydrate substrates," it is now known from the work of Breusch,[59] Buchanan *et al.*,[60] and Lehninger[61] that fatty acids are also oxidized via the Krebs cycle. Fatty acid oxidation may be visualized as proceeding in two stages. In the first stage, the fatty acid molecule is oxidatively degraded into 2-carbon units, which are now identified as acetyl-coenzyme A. These units may then condense with oxalacetate, to form citrate, as shown in Fig. 1, and thus enter the Krebs cycle. A second reaction path is recombination to form acetoacetate, but, since the ultimate fate of acetoacetate in tissues which possess the ability to oxidize this substance is likewise citrate formation,[59] the ultimate terminal oxidation of fatty acids is the same as that of pyruvate. The P/O ratios observed for the Krebs cycle probably may be assumed to hold for the terminal oxidation of the 2-carbon units derived from fatty acids, with account being taken of the fact that one oxidative step (the oxidation of pyruvate) is absent.

Very little is known about the oxidative reactions which lead to the production of acetyl-coenzyme A from fatty acids in animal tissues, but, since it is likely that these oxidations utilize the same electron transfer system as Krebs cycle oxidations, and since fatty acid oxidation takes place in the same enzyme preparations which are used for the study of oxidative phosphorylation, it would seem likely on *a priori* grounds that the over-all P/O ratio for fatty acid oxidation must be similar to that observed for the complete oxidation of pyruvate.

Direct measurements of phosphorylation coupled to fatty acid oxidation have been reported by Kennedy and Lehninger,[61a] who showed that the incorporation of radioactive inorganic phosphate into ATP could be linked to octanoate oxidation. Interest in oxidative phosphorylation and fatty

[59] F. L. Breusch, *Enzymologia* **11,** 169 (1944).

[60] J. Buchanan, W. Sakami, S. Gurin, and D. W. Wilson, *J. Biol. Chem.* **159,** 695 (1945).

[61] A. L. Lehninger, *J. Biol. Chem.* **161,** 413 (1945).

[61a] E. P. Kennedy and A. L. Lehninger, *J. Biol. Chem.* **179,** 957 (1949); *ibid.* **190,** 361 (1951).

acid oxidation is stimulated by the finding that apparently phosphorylation is not only possible but obligatory for fatty acid oxidation. Thus fatty acid oxidation cannot be de-coupled by the action of dinitrophenol, as can oxidations of the Krebs cycle; the addition of low concentrations of dinitrophenol completely abolishes fatty acid oxidation as well as phosphorylation. Again, aged enzyme preparations which retain for a long time the ability to oxidize substrates such as succinate lose fatty acid oxidase activity at about the same rate as oxidative phosphorylating capacity. These facts and the requirement of some enzyme preparations for a "sparking oxidation" of small amounts of a Krebs cycle intermediate or of DPN_{red} point to an intimate intermeshing of fatty acid oxidation and phosphorylation.

7. Role of Mitochondria in Oxidative Phosphorylation

Our understanding of the extremely important reactions of oxidative phosphorylation in animal tissues has been severely handicapped by technical difficulties, which have as yet by no means been overcome. These difficulties arise from the fact that the enzymes exist in animal tissues in insoluble, particulate form. In 1913, Warburg found that the ability of macerated liver tissue to take up oxygen was found largely in "Koernchen" or insoluble particles which were readily removed from extracts by centrifugation or by filtration. Many other studies on enzymes concerned with terminal respiration have indicated their insoluble or structure-bound character. Particular attention has been devoted to studies of succinoxidase and cytochrome oxidase. Indeed, these enzymes have come to be considered examples *par excellence* of "insoluble" enzymes.

In more recent years, a clearer understanding of the biological importance of structure in enzyme systems has been gained, largely as the result of investigations in the domain shared by cytology and enzymology. Cytologists have long recognized the occurrence of microscopically visible granules, either round or with thread-like structure, in the cytoplasm of nearly every kind of living cell. The importance of such cytoplasmic granules was suggested even in the nineteenth century by Altmann.[62] These intracellular bodies, now called mitochondria, have been the subject of a very large and for the most part somewhat speculative literature.

In 1934, Bensley and Hoerr[63] developed a technique for the isolation of cell fractions by means of differential centrifugation. With this technique it became possible for the first time to obtain sufficient quantities of these subcellular fractions for chemical and enzymological investigations. In extension of this work along enzymological lines, Claude[64, 65] discovered

[62] R. Altman, Die Elementarorganismen, Veit & Co., Leipzig, 1890.

[63] R. R. Bensley and N. L. Hoerr, *Anat. Rec.* **60,** 251 (1934).

[64] A. Claude, *Proc. Soc. Exptl. Biol. Med.* **39,** 398 (1938).

[65] A. Claude, *J. Exptl. Med.* **85,** 51, 61 (1946).

the existence of the submicroscopic particulate fraction, the microsomes. Claude also made the important finding that mitochondria and microsomes are composed in part of nucleoprotein, with the nucleic acid being exclusively of the pentose nucleic acid type. Since nucleoprotein is considered one of the fundamental constituents of living matter, found in similar particles of great biological interest, such as viruses and bacteriophage, this discovery suggested that the cytoplasmic granules were essential cellular units.

Evidence to substantiate this concept was immediately forthcoming. Investigations by Claude and by Schneider and their collaborators[66, 67, 68] indicated that the succinoxidase and cytochrome oxidase activities of liver cells are largely concentrated in the mitochondria, a finding which explained the results of earlier studies on these "insoluble" enzymes. These workers also developed a technique for the fractionation of tissue extracts in hypertonic sucrose solution, a technical advance of considerable value, since previous work with distilled water or buffered saline extracts did not preserve the morphological and enzymological properties of the fractions satisfactorily. With this method, it was possible to isolate mitochondria from liver tissue in a state approximating that found in the intact cell, with the full preservation of many labile enzyme activities which it had been impossible to study by previous methods.

Working with hypertonic sucrose homogenates, Kennedy and Lehninger [61a] found that isolated washed mitochondria of rat liver tissue oxidized not only succinate but also other key intermediates of the Krebs cycle. They also reported that fatty acid oxidation was localized in the mitochondria and furthermore demonstrated that phosphorylations coupled to these oxidations took place in these preparations, as could be shown with tracer phosphate experiments.

A washed-residue type of enzyme preparation capable of carrying out the reactions found in mitochondrial preparations has been intensively studied by D. E. Green and his school. Green has recently summarized some of the results of these valuable investigations.[69] Green has given the name "cyclophorase" to this enzyme system and has emphasized differences which are observed between enzymes bound to the cyclophorase entity, and the same enzymes studied in the solubilized "classical" form. From this point of view, he has stressed specific organization in space of the

[66] G. H. Hogeboom, A. Claude, and R. D. Hotchkiss, *J. Biol. Chem.* **165**, 615 (1946).

[67] W. C. Schneider, A. Claude, and G. H. Hogeboom, *J. Biol. Chem.* **172**, 451 (1948).

[68] G. H. Hogeboom, W. C. Schneider, and G. E. Pallade, *J. Biol. Chem.* **172**, 619 (1948).

[69] D. E. Green, *in* Edsall, Enzymes and Enzyme Systems, Harvard University Press, Cambridge, 1951, p. 15.

enzyme complex and has shown that considerable amounts of coenzymes are bound into the structure. Recently, Harmon[69a] has correlated the specific properties of "cyclophorase" preparations with their content of mitochondria and has shown that the mitochondria are the essential elements in the enzyme reactions concerned, although the situation is perhaps complicated by the presence of other cell components.

The knowledge that the enzymes which catalyze the oxidative phosphorylation reactions are organized into structural units in the living cell gives us some insight into the manner by which the cell "compartmentalizes" or effects a division of labor in its metabolic activities. For example, the breakdown of glucose in the liver cell, to the stage of pyruvic acid, does not take place in isolated mitochondria, because the complete sequence of glycolytic enzymes is not present in these particles but is found in the soluble supernatant. Anaerobic glycolysis, then, in the liver cell, must occur in the cytoplasm outside the mitochondria, and the pyruvic acid which results from this process must be subsequently taken up by the mitochondria which possess the complete apparatus of enzymes needed for oxidation of this substrate to carbon dioxide and water. The free energy released by this oxidative process is conserved by the mitochondria in the form of ATP, which may then diffuse back out of the mitochondria and be made available to the cell for synthesis or function. An alternative possibility, as yet inadequately explored, is the direct building up of cell constituents by the mitochondria and release of the finished products into the cell.

This grossly oversimplified picture of the interrelationships of the mitochondria and soluble enzymes does not rule out the presence in the soluble cytoplasm of certain of the enzymes involved in Krebs cycle reactions, nor does it mean that the mitochondria are completely lacking in all the glycolytic enzymes. What is important is the occurrence of all the complicated enzymatic machinery or processes like oxidative phosphorylation or fatty acid oxidation *as organized systems* in cytoplasmic structures, and the dependence of these processes on integrity of structure. It seems likely that the organization of such enzyme systems into compact units effects a considerable economy for the organism, since the effective concentration of substrates and coenzymes, as well as of enzymes, is enormously greater when compacted into these structures rather than freely diffused throughout the entire cell volume. It is also possible that structural organization of this sort may be a control mechanism, which channels substrates through a complete assembly line of enzymes, once a given process has been initiated.

The puzzling properties of many of the enzyme systems previously studied in crude washed residues are much more easily understood when it

[69a] J. Harman, *J. Exptl. Cell Research* **1**, 382, 394 (1950).

is realized that they are intimately bound up with mitochondrial structure and in many cases cannot survive treatment which affects the structural integrity of the mitochondria. Fatty acid oxidation and oxidative phosphorylation are examples of enzyme systems which cannot withstand treatment with distilled water or markedly hypotonic solutions. It is known that, when subjected to hypotonic media, mitochondria swell up and eventually disintegrate. The loss of these enzyme activities is seen to parallel loss of structural organization of the mitochondria. Results of treatment such as freezing and thawing, shaking with capryl alcohol, or drying may similarly be explained.

The realization that the enzymologist is still confronted with problems of structure in these completely cell-free preparations is in a way disheartening. The problem of working out these reactions is like a puzzle in which one box is opened and is found to contain another box more difficult to open, and this in its turn contains yet another box, and so on. In situations where structure is so important to the observed sequence of enzyme reactions, it may be necessary to reconstruct not only the sequence of reactions, but also their peculiar alignment in cell geometry, and to reproduce the localized conditions under which coenzymes and substrates are fitted into the particle. At present, this ideal reconstruction must be considered to be quite beyond the scope of our methods.

An alternative possibility remains, which suggests that appropriate techniques may be found for the solubilization of all of these enzymes, so that they may be brought into a form susceptible to purification and study, and that when more is known about the individual reactions we may be able to sum up this information into a coherent pattern quite apart from organized structure. It is possible that, by selecting the proper biological material for study, some of the enzyme systems may be found in a condition not closely tied up with cytoplasmic structure. A pertinent case is that of the fatty acid oxidase complex, which in mammalian tissues is extremely labile and cannot be studied under conditions which do not permit maintenance of mitochondrial structure. Barker and his co-workers[70] have found that the fatty acid oxidizing and synthesizing systems of the microorganism *Cl. kluyveri* can be readily extracted from dried cell preparations and studied in detail in completely soluble form. If our faith in the fundamental unity of major metabolic pathways is justified, it can be expected that the essential reactions of this type of fatty acid metabolism will be closely related to analogous reactions in animal tissues. Many points of similarity have in fact already been found. This indirect approach may permit a synthesis of data derived from many kinds of cells which may contribute

[70] E. R. Stadtman and H. A. Barker, *J. Biol. Chem.* **180**, 1085, 1095, 1117, 1169 (1949).

to the solution of problems which cannot now be directly attacked by the study of isolated mitochondria.

VI. Utilization of Metabolic Energy for Biosyntheses

1. Energetics of Biosynthetic Reactions

In the preceding pages, our attention has been directed toward the exergonic reactions by which foodstuffs are degraded in the animal body, with specific emphasis on the conservation in the form of high-energy bonds of the free energy liberated. The other half of the balance sheet, the utilization of metabolic energy for synthetic reactions and for function, is much more difficult to describe in detail, since in most cases only limited information is at hand concerning these synthetic processes.

The major constituents of the animal body—fats, proteins, and carbohydrates—are constantly being broken down and rebuilt from simple building blocks. Free energy is required for the task of building up cell constituents, not only because of the entropy change implicit in imposing a complex, ordered pattern on more or less randomized precursors of the cell, but also because the chemical reactions involved are themselves endergonic under physiological conditions.

Lipmann[8] has pointed out that a large number of synthetic reactions have one feature in common: formation of more complicated structural units by mechanisms involving the splitting out of water. A few such reactions, which may be considered typical are shown in the following equations.

$$3CH_3(CH_2)_nCOOH + \begin{array}{c} HOCH_2 \\ | \\ HOCH \\ | \\ HOCH_2 \end{array} \rightarrow \begin{array}{c} CH_3(CH_2)_nCO\text{—}OCH_2 \\ | \\ CH_3(CH_2)_nCO\text{—}OCH \\ | \\ CH_3(CH_2)_nCO\text{—}OCH_2 \end{array} + 3H_2O \qquad (32)$$

$$RCH(NH_2)COOH + R'CH(NH_2)COOH \rightarrow RCH(NH_2)CO\text{—}NHCH(R')COOH + H_2O \qquad (33)$$

$$2CH_3COOH \rightarrow CH_3COCH_2COOH + H_2O \qquad (34)$$

All the above reactions, as written, are endergonic, and consequently cannot spontaneously proceed sufficiently far to the right to give reasonable yields of product, unless coupled to some exergonic reaction in such a fashion that the over-all net reaction leads to a negative free energy change.

The free energy change for the production of an ester bond, or a simple peptide bond, such as are shown in equations 32 and 33, is approximately 3000 cal. per bond. This value is obtained when the free energy change is calculated with the standard state of water taken as pure water, or 55.6 molal. When the standard state of water is taken as 1 molal (*i.e.*, when it is considered in identical fashion with other reactants), the free energy change is about zero. This means that the high concentration of water in aqueous systems such as are essential for biosynthetic reactions forces the reactions to the left; free energy must be provided to overcome this factor. It will be recalled that many esterification reactions in ordinary organic chemistry can be caused to proceed with high yields of ester simply by providing for the continuous removal of water by some suitable device. This expedient is not available to living organisms, carrying out enzymatic syntheses in dilute aqueous media.

The free energy change in reaction 34 is much higher; Lipmann[8] has calculated it to be about 16,000 cal. In this reaction, not only is water split out, but a C—C bond of a β-keto acid is formed.

Since it is known that a large portion of the total energy available to the organism is channeled into high-energy phosphate bonds, through the medium of ATP, it is reasonable to suggest that phosphorylated intermediates may take part in coupled reactions to overcome the energy barrier and cause the reactions to proceed in the direction of synthesis. A series of reactions involving phosphorylated intermediates may be written in place of those above, which involve, not the splitting out of water, but rather of the elements of phosphoric acid.

$$3CH_3(CH_2)_nCO{\sim}P + \begin{array}{c} HOCH_2 \\ | \\ HOCH \\ | \\ HOCH_2 \end{array} \rightarrow \begin{array}{l} CH_3(CH_2)_nCO—OCH_2 \\ \qquad\qquad\qquad\quad | \\ CH_3(CH_2)_nCO—OCH \\ \qquad\qquad\qquad\quad | \\ CH_3(CH_2)_nCO—OCH_2 \end{array} + 3P_i \tag{35}$$

$$RCHNH_2CO{\sim}P \rightarrow R'CHNH_2COOH \rightarrow RCH(NH_2)CO—NHCH(R')COOH + P_i \tag{36}$$

$$CH_3CO{\sim}P + CH_3COOH \rightarrow CH_3COCH_2COOH + P_i \tag{37}$$

Reactions 35 and 36 would be strongly exergonic as written, since formation of the ester and peptide linkages is in effect coupled to the breakdown of the high-energy carboxyl phosphate ($\Delta F = -15{,}000$ cal.). Reaction 37 would no longer be highly endergonic, and appreciable synthesis of acetoacetate could be expected at reasonable concentrations of acetate and acetyl phosphate.

2. Role of Coenzyme A in Synthetic Reactions

Although the postulation of compounds such as acetyl phosphate and the acyl phosphates of the higher fatty acids as intermediates in synthetic reactions is very attractive on thermodynamic grounds, a considerable amount of evidence has been obtained which indicates that these phosphorylated compounds do not themselves play an important physiological part in mammalian tissues. When acetyl phosphate or the higher acyl phosphates are tested in mammalian enzyme systems, these substances do not exhibit the properties of the "activated" state but instead are rapidly destroyed by phosphatase action. On the other hand, it is now apparent from work of the laboratories of Lipmann, Ochoa, Stadtman, Lynen, and others that coenzyme A may take the place of phosphate in forming high-energy addition products.

An important advance in this area was made by Snell *et al.*,[71] who demonstrated that the sulfur-containing moiety of coenzyme A was β-thiolethylamine. Lynen[34] has summarized the evidence which has led to the conclusion that the sulfhydryl group of this portion of the coenzyme is the active site for high-energy group transfer. He has also reported the isolation in partly purified form of the active acetyl conjugate of coenzyme A, which is designated in his notation, $Co\bar{A}—S—COCH_3$. The tentative structure of this compound has been shown in Fig. 1.

Stadtman and Barker[21] had previously discovered in bacterial extracts an enzyme (phosphotransacetylase) which is now believed to carry out the following reaction:

$$CH_3CO{\sim}P + Co\bar{A}—SH \rightleftharpoons Co\bar{A}—S—COCH_3 + P_i \quad (38)$$

The failure of acetyl phosphate to substitute for "activated acetate" in animal tissues is due to the fact that phosphotransacetylase does not occur in higher animals. Instead, other pathways exist for the production of $Co\bar{A}—S—COCH_3$.

$$ATP + CH_3COOH + Co\bar{A}—SH \rightarrow Co\bar{A}—S—COCH_3 + ADP + P_i \quad (39)$$

Whereas in reaction 39 the ultimate source of the high-energy acetyl compound is ATP, it is significant to note that this bond can also be directly generated from metabolic energy liberated upon the oxidation of pyruvate, by a reaction not requiring the utilization of ATP.

$$CH_3COCOOH + Co\bar{A}—SH \xrightarrow{\text{"pyruvic oxidase"}} Co\bar{A}—S—COCH_3 + CO_2 + 2H^+ + 2e \quad (40)$$

The actual reaction summarized in equation 40 is very complex and may require the presence of unidentified cofactors as well as coenzyme A and cocarboxylase. Reaction 40 is an example of the trapping of metabolic energy in a form useful to the organism without direct participation of ATP. The generation of high-energy succinyl-coenzyme A has been described above in equations 28 and 29. It is possible that other reactions also exist by which coenzyme A can trap metabolic energy directly in a form suitable for synthetic uses. In this connection, a possible role for

[71] E. E. Snell, G. M. Brown, V. J. Peters, J. A. Craig, E. L. Wittle, J. A. Moore, V. M. McGlohon, and C. D. Bird, *J. Am. Chem. Soc.* **72**, 5349 (1950).

coenzyme A or a similar sulfhydryl compound in oxidative phosphorylation may be suspected.

The number of synthetic reactions known to involve coenzyme A is already large and is rapidly growing. Some of these are shown in Fig. 1. The implication of this coenzyme in the direct formation of high-energy addition products by pathways not involving the breakdown of ATP points to the quantitative importance of such mechanisms in the general picture of metabolic energetics.

3. Synthesis of Fatty Acids

Isotope tracer studies have proved that long-chain fatty acids may be built up in mammalian tissues entirely from 2-carbon units which are presumably identical with $Co\bar{A}{-}S{-}COCH_3$. As is the case with protein synthesis, it has been found very difficult to study fatty acid synthesis in the mammal in cell-free preparations. However, as discussed above, Stadtman and Barker have succeeded in carrying out both fatty acid oxidation and fatty acid synthesis in cell-free soluble enzyme preparations of *Cl. kluyveri*. The data obtained in these investigations fit in well with the indirect evidence from tracer experiments concerning fatty acid synthesis in the mammal. Kennedy and Barker[72] have suggested a scheme for the oxidation and synthesis of butyrate in *Cl. kluyveri*, based on the assumption that free intermediates do not occur as such in this process, but rather as conjugates with an unidentified coenzyme, probably coenzyme A.

$$CH_3CH_2CH_2COOH + Co\bar{A}{-}SH \rightleftharpoons CH_3CH_2CH_2CO{-}S{-}Co\bar{A}$$

$$\pm 2H \downarrow\uparrow$$

$$CH_2{:}CHCH_2COOH + Co\bar{A}{-}SH \rightleftharpoons CH_2{:}CHCH_2CO{-}S{-}Co\bar{A}$$

$$\pm H_2O \downarrow\uparrow \quad \pm 2H \downarrow\uparrow$$

$$CH_3COCH_2COOH + Co\bar{A}{-}SH \rightleftharpoons CH_3COCH_2CO{-}S{-}Co\bar{A}$$

$$\pm P_i \downarrow\uparrow$$

$$CH_3COOH + CH_3CO{\sim}P + Co\bar{A}{-}SH$$

In this scheme, the energy required for the condensation of 2-carbon units to form acetoacetyl-coenzyme A is furnished by the breakdown of acetyl phosphate, in a complex series of reactions probably requiring phosphotransacetylase. In animal tissues, which lack this enzyme, the energy would be furnished by reaction 39 or 40, or some similar reaction.

Stadtman *et al.*[73] have studied the synthesis of acetoacetate by an enzyme from liver in the presence of added phosphotransacetylase of bacterial origin, using isotopically labeled acetyl phosphate. These workers found that 2 moles of acetyl phosphate were used up for each mole of acetoacetate synthesized.

[72] E. P. Kennedy and H. A. Barker, *J. Biol. Chem.* **191**, 419 1951).
[73] E. R. Stadtman, M. Doudoroff, and F. Lipmann, *J. Biol. Chem.* **191**, 377 (1951).

$$2CH_3CO—S—Co\bar{A} + H_2O \rightarrow CH_3COCH_2COOH + 2Co\bar{A}—SH \quad (41)$$

Reaction 41 as written is highly exergonic, since only 16,000 cal. are required for the synthesis of acetoacetate. One mole of acetyl phosphate should furnish sufficient energy to allow the reaction to proceed. Here two high-energy bonds are dissipated, which may seem wasteful from a teleological point of view. However, it is possible that mechanisms exist in the intact tissues for the recovery of this energy by some type of feedback mechanism.

The elongation of long-chain fatty acids in mammals has certain aspects of similarity with the synthesis of acetoacetate. Zabin[74] has studied this process and has found that 2-carbon units are added to the carboxyl group of the homologous fatty acid with 2 fewer carbon atoms in a manner precisely analogous to the short-chain fatty acid synthesis described by Barker, Kamen, and Bornstein.[75] Although no direct evidence is at hand for the reaction in animal tissues, the following reaction sequence seems likely:

$$CH_3(CH_2)_nCO—S—Co\bar{A} + CH_3CO—S—Co\bar{A} \rightarrow CH_3(CH_2)_nCOCH_2CO—S—Co\bar{A} + Co\bar{A}—SH \quad (42)$$

$$CH_3(CH_2)_nCOCH_2CO—S—Co\bar{A} + 4H \rightarrow CH_3(CH_2)_{n+2}COOH + CoA—SH \quad (43)$$

The formation of fatty acid-coenzyme A complexes has also been suggested by Lynen.[34] It is possible that such mechanisms may also furnish the free energy necessary for the esterification of glycerol to form neutral fat or phospholipids.

4. Synthesis of Protein

The synthesis of proteins requires the formation of high-molecular-weight polymers of amino acids, linked through peptide bonds, and possessing highly specific spatial configurations. The synthesis of the peptide bond shown in equation 33 requires about 3000 cal. per mole. It is clear from this fact alone that previously suggested mechanisms for protein synthesis involving the reversal of reactions catalyzed by proteolytic enzymes are rather unlikely. Protein synthesis must be geared in some fashion to catabolic processes which can furnish the required energy. Although it has not been possible to study the synthesis of the extremely complex protein structure in simple enzyme systems, experiments with tissue slices have demonstrated the requirement for a vigorous respiration in order to obtain maximum incorporation of labeled amino acids into tissue protein.[76] The addition of uncoupling agents such as 2,4-dinitrophenol, which permit

[74] I. Zabin, *J. Biol. Chem.* **189**, 355 (1951).

[75] H. A. Barker, M. D. Kamen, and B. T. Bornstein, *Proc. Natl. Acad. Sci. U. S.* **31**, 373 (1945).

[76] T. Winnick, F. Friedberg, and D. M. Greenberg, *Arch. Biochem.* **15**, 160 (1947).

respiration to continue while preventing the formation of ATP, reduces incorporation of labeled amino acids to a very low level.[77] This suggests that oxidative phosphorylation processes are needed in order to activate the amino acids for peptide synthesis, but there is at present no direct evidence for the formation of phosphorylated derivatives such as those shown in reaction 36.

Although it would seem likely that the ultimate donor of energy to cause these reactions to proceed may be ATP, or some compound in equilibrium with ATP, this does not exclude the occurrence of exchange reactions on a lower energy level. Glutathione has been reported by Hanes[78] to undergo a variety of transpeptidation reactions, and it may be that this substance and other polypeptides may function to some degree as reservoirs of peptide bond energy.

Although it has not yet been possible to study the synthesis of proteins in purified enzyme systems, a considerable number of model reactions which involve the formation of peptide or amide linkages have been investigated. These include the synthesis of glutamine,[79] hippuric acid,[80] acetylated amines,[81] glutathione,[82] and urea.[83] It is of significance that every one of these model reactions, studied in isolated enzyme systems, has been shown to require ATP. In addition, two of these reactions, the formation of hippuric acid and the acetylation of amines, have been shown to require coenzyme A.

The requirement of coenzyme A for hippuric acid synthesis may be taken as evidence for a benzoyl-coenzyme A complex, similar to the acetyl-coenzyme A compound for which good evidence has been obtained. It is possible that coenzyme A-amino acid addition products are the real intermediates in protein synthesis, rather than the acyl phosphates shown in equation 37.

Another model enzyme system of interest in connection with protein synthesis is glutamyltransphorase, studied in bacteria by Waelsch *et al.*[84] and in plant extracts by Stumpf.[85] There is no net synthesis of amide linkages brought about by this enzyme, which exchanges the amide group of glutamine with added ammonia, or with hydroxylamine. Stumpf has

[77] L. D. Frantz, Jr., P. C. Zamecnik, J. W. Reese, and M. L. Stephenson, *J. Biol. Chem.* **174,** 773 (1948).
[78] C. S. Hanes, F. J. R. Hird, and F. A. Isherwood, *Nature* **166,** 288 (1950).
[79] J. F. Speck, *J. Biol. Chem.* **168,** 402 (1947).
[80] P. P. Cohen and R. W. McGilvery, *J. Biol. Chem.* **171,** 121 (1947).
[81] F. Lipmann, *Federation Proc.* **8,** 597 (1949).
[82] R. B. Johnston and K. Bloch, *J. Biol. Chem.* **188,** 221 (1951).
[83] P. P. Cohen and M. Hayano, *J. Biol. Chem.* **172,** 405 (1948).
[84] H. Waelsch, P. Owades, E. Borek, N. Grossowicz, and M. Schou, *Arch. Biochem.* **27,** 237 (1950).

shown that this enzyme also requires ADP or ATP, although at catalytic concentrations. This is to be contrasted with the enzyme studied by Speck,[79] which brought about a net synthesis of glutamine with a corresponding stoichiometric utilization of ATP.

5. The Synthesis of Carbohydrate

The synthesis of glucose or of glycogen by reversal of the anaerobic glycolytic pathway was suggested by the early work of Meyerhof and has been confirmed by more recent experiments with isotopic tracers. Since two phosphorylations must take place for each mole of triose phosphate formed from lactate, a total of 4 moles of ATP must be used up for each mole of carbohydrate resynthesized.

The early work of Meyerhof showed that there was an intimate connection between resynthesis of carbohydrate and aerobic metabolism. When the production of lactic acid by tissues under anaerobic conditions was compared with the production of lactic acid under aerobic conditions, it was found that the uptake of 1 mole of respired oxygen inhibited the formation of 1 to 2 moles of lactic acid; i.e., three to six times as much lactic acid as could have been oxidized by the observed oxygen uptake. This phenomenon of the suppression of glycolysis under aerobiosis has been called by Warburg the Pasteur effect.

There has been much controversy concerning the mechanism of the Pasteur effect, which it would be unprofitable to review here. One suggestion, put forth by Johnson,[86] is worth of mention, however, since it bears intimately on the channels of energy transfer and utilization. Johnson has pointed out that under aerobic conditions the much more efficient phosphorylation of ADP to ATP by oxidative phosphorylation diminishes the amount of inorganic phosphate available and increases the ratio of ATP to ADP. In effect, the anaerobic glycolytic mechanism must compete with the aerobic oxidative phosphorylation system for available adenine nucleotide and inorganic phosphate. Since the aerobic mechanisms are so much more efficient, the rate of anaerobic glycolysis is therefore sharply reduced under aerobic conditions. Meyerhof[87] has presented data which show that the presence of acceptor systems is rate-limiting in certain types of glycolyzing extracts, under conditions where considerable stimulation is brought about by adding ATP-ase to regenerate ADP and inorganic phosphate. This and other evidence can be taken to support the Johnson hypothesis, which in all events has the merit of focusing attention on the adenine nucleotide system as the common denominator of both types of metabolism.

[85] P. K. Stumpf, W. D. Loomis, and C. Michelson, *Arch. Biochem.* **30,** 126 (1951).
[86] M. T. Johnson, *Science* **94,** 200 (1951).
[87] O. Meyerhof, *J. Biol. Chem.* **157,** 105 (1945).

CHAPTER 17

Hydrolytic and Phosphorolytic Enzymes

WARREN H. YUDKIN

Department of Chemistry, Northwestern University, Evanston, Illinois

Chapter 17

I. Introduction

Hydrolytic enzymes catalyze reactions that involve the insertion of the elements of water into a molecule in such a way as to yield two different functional groups. The general reaction is:

$$R\text{—}R' + HOH \rightarrow ROH + R'H$$

The usual terms will be used in this discussion, but, for the sake of clarity, it might be well to review briefly the general properties of all enzymes. Enzymes are catalysts and, as such, they lower the free energy of activation of reactive molecules and thus shorten the time required for a reaction to reach equilibrium. Consequently, an enzyme, like a catalyst, does not favor any equilibrium position but rather it accelerates equally the reactions in both directions in a system approaching equilibrium. All enzymes thus far characterized show the properties of proteins, and all enzymes that have been crystallized are definitely proteins. Enzymes, therefore, are subject to denaturation by heat and other physical and chemical agents, have isoelectric points, and are synthesized only by a living cell.

Since the hydrolysis of most substrates is a highly exergonic reaction, the position at equilibrium is greatly in favor of the products of the split. Furthermore, the presence of a great excess of water, common in most biological systems, favors the hydrolytic reaction as opposed to the synthetic reaction. Some esterases, however, catalyze reactions with a free energy close to zero, and so the equilibrium mixture will contain appreciable quantities of reactant. If, on the other hand, the elements of phosphoric acid, rather than water, are inserted, then the energy released upon scission is lowered considerably and the equilibrium mixture contains a much higher proportion of reactant than in hydrolysis:

$$R—R' + HOPO_3H_2 \rightleftharpoons ROPO_3H_2 + R'H$$

This reaction is termed phosphorolysis.

Because the free energy of hydrolysis in most biological systems is a good deal less than zero, the hydrolytic enzymes generally act in a catabolic or digestive capacity and have been difficult to implicate in the biosynthesis of their substrates by a simple reversal of the hydrolytic reaction. It is becoming increasingly evident, however, that many hydrolytic enzymes, such as proteinases and phosphatases, can catalyze transfer or exchange reactions. Much of the exposition in this chapter will be devoted to the synthetic and transferring activities of enzymes considered hydrolytic. No attempt will be made to cover the original literature, and consequently the reader will be referred mainly to recent work and review articles.

1. Classification

Hydrolytic enzymes are classed roughly according to the nature of the substrate acted upon. Proteinases and peptidases catalyze the following reaction:

$$\cdots NH\underset{\substack{|\\ R}}{C}HCO—NH\underset{\substack{|\\ R'}}{C}HCO\cdots + HOH \rightarrow \cdots NH\underset{\substack{|\\ R}}{C}HCOOH + H_2N\underset{\substack{|\\ R'}}{C}HCO\cdots$$

Esterases catalyze the hydrolysis of esters, including those of fatty acids:

$$RCOOR' + HOH \rightarrow RCOOH + HOR'$$

Carbohydrases catalyze the scission of glycosides:

$$\overline{\qquad\qquad\quad} O \qquad\qquad\qquad\qquad \overline{\qquad\qquad\quad} O$$
$$—CH(OH)—\overset{|}{C}H—O—R' + HOH \rightarrow —CH(OH)—\overset{|}{C}H—OH + HOR'$$

Amidases are a rather ill-defined class of enzymes which may be said to catalyze the hydrolysis of C—N linkages other than a strict peptide link between two amino acids. This large class may be subdivided into the deaminases, amidases, and miscellaneous amidases. Deaminases act on free amino groups:

$$R—NH_2 + HOH \rightarrow R—OH + NH_3$$

Amidases act principally on amide linkages,

$$R\underset{\substack{\|\\ O}}{C}—NH_2 + HOH \rightarrow R\underset{\substack{\|\\ O}}{C}—OH + NH_3$$

but amidine groups also may be attacked.[1] The amidases are overlapping since many proteolytic enzymes hydrolyze amino acid amides. In addition there are other enzymes in the larger class of amidases which fit in neither of these subdivisions. Phosphatases catalyze the hydrolysis of esters of phosphoric acid:

$$RO{-}PO_3H_2 + HOH \rightarrow ROH + HOPO_3H_2$$

In a general sense they are also esterases.

The phosphorolytic enzymes at present are usually classed together. As noted above, these do not insert the elements of water but phosphoric acid instead. So far, no phosphates other than sugar phosphates have been the products of phosphorylase action which may be formulated by analogy with the carbohydrases:

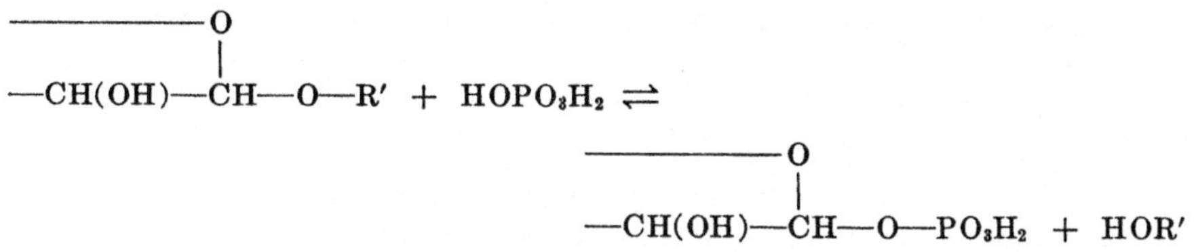

The free energy change in phosphorolysis compared to hydrolysis is much closer to zero, and thus the known phosphorolytic enzymes catalyze biosynthetic reactions as well as catabolic changes. They are implicated chiefly in the synthesis of starch by plants[2,3] and glycogen[4] by animals. Recent work by Kornberg and his associates[5,6] has pointed to a "pyrophosphorylase" in yeast and hog liver that catalyzes the following general reaction:

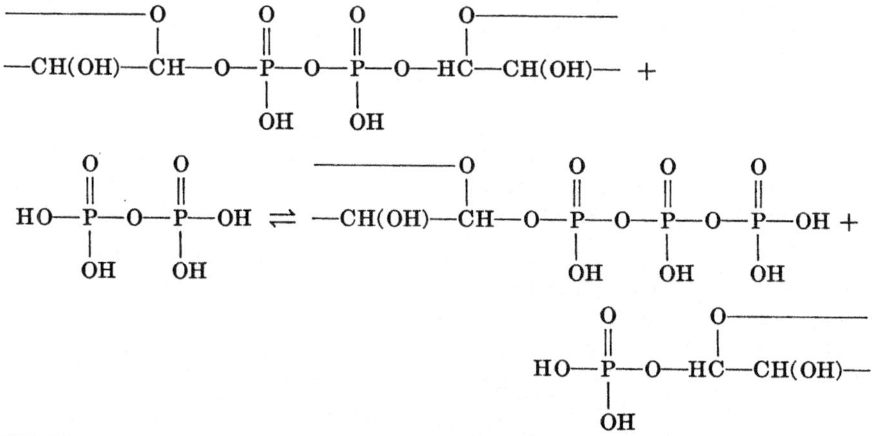

[1] V. A. Knivett, *Biochem. J.* **50,** xxx (1952).
[2] C. S. Hanes, *Proc. Roy. Soc.* (*London*) **B128,** 421 (1940).
[3] C. S. Hanes, *Proc. Roy. Soc.* (*London*) **B129,** 174 (1940).
[4] G. T. Cori and C. F. Cori, *J. Biol. Chem.* **131,** 397 (1939).
[5] A. Kornberg, *J. Biol. Chem.* **182,** 779 (1950).
[6] A. W. Schrecker and A. Kornberg, *J. Biol. Chem.* **182,** 795 (1950).

These reactions lead to a synthesis of cozymase or flavine -adenine dinucleotide from adenosine triphosphate with nicotinamide-ribofuranoside-5-phosphoric acid or riboflavine phosphate respectively.

The classification outlined here is presented mainly for convenience in arranging and relating a wealth of material. Enzymes display an amazing specificity in the reactions each will catalyze. Despite this, these classifications do not represent absolutes in specificity. Some proteinases will, for instance, catalyze the hydrolysis of amino acid esters and amides as well as peptides.

2. Kinetics

a. Kinetics of Non-Enzymatic Hydrolysis. The hydrolytic reaction is a bimolecular reaction. Under most circumstances hydrolysis is carried out in an aqueous system where the concentration of water is in such huge excess compared to the substrate that the water is practically constant. Under these circumstances the velocity, v, of the reaction, that is, the rate of disappearance of the substrate, $-d\,[S]/dt$ would be expected to depend only on the concentration of the substrate, $[S]$. This relationship is given by the equation:

$$v = -\frac{d[S]}{dt} = K_1[S] \tag{1}$$

The substrate concentration, $[S]$, at any time, t, will be equal to the amount of substrate, a, originally present, less the amount, x, that has been converted in the time, t. Then, integrating over the course of the reaction from a to $(a - x)$, there is obtained:

$$K_1 = \frac{1}{t}\cdot\ln\frac{a}{(a - x)} \tag{2}$$

This is the equation expressing the kinetics of a first-order reaction, a theoretical monomolecular reaction, and defining the first-order velocity constant, K_1. In this case, however, it has been called a pseudo monomolecular reaction, since there is actually a second molecule, water, entering the reaction. The decrease in the concentration of the water is so slight relative to the amount present as to have no influence ordinarily.

In an unhindered hydrolytic reaction not catalyzed by enzymes we would expect the relation shown in Fig. 1 to hold in view of equation 1.

b. Kinetics of Enzymatic Hydrolysis. Enzymatic hydrolysis exhibits a different relationship than that shown in Fig. 1. The enzymes urease[7] and invertase,[8] among many others tested, have, at constant concentration, a limiting initial velocity which is not affected by increasing the substrate concentration. This is expressed graphically in Fig. 2.

[7] D. D. Van Slyke and G. E. Cullen, *J. Biol. Chem.* **19,** 141, 211 (1914).

[8] A. J. Brown, *J. Chem. Soc. Trans.* **81,** 373 (1902).

In order to explain this phenomenon in enzymatic catalysis, Michaelis and Menten[9] and subsequent investigators[10, 11, 12] who have extended the theory postulated the reversible formation of a complex, ES, between the enzyme, E, and the substrate, S, prerequisite to the breakdown to the

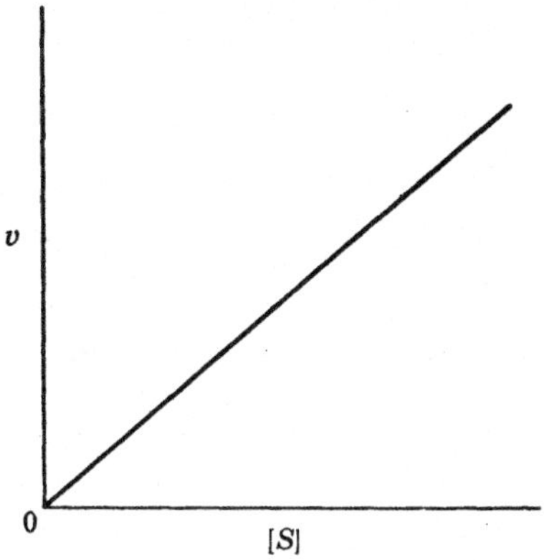

FIG. 1. The effect of increasing substrate concentration on the velocity of a monomolecular reaction.

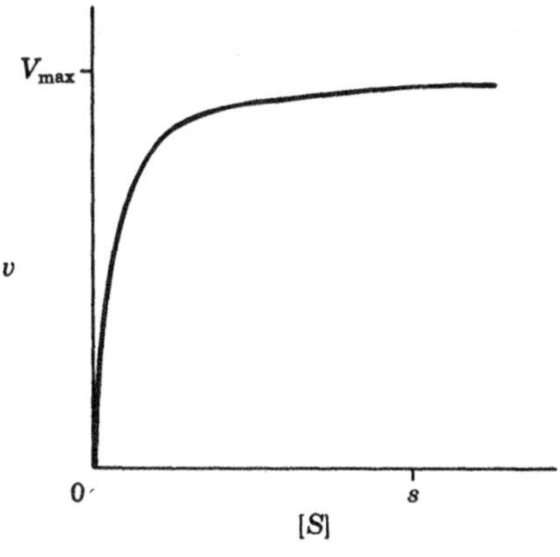

FIG. 2. The effect of increasing substrate concentration on the velocity of a hydrolytic reaction catalyzed by an enzyme.

products, P, of enzymatic catalysis and the simultaneous release of the enzyme:

$$E + S \underset{k_2}{\overset{k_1}{\rightleftharpoons}} ES \xrightarrow{k_3} P + E \tag{3}$$

Here k_1, k_2, and k_3 are velocity constants for the reactions indicated. Im-

[9] L. Michaelis and M. L. Menten, *Biochem. Z.* **49**, 333 (1913).
[10] G. E. Briggs and J. B. S. Haldane, *Biochem. J.* **19**, 338 (1925).
[11] D. D. Van Slyke, *Advances in Enzymol.* **2**, 33 (1942).
[12] A. C. Walker and C. L. A. Schmidt, *Arch. Biochem.* **5**, 445 (1944).

plicit in this theory is the picture of an enzyme molecule in a medium in which the substrate can be at such a high concentration that, as soon as any ES decomposes to yield the products of hydrolysis, the active enzyme liberated immediately combines with substrate to form more ES. Thus, at this substrate concentration (s in Fig. 2), the enzyme is said to be saturated and the velocity is at its maximum (V_{max} in Fig. 2) determined by the rate of the decomposition of ES.

The concentration of the free enzyme, $[E]$, is the amount of enzyme, $[E_0]$, originally present less the part that has gone to form ES:

$$[E] = [E_0] - [ES] \tag{4}$$

For the equilibrium of equation 3 and substituting for $[E]$ as defined by equation 4 the following expression for the equilibrium constant is derived:

$$K_m = \frac{[E]\cdot[S]}{[ES]} = \frac{([E_0] - [ES])[S]}{[ES]} \tag{5}$$

Rearrangement of equation 5 yields:

$$[ES] = \frac{[E_0]\cdot[S]}{K_m + [S]} \tag{6}$$

If it assumed that k_1/k_2 is very much greater than k_3 (equation 3), then the measured velocity of the enzyme reaction will depend on the velocity of the decomposition of ES into E and P:

$$v = k_3[ES] \tag{7}$$

One can then substitute for $[ES]$ defined in equation 6 and use the differential for the velocity as explained in equation 1:

$$v = -\frac{d[S]}{dt} = \frac{k_3\cdot[E_0]\cdot[S]}{K_m + [S]} \tag{8}$$

By employing the definitions of a and x as used in equation 2 the integrated form of equation 8 becomes:

$$k_3\cdot[E_0]\cdot t = K_m\cdot\ln\frac{a}{a - x} + x \tag{9}$$

This might be considered the general equation for the kinetics of hydrolytic enzymes behaving according to the Michaelis-Menten theory. The total enzyme concentration, $[E_0]$, has, for the purpose of this discussion, been kept constant.

In regions of low initial substrate concentrations, a, the term K_m ln $a/(a - x)$ predominates the right-hand member of equation 9 and the reaction kinetics will approach those of a first-order reaction (equation 2). This is apparent in the region not much below substrate concentration s

in Fig. 2. The actual value of s, the substrate concentration required to saturate the enzyme, depends upon the magnitude of K_m, for, as can be seen from equation 9, a large value for K_m will magnify the range of initial substrate concentration, a, over which first-order kinetics will hold. In regions of high initial substrate concentration the term $K_m \ln a/(a - x)$ will approach zero and equation 9 will reduce to:

$$k_3 \cdot [E_0] \cdot t = x \tag{10}$$

Since E_0 is held constant:

$$x = K_0 \cdot t \tag{11}$$

The reaction then follows zero-order kinetics; the quantity of substrate decomposed, x, depends only on time. If K_m is small, the term $K_m \ln a/(a - x)$ will approach zero more rapidly as a is increased and zero-order kinetics will occur at relatively low substrate concentrations.

The kinetics of an enzymatically catalyzed reaction are, therefore, a combination of first-order and zero-order kinetics as defined in equation 9. At low substrate concentrations first-order kinetics are evident and the kinetics become zero order as the substrate concentration is raised.

To evaluate K_m two approaches have been used.[9, 13] At the substrate concentration (s, Fig. 2) required for the enzymatic reaction to reach its maximum velocity (V_{max}, Fig. 2) practically all the active enzyme present is combined with the substrate. Thus at V_{max} (cf. equation 4),

$$[E_0] = [ES] \tag{12}$$

and according to conditions for Michaelis-Menten kinetics as in equation 7:

$$V_{max} = [ES] \cdot k_3 = [E_0] \cdot k_3 \tag{13}$$

Substitution of V_{max} in equation 8 gives:

$$v = \frac{V_{max} \cdot [S]}{K_m + [S]} \tag{14}$$

Rearranging this equation to

$$K_m = [S] \left(\frac{V_{max}}{v} - 1 \right) \tag{15}$$

describes the rectangular hyperbola in Fig. 2. When $\frac{1}{2}V_{max} = v$, equation 14 reduces to $K_m = [S]$. One method, therefore, for measuring K_m is to plot the velocity of the enzyme reaction against the substrate concentration which should reach high enough values to saturate the enzyme. The substrate concentration at one-half this maximal velocity is the K_m.[9]

[13] H. Lineweaver and D. Burk, *J. Am. Chem. Soc.* **56**, 658 (1934).

A more convenient and accurate method for estimating K_m is that of Lineweaver and Burk.[13] If the reciprocals of both sides of equation 14 are taken, the following equation is obtained:

$$\frac{1}{v} = \frac{1}{[S]}\left(\frac{K_m}{V_{\max}}\right) + \frac{1}{V_{\max}} \tag{16}$$

This is the equation for a straight line with slope $K_m/V_{\max}$ and ordinate intercept $1/V_{\max}$ when the reciprocal of the velocity is plotted against the reciprocal of non-saturating substrate concentrations as in Fig. 3. By extrapolation of the straight line the ordinate intercept $1/V_{\max}$ is found and $V_{\max}$ can be calculated. From the slope, $K_m/V_{\max}$ is found and K_m

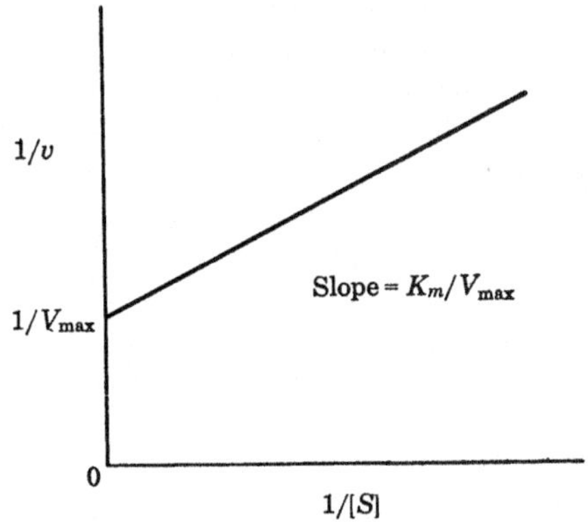

FIG. 3. The reciprocal of the substrate concentration plotted against the reciprocal of the velocity of an enzymatic reaction according to the method of Lineweaver and Burk.

can be calculated using $V_{\max}$. This method is simple because it does not require that the enzyme be saturated with substrate and thus K_m can be determined from velocity measurements at a minimum of two substrate concentrations. It also has the advantage that the data can be treated statistically so that the straight line may be defined by the method of least squares. Needless to say, this method is far more objective than drawing a curve that gives the "best fit" as often must be done if $[S]$ is plotted against v directly. It is advisable, of course, to use more than two points, for if further data deviates greatly from a straight line it is unlikely that the enzyme reaction is progressing according to Michaelis-Menten kinetics.

3. INHIBITION

Inhibitors of enzyme reactions are generally of two kinds, competitive and non-competitive. Competitive inhibitors compete with substrate molecules for the catalytically active sites on the enzyme molecules. Thus competitive inhibition can be expressed by,

$$E + I \rightleftharpoons EI \tag{17}$$

where E is the enzyme, I is the inhibitor, and EI is the enzyme-inhibitor complex which is unable to yield any products of reaction. The usual substrate combination (equation 3) is going on in the same system. The equilibrium constant, K_i, for the formation of EI is as follows:

$$K_i = \frac{[E][I]}{[EI]} \tag{18}$$

In the presence of a competitive inhibitor the maximum velocity remains unchanged. If the competitive inhibitor combines reversibly with the enzyme at the catalytically active site, then once ES is formed the inhibitor has no effect. If the substrate concentration is increased in the presence of a constant quantity of competitive inhibitor, the chances for forming ES rather than EI are greatly increased and reach finally the limiting concentration of ES to the exclusion of EI.

Although the concentration of free enzyme, $[E]$, must now be represented as

$$[E] = [E_0] - [ES] - [EI] \tag{19}$$

the relations expressed by equations 7 and 8 still hold true and thus the following effect on the Lineweaver-Burk plot can be derived:[14]

$$\frac{1}{v_i} = \frac{1}{[S]}\left(\frac{K_m}{V_{\max}}\right)\left(1 + \frac{[I]}{K_i}\right) + \frac{1}{V_{\max}} \tag{20}$$

The velocity in the presence of a competitive inhibitor is v_i. The slope of the line obtained by plotting $1/v_i$ against $1/[S]$ has increased by the factor $(1 + [I]/K_i)$ over the data from the uninhibited reaction (cf. equation 16). Thus, only the slope has changed and not the intercept. True competitive inhibition is characterized by an increase in slope in the Lineweaver-Burk plot, but by no change in intercept. Use of this method in demonstrating the competitive inhibition of carboxypeptidase by D-phenylalanine and D-histidine is well illustrated by the work of Elkins-Kaufman and Neurath.[15]

Non-competitive inhibition is the result of the combination of the inhibitor with non-catalytic sites on the enzyme molecule so as to render it inactive. In this case it is evident that $V_{\max}$ will be lowered to $V^i_{\max}$ since the concentration of active enzyme is effectively lowered. The final velocity

[14] P. W. Wilson, *in* Lardy, Respiratory Enzymes, rev. ed., Burgess Publishing Co., Minneapolis, 1949, p. 16.

[15] E. Elkins-Kaufman and H. Neurath, *J. Biol. Chem.* **175**, 893 (1948).

equation is:[14]

$$\frac{1}{v_i} = \left[\frac{1}{[S]}\left(\frac{K_m}{V^i_{\max}}\right) + \frac{1}{V^i_{\max}}\right]\left(1 + \frac{[I]}{K_i}\right) \tag{21}$$

Both the slope *and* the ordinate intercept are increased by the factor $(1 + [I]/K_i)$ in non-competitive inhibition.

4. Effect of Enzyme Concentration

In the previous sections the enzyme concentration was held constant for purposes of discussion. It is obvious from equation 9 that the velocity of an enzyme reaction is proportional to the amount of enzyme present. Van Slyke[11] has very well presented the methodology of measuring the concentration of hydrolytic enzymes. For studies on purification it is often the case that the initial velocity constants, proportional to the enzyme concentration, can be used. In situations where the substrate concentration is kept low or K_m is large, first-order kinetics predominate to the extent that K_1 (equation 2) is proportional to the enzyme concentration. In cases where the substrate concentration can be raised high enough to saturate the enzyme, that is, where K_m is small, the initial velocity of the reaction may be directly proportional to the enzyme concentration.[16]

Complications arise when a product of enzymatic reaction[17-20] or the substrate[21, 22, 23] itself, inhibits the enzyme reaction. In cases where inhibition is effected by the products of the reaction it is necessary to select as short a reaction time as possible in which to calculate the velocity. This type of effect by inorganic phosphate on the kinetics of prostatic acid phosphatase has been rigorously presented by Schønheyder.[24] Neurath and Schwert[25] have analyzed the kinetics of reactions competitively inhibited by reaction products.

When it is necessary to compare the activity of preparations of the same enzyme which may be inhibited during the course of reaction by accumulating reaction products, the inverse time-enzyme relation may be used.[11] Here the time required to hydrolyze a given fraction of the sub-

[16] G. W. Irving, Jr., J. S. Fruton, and M. Bergmann, *J. Biol. Chem.* **138,** 231 (1941).
[17] O. Bodansky, *J. Biol. Chem.* **120,** 555 (1937).
[18] R. V. MacAllister, K. M. Harmon, and C. Niemann, *J. Biol. Chem.* **177,** 767 (1949).
[19] K. M. Harmon and C. Niemann, *J. Biol. Chem.* **178,** 743 (1949).
[20] H. T. Huang and C. Niemann, *J. Am. Chem. Soc.* **73,** 1541 (1951).
[21] K. J. Laidler and J. P. Hoare, *J. Am. Chem. Soc.* **71,** 2699 (1949).
[22] G. A. Alles and R. C. Hawes, *J. Biol. Chem.* **133,** 375 (1940).
[23] K.-B. Augustinsson, *Arch. Biochem.* **23,** 111 (1949).
[24] F. Schønheyder, *Biochem. J.* **50,** 378 (1952).
[25] H. Neurath and G. W. Schwert, *Chem. Revs.* **46,** 69 (1950).

strate will vary inversely as the amount of enzyme present. For constant substrate concentrations the concentration of reaction products will be the same and thus a comparison can be made.

5. Effect of pH

Most hydrolytic enzymes show a more or less clearly defined region of hydrogen ion concentration at which the velocity is the greatest. This is the region of the pH optimum for activity. The pH-velocity curve may be symmetrical or not, but for a single enzyme it usually contains only one point of inflection. When experimental data lead to a pH-velocity curve with two peaks, it is often suspected that two enzymes are involved, each having a different pH optimum for activity. This has been aptly demonstrated in studies on certain phosphatases.[26] The reviews of Johnson[27] and of Neurath and Schwert[25] treat the theory of the effect of pH.

The shape of the pH-velocity curve and the pH optimum itself can be affected by activators or inhibitors. The shape of the pH-velocity curve of beef liver arginase, for instance, is altered by the activating ions, cobalt and nickel.[28]

The dependence of the velocity of an enzyme-catalyzed reaction on the hydrogen ion concentration indicates that all measurements, to be comparable, must be made at the same pH. This pH must be one at which the enzyme is active although not necessarily the optimum pH. Since the products of a reaction may differ markedly in acidity from the reactants, it is important to run all enzyme reactions in well-buffered systems. Moreover, the buffer should be of such a constitution as not to interfere with the activity of the enzyme, especially at the lower hydrogen ion concentrations, by complexing or precipitating metal activators. Certain buffers may also have an intrinsic effect, so it is practical to try several available at a given pH. Likewise, the concentration of buffers used at various pH values should be approximately the same ionic strength so that differences because of varying salt concentrations will be minimized.

An asymmetrical pH-velocity curve may be due to the instability of the enzyme at hydrogen ion concentrations very far from neutrality. The pH-stability curve may be determined by exposing the enzyme to varying hydrogen ion concentrations for a definite time interval and then measuring its activity at the optimal pH. In some cases the pH optimal for activity is in a range which inactivates the enzyme.

[26] J. Roche, *in* Sumner and Myrbäck, The Enzymes, Academic Press, New York, 1950, Vol. I, Part 1, p. 473.

[27] M. J. Johnson, *in* Lardy, Respiratory Enzymes, rev. ed., Burgess Publishing Co., Minneapolis, 1949, p. 51.

[28] M. S. Mohamed and D. M. Greenberg, *Arch. Biochem.* **8**, 349 (1945).

6. Effect of Temperature

Heat has two effects on enzyme systems: firstly, it accelerates the chemical reaction, and secondly, it inactivates the enzyme. The Arrhenius equation,

$$\frac{d \ln k'}{dt} = \frac{\mu}{RT^2} \tag{22}$$

relates the temperature, t, to the rate of reaction, k'. In this equation R is the gas constant (1.986 cal. per degree per mole), k' is the over-all reaction constant, T is the absolute temperature, and μ is the activation energy. Integration of equation 22 gives

$$\ln \frac{k_2'}{k_1'} = \frac{\mu}{R}\left(\frac{1}{T_1} - \frac{1}{T_2}\right) \tag{23}$$

where k_1' and k_2' may be any comparable reaction constants at absolute temperatures, T_1 and T_2, respectively. Thus a plot of log k against $1/T$ yields a straight line from the slope of which μ may be calculated (μ = slope/4.58).

In enzymatic reactions deviations from a straight line will occur at temperatures at which the enzyme begins to denature. The so-called optimum temperature for an enzyme reaction is a composite of the two opposing reactions, denaturation of the enzyme and acceleration of the chemical reaction. The activity of the enzyme should be measured as close to the initial velocity as possible so as to minimize the denaturation effect. The optimal temperature of an enzymatic reaction has little meaning unless it is considered in conjunction with the heat stability of the enzyme. Even though an enzyme may have its optimal temperature for activity when measured as initial velocity at say 40°, the rate of denaturation of the enzyme may be so high at this temperature that its effectiveness is greater at a lower temperature at which it can remain active for a longer time interval.

For each rise of 10 degrees in temperature in the range which does not inactivate the enzyme the velocity of an enzymatic reaction is increased by about two to three times. Between 20 and 30° a twofold rise in rate ($Q_{10} = 2$) corresponds to an activation energy of 12,300 cal. per mole, and a threefold increase ($Q_{10} = 3$) corresponds to an activation energy of 19,500 cal. per mole. In ordinary temperature ranges the activation energy of reactions catalyzed by hydrolytic enzymes falls roughly within this range.

Unless the catalytic surface of the enzyme is altered, an environmental change should theoretically have no effect on the activation energy.[29] Lipase, trypsin, and pepsin follow the Arrhenius equation, but below 0°

[29] I. W. Sizer, *Advances in Enzymol.* **3,** 35 (1943).

the activation energy becomes four or five times greater. It is not unusual to find that changes in pH, electrolyte concentration, and substrate concentration affect μ, but in these cases there is likely to be an effect on the catalytically active center. Changes in the nature of the substrate will cause a change in μ. The degree of purity of the enzyme has little effect on the activation energy, and in some cases it happens that enzymes from different sources but catalyzing the same reactions have the same activation energy.[29]

7. Zymogens

The extracellular proteinases, pepsin, trypsin, and chymotrypsin, are secreted as inactive precursors, zymogens, which are transformed into the active enzymes at the site of their action. The transformation of pepsinogen to pepsin is accomplished by acid or by pepsin and is accompanied by the appearance of 15% of the nitrogen in non-protein form.[30] The molecular weight of pepsinogen is about 4000 units greater than pepsin, but this may be within experimental error. It has been tempting to visualize the generation of pepsin from pepsinogen as an "unmasking" of the catalytically active site by a hydrolytic process catalyzed by acid or pepsin with the non-protein nitrogen representing in fact the remains of a low molecular weight "mask." Trypsin is generated from inactive trypsinogen by the action of enterokinase, a protein constituent of the secretion of the digestive glands of the small intestine, and by the action of trypsin itself. The action of enterokinase has the properties of an enzyme reaction. End-group assays with dinitrofluorobenzene[33a] and with carboxypeptidase[33b] have indicated that chymotrypsinogen and trypsinogen are cyclic peptides, but that α-chymotrypsin and trypsin are at least partly linear. Thus, the action of enterokinase and of trypsin may be to hydrolyze peptide bonds of these cyclic inactive precursors in one or more places to yield an active enzyme of approximately the same molecular weight as the zymogen.

The transformation of the respective zymogens to pepsin and trypsin follow, in uncomplicated systems, the kinetics of an autocatalytic reaction,

$$\frac{dA}{dt} = KA(A_e - A) \qquad (24)$$

where A is the activity at time, t, and A_e is the final activity. This is characteristic of a reaction that gives rise to a product which in turn catalyzes that reaction producing itself.

On the other hand, chymotrypsinogen is transformed into chymotrypsin only by the action of trypsin. In this case the reaction follows first-order kinetics. The enzymatic activation of chymotrypsinogen can proceed along

[30] R. M. Herriot, *J. Gen. Physiol.* **22**, 65 (1938).

alternative pathways to yield a variety of chymotrypsins[31] (denoted as α-, β-, γ-, π-, and δ-chymotrypsin) which differ in physical properties or specific activity but not in substrate specificities.[32, 33]

In the examples cited above, both zymogens and enzymes have been crystallized and therefore studied in relatively pure condition. There is evidence, however, that pancreatic carboxypeptidase is secreted as a zymogen, since fresh pancreatic extracts contain little or no carboxypeptidase activity. Upon incubation at 37° or treatment with trypsin the formation of carboxypeptidase occurs. This procarboxypeptidase has been partially purified.[34]

8. Cofactors

The activity of no hydrolytic enzyme has so far been shown to depend on the simultaneous presence of a specific organic compound of low molecular weight other than the substrate. This is in distinct contrast to many other enzymes such as the decarboxylases, the transaminases, and the oxidative enzymes. It has been alleged that the activity of pig pancreatic amylase is dependent on the presence of *m*-inositol,[35] but this claim has been contested by Fischer and Bernfeld,[36] who used purer preparations of the enzyme.

Many hydrolytic enzymes, however, do require certain metal ions[37] for activity, notably divalent cations in the range 12 to 30 in atomic number. In a few cases metal ions have a stabilizing effect rather than a true activation. The presence of halide ions are required for activity by the animal amylases. Some of the proteolytic enzymes and also muscle phosphorylase lose activity unless reducing agents such as cysteine or HCN are present. But, other than simple ions and reducing agents, there are no cofactors necessary in enzymatic hydrolysis.

Muscle phosphorylase, on the other hand, reaches its full activity only in the presence of adenylic acid.

9. Specificity

The concept of enzyme specificity has gradually changed during the years. Too often the infrequent cases of absolute specificity towards a

[31] C. F. Jacobsen, *Compt. rend. trav. lab. Carlsberg. Sér. chim.* **25**, 325 (1947).
[32] M. Kunitz, *J. Gen. Physiol.* **22**, 207 (1938).
[33] G. W. Schwert and S. Kaufman, *J. Biol. Chem.* **180**, 517 (1949).
[33a] P. Desnuelle, M. Rovery, and C. Fabre, *Compt. rend.* **233**, 1496 (1951).
[33b] J. A. Gladner and H. Neurath, *Biochim. et Biophys. Acta* **9**, 335 (1952); E. W. Davie and H. Neurath, *J. Am. Chem. Soc.* **74**, 6305 (1952).
[34] M. L. Anson, *J. Gen. Physiol.* **20**, 777 (1937).
[35] R. L. Lane and R. J. Williams, *Arch. Biochem.* **19**, 329 (1948).
[36] E. H. Fischer and P. Bernfeld, *Helv. Chim. Acta* **32**, 1146 (1949).
[37] A. L. Lehninger, *Physiol. Revs.* **30**, 393 (1950).

single molecule are given as examples when, in fact, the majority of enzymes show a relative specificity to an array of substrates. Of the hydrolytic enzymes only urease seems to show absolute specificity in that urea is the only molecule attacked.[38] The least attempt to modify the urea molecule results in inactivity as a substrate for urease.[39] Nevertheless, enzymes which will show relative specificity towards some groups on the substrate molecule may adopt an absolute specificity towards others. The tremendous quantity of work on β-glucosidase carried out by Helferich,[40] Pigman,[41] and Veibel,[42] has shown that this enzyme is absolutely specific with regard to carbon atoms 1, 2, and 3 on the sugar molecule and relatively specific with regard to carbons 4, 5, and 6 and also the aglycon.

The distinction between esterases and lipases is one of relative specificity. The lipases hydrolyze triglycerides more readily than they do esters of monohydric alcohols, whereas the reverse is true with regard to the esterases. The absolute specificity of esterases, therefore, applies only to the bond hydrolyzed, the ester, and not to vicinal groups. Relative specificity is influenced by the nature of the acid and alcohol moieties of the ester and by the stereochemistry of these components.[43] Substrates for the study of lipase action are most often chosen for experimental convenience, such as solubility in water, rather than as possible natural substrates of these enzymes. In such a situation it is natural that confusion arises.

II. Esterases

1. Pancreatic Lipase

Most of the investigations concerning this enzyme have been done with fairly crude glycerol extracts of pancreatic tissue. Bamann and Laeverenz[44] reported the preparation of crystalline pancreatic lipase, but confirmation of their results is lacking.[45] The kinetics of pancreatic lipase action have been treated by Schønheyder and Volqvartz[46] who used racemic 1-caprylyl

[38] J. B. Sumner, *in* Sumner and Myrbäck, The Enzymes, Academic Press, New York, 1951, Vol. I, Part 2, p. 873.

[39] The report of Shaw and Kistiakowski (*J. Am. Chem. Soc.* **72,** 634, 1950) that biuret is hydrolyzed by crystalline urease was later withdrawn (*ibid.* p. 2817) by these authors when they found that their preparation of biuret may very well have contained urea as an impurity.

[40] B. Helferich, *Ergeb. Enzymforsch.* **2,** 74 (1933); *ibid.* **7,** 83 (1938).

[41] W. W. Pigman, *Advances in Enzymol.* **4,** 41 (1944).

[42] S. Veibel, *in* Sumner and Myrbäck, The Enzymes, Academic Press, New York, 1951, Vol. I, Part 1, p. 583.

[43] E. Bamann and R. Ammon, *in* Bamann and Myrbäck, Die Methoden der Fermentforschung, Thieme, Leipzig, 1941; Academic Press, New York, 1945, p. 1704.

[44] E. Bamann and P. Laeverenz, *Z. physiol. Chem.* **223,** 1 (1934).

[45] R. A. Boissonnas, *Helv. Chim. Acta* **31,** 1577 (1948).

[46] F. Schønheyder and K. Volqvartz, *Biochim. et Biophys. Acta* **6,** 147 (1950).

glycerol as substrate. Again, however, the enzyme preparation was comparatively crude for such a detailed study. The investigations of Fodor[47] have pointed to the existence of two esterases in pancreatic extracts. The lipase which hydrolyzes triglycerides, such as olive oil, tributyrin, or tripropionin, is more resistant to treatment with heat and alkali than the esterase which hydrolyzes ethyl butyrate and other esters of monohydric alcohols. Although this evidence is indirect, Boissonnas[45] in purifying swine pancreatic lipase 17.2-fold interprets the varying activities of his fractions against the substrates, Tween 20, ethyl butyrate, ethyl acetate, and triacetin, as indicative of the presence of more than one enzyme. Until the esterase activities of pancreatic extracts are resolved, kinetic studies which are directed towards the elucidation of the mechanisms of lipolysis will be disputable.

The use of water-soluble substrates like Tween 20 (polyoxyethylene sorbitan monolaurate), triglycerides of volatile fatty acids, and esters of low molecular weight for studies of lipase action necessarily neglects the real problem of the availability of the predominant, relatively insoluble dietary triglycerides of C_{16} and C_{18} fatty acids to the lipase molecules acting in an aqueous medium. In these systems where a large portion of the substrate is in a phase separate from the enzyme, the extent and the stability of the dispersion of the fatty phase in the aqueous phase has an effect upon the rate of lipase action. Thus the optimum pH for pancreatic lipase action on the triglycerides of higher fatty acids may reflect the pH at which emulsification of the fat is the greatest and is therefore alkaline, pH 7 to 9. For Tween 20, a water-soluble substrate, the optimum pH is at 7.[48] Bile salts, soaps, and foreign proteins which activate lipase under alkaline conditions may act by emulsifying the substrate[49] rather than by directly affecting the enzyme protein. *In vivo* the rate-determining step may be the degree of emulsification of the fat rather than the velocity of the decomposition of the enzyme-substrate complex.

The ability of pancreatic lipase to synthesize as well as hydrolyze esters is well known.[50] The standard free energy of hydrolysis of ethyl acetate is small, about −400 cal.,[51] and this probably does not vary much for most esters.[52] The reaction is essentially reversible. In an aqueous system the

[47] P. J. Fodor, *Arch. Biochem.* **25**, 223 (1950); *ibid.* **26**, 307 (1950).

[48] R. M. Archibald, *J. Biol. Chem.* **165**, 443 (1946).

[49] K. Holwerda, *Biochem. Z.* **296**, 1 (1938).

[50] J. H. Kastle and A. S. Loevenhart, *J. Am. Chem. Soc.* **24**, 491 (1900). Also see the review by R. Ammon and M. Jaarma, *in* Sumner and Myrbäck, The Enzymes, Academic Press, New York, 1950, Vol. I, Part 1, p. 390.

[51] G. S. Parks and H. M. Huffman, Free Energies of Some Organic Compounds. Chemical Catalog Co., New York, 1932.

[52] L. P. Hammett, Physical Organic Chemistry, McGraw-Hill Book Co., New York, 1940.

free energy of hydrolysis will be influenced by the concentration of water and the pH. The solubility of the triglyceride, however, renders the effect of water constant. At low hydrogen ion concentrations the hydrolytic reaction will be more exergonic because the liberated fatty acid will be converted into soap and thus cannot take part in the synthetic reaction. The degree to which this effect occurs will vary with the acid strength of the fatty acids.

During the course of lipolysis the liberation of free fatty acid lowers the pH. This necessitates large buffer capacities in *in vitro* experiments. In the alkaline range especially, the effect of pH on esterase or lipase action may be a combined effect upon the reactants, the products, and the enzyme protein.

2. Tissue Esterases

The hydrolysis of ethyl butyrate is a ubiquitous phenomenon occurring in animal and plant tissues of many types. The enzyme responsible for the catalysis of this hydrolysis has been purified.[53] In general, tissues contain one or more aliphatic esterases (or "ali-esterases") which have a broad specificity for simple esters, possibly extending to esters of cholesterol and vitamin A.

a. Cholinesterases. Esters of choline are rapidly hydrolyzed by enzymes in pancreas, liver, serum, red blood corpuscles, and conducting tissues. The activity of serum cholinesterase is highest for the hydrolysis of *n*-butyrylcholine,[54] but non-choline esters are also split.[55] The present situation concerning the specificity of cholinesterases has been reviewed with impartiality by Whittaker.[56]

It is necessary to distinguish among three types of esterases: (1) The aliphatic esterases which are tissue esterases of broad specificity, highest for aliphatic esters but extending to choline esters. (2) Cholinesterases which occur mainly in pancreas, liver, and serum, the activity of which is greatest for choline esters. (3) Acetylcholinesterase,[57] present mainly in conducting tissue and red blood corpuscles, which has a relative specificity towards choline esters but is most active in splitting acetylcholine. It is to be noted, however, that the cholinesterases and acetylcholinesterases will to a lesser degree catalyze the hydrolysis of esters of aliphatic alcohols.[58, 59] In the collection of data on cholinesterases it is important that purified

[53] W. M. Connors, A. Pihl, A. L. Dounce, and E. Stotz, *J. Biol. Chem.*, **184**, 29 (1950).

[54] E. Stedman, E. Stedman, and L. H. Easson, *Biochem. J.* **26**, 2056 (1932).

[55] B. Vahlquist, *Skand. Arch. Physiol.* **72**, 133 (1935).

[56] V. P. Whittaker, *Physiol. Revs.* **31**, 312 (1951).

[57] K.-B. Augustinsson and D. Nachmansohn, *Science* **110**, 98 (1949).

[58] D. H. Adams, *Biochim. et Biophys. Acta* **3**, 1 (1949).

[59] L. M. Sturge and V. P. Whittaker, *Biochem. J.* **47**, 518 (1950).

preparations be used that are not contaminated with aliphatic esterases. Specific substrates, acetyl-β-methylcholine and benzoylcholine, have been used to differentiate between acetylcholinesterase (true cholinesterase)[60] and cholinesterase (pseudocholinesterase) respectively.[61]

One of the pillars upon which rests the prevailing theory of the chemical mediation of nerve impulses[62] is the uniqueness, in conducting tissue, of the enzyme that hydrolyzes acetylcholine. The characteristics of acetylcholinesterase that distinguish it from the other cholinesterases are as follows: (1) A small K_m when acetylcholine is the substrate. (2) Inhibition of the hydrolysis of acetylcholine by the substrate so that when the velocity is plotted against substrate concentration a bell-shaped curve results.[23] (3) A rate of hydrolysis that is greatest with acetylcholine, less with propionylcholine, and the least with butyrylcholine. None of these properties is shared by the other cholinesterases.[63] Acetylcholinesterase occurs, however, not only in conducting tissue but also in erythrocytes and cobra venom. The distribution of cholinesterases has been reviewed by Augustinsson.[64]

The mechanism of the action of acetylcholinesterase purified from the electric organs of *Electrophorus electricus* involves the attraction of the positively charged nitrogen of acetylcholine to an anionic site on the enzyme and cleavage of the substrate at an "esteratic" site of a nucleophilic character.[63] The irreversible inhibition by the alkyl phosphates, tetraethyl pyrophosphate (TEPP) and diisopropylfluorophosphate (DFP) may be due to phosphorylation of the nucleophilic esteratic site.[65] The phosphorylation by DFP of the phenolic hydroxyl group of free tyrosine has been demonstrated by Ashbolt and Rydon.[66] Chymotrypsin and citrus fruit acetylesterase are also inhibited by DFP and TEPP.[67]

The synthesis of acetylcholine catalyzed by acetylcholinesterase, as might be expected from the preceding considerations on the reversal of lipase action, will proceed at pH 5.1 in contrast to the hydrolysis which is most rapid at pH 8 to 9.[68] From equilibrium constants Hestrin[68] has calculated the $\Delta F°$ of the hydrolysis of acetylcholine to be about −3200 cal. with an estimated error of 10%.

b. Cholesterol Esterase. The position of this enzyme is controversial. Esterification of cholesterol and hydrolysis of cholesteryl esters take place in serum,[69] pancreas, liver, spleen, intestinal mucosa,[70] and other tissues. The purification of the enzyme has not progressed, and at this point it is

[60] B. Mendel and H. Rudney, *Biochem. J.* **37,** 59 (1943).
[61] B. Mendel, D. B. Mundell, and H. Rudney, *Biochem. J.* **37,** 473 (1943).
[62] D. Nachmansohn, *in* Pincus and Thimann, The Hormones, Academic Press, New York, 1951, Vol. II, p. 515.
[63] D. Nachmansohn and I. B. Wilson, *Advances in Enzymol.* **12,** 259 (1952).
[64] K.-B. Augustinsson, *Acta Physiol. Scand.* **15,** *Suppl.* 52 (1948).
[65] A. S. V. Burgen, *Brit. J. Pharmacol.* **4,** 219 (1949).
[66] R. F. Ashbolt and H. N. Rydon, *J. Am. Chem. Soc.* **74,** 1865 (1952).
[67] E. F. Jansen, M. D. Fellows-Nutting, and A. K. Balls, *J. Biol. Chem.* **175,** 975 (1948); *ibid.* **179,** 201 (1949).
[68] S. Hestrin, *Biochim. et Biophys. Acta* **4,** 310 (1950).
[69] W. M. Sperry, *J. Biol. Chem.* **113,** 599 (1936).
[70] W. Klein, *Z. physiol. Chem.* **254,** 1 (1938).

difficult to discern whether the esterification of cholesterol is due to a separate enzyme or is merely a manifestation of the broad specificity of a general esterase. Fodor[71] has found that the capacities of pancreatic extract to hydrolyze methyl buyrate and cholesteryl butyrate decline in a parallel fashion on treatment with heat.

The esterification of cholesterol is considered by Nieft and Deuel[72] and by Saviano and Baccari[73] to be due to an enzymatic activity separate from that which hydrolyzes cholesteryl esters. With but one exception the hydrolysis of cholesteryl esters has taken place in the same extract as that in which esterification of cholesterol occurred, and at a higher pH than esterification.[71,72,74] It is difficult to reconcile the concept of the true catalytic function of enzymes with the idea of completely separate enzymes involved in esterification and hydrolytic reactions.

The formation and hydrolysis of fatty acid esters of cholesterol play an important part in the "fatty liver" syndrome.[75,76,77] Pathological fatty livers contain an inordinate proportion of cholesteryl esters of fatty acids. Under treatment with lipotropic agents these esters are the most resistant and tend to leave the liver last.

3. Phospholipases

The phospholipases hydrolyze lecithin, cephalin, and other phospholipids. The products of the hydrolysis vary according to the bonds which are attacked, but only two phospholipases will be discussed here. The reader is referred to Zeller's[78] review for a complete discussion of these enzymes.

Phospholipase A[79] hydrolyzes the ester linkage of the unsaturated fatty acid in an α position on the glycerol moiety of lecithin to yield lysolecithin, a powerful hemolytic agent.[78] The enzyme occurs in snake venom and in tissue extracts including the pancreas. Both lecithins and cephalins are split, but triglycerides and sphingomyelins are not attacked.[80,81]

Phospholipase B hydrolyzes the ester linkage of the saturated fatty acid in the β position on the glycerol moiety of lysolecithin. Thus the

[71] P. J. Fodor, *Arch. Biochem.* **26,** 331 (1950).
[72] M. L. Nieft and H. J. Deuel, Jr., *J. Biol. Chem.* **177,** 143 (1949).
[73] M. Saviano and V. Baccari, *Arch. sci. biol.* (*Italy*) **31,** 22 (1946).
[74] L. Swell, J. E. Byron, and C. R. Treadwell, *J. Biol. Chem.* **186,** 543 (1950).
[75] J. P. Peters and D. D. Van Slyke, Quantitative Clinical Chemistry, Interpretations, The Williams and Wilkins Co., Baltimore, 1946, Vol. I, p. 423.
[76] I. L. Chaikoff and C. Entenman, *Advances in Enzymol.* **8,** 171 (1948).
[77] E. W. McHenry and J. M. Patterson, *Physiol. Revs.* **24,** 128 (1944).
[78] E. A. Zeller, *in* Sumner and Myrbäck, The Enzymes, Academic Press, New York, 1951, Vol. I, Part 2, p. 986.
[79] S. Belfanti, A. Contardi, and A. Ercoli, *Ergeb. Enzymforsch.* **5,** 213 (1936).
[80] D. Fairbairn, *J. Biol. Chem.* **157,** 633 (1945).
[81] S. J. Thannhauser and N. F. Boncoddo, *J. Biol. Chem.* **172,** 141 (1948).

hydrolysis of lecithin is carried to glycerylphosphorylcholine and free fatty acid. This enzyme is present in pancreas and other tissues but is absent from snake venom.

III. Proteolytic Enzymes

1. Proteinases

Proteinases catalyze the hydrolysis of peptide linkages in proteins. Those proteinases which have been purified have also been shown to have peptidase activity, that is, they catalyze the hydrolysis of synthetic polypeptides of known structure. Because the proteinases split proteins to yield lower molecular weight peptones, proteoses, and peptides they have been termed endopeptidases; they are capable of hydrolyzing peptide bonds that are not terminal. Exopeptidases can hydrolyze only peptide bonds that involve terminal amino acids. The synthesis of peptide substrates has been authoritatively reviewed by Fruton.[82]

The definition of the scope of the specificity of proteolytic enzymes with regard to the chemical configurations in the vicinity of the hydrolyzable peptide bond has necessarily been restricted to their peptidase activity in order that substrates of known chemical structure can be used. The familiar classification of proteinases on the basis of specificity towards peptides as model substrates[25, 83, 84, 85] has yet to meet rigorous testing on those rare polypeptides of high molecular weight and known constitution. That this is now beginning to come into the range of experimental possibility has been demonstrated by Sanger and Tuppy[86] in their investigations on the products of the enzymatic hydrolysis of polypeptide chains from insulin. Additional information on this question has come from the work of Katchalski[87] on the enzymatic hydrolysis of synthetic poly-L-lysine.

a. Pepsin. Pepsin is the proteinase secreted as pepsinogen by the chief cells of the gastric mucosa. The action of crystalline pepsin results in the hydrolysis of a peptide bond involving the amino group of tyrosine, phenylalanine,[88] diiodotyrosine, or cysteine.[89] The amino acid contributing the carbonyl group to the hydrolyzable bond can be glutamic acid (α-carboxyl), glycine,[88] cystine, cysteine,[90] methionine,[91] phenylalanine, or tyrosine.[89] As

[82] J. S. Fruton, *Advances in Protein Chem.* **5,** 1 (1949).
[83] M. Bergmann and J. S. Fruton, *Advances in Enzymol.* **1,** 63 (1941).
[84] M. Bergmann, *Advances in Enzymol.* **2,** 49 (1942).
[85] J. S. Fruton, *in* Green, Currents in Biochemical Research, Interscience Publishers, New York, 1946, p. 123.
[86] F. Sanger and H. Tuppy, *Biochem. J.* **49,** 488 (1951).
[87] E. Katchalski, *Advances in Protein Chem.* **6,** 123 (1951).
[88] J. S. Fruton and M. Bergmann, *J. Biol. Chem.* **127,** 627 (1939).
[89] L. E. Baker, *J. Biol. Chem.* **193,** 809 (1951).
[90] C. R. Harington and R. V. Pitt-Rivers, *Biochem. J.* **38,** 417 (1944).
[91] C. A. Dekker, S. P. Taylor, Jr., and J. S. Fruton, *J. Biol. Chem.* **180,** 155 (1949).

yet, however, no complete study of the specificity of the peptidase action of pepsin has been reported. That such an investigation is timely is indicated by the study of Sanger and Tuppy[86] on the peptic degradation of fraction B of insulin. They found that the major sites of hydrolysis were the peptide bonds between the following amino acid residues: leucylvalyl, tyrosylleucyl, phenylalanylphenylalanyl, and phenylalanyltyrosyl. Polylysine is not attacked by pepsin.[87]

For hydrolytic action on proteins pepsin works optimally at pH 1.8 to 2. Earlier experiments[88] indicated that pepsin displayed its optimum peptidase activity at pH 4, but recently Baker[89] has shown that, for many peptides, hydrolysis by pepsin takes place most rapidly at pH 2.

b. Trypsin. The structural requirements for peptide substrates of crystalline trypsin have been summarized by Neurath and Schwert[25] partly as follows:

"1. A positive charge separated from the susceptible bond by a chain not shorter than that of arginine or lysine. The maximum distance of separation has not been established.

"2. A polar group in the α position to the carbonyl carbon atom of the susceptible bond.

"3. The susceptible peptide, amide, or ester bond."

Trypsin, chymotrypsin, and carboxypeptidase, long thought to be specific only for the hydrolysis of peptide bonds also display a specific esterase activity.[92] Trypsin, for example, splits ammonia from benzoylargininamide and as well splits ethanol from benzoylarginine ethyl ester. No activity, however, is shown toward ethyl butyrate, a substrate for an aliphatic esterase.[92]

In accord with the structural requirements for tryptic hydrolysis quoted above, the sites of trypsin activity towards insulin fraction B was the peptide bond between lysine and alanine.[86] No other sites were attacked, nor were there present any other sites having the required structure. Synthetic polylysine having an average molecular weight of 4100 was split almost quantitatively to lysyllysine.[87] At this time no data are available to determine whether in this case trypsin acts like α-amylase on amylose, by dividing the polymer until the dipeptide stage is reached, or like β-amylase, by "chewing" off dipeptides one at a time from the end of the polymer. No activity by trypsin is shown towards polyglycine or poly(glycine + leucine).[93]

It has recently been reported from Neurath's laboratory[94] that the

[92] G. W. Schwert, H. Neurath, S. Kaufman, and J. E. Snoke, *J. Biol. Chem.* **172**, 221 (1948).

[93] V. Go and H. Tani, *Bull. Chem. Soc. Japan*, **14**, 510 (1939).

[94] M. M. Green, J. A. Gladner, L. W. Cunningham, Jr., and H. Neurath, *J. Am. Chem. Soc.* **74**, 2122 (1952).

activity of dialyzed trypsin is somewhat increased by the presence of divalent ions of calcium, manganese, cadmium, and cobalt. Since the activity of the dialyzed trypsin in the presence of 0.001 M Ca^{++} was increased only about 1.3-fold, the authors do not regard it as typical metal activation.

c. Chymotrypsin. Both chymotrypsin and trypsin are secreted as zymogens in the pancreatic juice. Neurath[95] has reviewed the properties of chymotrypsin. The structural requirements of crystalline chymotrypsin for its more active peptide substrates are as follows: (1) A hydrolyzable peptide bond. (Chymotrypsin will also hydrolyze an ester linkage.) (2) An aromatic amino acid, tyrosine, phenylalanine, or tryptophane (and, to a lesser extent, methionine), in L-configuration as the donor of the carbonyl group to the hydrolyzable bond. (3) A "secondary peptide" group formed by the amino group of the above amino acid and an acyl group which can be benzoyl, nicotinyl, acetyl, carbobenzoxyglycyl, or glycyl.

Typical peptide substrates for chymotrypsin are benzoyl-L-tyrosylglycinamide[96] and nicotinyl-L-tryptophanamide.[20] The character of its pH-velocity curve is much the same as for trypsin. In accord with the above observations the major action of chymotrypsin on fraction B of insulin is to split the residues, tyrosylleucyl, phenylalanyltyrosyl, and tyrosylthreonyl. A central phenylalanylphenylalanyl pair remains intact, however.[86] Polylysine is split very slowly; 25% of the peptide bonds present are cleaved in about 3 days at 37°. No free lysine was detected.[87] α-Chymotrypsin can be activated about 1.5-fold by the presence of 0.01 M Ca^{++}.[94]

d. Cathepsin. The intracellular proteinase activity of tissues, called cathepsin,[97] has so far been separated into activities classified now as cathepsins A, B, and C.[98] In addition, there is a variety of intracellular peptidases. Cathepsin A, formerly cathepsin I,[98a] hydrolyzes carbobenzoxy-L-glutamyl-L-tyrosine and is therefore considered to have a specificity similar to that of pepsin. On considering that the specificity of pepsin is comparatively ill-defined and that cathepsin A has not yet been purified to any great degree, it is difficult to determine whether the hydrolytic activity described as cathepsin A is actually a proteinase or is merely one of the numerous peptidases of animal tissue.[98]

Cathepsin B hydrolyzes benzoyl-L-argininamide and in other ways bears a specificity pattern similar to trypsin.[98, 98a] In contrast to cathepsin A, cysteine is required for cathepsin B to demonstrate full activity. No com-

[95] H. Neurath, *in* Barron, Modern Trends in Physiology and Biochemistry, Academic Press, New York, 1952, p. 453.

[96] M. Bergmann and J. S. Fruton, *J. Biol. Chem.* **124**, 321 (1938).

[97] R. Willstätter and E. Bamann, *Z. physiol. Chem.* **180**, 127 (1928).

[98] H. H. Tallan, M. E. Jones, and J. S. Fruton, *J. Biol. Chem.* **194**, 793 (1952).

[98a] J. S. Fruton, G. W. Irving, Jr., and M. Bergmann, *J. Biol. Chem.* **141**, 763 (1941).

plete purification of cathepsin B has been achieved, and thus no evidence for its proteinase activity is available.

Beef spleen cathepsin C has been partially purified.[98] Preparations having activity towards glycyl-L-phenylalaninamide some one hundred times greater than crude aqueous extracts have been shown, upon activation by cysteine, to hydrolyze glycyl-L-tyrosinamide and α-L-glutamyl-L-tyrosine ethyl ester. No activity was shown towards carbobenzoxy-L-glutamyl-L-tyrosine and benzoyl-L-argininamide. Cathepsin C, therefore, has substrate requirements similar to those of chymotrypsin. These purified preparations cleaved hemoglobin and serum albumin. The rate of hydrolysis of hemoglobin was enhanced by the addition of cysteine, but the hydrolysis of serum albumin was completely dependent on it. Measured at pH 5, the region of hydrogen ion concentration optimal for the hydrolysis of glycyl-L-phenylalaninamide, the activity in the presence of cysteine toward hemoglobin was increased some sixfold and, towards the peptide substrate, about ninefold.[98]

At best, the evidence is indirect that cathepsins which toward peptides exhibit specificity patterns similar to those of known proteinases can, in fact, hydrolyze proteins.[99] A satisfactory answer to this question must await the purification of the cathepsins to the degree of the extracellular proteinases. The main differences between the extracellular proteinases and cathepsins are as follows: (1) The presence of a reducing agent is essential for the activity of cathepsins B and C but not for trypsin or chymotrypsin. (2) Crystalline tyrpsin inhibitor has no effect on cathepsin B.[99] (3) The pH optima for cathepsins B and C lie between pH 3.5 and 6 compared to pH 8 to 9 favorable for the action of trypsin and chymotrypsin.

2. Peptidases

Numerous peptidases with varying specificities have been described. Peptidases may be defined as enzymes that split peptide bonds of only terminal amino acids and thus are exopeptidases. Aminopeptidases hydrolyze peptide bonds involving acids with a free α-amino group; carboxypeptidase requires that the α-carboxyl group be free. Dipeptidases and tripeptidases will split substrates in which there is only one peptide bond or only two peptide bonds, respectively. Peptidases such as prolinase, prolidase, leucineaminopeptidase, etc., exhibit side-chain specificity requirements as well as "backbone" requirements. Comprehensive reviews of peptidase activity have recently been written by Smith.[100, 101, 102]

Complete activation by divalent metal ions is very common to the pep-

[99] H. R. Gutmann and J. S. Fruton, *J. Biol. Chem.* **174,** 851 (1948).

[100] E. L. Smith, *in* Sumner and Myrbäck, The Enzymes, Academic Press, New York, 1951, Vol. I, Part 2, p. 793.

[101] E. L. Smith, *Advances in Enzymol.* **12,** 191 (1951).

[102] E. L. Smith, *in* Edsall, Enzymes and Enzyme Systems, Harvard University Press, Cambridge, 1951, p. 47.

tidases. Smith and Hanson[103] have interpreted their experiments with certain metal poisons to indicate that pancreatic carboxypeptidase is a magnesium protein. Doubt is cast on this hypothesis by the work of Neurath and De Maria,[104] who find that orthophosphate, pyrophosphate, oxalate, citrate, and cyanide have no effect on the *initial* velocity of the hydrolysis of carbobenzoxyglycyl-L-phenylalanine by crystalline pancreatic carboxypeptidase. This enzyme, however, is competitively inhibited by L-phenylalanine, a product of the hydrolysis of the peptide substrate, and thus the effect of these anions, demonstrated with orthophosphate, was to alter the enzyme-phenylalanine dissociation constant, K_i. Intracellular exopeptidases when activated by sulfhydryl compounds are considered to be cathepsins. In the present discussion the term cathepsin was limited to those intracellular enzymes which are homospecific[84] with extracellular proteinases.

Carboxypeptidase will attack proteins, presumably by splitting off one amino acid at a time from the carboxyl end of the peptide chain. This specificity is responsible for the utilization of carboxypeptidase to determine carboxy terminal amino acids in insulin,[105] fractions from oxidized insulin, lysozyme,[105a] tobacco mosaic virus,[105b] chymotrypsinogen, DFP-chymotrypsin, trypsinogen, and DFP-trypsin.[33b] The rate of hydrolysis of the carboxy terminal amino acid depends on the amino acid split off and the one adjacent to it, as has been demonstrated by the use of peptide substrates.[25] The reaction of carboxypeptidase with proteins containing carboxy terminal amino acids has so far clearly indicated, by the rapid liberation of one specific amino acid, which one is carboxy terminal. If, however, the peptide resulting from the cleavage now bears a structure more easily hydrolyzed by the enzyme, either because of the specificity requirements of the enzyme or of steric disposition, the results of the hydrolysis can be difficult to interpret. An amino acid liberated rapidly is considered to be carboxy terminal, especially when there is only one amino terminal end group. Amino acids liberated subsequently, however, may have been carboxy penultimate on the peptide chain attacked or they may have been carboxy terminal on another chain. When end-group assays reveal more than one amino terminal amino acid, this situation may obtain.

3. Protein Synthesis and Proteolytic Enzymes

The formation of a peptide bond by the extraction of the elements of water is an endergonic reaction. It has been estimated that the standard

[103] E. L. Smith and H. T. Hanson, *J. Biol. Chem.* **176,** 997 (1948); *ibid.* **179,** 803 (1949).

[104] H. Neurath and G. DeMaria, *J. Biol. Chem.* **186,** 653 (1950).

[105] J. Lens, *Biochim. et Biophys. Acta* **3,** 367 (1949); A. R. Thompson, *Nature* **169,** 495 (1952).

[105a] J. I. Harris, *J. Am. Chem. Soc.* **74,** 2944 (1952).

[105b] J. I. Harris and C. A. Knight, *Nature* **170,** 613 (1952).

free energy of formation of the peptide bond is in the neighborhood of 3000 cal.[106] Applied to the formation of hippuric acid in solution under physiological conditions, this value changes little.[107] This apparently large free energy of formation must therefore be supplied by a coupled exergonic reaction if peptide synthesis is to take place by a reversal of the hydrolytic reaction. The rich array of intracellular proteinases and peptidases, each with its own specific structural requirements for hydrolysis and presumably for synthesis, has been ever present to serve as catalysts both for reactions and for discussion.[108, 109]

There is reason to believe that the flat value of 3000 cal. for the formation of *any* peptide bond may be misleading.[109a] Employing a tracer technique, Dobry, Fruton, and Sturtevant[110] have provided evidence that the standard free energy of hydrolysis of benzoyl-L-tyrosylglycinamide to yield benzoyl-L-tyrosine and glycinamide at pH 7.9, 25°, in aqueous solution is -420 ± 50 cal. per mole. Chymotrypsin was used to catalyze this reaction. Heat measurements on the enzymatic hydrolysis of the amide linkage in benzoyl-L-tyrosinamide by chymotrypsin at neutral pH and 25°[111] imply that the free energy of hydrolysis for this amide bond can be ten times greater than the free energy of hydrolysis of benzoyl-L-tyrosylglycinamide described above. Consequently, it does not appear that all peptide linkages can be treated equally when basing arguments on the energy required for their formation.

That proteolytic enzymes can indeed catalyze the reversal of proteolysis was shown by Bergmann and Fraenkel-Conrat[112] in 1937 in their study of the synthesis of insoluble amides and anilides catalyzed by papain and bromelin. These studies have been extended recently by other investigators.[113, 114] The enzymes displayed the same specificity in synthesis as they did in hydrolysis. Moreover, when the free energy of the reaction was lowered by the subseqeunt hydrolysis of the product rather than by precipitation, the synthetic reaction also occurred but the net result of the reaction was a hydrolysis.[115] These experiments serve only as models since *in vivo* such insoluble products as the anilides are not formed nor would protein synthesis result in net hydrolysis.

[106] H. Borsook and H. H. Huffman, *in* Schmidt, Chemistry of Amino Acids and Proteins, Charles C Thomas, Springfield, Ill., 1938, p. 822.

[107] H. Borsook and J. W. Dubnoff, *J. Biol. Chem.* **132,** 307 (1940).

[108] H. Borsook and C. L. Deasy, *Ann. Rev. Biochem.* **20,** 209 (1951).

[109] M. Bergmann and J. S. Fruton, *Ann. N. Y. Acad. Sci.* **45,** 409 (1944).

[109a] J. W. Breitenbach, J. Derkosch, and F. Wessely, *Nature* **169,** 922 (1952).

[110] A. Dobry, J. S. Fruton, and J. M. Sturtevant, *J. Biol. Chem.* **195,** 149 (1952).

[111] A. Dobry and J. M. Sturtevant, *J. Biol. Chem.* **195,** 141 (1952).

[112] M. Bergmann and H. Fraenkel-Conrat, *J. Biol. Chem.* **119,** 707 (1937).

[113] E. Waldschmidt-Leitz and K. Kühn, *Z. physiol. Chem.* **285,** 23 (1950).

[114] T. Yoneya, *J. Biochem.* (*Japan*) **38,** 343 (1951).

[115] O. K. Behrens and M. Bergmann, *J. Biol. Chem.* **129,** 587 (1939).

The syntheses of glutamine[116, 117] and of glutathione[118] are the only analyzed examples, to date, of peptide bond synthesis with the formation of a physiological substrate. In both cases a new peptide bond involving the γ-carboxyl group of glutamic acid is formed, and in both cases the simultaneous decomposition of ATP is obligatory for peptide synthesis. These syntheses, it should be noted, are not catalyzed by the enzymes responsible for the hydrolysis of glutamine or glutathione.

The exchange in peptides of one amino acid or ammonia (as in an amide) for another amino acid or peptide is a reaction catalyzed by proteolytic enzymes and has been termed "transpeptidation."[119] In transpeptidation the total number of peptide bonds remains the same; there is no net hydrolysis or synthesis.

A simple case of the general transpeptidation reaction was the transamidation resulting in the formation of hippuric anilide from aniline and hippuric amide in the presence of papain.[112] Since this reaction proceeded much faster than the enzymatic synthesis of hippuric anilide from hippuric acid and aniline, it seems reasonable to infer that exchange, in the former reaction, took place between the aniline and ammonia. Waley and Watson[120] subjected L-lysyl-L-tyrosyl-L-lysine and L-lysyl-L-tyrosyl-L-leucine to treatment with chymotrypsin and trypsin at pH 7.8. In the hydrolysis mixture of either of these substrates they were able to identify lysyllysine which could have arisen only by rearrangement of the amino acids in peptide bond. The peptide may have reacted with the lysine liberated by hydrolysis:

$$\text{Lysine} + \text{lysyltyrosyllysine} \rightarrow \text{Lysyllysine} + \text{tyrosyllysine}$$

In a similar manner there occurs the formation of γ-glutamyl peptides by sheep kidney extracts from glutathione and free amino acids.[121] The activity of esters as substrates is shown by methionine isopropyl ester condensing to form methionylmethionine and methionylmethionylmethionine catalyzed by trypsin or by chymotrypsin.[122, 123] A similar condensation has been noted for threonine isopropyl ester.[124]

Experiments in Fruton's laboratory with proteolytic enzymes have demonstrated the enzymatic exchange of labeled ammonia[125] and amino

[116] J. F. Speck, *J. Biol. Chem.* **179,** 1405 (1949).
[117] W. H. Elliott, *Nature* **161,** 128 (1948).
[118] R. B. Johnston and K. Bloch, *J. Biol. Chem.* **188,** 221 (1951).
[119] J. S. Fruton, *Yale J. Biol. and Med.* **22,** 263 (1950).
[120] S. G. Waley and J. Watson, *Nature* **167,** 360 (1951).
[121] C. S. Hanes, F. J. R. Hird, and F. A. Isherwood, *Nature* **166,** 288 (1950).
[122] M. Brenner, H. R. Muller, and R. W. Pfister, *Helv. Chim. Acta* **33,** 568 (1950).
[123] M. Brenner and R. W. Pfister, *Helv. Chim. Acta* **34,** 2085 (1951).
[124] M. Brenner, E. Sailer, and K. Rufenacht, *Helv. Chim. Acta* **34,** 2096 (1951).
[125] R. B. Johnston, M. J. Mycek, and J. S. Fruton, *J. Biol. Chem.* **185,** 629 (1950).

acid amides[126] with their counterparts in peptide amides. Finally, the elongation of peptide chains by the process of transpeptidation was shown to occur catalyzed by chymotrypsin or papain.[127] In all cases cited the greatest degree of transpeptidation occurred at hydrogen ion concentrations equal to or less than that optimal for hydrolysis. Chymotrypsin catalyzed the following reaction at pH 7.9:

Benzoyltyrosinamide + glycinamide → Benzoyltyrosylglycinamide + ammonia

Simultaneous with transpeptidation occurs the enzymatic deamidation of benzoyltyrosinamide and the hydrolysis of benzoyltyrosylglycinamide. But, by the time 40% of the benzoyltyrosinamide had been deamidated at least 14% of this amount had been converted to benzoyltyrosylglycinamide.[127] Experiments with partially purified cathepsin C show that at about pH 7 replacement of the amide $—NH_2$ group of glycylphenylalaninamide by argininamide or by another molecule of glycylphenylalaninamide will take place. Serving as both substrate and replacement agent glycylphenylalaninamide will polymerize to form an octapeptide and ammonia. A mixture of cathepsin C and papain at pH 7.4 gave evidence of performing the following transpeptidation when the proper substrates were added:

Carbobenzoxyisoglutamine + phenylalaninamide + argininamide →

Carbobenzoxy-α-glutamylphenylalanylargininamide + $2NH_3$

This coupled transpeptidation is the result of the formation of the glutamylphenylalanyl link by the papain and the phenylalanylarginyl link by cathepsin C.[128] (cf. vol. I p. 205–211)

Transpeptidation, obviously, does not explain the synthesis of new peptide bonds. It is to be expected that glutathione and glutamine will not remain the only examples of the synthesis of new peptide bonds. Transpeptidation by proteinases, each with its own specific requirements, however, may operate to provide redistribution of peptide bonds so that the manifold specificities of proteins are effected.[119, 127]

IV. Carbohydrases

1. Depolymerization of Polysaccharides

The exogenous polysaccharides most often encountered by heterotrophic organisms are starch and cellulose. Starch is the name given to the inhomogeneous reserve polysaccharide in plants consisting of D-glucose residues linked in α configuration at C_1 to either C_4' or C_6' on the next unit.

[126] R. B. Johnston, M. J. Mycek, and J. S. Fruton, *J. Biol. Chem.* **187**, 205 (1950).
[127] J. S. Fruton, R. B. Johnston, and M. Fried, *J. Biol. Chem.* **190**, 39 (1951).
[128] M. E. Jones, W. R. Hearn, M. Fried, and J. S. Fruton, *J. Biol. Chem.* **195**, 645 (1952).

Polymeric and weakly aggregated in solution, starch has no definite molecular weight. Two components, amylose and amylopectin, are recognized and can be separated. Amylose is a straight-chain dextran consisting of 200 to 300 glucose residues all linked 1,4′-α.[129] Amylopectin is a branched polysaccharide. The most frequent glucosidic linkage is 1,4′-α, but at intervals branching occurs by formation of 1,6′-α-glucosidic links. Methylation studies (end-group assay) reveal that the *average* chain length of each branch is 24 to 30 glucose units,[130, 131] that is, for every 24 to 30 1,4′-α links there is a 1,6′-α link. However, if the molecule is regarded as an irregularly branched bush, the branches nearer to the reducing end of the molecule (corresponding to the root of the bush) are shorter than the outside branches.[132] Iodine yields a deep blue color with amylose and a reddish-purple color with pure amylopectin. The blue iodine test is given by a chain longer than 35 glucose units and, by degrees, changes to red at 13 units and finally becomes colorless at 6 units.[133] Glycogen, the reserve polysaccharide of animals, resembles amylopectin. It is more highly branched, however, the average chain length being about 10 units.[134] Glycogen yields a reddish color with iodine. Although the degree of branching in glycogen is greater than in most amylopectins, polysaccharides from different sources display different properties. Thus the amylopectin from corn (*Zea mays*) shows all the properties of glycogen.[135, 136] Rabbit liver glycogen[136a] and corn amylopectin[136b] have been shown to contain fructose residues. Since the disaccharide that was isolated was maltulose, it is possible that the fructose is linked α-1,4. How this will affect the interpretation of the action of glycogenolytic enzymes remains to be seen.

The state of aggregation of starch and of glycogen are in question.[137] Molecular weights vary in amylose and in amylopectin from a single source, depending on the method of fractionation. In addition to dealing with association effects one has to be satisfied with an average molecular weight for the polymer. By end-group assay, amylose has a minimum molecular weight of about 32,000 (for 200 hexose units) to 48,000 (for 300 hexose units). Molecular weight measurements show that amylose may consist of one to three of these chains associated. The minimum molecular weight for

[129] K. H. Meyer, M. Wertheim, and P. Bernfeld, *Helv. Chim. Acta* **23,** 865 (1940); *ibid.* **24,** 378 (1941).
[130] K. Freudenberg and H. Boppel, *Ber.* **73,** 609 (1940).
[131] C. C. Barker, E. L. Hirst, and G. T. Young, *Nature* **147,** 296 (1941).
[132] K. Myrbäck, *Advances in Carbohydrate Chem.* **3,** 251 (1948).
[133] M. A. Swanson, *J. Biol. Chem.* **172,** 825 (1948).
[134] T. G. Halsall, E. L. Hirst, and J. K. N. Jones, *J. Chem. Soc.* **1947,** 1399.
[135] D. L. Morris and C. T. Morris, *J. Biol. Chem.* **130,** 535 (1939).
[136] C. T. Morris, *J. Biol. Chem.* **154,** 503 (1944).
[136a] S. Peat, P. J. P. Roberts, and W. J. Whelan, *Biochem. J.* **51,** xvii (1952).
[136b] W. J. Whelan and P. J. P. Roberts, *Nature* **170,** 748 (1952).

an amylopectin branch is about 3500. A molecule of amylopectin, on the basis of molecular weight determinations, may contain at least 15 branches. The determination of the molecular weight of glycogen is beset by somewhat the same problems. The picture of the glycogen molecule is that of a more highly branched polysaccharide with the number of branches ranging from about 20, upward. The outline of polysaccharide structure given here is necessarily curtailed but sufficient to permit a discussion of the action of amylolytic enzymes. For more detailed accounts the reviews of Meyer[138, 139] and of Myrbäck[132] are recommended.

a. Amylases. Both animals and plants contain amylases which to some degree can hydrolyze starch or glycogen to maltose. Crude preparations of amylases are often referred to as diastases.

(1) *α-Amylase Action.* On treatment with α-amylase amylose is finally hydrolyzed to a mixture of α-maltose (87%) and glucose (13%). Random attack at non-terminal 1,4′-α-glucosidic linkages yields intermediate dextrins of decreasing chain length and, as hydrolysis continues, a mixture of α-maltose and α-maltotriose (4-[α-maltosyl]-D-glucose).[140] Maltotriose is split very slowly to α-maltose and α-glucose.[141] The 1,4′-α linkages of amylopectin are also split randomly to yield a series of dextrins, and finally, after 20 hr. at 20°, a mixture of 19% glucose, 73% maltose, and 8% isomaltose results from the degradation of an amylopectin with 4% end groups.[142] These proportions will, of course, vary with the degree of branching of the amylopectin. The evidence is dubious that α-amylase can hydrolyze 1,6′-α-linkages which occur at branch points in amylopectin. Isomaltose is believed to be 6-(α-D-glucopyranosyl)-D-glucose.[143] Whelan and Roberts [136b] have modified this conception of the action of α-amylase. Salivary α-amylase cannot hydrolyze maltotriose,[143a] and thus any but terminal α-1,4-glucosidic links are attacked. Exhaustive treatment of rabbit liver glycogen with the same enzyme yielded a mixture of dextrins, the lowest of which was a pentasaccharide. These investigators[136b] have postulated, therefore, that each of the three α-1,4 links adjacent to the branch link is resistant to hydrolysis by α-amylase.

If the structure of amylopectin involves 1,6′-α-glucosidic links at branch points and the action of amylase is to hydrolyze only 1,4′-α-glucosidic links, then a portion of the final products of the enzymatic hydrolysis of amylopectin should be the di-

[137] K. H. Meyer, *Advances in Colloid Sci.* **1,** 143 (1942).
[138] K. H. Meyer, *Advances in Enzymol.* **3,** 109 (1943).
[139] K. H. Meyer and G. C. Gibbons, *Advances in Enzymol,* **12,** 341 (1951).
[140] J. M. Sugihara and M. L. Wolfrom, *J. Am. Chem. Soc.* **71,** 3357 (1949).
[141] K. H. Meyer and P. Bernfeld, *Helv. Chim. Acta,* **24,** 359E (1941).
[142] K. H. Meyer and W. F. Gonon, *Helv. Chim. Acta,* **34,** 308 (1951).
[143] K. Myrbäck, *Svensk. Kem. Tid.* **53,** 264 (1941).
[143a] P. J. P. Roberts and W. J. Whelan, *Biochem. J.* **51,** xviii (1952).

saccharide, isomaltose. Until the work of Wolfrom and his associates,[144, 145] in characterizing the disaccharide obtained from the acid partial hydrolysis of the 1,6′-α-linked dextran from *Leuconostoc dextranicum*, the properties and derivatives of isomaltose were indefinite. In some cases[142] isomaltose obtained from the hydrolysis of amylopectin by purified α-amylase was characterized by such properties as not being fermentable by yeast, exhibiting a degree of polymerization close to two, and a correct reducing value. Despite the strength of these implications, proof had long been lacking for the presence of isomaltose among the products of enzymatic hydrolysis. True isomaltose has been isolated from the mixture resulting from the acid partial hydrolysis of amylopectin,[146] but application of the same technique of isolation has failed to reveal isomaltose as a product of the degradation of the same substrate by diastase preparations from barley malt.[144, 147] In such barley malt preparations it is highly probable that an amylo-1,6-glucosidase is also present which might catalyze the cleavage of the 1,6′-α-glucosidic branch points.[148]

Hydrolysis of amylopectin by takadiastase preparations from *Aspergillus oryzae* gave the expected quantity, judging from the degree of branching of the substrate, of true isomaltose, well characterized.[149] But it has since been shown that *A. oryzae* extracts contain a very interesting transglucosidase which can lead to the synthesis, among other oligosaccharides, of isomaltose from maltose.[150]

It appears that in experiments where good characterization of isomaltose has been achieved, heterogeneous enzyme preparations have been employed, and in experiments where pure enzyme preparations have been used, the isomaltose has been inadequately characterized. In both cases the conclusions concerning the action of amylase tend to be obscure.

The random attack of α-amylase on starch or glycogen results in the initial production of dextrins. It is seldom that time is allowed for the hydrolysis to proceed all the way to maltose and thus α-amylases are sometimes called dextrinogenic amylases. Reducing sugars and dextrins are released as α-anomers, hence the name α-amylase.

(2) *β-Amylase Action.* Amylose is hydrolyzed to β-maltose by β-amylase. From the initiation of amylolytic action maltose appears, unlike the initial action of α-amylase which produces only dextrins. The picture of β-amylase action is that it continuously hydrolyzes only the penultimate 1,4′-α-glucosidic link from the non-reducing end of the amylose chain, so that in effect one maltose unit at a time is cleaved from the polysaccharide until all the amylose molecule is converted to maltose.[151]

[144] M. L. Wolfrom, L. W. Georges, and I. L. Miller, *J. Am. Chem. Soc.* **71**, 125 (1949).

[145] L. W. Georges, I. L. Miller, and M. L. Wolfrom, *J. Am. Chem. Soc.* **69**, 473 (1947).

[146] M. L. Wolfrom, J. T. Tyree, T. T. Galkowski, and A. N. O'Neill, *J. Am. Chem. Soc.* **73**, 4927 (1951); *ibid.* **72**, 1427 (1950).

[147] M. L. Wolfrom, L. W. Georges, A. Thompson, and I. L. Miller, *J. Am. Chem. Soc.* **71**, 2873 (1949).

[148] E. Kneen and J. M. Spoerl, *Am. Soc. Brewing Chemists, Proc.* 20 (1948).

[149] E. M. Montgomery, F. B. Weakley, and G. E. Hilbert, *J. Am. Chem. Soc.* **71**, 1682 (1949); *ibid.* **69**, 2249 (1947).

[150] J. H. Pazur and D. French, *J. Biol. Chem.* **196**, 265 (1952).

[151] P. Bernfeld, *Advances in Enzymol.* **12**, 379 (1951).

Two theories of β-amylase action on amylose may be mentioned. The "single-chain" theory[152, 153, 154] considers that a single chain of amylose is degraded completely by a molecule of the β-amylase before another amylose chain is attacked. The "multi-chain" theory[155, 156] states that degradation of one amylose chain by a molecule of β-amylase does not necessarily have to be complete before the same enzyme molecule begins to cleave another chain. The molecular sizes and the proportions of the intermediate dextrins during the course of the reaction, as measured by maltose liberation, is the method used to test the validity of these theories.

It is worth while noting, however, that Thomas, Whelan, and Peat[156a] found that sweet potato, crystalline β-amylase or soybean, purified β-amylase will cleave pure amylose only 70% to completion. Cruder soybean preparations contain an additional factor ("Z enzyme") which when added to the purified amylase will allow hydrolysis of the amylose to proceed to completion. The Z enzyme has been purified and found to have all the properties of a β-glucosidase, and, in fact, emulsin will replace it. Consequently it appears that amylose contains one or more β-linked glucose residues in the chain, or more likely attached to the chain at the free 2, 3, 5, or 6 position of an intra-chain glucose residue.[156b] Since it would probably take only one such β-linked residue per chain to interrupt the progress of β-amylase, it is not unusual that methylation studies have not detected it. The structure outlined previously for amylose may be modified somewhat to include these observations.

The action of β-amylase on amylopectin or glycogen results in incomplete depolymerization of the polysaccharide. The branch points consisting of 1,6′-α-glucosidic links rather than 1,4′-α are effective blocks to the continuance of β-amylase action. The resulting resistant dextrin is called the "limit dextrin," and its properties are exceedingly important in the elucidation of amylopectin and glycogen structure. The β-maltose produced by β-amylase action on amylopectin is the result of the cleavage of only the outer 1,4′-α-linked chains. The limit dextrin, therefore, is the polysaccharide molecule denuded of its *outer* branches only. A glucose or a maltose residue (average is 1.5 glucose residues) may be left on the outer portion of the limit dextrin proximal to the 1,6′ branch point; they remain as nubs of the former outer branches. The properties of this limit dextrin

[152] M. A. Swanson, *J. Biol. Chem.* **172,** 805 (1948).
[153] R. W. Kerr, *Nature* **164,** 757 (1949).
[154] R. W. Kerr and F. C. Cleveland, *J. Am. Chem. Soc.* **73,** 2421 (1951).
[155] R. H. Hopkins and B. Jelinek, *Nature* **164,** 955 (1949).
[156] E. J. Bourne and W. J. Whelan, *Nature* **166,** 258 (1950).
[156a] G. J. Thomas, W. J. Whelan, and S. Peat, *Biochem. J.* **47,** xl (1950); S. Peat, W. J. Whelan, and S. J. Pirt, *Nature* **164,** 499 (1949).
[156b] S. Peat, S. J. Pirt, and W. J. Whelan, *J. Chem. Soc.* **1952,** 705, 714, 722.

have led Meyer, Gürtler, and Bernfeld,[157] to the conclusion that amylopectin and glycogen have an irregularly branched, bush-like structure.[132]

When about 60% of an amylopectin molecule is transformed by β-amylase into maltose, the hydrolysis stops. Since for glycogen this conversion limit is reached when only about 47% is degraded to maltose, it is evident that the outer branches in glycogen constitute a smaller proportion of the molecule. The details of the interior structure of glycogen have been revealed by treating the molecule alternately with phosphorylase and amylo-1,6-glucosidase.[158]

The immediate production of maltose by β-amylase has led to its being called maltogenic amylase. In all cases β-amylase produces only β-maltose and none of the α-anomers. The mechanism of the inversion of the anomeric carbon atom during cleavage by the enzyme is unknown. The occurrences of β-amylase activity is restricted to higher plants.

(3) *Individual Amylases.* All animal amylases that have so far been studied have been α-amylases. Human salivary and pancreatic amylase and pig pancreatic amylase require the presence of certain anions for full activity. Dialysis of both enzyme and substrate reduce the amylolytic activity to 15% of its maximum value. Addition of 0.001 *M* chloride ion, whether as NaCl, KCl, NH_4Cl, $MgCl_2$, or $CaCl_2$, will restore the activity to 80% of its maximum value. The same concentration of bromide, iodide, or nitrate ion is not as effective. Sodium chloride at a concentration of 0.01 *M* restores complete activity to the enzyme.[159] The optimum pH for salivary and pancreatic amylase action is pH 6.9.

Of the animal amylases only the three discussed above have been obtained in crystalline form. Amylase activity, however, is widespread among the animal kingdom.[160] It is also present in human blood plasma and in urine.[161] Molds and bacteria contain α-amylases. *Aspergillus oryzae* is the source of commericial takadiastase. The reviews of Myrbäck and Neumüller[162] and of Bernfeld[151] should be consulted for additional information on amylases.

b. Isoamylase and Amyloglucosidase. The hydrolysis of 1,6′-α-glucosidic linkages in polysaccharides is attributable to at least two types of enzymes separate from the amylases proper. Hobson, Whelan, and Peat[163] have described an "R enzyme" in bean and potato extracts which serves to make amylopectin more completely hydrolyzable by β-amylase, appar-

[157] K. H. Meyer, P. Gürtler, and P. Bernfeld, *Nature* **160,** 900 (1947).

[158] G. T. Cori and J. Larner, *J. Biol. Chem.* **188,** 17 (1951).

[159] P. Bernfeld, A. Staub, and E. H. Fischer, *Helv. Chim. Acta* **31,** 2165 (1948).

[160] C. L. Prosser, *in* Prosser, Comparative Animal Physiology, W. B. Saunders, Philadelphia, 1950, p. 144.

[161] M. Somogyi, *J. Biol. Chem.* **134,** 301 (1940).

[162] K. Myrbäck and G. Neumüller, *in* Sumner and Myrbäck, The Enzymes, Academic Press, New York, 1950, Vol. I, Part 1, p. 653.

[163] P. N. Hobson, W. J. Whelan, and S. Peat, *Biochem. J.* **47,** xxxix (1950).

ently by cleaving the 1,6′-α branch points. The isoamylase of Maruo and Kobayashi[164] may be identical with this debranching R enzyme. In addition to these amylases acting on the 1,6′-α-glucosidic links of amylopectin there also exist amylo-1,6-glucosidases which hydrolyze the same bonds in limit dextrins of branched polysaccharides treated with β-amylase. An amylo-1,6-glucosidase occurs in yeast[165] and in barley malt.[148] It is evident that crude diastase preparations may contain varying quantities of the enzymes involved in the scission of 1,6′-α-glucosidic links.[162] For this reason the purest enzymes available should be used for the determination of the end products of amylase activity.

Muscle also contains an amylo-1,6-glucosidase which hydrolyzes the 1,6′-α-glucoside links of the limit dextrin obtained from phosphorylase action on glycogen or amylopectin.[158] The products of the reaction are glucose and a dextrin which can further be partially phosphorolyzed by the phosphorylase.

c. Cellulase. Cellulose is the principal structural polysaccharide of plants. The only animal occurrence of this polysaccharide is in the tests of tunicates, *Phallusia* and *Molgula*. The repeating structure of cellulose is similar to amylose with one critical exception, namely, the D-glucose residues are bonded 1,4′-β rather than 1,4′-α. Cellulose, compared to amylose, is much more insoluble, is resistant to acid hydrolysis, and is completely inactive as a substrate for amylase. The markedly different physical and chemical properties of amylose compared to cellulose is due simply to the configuration of the glycosidic carbon atom. Because of this there are far-reaching changes in the aspect the polymer presents.[166] Cellulose is heterogeneous with regard to its molecular dispersion. Association among molecules is important, and this is greatly affected by traces of oxygen. Methylation is difficult. What determinations of molecular weight have been done indicate that native cellulose has upward of 3500 hexose units.[167]

Enzymes that can catalyze the hydrolysis of cellulose occur mainly in plants and microorganisms. Among the Metazoa, clams (*Mactra* and *Mya*), snails (*Helix* and *Limnaea*), shipworms (*Teredo*), and certain wood insects secrete cellulase in their digestive juices. Cellulose digestion by termites, wood cockroaches (*Cryptocercus*), and ruminants is carried out by symbiotic protozoa and bacteria which live in the gut or the rumen. Even in non-ruminants, however, the caecum or the colon may contain bacteria which are able to digest cellulose. Whether the products of this digestion are absorbed is not known.[160]

[164] B. Maruo and T. Kobayashi, *Nature* **167,** 606 (1951).
[165] K. H. Meyer and P. Bernfeld, *Helv. Chim. Acta* **23,** 875 (1940).
[166] O. Hassel and B. Ottar, *Acta Chem. Scand.* **1,** 929 (1947).
[167] W. W. Pigman and R. M. Goepp, Jr., Chemistry of the Carbohydrates, Academic Press, New York, 1948.

The immediate product of cellulose hydrolysis by cellulase is believed to be cellobiose,[168] but in crude extracts a cellobiase is generally present and thus D-glucose is the product. The successful separation of cellobiase from cellulase indicates that the two enzymes are not identical. Pigman has reviewed the literature on cellulases.[169]

d. Mucopolysaccharidases. Mucopolysaccharides are roughly defined as polysaccharides containing residues of hexoseamine as components.

(1) *Chitinase.* The organic skeletal substance of crustaceans, insects, and fungi is composed of chitin, a chain polymer of N-acetylglucosamine (N-acetyl-2-amino-2-desoxy-D-glucose) residues connected through 1,4′-β-glucosidic links. Chitin is usually associated with protein.[170]

Helix pomatia contains an enzyme capable of splitting chitin to N-acetylglucosamine.[171] The work of Zechmeister, Tóth, and Vajda[172] indicates the possibility that chitobiose (4-glucosamine D-glucosaminide) may be an intermediate because *Helix* digestive tracts also contain a chitobiase.

(2) *Hyaluronidases.* The properties of these enzymes and other mucopolysaccharidases have been reviewed by Fishman.[173] Since these enzymes are involved in the hydrolysis of mucus, their function in fertilization and other types of cytolysis is extremely important. The significance of these enzymes in nutrition has been considered by Duran-Reynals.[174]

2. Disaccharidases

The specificity of disaccharidases has been an object of study for many years by Helferich[175] and his school. Reference to some of these studies has already been made,[40] and additional information can be found in the reviews by Pigman,[41] Veibel,[42] and Gottschalk.[176] The implications of these studies are that the degree of variation in the aglycon of a hydrolyzable glycoside may be great but that the requirements of the enzyme towards the glycosyl radical are considerably stricter. On the glycosyl group the anomeric carbon atom must be of the correct configuration and so to a degree the other asymmetric carbon atoms. Combination of the glycoside with the enzyme is therefore considered to take place at the glycosyl radical

[168] H. Pringsheim, *Z. physiol. Chem.* **78,** 266 (1912).
[169] W. W. Pigman, *in* Sumner and Myrbäck, The Enzymes, Academic Press, New York, 1951, Vol. I, Part 2, p. 725.
[170] A. R. Trim, *Biochem. J.* **35,** 1088 (1941).
[171] P. Karrer and A. Hofmann, *Helv. Chim. Acta* **12,** 616 (1929).
[172] L. Zechmeister, G. Tóth, and E. Vajda, *Enzymologia* **7,** 170 (1939).
[173] W. H. Fishman, *in* Sumner and Myrbäck, The Enzymes, Academic Press, New York, 1951, Vol. I, Part 2, p. 769.
[174] F. Duran-Reynals, *Bact. Revs.* **6,** 197 (1942).
[175] B. Helferich, *in* Sumner and Myrbäck, The Enzymes, Academic Press, New York, 1950, Vol. I, Part 1, p. 79.
[176] A. Gottschalk, *Advances in Carbohydrate Chem.* **5,** 49 (1950).

with such configurations as must be fixed. The general reactions of the disaccharidases are those of glycosidases; however, here the emphasis will be placed on their role in hydrolyzing disaccharides.

a. Maltase. Maltase is a general α-D-glucosidase occurring widespread throughout the animal and plant kingdoms.[176a] Its physiological substrates are maltose and sucrose. The specificity of maltases from different sources towards the aglycon residue of α-D-glucosides may vary a great deal. For this reason there has been considerable controversy over the hydrolysis of sucrose by maltase. Weidenhagen[177] maintains that all α-D-glucosidic linkages are split by the same type of α-D-glucosidase on the basis that his purified maltase from yeast hydrolyzes maltose, sucrose, and melezitose (3-[α-D-glucopyranosyl-β-D-fructofuranosyl]-α-D-glucopyranoside). It is known, however, that maltases from many other sources have no effect on sucrose, but this may be only another example of difference in specificity of maltase from different sources. Since Weidenhagen's preparations were free of invertase, it does appear that sucrose can be split by an enzyme with specific requirements for the α-D-glucopyranose residue. This is further supported by the fact that melezitose, which is not cleaved by invertase, is hydrolyzed by Weidenhagen's yeast maltase. It remains to be seen whether Weidenhagen's[177] preparation of yeast maltase is not a mixture of maltases with varying specificity requirements.

b. Invertase. Invertase hydrolyzes sucrose to yield invert sugar, an equimolar mixture of glucose and fructose. Sucrose, the only ubiquitous non-reducing disaccharide, has two glycosyl radicals; one is α-D-glucopyranosyl, and the other is β-D-fructofuranosyl. Consequently sucrose may be cleaved by two enzymes, one that is specific for α-D-glucopyranosides, and one that is specific for β-fructofuranosides. The α-glucosidase, if separate from maltose, that exhibits invertase activity is called glucoinvertase. Frequently invertase specific for β-fructofuranosides is called β-*h*-fructosidase in order to differentiate it from glucoinvertase, but a more modern name is β-fructofuranosidase. Raffinose (1-[6-(α-D-galactopyranosyl)-α-D-glucopyranosyl]-β-D-fructofuranoside) is often used as a substrate to differentiate between β-fructofuranosidase, through which it will be hydrolyzed to yield fructose and mellibiose, and glucoinvertase, which will not catalyze its hydrolysis. This subject is discussed in detail by Neuberg and Mandl.[178] Animal invertase is possibly all glucoinvertase, whereas yeasts, molds, and higher plants contain β-fructofuranosidase or a mixture of both invertases.

[176a] A. Gottschalk, *in* Sumner and Myrbäck, The Enzymes, Academic Press, New York, 1950, Vol. I, Part 1, p. 551.

[177] R. Weidenhagen, *Fermentforschung*, **11**, 155 (1930).

[178] C. Neuberg and I. Mandl, *in* Sumner and Myrbäck, The Enzymes, Academic Press, New York, 1950, Vol. I, Part 1, p. 527.

Invertase has generally been regarded as an enzyme purely hydrolytic in action. Evidence is accumulating, however, for the role of invertase as a transglycosidase because of the appearance of several oligosaccharides in invertase-sucrose mixtures.[179, 180, 181]

c. β-Glucosidase. β-Glucosidase is present in many higher plants, molds, yeasts, and animals.[42] The hydrolysis of β-galactosides such as lactose may be due to β-glucosidase, since the specificity requirement at carbon atom 4 for the glycosyl radical of substrates may be relative.

Other enzymes attacking disaccharides are α-galactosidase, α-mannosidase, and β-mannosidase.[182]

V. Nucleolytic Enzymes

1. Depolymerization of Nucleic Acids

Nucleic acids are polymerized nucleotides which occur associated with basic proteins as nucleoproteins. Nucleotides, as they are isolated from hydrolyzates of nucleic acids, are phosphoric esters of nucleosides. Nucleosides are β-D-ribofuranosides or β-D-2-desoxyribofuranosides of purines or pyrimidines linked at N_9 or N_3, respectively. Ribonucleic acids (RNA) occur universally and contain, as nitrogenous bases, adenine, guanine, cytosine, and uracil. Desoxyribonucleic acids (DNA) occur mainly in cell nuclei and animal viruses. DNA contain the same nitrogenous bases as RNA with the important exception that uracil is replaced by thymine. Samples of RNA or DNA from various sources differ in molecular weight and in the proportions of the four nitrogenous bases they contain.

RNA is easily hydrolyzed to mononucleotides by a weak solution of a strong base. When chromatographed on an ion-exchange resin the hydrolysate has been shown to contain two isomers, termed *a* and *b*, of each of the four mononucleotides present.[182a] Adenylic acid *a* is phosphorylated at C_2' and adenylic acid *b* is phosphorylated at C_3'.[182b] Enzymatic hydrolysis of RNA by phosphodiesterase from snake venom, separated from the accompanying 5′-nucleotidase results in the formation of large quantities of nucleotides phosphorylated at C_5'.[182c] The "backbone," then, of RNA consists of nucleosides joined to each other through a single phosphodiester linkage from either C_2' or C_3' of one ribosyl radical to C_5' of the next.

There is evidence from titration curves[183] and methylation experiments[184]

[179] J. S. D. Bacon and J. Edelman, *Biochem. J.* **49,** 529 (1951).

[180] E. H. Fischer, L. Kohtès, and J. Fellig, *Helv. Chim. Acta* **34,** 1132 (1951).

[181] L. M. White and G. E. Secor, *Arch. Biochem.* **36,** 490 (1952).

[182] S. Veibel, *in* Sumner and Myrbäck, The Enzymes, Academic Press, New York, 1950, Vol. I, Part 1, p. 621.

[182a] C. E. Carter and W. E. Cohn, *Fed. Proc.* **8,** 190 (1949).

[182b] J. X. Khym, D. G. Doherty, E. G. Volkin, and W. E. Cohn, *J. Am. Chem. Soc.* **75,** 1262 (1953).

[182c] W. E. Cohn and E. Volkin, *Arch. Biochem.* **35,** 465 (1952).

that a considerable degree of branching may occur. Jordan[183] interprets the titration data on RNA to indicate that 1 of every 4 phosphorus atoms is triply bonded. Therefore 1 in every 4 is monoesterified and the other 2 are involved in diester formation. Methylation experiments[184] reveal ratios of ribose to monomethylribose to dimethylribose of such an order that can allow a branch every 2.4 to 3.8 nucleotides along a straight chain. Branching would be considered to take place by a phosphodiester link to the position (C_2' or C_3') not occupied in backbone formation on the backbone ribosyl radical. There is also evidence of branching occurring at triply esterified phosphate linkages in the backbone. At present there are no indications concerning the exact carbon atoms of the ribosyl radical which are involved in the diester links of the backbone. It is assumed by some that the link is most probably from C_3' to C_5' in RNA, because if DNA has the same basic structure, involvement of C_2' is eliminated. Nevertheless, in RNA the choice between C_2' and C_3' is available. Eventually the answer to this may be found in a study of the monomethylribose arising from mild methylation and subsequent hydrolysis of RNA.[184]

The ease of base hydrolysis of RNA is attributed to the preliminary formation of cyclic 2′,3′-phosphotriesters which then cleave easily at P—O—C_5', since the phosphate is now triply bonded, to give cyclic nucleoside-2′,3′-phosphates.[184a] On further hydrolysis the cyclic nucleoside-2′,3′-phosphate yields a mixture of nucleoside-2′ and -3′ phosphates.

DNA is not easily hydrolyzed to nucleotides by acid or base. This can now be explained as due to the impossibility of forming cyclic triesters before hydrolysis, since there is no hydroxyl group on C_2'. Likewise C_2' is not available for internucleoside links in DNA. The results of titrating DNA can be explained by the dissociation of 1 secondary phosphoryl group for every 10 to 20 nucleotides in addition to the amino and —NH—CO— dissociations of the purines and pyrimidines. Thus there is 1 triester and 1 monoester of phosphoric acid in 10 to 20 diesters.[184b] Branching is at a minimum. Enzymatic hydrolysis results in nucleoside-5′-phosphates.[184c] Supported by physical data, the concept of the DNA molecule is that of a long, slightly branched chain structure of C_3' nucleotides linked to each other by a phosphoester bond at C_5'.

The molecular weights of nucleic acids are subject to numerous errors,

[183] D. O. Jordan, *in* Butler and Randall, Progress in Biophysics and Biophysical Chemistry, Academic Press, New York, 1951, Vol. II, p. 51.

[184] A. S. Anderson, G. R. Barker, J. M. Gulland, and M. V. Lock, *J. Chem. Soc.* **1952,** 369.

[184a] D. M. Brown and A. R. Todd, *J. Chem. Soc.* **1952,** 52; D. M. Brown, D. I. Magath, and A. R. Todd, *ibid.* 2708.

[184b] D. O. Jordan, *Ann. Rev. Biochem.* **21,** 209 (1952).

[184c] E. Volkin, J. X. Khym, and W. E. Cohn, *J. Am. Chem. Soc.* **73,** 1533 (1951).

one of which can result from degradation during isolation. This decomposition may occur as the result of harsh treatment or of enzymes accompanying the nucleic acids. Contamination by degradative enzymes may also account for some of the large differences in guanine content of pancreas RNA and yeast RNA.[185] Considerable disagreement exists as to the molecular weight of RNA from cells, since it is quite probable that no one has yet obtained an intact molecule. Desreux's[186] value of 50,000 is the highest recorded, but RNA from tobacco mosaic virus[187] has a molecular weight of about 10^5. DNA tends to aggregate in solution by comparatively weak bonds, and thus molecular weights have little meaning. Doty and Bunce[188] found values of 4.0, 6.7, and 4.0×10^6 for molecular weights of 3 different samples of DNA. In general, preparations of DNA show the physical properties of rod-like particles about 6000 Å long.

The sequence of the four kinds of nucleotides that go to make up a nucleic acid is unknown. The molar proportions of the nitrogenous bases present in nucleic acid are varied, and even a statistical tetranucleotide is excluded. DNA, more than RNA, approaches but does not reach equimolar quantities of nitrogenous bases. Astbury[189] believes that the fiber diagram of thymus DNA shows that there is a definite order to the four nucleotides, at least in the more crystalline parts of the structure. This is about the only evidence for an orderly arrangement. The high molecular weight of nucleic acids and the possibilities of altering the proportions and the order of nucleotides lead to numerous isomers. Thus these biological polymers can bear a degree of specificity similar to that of proteins and polysaccharides. The transforming agent for pneumococcal types is believed by McCarty[190] to be highly polymerized DNA. Whether the specificity exhibited by this transforming principle is due to a small amount of protein rather than due to the DNA alone is not settled.[191] How much of the specificity of gene action is due to DNA and how much is due to the protein in nucleoprotein is an outstanding question.

Several excellent reviews of nucleic acid structure are available.[183, 192, 193, 194]

a. Ribonuclease. Ribonuclease cleaves specifically ribonucleic acid into

[185] J. E. Bacher and F. W. Allen, *J. Biol. Chem.* **183,** 641 (1950).
[186] V. Desreux, *Trans. Faraday Soc.* **46,** 790 (1950).
[187] S. S. Cohen and W. M. Stanley, *J. Biol. Chem.* **144,** 589 (1942).
[188] P. Doty and B. H. Bunce, *J. Am. Chem. Soc.* **74,** 5029 (1952).
[189] W. T. Astbury, *Symposia Soc. Exptl. Biol.* **1,** 66 (1947).
[190] M. McCarty, *Bact. Revs.* **10,** 63 (1946).
[191] A. E. Mirsky and A. W. Pollister, *J. Gen. Physiol.* **30,** 117 (1946).
[192] R. S. Tipson, *Advances in Carbohydrate Chem.* **1,** 193 (1945).
[193] J. N. Davidson, The Biochemistry of the Nucleic Acids, John Wiley and Sons, New York, 1950.
[194] F. Schlenk, *Advances in Enzymol.* **9,** 455 (1949).

smaller molecules with no release of free inorganic phosphate. Kunitz[195] obtained crystalline ribonuclease from beef pancreas; McDonald[196] achieved a preparation free from contaminating proteolytic enzymes. The properties of crystalline ribonuclease have been reviewed.[197]

The action of ribonuclease on RNA is not clear. Free acid groups are formed without the liberation of inorganic phosphate. Some of the split products are diffusible through collodion membranes and probably contain all four nitrogenous bases although it is unlikely that all are free mononucleotides.[198] When yeast RNA is used as the substrate, the diffusible products are heterogeneous but show a lower purine/pyrimidine ratio than the substrate.[199] Schmidt and his associates[200, 201, 202] have developed a procedure for the measurement specifically of free mononucleotides by analysis of the inorganic phosphate released by the action of prostatic phosphatase. They find that the action of ribonuclease on yeast RNA releases, mainly as pyrimidine mononucleotides, about half of the total phosphorus. The purines are retained in a higher molecular weight fraction precipitatable by uranium salts. They visualize the sequence of nucleotides in yeast RNA to involve regions of two to three pyrimidine nucleotides in succession, the internucleotide linkages of which are cleaved by ribonuclease.[202] By another analytical technique Schmidt's experimental results have been confirmed.[203]

Despite large excesses of the enzyme there is always left a large molecular weight, apparently non-diffusible residue amounting to 15 to 50% of the initial substrate. The variation in the residual fraction, which can be called the "limit polynucleotides," may be due to the differences in RNA substrates, or it may be due to experimental technique. Markham and Smith[204] have shown that the diffusibility of the limit polynucleotides is influenced strongly by the salt concentration of the medium rather than merely by the size of the molecules. By the development of paper chromatographic and electrophoretic techniques for nucleotides these authors have gained considerable insight into the structure of RNA and the action of ribonuclease. Their conclusions are somewhat similar to those of Schmidt and his associates[202] and are as follows. Ribonuclease specifically catalyzes the

[195] M. Kunitz, *J. Gen. Physiol.* **24,** 15 (1940).

[196] M. R. McDonald, *J. Gen. Physiol.* **32,** 39 (1948).

[197] J. H. Northrop, M. Kunitz, and R. M. Herriott, Crystalline Enzymes, 2nd ed., Columbia University Press, New York, 1948, p. 171.

[198] H. S. Loring and F. H. Carpenter, *J. Biol. Chem.* **150,** 381 (1943).

[199] J. E. Bacher and F. W. Allen, *J. Biol. Chem.* **183,** 633 (1950).

[200] G. Schmidt, R. Cubiles, B. H. Swartz, and S. J. Thannhauser, *J. Biol. Chem.* **170,** 759 (1947).

[201] G. Schmidt, N. Zöllner, L. Hecht, R. Cubiles, M. Wargon, and S. J. Thannhauser, *Federation Proc.* **9,** 224 (1950).

[202] G. Schmidt, R. Cubiles, and S. J. Thannhauser, *J. Cellular Comp. Physiol.* **38,** *Suppl.* **1,** 61 (1951).

hydrolysis of secondary phosphate esters of pyrimidine ribonucleoside-3′-phosphates. An RNA chain consists of nucleotides linked as previously described, but so arranged that there are regions of up to about five purine or pyrimidine nucleosides in succession. Treatment with ribonuclease liberates, as cyclic mononucleotides, all pyrimidine nucleosides bonded 3′-5′ to a purine or pyrimidine nucleoside and 5′-3′ to another pyrimidine nucleoside. Resistant to ribonuclease are bonds between purine nucleotides, purine nucleotide-3′-pyrimidine nucleotide-5′-links, and cyclic purine nucleotides. Consequently a limit polynucleotide will consist of purine nucleotides with always one terminal pyrimidine nucleotide linked by C_5' to the purine nucleotides and having a monoesterified phosphate group at C_3'. The pyrimidine cyclic mononucleotides liberated are further attacked by the enzyme to yield only the 3′-nucleoside phosphate.[205] The limit of ribonuclease action will depend on the number of purine nucleotides in the RNA. The size of the limit polynucleotides will depend on the number of purine nucleotides linked to each other in any region of the chain. Thereby it becomes unnecessary to postulate any type of branching of RNA to explain the action of ribonuclease.

b. Desoxyribonuclease. Desoxyribonuclease from beef pancreas was crystallized by Kunitz.[206] He believes that the molecular weight of the products of desoxyribonuclease action approach that of tetranucleotides.[207] On the other hand, Overend and Webb[208] have concluded that the products are heterogeneous in molecular weight and, like the non-diffusible fraction in ribonuclease-RNA mixtures, the non-diffusible fraction of DNA after treatment with desoxyribonuclease is richer in purines than undegraded DNA. A desoxyribonuclease has also been isolated from yeast.[209] Desoxyribonuclease requires activation by magnesium or manganese ions, whereas ribonuclease is inhibited by magnesium.[210] Separate protein inhibitors for both yeast[209] and pancreatic enzymes[211] are known.

The probability is very strong that complete depolymerization of nucleic acids to the stage of nucleotides requires the intervention of more than one enzyme. Studies on the distribution of nucleolytic activity have shown it to be common among diverse cells.

[203] C. E. Carter and W. E. Cohn, *J. Am. Chem. Soc.* **72,** 2604 (1950).
[204] R. Markham and J. D. Smith, *Biochem. J.* **52,** 552, 558, 565 (1952).
[205] D. M. Brown, C. A. Dekker, and A. R. Todd, *J. Chem. Soc.* **1952,** 2715; J. X. Khym, D. G. Doherty, E. Volkin, and W. E. Cohn, *J. Am. Chem. Soc.* **75,** 1262 (1953).
[206] M. Kunitz, *J. Gen. Physiol.* **33,** 349 (1950).
[207] M. Kunitz, *J. Gen. Physiol.* **33,** 363 (1950).
[208] W. G. Overend and M. Webb, *J. Chem. Soc.* **1950,** 2746.
[209] S. Zamenhof and E. Chargaff, *J. Biol. Chem.* **180,** 727 (1950).
[210] C. Lamanna and M. F. Mallette, *Arch. Biochem.* **24,** 451 (1949).
[211] W. Dabrowska, E. J. Cooper, and M. Laskowski, *J. Biol. Chem.* **177,** 991 (1949).

2. Nucleotidases

The two hydrolyzable bonds in nucleotides are the N-ribosidic link and the phosphoester link at C_3' or at C_5' in muscle adenylic acid. Repeated work has shown that phosphatases can split the phosphate from free nucleotides. It is a controversial issue, however, whether there exist phosphatases which can cleave free phosphate from nucleic acids without the preliminary intervention of a depolymerizing enzyme.[212]

Nucleotide-N-ribosidase is discussed by Schlenk (ref. 194, p. 508). There is doubt that the ribosidic bond of nucleotides can be enzymatically hydrolyzed.

3. Nucleosidases

Enzymes that catalyze the hydrolysis of the ribosidic linkage in nucleosides have been frequently described as enzymes requiring phosphate ion for activation. The discovery by Kalckar[213] of purine nucleoside phosphorylase has made a reinvestigation of nucleosidases necessary. At present it is apparent that many of the so-called nucleosidases are indeed phosphorylases. An exception to this, however, is the uridine nucleosidase from yeast, partially purified by Carter.[214] Here the phosphorolytic mechanism does not operate, and the products, uracil and ribose, arise by simple hydrolysis of the nucleoside. There is no reason to think that more of such hydrolytic nucleosidases will not be discovered. Even if it is not likely that a new ribosidic link can be synthesized by this type of enzyme one might speculate that it functions in the synthesis of some nucleosides by a process of transglycosidation in which phosphate is not involved:

$$\text{Adenine} + \text{uridine} \rightarrow \text{Adenosine} + \text{uracil}$$

VI. The Chemical Process of Digestion

Digestion in higher animals involves four processes, chemical, mechanical, nervous, and hormonal. In lower animals the hormonal and nervous factors are comparatively unknown, and in unicellular organisms the mechanical process of digestion is necessarily diminished. Mechanical processes bring particulate nutrients into contact with enzymes, the secretion of which is controlled by nervous and hormonal mechanisms. Common to all heterotrophic organisms, however, is the chemical process of digestion which is essentially hydrolytic. The high molecular weight substances, carbohydrates, proteins, and nucleic acids, are hydrolyzed to yield smaller molecules which are absorbed and assimilated. Assimilation is defined as the incor-

[212] G. Schmidt, R. Cubiles, and S. J. Thannhauser, *Cold Spring Harbor Symposia Quant. Biol.* **12,** 161 (1947).

[213] H. M. Kalckar, *Federation Proc.* **4,** 248 (1945).

[214] C. E. Carter, *J. Am. Chem. Soc.* **73,** 1508 (1950).

poration of material into cellular substance which can be active, or reserve metabolites. Fats are at least partially hydrolyzed to yield stable emulsions in the presence of bile salts at pH 5 to 8. In this discussion the part of nucleic acids is minimized because little is known about the extent to which they are catabolized. Digestion, then, is a hydrolytic process by which exogenous metabolites lose their specificity and become available for conversion by any cell.

The properties of each polysaccharide and protein depend on its source. These properties, both chemical and physical, reflect the nature, the order, and the mode of linkage of the respective monosaccharides and amino acids. The digestive process reduces the legion properties of these polymers to those of their comparatively simple components. Thus, gluten from wheat and casein from milk, each having specific properties, are hydrolyzed to practically the same amino acids which in turn can be recondensed by the cell to form its own specific protein. The specificity of polysaccharides and proteins are further attested by their ability to act as antigens; simple monosaccharides, amino acids, or polypeptides have no antigenic properties. It is noteworthy that triglycerides, which may be only partially hydrolyzed, have limits to their specificity. It is well known that fat characteristic of a species may be substituted for by fat that is not characteristic, if fatty acids normally available to the animal are withheld and the diet radically changed. Moreover, triglycerides have not been found to act as antigens.

The "availability" of a substance to a cell is a vague term. Whether or not a cell can metabolize a substance depends on the permeability of the cell to that substance and upon the complement of enzymes in the cell. Thus, it is apparent that muscle cells have the enzymes required to metabolize glucose-6-phosphate, yet the comparative impermeability of these cells to even inorganic phosphate[215] makes glucose-6-phosphate less available than glucose. A mutant of *Escherichia coli*, unable to utilize added glucose, has been found to be capable of using added maltose.[216] The maltose is used as a substrate by a transglucosidase to form a polysaccharide according to the following equation:

$$\text{Maltose} \rightarrow \text{Polysaccharide} + \text{glucose}$$

During this process glucose does not accumulate in the intact cell despite the failure of the organism to use exogenous glucose. The polysaccharide is phosphorolyzed in the usual way to glucose-1-phosphate, which is probably fermented by the Embden-Meyerhof pathway. No information is available concerning the fate of the glucose produced by the transglucosi-

[215] C. Hevesy and O. Rebbe, *Acta Physiol. Scand.* **1**, 171 (1940).

[216] M. Doudoroff, W. Z. Hassid, E. W. Putnam, A. L. Potter, and J. Lederberg, *J. Biol. Chem.* **179**, 921 (1949).

dase reaction, but hexokinase is known to be absent. This study[216] serves to show that the monomeric type as the product of hydrolysis is not always the required metabolite. Why exogenous glucose is not available to the cell, whereas endogenous free glucose appears to be available, is not apparent.

The site of digestion varies. Some unicellular organisms secrete digestive enzymes into a food vacuole in which a food particle is trapped. Organisms with no mechanism for trapping particulate food must rely on dissolved nutrients and are therefore called compulsory osmotrophs.[217] Bacteria fit into this category. Some osmotrophs secrete enzymes into the medium in which they live in order, presumably, to solubilize the insoluble nutrients in the medium. Thus, digestion has been considered to take place intracellularly in the phagotrophs and extracellularly in the osmotrophs. It is debatable whether a food vacuole is really of the internal environment of the cell. It is more likely to be a piece of the external environment temporarily isolated by the cell which prevents the diffusion of its digestive effort. The question of intracellular versus extracellular digestion is more a question of whether or not a phagocytic mechanism is involved.[218]

Penicillium chrysogenum produces an adaptive pectinase in its medium under the proper conditions. This enzyme complex will, among other things, hydrolyze the D-glycosidic linkages between the galacturonic acid residues in pectin by virtue of a polygalacturonidase.[219] It might be argued that the enzyme by producing soluble galacturonic acid from the polyuronide has rendered it into molecules small enough to be absorbed through the cell membrane. If this is so, however, it is also worthwhile considering how the high molecular weight enzyme permeated that same membrane. If it is assumed that the enzyme is formed within a cell and that it is liberated by a live cell, then the question of the passage of large molecules becomes a matter of numbers. Only a comparatively small number of enzyme molecules need be passed in order to hydrolyze large quantities of substrate.

The end point of digestion is a matter of definition. Previously it was pointed out that the result of polysaccharide or protein digestion must involve a loss of specificity. In addition, the end product must be "available" to the cells. It is possible that the cells may further hydrolyze, and in that sense, digest, absorbed metabolites before assimilating or catabolizing them. The extracellular proteinases of the mammalian gut act on proteins to the extent of hydrolyzing them to a mixture composed of a high proportion of peptides compared to amino acids.[220] Most of the work on intestinal

[217] A. Lwoff, Biochemistry and Physiology of Protozoa, Academic Press, New York, 1951, Vol. I, p. 7.

[218] C. M. Yonge, *Biol. Revs. Cambridge Phil. Soc.* **12,** 87 (1937).

[219] J. Phaff, *Arch. Biochem.* **13,** 67 (1947).

[220] E. S. London and N. Kotschneff, *Z. physiol. Chem.* **228,** 235 (1934).

peptidases has been done on the peptidases of the mucosal cells rather than of the intestinal juice itself. Some of the peptides absorbed by the mucosal cells may be hydrolyzed in those cells before entering the portal circulation.

The extent of digestion of triglycerides in the intestine of the mammal is a controversial subject.[221] Frazer[222] asserts that the hydrolytic activity of lipase in the small intestine is not great enough to account for complete hydrolysis of all the triglycerides absorbed. At the hydrogen ion concentrations in the intestine, stable emulsions of triglycerides with a particle size of 0.5 μ or less can be formed, providing there is present at the same time a mixture of bile salts, fatty acids, and monoglycerides. The function of lipase, therefore, is to provide some fatty acids and monoglycerides from the hydrolysis of part of the triglycerides. Particles of neutral fat 0.5 μ in diameter are capable, according to Frazer,[222] of being absorbed by mucosal cells. Verzár,[221, 223] on the other hand, maintains that all the absorbed triglycerides were hydrolyzed completely by pancreatic lipase.

VII. Amidases

1. Arginase

The commonest enzymatic reaction known which produces urea is the hydrolysis of arginine catalyzed by the enzyme arginase. Ornithine, the other product of arginase action, is converted to arginine by a series of reactions, the Krebs-Henseleit cycle, involving citrulline as an intermediate.[224] Arginase is present in the livers of all ureotelic animals but absent from those of uricotelic animals. The foregoing statement, known as Clementi's rule, has been verified repeatedly.[225] The liver is generally recognized as the site of urea synthesis[226] and usually contains the greatest quantity of arginase as well as the enzyme system involved in the synthesis of arginine from ornithine according to the general scheme of Krebs and Henseleit. Selachian fishes, however, have a very high extrahepatic arginase activity[227, 228] which may be associated with their constitutional, high uremia. The Krebs-Henseleit cycle has not been investigated in selachian muscle.

Arginase has not yet been crystallized. The arginase crystals of Bach[229]

221 F. Verzár, *Arch. sci. physiol.* **2**, 43 (1948).

222 A. C. Frazer, *Arch. sci. physiol.* **2**, 15 (1948).

223 F. Verzár and E. J. McDougall, Absorption from the Intestines, Longmans, Green and Co., London, 1936.

224 S. Ratner, *in* McElroy and Glass, Phosphorous Metabolism, The Johns Hopkins Press, Baltimore, 1951, Vol. I, p. 601; S. Grisolia, *ibid.* p. 619.

225 E. Baldwin, *Biol. Revs. Cambridge Phil. Soc.* **11**, 247 (1936).

226 J. L. Bollman, F. C. Mann, and T. B. Magath, *Am. J. Physiol.* **69**, 371 (1924).

227 A. Hunter, and J. A. Dauphinee, *Proc. Roy. Soc. (London)* **B97**, 227 (1924).

228 A. Hunter, *J. Biol. Chem.* **81**, 505 (1929).

229 S. J. Bach, *Nature* **158**, 376 (1948).

have been shown to be controvertible.[230] Purified preparations are highly specific for L-arginine. The presence of a free amino group is not essential for arginase activity, and such compounds as octopine[231] and arginic acid are degraded to urea. The guanidine and carboxyl group must both be free. Richards and Hellerman,[232] however, report that agmatine is hydrolyzed. Canavanine is also split to yield urea.[233]

The course of enzymatic hydrolysis of arginine has been found to follow Michaelis-Menten kinetics.[234] Competitive inhibition of arginase is effected by ornithine and lysine, but most other α-amino acids inhibit non-competitively.[235] Arginase is activated by Co^{++}, Mn^{++}, or Ni^{++}. The characteristics of this activation has led Hellerman[236] to propose that the role of the activating ion is to form coordinate bonds with both substrate and enzyme. Smith[102] has proposed a similar mechanism operating in metal-activated peptidases.

2. Glutaminase and Asparaginase

The enzymatic hydrolysis of these important ω-amides of glutamic and aspartic acids is widespread in organisms. These enzymes have been reviewed by Zittle,[237] and little more can be added here. Except in the renal production of urinary ammonia[238] the hydrolytic cleavage of glutamine is probably an unimportant reaction, since uncoupled hydrolysis of these amides would involve a considerable loss of free energy as heat.

Protein breakdown in seedlings and leaves in the absence of light is accompanied by the formation of asparagine or glutamine.[239] It is improbable that the formation of these amides from ammonia is due to the reversal of hydrolysis, but rather it is likely that it proceeds according to a mechanism involving phosphorylation by adenosine triphosphate.[116, 117] Continued existence in the dark results in an accumulation of ammonia in the plant due to the hydrolysis of the amide. If, however, the seedling or leaf is exposed to light, protein synthesis takes place with the concommitant disappearance of the amide.[239] Whether the amidases are important in this transformation is not known. The synthesis of amino acids from glutamine or asparagine may possibly occur without the intermediary formation of ammonia. The hydrolytic degradation of asparagine and glutamine, therefore, seems to be a sign of acute starvation in plants.

---

[230] M. S. Mohamed, *Acta Chem. Scand.* **3,** 56 (1949).

[231] S. Akasi, *J. Biochem.* (*Japan*) **26,** 129 (1937).

[232] M. M. Richards and L. Hellerman, *J. Biol. Chem.* **134,** 237 (1940).

[233] M. Damodaran and K. G. A. Narayanan, *Biochem. J.* **34,** 1449 (1940).

[234] D. M. Greenberg and M. S. Mohamed, *Arch. Biochem.* **8,** 365 (1945).

[235] A. Hunter and C. E. Downs, *J. Biol. Chem.* **157,** 427 (1945).

[236] L. Hellerman, *Physiol. Revs.* **17,** 454 (1937).

[237] C. A. Zittle, *in* Sumner and Myrbäck, The Enzymes, Academic Press, New York, 1951, Vol. I, Part 2, p. 922.

[238] R. M. Archibald, *Chem. Revs.* **37,** 161 (1945).

[239] A. C. Chibnall, Protein Metabolism in the Plant, Yale University Press, New Haven, 1939.

3. Urease

Urease catalyzes the hydrolysis of urea to yield carbon dioxide and ammonia. Urease activity is common in higher plants, bacteria, and molds but is of limited distribution in animals.[38] In no case has ureolytic activity been reported in the liver of vertebrates, although urea may be hydrolyzed by gastric mucosa and erythrocytes. Since urease was the first enzyme to be isolated in crystalline form,[240] it has been the object of considerable study.

Although the jack bean (*Canavalia ensiformis*) contains arginase,[233] there is no evidence, as there is for *Neurospora*,[241] for the operation of the Krebs-Henseleit cycle in higher plants. There exists simultaneously in the jack bean free canavanine[241a] and the enzymes, arginase[233] and urease,[240] which have the power to hydrolyze it to canaline, ammonia, and carbon dioxide. Furthermore, both urea and urease have been found to coexist in several plants.[242] The *raison d'être* of an unemployed enzyme can be a source of enjoyable speculation, but to postulate that it exists to prevent the accumulation of its substrate if ever the substrate *should* appear on the scene is to beg the question. Urease may be an exceedingly important link in the nitrogen cycle whereby animal excreta is transformed to ammonia and made available for protein synthesis by autotrophs.[243] The synthesis of urea from the amino nitrogen of amino acids is an endergonic process requiring, in animals, the participation of adenosine triphosphate.[224] If urea in plants arises by the same mechanism, its hydrolysis by urease would accomplish little.

A large excess of substrate inhibits the action of urease. This inhibition is interpreted by assuming that there are two neighboring sites on the enzyme, one for urea and one for water. Inhibition occurs when urea is in such excess that it occupies the water site as well.[21] It is of compelling interest that phosphate inhibits urease activity competitively.[244]

The presence of urease in several ammonotelic invertebrates has been shown by many investigators.[245] In these animals urea arises from the degradation of purines.

4. Deaminases of Purines, Pyrimidines, Nucleosides, and Nucleotides

The pathway of purine catabolism involves the hydrolysis of amino purines to oxypurines and ammonia. From the available evidence, this

[240] J. B. Sumner, *J. Biol. Chem.* **69,** 435 (1926).
[241] A. M. Srb and N. H. Horowitz, *J. Biol. Chem.* **154,** 129 (1944).
[241a] M. Kitagawa and T. Tomita, *Proc. Imp. Acad.* (*Tokyo*) **5,** 380 (1929).
[242] R. Fosse, *Ann. inst. Pasteur* **30,** 739 (1916).
[243] J. B. Sumner and G. F. Somers, Chemistry and Methods of Enzymes. 2nd ed., Academic Press, New York, 1947, p. 156.
[244] K. M. Harmon and C. Niemann, *J. Biol. Chem.* **177,** 601 (1949).
[245] M. Florkin and G. Duchateau, *Arch. intern. physiol.* **53,** 267 (1943).

deamination may occur when the purine is free or bound in nucleosides and nucleotides. Each substrate, however, appears to be hydrolyzed by a separate enzyme. Little is known concerning pyrimidine catabolism except for the enzyme, cytosine deaminase, in *Escherichia coli* and yeast.[246]

The free amino purines, guanine and adenine, are hydrolyzed by guanase and adenase to xanthine and hypoxanthine, respectively. Adenase is of limited distribution compared to guanase, but the role of either enzyme is unknown since deamination of the purine can also occur while it is in riboside linkage. The pig, which can suffer from guanine gout, is one of the few animals that has no hepatic guanase.[247] Nevertheless, the deamination of guanosine and guanylic acid may occur.

Evidence exists for specific enzymes which will deaminate 3-adenylic acid, 5-adenylic acid, guanylic acid, adenosine, guanosine, and cytidine.[248] Little is known concerning their distribution.

5. Histidase

Edlbacher[249] has shown that the enzymatic hydrolysis of histidine proceeds according to the following equation:

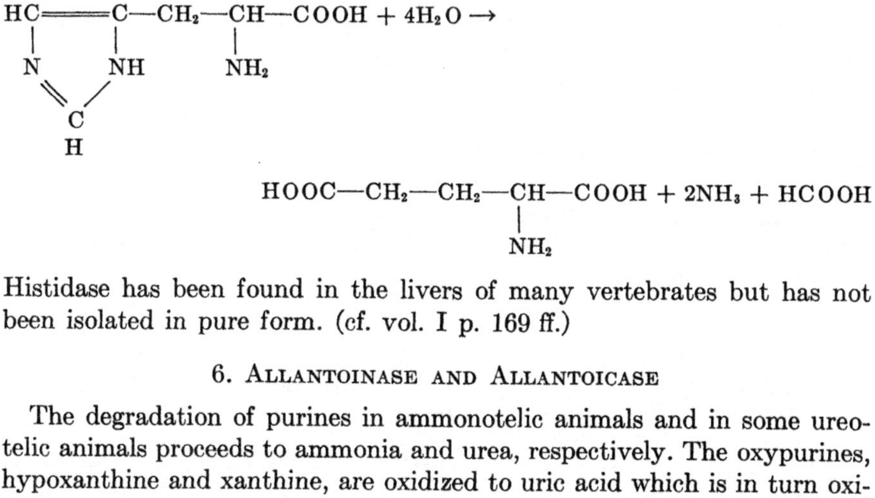

Histidase has been found in the livers of many vertebrates but has not been isolated in pure form. (cf. vol. I p. 169 ff.)

6. Allantoinase and Allantoicase

The degradation of purines in ammonotelic animals and in some ureotelic animals proceeds to ammonia and urea, respectively. The oxypurines, hypoxanthine and xanthine, are oxidized to uric acid which is in turn oxidized, under the influence of uricase, to allantoin. Hydrolysis of allantoin proceeds in two stages to yield allantoic acid, and finally urea and glyoxylic acid. In ammonotelic animals urease cleaves the urea into carbon dioxide and ammonia.

[246] E. Chargaff and J. Kream, *J. Biol. Chem.* **175,** 993 (1948).

[247] A. Schittenhelm and J. Schmid, *Z. exptl. Path.* **4,** 432 (1907).

[248] M. Laskowski, *in* Sumner and Myrbäck, The Enzymes, Academic Press, New York, 1951, Vol. I, Part 2, p. 956.

[249] S. Edlbacher, *Ergeb. Enzymforsch.* **9,** 131 (1943).

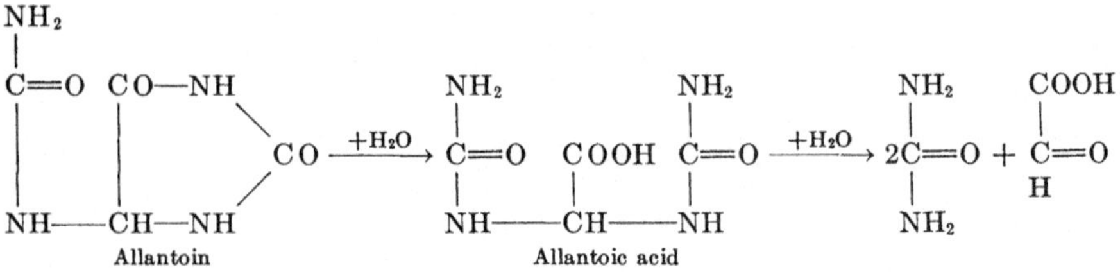

A comprehensive review of these enzymes has been presented by Przylecki and Truszkowski.[250] Since no pure preparations of these enzymes are available, little is known about them. Their distribution is discussed by Florkin and Duchateau.[245]

VIII. Nitrogen Excretion

The main nitrogenous end products of protein metabolism are amino acids, ammonia, uric acid, and urea. Much has been written concerning the excretion of these substances by various animals and the relationship this bears to their evolution, habitat, and mode of reproduction.[251-256] The excretion of guanine by spiders[257] and trimethylamine oxide by marine teleosts[258] are important exceptions. Guanine deposited in fish scales and pterines in butterfly pigments might also be regarded as excretory products if one were also to consider as such the keratin of mammalian hair.

The origin of trimethylamine oxide in marine teleosts is open to serious question. The experiments of Benoit and Norris[259] indicate strongly that trimethylamine oxide is of dietary origin. Fingerling Chinook salmon (*Oncorhynchus tschawytscha*) born in fresh water contain little or no trimethylamine oxide, but their marine parents have about 8 micromoles per gram of muscle. Fingerlings were kept on a diet of liver, free from trimethylamine oxide, for about 6 weeks during which the salinity of their medium was raised gradually to correspond with the changes that would normally occur in their run to the sea. At the end of this period the fish were living

[250] S. J. von Przylecki and R. Truszkowski, *in* Bamann and Myrbäck, Die Methoden der Fermentforschung, Vol. III, Thieme, Leipzig, 1941; Academic Press, New York, 1945, p. 2489.

[251] J. Needham, Biochemistry and Morphogenesis, Cambridge University Press, Cambridge, 1942.

[252] E. Baldwin, An Introduction to Comparative Biochemistry, 3rd ed., Cambridge University Press, Cambridge, 1948.

[253] M. Florkin, L'Evolution Biochimique, Desoer, Liege, 1944.

[254] C. Heidermanns, *Naturwissenschaften*, **26**, 263, 279 (1938).

[255] E. I. B. Dresel and V. Moyle, *J. Exptl. Biol.* **27**, 210 (1950).

[256] V. Moyle, *Biochem. J.* **44**, 581 (1949).

[257] K. Vajropala, *Nature* **136**, 145 (1935).

[258] A. Grollman, *J. Biol. Chem.* **81**, 267 (1929).

[259] G. J. Benoit, Jr. and E. R. Norris, *J. Biol Chem.* **158**, 439 (1945).

in salt water and, but for their diet, were marine. Analysis of their muscle during all this time showed it to contain the same negligible amount of trimethylamine oxide that was present in the fingerlings. But when scallop muscle, which contains large quantities of trimethylamine oxide, composed half of their diet, the trimethylamine oxide content of the salmons' muscle rose to values of wild adult fish. Shewan[260] has assembled data from various sources concerning the content of trimethylamine oxide in fish and other animals living in various fresh water and marine environments. Several fresh-water teleosts contain greater quantities of the oxide than do a few of the marine teleosts. Though the work cited above has been done on the trimethylamine oxide content of fish muscle rather than urine, it definitely suggests that trimethylamine oxide in fish is exogenous and does not represent a nitrogenous end product of protein metabolism in these animals.

Furthermore, it is at least of teleological interest to consider the efficiency of nitrogenous end products. In the order of decreasing nitrogen content, they are ammonia (82% N), urea (47% N), and uric acid (33% N). Trimethylamine oxide contains 19% nitrogen. Germane to this discussion is the possibility that the methyl groups in trimethylamine oxide may be "labile" methyl groups necessary to the metabolic economy of many animals.[261] A wholesale synthesis and excretion of trimethylamine oxide would serve more effectively to deplete the fish of methyl groups rather than nitrogen.

The end products of purine metabolism are uric acid in primates and uricotelic animals, allantoin in mammals other than primates, urea in the rest of the ureotelic animals, and ammonia in many of the ammonotelic animals.

IX. Phosphatases

The phosphatases may be considered as a special type of esterase which catalyzes the hydrolysis of esters (or anhydrides) of phosphoric acid. Common natural substrates for these enzymes include nucleotides, inorganic and organic pyrophosphate linkages, sugar phosphates, phytic acid, and glycerylphosphorylcholine. Except for inorganic pyrophosphatase from yeast,[262] no phosphatase has been obtained in crystalline form. Phosphatases have been classified according to the substrate hydrolyzed, the optimum pH for hydrolysis, and the characteristics of activation by cations. Authors often do not take pains to test the specificity of their crude enzyme preparations and thus a bewildering legion of phosphatases has been described. No

[260] J. M. Shewan, *in* Williams, The Biochemistry of Fish, Biochemical Society Symposia No. 6, Cambridge University Press, Cambridge, 1951, p. 28.
[261] H. K. Barrenscheen and M. Pantlitschko, *Z. physiol. Chem.* **284,** 250 (1949).
[262] M. Kunitz, *J. Gen. Physiol.* **35,** 423 (1952).

attempt will be made here to dwell on this aspect, since many good reviews on the subject are available.[26, 263, 264, 265]

The hydrolysis of organic phosphates in cellular metabolism can be a prelude to the utilization of the phosphate, as in calcification, or the organic part, as in glucose transport. The hydrolysis of glucose-6-phosphate to produce blood glucose takes place in the liver catalyzed by what may be a specific phosphatase.[266, 267] For every organic phosphate found to form in the course of intermediary metabolism there is probably some phosphatase which will hydrolyze it. Whether this hydrolysis proceeds seems to depend on cellular organization and the concentration of orthophosphate, phosphate ester, phosphatase, and hydrogen ion. Many phosphatases are inhibited competitively by orthophosphate. Activation by magnesium ion or other divalent cations frequently occurs and may be used as a means for differentiating prostatic phosphatase from erythrocyte phosphatase, both of which have optimal activity in the acid range pH 5 to 6.

The hydrolysis of glucose-1-phosphate by acid, prostatic acid phosphatase, and intestinal alkaline phosphatase has been studied by Cohn[268] who used H_2O^{18} in the medium. Her fundamental work has shown that upon acid hydrolysis the C—O bond in glucose-1-phosphate is ruptured, but on enzymatic hydrolysis it is the O—P bond that is cleaved. In the first case no O^{18} was detected in the inorganic phosphate formed, but in the second case the O^{18} was taken up.

The reversibility of the hydrolysis catalyzed by phosphatase has been demonstrated on many occasions by the synthesis of various organic phosphates and inorganic pyrophosphate. In the face of a high concentration of water, however, the synthesis of organic phosphates would not be at all favored unless the phosphorylated product were removed. Enzymatic phosphorylation, consequently, is usually accomplished by adenosine triphosphate (ATP) and an appropriate enzyme. The energy required to form the phosphoester linkage is supplied by the relatively higher-energy content of ATP compared to inorganic phosphate.

Phosphorylation by ATP to form a low-energy compound is essentially irreversible. The mechanism of dephosphorylation is not accompanied by a synthesis of ATP, but instead it is a simple hydrolytic cleavage mediated by phosphatase.

[263] J. Roche and Nguyen-van-Thoai, *Advances in Enzymol.* **10,** 83 (1950).

[264] F. Moog, *Biol. Revs. Cambridge Phil. Soc.* **21,** 41 (1946).

[265] E. Bamann and M. Meisenheimer, *in* Bamann and Myrbäck, Die Methoden der Fermentforschung, Thieme, Leipzig, 1941; Academic Press, New York, 1945, p. 1620; R. Robison, *ibid.* p. 1655; A. Schnäffner, *ibid.* p. 1663.

[266] C. de Duve, J. Berthet, H. G. Hers, and L. Dupret, *Bull. soc. chim. biol.* **31,** 1242 (1949).

[267] M. A. Swanson, *J. Biol. Chem.* **184,** 647 (1950).

[268] M. Cohn, *J. Biol. Chem.* **180,** 771 (1949).

The enzymatic transfer of a phosphate from one compound to another through the mediation of the adenylpyrophosphates is well known. Enzymes which catalyze the equilibrium between ATP and high-energy compounds are known as transphosphorylases. A more systematic nomenclature is that of Dixon,[269] who calls a phosphokinase any enzyme that catalyzes the transfer of a phosphate from ATP. This would include the hexokinases as well as the transphosphorylases.

Recently it has been shown that phosphatase can effect transfer of phosphate groups without the intervention of adenine nucleotides. Axelrod[270] first demonstrated this phenomenon in phosphatase preparations from citrus fruits; Appleyard[271] followed with preparations from prostate glands. Meyerhof and Green,[272] using purified intestinal phosphatase, demonstrated phosphate transfer from creatine phosphate, glucose-1-phosphate, or phosphopyruvate to glucose, fructose, and glycerol. As in transpeptidation and transglycosidation, the net hydrolysis is zero. That free phosphate from the phosphate donors is involved was eliminated because synthesis will not take place under the experimental conditions if the phosphate donor is replaced by inorganic phosphate.

The physiological roles of the phosphatases have been discussed by Roche.[26] According to the theory of Robison and Fell, in the process of calcification the function of phosphatase is to free organically bound phosphate by hydrolysis. This enables the formation of calcium phosphate at osteogenetic sites. The high concentration of alkaline phosphatases in the cells of kidney tubules and intestinal mucosa and at other sites of absorption has led to their involvement in the phenomenon of active absorption. Aside from histological studies, slim evidence exists for phosphorylation mediated by phosphatases accounting for active absorption.

1. Adenosinetriphosphatase and Apyrase

There are at least three enzymes which will hydrolyze adenosine triphosphate. Myosin, one of the active proteins of muscle, has very strongly associated with it an adenosinetriphosphatase which is activated by calcium ions and inhibited by magnesium.[273] Cleavage of only the terminal phosphate of ATP occurs, and inosine triphosphate is the only other substrate attacked. Not associated with myosin is a water-soluble adenosinetriphosphatase described by Kielley and Meyerhof.[274] It is strongly

[269] M. Dixon, Multi-Enzyme Systems, Cambridge University Press, Cambridge, 1949.

[270] B. Axelrod, *J. Biol. Chem.* **172,** 1 (1948); *ibid.* **176,** 295 (1948).

[271] J. Appleyard, *Biochem. J.* **42,** 596 (1948).

[272] O. Meyerhof and H. Green, *J. Biol. Chem.* **183,** 377 (1950).

[273] V. A. Engelhardt, *Advances in Enzymol.* **6,** 147 (1946).

[274] W. W. Kielley and O. Meyerhof, *J. Biol. Chem.* **176,** 591 (1948); *ibid.* **183,** 391 (1950).

activated by magnesium and inhibited by calcium in the presence of magnesium. Phosphatases which act on the terminal phosphate of either ATP or adenosine diphosphate have been termed apyrases by Meyerhof.[275] Apyrase from potatoes[276, 277] and from yeast[275] has been purified. Calcium activates potato apyrase but depresses the enzyme from yeast. Most cells contain one or more of these enzymes which can hydrolyze ATP. Difficulty arises in the differentiation between true adenosinetriphosphatase and apyrase because mammalian muscle, at least, contains myokinase which will dismute adenosine diphosphate into adenylic acid and ATP.[276] A combined effect of a true adenosinetriphosphatase and myokinase will then appear to be that of an apyrase. In the earlier work of Humphrey[278] on the occurrence of adenosinetriphosphatase in invertebrate muscle, little differentiation was done, and it is possible that ATP hydrolysis is due to many enzymes, including an apyrase. *Saxostrea commercialis*, however, seems to have a true adenosinetriphosphatase.[279]

The exact role of adenosinetriphosphatase in muscle contraction is largely conjectural. Upon the hydrolysis of ATP heat is released, but how the energy of hydrolysis is converted to mechanical energy is not known.[280] The function of the adenylpyrophosphatases in yeast[275] and in brain[281] metabolism has been treated by Meyerhof and his associates.

X. Phosphorylases

Phosphorolysis is the term introduced by Parnas[282] to describe the cleavage of substrates by phosphoric acid. The relation of phosphorolysis to hydrolysis is explained in Section I of this chapter. The phosphorylases have been reviewed by Hassid, Doudoroff, and Barker[283] Detailed information can be obtained from this authoritative review.

Phosphorylases have been described for the polysaccharides, glycogen and amylose, and for sucrose. Phosphorylases from both muscle and plants are able to cleave the 1,4′-α linkages of glycogen and starch to yield glucose-1-phosphate. Sucrose phosphorylase is a separate enzyme obtained from bacteria which produces glucose-1-phosphate and fructose from sucrose. Phosphorolysis has also been involved in the cleavage of the pentosidic

[275] O. Meyerhof, *J. Biol. Chem.* **157**, 105 (1945).
[276] H. M. Kalckar, *J. Biol. Chem.* **153**, 355 (1944).
[277] P. S. Krishnan, *Arch. Biochem.* **20**, 261 (1949).
[278] G. F. Humphrey, *Physiol. Comparata et Oecol.* **1**, 89 (1948).
[279] G. F. Humphrey, *Physiol. Comparata et Oecol.* **1**, 366 (1949).
[280] W. F. H. M. Mommaerts, Muscular Contraction, Interscience Publishers, New York, 1950.
[281] O. Meyerhof and J. R. Wilson, *Arch. Biochem.* **23**, 246 (1949).
[282] J. K. Parnas, *Ergeb. Enzymforsch.* **6**, 57 (1937).
[283] W. Z. Hassid, M. Doudoroff, and H. A. Barker, *in* Sumner and Myrbäck, The Enzymes, Academic Press, New York, 1951, Vol. I, Part 2, p. 1014.

bond of the following nucleosides: guanosine, inosine,[284] guanine desoxyribonucleoside, hypoxanthine desoxyribonucleoside,[285] thymidine,[286] uridine,[287] and cytidine.[288] In all these cases ribose-1-phosphate or desoxyribose-1-phosphate is probably formed. Pyrimidine nucleosides are not phosphorolyzed by the preparations active for purine nucleosides. Desoxyribosides and ribosides are probably cleaved by the same enzyme.

According to the views of Lynen,[289] coenzyme A (CoA) is acetylated at a free sulfhydryl group which then becomes —S—$COCH_3$. Transacetylase in *Escherichia coli* has been shown by Stadtman and his associates[290, 291] to catalyze the following equilibrium:

$$\text{Acetyl—CoA} + H_3PO_4 \rightleftharpoons \text{Acetyl—}PO_3H_2 + \text{CoA}$$

The emphasis has always been on acetyl transfer, but this might also be considered as an example, the only one as yet, of phosphorolysis of a linkage that is not glycosidic.

The concentration of inorganic phosphate is a limiting factor in phosphorolysis. Unlike water in hydrolysis, the phosphate is not in excess in most biological systems. The phosphate ester produced is of a higher energy content (roughly by 4800 cal. for glucose-1-phosphate) than the free sugar. The action of muscle and potato phosphorylase is specific for phosphate or arsenate as the glucose acceptor. Sucrose phosphorylase can operate with another monosaccharide as a glucose acceptor in the absence of phosphate. The reaction effectively exchanges the glucosidic linkage from sucrose to form a glucoside with the added sugar. This process is called transglycosidation,[292] and for the synthesis of polysaccharides it is catalyzed by enzymes apart from phosphorylases and amylases.[293]

Glycogen and amylopectin are not degraded completely by phosphorylase and phosphate alone, nor can phosphorylase synthesize a branched polysaccharide from glucose-1-phosphate. The 1,6′-α-glucosidic link is an effective block to phosphorylase which acts only on non-reducing terminal 1,4′ links. The scission of 1,6′ linkages is charged to hydrolysis by amylo-1,6-glucosidase[158] in muscle or R enzyme in plants.[163] Phosphorolysis of 1,6′

[284] H. M. Kalckar, *J. Biol. Chem.* **167,** 477 (1947).

[285] M. Friedkin and H. M. Kalckar, *J. Biol. Chem.* **184,** 437 (1950).

[286] L. A. Manson and J. O. Lampen, *Federation Proc.* **8,** 224 (1949).

[287] L. M. Paege and F. Schlenk, *Arch. Biochem.* **28,** 348 (1950).

[288] T. P. Wang, *Federation Proc.* **9,** 399 (1950).

[289] F. Lynen and E. Reichert, *Angew. Chem.* **63,** 47 (1951).

[290] E. R. Stadtman and H. A. Barker, *J. Biol. Chem.* **184,** 769 (1950).

[291] E. R. Stadtman, G. D. Novelli, and F. Lipmann, *J. Biol. Chem.* **191,** 365 (1951).

[292] M. Doudoroff, H. A. Barker, and W. Z. Hassid, *J. Biol. Chem.* **168,** 725 (1947).

[293] W. Z. Hassid, *in* McElroy and Glass, Phosphorus Metabolism, The Johns Hopkins Press, Baltimore, 1951, Vol. I, p. 11; E. J. Hehre, *ibid.* p. 48.

bonds by an "isophosphorylase" has been claimed,[294] but repetition of the work has not supported the operation of this mechanism.[295, 296]

The synthesis of amylose chains by phosphorylase requires the addition of a primer in the form of a polysaccharide. The enzyme cannot act to condense molecules of glucose-1-phosphate with each other, but rather the enzyme transfers the glucosidic link in glucose-1-phosphate to the 4 position of the glucose residue already on the non-reducing end of the chain.[297] Muscle phosphorylase requires primers of larger molecular weight than does potato phosphorylase which can be primed by maltotriose.[298] As well as the concentration of phosphate, the limiting factor in chain synthesis by phosphorylase is the length of the chain. Thus, with pure phosphorylases and priming agents, amyloses are synthesized.

The synthesis of 1,6′-α-glucosidic links in glycogen and starch is due to transglucosidation between two 1,4′ chains where a 1,4′ bond in one chain is transferred to make a 1,6′ bond to the other chain.

[294] P. Bernfeld and A. Meutémédian, *Helv. Chim. Acta* **31**, 1724 (1948); *ibid.* **31**, 1735 (1948).

[295] S. Nussenbaum and W. Z. Hassid, with a note by G. T. Cori and B. Illingworth, *J. Biol. Chem.* **190**, 673 (1951).

[296] J. M. Bailey and W. J. Whelan, *J. Chem. Soc.* **1950** 3573.

[297] G. T. Cori, M. F. Swanson, and C. F. Cori, *Federation Proc.* **4**, 234 (1945).

[298] C. Weibull and A. Tiselius, *Arkiv. Kemi. Mineral. Geol.* **19A** (19), 1 (1945).

CHAPTER 18

The Respiratory Enzymes

B. L. HORECKER AND A. KORNBERG

National Institute for Arthritis and Metabolic Diseases, Bethesda, Maryland

I. Introduction

A consideration of the respiratory enzymes should reasonably include not only reactions which directly involve electron transport but also those which alter the metabolites in other ways, and which are nonetheless essential links in the chain of events that result in respiration. Other chapters deal with these reactions, such as deaminations and transaminations, carboxylations and decarboxylations, transphosphorylations, etc. In this chapter we will be concerned only with those steps in the metabolic pathway which are directly concerned with electron transport. The rapid and extensive developments in this phase of biochemistry as well as its rich historical background have been covered in excellent reviews which have appeared in recent years.[1, 2, 3] Here we will attempt only to survey the field in a general way and to highlight some of the recent contributions that point to new directions.

Less than thirty years ago the nature of the processes which linked the removal of electrons from oxidizable substrates to the reduction of molecular oxygen was completely obscure. Some workers proposed that the activation of an oxygen molecule initiated the process, while others claimed that the activation of the substrate was the critical event. This controversy has been settled over the years in a way that validates both points of view. It is now established that a whole series of reactions is responsible for respiration, some of which deal with dehydrogenation of the substrate and a transfer of electrons in the complete absence of oxygen, and others which permit such activated substrates or carriers to react with an enzymatically altered oxygen molecule. The scheme in Fig. 1 summarizes a common

[1] J. M. Luck, *Annual Reviews of Biochemistry*, Vols. 15–20, Annual Reviews, Inc., Stanford, 1946–1951.
[2] J. B. Sumner and K. Myrbäck, The Enzymes, Academic Press, New York, 1951.
[3] H. A. Lardy, Respiratory Enzymes, Burgess Publishing Co., Minneapolis, 1949.

pattern of events in which a specific example might be the oxidation of glucose-6-phosphate to 6-phosphogluconate. The 2 electrons removed from the sugar are transferred by specific enzymes to the coenzyme triphosphopyridine nucleotide (TPN) or diphosphopyridine nucleotide (DPN) to form the corresponding reduced coenzymes. The latter may be reoxidized when linked with specific enzymes which bear the oxidized forms of either flavin adenine dinucleotide (FAD) or flavin mononucleotide (FMN). The reduced flavoproteins may in turn be oxidized by oxygen or by the intervention of another group of electron carriers, the cytochromes.

It has been appreciated for a long time that even in aerobic organisms certain vital functions are performed under relatively anaerobic conditions. For example, when the local oxygen supply is exhausted during vigorous muscular activity, a fermentative mechanism comes into play which can

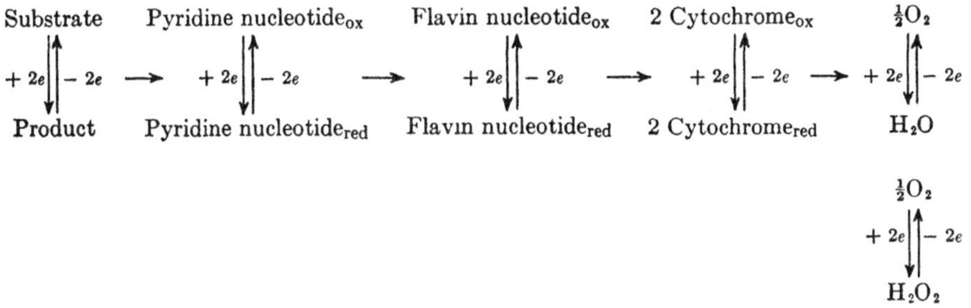

FIG. 1. Scheme of electron transport to oxygen.

provide energy for continued muscle contraction. Among microorganisms there is a wide assortment of prominent mechanisms which under rigidly anaerobic conditions provide energy for growth and function by means of oxidative reactions. Such oxidative steps are possible by linkage with reductive reactions, either by direct dismutation of the substrate or through the intervention of electron carriers. The point to be emphasized is that such fermentative reactions are as worthy of attention in an over-all consideration of respiratory enzymes as those steps which lie on a direct path to oxygen.

Unfortunately, many enzymes which fall clearly within the scope of oxidation-reduction catalysis will have to be neglected because of limitations of space and the inadequate amount of information available about them. In this group are the hydrogenases, the sulfur oxidases, the nitrogen oxidases, and a host of related enzymes found in autotrophic bacteria, as well as in plant and animal organisms.

II. Pyridine Proteins

1. Coenzymes

a. Properties. DPN was first recognized as a cofactor necessary for alcoholic fermentation in yeast. The studies of Warburg[4, 5] and Schlenk and von Euler[6, 7] led to the development of a structural formula (Fig. 2) which has been proved valid by subsequent studies.[8, 9, 10] DPN is a dinucleotide in which the constituent mononucleotides, adenosine-5′-phosphate (muscle adenylic acid) and nicotinamide ribose-5′-phosphate, are linked by a pyrophosphate bridge. Recent reports have shown the synthesis of this coenzyme to occur by a condensation of nicotinamide ribose-5′-phosphate

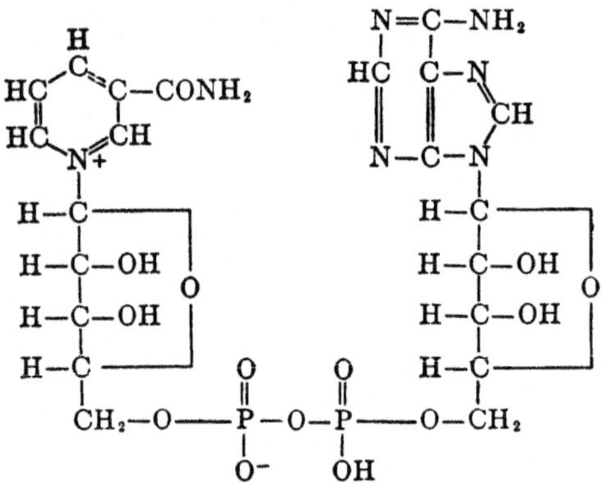

Diphosphopyridine nucleotide

Fig. 2

(NRP) and adenosine triphosphate (ATP) with the formation of inorganic pyrophosphate (PP).[10]

$$\text{NRP} + \text{ATP} \rightleftharpoons \text{DPN} + \text{PP}$$

The DPN molecule undergoes several additional types of enzymatic degradation.[9, 11]

TPN was isolated by Warburg and co-workers as the coenzyme required for the direct oxidation of glucose-6-phosphate to 6-phosphogluconate. For many years it was recognized that the structure of TPN closely resembled

[4] O. Warburg, W. Christian, and A. Griese, *Biochem. Z.* **282,** 157 (1935).
[5] O. Warburg and W. Christian, *Biochem. Z.* **287,** 291 (1936).
[6] H. von Euler, H. Albers, and F. Schlenk, *Z. physiol. Chem.* **240,** 113 (1936).
[7] F. Schlenk and H. von Euler, *Naturwissenschaften* **24,** 794 (1936).
[8] F. Schlenk, *J. Biol. Chem.* **146,** 619 (1942).
[9] A. Kornberg and W. E. Pricer, Jr., *J. Biol. Chem.* **182,** 763 (1950).
[10] A. Kornberg, *J. Biol. Chem.* **182,** 779 (1950).
[11] A. Nason, *Federation Proc.* **10,** 228 (1951).

that of DPN, differing only in the presence of one additional phosphate group. A common assumption was that the three phosphates were disposed in a triphosphate chain with nicotinamide nucleoside on one end and adenosine on the other. With the availability of improved enzymatic methods for degrading the molecule, it has been established that the extra phosphate group of TPN is attached to the adenosine part of the molecule and has the same configuration as adenylic acid "a."[12] Adenylic acid "a" has now been demonstrated to be adenosine-2′-phosphate, and adenylic acid "b" to be the 3′-phosphate.[12a] The structure of TPN is given in Fig. 3.

The properties of these coenzymes have been reviewed by Schlenk.[13]

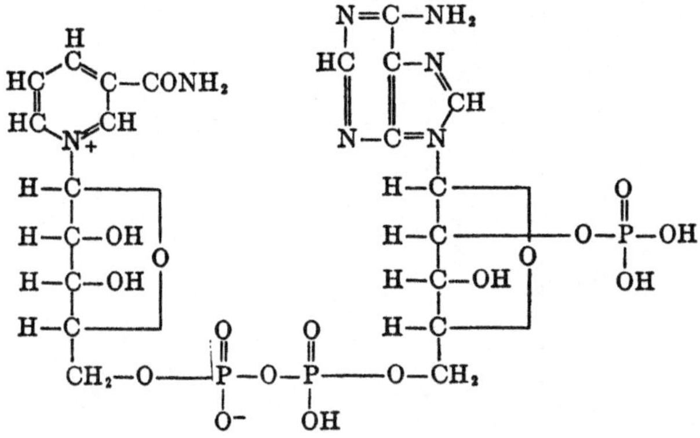

Triphosphopyridine nucleotide

FIG. 3

Their ultraviolet absorption spectra have been of especial interest and usefulness. As can be seen from Fig. 4 the spectrum of the oxidized form of DPN (and to the best of our knowledge the absorption characteristics of TPN are identical) displays a sharp maximum at 260 mμ and thus is similar to that of other adenine-containing compounds. There is no absorption above 300 mμ. Upon reduction of the pyridine ring there is a slight decrease in absorption at 260 mμ, but, most striking, there is the appearance of a new band at 340 mμ. These properties, first described by Warburg and his associates[4, 5] have been exploited by him and others as a basis for highly sensitive and specific analytic methods for a variety of enzymes, substrates, and other coenzymes. In this connection it is important to clarify a source of serious confusion regarding the correct values for the absorption coeffi-

[12] A. Kornberg and W. E. Pricer, Jr., *J. Biol. Chem.* **186**, 557 (1950).

[12a] J. X. Khym, D. G. Doherty, E. Volkin, and W. E. Cohn, *J. Am. Chem. Soc.* **75**, 1262 (1953).

[13] F. Schlenk, *in* Sumner and Myrbäck, The Enzymes, Academic Press, New York, 1951, Vol. II, Part 1, p. 250.

cients of DPN and TPN, especially for the 340-mμ bands of the reduced forms.

The literature contains numerous values which may differ by as much as 30%.[14] There are two prominent sources for the variation: one the assumption that impure preparations are pure, which leads to low values, and the other, the use of improper methods for reducing the coenzyme, which generally leads to high values. For example, the use of a chemical reducing agent such as hydrosulfite will result in the reduction of any coenzyme fragments containing the pyridinium linkage, all of which possess

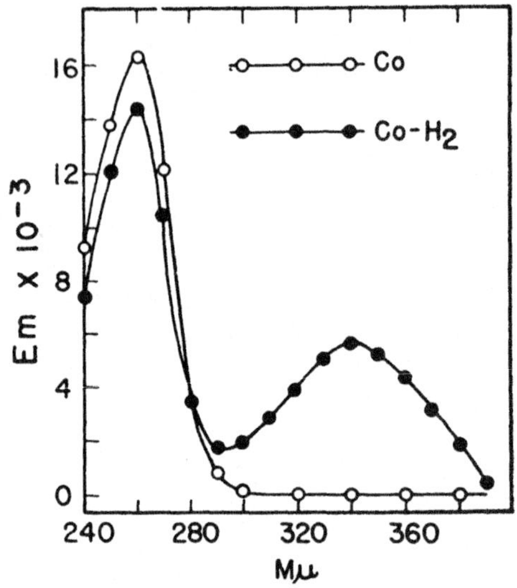

FIG. 4. Absorption spectra of the oxidized and reduced pyridine nucleotides (from Schlenk[13]). To show the correct extinction coefficients, the values in the ordinate are to be multiplied by 1.10.

an absorption band at 340 mμ. Since analytically pure samples of DPN and TPN were and still are unavailable, a method was developed which establishes the absorption coefficient in a way that is independent of the purity of the coenzyme. By selecting a simple enzymatic reaction in which *a pure substrate is completely oxidized* (*or reduced*) in the presence of an excess of coenzyme, the change in optical density observed represents the molar equivalent of the amount of substrate added. In this way a value of 6.22×10^6 sq. cm. $\times$ mole^{-1} was derived for the 340-mμ bands of both DPN and TPN[15] (Table 1).

b. Interconversion. A number of enzymatic mechanisms have been de-

[14] D. L. Drabkin, *J. Biol. Chem.* **157,** 563 (1945).
[15] B. L. Horecker and A. Kornberg, *J. Biol. Chem.* **175,** 385 (1948).

scribed which indicate that the pyridine coenzymes DPN and TPN can interact with one another.

(1) *Phosphate Transfer*. With a purified enzyme from yeast[17] the following reactions have been demonstrated:

$$DPN + ATP \rightarrow TPN + ADP$$

$$DPNH_2 + ATP \rightarrow TPNH_2 + ADP$$

There is also evidence for a similar reaction in pigeon liver extracts.[18]

TABLE 1

MOLECULAR EXTINCTION COEFFICIENTS FOR DPN AND TPN AT 340 mμ

	Density		Concentration of substrate, moles × cm.$^{-3}$ × 10^6	Molecular extinction coefficient, sq. cm. × mole^{-1} × 10^{-6}
	Initial*	Final		
Pyruvate-$DPNH_2$	0.684	0.389	47.0	6.28
Pyruvate-$DPNH_2$	0.683	0.440	38.5	6.31
Pyruvate-$DPNH_2$	0.648	0.190	73.3	6.25
Isocitrate-TPN	0.167	0.526	60.6	5.93
Pyruvate-$TPNH_2$	0.494	0.212	45.0	6.27
Acetaldehyde-$DPNH_2$	0.620	0.485	22.1	6.11
Acetaldehyde-$DPNH_2$	0.607	0.408	32.6	6.11
			AVERAGE	6.22†

* Corrected for dilution due to substrate addition.

† This average was obtained by excluding the result with isocitrate. With the value included the average is 6.18. This value has recently been confirmed with pure crystalline barium glucose-6-phosphate.[16]

The presence in potato,[9] kidney,[19] and other sources of a phosphatase which splits out the "third" phosphate of TPN has been established.

$$TPN + H_2O \rightarrow DPN + \text{phosphate}$$

It may be presumed that the same enzyme will convert $TPNH_2$ to $DPNH_2$.

(2) *Electron Transfer*. Another mechanism for the interaction of the pyridine nucleotides was described by Colowick *et al.*[20, 21, 22] They observed an enzyme in extracts of *Pseudomonas fluorescens*, *Azotobacter*, and, more

[16] W. A. Wood and B. L. Horecker, Biochem. Prepns., in press.

[17] A. Kornberg, *J. Biol. Chem.* **182,** 805 (1950).

[18] A. H. Mehler, A. Kornberg, S. Grisolia, and S. Ochoa, *J. Biol. Chem.* **174,** 961 (1948).

[19] D. R. Sanadi, *Arch. Biochem.* **35,** 268 (1952).

[20] S. P. Colowick, N. O. Kaplan, E. F. Neufeld, and M. M. Ciotti, *J. Biol. Chem.* **195,** 95 (1952).

[21] N. O. Kaplan, S. P. Colowick, and E. F. Neufeld, *J. Biol. Chem.* **195,** 107 (1952).

[22] N. O. Kaplan, S. P. Colowick, and E. F. Neufeld, *Federation Proc.* **11,** 238 (1952).

recently, animal tissues which carried out the following reaction:

$$TPNH_2 + DPN \rightleftharpoons DPNH_2 + TPN$$

These investigators were able to show that an electron rather than a phosphate transfer was involved. With desamino DPN (prepared by enzymatic hydrolysis of the 6-amino group of adenine), which can replace DPN in this reaction, the product in the case of electron transfer would be desamino $DPNH_2$, whereas in the case of phosphate transfer it would be desamino TPN. The occurrence of the reaction:

$$TPNH_2 + \text{desamino DPN} \rightarrow TPN + \text{desamino } DPNH_2$$

was established by the identification of desamino $DPNH_2$.

Colowick has pointed out[23] that this transhydrogenase system may provide a shuttle when and where necessary for linking DPN-specific dehy-

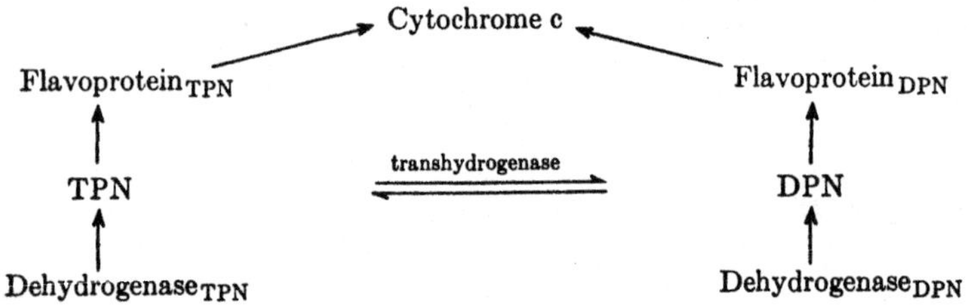

FIG. 5. The role of transhydrogenase in electron transfer.

drogenases with TPN-specific flavoproteins or vice versa, as shown in Fig. 5. Only indirect evidence is thus far available for the reduction of TPN by $DPNH_2$.[21]

c. Specificity. The availabity of refined spectrophotometric methods for the detection and assay of dehydrogenases, coupled with the more widespread interest in enzyme purification, has led to substantial revisions and expansion of our knowledge concerning the coenzyme specificity of dehydrogenases. It is apparent from current research activities that further tabulations of this subject will also be short-lived. It may be helpful to indicate the general categories.

1. Strict specificity toward a coenzyme regardless of the source material. Alcohol dehydrogenase isolated from yeast[24] and horse liver[25] reacts with DPN but not at all with TPN. Triosephosphate dehydrogenase might have been cited as another example of strict DPN specificity except that current

[23] S. P. Colowick, *in* McElroy and Glass, Phosphorus Metabolism, The Johns Hopkins Press, Baltimore, 1951, Vol. 1, p. 436.

[24] E. Racker, *J. Biol. Chem.* **184,** 318 (1950).

[25] R. K. Bonnichsen, *Acta Chem. Scand.* **4,** 714 (1950).

reports indicate the presence in green leaves of a triosephosphate dehydrogenase which reacts with TPN.[26, 27] It would seem that further investigations of a variety of plants and microorganisms may completely eliminate this category.

2. Strict specificity toward a coenzyme only in a given source material. The malic enzyme,[28] which oxidatively decarboxylates malic acid to pyruvic acid, is TPN specific in pigeon liver but DPN specific in *Lactobacillus arabinosus*.[29]

3. True non-specificity. The essentially equal reactivity of glutamic dehydrogenase of animal origin with either DPN or TPN was recognized at an early date.[30, 31] The slight but nonetheless significant non-specificity of other dehydrogenases escaped detection by the less sensitive methods which relied on methylene blue reduction. For example, re-examination of the specificity of lactic and malic dehydrogenases[18] previously regarded as DPN specific have revealed reactivity with TPN. As shown in Table 2, this applies to several source materials.

4. Non-specificity due to the presence of a DPN-specific and TPN-specific enzyme in the same source material. It was demonstrated that in yeast two distinct isocitric dehydrogenases could be isolated, each free of the other.[32] Both enzymes catalyze the same over-all oxidative decarboxylation, although some distinguishing features exist and will be discussed below. Even though the bakers' and brewers' yeast used as source material are not considered pure cultures, the fact that the activity of both enzymes was approximately equal in a variety of samples suggests that the two enzymes co-exist in the same cell.

d. Protein Binding. Unlike the flavin nucleotides, which are tightly bound to proteins and are readily observed as flavoprotein complexes, the pyridine nucleotides have been considered to be almost completely dissociated from their apodehydrogenases. There are now at least two important exceptions. The first of these is the demonstration that the crystalline triosephosphate dehydrogenase from muscle has firmly bound to it 2 molecules of DPN, assuming a minimum molecular weight of 100,000 for the protein.[33] The bound coenzyme is not removed by dialysis or recrystallization but can be

[26] M. Gibbs, personal communication.

[27] D. Arnon, personal communication.

[28] S. Ochoa, A. H. Mehler, and A. Kornberg, *J. Biol. Chem.* **174,** 979 (1948).

[29] S. Korkes, A. del Campillo, and S. Ochoa, *J. Biol. Chem.* **187,** 891 (1950).

[30] J. G. Dewan, *Biochem. J.* **32,** 1378 (1938).

[31] H. von Euler, E. Adler, G. Günther, and N. B. Das, *Z. physiol. Chem.* **254,** 61 (1938).

[32] A. Kornberg and W. E. Pricer, Jr., *J. Biol. Chem.* **189,** 123 (1951).

[33] J. F. Taylor, S. F. Velick, G. T. Cori, C. F. Cori, and M. W. Slein, *J. Biol. Chem.* **173,** 619 (1948).

removed by treatment with Norit or by alkaline phosphatase. The presence of the coenzyme is required in order for the enzyme to crystallize. Since the bound DPN can be separated from the enzyme by charcoal adsorption and since it also exchanges rapidly with P^{32}-labeled DPN,[34] it appears that the binding is not of the covalent type and that the bound DPN has a significant dissociation.

Another example of a pyridinoprotein is provided by Theorell and Bonnichsen,[35] who observed a shift in the reduced band of $DPNH_2$ to lower wavelengths in the presence of stoichiometric amounts of liver alcohol dehydrogenase. Attempts to observe a similar phenomenon with liver glutamic dehydrogenase have thus far been unsuccessful.[54]

e. Oxidative Phosphorylation. The knowledge that the function and growth of a cell are dependent on energy-yielding oxidative processes has been appreciated for a long time. However, the precise mechanisms whereby the energy released from exothermic oxidative reactions is converted into useful chemical forms are still largely obscure. The high-energy phosphate bond, with free energy changes of the order of 12,000 cal., has been the best-known device for trapping the energy of reactions. But most recently the high-energy character of the acyl-sulfur linkage and its interaction with high-energy phosphate compounds has become apparent.

In the discussions of triosephosphate, acetaldehyde, and α-keto acid dehydrogenases, it will be seen that there is a fairly close approach to an understanding of the energetic coupling in these oxidations. What is still a major mystery is the mechanism whereby the oxidation of the reduced pyridine nucleotides serves as the major energy source in most forms of aerobic metabolism.

2. Apodehydrogenases

The recent review of the pyridine proteins presented by Schlenk[13] makes it unnecessary to present this subject in great detail. Instead we will attempt only to touch on some of the current advances with the partially or highly purified pyridine nucleotide dehydrogenases. A brief summary of the reaction products, coenzyme specificity, source of the enzyme, and its state of purity is given in Table 2.

a. Glucose Dehydrogenase. Glucose dehydrogenase has been purified more extensively by Strecker and Korkes.[36] They demonstrated that TPN as well as DPN served as the coenzyme and that the oxidation was reversible starting with gluconolactone. The enzyme was also shown to be specific for the β form of glucose. The subsequent oxidation of gluconate by *Pseu-*

[34] C. F. Cori, S. F. Velick, and G. T. Cori, *Biochim. et Biophys. Acta* **4,** 160 (1950).
[35] H. Theorell and R. Bonnichsen, *Acta Chem. Scand.* **5,** 1105 (1951).
[36] H. J. Strecker and S. Korkes, *J. Biol. Chem.* **196,** 769 (1952).

TABLE 2

THE PYRIDINE NUCLEOTIDE-LINKED DEHYDROGENASES

Name of Dehydrogenase	Product	Coenzyme	Source	Purity of proteins	Reference
Glucose	Gluconolactone	TPN, DPN	Liver	Partial	36
Glucose-6-phosphate (Zwischenferment)	6-Phosphogluconolactone	TPN	Yeast	Partial	17, 37
6-Phosphogluconic	CO_2 + ribulose-5-phosphate	TPN	Yeast	Partial	38
3-Phosphoglyceraldehyde (Triose Phosphate)	1,3-Diphosphoglycerate	DPN	Yeast	Crystalline	39
		DPN	Muscle	Crystalline	33
α-Glycerophosphate	Dihydroxyacetone phosphate	DPN	Muscle	Crystalline	40
Lactic	Pyruvate	DPN, TPN	Muscle	Crystalline	41–44
Pyruvic	Acetyl-CoA + CO_2	DPN	Muscle	Partial	45, 46
		DPN	*E. coli*	Partial	47
Alcohol	Aldehyde	DPN	Yeast	Crystalline	24
		DPN	Liver	Crystalline	25
Acetaldehyde	Acetate	DPN	Liver	Partial	48
		DPN, TPN	Yeast	Partial	49
		TPN	Yeast	Partial	50
Isocitric	α-Ketoglutarate + CO_2	TPN	Muscle	Partial	51
		TPN	Yeast	Partial	32
		DPN	Yeast	Partial	32
α-Ketoglutaric	Succinyl-CoA + CO_2	DPN	Muscle	Partial	52, 53
Glutamic	α-Ketoglutarate + NH_3	DPN,TPN	Liver	Partial	54, 55
		TPN	*E. coli*, yeast	Partial	56, 57
		DPN	Plants	Partial	58
Malic	Oxaloacetate	DPN, TPN	Muscle	High	59
Malic enzyme	Pyruvate + CO_2	TPN	Liver	Partial	28
		DPN	*L. arabinosus*	Partial	29

[37] E. Negelein and W. Gerischer, *Biochem. Z.* **284,** 289 (1936).

[38] B. L. Horecker and P. Z. Smyrniotis, *J. Biol. Chem.* **193,** 303 (1951).

[39] O. Warburg and W. Christian, *Biochem. Z.* **303,** 40 (1939).

[40] T. Baronowski, *J. Biol. Chem.* **180,** 535 (1949).

[41] F. B. Straub, *Biochem. J.* **34,** 483 (1940).

[42] F. Kubowitz and P. Ott, *Biochem. Z.* **314,** 94 (1943).

[43] A. Kornberg and W. E. Pricer, Jr., *J. Biol. Chem.* **193,** 481 (1951).

[44] E. Racker, *J. Biol. Chem.* **196,** 347 (1952).

[45] R. S. Schweet, M. Fuld, K. Cheslock, and M. H. Paul, *in* McElroy and Glass, Phosphorus Metabolism, The Johns Hopkins Press, Baltimore, 1951, Vol. I, p. 246.

domonas fluorescens to yield 2-keto gluconate has thus far been studied only with intact cells.[60] Evidence of a gluconokinase in bacteria[61] and yeast[62] is found in recent reports.

b. Glucose-6-phosphate Dehydrogenase (Zwischenferment). Renewed interest in the direct oxidation of glucose-6-phosphate as an alternate metabolic pathway has come from several sources, such as studies on the mechanism of pentose synthesis,[38, 63, 64] on the elucidation of fermentative mechanisms with isotopically labeled glucose,[65] and on the fluoride-insensitive production of pyruvate and propionate.[66, 67] New facts may emerge from the study of the glucose-6-phosphate dehydrogenase in these varied biologic materials. A somewhat simplified preparation of the dehydrogenase from yeast has been described.[17]

c. Phosphogluconic Dehydrogenase. This enzyme has been purified[38] and shown to catalyze the reversible equilibrium:[68]

$$\text{6-Phosphogluconate} + \text{TPN} \rightleftharpoons CO_2 + \text{ribulose-5-phosphate} + \text{TPNH}_2$$

The presence in these preparations of a pentose phosphate isomerase results in the production of ribose-5-phosphate, the equilibrium favoring the aldose.

[46] S. Ochoa, *J. Biol. Chem.* **195,** 541 (1952).
[47] S. Korkes, A. del Campillo, I. C. Gunsalus, and S. Ochoa, *J. Biol. Chem.* **193,** 721 (1951).
[48] E. Racker, J. Biol. Chem. **177,** 883 (1949).
[49] S. Black, *Arch. Biochem.* **34,** 86 (1951).
[50] J. E. Seegmiller, *J. Biol. Chem.* in press.
[51] S. Ochoa, *J. Biol. Chem.* **174,** 115, 133 (1948); S. Ochoa and E. Weisz-Tabori, *ibid.*, **174,** 123 (1948).
[52] D. R. Sanadi and J. W. Littlefield, *J. Biol. Chem.* **193,** 683 (1951).
[53] S. Kaufman, *in* McElroy and Glass, Phosphorus Metabolism, The Johns Hopkins Press, Baltimore, 1951, Vol. I, p. 370.
[54] J. A. Olson and C. B. Anfinsen, *J. Biol. Chem.* in press.
[55] H. J. Strecker, *Arch. Biochem. Biophys.* **32,** 448 (1951).
[56] E. Adler, H. Hellström, G. Günther, and H. von Euler, *Z. physiol. Chem.* **255,** 14 (1938).
[57] E. Adler, G. Günther, and J. E. Everett, *Z. physiol. Chem.* **255,** 27 (1938).
[58] E. Adler, N. B. Das, H. von Euler, and U. Heyman, *Compt. rend. trav. lab. Carlsberg* **22,** 15 (1938).
[59] F. B. Straub, *Z. physiol. Chem.* **275,** 63 (1942).
[60] L. B. Lockwood, B. Tabenkin, and G. E. Ward, *J. Bact.* **42,** 51 (1941).
[61] S. S. Cohen, *J. Biol. Chem.* **189,** 617 (1951).
[62] H. Z. Sable and A. J. Guarino, *J. Biol. Chem.* **196,** 395 (1952).
[63] D. B. M. Scott and S. S. Cohen, *J. Biol. Chem.* **188,** 509 (1951).
[64] F. Dickens, *Biochem. J.* **32,** 1626, 1636 (1938).
[65] M. Gibbs and R. D. DeMoss, *Federation Proc.* **10,** 189 (1951).
[66] N. Entner and M. Doudoroff, *J. Biol. Chem.* **196,** 853 (1952).
[67] H. G. Wood and F. Leaver, *Federation Proc.* **11,** 313 (1952).
[68] B. L. Horecker and P. Z. Smyrniotis, *J. Biol. Chem.* **196,** 135 (1952).

Proof for the existence of a mechanism for the direct oxidation of pentose phosphate has not been provided as yet. However, it has been demonstrated that one pathway of pentose phosphate metabolism involves initially a cleavage to produce triose phosphate and a 2-carbon unit,[69, 70, 71] the nature of which is the subject of current investigation.

d. Phosphoglyceraldehyde Dehydrogenase. The classical report of Warburg and Christian on the crystallization and mode of action of this enzyme now stands as a major milestone rather than the final goal in the understanding of this reaction. Knowledge derived from the existence of high-energy acyl-sulfur linkages in related reactions has led to new hypotheses such as that formulated by Racker:[71]

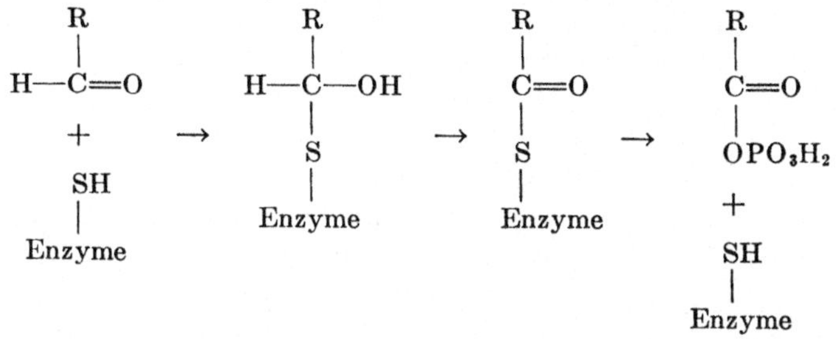

He has observed the presence of glutathione in the crystalline enzyme[72] and has thus strengthened the concept that an addition product of the aldehyde and sulfhydryl portion of the enzyme is oxidized and then cleaved by inorganic phosphate to yield 1,3-diphosphoglyceric acid. The inhibitory role of arsenate in this reaction is accounted for by the production of the highly unstable 1-arsenyl-3-phosphoglyceric acid.

e. Glycerophosphate Dehydrogenase. This enzyme was crystallized from myogen A by Baranowski,[40] and the equilibrium was shown to favor strongly the reduction of dihydroxyacetone phosphate. Nevertheless, under aerobic conditions the utilization of glycerol very likely proceeds by way of the triose phosphates after a preliminary phosphorylation of the glycerol by a kinase reaction. In view of recent evidence[73] that α-glycerophosphate

[69] Z. Dische, *Naturwissenschaften* **26,** 252 (1938).
[70] B. L. Horecker and P. Z. Smyrniotis, *Federation Proc.* **11,** 232 (1952).
[71] E. Racker, *in* McElroy and Glass, Phosphorus Metabolism, The Johns Hopkins Press, Baltimore, 1951, Vol. I, p. 145.
[72] I. Krimsky, and E. Racker, *J. Biol. Chem.* **198,** 721 (1952).
[72a] S. Korkes, A. Del Campillo, I. C. Gunsalus, and S. Ochoa, *J. Biol. Chem.* **193,** 721 (1952).
[72b] V. Jagannathan and R. S. Schweet, *J. Biol. Chem.* **196,** 551 (1952).
[72c] S. Kaufman, McElroy and Glass, *in* Phosphorous Metabolism, The Johns Hopkins Press Baltimore, 1951, p. 370.
[72d] D. R. Sanadi and J. W. Littlefield, Science **116,** 327 (1952).
[73] A. Kornberg and W. E. Pricer, Jr., *J. Am. Chem. Soc.* **74,** 1617 (1952).

is a precursor in the enzymatic formation of phosphorus-containing lipids, this enzyme may prove to be of great importance in fat metabolism.

f. Pyruvic Dehydrogenase and α-Ketoglutaric Dehydrogenase. Both enzymes can be considered together as examples of α-keto acid oxidation. Although it is premature to consider the exact mechanism of this type of oxidation, it is worth while to direct attention to some of the important advances that have been made and the hypotheses that are under serious consideration.[72a, 72b, 72c, 72d] The involvement of an acyl-coenzyme A intermediate is established:

$$\text{Pyruvate} + \text{DPN} + \text{CoA} \rightarrow \text{Acetyl-CoA} + \text{DPNH}_2 + \text{CO}_2$$

$$\alpha\text{-Ketoglutarate} + \text{DPN} + \text{CoA} \rightarrow \text{Succinyl-CoA} + \text{DPNH}_2 + \text{CO}_2$$

It remains to be determined what function thiamin pyrophosphate, an essential cofactor, serves in these reactions. The role of the pyruvate oxidation factor[74] in cell-free systems also remains to be clarified.*

g. Acetaldehyde Dehydrogenase. Several different enzymes which catalyze the oxidation of acetaldehyde have been described in the past few years in yeast. The enzymes discovered by Racker in liver[48] and Black[49] and Seegmiller[50] in yeast produce acetic acid without any detectable participation of phosphate or coenzyme A. The preparation of Stadtman and Burton[75] from *Cl. kluyveri* requires coenzyme A and converts acetaldehyde to acetyl-CoA.

$$\begin{array}{l} \text{Acetaldehyde} + \text{DPN} + \text{CoA} \rightleftharpoons \text{Acetyl—CoA} + \text{DPNH}_2 \\ \qquad\qquad\qquad\qquad\qquad\qquad\qquad\quad \updownarrow \\ \qquad\qquad\qquad\qquad\qquad\qquad\qquad \text{Acetyl phosphate} + \text{CoA} \end{array}$$

In the presence of inorganic phosphate and transacetylase (in crude extracts) acetyl phosphate is produced. Reversal of this reaction has been achieved. Since sulfhydryl activation of the liver enzyme has been noted,

* Pyruvate oxidation factor has been identified as α-lipoic acid,

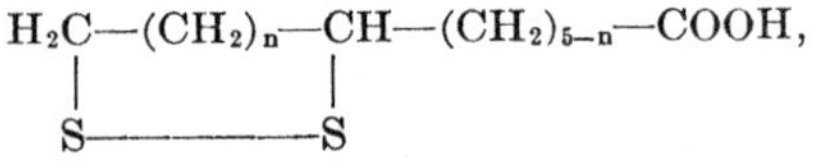

(J. A. Brockman, Jr., E. L. R. Stokstad, E. L. Patterson, J. V. Pierce, M. Macchi and F. P. Day, J. Am. Chem. Soc. **74,** 1868 (1952); C. S. Hornberger, Jr., R. F. Heitmiller, I. C. Gunsalus, G. H. Schnakenberg, and L. J. Reed, J. Am. Chem. Soc. **74,** 2382 (1952)) and it has been shown (L. J. Reed and B. G. De Busk, J. Biol. Chem. **199,** 873, 881 (1952)) that the coenzyme for pyruvate and α-keto-glutarate oxidation is the amide of α-lipoic acid and cocarboxylase, named lipothiamide pyrophosphate.

[74] I. C. Gunsalus, *in* McElroy and Glass, Phosphorus Metabolism, The Johns Hopkins Press, Baltimore, 1951, Vol. I, p. 289.

[75] R. M. Burton, *Federation Proc.* **11,** 193 (1952).

it may also prove to involve a mechanism analogous to the coenzyme A-requiring dehydrogenase of *Cl. kluyveri*.

h. Isocitric Dehydrogenase. In yeast[32] two enzymes were found which could be distinguished not only on the basis of their coenzyme specificity but also in other important respects. The TPN-specific dehydrogenase resembles closely a TPN-specific enzyme derived from acetone powders of mammalian tissue. As described by Ochoa[51] for the mammalian TPN-specific enzyme, the yeast enzyme catalyzes the decarboxylation of oxalosuccinate, the reductive carboxylation of α-ketoglutarate, and the reduction of oxalosuccinate. The DPN-specific enzyme fails to catalyze the reductive carboxylation of α-ketoglutarate and does not appear to have oxalosuccinate as an intermediate. As a further distinction from the TPN-specific enzyme, there is a specific and absolute requirement for adenosine-5-phosphate, the function of which is not clear.

III. Flavoprotein Enzymes

The flavoproteins, or yellow enzymes, have been shown to contain either alloxazine mononucleotide (flavin mononucleotide, FMN) or isoalloxazine adenine dinucleotide (flavin adenine dinucleotide, FAD) (Fig. 6). The

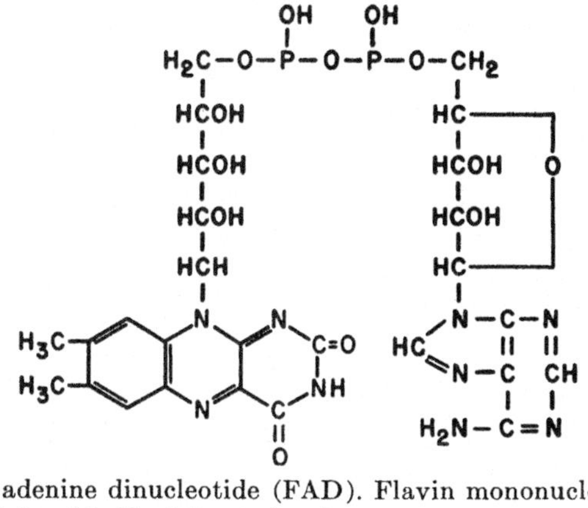

FIG. 6. Flavin adenine dinucleotide (FAD). Flavin mononucleotide (FMN) contains only the left-hand half of the molecule.

flavin prosthetic groups give these enzymes their yellow or orange-yellow color. In the oxidized form the flavoproteins have the characteristic two-banded absorption spectrum of riboflavin (Fig. 7) although the absorption maxima may be shifted toward longer wavelengths. On reduction, either with substrate or with reducing agents such as sodium hydrosulfite, the yellow color is more or less bleached with a corresponding disappearance of the absorption bands. In the reduced form one pair of conjugated double bonds is replaced by a single double bond (Fig. 8). The reduced forms of the free prosthetic groups are very rapidly oxidized by molecular oxygen, but this ability to react directly with oxygen (autoxidation) is greatly dimin-

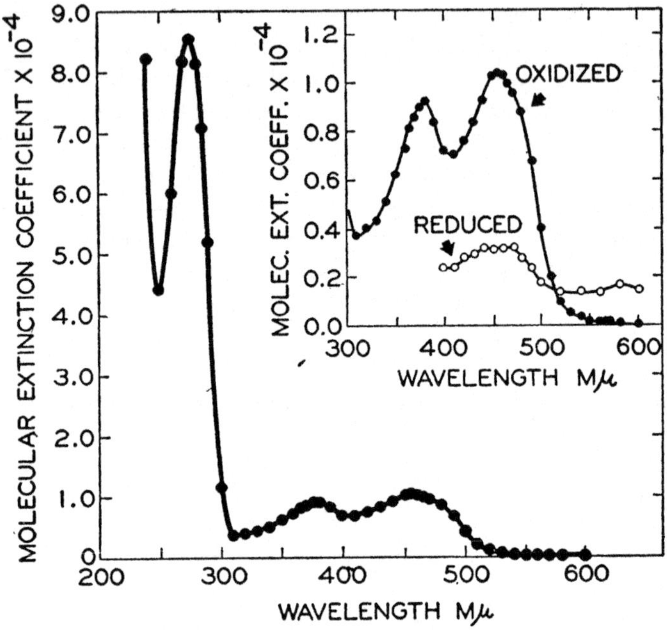

FIG. 7. The absorption spectrum of TPN-cytochrome c reductase isolated from liver. The reduced spectrum was obtained by reduction with $TPNH_2$. The insert is enlarged to show the details of the bands at 380 and 455 mμ. (From Horecker.[97])

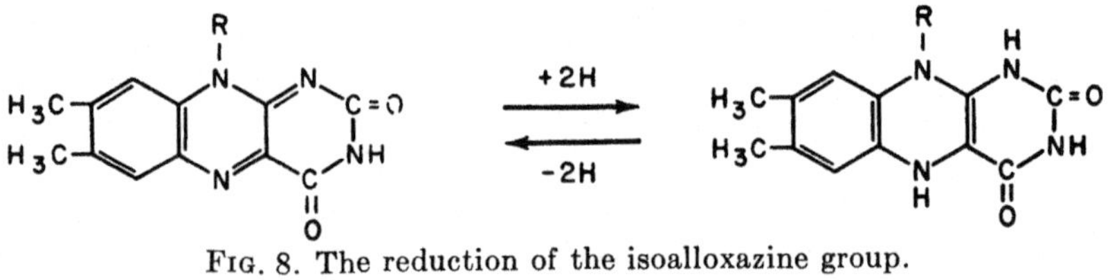

FIG. 8. The reduction of the isoalloxazine group.

ished in combination with the enzyme protein and in some cases is almost entirely absent. In the reaction with oxygen, hydrogen peroxide is produced (see Fig. 1).

1. The Oxidation of the Pyridine Nucleotides

Flavoproteins occur as "oxidases" which catalyze the direct oxidation of various substrates by oxygen. They appear to be of major importance, however, in the hydrogen transport system where they catalyze the oxidation of the reduced pyridine nucleotides. A number of flavin enzymes have been implicated in this system, but the true physiological role of many of these is not well understood. This is true of the "old" yellow enzyme of Warburg, which catalyzes the direct oxidation of pyridine nucleotides by oxygen, and of Haas' "new" yellow enzyme and diaphorase which react rapidly only with dyes such as methylene blue. On the other hand, the cytochrome c reductases, which bring about the reduction of cytochrome c by reduced pyridine nucleotides, appear to be in the direct path of hydrogen transport.

a. The "Old" Yellow Enzyme. Extracts of yeast cells show a respiration which, unlike that of the intact cells, is not sensitive to cyanide.[76] Cyanide insensitive respiration has also been demonstrated in anaerobic bacteria.[77] From extracts of dried brewer's yeast, Warburg and Christian[78] isolated a flavoprotein, later designated as the "old" yellow enzyme, which appeared to account for this cyanide-insensitive respiration.

Theorell[79] obtained the enzyme in essentially pure form by electrophoretic separation. The prosthetic group was FMN, and the absorption spectrum showed bands at 380 and 465 mμ which disappeared on reduction. From the flavin content the molecular weight was estimated to be about 80,000, a value which was confirmed by ultracentrifugal measurements.[80]

Theorell[79] succeeded in dissociating the prosthetic group of "old" yellow enzyme from the protein by means of dialysis against dilute acid. The active enzyme was regenerated when the two components were mixed. Later Warburg and Christian[81] introduced a more generally applicable method for this purpose which involved acidification to pH 1 to 3 in the presence of ammonium sulfate. The protein component is precipitated, and the flavin remains in solution. This procedure for the reversible separation of protein and prosthetic group has been successfully applied to many of the known flavoprotein enzymes.

The role of the "old" yellow enzyme in metabolism is not known. It is reduced by $TPNH_2$[81] or $DPNH_2$[82] and the reduced form is reoxidized by atmospheric oxygen. Its activity in this system, however, is too low to account for more than a fraction of yeast respiration. It is more rapidly reoxidized by methylene blue, but very slowly, if at all, by cytochrome c.

[76] O. Warburg and W. Christian, *Biochem. Z.* **231,** 493 (1931).
[77] O. Meyerhof and P. Finkle, *Chem. Zelle u. Gewebe* **12,** 157 (1925).
[78] O. Warburg and W. Christian, *Biochem. Z.* **254,** 438 (1932); **263,** 228 (1933).
[79] H. Theorell, *Biochem. Z.* **278,** 263 (1935).
[80] R. A. Kekwich and K. O. Pederson, *Biochem. J.* **30,** 2201 (1936).
[81] O. Warburg and W. Christian, *Biochem. Z.* **298,** 368 (1938).
[82] H. von Euler, E. Adler, and H. Hellström, *Svensk Kem. Tid.* **47,** 290 (1934).

"Old" yellow enzyme has been isolated only from yeast; a similar enzyme, however, has been shown to account for the oxygen uptake of anaerobic lactic acid bacteria,[83] which contain no demonstrable hemochromogens. The hydrogen peroxide produced in such respiration kills the organism.

Haas[84] has described the step-wise reduction and oxidation of "old" yellow enzyme, with evidence for a red free radical which has undergone a one-electron change. This concept of two-step oxidation and reduction involving free radical formation has been discussed at length by Michaelis and Schubert.[85]

The protein of "old" yellow enzyme also combines with FAD to produce a synthetic enzyme with almost identical catalytic properties.[81]

b. The "New" Yellow Enzyme. A number of other flavoproteins have been shown to catalyze the oxidation of the reduced pyridine nucleotides, but the physiological oxidant for these enzymes is not known. They are rapidly oxidized by dyes such as methylene blue, react sluggishly with oxygen, and not at all with cytochrome c. The wide distribution and great activity of these enzymes has stimulated an intensive search for auxiliary enzymes to link them to the cytochrome system, but none has yet been reported.

From dried brewer's yeast Haas[86] isolated a yellow enzyme differing from the "old" yellow enzyme in both protein and prosthetic group. The absorption maxima occur at 377 and 455 mμ, and the prosthetic group was identified as FAD. The molecular weight calculated from the flavin content was 65,000. The protein and prosthetic groups were separated by treatment with acid ammonium sulfate; the protein could be reactivated only with FAD. While the enzyme is reduced by $TPNH_2$ about three times more rapidly than is the "old" yellow enzyme, reoxidation with oxygen is about seven times slower. Reoxidation by methylene blue, on the other hand, is very much more rapid. No data on the activity of this enzyme with $DPNH_2$ have been reported; the relation between this enzyme and diaphorase has not been determined.

c. Diaphorase or Coenzyme Factor. The oxidation of reduced pyridine nucleotide by methylene blue was observed in animal tissues[87, 88, 89] and in bacteria and yeast.[90] This activity has been referred to as "diaphorase"

[83] H. von Euler, E. Adler, and H. Hellström, *Biochem. Z.* **260,** 499 (1933).
[84] E. Haas, *Biochem. Z.* **290,** 291 (1937).
[85] L. Michaelis and M. P. Schubert, *Chem. Revs.* **22,** 437 (1938).
[86] E. Haas, *Biochem. Z.* **298,** 378 (1938).
[87] E. Adler, H. von Euler, and H. Hellström, *Arkiv. Kemi Mineral. Geol.* **12 B,** No. 38 (1937).
[88] H. von Euler and H. Hellström, *Z. physiol. Chem.* **252,** 31 (1938).
[89] J. G. Dewan and D. E. Green, *Biochem. J.* **32,** 626 (1938).
[90] D. E. Green and J. G. Dewan, *Biochem. J.* **32,** 1200 (1938).

or coenzyme factor. In animal tissues it is associated with insoluble particles from which it can be liberated by treatment with dilute ammonium sulfate and ethanol at 40°. From the solution thus obtained Straub[91] isolated a flavoprotein containing FAD as the prosthetic group. The absorption maxima are found at 359 and 451 mμ, and the molecular weight is 70,000, about the same as that of previously isolated flavoproteins. This flavoprotein was shown[92] to possess the catalytic properties of coenzyme factor (diaphorase). The purified enzyme is reduced by $DPNH_2$; the activity with $TPNH_2$ was not reported, although this activity is present in crude enzyme preparations.[89] The leucoflavoprotein is rapidly reoxidized by methylene blue but reacts sluggishly with oxygen. No physiological oxidant was found.

Crude enzyme preparations from various animal tissues show varying ratios of activity with $DPNH_2$ and $TPNH_2$, suggesting the existence of at least two distinct diaphorases.[93]

d. Yeast TPN-Cytochrome c Reductase. In the purification of "new" yellow enzyme from yeast, Haas[86] made the observation that, although crude preparations had the ability to reduce cytochrome c, this activity was lost during the purification procedure. This enzyme was subsequently isolated by Haas, Horecker, and Hogness[94] who used an assay system in which the rate of cytochrome c reduction was measured spectrophotometrically. It was identified as a flavoprotein having FMN as the prosthetic group. The absorption spectrum showed maxima at 385 and 455 mμ, and the molecular weight calculated from the flavin content was 75,000. The preparation was shown by titration with cytochrome c to be about 87% pure. A small amount of hemoprotein contaminant was present, but this was later removed by an improved purification procedure[95] and shown to be unnecessary for the enzymatic activity. The product had a purity of 98%. Although the prosthetic group of yeast TPN-cytochrome c reductase is apparently identical with that of the "old" yellow enzyme, the two enzymes differ with respect to their catalytic activities. Cytochrome c reductase is more rapidly reduced by $TPNH_2$ and is rapidly reoxidized by cytochrome c; the latter reaction occurs slowly if at all with "old" yellow enzyme. Cytochrome c reductase, on the other hand, shows very little activity with oxygen.

The prosthetic group of yeast-cytochrome c reductase is reversibly dissociated from the protein by treatment with acid in the presence of

[91] F. B. Straub, *Biochem. J.* **33,** 787 (1939).
[92] H. S. Corran, D. E. Green, and F. B. Straub, *Biochem. J.* **33,** 793 (1939).
[93] E. Adler, H. von Euler, and G. Günther, *Nature* **143,** 641 (1939).
[94] E. Haas, B. L. Horecker, and T. R. Hogness, *J. Biol. Chem.* **136,** 747 (1940).
[95] E. Haas, C. J. Harrer, and T. R. Hogness, *J. Biol. Chem.* **143,** 341 (1942).

ammonium sulfate. The flavin obtained exchanges quantitatively with that of "old" yellow enzyme but is inactive in the D-amino acid oxidase test for FAD. The protein of cytochrome c reductase, however, is reactivated by FAD, although the catalytic activity obtained is about one-third less than with FMN.[96] This combination with FAD resembles that observed in the case of the "old" yellow enzyme. In each case a "synthetic" enzyme is formed where catalytic properties are largely determined by the nature of the protein component.

e. Liver TPN-Cytochrome c Reductase. In animal tissues TPN-cytochrome c reductase is associated with insoluble particles, together with diaphorase, and for many years the existence of a separate cytochrome c reductase was not established. The enzyme was obtained in solution by tryptic digestion of a homogenate prepared from pig liver acetone powder.[97] TPN-cytochrome c reductase isolated from these solutions was identified as a flavoprotein with FAD as the prosthetic group. The enzyme was dissociated into protein and prosthetic group by the method of Warburg and Christian.[81] The flavin was identified as FAD by its activity in the D-amino acid oxidase test and also by the increase in fluorescence[98] which occurs when FAD is hydrolyzed to FMN and adenylic acid by nucleotide pyrophosphatase from potato.[9] The absorption spectrum of the purified enzyme, with absorption bands at 380 and 455 mμ, is shown in Fig. 7. From the flavin content the molecular weight was estimated to be 68,000.

In spite of the difference in prosthetic groups, the catalytic activity of TPN-cytochrome c reductase is essentially identical with that of the yeast enzyme. In the presence of excess TPN, reduced with hexosemonophosphate and glucose-6-phosphate dehydrogenase (Zwischenferment) the turnover number was found to be 1140 moles of cytochrome per mole of flavoprotein per minute with the liver enzyme and 1300 with the yeast enzyme. Cytochrome c reductase from liver also resembles the enzyme from yeast in the behavior of the protein component with FAD and FMN (Table 3). In each case the enzyme reconstituted with FMN is more active than with FAD, despite the fact that the natural prosthetic group of the liver enzyme is FAD.

The TPN-cytochrome c reductases appear to function without the participation of accessory enzymes or factors. In the purified preparations, both from yeast and liver, the virtual absence of hemoprotein contaminants is shown by the absence of the Soret band. Although the introduction of such a factor with the impure Zwischenferment preparations has not been ruled out, evidence against this possibility is available in the case of liver

[96] E. Haas, B. L. Horecker, and T. R. Hogness, unpublished observation.
[97] B. L. Horecker, *J. Biol. Chem.* **183,** 593 (1950).
[98] H. B. Burch, O. A. Bessey, and O. H. Lowry, *J. Biol. Chem.* **175,** 457 (1948).

cytochrome c reductase. This was found to have identical activity in a test system in which the Zwischenferment-glucose-6-phosphate system for the reduction of TPN was replaced by isocitrate and isocitric dehydrogenase.

f. DPN-Cytochrome c Reductase. The reduction of cytochrome c by reduced DPN was observed by Lockhart and Potter[99] with suspensions of pig heart and a soluble DPN-cytochrome c reductase from yeast was reported by Altschul, Persky, and Hogness.[100] The enzyme was also purified by Heppel[101] from liver acetone powder suspensions, where it was found to occur together with TPN-cytochrome c reductase. However, although the DPN-cytochrome c reductase activity is twenty to thirty times as great in the liver preparations as is the corresponding activity with TPN, it is completely destroyed by the tryptic digestion which renders the TPN-reductase soluble. With milder digestion procedures a preparation is obtained which is considerably purified although it appears still to be associated with small particles.

TABLE 3

ACTIVITY OF RECONSTITUTED CYTOCHROME c REDUCTASES WITH FMN AND FAD

Source of Protein	Activity	
	FMN	FAD
Yeast	0.027	0.011
Liver	0.035	0.030

Recently Edelhoch, Hayaishi, and Tepley[102] have obtained soluble preparations of DPN-cytochrome c reductase from pigeon breast muscle or pig heart particles by freezing at pH 4 or treatment with dilute ethanol. The enzyme, purified from such extracts by Mahler, Vernon, and Sarkar,[103] has the properties of a flavoprotein, although the prosthetic group has not yet been identified with either FMN or FAD.

Slater[104] has reported evidence that in heart muscle the oxidation of $DPNH_2$ depends on the presence of a component which has been since referred to as "Slater's factor." The evidence is based largely on the effect of BAL (British Anti-Lewisite, 2,3-dimercaptopropanol). Treatment of heart muscle preparations with this substance leads to an irreversible de-

[99] E. E. Lockhart and V. R. Potter, *J. Biol. Chem.* **137,** 1 (1941).

[100] A. M. Altschul, H. Persky, and T. R. Hogness, *Science* **94,** 349 (1941).

[101] L. A. Heppel, *Federation Proc.* **8,** 205 (1949).

[102] H. Edelhoch, O. Hayaishi, and L. J. Tepley, *J. Biol. Chem.* in press.

[103] H. R. Mahler, L. P. Vernon, and N. K. Sarkar, *Federation Proc.* **11,** 253 (1952).

[104] E. C. Slater, *Biochem. J.* **46,** 484, 499 (1950).

struction of the ability of these enzymes to oxidize $DPNH_2$ with cytochrome c, without affecting the reaction of $DPNH_2$ with methylene blue. The participation of such a factor in purified systems has not been demonstrated.

2. The Flavoprotein Oxidases

a. D-Amino Acid Oxidase. A number of flavoprotein enzymes have been isolated which catalyze the direct oxidation of substrate by oxygen, with no requirement for a pyridine nucleotide coenzyme. Among these is D-amino acid oxidase, described in pig kidney acetone powder extracts by Krebs.[105] The flavoprotein nature of this enzyme was established in Warburg's laboratory, where Warburg and Christian[106] isolated the prosthetic group and Negelein and Brömel[107] the protein component, following separation of these by treatment with acid ammonium sulfate. The prosthetic group purified by extraction with phenol and precipitation of the metal salts was identified as a dinucleotide containing riboflavin phosphate and adenylic acid (Fig. 6). This was the first identification of FAD as a flavoprotein prosthetic group. The assay method was based on the oxygen uptake with D-alanine as a substrate in the presence of an excess of the protein component. This method is highly sensitive and specific for FAD and is most widely used for the determination of this substance. FMN is completely inactive. FAD can also be assayed by the fluorimetric method previously described.[9, 98] The protein component was obtained in highly purified form with the aid of the same assay procedure, except that FAD was added in excess.

In the presence of catalase the oxidation of alanine leads to the formation of pyruvic acid and ammonia:

$$CH_3 \cdot CHNH_2 \cdot COOH + \tfrac{1}{2}O_2 \rightarrow CH_3 \cdot CO \cdot COOH + NH_3$$

However, hydrogen peroxide is a primary product of the reaction, and in purified preparation free of catalase this reacts with pyruvic acid to form acetic acid.[107]

$$CH_3 \cdot CHNH_2 \cdot COOH + O_2 \rightarrow CH_3 \cdot COOH + NH_3 + H_2O$$

Enzymes which catalyze the oxidation of D-amino acids have also been described in molds,[108, 109] but these have not been identified as flavoproteins.

b. L-Amino Acid Oxidases. From rat liver a flavoprotein has been iso-

105 H. A. Krebs, *Biochem. J.* **29,** 1620 (1935).
106 O. Warburg and W. Christian, *Biochem. Z.* **296,** 294; **298,** 150 (1938).
107 E. Negelein and H. Brömel, *Biochem. Z.* **300,** 225 (1939).
108 N. H. Horowitz, *J. Biol. Chem.* **154,** 141 (1944).
109 R. L. Emerson, M. Pugiss, and S. G. Knight, *Arch. Biochem.* **25,** 299 (1950).

lated[110] which catalyzes the oxidation of L-amino acids. The prosthetic group contained 1 mole of phosphorus per mole of flavin and showed no extra absorption at 260 mμ due to adenylic acid. These properties are consistent with FMN. The molecular weight was 120,000, and the absorption maxima occurred at 360 and 460 mμ. This enzyme exhibits a very low catalytic activity; with L-leucine as the substrate the turnover number was found to be 6 moles of substrate oxidized per minute per flavin equivalent. α-Hydroxy acids were oxidized at a rate which was about three times the rate with L-leucine.[110, 111]

A much more active L-amino acid oxidase, discovered in snake venom by Zeller,[112] has been isolated by Singer and Kearney[113] and identified as a flavoprotein with FAD as the prosthetic group. The enzyme has a molecular weight of about 62,000, and the spectrum shows absorption bands at 389 and 465 mμ. With L-leucine as the substrate, the turnover number at 38° is 3100 moles of substrate oxidized per minute per mole of flavin; even crude venom has higher activity than the purified rat liver enzyme. Snake venom L-amino acid oxidase exists in active and inactive forms which are interconvertible under the influence of pH and inorganic ions.

c. Glycine Oxidase. Pig heart acetone powder extracts contain an enzyme[114] which oxidizes glycine or sarcosine to glyoxylic acid:

$$CH_2NH_2\cdot COOH + \tfrac{1}{2}O_2 \rightarrow CHO\cdot COOH + NH_3$$

With sarcosine methylamine is produced instead of ammonia. Although the enzyme has not been obtained in highly purified form it is inactivated by precipitation with ammonium sulfate, and the apoenzyme thus obtained is reactivated by the addition of FAD. The failure to observe hydrogen peroxide formation during substrate oxidation is probably due to the presence of catalase.

d. Xanthine Oxidase. Two widely different enzymatic activities in milk, the oxidation of aldehydes discovered by Schardinger[115] and the oxidation of purines described by Morgan, Stewart, and Hopkins[116] now appear to be due to the same enzyme, xanthine oxidase. The enzyme was isolated by Ball[117] and characterized as a flavoprotein. The nature of the prosthetic

[110] M. Blanchard, D. E. Green, V. Nocito, and S. Ratner, *J. Biol. Chem.* **155,** 421 (1944); **161,** 583 (1945).

[111] B. Iselin and E. A. Zeller, *Helv. Chim. Acta* **29,** 1508 (1940).

[112] E. A. Zeller, *Helv. Physiol. et Pharmacol. Acta* **2,** C33 (1944).

[113] T. Singer and E. B. Kearney, *Arch. Biochem.* **27,** 348; **29,** 190 (1950).

[114] S. Ratner, V. Nocito, and D. E. Green, *J. Biol. Chem.* **152,** 119 (1944).

[115] F. Schardinger, *Z. Untersuch. Nahr. u. Genussm.* **5,** 1113 (1902).

[116] E. J. Morgan, C. P. Stewart, and F. G. Hopkins, *Proc. Roy. Soc.* (*London*) **B94,** 109 (1922).

[117] E. Ball, *J. Biol. Chem.* **128,** 51 (1939).

group is still in doubt. It can be demonstrated that FAD is present in the enzyme, but all efforts to reactivate the split enzyme with FMN have failed. Ball reported the activation of a dialyzed preparation by a heat-stable factor which could not be replaced by FAD. However, this reactivation has been attributed to a non-specific reactivation which could be obtained with several metal binding agents.[118] Evidence for a relation of FAD and enzymatic activity has been obtained by Corran *et al.*[119] and by Horecker and Heppel,[42] who found the catalytic activity to be proportional to the FAD content throughout the purification procedure. The purified preparations obtained by Ball and by Corran *et al.* did not show the typical flavoprotein absorption spectrum. An additional component was present which contributed to the absorption below 400 mμ and obscured the flavin bands. A more active preparation of milk xanthine oxidase obtained by Horecker and Heppel[120] was estimated from its FAD content to be about 62% pure. The absorption spectrum, however, was essentially the same as that shown by Ball's preparation. The absorption bands were removed on addition of hypoxanthine, providing further evidence for the participation of flavin in the enzyme action.*

The reduced form of xanthine oxidase is oxidized by molecular oxygen or by methylene blue. It will also react with cytochrome c at a rate which is about one-half the rate with oxygen.[120] However, the reduction of cytochrome c is dependent on the presence of oxygen, and very little reaction occurs under anaerobic conditions. This requirement of oxygen for the reduction of cytochrome c by xanthine oxidase has been explained in terms of the formation of a free radical in the oxidation of the leucoflavoprotein. Oxygen would be necessary to form the free radical from the leucoform, but the free radical could react with either oxygen or cytochrome c.†

Milk xanthine oxidase, in addition to its activity with purines and aldehydes, also catalyzes the oxidation of $DPNH_2$ with oxygen or methylene blue. This activity was first associated with a milk flavoprotein isolated by Corran and Green[121] which was later reported to be identical with xanthine

* D. B. Morell, Biochem. J. **51,** 657 (1952) has recently reported results from which it is concluded that the prosthetic group of xanthine oxidase contains no colored components other than FAD.

† D. B. Morell, Biochem. J. **51,** 666 (1952), using somewhat different test conditions, has been unable to observe the requirement of oxygen for cytochrome c reduction.

[118] H. M. Kalckar, N. O. Kjeldgaard, and H. Klenow, *Biochem. et Biophys. Acta* **5,** 575 (1950).

[119] H. S. Corran, J. G. Dewan, A. H. Gordon, and D. E. Green, *Biochem. J.* **33,** 1694 (1939).

[120] B. L. Horecker and L. A. Heppel, *J. Biol. Chem.* **178,** 683 (1949).

[121] H. S. Corran and D. E. Green, *Biochem. J.* **32,** 2231 (1938).

oxidase.[119, 122] The activity with reduced coenzyme was much less than with hypoxanthine.

Xanthine oxidase has been partially purified from liver.[119] The purified preparations catalyze the oxidation of aldehydes, purines, and $DPNH_2$, much as does the milk enzyme. Although flavoprotein was present in the purified preparations, it was not shown to be directly related to the enzyme activity.

e. Glucose Oxidase, Notatin. Notatin (or penicillin B), isolated on the basis of its antibiotic properties,[123, 124] was identified as a flavoprotein which was identical with the glucose oxidase discovered in molds by Müller.[125] It had been suggested by Franke and Deffner[126] that this enzyme was a flavoprotein. Notatin has absorption bands at 377 and 455 mμ, and the prosthetic group was identified as FAD.[127] However, the process of splitting the enzyme appears to denature the protein, and it has not been possible to establish that FAD will restore the catalytic activity. The molecular weight has been determined by sedimentation measurements to be 152,000,[128] which is twice the equivalent weight calculated from the FAD content. Thus 1 mole of enzyme contains 2 moles of FAD. The purified enzyme has a high degree of specificity for glucose[129] and has been employed for the assay of this sugar in biological materials. The antibiotic activity of notatin was shown to be due to the formation of hydrogen peroxide during the catalytic oxidation of glucose.[123, 124] Xanthine oxidase was shown to have similar antibiotic activity,[130] and presumably this would be true of other flavoproteins which react with oxygen, unless catalase is also present.

Notatin has been shown to be specific for the β-D-pyranose form of glucose[129] and to produce D-glucono-δ-lactone[131] as the primary oxidation product.

f. Fumaric Hydrogenase. This flavoprotein, isolated from yeast by Fischer, Roedig, and Rauch,[132] appears to have the unique property (for a flavoprotein) of catalyzing the reduction of a substrate rather than its

[122] E. G. Ball and P. A. Ramsdell, *J. Biol. Chem.* **131,** 767 (1939).

[123] C. E. Coulthard, R. Michaelis, W. F. Short, G. Sykes, G. E. H. Skrimshire, A. F. B. Standfast, J. H. Brikinshaw, and H. Raistrick *Biochem. J.* **39,** 24 (1945).

[124] J. T. Van Braggen, F. J. Reitherl, C. K. Cain, P. A. Katzman, E. A. Doisy, R. D. Muir, E. C. Roberts, W. L. Gaby, D. M. Homan, and L. R. Jones, *J. Biol. Chem.* **148,** 365 (1943).

[125] D. Müller, *Biochem. Z.* **199,** 136 (1928).

[126] W. Franke and M. Deffner, *Ann.* **541,** 117 (1939).

[127] D. Keilin and E. F. Hartree, *Biochem. J.* **42,** 221 (1948).

[128] R. Cecil and A. G. Ogston, *Biochem. J.* **42,** 229 (1948).

[129] D. Keilin and E. F. Hartree, *Biochem. J.* **50,** 331 (1951).

[130] D. E. Green and R. Pauli, *Proc. Soc. Exptl. Biol. Med.* **54,** 148 (1943).

[131] R. Bentley and A. Neuberger, *Biochem. J.* **45,** 584 (1949).

[132] F. G. Fischer, A. Roedig, and K. Rauch, *Ann.* **552,** 203 (1942).

oxidation. In the presence of dyes such as methylene violet, fumarate is reduced to succinate:

Leuco dye + fumarate → Oxidized dye + succinate

The prosthetic group was readily removed from the protein by dialysis, charcoal treatment, or exposure to acid ammonium sulfate. The activity of of the protein component was restored by the addition of FAD. The physiological reducing agent is unknown.

IV. The Hemoprotein Enzymes

The importance of iron enzymes in respiration has been thoroughly established by the work of Warburg, Keilin, and Theorell. Whenever isola-

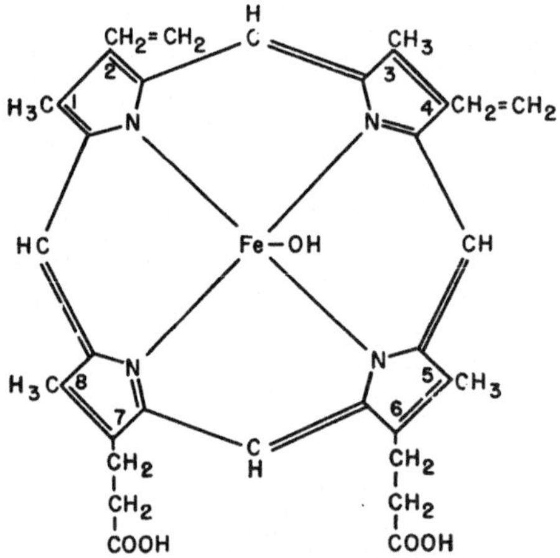

FIG. 9 Protohematin IX.

tion and characterization of these enzymes has been possible, they have proved to contain divalent or trivalent iron in the form of a porphyrin complex, or heme. In most cases the prosthetic group is identical with that of hemoglobin, protohematin IX (Fig. 9), although evidence for different porphyrins is available in some cases. A general method for the removal of the prosthetic group by treatment with acid acetone is available, and in some cases this separation of the prosthetic group from the protein component is reversible. The hemoprotein enzymes include the cytochromes, which are concerned with the activation of oxygen and electron transport, as well as the catalases and peroxidases, which catalyze oxidations by hydrogen peroxide.

1. The Cytochromes

The pigments now known as the cytochromes were first described by MacMunn[133] in 1886 and called histohematin. They were rediscovered and given the name cytochromes in 1925 by Keilin[134] who obtained evidence for the presence of three components and confirmed MacMunn's suggestion that these substances function as respiratory catalysts in the intact cell. In the reduced form these substances give rise in vertebrate and invertebrate muscle, yeast, bacteria, and plant cells to a number of absorption bands. These bands occur at or near the following wavelengths, depending on the tissue examined.[135]

Band	a	b	c	d
Wavelength (mμ)	605	565	550	520

The oxidized forms show diffuse bands at 567 and 529 mμ. The a, b, and c bands are the α bands of cytochromes a, b, and c, respectively, while the d band is formed by the fused β bands of cytochromes b and c. The region between 415 and 455 mμ contains the very intense γ, or Soret, bands which are not readily observed with the visual spectroscope. Under some conditions the a band can be shown to be due to two components, one of which forms a complex with carbon monoxide. This has been designated by Keilin and Hartree as cytochrome a_3[135]. Only one of the cytochromes, cytochrome c, has been obtained in soluble form and purified. The structure and function of the a and b cytochromes is not completely understood. A convenient source for the study of the cytochromes is the heart muscle preparation of Keilin and Hartree[136] which consists of a suspension of fine particles prepared from washed minced heart muscle.

a. Cytochrome c. Methods have been developed for the isolation of cytochrome c from yeast[137] and heart muscle.[138, 139] The procedures depend on the stability of cytochrome c in acid solution and its solubility in trichloracetic acid and ammonium sulfate solutions. As a result of these properties it is easily separated from the bulk of the tissue proteins. The early methods all yielded an apparently homogeneous product which contained 0.34% Fe, corresponding to a molecular weight of about 16,000 However, when this product was subjected to electrophoresis at pH 10.5[140]

[133] C. A. MacMunn, *Phil. Trans. Roy. Soc. (London)* **177,** 267 (1886).
[134] D. Keilin, *Proc. Roy. Soc. (London)* **B98,** 312 (1925).
[135] D. Keilin and E. F. Hartree, *Proc. Roy. Soc. (London)* **B127,** 167 (1939).
[136] D. Keilin and E. F. Hartree, *Proc. Roy. Soc. (London)* **B125,** 171 (1938).
[137] D. Keilin, *Proc. Roy. Soc. (London)* **B106,** 418 (1930).
[138] H. Theorell, *Biochem. Z.* **285,** 207 (1936).
[139] D. Keilin and E. F. Hartree, *Proc. Roy. Soc. (London)* **B122,** 298 (1937).
[140] H. Theorell and A. Akesson, *Science* **90,** 67 (1939).

(above the isoelectric point) a product containing 0.43% Fe resulted. A similar purification was achieved by fractionation with ammonium sulfate at pH 10,[141] by means of which a colorless fraction was separated from the Fe-containing portion. The molecular weight calculated from the Fe content was 13,000.

Isolated cytochrome c in neutral solution shows absorption bands at 408, 530, and 695 mμ in the oxidized form[142] and at 415, 520, and 550 mμ in the reduced form.[138] The heme group is not readily removed by treatment with acid acetone. Following hydrolysis with hydrochloric acid a porphyrin can be isolated which is referred to as porphyrin c.[143] This porphyrin differs from protoporphyrin IX in the nature of side chains 2 and 4 in which the vinyl groups are replaced by thioether bridges with cysteine (Fig. 10).

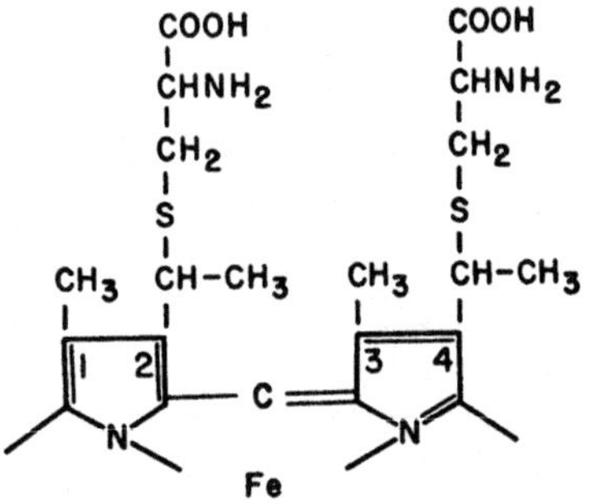

FIG. 10. Cysteine linkages in porphyrin c. The rest of the structure is identical with that of protoporphyrin IX.

Although such linkages may arise secondarily by the addition of cysteine to protoporphyrin under the conditions of hydrolysis,[144] they also are present where such recombination is excluded[145] and are now considered to exist in the intact cytochrome c molecule. The binding of these groups to protein may account for the difficulty with which the prosthetic group is removed from the protein. Treatment of cytochrome c with hydrogen bromide and glacial acetic acid results in the hydrolysis of these thioether linkages with the resultant formation of hematoporphyrin.[146]

Oxidized cytochrome c reacts slowly with cyanide to form a stable complex[147, 148] in which the absorption band at 695 mμ disappears and the band

[141] D. Keilin and E. F. Hartree, *Biochem. J.* **39,** 289 (1945).
[142] H. Theorell and A. Akesson, *J. Am. Chem. Soc.* **63,** 1804 (1941).
[143] H. Theorell, *Enzymologia* **6,** 88 (1939).
[144] H. Theorell, *Biochem. Z.* **301,** 201 (1939).
[145] K. Zeile and H. Meyer, *Z. physiol. Chem.* **262,** 178 (1939).
[146] R. Hill and D. Keilin, *Proc. Roy. Soc.* (*London*) **B107,** 286 (1930).
[147] B. L. Horecker and A. Kornberg, *J. Biol. Chem.*, **165,** 11 (1946).
[148] P. George and C. L. Tsou, *Biochem. J.* **50,** 440 (1952).

at 530 mμ is shifted to 535 mμ.[149] A similar compound is formed with azide at high concentrations of this substance.[150] Ferricytochrome c also forms an unstable complex with hydrocyanic acid.[148]

The catalytic activity of cytochrome c is due to the alternate oxidation and reduction of the iron in the prosthetic group. It has been shown that in intact yeast cells at least 95% of the oxygen uptake occurs by way of cytochrome c.[151] The oxidized form is reduced by enzymes such as the TPN- and DPN-cytochrome c reductases, succinic dehydrogenase, and others including α-glycerophosphate dehydrogenase.[152] Except for the TPN-cytochrome c reductases these enzymes have not yet been isolated in pure form, and they may contain more than one component necessary for the over-all activity. In particular the role of cytochrome b in cytochrome c reduction remains obscure. Ball[153] has shown this substance to have a lower oxidation-reduction potential than cytochrome c, and it appears to participate in electron transport between certain substrates and cytochrome c. In the cell cytochrome c is associated with particulate fractions (i.e., the mitochondria and microsomes of liver cells[154]) from which it is readily removed by treatment with water. There is evidence that the bound form is different from isolated cytochrome c. Thus endogenous cytochrome c does not form a cyanide complex[155] and has a catalytic activity which is 100 to 1000 times greater than that of added cytochrome c.[155-158] At neutral pH ferrocytochrome c does not react with oxygen but becomes autoxidizable below pH 4.[139] However, nearly all aerobic cells contain an enzyme which catalyzes the very rapid oxidation of ferrocytochrome c by molecular oxygen.

b. Cytochrome Oxidase, Cytochrome a and Atmungsferment. The oxidation of cytochrome c was first attributed to an enzyme known as indophenol oxidase, which catalyzed the oxidation of the "nadi" reagent (α-naphthol and dimethyl *p*-phenylenediamine) to form indophenol. This activity was later shown to be due to the presence of cytochrome c in the oxidase preparations.[159, 160] The enzyme is highly specific for cytochrome c and was there-

[149] V. R. Potter, *J. Biol. Chem.* **137,** 13 (1941).
[150] B. L. Horecker and J. N. Stannard, *J. Biol. Chem.* **172,** 589 (1948).
[151] E. Haas, *Naturwissenschaften* **22,** 207 (1934).
[152] D. E. Green, *Biochem. J.* **30,** 629 (1936).
[153] E. Ball, *Biochem. Z.* **295,** 262 (1938).
[154] W. C. Schneider, A. Claude, and G. H. Hogeboom, *J. Biol. Chem.* **172,** 451 (1948).
[155] C. L. Tsou, *Biochem. J.* **50,** 493 (1952).
[156] D. Keilin and E. F. Hartree, *Proc. Roy. Soc. (London)* **B129,** 227 (1940).
[157] E. C. Slater, *Biochem. J.* **46,** 499 (1950).
[158] E. C. Slater, *Biochem. J.* **44,** 305 (1949).
[159] D. Keilin and E. F. Hartree, *Proc. Roy. Soc. (London)* **B125,** 171 (1938).
[160] E. Stotz, A. E. Sidwell, and T. R. Hogness, *J. Biol. Chem.* **124,** 733 (1938).

fore renamed cytochrome oxidase. Since the activity is associated with insoluble particles, its isolation has not yet been accomplished, but from its behavior with metal-binding agents it has been characterized as a hemoprotein. Preparations of cytochrome oxidase always show the band of cytochrome a which appears to be related to the enzymatic activity. The reduced band of cytochrome a in heart muscle preparations appears along with the reduced bands of cytochromes b and c when these preparations are treated with succinate. The a band has been shown[135] to be due to two components (Fig. 11). One of these has reduced bands at 605 and 452 mμ. The second component has bands at 600 and 448 mμ and forms a carbon monoxide complex in which the bands are shifted to 590 and 432 mμ. This component, called cytochrome a_3, to distinguish it from cytochromes a_1 and a_2, reported to occur in bacteria, has properties which are consistent with those of cytochrome oxidase. It also appears to be identical with Atmungsferment described by Warburg and his co-workers.[161] Cytochrome oxidase, Atmungsferment, and cytochrome a_3 are now generally considered to be the same enzyme.

Atmungsferment was the name given by Warburg to the enzyme which catalyzes the primary reduction of oxygen and thus forms the end of the electron transport chain. The absorption spectrum of the carbon monoxide compound of this substance was deduced from a study of the effect of carbon monoxide and light on the respiration of intact yeast cells. It had been shown[162] that the carbon monoxide compound of hemoglobin was dissociated by light. The observation that the respiration of yeast cells was inhibited by carbon monoxide and that this inhibition was greater in the dark than in the light[163] was in support of Warburg's suggestions concerning the role of iron catalysts in the activation of oxygen. The degree of inhibition by carbon monoxide was dependent on the oxygen tension as well as the carbon monoxide tension, and this was interpreted in terms of a competition between oxygen and carbon monoxide for the active site on the enzyme. Since this enzyme combines with oxygen, it was postulated that it must be responsible for the primary activation of oxygen. It was further observed that the effect of light on the inhibition of respiration by carbon monoxide was dependent upon the wavelength of light used. With the assumption that only light which was absorbed was effective, the absorption spectrum was determined from the extent of reversal of inhibition of respiration by carbon monoxide, using a series of highly monochromatic light sources and measuring respiration in the presence of carbon monoxide both in the light and in the dark (Fig. 12). The spectrum obtained in this

[161] O. Warburg, Heavy Metal Prosthetic Groups, Clarendon Press, Oxford, 1949.
[162] J. B. S. Haldane and J. L. Smith, *J. Physiol.* **20**, 497 (1896).
[163] O. Warburg, *Biochem. Z.* **177**, 471 (1926).

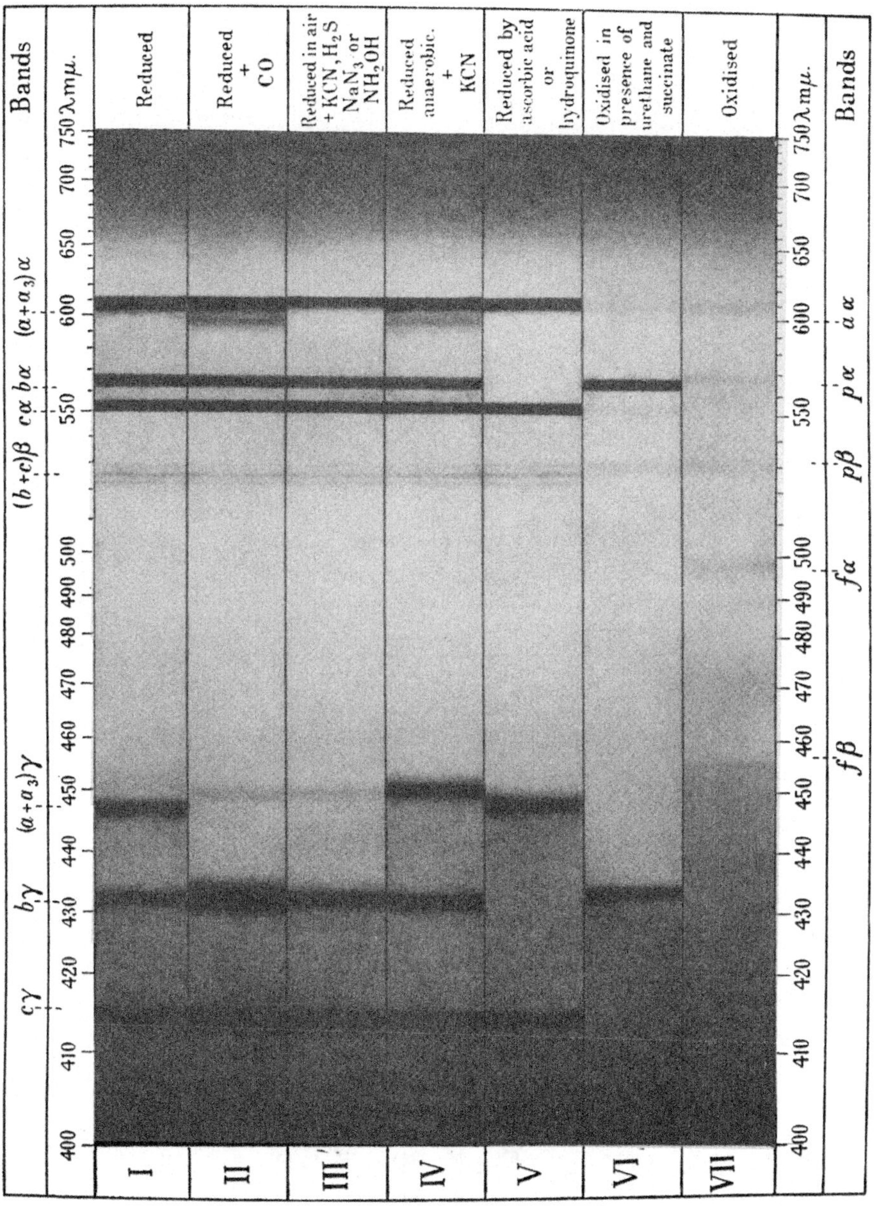

FIG. 11. The absorption bands of the cytochromes. (From Keilin and Hartree.)

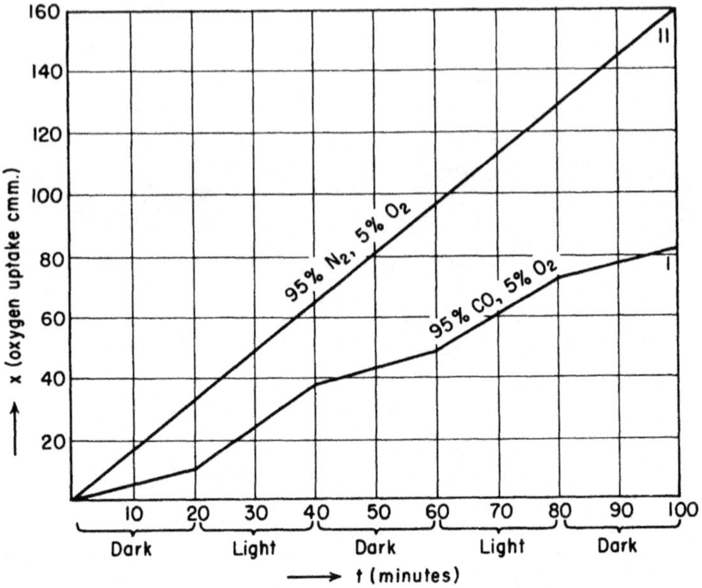

FIG. 12. The action of light on the carbon monoxide inhibition of respiration. (From Warburg.)

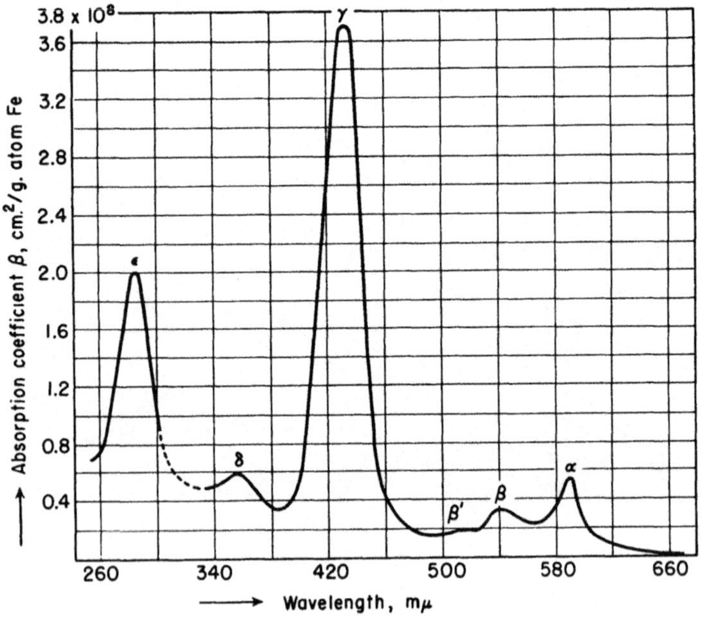

FIG. 13. Carbon monoxide compound of the oxygen-transporting respiratory enzyme. (From Warburg.)

manner showed a very intense Soret band at 430 mμ and a strong band at 590 mμ, with weak bands at 520 and 540 mμ.[163, 164] The bands of the cytochrome a_3-carbon monoxide complex coincide with the strong bands of Atmungsferment. The general character of the absorption spectrum confirms the hemoprotein nature of this enzyme (Fig. 13).

Although the enzyme itself has not been isolated, heme compounds with the spectroscopic properties of Atmungsferment have been isolated from various sources, including the snail *Spirographis spallanzanii*,[165] heart muscle,[166, 167] and *Corynebacterium diphtheriae*.[167] *Spirographis* hemin differs from protohemin in the presence of a formyl group instead of the vinyl group in position 2.[168] Heart muscle and *Corynebacterium* hemin a also possess an aldehyde side chain which has not been further identified.

The conventional method for the study of cytochrome oxidase is based on the rate of oxygen uptake with a suitable substrate present to reduce cytochrome c. Hydroquinine, *p*-phenylenediamine, and ascorbic acid have been used for this purpose. These methods have been critically evaluated by Slater.[158] Cytochrome oxidase activity can be followed spectrophotometrically by measurement of the rate of disappearance of the reduced band of cytochrome c at 550 mμ.[169, 170]

c. Cytochrome b, Succinic Dehydrogenase, Yeast Lactic Dehydrogenase. In heart muscle preparations the reduced bands of cytochrome b appear, together with those of cytochromes a and c, on addition of succinate.[153] This would indicate that all the cytochrome components are involved in succinate oxidation, and the entire complex is therefore known as succinoxidase. For the reduction of methylene blue by succinate the cytochromes, with the possible exception of cytochrome b, are not required. This activity is referred to as succinic dehydrogenase. The coupling between succinic dehydrogenase and cytochrome oxidase has been shown to require a colloidal framework such as is present in the particulate suspension or which can be provided by aluminum hydroxide gel[171] or by denatured globin.[172] Evidence for an enzymatic component between cytochrome b and cytochrome c has been obtained by Slater.[173] With heart muscle preparations treated with BAL, succinate addition will result in the reduction of methyl-

[164] F. Kubowitz and E. Haas, *Biochem. Z.* **255,** 247 (1932).
[165] H. M. Fox, *Proc. Roy. Soc.* (*London*) **B99,** 199 (1926).
[166] E. Negelein, *Biochem. Z.* **266,** 412 (1933).
[167] W. A. Rawlinson and J. H. Hale, *Biochem. J.* **45,** 247 (1949).
[168] H. Fischer and C. v. Seemann, *Z. physiol. Chem.* **242,** 133 (1936).
[169] H. G. Albaum, J. Tepperman, and O. Bodansky, *J. Biol. Chem.* **163,** 641 (1946).
[170] J. N. Stannard and B. L. Horecker, *J. Biol. Chem.* **172,** 599 (1948).
[171] B. L. Horecker, E. Stotz, and T. R. Hogness, *J. Biol. Chem.* **128,** 251 (1939).
[172] D. Keilin and E. F. Hartree, *Biochem. J.* **44,** 205 (1949).
[173] E. C. Slater, *Biochem. J.* **45,** 14 (1949).

ene blue but the activity with oxygen is absent. Cytochrome oxidase activity in such preparations is still intact. Spectroscopic examination shows that only cytochrome b, but not cytochrome c or cytochrome a, is reduced in BAL-treated preparations. These results suggest that a factor has been destroyed which catalyzes the reaction between cytochrome b and cytochrome c. The same factor is required for the reduction of cytochrome c by $DPNH_2$ (see p. 305) except that in this case cytochrome b is not reduced. On this basis cytochrome b is assigned a special role in the oxidation of succinate. Cytochrome b and succinic dehydrogenase may be identical, although some evidence for their separation is available.[174]

Lactic dehydrogenase in yeast differs from the muscle enzyme in the lack of a requirement for, or reaction with, DPN. The yeast enzyme has been purified by Bach, Dixon, and Zerfas[175] and found to be associated with a hemoprotein with reduced bands at 530 and 556 mμ which has been designated as cytochrome b_2. The reduced bands appear instantly on addition of lactate, but, as in the case of heart muscle succinic dehydrogenase, the identity of the hemoprotein with lactic dehydrogenase has not been fully established. Crude preparations will reduce cytochrome c, but this activity is diminished on purification. An additional factor is thought to be required for the reaction with cytochrome c. The reduced cytochrome b_2 is slowly reoxidized by air.

In *Corynebacterium diphtheriae* a hemoprotein which resembles cytochrome b is formed[176] in large quantities when the organisms are grown in the presence of high concentrations of iron. The formation of hemoprotein parallels a decrease in the elaboration of toxin, and this has led to the suggestion that the toxin is the protein part of cytochrome b, or a precursor of it. The hemoprotein, obtained by sonic disruption of the bacterial cells, resembles heart muscle cytochrome b in its reduction by succinate. The reduced form reacts slowly with oxygen but is not oxidized by cytochrome c.

d. The Hemoprotein Enzymes in Electron Transport. The path of electrons in substrate oxidation may tentatively be summarized by the scheme on the following page.

In this scheme no function has been assigned to cytochrome a, which may be a component of cytochrome oxidase. The role of diaphorase and of cytochrome b still requires clarification. Further evidence for the existence of the enzyme referred to as Slater's factor is required.

2. Enzymes Utilizing Hydrogen Peroxide

a. Catalase. Nearly all plant and animal tissues and most microorganisms possess the ability to decompose hydrogen peroxide, owing to the

[174] C. Widmer, Jr. and E. Stotz, *Federation Proc.* **11,** 310 (1952).
[175] S. J. Bach, M. Dixon, and L. G. Zerfas, *Biochem. J.* **40,** 229 (1940).
[176] A. M. Pappenheimer, Jr., *J. Biol. Chem.* **167,** 251 (1947).

presence of enzymes known as catalases. Zeile and Hellström[177] established that hematin is a component of catalase. The enzyme was first crystallized from beef liver by Sumner and Dounce[178] and has since been crystallized from blood, kidney, and bacteria. Catalase activity is generally determined by the method of Euler and Josephson[179] in which residual peroxide is titrated with permanganate. For kinetic studies this method is of limited applicability, since enzymatic activity decreases during the determination and the true initial activity cannot be precisely determined. This difficulty is met in the rapid titration method of Bonnichsen, Chance, and Theorell[180] which permits the use of high catalase concentrations, since the assay is completed within the first minute. Results obtained with this method at

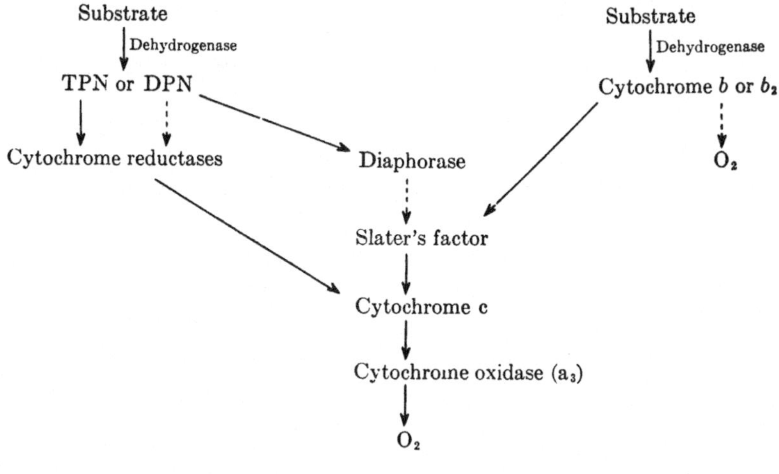

higher concentrations of catalase and hydrogen peroxide show much higher activity than with the classical methods. A rapid manometric method has also been introduced by Appleman.[181]

Catalase contains protoporphyrin IX and in addition variable quantities of biliverdin,[182] depending on the method of preparation.[183] Catalase containing biliverdin is still catalytically active. The molecular weight is about 250,000,[184, 185] and the iron content is about 0.09%. Each molecule therefore

[177] K. Zeile and H. Hellström, *Z. physiol. Chem.* **192,** 171 (1930).
[178] J. B. Sumner and A. L. Dounce, *J. Biol. Chem.* **121,** 417 (1937).
[179] H. v. Euler and K. Josephson, *Ann.* **455,** 9 (1927).
[180] R. Bonnichsen, B. Chance, and H. Theorell, *Acta Chem. Scand.* **1,** 685 (1947).
[181] D. Appleman, *Analytical Chem.* **23,** 1627 (1951).
[182] R. Lemberg and J. W. Legge, *Biochem. J.* **37,** 117 (1943).
[183] R. Bonnichsen, *Acta Chem. Scand.* **2,** 561 (1948).
[184] J. B. Sumner and N. Gralen, *J. Biol. Chem.* **125,** 33 (1938).
[185] K. Agner, *Biochem. J.* **32,** 1702 (1938).

contains 4 atoms of iron. This is present in the ferric form and is reduced with great difficulty, in contrast with other hemoproteins. The absorption spectrum shows bands at 400, 536, and 623 mμ.[185] These bands are displaced in the presence of cyanide, azide or fluoride, indicating compound formation with these substances.[186]

A new role for catalase, other than the decomposition of hydrogen peroxide, was discovered by Keilin and Hartree,[187] who observed that in the presence of high concentrations of catalase and a suitable substrate, such as ethanol, hydrogen peroxide formed by enzymatic action was not decomposed but instead was used for the coupled oxidation of the substrate:

$$\text{Xanthine} + O_2 \xrightarrow{\text{xanthine oxidase}} \text{Uric acid} + H_2O_2$$

$$\text{Ethanol} + H_2O_2 \xrightarrow{\text{catalase}} \text{Acetaldehyde} + 2H_2O$$

Since added hydrogen peroxide was not as efficiently utilized as hydrogen peroxide generated *in situ*, this reaction was attributed to a nascent form of hydrogen peroxide which was formed enzymatically. The mechanism of this reaction was clarified by Chance,[188] who showed the "peroxidative" activity of catalase to be not essentially different from the "catalatic" (hydrogen peroxide decomposing) activity. By a study of the spectral absorption changes in catalase-hydrogen peroxide mixtures with a modification of the rapid-flow apparatus of Roughton and Millikan,[189] Chance obtained evidence for primary and secondary complexes of catalase with hydrogen peroxide. In terms of the formation of these complexes the peroxidative and catalatic activities of catalase may be represented as follows:

$$E + H_2O_2 \rightarrow E \cdot H_2O_2 \quad (1)$$

$$E \cdot H_2O_2 + H_2O_2 \rightarrow E + 2H_2O + O_2 \quad (2)$$

$$E \cdot H_2O_2 + A \rightarrow E + H_2O + AO \quad (3)$$

where A is the substrate which is undergoing coupled oxidation, such as ethanol. Chance showed reaction 2 to be very rapid and therefore to predominate when the hydrogen peroxide concentration is high relative to that of catalase. Reaction 3 is found to occur when the concentration of acceptor is high relative to the concentration of peroxide. When all the peroxide is bound as in equation 1, then none is free to react as in equation 2, and it is utilized quantitatively for reaction 3. This is the situation in the experiments of Keilin and Hartree in which endogenously formed hydrogen peroxide in the presence of high catalase concentrations is quantitatively used

[186] D. Keilin and E. F. Hartree, *Proc. Roy. Soc.* (*London*) **B121**, 173 (1936).
[187] D. Keilin and E. F. Hartree, *Proc. Roy. Soc.* (*London*) **B119**, 141 (1936).
[188] B. Chance, *Biochem. J.* **46**, 387 (1950).
[189] F. J. W. Roughton and G. A. Millikan, *Proc. Roy. Soc.* (*London*) **A155**, 258 (1936).

for the coupled oxidation. Added peroxide, on the other hand, is decomposed by the very rapid reaction 2 before much of it can be used in reaction 3. The concentration of catalase in tissues is much higher than is necessary for the simple decomposition of hydrogen peroxide, indicating that the major function of this enzyme is peroxidation.

b. Peroxidases. Like the catalases, the peroxidases contain hematin as the prosthetic group and activate hydrogen peroxide for oxidation reactions. They differ from the catalases in the absence of catalatic activity; they do not catalyze reaction 2, which represents the oxidation of one molecule of hydrogen peroxide by another. Chance[190] has shown that peroxidase activity may be represented by equations 1 and 3. In general, peroxidases are active with a wide variety of substrates including phenolic compounds, amines, and leuco dyes, whereas the peroxidative activity of the catalases is limited to such hydroxyl compounds as alcohols[191] or nitrite.[192] The classical method for the determination of peroxidase activity is based on the formation of purpurogallin in the oxidation of pyrogallol.[193]

Hemoproteins all possess some peroxidative activity, and in most animal tissues hemoglobin, myoglobin, and the cytochromes account for the low peroxidative activity present. Most plant tissues, however, contain hemoproteins with much greater activity. These are designated as peroxidases. Of these the best known is horseradish peroxidase which has been isolated in crystalline form by Theorell.[194] The absorption spectrum resembles that of acid methemoglobin, with maxima at 402, 500, and 640 mμ.[195] The iron is present in the trivalent form. Treatment with cold acid acetone dissociates horseradish peroxidase into the protein and hemin components. When these are recombined the spectrum and peroxidase activity are restored.[196,197] The prosthetic group of horseradish peroxidase is protohematin IX, but some activity is obtained with mesohematin (with the vinyl groups reduced to ethyl groups) as well as with deuterohematin (with the vinyl groups removed and replaced by hydrogen).

In addition to peroxidase activity, horseradish peroxidase catalyzes the uptake of oxygen with dihydroxymaleic acid as the substrate.[198] Traces of hydrogen peroxide are essential for this activity, and the reaction is inhib-

[190] B. Chance, *J. Biol. Chem.* **151,** 553 (1943).
[191] D. Keilin and E. F. Hartree, *Biochem. J.* **39,** 293 (1945).
[192] L. A. Heppel and V. T. Porterfield, *J. Biol. Chem.* **178,** 549 (1949).
[193] R. W. Willstätter and A. Pollinger, *Ann.* **430,** 269 (1923).
[194] H. Theorell, *Arkiv Kemi Mineral. Geol.* **16A,** No. 2 (1942).
[195] H. Theorell, *Enzymologia* **10,** 3 (1942).
[196] H. Theorell, *Arkiv Kemi Mineral. Geol.* **14B,** No. 20 (1940).
[197] H. Theorell and A. Maehly, *Acta Chem. Scand.* **4,** 422 (1950).
[198] H. Theorell and B. Swedin, *Naturwissenschaften* **27,** 95 (1939); *Nature* **145,** 71 (1940).

ited by carbon monoxide. This can be interpreted in terms of a two-step reaction in which one step involves the ferrous form of the peroxidase.

Other plant peroxidases resemble horseradish peroxidase in absorption spectra and the prosthetic group. Peroxidases which differ in color and presumably in the nature of the prosthetic group have been found in animal tissues. They are green in color and often referred to as verdoperoxidases. One has been purified from leucocytes by Agner,[199] and a second, called lactoperoxidase, has been isolated from milk by Theorell and Paul.[200]

Yeast contains a peroxidase which is highly specific for reduced cytochrome c.[201] It is a hemoprotein with protohematin IX as the prosthetic group.

In animal tissues[202] and bacteria[203] the oxidation of tryptophan to formylkynurenine is catalyzed by the adaptive enzyme, tryptophan peroxidase. The oxidation requires traces of hydrogen peroxide and appears to proceed in two stages, the first of which is an oxidation of tryptophan by hydrogen peroxide and the second the oxidation of the product by oxygen in which hydrogen peroxide is regenerated. The net result is the utilization of one mole of oxygen. The intermediate product has not been isolated. From inhibition studies both the ferrous and ferric forms of the enzyme (or enzymes) appear to participate in the catalytic action.

$$\text{Tryptophan} + H_2O_2 \xrightarrow{\text{peroxidase}} \text{Intermediate}$$

$$\text{Intermediate} + O_2 \xrightarrow{\text{oxidase}} \text{Formylkynurenine} + H_2O_2$$

V. The Copper Enzymes

A widely distributed group of enzymes known as the tyrosinases or polyphenol oxidases catalyzes the oxidation of phenolic substances by oxygen. Where these have been isolated, they have been shown to be copper proteins. These enzymes are particularly abundant in plant tissues where they may function as terminal oxidases in place of the cytochrome system; the relative importance of the phenol oxidases in plant cell respiration, however, has not yet been determined.[204, 205] The oxidation of ascorbic acid in plant tissues is also due to the presence of a copper enzyme.

[199] K. Agner, *Acta Physiol. Scand.* **2,** *Suppl.* 8 (1941).
[200] H. Theorell and K. G. Paul, *Arkiv Kemi Mineral. Geol.* **18A,** No. 12 (1944).
[201] A. M. Altschul, R. Abrams, and T. R. Hogness, *J. Biol. Chem.* **136,** 777 (1940).
[202] W. E. Knox and A. H. Mehler, *J. Biol. Chem.* **178,** 419 (1950).
[203] O. Hayaishi and R. Y. Stanier, *J. Bact.* **62,** 691 (1951).
[204] J. M. Nelson, *in* McElroy and Glass, Copper Metabolism, The Johns Hopkins Press, Baltimore, 1950, p. 76.
[205] D. I. Arnon, *in* McElroy and Glass, Copper Metabolism, The Johns Hopkins Press, Baltimore, 1950, p. 89.

1. Phenol Oxidases

a. Potato Polyphenol Oxidase. The first of the phenol oxidases to be isloated was the polyphenol oxidase of the white potato which Kubowitz[206] demonstrated to be a copper protein. During the purification procedure the enzymatic activity was found to be proportional to the copper content and the final product contained 0.12% copper. The copper could be removed by dialyzing the enzyme against a cyanide solution, leaving the inactive protein which could be reactivated by the addition of copper salts. Other metals were inactive. Since the enzymatic activity was inhibited by carbon monoxide as well as other metal binding agents such as cyanide and diethyldithiocarbamate, Kubowitz concluded that both the cuprous and cupric forms of the metal participated in the enzyme action and postulated that copper was alternately reduced and oxidized. Kubowitz employed a manometric assay method in which catechol was oxidized to benzoquinone. However, since this product inactivated the enzyme, the substrate catechol was added in catalytic amount together with hexosemonophosphate, Zwischenferment, and TPN. In this system glucose-6-phosphate served as the ultimate substrate, since the benzoquinine formed was rapidly reduced to catechol by $TPNH_2$:

$$\text{Glucose-6-phosphate} + \text{TPN} \rightarrow \text{6-Phosphogluconate} + \text{TPNH}_2$$

$$\text{Benzoquinone} + \text{TPNH}_2 \rightarrow \text{Catechol} + \text{TPN}$$

$$\text{Catechol} + \tfrac{1}{2}\text{O}_2 \rightarrow \text{Benzoquinone} + \text{H}_2\text{O}$$

$$\text{Sum: Glucose-6-phosphate} + \tfrac{1}{2}\text{O}_2 \rightarrow \text{6-Phosphogluconate} + \text{H}_2\text{O}$$

These reactions serve as a model for the oxidation of metabolites by way of the pyridine nucleotides and the phenol oxidases. Dihydroxyphenylalanine (dopa) formed in the oxidation of tyrosine[207] might serve as the natural phenolic carrier for this type of respiration.[208]

b. Mushroom Polyphenol Oxidase. Another of the phenol oxidases which has been the subject of considerable study is mushroom polyphenol oxidase, isolated from the edible mushroom by Keilin and Mann.[209] The product contained 0.3% copper. Crude preparations showed little relation between copper content and enzymatic activity, because of the presence of much inactive copper. In the more purified preparations, however, the activity clearly depended on the copper content. From a wild mushroom, Dalton and Nelson[210] obtained a preparation containing 0.25% of copper. It is

[206] F. Kubowitz, *Biochem. Z.* **292,** 221 (1937); **229,** 32 (1939).
[207] H. S. Raper, *Biochem. J.* **20,** 735 (1926).
[208] E. S. Robinson and J. M. Nelson, *Arch. Biochem.* **4,** 111 (1944).
[209] D. Keilin and T. Mann, *Proc. Roy. Soc.* (*London*) **B125,** 187 (1938).
[210] H. R. Dalton and J. M. Nelson, *J. Am. Chem. Soc.* **61,** 2946 (1939).

not known whether the low copper content of Kubowitz's preparation is due to the presence of copper-free impurities or whether the mushroom preparations contain inert copper proteins. None of the preparations thus far obtained has been shown to consist of a homogeneous protein.

The absence of evidence for homogeneity has prevented a clarification of the dual specificity of the phenol oxidases. Most preparations oxidize both monophenols and *o*-diphenols. With the monophenols, however, the maximum oxidation rate with purified enzymes is always preceded by an induction period which can be eliminated by the addition of traces of diphenol.[211] This autocatalytic behavior with monophenols and the priming effect of diphenols may be due to a catalytic role of diphenols formed in the oxidation of the monophenolic substrate.[212] Thus the oxidation of monophenols appears to follow the pattern:

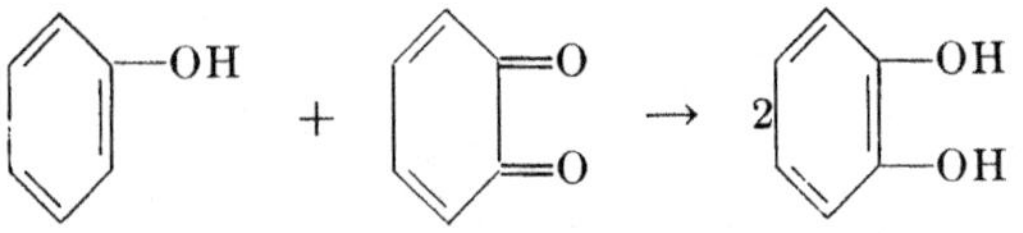

(1)

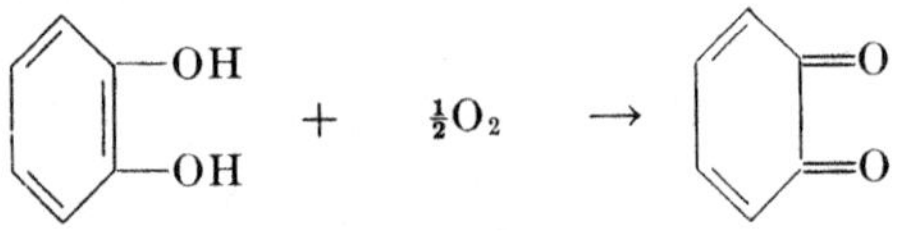

(2)

Reaction 1, as well as reaction 2, is enzymatically catalyzed, since no reaction between *o*-quinone and monophenol occurs in the absence of enzyme.[213] The *o*-quinones are not end products of enzymatic action but undergo further oxidations and condensations with the ultimate formation of insoluble polymers like melanin. Among the phenolic substances oxidized by these enzymes are some which occur naturally, such as tyrosine (I),[207] dopa (II),[208] and urushiol (III),[214] a phenolic substance which occurs in plants of the genus Rhus.

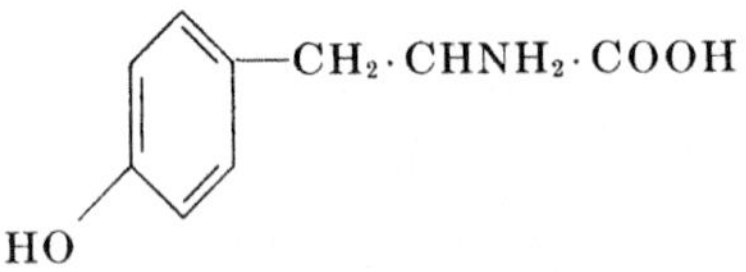

I

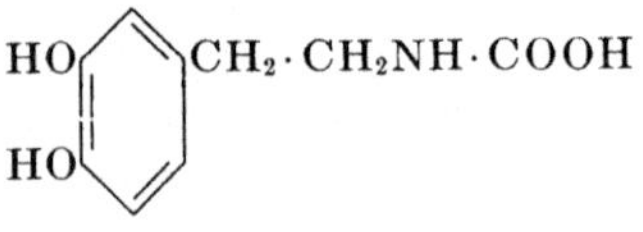

II

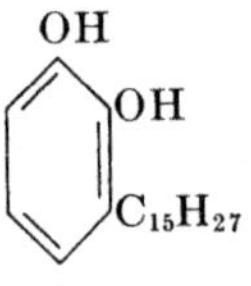

III

c. Laccase. This enzyme, isolated from the latex of the lacquer tree,[215]

[211] J. M. Nelson and C. R. Dawson, *Advances in Enzymol.* **4,** 99 (1949).
[212] A. B. Lerner, T. B. Fitzpatrick, E. Calkins, and W. H. Summerson, *J. Biol. Chem.* **178,** 185 (1949).
[213] C. A. Bordner and J. M. Nelson, *J. Am. Chem. Soc.* **61,** 1507 (1939).
[214] R. Majima, *Ber.* **55,** 172 (1922).
[215] D. Keilin and T. Mann, *Nature* **143,** 23 (1939); **145,** 304 (1940).

is a blue protein containing 0.15% of copper. On addition of a substrate solutions of laccase are decolorized but turn blue again on shaking with air. The color is thus shown to be associated with enzymatic activity. The colored component could, however, be obtained free of copper. In the case of laccase the copper present does not seem to undergo reversible oxidation and reduction, since the activity is not inhibited by carbon monoxide.

d. Mammalian Tyrosinase. In mammalian tissues tyrosinase activity is associated with insoluble particles in skin[216] and melonomas.[217] This enzyme is often referred to as dopa oxidase, since dopa is thought to be the natural substrate. However, tyrosine is also oxidized.

2. L-Ascorbic Acid Oxidase

Extract of plant tissues catalyze the oxidation of L-ascorbic acid by oxygen:

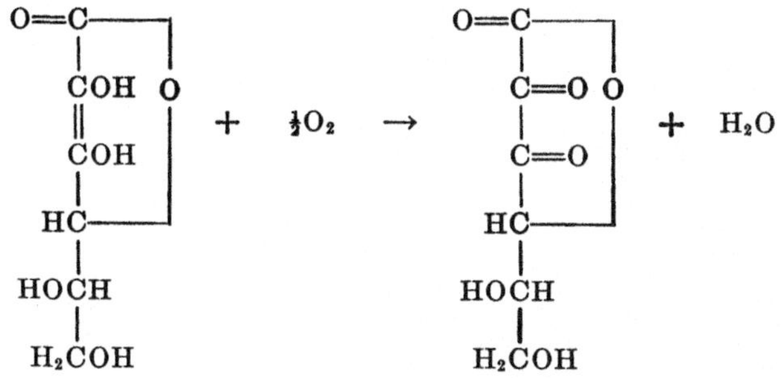

From extracts of the crook-neck squash[218] and cucumber[219] purified preparations containing 0.15 to 0.25% of copper have been obtained. Since ascorbic acid oxidase does not attack mono- or diphenols, it is distinct from the phenol oxidases. A number of synthetic copper protein complexes have been shown to have ascorbic oxidase activity.[220] This activity, however, is much less than that of the isolated enzyme. These complexes form hydrogen peroxide which is not produced in the oxidation catalyzed by ascorbic acid oxidase.

Ascorbic acid oxidase is inhibited by agents which bind copper, including cyanide, diethyldithiocarbamate, and 8-hydroxyquinoline. Purified solutions are blue-green in color. The biological function of this enzyme, like that of the phenol oxidases, is not clear.

[216] B. Block, *Am. J. Med. Sci.* **177,** 609 (1929).
[217] G. H. Hogeboom and M. H. Adams, *J. Biol. Chem.* **145,** 273 (1942).
[218] H. Tauber, *Ergeb. Enzymforsch.* **7,** 301 (1938).
[219] E. Stotz, *J. Biol. Chem.* **133,** c (1940).
[220] J. F. McCarthy, L. F. Green, and C. G. King, *J. Biol. Chem.* **128,** 455 (1939).

CHAPTER 19

Coenzymes

G. DAVID NOVELLI AND MORRIS SOODAK

Biochemical Research Laboratory and Robert W. Lovett Memorial Laboratory Massachusetts General Hospital, Boston, Massachusetts

I. Introduction

1. The Concept and Definition of a Coenzyme

Our present concept of coenzymes derives from the observation of Harden and Young[1] that Buchner's yeast juice could be separated by dialysis into a crystalloidal and a colloidal fraction. Neither fraction was active alone, but activity was restored by recombination of the two fractions. This crystalloidal fraction subsequently came to be known as the coenzyme for fermentation.

Ideas concerning the nature and definition of coenzymes have undergone considerable change since the first experimental demonstration of a coenzyme. At first anything non-protein in nature which was heat-stable, dialyzable, and capable of "activating" or "reactivating" an inactive enzyme preparation was considered a coenzyme. As this dialyzable portion of yeast juice was fractionated and further investigated, it became apparent that materials capable of activating enzymes could be classified into two

[1] A. Harden and W. J. Young, *J. Physiol.* (*London*) **32,** *Proc.* Nov. (1905).

large groups:

1. Inorganic cations and anions.
2. Low molecular weight, heat-stable organic molecules.

With the isolation and crystallization of enzymes, the process of enzymatic catalysis was studied in great detail. From these studies it was soon evident that many active enzymes exist as a complex of a protein associated with a non-protein prosthetic group. Such an enzyme complex is known as an "holoenzyme." When the prosthetic group is removed from the protein, the inactive enzyme is known as an "apoenzyme," viz:

$$\text{Holoenzyme} = \text{Apoenzyme} + \text{prosthetic group}$$

The prosthetic group is loosely referred to as the coenzyme. But, as was mentioned earlier, certain metals and inorganic ions may also function as a prosthetic group. It is customary to reserve the designation "coenzyme" for the prosthetic groups of an organic nature. Therefore, although recognizing the apparent impossibility of fitting nature to the rigid mold of a man-made definition without modifying statements and subsequent exceptions, for the purposes of the present discussion a coenzyme may be defined as *a small organic molecule which is relatively heat-stable and whose presence is absolutely essential for enzyme action.*

2. General Properties of the Coenzymes

a. Dissociation from the apoenzyme. The coenzymes exhibit all degrees of dissociation ranging from the relatively freely dissociable diphosphopyridine nucleotide (DPN) and the difficulty dissociable flavin adenine dinucleotide (FAD), to the apparently undissociable metalloporphyrins. The degree of dissociation of a single coenzyme will vary with the protein to which it is bound, as well as the state of the coenzyme. For example, DPN has a dissociation constant from alcohol dehydrogenase[2] of 9×10^{-5}, and the reduced coenzyme has a dissociation constant of 3×10^{-5}. This is compared with the constant of the same coenzyme with another enzyme, i.e., triosephosphate dehydrogenase,[3] which is 4×10^{-8}. FAD is split from its enzyme by using drastic procedures, such as acid ammonium sulfate,[4] and cocarboxylase is removed from its enzyme only by treatment in acid or alkaline solutions.[5]

In many cases when a coenzyme has been dissociated from its enzyme, recombination is difficult and frequently requires a high concentration of the

[2] E. Negelein and H. J. Wulff, *Biochem. Z.* **293,** 351 (1937).
[3] C. Cori, S. F. Velick, and G. T. Cori, *Biochim. et Biophys. Acta* **4,** 160 (1950).
[4] F. Lipmann, *Cold Spring Harbor Symposia Quant. Biol.* **7,** 248 (1939).
[5] K. Lohmann and P. Schuster, *Biochem. Z.* **294,** 112 (1934).

coenzyme. After FAD is split from certain flavo-proteins, it requires a 1000-fold excess of FAD to restore the enzyme to its original degree of reactivity.[4]

Very little is known or understood concerning the dissociation of a coenzyme or the forces and groups which participate in binding the coenzyme to the enzyme. In yeast extract, DPN is apparently freely dissociable from its enzymes; thus it shuttles back and forth between triosephosphate dehydrogenase and alcohol dehydrogenase. In mitochondria, on the other hand, the coenzymes are tightly bound to their proteins for they are not removed by repeated washing of these particles.[6]

The coenzymes range in molecular weight from approximately 200 to 1000. They are all of a size small enough to render them rather freely dialyzable through cellophane membranes, provided that they dissociate from their respective proteins. The charge on the molecule as well as its size will influence the rate of its dialyzability.

All the known coenzymes are composed of at least one or more of the following materials:

1. One of the B vitamins.
2. A purine or pyrimidine base.
3. A sugar or sugar alcohol with 5 or 6 carbons.
4. Phosphate.

The function of these various constituents of the coenzymes in many cases is completely unknown. Phosphate is believed to be the group through which the coenzyme attaches itself to the protein, but the fact that riboflavin can, in part, reactivate the Warburg and Christian yellow enzyme makes this theory a little uncertain.[7]

From a nutritional standpoint, it is significant that five of the B-complex vitamins (riboflavin, nicotinamide, thiamine, vitamin B_6, and pantothenic acid) have been shown to be constituents of the coenzymes. The nutritional requirement of these vitamins is explained on the basis of their coenzyme function. In all cases the coenzyme form appears to be the sole bound form of the vitamin, and this then becomes the only metabolically active form for these particular vitamins.

It is a curious coincidence that the first isolation of a crystalline vitamin from natural sources[8] corresponded in time to the first isolation of a crystalline enzyme.[9] The close relationship between the vitamins and the coenzymes was clearly demonstrated in the 1930's when in rapid succession riboflavin was shown to be a constituent of the coenzyme FAD,[10] and cocar-

[6] D. E. Green, *Biol. Bull.* **26,** 410 (1951).

[7] O. Warburg and W. Christian, *Biochem. Z.* **254,** 438 (1932).

[8] B. C. P. Jansen and W. F. Donath, *Chem. Weekblad* **23,** 201 (1926).

[9] J. B. Sumner, *J. Biol. Chem.* **69,** 435 (1926).

[10] O. Warburg and W. Christian, *Biochem. Z.* **254,** 428 (1932).

boxylase was shown to be thiaminepyrophosphate.[5] It is an interesting fact that TPN[11] was isolated and studied as a coenzyme a few years before nicotinamide was shown to be a vitamin in animal nutrition.[12] In the 1940's vitamin B_6[13] and pantothenic acid[14] were added to the list of B vitamins with a coenzyme function. This identification of some of the B vitamins as coenzymes led Green[15] to propose in 1941 the enzyme-trace substance theory, in which the prediction is made that any substance necessary in the diet in trace amounts must be an essential component of an enzyme system. This notion, particularly with respect to the B vitamins, is well implanted in biochemical thinking, so that the announcement of the coenzyme function of biotin, folic acid, B_{12}, etc., is anticipated by most biochemists.

b. Specificity. A property shared by the coenzymes is their specificity. The specificity of coenzymes may be viewed somewhat differently from the type of specificity with which we generally regard the enzymes. In general (and there are notable exceptions), enzyme specificity may be regarded as substrate specificity extending to the molecule as a whole. That is, a given enzyme will be specific for a single organic compound or closely related substances; i.e., succinoxidase will oxidize only succinic acid. Coenzymes, on the other hand, have what shall be called "functional group specificity." That is, the coenzyme is specific for a particular grouping in an organic molecule and will generally be seen to function with a large variety of substrate molecules which have only this functional group in common. The coenzyme, furthermore, is usually involved in the transfer of the group for which it is specific. Therefore, coenzyme reactions may be viewed as "group transfer" reactions. It is the functional group specificity of the coenzymes which permits them to operate with a variety of apoenzymes and their corresponding substrates.

3. Classification of Coenzymes

There are several ways in which one can classify coenzymes. They may be classified on the basis of their vitamin content, on the basis of their structure, or on the basis of the enzymes with which they function. However, a convenient grouping of the coenzymes can be made on the basis of

[11] O. Warburg and W. Christian, *Biochem. Z.* **275,** 112 (1934).

[12] C. A. Elvehjem, R. J. Madden, F. M. Strong, and D. W. Wooley, *J. Am. Chem. Soc.* **59,** 1767 (1937).

[13] I. C. Gunsalus, W. D. Bellamy, and W. W. Umbreit, *J. Biol. Chem.* **155,** 685 (1944).

[14] F. Lipmann, N. O. Kaplan, G. D. Novelli, L. C. Tuttle, and B. M. Guirard, *J. Biol. Chem.* **167,** 869 (1947).

[15] D. E. Green, *Advances in Enzymol.* **1,** 177 (1941).

their functional group specificity. Such a classification has the advantage that it focuses attention on the group transfer reaction of the coenzyme. If only the well-defined coenzymes are considered and these are classified on a functional basis, the following scheme emerges:

a. H atom (or electron) transfer.
   1. Diphosphopyridine nucleotide (DPN).
   2. Triphosphopyridine nucleotide (TPN).
   3. Flavin mononucleotide (FMN).
   4. Flavin adenine dinucleotide (FAD).

b. Phosphate transfer.
   1. The adenylic system (AMP, ADP, ATP).
   2. Glucose diphosphate.
   3. Diphosphoglyceric acid.

c. CO_2-decarboxylations.
   1. Thiamine pyrophosphate (TPP) (cocarboxylase).
   2. Pyridoxal phosphate.

d. Acyl transport.
   1. Coenzyme A. (CoA)

e. NH_2 transport.
   1. Pyridoxal phosphate.

f. Miscellaneous.
   1. Uridine diphosphate glucose (UDPGlucose).
   2. Glutathione (GSH, GSSG).
   3. Pyruvate oxidation factor (POF).
   4. Biotin.

II. Diphosphopyridine and Triphosphopyridine Nucleotides

1. Introduction

DPN and TPN are the major coenzymes involved in hydrogen transport at the substrate level. Their main function appears to be to remove hydrogen from certain substrates (in cooperation with the proper dehydrogenases) and to transfer the hydrogen (or electron) to another coenzyme in the hydrogen transport series, or to another substrate, which is accordingly reduced.

These two coenzymes, along with their apoenzymes, have recently been reviewed in considerable detail in an excellent article by Schlenk.[16] Certain aspects of their metabolism have been presented by Kornberg.[17]

[16] F. Schlenk, *in* Sumner and Myrbäck, The Enzymes, Academic Press, New York, 1950, Vol. II, Part 1, p. 250.

[17] A. Kornberg, *in* McElroy and Glass, Phosphorus Metabolism, The Johns Hopkins Press Baltimore, 1950, Vol. I, p. 392.

2. Historical

DPN and TPN are of considerable interest, not only because of their widespread occurrence as coenzymes for a large number of metabolically important dehydrogenases, but because they were among the first factors to be recognized as coenzymes and were also among the first coenzymes to be isolated in relatively pure state.

Interest in DPN dates back to the observation of Harden and Young[1] on the cofactor for fermentation. This observation led von Euler and Myrbäck[18] to coin the term "cozymase" for this cofactor. It has since been shown that "cozymase" is actually a complex system containing three or more coenzymes necessary for alcoholic fermentation.

Although DPN is the older of the two coenzymes in point of historical recognition, its partner, TPN, was the first of the pair to be isolated, in the classical investigations of Warburg and his co-workers.[19] Shortly thereafter, DPN was isolated by von Euler[20] and his school. It is interesting to note that nicotinamide was recognized as a component of these coenzymes before the acid of this amide was discovered to be a nutritionally important vitamin. This represents the only case in which the coenzyme form of a vitamin was recognized before the vitamin function was discovered, a point which further emphasizes the close correlation between the fields of nutrition and of enzyme chemistry.

3. Structure and Chemistry

The structures of DPN and TPN are given in Fig. 1. These dinucleotides consist of the elements nicotinamide, ribose, phosphate, and adenine. TPN has one more phosphate group than DPN. Neither of these structures has been confirmed by total organic synthesis, but the evidence for these formulations, which is too lengthy to be presented here in detail, is sufficiently strong to permit their acceptance (see Schlenk[16] and Kornberg[17]). Whether the phosphomonoester group of TPN is esterified at the 2′- or 3′-hydroxyl of the adenosine moiety is not yet clearly established and awaits the decision on the location of the phosphate group in adenylic acid "a" and "b." In any case, Kornberg[17] and Kaplan, Colowick, and Stolzenbach[21] have presented evidence that the adenylic acid portion of TPN is identical to adenylic acid "a."

[18] H. von Euler and K. Myrbäck, *Z. physiol. Chem.* **131,** 179 (1923).

[19] O. Warburg, W. Christian, and A. Griese, *Biochem. Z.* **282,** 157 (1935).

[20] F. Schlenk and H. von Euler, *Naturwissenschaften* **24,** 794 (1936).

[21] N. O. Kaplan, S. Colowick, and F. E. Stolzenback, *Federation Proc.* **10,** 204 (1951).

* The position of the phosphate group in adenylic acid "a" and "b" has been established as 2′, & 3′ respectively.[21a]

[21a] Khym, J. X., D. G. Doherty, E. Volkin, W. E. Cohn, *J. Am. Chem. Soc.*, **74:** 3204 (1952).

The structure of TPN was the more difficult to determine, but the fact that the coenzymes could be interconverted enzymatically as shown below:

$$DPN + ATP \rightarrow TPN + ADP \quad (1)$$

$$TPN + \text{acid phosphatase} \rightarrow DPN + \text{phosphate} \quad (2)$$

and the elegant studies of Kornberg[22] in which the cleavage of TPN by nucleotide pyrophosphatase gave the products nicotinamide mononucleotide (NMN) and adenosine diphosphate, viz:

$$TPN \xrightarrow{\text{nucleotide pyrophosphatase}} \text{Nicotinamide mononucleotide} + \text{adenosine diphosphate} \quad (3)$$

is strong evidence for this formulation.

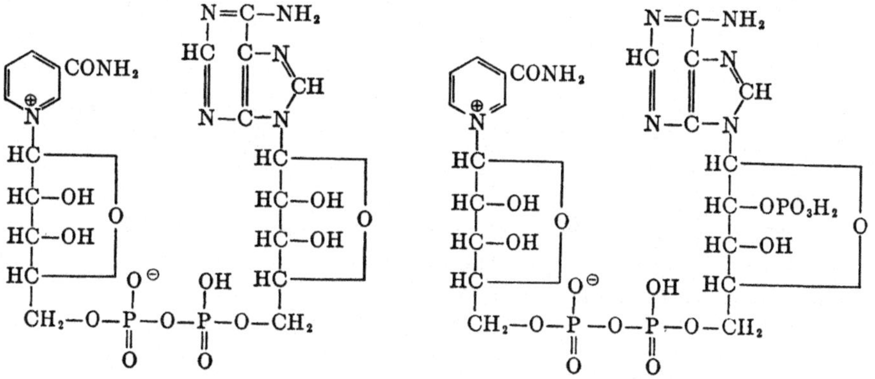

Fig. 1

By using enzymatic methods for the breakdown and resynthesis of these coenzymes, Kornberg[17] has confirmed the structure of DPN as first postulated by Schlenk and von Euler[20] and has elucidated very convincing data for the validity of the structure of TPN.

The structure of these compounds was investigated by noting the products of hydrolysis.

1 Acid hydrolysis of the coenzymes splits the glycosidic bond to the bases, as well as the pyrophosphate bridge. The products are thus nicotinamide, ribose-5-phosphate,[23] and adenine.

2. Weak alkaline hydrolysis at low temperature splits off only the base, nicotinamide, while at higher temperatures adenosine diphosphate is obtained.[24]

[22] A. Kornberg and W. E. Pricer, Jr., *J. Biol. Chem.* **186,** 557 (1950).

[23] F. Schlenk, *J. Biol. Chem.* **146,** 619 (1942).

[24] R. Vestin, F. Schlenk, and H. von Euler, *Ber.* **70,** 1369 (1937).

In the oxidized state the coenzymes are rapidly inactivated at room temperature by 0.1 *N* alkali (presumably due to splitting off nicotinamide) but are considerably more stable to dilute acid. The reduced coenzymes on the other hand are very labile in dilute acid but are stable in dilute alkali.[25] This rapid destruction by acid of the reduced coenzymes is considered to be due to the addition of a molecule of acid to the reduced pyridine ring system. The alkali stability of the reduced molecules is explained by the fact that the nitrogen is trivalent in the reduced coenzymes and thus has a stability similar to the glucosidic-type link in purine nucleosides. The alkali lability of the oxidized molecules is therefore due to the quaternary nitrogen ion in glucosidic linkage.[26]

4. Isolation of the Coenzymes

a. TPN. The original isolation by Warburg[19] was from 250 l. of horse blood. After separation of the erythrocytes, the cells were hemolyzed and the stromata removed by acetone. By a series of steps involving fractionation with mercuric acetate and barium hydroxide, precipitation from acid methanol with ethyl acetate and fractional precipitation of the lead salt with ethanol, about 1 g. of pure TPN was obtained.

b. DPN. The first pure preparation of DPN was made by von Euler[18] and his co-workers by a hot-water extraction of yeast, removal of impurities with lead acetate followed by the precipitation in successive steps as Hg, Ag, and Cu salts, then treatment with Ba and Pb followed by alcohol fractionation. Final purification was effected with aluminum oxide. Both of these original procedures for the isolation of these coenzymes were tedious, were fraught with difficulties, and gave poor yields. Therefore, the methods have undergone extensive revisions. The most recent methods (Kornberg,[27] Hogeboom,[28] Mueller,[29] and Neilands[30]) involve the use of adsorption and elution from charcoal, countercurrent distribution with immiscible solvents, and adsorption and elution from various resins. These newer methods have resulted in a great simplification of the procedure and the yields and purifications have been greatly improved.

5. Distribution

The distribution of the coenzymes follows the distribution of their vitamin component, niacin. They have been found in all cells which have

[25] H. von Euler, H. Albers, and F. Schlenk, *Z. physiol. Chem.* **240,** 113 (1936).
[26] E. Haas, *Biochem. Z.* **288,** 123 (1936).
[27] A. Kornberg and B. Horecker, unpublished method.
[28] G. H. Hogeboom and G. T. Barry, *J. Biol. Chem.* **176,** 935 (1948).
[29] G. A. LePage and G. C. Mueller, *J. Biol. Chem.* **180,** 975 (1949).
[30] J. B. Nielands and A. Akeson, *J. Biol. Chem.* **188,** 307 (1951).

been examined for their presence. In view of the essential nature of the reactions catalyzed by these coenzymes, it would be exceedingly surprising if they should be found to be absent from some type of cell. The quantity of the two coenzymes, as well as the ratio between the two, varies considerably with different cells,[25] but the quantity of DPN is usually much larger than that of TPN. A notable exception is the green leaf, in which the TPN content is much higher than that of DPN. This is now explained on the basis of a TPN-linked triosephosphate dehydrogenase in the green part of the plant, in contrast to the situation in the root and seed.[31] The state of oxidation or reduction of the coenzyme will depend on the oxidation-reduction potential of the cell. In most cells about 35% of the total coenzymes are in the reduced state. In certain malignant cells this quantity is increased, presumably due to the lower oxidation-reduction potentials maintained by such cells.

6. Methods for Assay

A variety of methods (chemical, microbiological, and enzymatic) are available for the determination of these coenzymes.

1. The classical method for the measurement of DPN has been to measure the rate of fermentation, as followed by CO_2 evolution of an apozymase preparation.[32] The apozymase is easily prepared by repeated extraction of dried brewer's yeast and upon the addition of DPN is capable of causing alcohol fermentation.

2. The original method for the measurement of TPN was described by Warburg.[33] Here advantage is taken of the fact that glucose-6-phosphate dehydrogenase (*Zwischenferment*) is specific for TPN. This enzyme catalyzes the following reaction:

$$\text{Glucose-6-phosphate} + \text{TPN} \rightarrow \text{6-Phosphogluconate} + \text{TPNH}_2 \qquad (4)$$

The reaction can be followed in several ways In the original method, Zwischenferment was coupled to the yellow enzyme by which $TPNH_2$ was oxidized by the flavin enzyme, which in turn was oxidized by molecular oxygen, and thus the reaction could be followed manometrically.

After absorption of the reduced coenzymes at 340 mμ was discovered, the reaction was followed spectrophotometrically. Zwischenferment is usually used for the measurement of TPN because of its absolute specificity and because of its ease of preparation. A very sensitive method has been

[31] M. Gibbs, *Nature* 1952, in press.
[32] K. Myrbäck, *Z. physiol. Chem.* **177,** 158 (1928).
[33] O. Warburg, W. Christian, and A. Griese, *Biochem. Z.* **279,** 142 (1935).

described by Haas[34] which couples Zwischenferment with the reduction of cytochrome c. By following the rate of reduction of cytochrome c spectrophotometrically, quantities as small as 0.02 μg. of TPN can be determined.

3. Since the recent description by Racker[35] and Bonnichsen[36] of a simplified method for the preparation of crystalline alcohol dehydrogenase from yeast, this enzyme has been used to measure DPN. The reaction

$$\text{Ethanol} + \text{DPN} \rightleftharpoons \text{Acetaldehyde} + \text{DPNH}_2 \quad (5)$$

leads to the reduction of DPN which can be measured spectrophotometrically at 340 mμ.

A chemical method using hydrosulfite[37] has been used for determining both coenzymes simultaneously, or total pyridine nucleotides. It has the advantage that it is rapid and convenient, but it does not distinguish between DPN, TPN, or NMN. This method is used primarily for the determination of total pyridine nucleotides or in the preparation of the reduced form of the individual coenzymes.

7. Mechanism of Action

The mechanism of action of the coenzymes was discovered by Warburg and his co-workers in 1935.[33] They studied the oxidation of glucose-6-phosphate in yeast extracts. This oxidation is caused by the combination of two enzymes, glucose-6-phosphate dehydrogenase (Zwischenferment) and the "old" yellow enzyme.

$$\text{Glucose-6-phosphate} + \text{TPN} \xrightarrow{\text{Zwischenferment}} \text{6-Phosphogluconate} + \text{TPNH}_2 \quad (6)$$

$$\text{TPNH}_2 + \text{flavoprotein} \rightarrow \text{TPN} + \text{flavoprotein H}_2 \quad (7)$$

$$\text{Flavoprotein H}_2 + \text{O}_2 \rightarrow \text{Flavoprotein} + \text{H}_2\text{O}_2 \quad (8)$$

They followed the course of reaction 6 spectrophotometrically and observed that the reduced coenzyme showed a characteristic absorption maximum at 340 mμ, whereas the oxidized coenzyme has no absorption at this wavelength. Von Euler, Adler, and Hellstrom[38] somewhat later made the same observation with DPN in the oxidation of ethyl alcohol by alcohol dehydrogenase (equation 5). Warburg discovered that these two coenzymes could be chemically reduced by sodium hydrosulfite,[19] and Ohlmeyer[39] has used this method for the preparation of relatively pure reduced DPN.

[34] E. Haas, C. J. Harper, and T. R. Hogness, *J. Biol. Chem.* **142,** 835 (1942).
[35] E. Racker, *J. Biol. Chem.* **184,** 313 (1950).
[36] R. K. Bonnichsen, *Acta Chem. Scand.* **4,** 715 (1950).
[37] D. E. Green and J. G. Dewan, *Biochem. J.* **31,** 1069 (1937).
[38] H. von Euler, E. Adler, and H. Hellstrom, *Z. physiol. Chem.* **241,** 239 (1936).
[39] P. Ohlmeyer, *Biochem. Z.* **297,** 66 (1938).

Warburg[19] pointed out that the site of reduction is the pyridinium ring which gives rise to the characteristic absorption at 340 mμ. The reaction is formulated as follows:

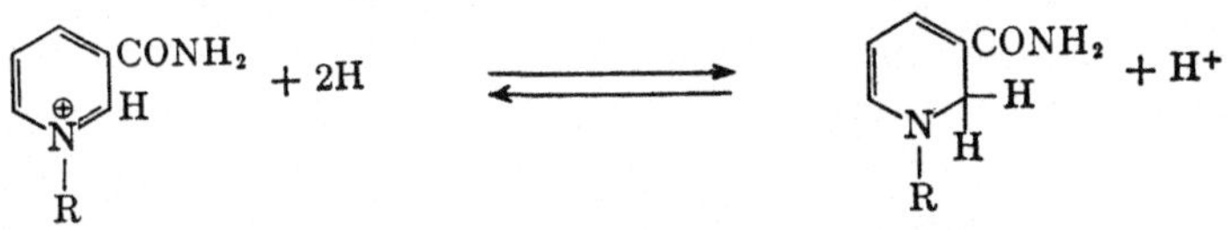

FIG. 2. Reduction of the pyridinium ring.*

In the oxidized state the nicotinamide nucleus exists as a quarternary pyridinium ion which forms a salt with one of the ionized OH groups of the pyrophosphate bridge. When the coenzyme undergoes reduction, 2 electrons add across the double bond, converting the heterocyclic nitrogen to a trivalent weakly basic nitrogen atom. As a result, the reduced coenzyme behaves as a dibasic acid. Therefore, during the reduction of the coenzyme an acid equivalent is formed. Haas[40] has pointed out that in the chemical reduction of the coenzyme the liberation of acid can be followed manometrically by conducting the reaction in a bicarbonate medium.

It is the reduction of the pyridinium ring system that causes the reduced coenzymes to show an absorption maximum at 320 to 360 mμ. At 340 mμ the molar extinction coefficient for the reduced coenzymes is 6.3×10^3. Recently Theorell and Bonnichsen,[41] working with horse liver alcohol dehydrogenase have shown that as the apoenzyme concentration approaches that of $DPNH_2$ the 340 band shifts toward shorter wavelengths, and when the $DPNH_2$ and apoenzyme are in equivalent concentration the absorption peak is at 325 mμ. $DPNH_2$ is bound two hundred times more strongly than DPN under these conditions. Addition of iodoacetate which inhibits the enzyme causes an immediate shift of the 325 peak back to 340, and this is taken to indicate that an —SH group of the apoenzyme is involved in binding the $DPNH_2$ to form the holoenzyme.

Nicotinamide does not undergo hydrosulfite reduction. In order to undergo reduction, the ring nitrogen must be pentavalent. Nicotinamide riboside as well as nicotinamide mononucleotide will undergo reduction when treated with hydrosulfite, but these compounds are incapable of acting biologically in place of the coenzyme. For full biological activity the coenzymes must be intact. Desamino DPN with the 6-amino group of

* M. E. Pullman [*Federation Proc.* **12,** 255 (1953)] has recently demonstrated that in the reduction of DPN the electrons add in a position para to the nitrogen rather than ortho.

[40] E. Haas, *Biochem. Z.* **285,** 368 (1936).

[41] H. Theorell and R. Bonnichsene, *Acta Chem. Scand.* **5,** 1105 (1951).

adenine removed, however, has some activity with alcohol dehydrogenase, although much less than DPN.[42]

Michaelis[43] has suggested that the hydrogen transfer takes place in steps, i.e., one electron at a time, leading to an intermediate semiquinoid radical (monohydropyridine nucleotide). This would facilitate the establishment of an equilibrium between the oxidized and reduced states.

There have been many unsuccessful attempts to link the nicotinamide coenzymes with phosphate transport. This problem has been reopened by the investigations on aerobic phosphorylation during which the generation of energy-rich phosphate bonds is coupled to electron transport. Green[44] and his collaborators postulated that the pyridine nucleotides might function in this capacity. Interest was further aroused by the observation of Friedkin and Lehninger[45] that the oxidation of reduced DPN could support the generation of energy-rich phosphate bonds. However, to date no conclusive evidence has been presented which would directly implicate these coenzymes in the phosphorylation reaction itself.

Starting from the original observation of Meyerhof[46] that DPN reacts with cyanide and bisulfite to form complexes with absorption spectra resembling that of the enzymatically reduced coenzymes, Kaplan[47] and Colowick have made some attempt to link the mechanism of action of DPN to the generation of energy-rich phosphate bonds during hydrogen transport. These workers satisfied themselves that cyanide, bisulfite, and hydroxyl ion add across the double bond in the pyridinium ring much like hydrogen during the enzymatic reduction of the coenzyme. It is reasoned that such behavior is indicative of the carbonyl properties of this double bond. Therefore, a possible mechanism for the generation of an energy-rich phosphate bond is visualized involving the addition of phosphate to the double bond, followed by dehydrogenation to give a DPN pyridone. DPN is again regenerated by reaction of the pyridone with $DPNH_2$ to give 2 molecules of DPN.

8. Reactions Catalyzed

The reactions catalyzed by DPN and TPN are: (*a*) specifically reversible dehydrogenations of aldehydes, (*b*) primary and secondary alcohols, (*c*)

[42] M. E. Pullman, S. P. Colowick, and N. O. Kaplan, *J. Biol. Chem.* **194,** 593 (1952).

[43] L. Michaelis, *in* Sumner and Myrbäck, The Enzymes, Academic Press, New York, 1951, Vol. II, Part 1, p. 1.

[44] R. J. Cross, J. V. Taggart, G. A. Covo, and D. E. Green, *J. Biol. Chem.* **177,** 655 (1949).

[45] M. Friedkin and A. L. Lehninger, *J. Biol. Chem.* **178,** 611 (1949).

[46] O. Meyerhof, P. Ohlmeyer, and W. Mohle, *Biochem. Z.* **297,** 113 (1938).

[47] N. O. Kaplan, *in* McElroy and Glass, Phosphorus Metabolism, The Johns Hopkins Press, Baltimore, 1951, Vol. I, p. 428.

and amines. These systems are depicted in the following general schemes:

$$\mathrm{R{-}\overset{H}{C}{=}O + DPN \rightleftharpoons R{-}\overset{O}{\overset{\|}{C}}{-}OH + DPNH_2} \quad (a)$$

$$\mathrm{R{-}CH_2OH + DPN \rightleftharpoons R{-}CHO + DPNH_2} \quad (b)$$

$$\mathrm{R{-}NH_2 + DPN \rightleftharpoons R{=}NH + DPNH_2} \quad (c)$$

The reactions as written actually do not depict DPN and TPN acting as coenzymes, but since they are only partical reactions the coenzymes are, in reality, acting as substrates. They behave catalytically as coenzymes only when they can couple in an oxidation-reduction sequence. Hence they have been called "mobile coenzymes"; i.e., they undergo reduction while on a given apoenzyme, leave this protein, and attach to another where they become oxidized. This has led to a consideration of these coenzymes as the classical type of dissociable coenzyme which shuttles about among a large variety of dehydrogenases. In the light of the recent observations from the Cori laboratory[3] such a generalization should be revised. When triosephosphate dehydrogenase was recrystallized, it was found to contain 1 mole of DPN per mole of enzyme. Extensive dialysis or dilution could not remove this bound DPN. The dissociation constant was found to be at least 1×10^{-11} or even smaller. This is in contrast to the usual value of 4×10^{-8} observed with *added* DPN. This led the Cori group to postulate two sites of attachment of DPN to the enzyme. The fact that the bound DPN does, in fact, dissociate, is evident from equilibration of this bound DPN with added radioactive DPN. Furthermore, the bound DPN was shown to interact with lactic dehydrogenase, but the rate was too fast to be accounted for by a dissociation of $DPNH_2$ from the triosephosphate dehydrogenase. Thus it appears possible that in this case there is an enzyme-enzyme interaction.

The coupled reactions which these coenzymes undergo are of two types.

a. Oxidation-Reduction Coupled to Substrate. In this type of reaction the coenzyme which is reduced in the first step is reoxidized in the second step by another substrate. An example of this type of reaction is the following set of equations:

$$\text{3-Phosphoglyceraldehyde} + \mathrm{DPN} \rightleftharpoons \text{3-Phosphoglyceric acid} + \mathrm{DPNH_2} \quad (9)$$

$$\text{Pyruvate} + \mathrm{DPNH_2} \rightleftharpoons \text{Lactate} + \mathrm{DPN} \quad (10)$$

In this reaction the reduction of DPN by glyceraldehyde is coupled to the oxidation of reduced DPN by the reduction of pyruvate to lactate. The over-all reaction would be written as follows:

$$\text{3-Phosphoglyceraldehyde} + \text{pyruvate} \rightleftharpoons \text{3-Phosphoglyceric acid} + \text{lactate} \quad (11)$$

Thus a small amount of DPN acting catalytically could effect the transformation of large amounts of substrate. The essence of this type of linked reaction is that the oxidation of reduced DPN is effected by *another substrate*, in the example cited, pyruvate, and the reactions are therefore anaerobic and require the presence of a large amount of hydrogen accepting substrate. This type of linked oxidation-reduction occurs during glycolysis where the situation is ideal for reactions of this type. Here we have the continual formation of pyruvate which can act as hydrogen acceptor for the DPN, reduced in the oxidation step. Usually the formation of pyruvate just keeps pace with the oxidation of glyceraldehydephosphate and the reaction is thereby allowed to go to completion.

b. Oxidation-Reduction Coupled to Electron Transport. This type of reaction has an initial step similar to the previously considered reaction but differs in that the oxidation of the reduced pyridine nucleotide is carried out by a second hydrogen-carrying system, rather than by a substrate. An example of this type is shown below:

$$\text{Glucose-6-phosphate} + \text{TPN} \rightarrow \text{6-Phosphogluconate} + \text{TPNH}_2 \tag{12}$$

$$\text{TPNH}_2 + \text{flavoprotein} \rightarrow \text{TPN} + \text{flavoprotein H}_2 \tag{13}$$

$$\text{Flavoprotein H}_2 + \text{O}_2 \rightarrow \text{Flavoprotein} + \text{H}_2\text{O}_2 \tag{14}$$

In this sequence of reactions the reduced TPN is oxidized by a flavin-containing enzyme, and the reduced flavin is subsequently oxidized by molecular oxygen. In other cases, the reduced flavin can be oxidized by the cytochrome series of enzymes (see Horecker and Kornberg, Chapter 18).

The essence of this type of reaction is that the hydrogens are removed from the reduced coenzymes, passed along a chain of hydrogen-carrying systems, and eventually used to reduce oxygen to the state of water. These reactions, then, are essentially aerobic and function in the respiration of the cell.

III. Flavin Mononucleotide and Flavin Adenine Dinucleotide

1. Introduction

The flavin-containing coenzymes are similar in importance to the pyridine coenzymes as functioning in hydrogen transport. In general the flavin coenzymes mediate hydrogen transport between coenyzmes, rather than between substrates, although their function in anaerobic bacteria and certain other tissues indicates that they may also play an important role in hydrogen transport at the substrate level.

2. Historical

A class of yellow pigments named "flavins" was recognized in 1879 by Blyth,[48] who isolated an impure flavin from whey. Bleyer and Kallman[49] in 1925 described a yellow substance from milk which they called "lactochrom." The year 1932 brought the riboflavin problem into sharp focus. Szent-Györgyi and associates,[50] working with heart muscle, obtained a "cytoflav" preparation which had the property of losing its yellow color on reduction and regaining it upon oxidation and suggested a possible role in cellular oxidation. Warburg and Christian[51] reported on the "yellow enzyme" from yeast, which contained a yellow pigment attached to a high molecular weight substance thought to be polysaccharide in nature. This enzyme mediated the oxidation of glucose-6-phosphate by oxygen in the presence of Zwischenferment (glucose-6-phosphate dehydrogenase) and a coenzyme later shown to be TPN. These workers split the yellow pigment from the complex by treatment with acidic methanol and upon irradiation of the pigment in alkaline solution obtained the first pure flavin derivative, "lumiflavin," which was soluble in chloroform. In 1933 two groups[52] isolated a yellow compound from eggs and milk which proved to be identical with vitamin B_2. Its spectrum was similar to that of Warburg and Christian's lumiflav and was thought to be lumiflavin plus a pentose of some sort. The use of spectrographic methods enable Stern and Holiday[53] to establish the presence of an alloxazine structure in Warburg's coenzyme. The groups of Kuhn[54] and Karrer[55] elucidated and proved the structure of riboflavin, 6,7-dimethyl-9-ribityl-isoalloxazine.

In one of the earliest applications of the electrophoretic techniques developed by Tiselius, Theorell[56] (1934–1937) was able to prepare a homogeneous chromoprotein with the properties of the yellow enzyme. The polysaccharide contaminant which prevented Warburg from isolating the enzyme was easily removed by this new procedure. The yellow holoenzyme was dissociated by dialyzing against an acidic solution. The apoenzyme, after dialysis against water, could be reconstituted by adding the chromogen

[48] A. W. Blyth, *J. Chem. Soc.* **35,** 530 (1879).

[49] B. Bleyer and A. Kallman, *Biochem. Z.* **155,** 54 (1924).

[50] I. Banga, A. Szent-Györgyi, and L. Vargha, *Z. physiol. Chem.* **210,** 228 (1932); A. Szent-Györgyi and I. Banga, *Biochem. Z.* **246,** 203 (1932).

[51] O. Warburg and W. Christian, *Biochem. Z.* **257,** 492 (1933).

[52] R. Kuhn, P. Györgyi, and T. Wagner-Jauregg, *Ber.* **66,** 317 (1933); P. Ellinger and W. Koschara, *Ber.* **66,** 315 (1933).

[53] K. G. Stern and E. R. Holiday, *Ber.* **67,** 1104, 1442 (1934).

[54] R. Kuhn, K. Reinemund, H. Kaltschmitt, R. Strobele, and H. Trischman, *Naturwissenschaften* **23,** 260 (1935).

[55] P. Karrer, K. Schopp, and F. Benz, *Helv. Chim. Acta* **18,** 426 (1935).

[56] H. Theorell, *Biochem. Z.* **272,** 155 (1934); **278,** 263 (1935); **290,** 293 (1937).

to yield the active holoenzyme. The chromogen, the free prosthetic group, was found to migrate toward the anode, while riboflavin moved toward the cathode. This implicated an acid group in the former. One mole of phosphate was found per mole of riboflavin. Isolation of a crystalline calcium salt led to the elucidation of the structure as flavin monophosphoric acid (flavin mononucleotide, FMN), in which the phosphoric acid is attached to carbon 5 of the ribityl side chain of riboflavin.[57]

Straub[58] and Warburg and Christian[59] found that the prosthetic group of *d*-amino acid oxidase was a flavin derivative, not identical with FMN. The latter group isolated the coenzyme and showed it to be flavin adenine-dinucleotide (FAD).

3. Structure and Chemistry

The structures for riboflavin monophosphate (FMN), and flavin adenine-dinucleotide (FAD) are as follows:

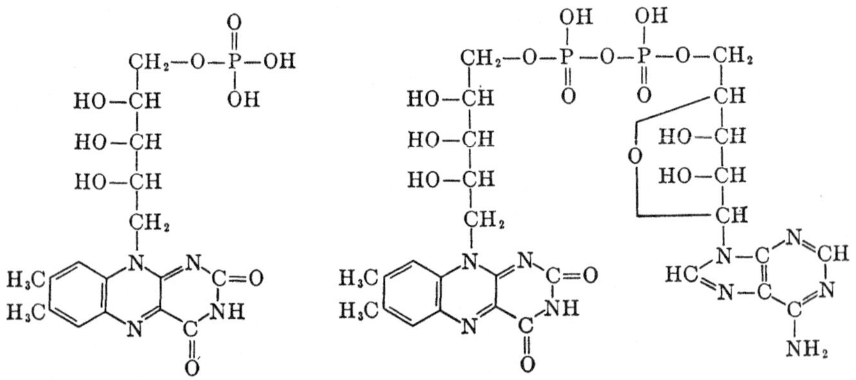

Riboflavin monophosphate (FMN)

Flavin adenine dinucleotide (FAD)

Fig. 3

The flavin or isoalloxazine moiety is a polycyclic structure comprised of a benzene, a pyrazine, and a pyrimidine ring. It is of interest to note that the latter two rings are found in the pteridin portion of folic acid and that this pteridin structure was first found in certain pigments isolated from butterfly wings.

Treating FAD with *N* HCl at 100° destroys biological activity and

[57] R. Kuhn, H. Rudy, and F. Weygand, *Ber.* **69,** 1543 (1936).

[58] F. S. Straub, *Nature* **141,** 603 (1938).

[59] O. Warburg and W. Christian, *Biochem. Z.* **295,** 261; **296,** 294; **297,** 417; **298,** 150 (1938).

yields FMN and adenosine-5-phosphate.[60] Trichloroacetic acid extracts kept at 37° overnight will affect this hydrolysis.[61] Biological activity is also destroyed by treating with *N* alkali at room temperature. Forrest and Todd[62] have shown that the action of weak alkali splits the molecule to yield adenylic acid and an alloxazine mononucleotide containing a cyclic phosphate esterified at positions 4′ and 5′ of the ribityl residue. The riboflavins are also sensitive to irradiation. Warburg and Christian,[51] as mentioned above, irradiated FMN in alkaline solution and isolated lumiflavin in which 4 carbons of the ribityl side chain are split off. Karrer and coworkers[63] were able to isolate "lumichrom," the 6,7-dimethylisoalloxazine, in which the ribityl moiety is completely removed, after irradiation of riboflavin in neutral solution in the presence of air.

a. Absorption Spectra and Fluorescent Properties. Studies on the absorption spectra and the fluorescent properties of the biologically active flavins have proved to be valuable in understanding the more detailed chemistry of FAD, as well as to give clues as to the binding of FMN and FAD to their apoenzymes.

Riboflavin, greenish-yellow in color, has three absorption maxima at 445, 375, and 260 mμ.[59] In FAD, reddish-yellow in color, there is a shift of the 445 band to 450 mμ. This same shift has been observed by Weber[64] when the spectrum of riboflavin is taken in the presence of caffeine or adenosine, which are shown to form complexes with riboflavin. FMN, greenish-yellow, has maxima at 445, 375, and 260 mμ.[56] The "old" yellow enzyme is a pure yellow, and this is due to a shift toward longer wavelengths in the entire spectral range, i.e., maxima at 465, 380, and 265 mμ. The case of xanthine oxidase is an even more striking example of the shift toward the red. Purified xanthine oxidase is red in color. Until very recently this red coloration was thought to indicate the presence of a second pigment in the purified enzyme. Morell[65] has rectified this concept by proving that FAD is the only prosthetic group involved and that the red coloration is due to an even more marked shift to the red which results from the fact that the bound FAD absorbs 67% higher at 450 mμ than does the free FAD (which can be split from the enzyme by treatment with high concentrations of divalent salts).

The fluorescence of riboflavin and its biological derivatives confirms the interesting changes in physical properties (and probably therefore

[60] E. P. Abraham, *Biochem. J.* **33,** 543 (1939).

[61] O. A. Bessey, O. H. Lowry, and R. Love, *J. Biol. Chem.* **180,** 755 (1949).

[62] H. S. Forrest and A. R. Todd, *J. Chem. Soc.* **1950,** 3295.

[63] P. Karrer, H. Salomon, K. Schopp, E. Schlitter, and H. Fritzsche, *Helv. Chim. Acta* **17,** 1010 (1934).

[64] G. Weber, *Biochem. J.* **47,** 114 (1950).

[65] D. B. Morell, *Biochem. J.* **50,** vii (1951).

chemical reactivities) noted above. Although riboflavin shows a strong yellow fluorescence at neutral pH, FAD fluoresces only weakly, and the "old" yellow enzyme and other flavoproteins with firmly bound coenzymes do not fluoresce at all under these conditions. The shift to the red in the absorption spectrum is here confirmed in a more striking manner by the weakening and disappearance of fluorescence. An analysis of the effects of pH on the fluorescence of these compounds led Bessey, Lowry, and Love[61] to conclude that the lack of fluorescence of FAD compared with that of riboflavin is due to an internal complex between the flavin and adenine moieties of the FAD which results in quenching. That this is the case has been demonstrated by Weber,[64] who found that the quenching of the fluorescence of riboflavin by purines is due to complex formation and that within the FAD molecule such a complex also exists.

It should also be mentioned that the oxidation-reduction potentials of free and bound FMN are different.[66]

The isolation of FAD, according to Warburg and Christian,[59] involves essentially the following steps. A boiled extract of yeast is filtered and brought to 66% saturation with ammonium sulfate and the coenzyme extracted into phenol. Water and ether are added to the phenol solution, and the coenzyme is driven into the aqueous phase. The dinucleotide is precipitated at pH 2 as the silver salt.

Chemical syntheses of FMN from riboflavin by phosphorylating agents have recently been made available.[62, 67]

4. Mechanism of Action

FMN and FAD, in a manner analogous to DPN and TPN, undergo alternate oxidation and reduction. In the flavin series the hydrogens add across the conjugated double bond system made up of the 1-nitrogen (pyrimidine ring), 2 carbons, and the 10-nitrogen (pyrazine ring). Whereas the oxidized form is yellow, the reduced or leuco form is almost colorless. The semiquinoid form, the monohydroflavin, has been demonstrated in chemical systems by several workers[67, 68] and in an enzyme system by Haas.[69] The Haas intermediate is thought to be part of a complex consisting of the three components, FMN-apoenzyme-TPNH. The transformation of the oxidized form to the leuco form may be visualized as follows:

[66] R. Kuhn and P. Boulanger, *Ber.* **69,** 1557 (1936).

[67] R. Kuhn and T. Wagner-Jauregg, *Ber.* **67,** 351 (1934).

[68] L. Michaelis, M. P. Schubert, and C. V. Smythe, *J. Biol. Chem.* **116,** 587 (1936).

[69] E. Haas, *Biochem. Z.* **290,** 291 (1937).

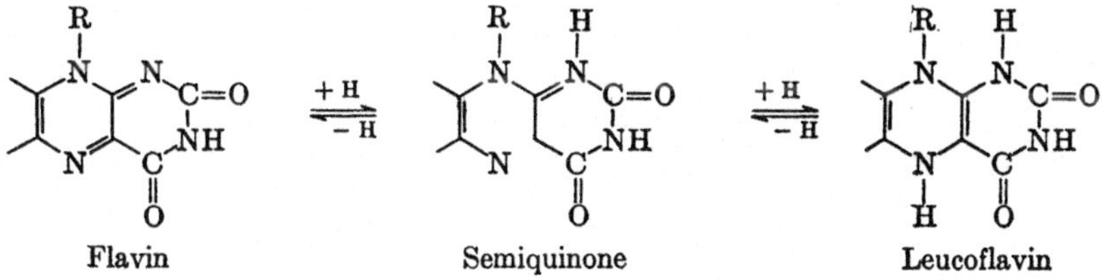

FIG. 4

For a discussion of this and other problems concerning the mechanism of oxidation reduction reactions, the article of Leonor Michaelis,[43] published after his death, is recommended.

Chemically the oxidized form of the flavins can be reduced by hydrosulfite to the leuco form, and the latter is easily reoxidized back by oxygen. Singer and Kearney[70] have demonstrated that the flavins could mediate a nonenzymatic reduction of cytochrome c by reduced pyridine nucleotides This chemical system may serve as a model for the enzymatic reactions in which flavins function in this same capacity.

5. METHOD OF ESTIMATION

a. Chemical. Bessey, Lowry, and Love[61] have developed what is probably the method of choice for the determination of the various forms of riboflavin found in tissues. The following facts form the basis for the assay. Whereas riboflavin and FMN fluoresce strongly at neutral pH, FAD fluoresces to an extent equivalent to about 10% of the former compounds. Riboflavin is soluble in benzyl alcohol, but FMN, because of its phosphate group, is not. FAD likewise is insoluble in organic solvents. As the pH of the extract is lowered, the fluorescence of the riboflavin and FMN falls while that of the FAD rises, so that at pH 2.9 the latter reaches a maximum equivalent to 70% of that of riboflavin and FMN at that pH. Blanks for the extracts are obtained by reducing the flavins with hydrosulfite. The leucoflavins do not fluoresce.

b. Enzymatic. The methods of Warburg and Christian,[71] Negelein and Bromël,[71a] or the modification by Krebs[72] is usually employed for the quantitative assay of FAD for which it is specific. An extract of an acetone-dried hog kidney preparation is the source of the crude D-amino acid oxidase used in this procedure. The crude extract is treated with acid ammonium sulfate which results in the precipitation of the apoenzyme, the FAD remaining in the supernatant fluid. The oxidative deamination of a D-amino

[70] T. P. Singer and E. B. Kearney, *J. Biol. Chem.* **183,** 409 (1950).
[71] O. Warburg and W. Christian, *Biochem. Z.* **298,** 368 (1938).
[71a] E. Negelein and H. Bromël, *Biochem. Z.* **300,** 225 (1939).
[72] H. A. Krebs, *Enzymologia* **7,** 53 (1939).

acid, with *dl*-alanine usually employed as substrate, is followed either manometrically by measuring the oxygen uptake or chemically by a determination of the NH_3 or keto acid liberated and is compared against rates obtained with known amounts of FAD when added to the apoenzyme solution.

6. Flavin Content of Tissues

Free riboflavin appears to be present to a significant extent in the retina of certain animals and fishes, in milk, and in the urine. The findings of Bessey, Lowry, and Love[61] indicate that riboflavin, like the other B vitamins, occurs mostly in bound forms, FAD being the major constituent. The FAD content varies from 90% of the total riboflavin content of the muscle to 70% in the kidney. The remainder occurs mainly as FMN, with free riboflavin being present in only trace amounts.

7. Metabolism of Flavin Coenzymes

Klein and Kohn[73] found that riboflavin can be converted to FAD by red blood cells, both *in vitro* and *in vivo.* Trufanov[74] reported the synthesis of FAD by rat liver slices. Recently, the mechanism of the over-all conversion has been elucidated. Kearney and Englard[75] have isolated a kinase from brewer's yeast which catalyzes the following reaction:

$$\text{ATP} + \text{riboflavin} \rightarrow \text{FMN} + \text{ADP} \tag{15}$$

Kornberg and his co-workers have shown that an enzyme in yeast and liver[76, 77] can synthesize FAD from FMN and ATP according to the following scheme:

$$\text{FMN} + \text{ATP} \rightleftharpoons \text{FAD} + \text{pyrophosphate} \tag{16}$$

8. Function

Theorell[78] has recently written an excellent review article on the flavoproteins. Only a short discussion of their function will be presented here.

The flavoproteins have been classified in two ways: (1) on the basis of the nature of the substance donating 2H to the flavin coenzyme—(*a*) a substrate molecule or (*b*) a reduced coenzyme; and (2) on the basis of the substance to which the leucoflavin transfers its hydrogens—(*a*) molecular

[73] J. R. Klein and H. I. Kohn, *J. Biol. Chem.* **139,** 177 (1940).
[74] A. V. Trufanov, *C. A.* **38,** 131 (1944).
[75] E. B. Kearney and S. Englard, *J. Biol. Chem.* **193,** 821 (1951).
[76] A. Kornberg and W. E. Pricer, Jr., *J. Biol. Chem.* **182,** 763 (1950).
[77] A. W. Schrecker and A. Kornberg, *J. Biol. Chem.* **182,** 795 (1950).
[78] H. Theorell, *in* Sumner and Myrbäck, The Enzymes, Academic Press, New York, 1951, Vol. II, Part 1, p. 329.

oxygen, (*b*) cytochrome c, (*c*) fumaric acid, and (*d*) dyes. The various combinations possible between these two groups have been demonstrated.

Although most of the flavoproteins contain FAD as their prosthetic group, Warburg's "old" yellow enzyme, a cytochrome c reductase, and the *l*-amino acid oxidase contain FMN as coenzyme.

The diaphorases are flavoproteins mediating the transfer of hydrogens from the reduced pyridine nucleotides (DPN and TPN) to dyes. The cytochrome c reductases mediate this transfer to cytochrome c, the cytochrome oxidase systems taking the hydrogens to oxygen. These systems have been demonstrated in plant and animal tissues and are discussed in the chapter 18. The flavin coenzymes thus play a central role in the transfer of hydrogens from the substrate level to oxygen via the cytochrome system.

The flavoproteins are thought to assume an even more significant role in organisms and tissues in which the cytochrome system is either nonexistent or has been poisoned by cyanide. The anaerobic bacteria, many facultative anaerobes, carrot leaves, and avocado fruit, as well as the helminths, contain no cytochrome system. Needham[79] has reviewed the problem of the respiratory systems of many animal species and gives many examples of significant non-ferrous respiration. It is also known, that some respiration remains even after poisoning the metal-containing cytochrome system by cyanide. In all these cases it is felt that the flavoproteins serve a role in the terminal transfer of hydrogens either directly to free oxygen in aerobic systems or to reducible substances of various kinds under anaerobic conditions.

Several reduced flavoproteins react directly with oxygen to give hydrogen peroxide. The latter is usually metabolized by a catalase. It is now generally felt that the catalases function primarily in a peroxidative capacity. Seeley and Vandenmark[80] have evidence for concluding that flavoproteins in *Streptococcus faecalis* serve as peroxidases. This is another example of the flavoprotein mediating a reaction usually attributed to a hematin-containing enzyme.

The following enzymes have been shown to be flavoproteins: D-amino acid oxidases, *l*-amino acid oxidases, xanthine oxidases, aldehyde oxidases, the glucose dehydrogenase, notatin, the diamine oxidase, and fumaric hydrogenase. Details concerning these enzymes may be found in the review articles of Theorell[78] and Krebs.[81]

[79] J. Needham, Biochemistry and Morphogenesis, The Macmillan Company, New York, 1942.

[80] H. W. Seeley and P. J. Vandenmark, *J. Bact.* **61**, 27 (1951).

[81] H. A. Krebs, *in* Sumner and Myrbäck, The Enzymes, Academic Press, New York, 1951, Vol. II, Part 1, p. 499.

IV. Adenine Nucleotides

1. Introduction

The adenine nucleotides (AMP, ADP, and ATP) comprise a family of cofactors which are of prime importance in the transport of phosphate. The importance of phosphate as a means of transforming chemical potential and oxidation energy into metabolically active forms has been amply discussed by Lipmann[82] and by Kalckar[83] and others and needs no elaboration here. The major role of the adenine nucleotide system is the transport and storage of phosphate bond energy.

In a recent review, Colowick[84] has admirably discussed the role of the adenine nucleotides in various transphosphorylation reactions, and a thorough consideration is given to the enzymes involved. Only the more general aspects of the problem will therefore be discussed.

2. Structure and Chemistry

The adenine nucleotide system consists of the following members: adenosine 5′-phosphoric acid (AMP), adenosine diphosphoric acid (ADP), and adenosine triphosphoric acid (ATP). The structures of these compounds are given in Fig. 5. These structures have been confirmed by the classical chemical synthesis of Todd and his co-workers.[85]

It is to be noted that these compounds differ from each other only in the number of phosphate groups. The phosphate group of AMP is bound in a simple ester linkage, whereas the terminal phosphate of ADP and the two terminal phosphate groups of ATP are bound in anhydride linkage and are, in Lipmann's nomenclature,[82] "energy-rich" linkages. These are indicated by the symbol (∼). That is, these linkages liberate around 10,000 cal. per mole on hydrolysis, in contrast to around 3000 cal. per mole liberated by the low-energy, simple ester bonds. The energy-rich bonds of ADP and ATP are characterized by their acid lability; i.e., they are hydrolyzable in 10 min. by 1 *N* acid at 100°, conditions under which the usual low-energy bond (for example the hexose-6-phosphates) are stable.

The nucleotides are interconvertible by the enzyme myokinase[86] which catalyzes the following transformation:

$$2\text{ADP} \rightleftharpoons \text{AMP} + \text{ATP} \qquad (17)$$

At equilibrium, about two-thirds of the ADP is converted and the same

[82] F. Lipmann, *Advances in Enzymol.* **1,** 99 (1941).

[83] H. M. Kalckar, *Chem. Revs.* **28,** 71 (1941).

[84] S. P. Colowick, *in* Sumner and Myrbäck, The Enzymes, Academic Press, New York, 1951, Vol. II, Part 1, p. 114.

[85] J. Baddiley and A. R. Todd, *J. Chem. Soc.* **1947,** 648.

[86] S. P. Colowick and H. M. Kalckar, *J. Biol. Chem.* **148,** 117 (1943).

equilibrium is attained starting from either side. The importance of this enzyme will be considered in a later section.

3. Distribution and Method of Measurement

The adenine nucleotides are widely distributed in living materials, having been reported for all types of animal, plant, and microbial tissue. There is one report of the isolation from *Thiobacillus thiooxidans*[87] of ATP with the phosphate esterified in the 3 position of ribose rather than in the 5 position.

Chemical, manometric, and spectophotometric methods, as well as a

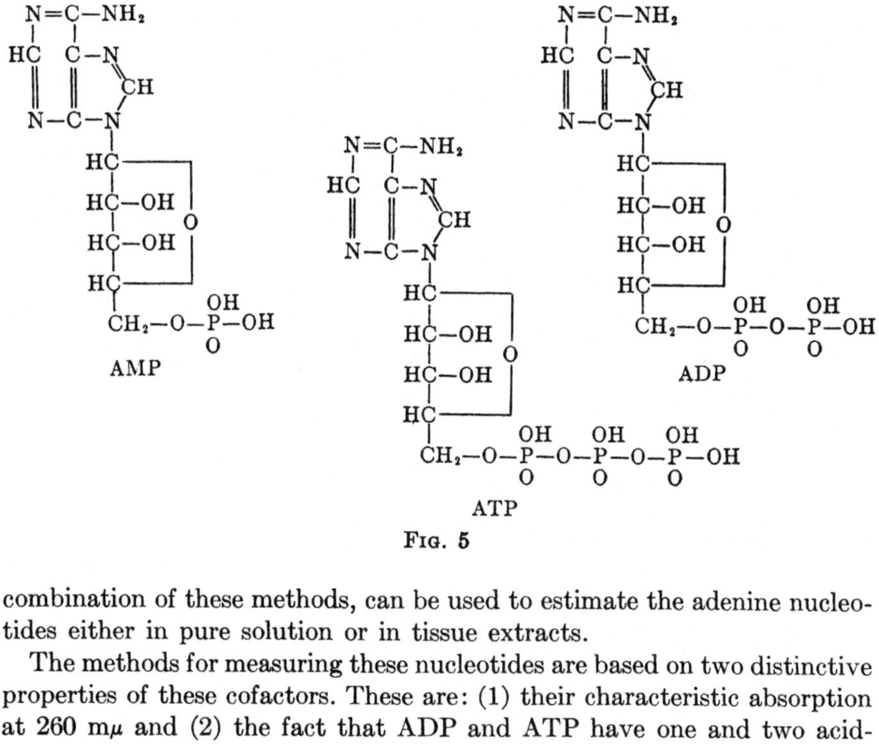

Fig. 5

combination of these methods, can be used to estimate the adenine nucleotides either in pure solution or in tissue extracts.

The methods for measuring these nucleotides are based on two distinctive properties of these cofactors. These are: (1) their characteristic absorption at 260 mμ and (2) the fact that ADP and ATP have one and two acid-labile phosphate groups, respectively. For example, the pyrophosphate linkages of ATP are completely hydrolyzed to orthophosphate in 10 min. at 100° (1 *N* acid), whereas the hexose-6-phosphates are not appreciably hydrolyzed by this treatment.

a. Adenylic acid (AMP) can be measured quantitatively by using Schmidt's[88] muscle adenylic acid deaminase and measuring the change in

[87] G. A. LePage and W. W. Umbreit, *J. Biol. Chem.* **148,** 255 (1943).
[88] G. Schmidt, *Z. physiol. Chem.* **179,** 243 (1928).

absorption at 260 mμ. This method was developed by Kalckar.[89] When AMP is deaminated to inosinic acid by this method, there is a proportional decrease in absorption at 260 mμ and a corresponding increase at 240 mμ. By noting this change, AMP can be quantitatively determined.

b. ATP can be estimated by noting the disappearance of acid-labile phosphate when hexokinase[86] is allowed to transfer the terminal phosphate of ATP to glucose.

$$\text{ATP} + \text{glucose} \rightarrow \text{Glucose-6-phosphate} + \text{ADP} \quad (18)$$

From this equation, it can be seen that when the reaction is allowed to go to completion there will result a disappearance of 50% of the acid-labile phosphate. Because of the fact that in the above reaction an acid equivalent is liberated, owing to the strongly acid character of the new OH group of ADP, Colowick and Kalckar[86] have developed a manometric method for following the course of this reaction. If the reaction is conducted in a bicarbonate buffer, the evolution of CO_2 can be used to estimate the transfer of phosphate and thus the original ATP concentration.

Kornberg[90] has coupled the hexokinase reaction with Zwischenferment and TPN and has developed a spectrophotometric method for measuring ATP. After glucose-6-phosphate is formed by the hexokinase reaction, Zwischenferment will oxidize the glucose-6-phosphate to phosphogluconic acid with a concomitant reduction of TPN. The reduction of TPN then reflects the ATP concentration in the solution under test.

c. ADP can be specifically measured by using pyruvate kinase. The reaction utilized is the following:

$$\text{Phosphoenolpyruvate} + \text{ADP} \rightleftharpoons \text{Pyruvate} + \text{ATP} \quad (19)$$

If this reaction is coupled with lactic dehydrogenase and reduced DPN, the pyruvate formed will be immediately reduced to lactate, with a corresponding oxidation of the reduced DPN. The oxidation of $DPNH_2$ is then a direct measure of the ADP present in the original solution.[90]

4. Function

The functions performed by the adenine nucleotides can be classified into three general types, as follows: (*a*) transport of phosphate on the energy-rich level; (*b*) transport from high energy to low energy; and (*c*) activation of compounds for biosynthetic reactions.

a. Transport of Phosphate at the High-Energy Level. In this class of reactions four types of processes can be distinguished, namely: (1) the uptake of energy-rich phosphate bonds created at the substrate level, (2)

[89] H. M. Kalckar, *J. Biol. Chem.* **167**, 445 (1947).

[90] A. Kornberg, personal communication.

the uptake of energy-rich phosphate bonds created during electron transport, (3) storage of energy-rich bonds, and (4) intermolecular rearrangement of energy-rich bonds.

(1) *The Uptake of Energy-Rich Phosphate Bonds Created at the Substrate Level.* The two general reactions by which energy-rich phosphate bonds generated at the substrate level are transferred to the adenine nucleotides are given below:

$$R-\underset{\underset{O}{\|}}{C}-O-PO_3H_2 + ADP \rightleftharpoons R-\underset{\underset{O}{\|}}{C}-OH + ATP \qquad (20)$$

$$R = \underset{\underset{COOH}{|}}{C}-O-PO_3H_2 + ADP \rightleftharpoons R-\overset{\overset{COOH}{|}}{\underset{\underset{O}{\|}}{C}} + ATP \qquad (21)$$

In reaction 20 an acyl phosphate transfers its phosphate group to ADP, giving rise to the formation of ATP. Examples of this type are the transfer of phosphate from phosphoglyceryl phosphate[91] and from acetyl phosphate.[92] Phosphoglyceryl phosphate arises from the oxidation of phosphoglyceraldehyde during the oxidation reduction step of glycolysis. Acetyl phosphate may arise from the oxidation of fatty acids,[93] from the oxidation of acetaldehyde,[94] during the metabolism of pyruvate by bacteria, i.e., phosphoroclastic split,[95] fermentation,[96] dismutation,[97] and oxidation.[98]

In reaction 21 an enol phosphate is transferred to ADP. An example of this type of reaction is the phosphate transfer from phosphoenol pyruvate, already considered.

(2) *The Uptake of Energy-Rich Phosphate Bonds Generated during Electron Transport.* Very little is understood regarding the mechanism of generation of energy-rich phosphate bonds during electron transport, but, whatever the mechanism of formation, it is clear that the utlimate acceptor is the adenine nucleotide system.

(3) *Storage of Energy-Rich Phosphate Bonds.* After the energy-rich bonds are generated during the various metabolic processes, they are transferred to ADP in the formation of ATP. From this point a variety of things can

[91] T. Bucher, *Biochim. et Biophys. Acta* **1,** 292 (1947).
[92] F. Lipmann, *J. Biol. Chem.* **155,** 55 (1944).
[93] E. R. Stadtman and H. A. Barker, *J. Biol. Chem.* **180,** 1095 (1949).
[94] E. R. Stadtman and H. A. Barker, *J. Biol. Chem.* **180,** 1117 (1949).
[95] M. F. Utter, F. Lipmann, and C. H. Werkman, *J. Biol. Chem.* **158,** 521 (1945).
[96] H. J. Koepsall and M. Johnson, *J. Biol. Chem.* **145,** 379 (1942).
[97] S. Korkes, A. del Campillo, I. C. Gunsalus, and S. Ochoa, *J. Biol. Chem.* **193,** 721 (1951).
[98] F. Lipmann, *Cold Spring Harbor Symposia Quant. Biol.* **7,** 248 (1939).

happen. For example, the energy may be tucked away in storage until such time as it is needed to drive some vital process. In vertebrates, one of the largest requirements for energy is muscular activity. It is not surprising to find in muscle a storehouse of energy, since in the course of existence the muscles may be called upon rather abruptly to do work of a vital nature which requires the sudden expenditure of large amounts of energy. This storehouse of energy in muscle is phosphocreatine. ATP reacts with creatine under the influence of the enzyme creatine kinase to form phosphocreatine.[99]

$$\text{Phosphocreatine} + \text{ADP} \rightleftharpoons \text{Creatine} + \text{ATP} \quad (22)$$

In invertebrates, the storage function is taken over in an analogous manner by arginine.

$$\text{Phosphoarginine} + \text{ADP} \rightleftharpoons \text{Arginine} + \text{ATP} \quad (23)$$

The ready reversibility of these reactions, since there is relatively no free energy change involved, makes it easy for the storage compounds, phosphocreatine and phosphoarginine, to deliver energy-rich phosphate to ADP in the formation of ATP as the demand of the organism requires.

(4) *Intermolecular Rearrangements.* In the uptake of energy-rich phosphate generated either at the substrate level or by electron transport, it was noted that ADP is the primary acceptor. Since the cell does not have an unlimited amount of adenine nucleotide, ADP must be regenerated from ATP in order to have enough phosphate acceptor available to take up the energy-rich phosphate as fast as it is formed. In the previous section, one method of regeneration of ADP was illustrated in the storage of energy as phosphocreatine or phosphoarginine. A second method by which ADP may be regenerated is in the utilization of the ATP energy for biosynthetic purposes. This reaction will be dealt with below.

Living systems have still another mechanism for making ADP available as a phosphate acceptor and that is the myokinase reaction[86] previously mentioned whereby 1 mole of ATP may react with 1 mole of AMP to yield 2 moles of ADP. Such a device then serves the purpose of regenerating ADP from ATP as long as a source of AMP is available. This enzyme will serve to keep the proper adenine nucleotides in desirable concentration as the needs of the cell dictate.

b. Transport of Phosphate from High-Energy to Low-Energy Levels. The utilization of pyrophosphate bond energy of ATP for the formation of low-energy bonds, such as glucose-6-phosphate from glucose, is accompanied by the release of a large amount of free energy. For this reason these reactions have been considered to be essentially irreversible.

In addition to glucose and fructose-6-phosphates, a wide variety of

[99] K. Lohmann, *Biochem. Z.* **271,** 264 (1934).

compounds with alcoholic hydroxyl groups have been shown to accept phosphate from ATP. Some of these compounds are various hexoses, pentoses, glycerol, adenosine, and DPN. The type reaction is represented by the equation:

$$ATP + R{-}OH \rightarrow R{-}O{-}PO_3H_2 + ADP \qquad (24)$$

The best-studied example of this type of reaction is the conversion of glucose to glucose-6-phosphate by the enzyme hexokinase. The purpose of this reaction appears to be to mold the glucose molecule in such a fashion that during the subsequent fermentation this low-energy phosphate is again raised to the high-energy state. Starting with glucose, the fermentation requires the expenditure of two high-energy bonds, but four energy-rich bonds are produced, which give an over-all net gain of two energy-rich bonds.

c. Transport of Energy-Rich Bonds for the Activation of Compounds for Biosynthetic Purposes. This group of reactions requires the utilization of the energy of the pyrophosphate bonds of ATP for various syntheses.

The utilization of ATP energy in the synthesis of fatty acids, sterols, carbohydrates, and proteins has been demonstrated. In some cases a direct demonstration has been possible; in other cases, viz., protein synthesis, only an indirect approach has been possible.

Since various synthetic reactions will be considered in detail in other chapters of this volume, only the more general aspects of the problem will be considered here.

The classical example of the utilization of ATP energy for synthetic purposes is the synthesis of glycogen by reversal of the reactions of glycolysis. Since all the enzymatic steps in glycolysis have been shown to be reversible, it is possible to synthesize glycogen from pyruvate by reversing the entire reaction. This would require the net expenditure of 2 moles of energy-rich phosphate per mole of pyruvate converted to glucose.

Other reactions in which ATP energy is utilized for biosynthetic reactions are less well understood. In general, one can visualize three possible mechanisms for the activation of molecules for synthetic reactions. These are: (1) direct phosphorylation of the substrate, (2) phosphorylation of a coenzyme, and (3) phosphorylation of the enzyme.

(1) *Direct Phosphorylation of the Substrate.* There are several examples of this type of reaction. In the reversal of glycolysis pyruvate is phosphorylated by ATP to give phosphoenol pyruvate, and the phosphorylation of 3-phosphoglyceric acid occurs prior to its reduction to the aldehyde.

Another reaction of this type is the phosphorylation of acetate and of butyrate, yielding the respective acyl phosphates.[92]

$$\text{Acetate} + \text{ATP} \rightleftharpoons \text{Acetyl phosphate} + \text{ADP} \qquad (25)$$

However, it must be admitted that, of all the synthetic reactions in which ATP is involved, the observation of a phosphorylated substrate is relatively rare. In the majority of the ATP-driven reactions attempts to demonstrate the intermediate formation of a phosphorylated substrate molecule have been unsuccessful. Outstanding examples of this are (1) synthesis of glutamine from glutamic acid, ATP, and ammonia and (2) protein synthesis.

*(2) *The Intermediate Phosphorylation of a Coenzyme.* For a long time the reaction between ATP, acetate, and CoA leading to the formation of acetyl-CoA was another example of a reaction in which the demonstration of a phosphorylated intermediate was unsuccessful. The transfer of phosphate from ATP to acetate to form acetyl phosphate was excluded, since the addition of acetyl phosphate in place of acetate + ATP could not support the subsequent synthetic reaction.

The recent elucidation by Lipmann[100, 101] and his co-workers of the mechanism of this reaction has now added a new type of phosphate transfer by ATP. In this reaction the entire pyrophosphate group of ATP is transferred to the coenzyme. The reaction is:

$$\text{ATP} + \text{CoA} \rightleftharpoons \text{CoA—P—P} + \text{AMP} \tag{26}$$

$$\text{CoA—P—P} + \text{acetate} \rightleftharpoons \text{Acetyl—CoA} + \text{P—P} \tag{27}$$

Reaction 26 shows that the pyrophosphate group of ATP is transferred to CoA to form pyrophosphoryl-CoA and AMP. And in reaction 27 acetate displaces the pyrophosphate group from CoA to form acetyl-CoA and inorganic pyrophosphate. By reversal of these reactions a new mechanism for the formation of ATP from inorganic pyrophosphate and AMP becomes available.

A somewhat similar reaction for the formation of ATP from inorganic pyrophosphate was described by Kornberg.[102] This reaction differs from the CoA reaction in that ATP is formed by the pyrophosphorolysis of DPN. The reaction is described by the equation:

$$\text{DPN} + \text{P—P} \rightleftharpoons \text{ATP} + \text{NMN} \tag{28}$$

Here, then, are two examples of the formation of ATP from inorganic pyrophosphate. The intermediate pyrophosphorylation of CoA is the first example of the activation of a coenzyme by ATP in which the energy-rich bond is maintained so that it can be utilized to activate a substrate molecule for synthetic purposes.

(3) *The Intermediate Phosphorylation of Enzyme.* There are no known

* See Addendum at end of chapter.

[100] S. Black, M. E. Jones, R. M. Flynn, and F. Lipmann, *Federation Proc.* **11,** 189 (1952).

[101] F. Lipmann, M. E. Jones, S. Black, and R. M. Flynn, *J. Am. Chem. Soc.* **74,** 2384 (1952).

[102] A. Kornberg, *J. Biol. Chem.* **182,** 779 (1950).

reactions of this type, but this mechanism is proposed as a possible explanation for these reactions involving the utilization of ATP energy in which no phosphorylated intermediate can be detected (glutamine synthesis for example). The reaction could be visualized to follow this course:

$$E + \text{ATP} \rightleftharpoons E\text{—}P + \text{ADP} \tag{29}$$

$$E\text{—}P + \text{A} \rightleftharpoons E\text{—A} + P \tag{30}$$

$$E\text{—A} + \text{B} \rightleftharpoons \text{A—B} + E \tag{31}$$

In reaction 29 ATP may react with some site on the enzyme to form an enzyme phosphate (or pyrophosphate) in which the energy-rich character of the bond is maintained. Reaction 30 depicts the exchange of the phosphate group with substrate molecule A which now becomes activated. In equation 31 is depicted the reaction between the activated substrate enzyme and an acceptor molecule B to complete the condensation reaction and liberate free enzyme. The essence of this scheme is that no phosphorylated substrate molecules appear during the reaction.

There are of course other reactions which are theoretically possible in which no phosphorylated substrate would appear. For example, the first could involve a complex between enzyme and ATP, viz.,

$$E + \text{ATP} \rightleftharpoons E\text{—ADP} + P \tag{32}$$

and then the same sequence of reactions as shown above would achieve the same purpose, i.e., the activation of the substrate without its direct phosphorylation.

It appears likely that this type of reaction may eventually be found as an explanation for these reactions involving ATP and in which a phosphorylated substrate cannot be demonstrated.

d. Miscellaneous Reactions. There are a few reactions in which the adenylic system appears to be involved, but as yet the mechanism of these reactions is not known. Adenylic acid[103] has been shown to accelerate the activity of phosphorylase b, and it is found in the crystalline enzyme. Phosphorylase a, however, seems not to require AMP.

Stumpf[104] has described the exchange of isotopic ammonia with the amide group of glutamine as requiring a catalytic amount of ATP. As yet no reason for such a requirement has been postulated.

V. Glucose Diphosphate and Diphosphoglyceric Acid

1. Introduction

The coenzymes of the phosphomutases (glucose diphosphate and diphosphoglyceric acid) are a recently discovered group of coenzymes which

[103] G. T. Cori and A. A. Green, *J. Biol. Chem.* **151,** 31 (1943).
[104] P. K. Stumpf and W. D. Loomis, *Arch. Biochem.* **25,** 451 (1950).

play a role in the transport of phosphate at the low energy level. It has already been seen that the adenylic nucleotides transport phosphate between molecules at the high-energy level or from a high level to a low level. The phosphomutases are enzymes which transfer a phosphate group from one position in a molecule to another in the same molecule. Two such enzymes are found in the glycolytic scheme. Phosphoglucomutase, in the presence of glucose-1-6-diphosphate catalyzes the interconversion of glucose-1-phosphate to glucose-6-phosphate, and phosphoglyceromutase, in the presence of 2,3-phosphoglyceric acid, equilibrates 3-phosphoglyceric acid and 2-phosphoglyceric acid. This section will deal almost exclusively with the coenzyme glucose diphosphate.

2. Historical

Cori, Colowick, and Cori[105] were the first to study the phosphoglucomutase reaction, and they demonstrated the widespread occurrence of this enzyme in animal tissues and in yeast. The muscle enzyme was finally crystallized by Najjar.[106] Leloir and his co-workers[107, 108, 109] proved the necessity of a coenzyme for the reaction and showed that the coenzyme was glucose-1-6-phosphate. For many years the coenzyme had escaped detection because the glucose-1-phosphate used as the substrate for the reaction was contaminated with minute amounts of the glucosediphosphate.

3. Structure and Chemistry of Glucose Diphosphate[109]

The structure of the coenzyme is given in Fig. 6. The coenzyme is a nonreducing substance. Heating in 0.1 *N* hydrochloric acid at 100° for 10 min. hydrolyzes the labile 1-phosphate group and destroys biological activity. One equivalent of aldose is formed per mole of phosphate split off. Glucose-6-phosphate was shown to be the remaining product.

a. Isolation. The isolation of the coenzyme was carried out by the following procedure: Fresh yeast is incubated with sugars and phosphate in the presence of ether. This raises the coenzyme content some 10- to 100-fold. Starting with a boiled juice, a lead salt is precipitated, and then decomposed with H_2S. Contaminating fructose diphosphate, which accumulates along with the coenzyme, is selectively destroyed by alkali treatment.

[105] G. T. Cori, S. P. Colowick, and C. F. Cori, *J. Biol. Chem.* **124,** 543 (1938).
[106] V. A. Najjar, *J. Biol. Chem.* **175,** 281 (1948).
[107] R. Caputto, L. F. Leloir, R. E. Trucco, C. E. Cardini, and A. Paladini, *Arch. Biochem.* **18,** 201 (1948).
[108] L. F. Leloir, R. E. Trucco, C. E. Cardini, A. Paladini, and R. Caputto, *Arch. Biochem.* **19,** 339 (1948).
[109] C. E. Cardini, A. Paladini, R. Caputto, L. F. Leloir, and R. E. Turcco, *Arch. Biochem.* **22,** 87 (1949),

After a second lead precipitation, a barium precipitation is followed by an acetone precipitation of the barium-free product. The coenzyme is finally isolated as a water-insoluble barium salt at pH 8.

The coenzyme is the α isomer of glucose diphosphate. Its synthesis has been achieved by Posternak[110] in the following manner: β-1-Bromo-2,3,4-triacetyl-6-diphenylphosphonoglucose is reacted with silver diphenylphosphate. The product, upon removal of the phenyl and acetyl groups, yields the coenzyme.

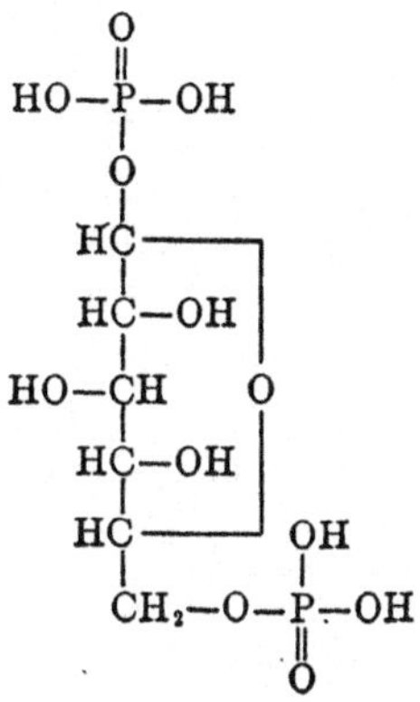

Glucose-1-6-diphosphate

FIG. 6

4. MECHANISM OF ACTION

Before knowledge of the coenzyme existed, the transformation of glucose-1-phosphate to the -6-phosphate was considered to be an intramolecular reaction, since neither free glucose nor free phosphate entered into exchange with the glucose phosphates as determined by isotopic experiments. However, after the discovery of the coenzyme, Leloir[108] postulated the following intermolecular reaction. The -1-phosphate group of the coenzymes is transferred, a transphosphorylation reaction, to position 6 of a glucose-1-phosphate molecule, forming a new coenzyme molecule. The original coenzyme molecule becomes, therefore, a glucose-6-phosphate molecule. This is shown in the following scheme:

glucose-1-Phosphate + glucose-1-6-diphosphate ⇌

glucose-1-6-Diphosphate + glucose-6-phosphate (33)

The coenzyme is thus being converted to substrate and is simultaneously being regenerated from substrate. This mechanism was proved by Suther-

[110] T. Posternak, *J. Biol. Chem.* **180**, 1269 (1949).

land and co-workers,[111] who incubated glucose-1-phosphate, labeled with C^{14} and P^{32} in the presence of unlabeled coenzyme. They found both of the isotopic markers in the coenzyme isolated at the end of the reaction. This was the first demonstration of such a mechanism in enzyme chemistry.

Sutherland and co-workers[112] also found the same mechanism to apply to the phosphoglyceromutase reaction mediated by 2,3-phosphoglyceric acid.

$$\text{3-Phospho}\textit{glycerate} + \text{2,3-diphosphoglycerate} \rightleftharpoons \text{2,3-Diphospho}\textit{glycerate} + \text{2-phosphoglycerate} \quad (34)$$

5. Function

These coenzymes serve the specific function of mediating a particular transphosphorylation at the low-energy level. It is interesting that Greenwald[113] had found in 1924 that the erythrocytes of certain mammals contained large amounts of 2,3-phosphoglyceric acid. The function of this substance is only now understood. Leloir[114] has also found that the phosphoglucomutase of yeast and muscle can convert mannose-1-phosphate to mannose-6-phosphate at a rate forty times slower than the glucose phosphate conversion. Some mannose-1-6-diphosphate was also detected. He has suggested that the 1,6-diphosphates of other hexoses and the 1,5-diphosphates of pentoses will probably prove to be coenzymes for their respective phosphomutases.

6. Enzymatic Estimation and Distribution of Glucose Diphosphate

An extract of dried brewer's yeast serves as the source of the phosphoglucomutase enzyme. Glucose-1-phosphate (non-reducing) in the presence of magnesium and the enzyme are incubated for 10 min. at 37°. The glucose-6-phosphate (reducing) formed is measured by a copper reduction method and compared against a curve prepared using known amounts of the diphosphate. In the organs of the rat, the following values in micromoles per gram were observed: muscle, 0.006 to 0.1; heart, 0.026; brain, 0.027; liver, 0.017; kidney, 0.008; intestine, 0.007; and blood, 0.013 to 0.130. The coenzyme is probably present wherever phosphoglucomutase is found. In the blood, only the red blood cells contain the coenzyme. Increases of about 100% in coenzyme content of muscle were found after injection of glucose, adrenalin, or insulin.

[111] E. W. Sutherland, M. Cohn, T. Posternak, and C. F. Cori, *J. Biol. Chem.* **180,** 1285 (1949).

[112] E. W. Sutherland, T. Posternak, and C. F. Cori, *J. Biol. Chem.* **179,** 501 (1949).

[113] I. Greenwald, *J. Biol. Chem.* **63,** 339 (1925).

[114] C. F. Leloir, *in* McElroy and Glass, Phosphorus Metabolism, The Johns Hopkins Press, Baltimore, 1951, Vol. I, p. 67.

7. Metabolism of Glucose Diphosphate

The biochemical formation of the coenzyme has been found to occur by two possible mechanisms, both demonstrated by the Argentina group. Muscle and yeast contain an enzyme,[115] a phosphoglucokinase, which mediates the following reaction:

ATP + glucose-1-phosphate → Glucose-1-6-phosphate + ADP (35)

Escherichia coli contains an enzyme[116] which carries out a transphosphorylation between 2 moles of glucose-1-phosphate to yield glucose diphosphate.

2 Glucose-1-phosphate ⇌ Glucose-1-6-diphosphate + glucose (36)

VI. Thiamine Pyrophosphate

1. Introduction

Thiamine pyrophosphate (TPP) is the active form of thiamine in the cells of all organisms and, as such, is important as a coenzyme in the decarboxylation of α-keto acids, as well as in the biosynthesis of certain acyloins. In view of the recent findings of Reed and De Busk (see section on pyruvate oxidation factor) TPP may form a complex with α-lipoic acid, which may be the coenzyme for the oxidative decarboxylation of α-keto acids.

2. Historical

Neuberg and Karczag[117] in 1911 were the first to demonstrate an enzymatic cleavage of a carbon-to-carbon bond when they found that an enzyme in yeast, carboxylase, mediated the decarboxylation of pyruvate to acetaldehyde and CO_2. In 1932 Auhagen[118] found that alkaline-washed yeast carboxylase preparations were inactive unless the washings were added back to the inactive preparations. Magnesium and a heat-stable organic factor called cocarboxylase were shown to be the active factors in the washings. The cofactor was found to be present in all tissues. In the same year Simola[119] provided evidence for the relationship to thiamine. He demonstrated the presence of thiamine in cocarboxylase preparations and found that thiamine-deficient animals contained less cocarboxylase than did tissues of normal animals.

Although thiamine was isolated in pure form in 1926,[8] the structure and

[115] A. Paladini, R. Caputto, L. F. Leloir, R. E. Trucco, and C. E. Cardini, *Arch. Biochem.* **23,** 55 (1949).

[116] L. F. Leloir, R. E. Trucco, C. E. Cardini, A. Paladini, and R. Caputto, *Arch. Biochem.* **24,** 65 (1949).

[117] C. Neuberg and L. Karczag, *Biochem. Z.* **36,** 60, 68 (1911).

[118] E. Auhagen, *Z. physiol. Chem.* **204,** 149 (1932); **209,** 20 (1932); *Biochem. Z.* **258,** 330 (1933).

[119] P. E. Simola, *Biochem. Z.* **254,** 229 (1932).

synthesis of the vitamin were made known in 1936. In the following year Lohmann and Schuster[5] isolated the crystalline coenzyme and showed it to be the pyrophosphate ester of thiamine. The coenzyme has been variously referred to as cocarboxylase, diphosphothiamine, aneurindephosphate, and thiamine pyrophosphate (TPP). In this chapter the terms cocarboxylase and thiamine pyrophosphate will be used.

Although Barron[120] had shown as early as 1932 that gonococci could carry out an oxidative breakdown of pyruvate to acetate and CO_2, it was the work of Peters and his group[121] on thiamine-deficient rats and pigeons which indicated that thiamine was necessary for the oxidative breakdown of pyruvate which occurred in animal tissues. Shortly after it was shown by many investigators that pyruvate breakdown in animal tissues, as well as in many bacteria, can occur only in the presence of suitable hydrogen acceptors to yield acetate and CO_2 and that free acetaldehyde was not an intermediate.

Lipmann's[4] discovery in 1939 that acetyl phosphate results from the oxidative decarboxylation of pyruvate foretold the fundamental role of TPP in initiating reactions leading to the production of an activated 2-carbon fragment.

3. Structure and Chemistry

The structure of TPP is generally written as follows:

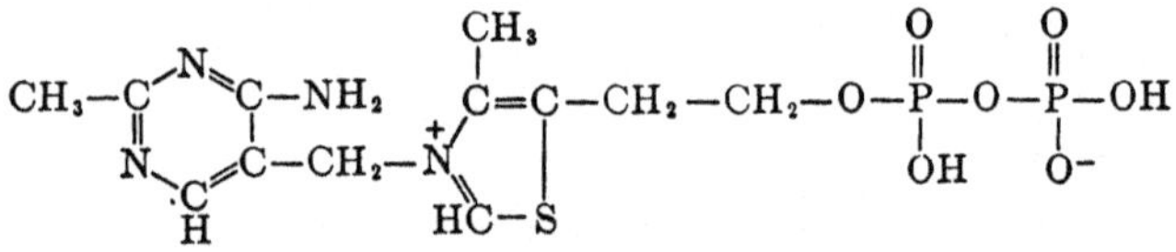

Thiamine pyrophosphate (TPP)

Fig. 7

Thiamine consists of a pyrimidine moiety linked by means of a methylene bridge to a thiazole fragment. The pyrophosphate is esterified on the hydroxy ethyl group of the thiazole moiety. It should be noted that only one of the phosphate groups is of a high-energy type. The nitrogen of the thiazole ring is a quaternary nitrogen.

TPP is inactivated by treatment with 1 *N* hydrochloric acid at 100° for 15 min. owing to the hydrolysis of the pyrophosphate linkage and yields orthophosphate and thiamine monophosphate. Treatment of TPP with alkali splits off pyrophosphate and gives free thiamine. TPP, like thiamine,

[120] E. S. G. Barron and P. C. Miller, *J. Biol. Chem.* **97**, 691 (1932).
[121] R. A. Peters, *Lancet* **230**, 1161 (1936).

is cleaved by sulfite to form the sulfuric acid derivative of the pyrimidine moiety and the thiazole pyrophosphate moiety.[122]

Upon addition of 2 moles of alkali to thiamine hydrochloride the quaternary thiazole ring opens to form an acidic sulfhydryl group. This isomeric sulfhydryl form, as well as the disulfide form of both thiamine[123] and TPP,[124] have been prepared. The formula of the sulfhydryl form of TPP is the following:

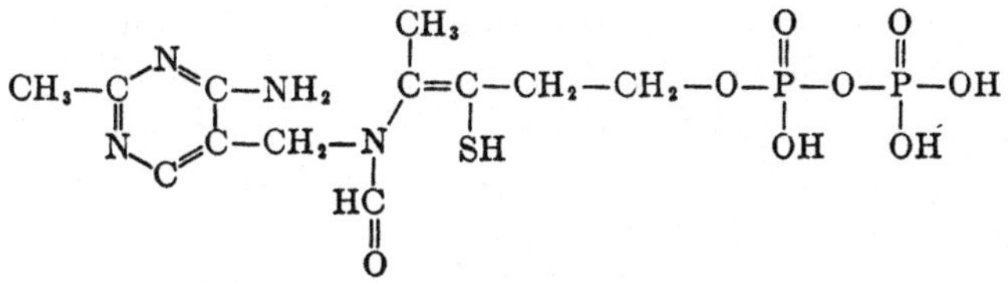

"Thiol" form of thiamine pyrophosphate

FIG. 8

Mild oxidation with hydrogen peroxide at pH 7.5 forms the disulfide, while the disulfide can be reduced back to the sulfhydryl form by cysteine or reduced glutathione.

The synthesis of TPP from thiamine has been carried out by several pyrophosphorylation procedures. The usual procedure involves the reaction of thiamine and polyphosphoric acid obtained by heating orthophosphoric acid. Velluz[125] pointed out that the primary product in this procedure was thiamine triphosphoric acid. Viscontini, Bonetti, and Karrer[126] worked out the details for obtaining thiamine tri-, pyro- and monophosphates in good yield.

The isolation of the pure crystalline TPP from brewer's yeast extract, as carried out by Lohmann and Schuster,[5] involved various precipitations as barium, picric acid, and phosphotungstic acid salts, fractional precipitation with solvents, and an adsorption with fuller's earth. The method has no practical application.

4. Mechanism of Action

The mechanism of action of TPP is still unknown. However, an analysis of the reactions carried out by TPP enzymes and of the structural require-

[122] J. Weijlard and H. Tauber, *J. Am. Chem. Soc.* **60**, 2263 (1938).

[123] O. Zima and R. R. Williams, *Ber.* **730**, 941 (1940); O. Zima, K. Ritsert, and T. Moll. *Z. physiol. Chem.* **267**, 210 (1941).

[124] P. Karrer and M. Viscontini, *Helv. Chim. Acta* **29**, 711, 1981 (1946).

[125] L. Velluz, G. Amiard, and J. Bartos, *Bull. soc. chim.* **15**, 871 (1948).

[126] M. Viscontini, G. Bonnetti, and P. Karrer, *Helv. Chim. Acta* **32**, 1478 (1949).

ments for coenzyme activity leads to some conclusions as to the possible mode of action.

In the yeast carboxylase system thiamine and thiamine monophosphate are inactive. Thiamine pyrophosphate reactivates the apoenzyme maximally. It has recently been shown that thiamine triphosphate at a concentration four to five times that of TPP can saturate the apoenzyme giving an activity equivalent to 80% of the TPP reconstituted system.[127] Since thiamine and thiamine monophosphate are inactive, it would appear that a pyrophosphate linkage is necessary for activity. This linkage is assumed to be the site of attachment to the apoenzyme.

The metal requirement of the carboxylase system is not so specific. Although magnesium and manganese are the best activators, other divalent cations in higher concentrations may replace the former. The metal requirement is not clearly understood but is assumed to play a role in binding the coenzyme to the apoenzyme. Highly purified preparations of yeast carboxylase[128] contain 1 gram-atom of magnesium to 1 gram-mole of TPP to 75,000 grams of protein.

The essential nature of the amino group of the pyrimidine moiety of the molecule is dramatically demonstrated by the fact that desaminothiamine, oxythiamine, is a potent antimetabolite *in vivo*.[129] In the yeast carboxylase system both oxythiamine pyrophosphate[130] and oxythiamine triphosphate[131] are competitive inhibitors of TPP. Neither oxythiamine nor its phosphate are active in this respect. The function of the amino group has been the subject of much speculation since 1933 when Langenbeck,[132] three years before the structure of thiamine was known, predicted that the vitamin would contain an amino group on the basis of work indicating that certain primary amines catalyzed the decarboxylation of α-keto acids in non-aqueous media. A Schiff base intermediate (see section on pyridoxal phosphate) was postulated. Stern and Melnick,[133] however, showed that thiamine does not catalyze the decarboxylation of pyruvic acid in non-aqueous media and that the amino group does not behave as a primary amine as evidenced by its very sluggish behavior towards nitrous acid and the complete absence of reactivity towards ketene, a good acetylating agent. Recent work on the infrared spectrum of amino pyrimidines[134] provides evidence that these amino groups show the spectral properties of amide groups.

[127] L. Velluz, G. Amiard, and J. Bartos, *J. Biol. Chem.* **180,** 1137 (1949).

[128] F. Kubowitz and W. Luttgens, *Biochem. Z.* **307,** 170 (1941).

[129] L. R. Cerecedo, M. Soodak, and A. J. Eusebi, *J. Biol. Chem.* **189,** 293 (1951).

[130] A. J. Eusebi and L. R. Cerecedo, *Abstracts* 116th Meeting, American Chemical Society, September, 1949, p. 61c.

[131] L. Velluz and M. Herbain, *J. Biol. Chem.* **190,** 241 (1951).

[132] W. Langenbeck, *Ergeb. Enzymforsch.* **2,** 314 (1933).

[133] K. G. Stern and J. L. Melnick, *J. Biol. Chem.* **131,** 597 (1939).

[134] L. N. Short and H. W. Thompson, *J. Chem. Soc.* **1952,** 168.

The function of the amino group in TPP is, therefore, still an open question. However Reed and De Busk[134a, b] suggest that α-lipoic acid is connected to thiamine through a peptidic linkage between the carboxyl of α-lipoic acid and the amino group of thiamine to form lipothiamide. This finding would account for the essential nature of the amino group as far as oxidative decarboxylation is concerned.

Recent evidence suggests that the thiazole ring and, in particular, the sulfur atom of that ring may be the center involved in the primary attack on pyruvate. Thiamine disulfide is as active as thiamine *in vivo*,[123] and Karrer and Viscontini[124] have found that the thiol form of TPP is as active as TPP in the yeast carboxylase assay. Evidence for the open-ring form of thiamine in nature also is at hand. Approximately 50% of the thiamine in milk has long been known to be present in a form which is released only upon treatment with a proper proteolytic enzyme.[135] In the light of present-day knowledge this bound form is suggestive of a thiamine-protein-S-S complex or of lipothiamide. Bonvicino and Hennessey[136] have prepared a complex of the former type and found it to be biologically active to the extent of 90%. Myrbäck and his co-workers[137] have evidence to the effect that well-aerated baker's yeast contains what is probably the disulfide form of TPP.

It is of interest to note that the thiol form of TPP contains the β-thiolethylamine configuration present in coenzyme A. The fact that glutathione, coenzyme A, and the pyruvate oxidation factor all are sulfur-containing compounds is highly suggestive of the possibility that TPP is also active in the thiol form. Cavallini[138] has recently shown that *in vitro* at pH 7.4, the oxidation of GSH by oxygen in the presence of traces of copper or cytochrome c may be coupled with a simultaneous oxidation of pyruvate to acetate and CO_2. The reaction probably depends on the intermediate formation of a GSH-pyruvate complex and is suggested as a possible model for the action of TPP.

The functional group of TPP involved in the attack on α-keto acids still, therefore, remains in doubt. The nature of the primary product formed upon decarboxylation of the keto acid also remains to be unequivocally elucidated. Rather good evidence exists, however, which indicates that pyruvate is attacked by TPP enzymes to yield CO_2 and a 2-carbon fragment-TPP complex in which the C_2 fragment is at the aldehyde level of oxidation. This primary product will here be called the "acetaldehyde-

[134a] L. J. Reed and B. G. De Busk, *J. Biol. Chem.*, **199,** 873 (1952).

[134b] L. J. Reed and B. G. De Busk, *J. Biol. Chem.* **199,** 881 (1952).

[135] N. Halliday and H. J. Deuel, Jr., *J. Biol. Chem.* **140,** 555 (1941).

[136] G. E. Bonvicino and D. J. Hennessy, *Abstracts* 121st Meeting, American Chemical Society, March, 1952, p. 4c.

[137] K. Myrbäck, I. Wallin, and I. Magnell, *Svensk. Kem. Tid.* **57,** 124 (1945).

[138] D. Cavallini, *Biochem. J.* **49,** 1 (1951).

TPP" complex. It will be seen that by assuming such a complex the TPP catalyzed reactions fall into a logical pattern.

Several reviews[139, 140] are available which provide a detailed account of the reactions mediated by TPP. The discussion to follow will center about a scheme for understanding how these reactions may give rise to a variety of products. Weil-Malherbe[141] and Martius,[142] in the early forties, presented hypotheses to account for these products. More recent knowledge has provided the basis for the modification of these earlier schemes.

The action of TPP-containing enzymes on pyruvate falls into two groups: (1) decarboxylations of the non-oxidative type which lead to products at the aldehyde level of oxidation, namely, acetaldehyde and the acyloins, and (2) decarboxylations of the oxidative type which lead to substances at the acid level of oxidation, such as acetic acid, acetyl phosphate, and acetyl-CoA.

The postulated common intermediate involved in the formation of these products is the acetaldehyde-TPP complex which arises from the non-oxidative decarboxylation of pyruvate.

$$\text{Pyruvate} + \text{TPP-enz.} \rightarrow \text{Acetaldehyde-TPP-enz.} + CO_2 \qquad (37)$$

The acetaldehyde-TPP may then (1) split to yield acetaldehyde, (2) react with itself or a mole of free acetaldehyde to give the acyloin, acetylmethylcarbinol (AMC), or (3) pass on the activated acetaldehyde molecule to a system which can oxidize it to a substance at the acid level of oxidation. In each case the TPP is made available for another catalytic cycle. The acetaldehyde-TPP intermediate has also been shown to be derived, in special instances, from free acetaldehyde[143] and diacetyl.[144] In some bacteria, the intermediate, derived from 1 mole of pyruvate, donates the acetaldehyde to a second mole of pyruvate to form acetolactic acid.[145] The latter upon decarboxylation yields AMC. These reactions are pre-

[139] F. Lipmann, *Cold Spring Harbor Symposia Quant. Biol.* **7,** 248 (1939); S. Ochoa, *in* Biological Action of Vitamins, University of Chicago Press, 1942, p. 17.

[140] S. Ochoa, *Physiol. Revs.* **31,** 56 (1951); E. Stotz, *Advances in Enzymol.* **5,** 145 (1945); B. C. P. Jansen, *Vitamins and Hormones* **7,** 83 (1949); B. Vennesland, *in* Sumner and Myrbäck, The Enzymes, Academic Press, New York, 1951, Vol. II, Part 1, p. 49.

[141] H. Weil-Malherbe, *Nature* **145,** 106 (1940).

[142] C. Martius, *Z. physiol. Chem.* **279,** 96 (1943).

[143] T. P. Singer, *Biochim. et Biophys. Acta* **8,** 108 (1952).

[144] R. S. Schweet, M. Fuld, K. Cheslock, and M. H. Paul, *in* McElroy and Glass, Phosphorus Metabolism, The Johns Hopkins Press, Baltimore, 1951, Vol. I, p. 246.

[145] D. Watt and L. O. Krampitz. *Federation Proc.* **6,** 301 (1947); L. O. Krampitz, *Arch. Biochem.* **17,** 81 (1948).

sented in the following scheme:

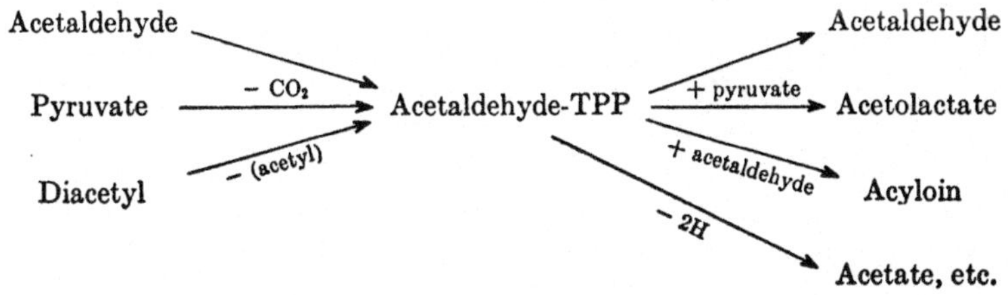

FIG. 9

a. Products of the Aldehyde Level. Carboxylase carries out the simple, non-oxidative decarboxylation of pyruvate to acetaldehyde and CO_2. The enzyme is present in yeast, molds, some bacteria, and some plant tissues, but it does not occur in animal tissues.

The formation of acetoin or acetyl methyl carbinol (AMC) is mediated by some TPP-containing enzymes of the non-oxidative as well as of the oxidative type. It is a product at the aldehyde level of oxidation and may be looked upon as a condensation of 2 moles of acetaldehyde, i.e., an acyloin condensation.

$$CH_3—CHO + OCH—CH_3 \rightarrow CH_3—CO—C(HOH)—CH_3 \qquad (38)$$

Three mechanisms, at least, exist for the formation of AMC enzymatically. These are: (1) from 2 moles of acetaldehyde, (2) from 1 mole of pyruvate and 1 mole of acetaldehyde, and (3) from 2 moles of pyruvate.

1. The highly purified wheat germ carboxylase of Singer[143] carries out the simple decarboxylation of pyruvate to acetaldehyde and CO_2. It can also synthesize AMC from pyruvate and acetaldehyde or form 2 moles of acetaldehyde, at a slower rate. The fact that this enzyme can form AMC from acetaldehyde alone indicates that this TPP-enzyme complex can activate acetaldehyde directly. This strongly suggests the formation of an acetaldehyde-TPP intermediate directly from acetaldehyde, which has the same capacity as that derived from the decarboxylation of pyruvate, to condense with a free acetaldehyde molecule to form AMC.

2. AMC is usually formed from 1 mole of pyruvate and 1 mole of acetaldehyde. Isotope experiments with labeled pyruvate indicate that the acetyl portion of the AMC is always derived from the pyruvate. In other words, the active acetaldehyde is derived from the pyruvate. The simple carboxylases of yeast and wheat germ carry out this reaction.[143]

The pyruvic oxidases of animal tissues also form AMC. Jagannathan and Schweet[144, 146] have described a homogeneous pyruvic oxidase prepara-

[146] V. Jagannathen and R. S. Schwett, *J. Biol. Chem.* **196**, 551 (1952).

tion from pigeon breast muscle which can carry out the following reactions: (*a*) the oxidative decarboxylation of pyruvate to acetic acid, (*b*) the formation of AMC from pyruvate which can be increased by adding acetaldehyde, and (*c*) the formation of AMC and acetate from diacetyl ($CH_3COCOCH_3$). Like all pyruvic oxidase systems, free acetaldehyde is not an intermediate. The ratio of acetate to AMC formed by this preparation in the presence of pyruvate and acetaldehyde is changed by various procedures. If the concentration of acetaldehyde is raised, then more AMC and less acetate results. Conversely, by increasing the oxidizing potential of the system, more acetate and less AMC is formed. This is strong evidence for the hypothesis that an acetaldehyde-TPP intermediate is the primary product formed from pyruvate upon decarboxylation. The intermediate is then trapped at the aldehyde level in the formation of AMC by reacting with another like itself or with a mole of added acetaldehyde or is oxidized to acetate. Similar evidence is derived from the diacetyl reaction, or the diacetyl mutase system. Diacetyl behaves as an alternate substrate for pyruvic oxidase. Anaerobically it forms one C_2 fragment at the aldehyde level, the acetaldehyde-TPP intermediate, and one at the acetate level. The acetaldehyde-TPP intermediate acts exactly as that derived from pyruvate.

3. In the case of certain bacteria 2 moles of pyruvate are necessary for the formation of a mole of AMC. Krampitz[145, 147] and his co-workers have elucidated the mechanism involved. One enzyme, which requires TPP, forms acetolactate from 2 moles of pyruvate, the acetaldehyde-TPP intermediate from 1 mole of pyruvate condensing with the second mole of pyruvate. The second enzyme decarboxylates the acetolactate to form AMC.

$$CH_3COCOOH + TPP\text{—}E_1 \rightarrow CH_3CHO\text{—}TPP\text{—}E_1 + CO_2 \qquad (39)$$

$$CH_3CHO\text{—}TPP\text{—}E_1 + CH_3COCOOH \rightarrow CH_3CO\text{—}\underset{\displaystyle COOH}{\overset{\displaystyle HO}{\overset{|}{\underset{|}{C}}}}\text{—}CH_3 + TPP\text{—}E_1 \qquad (40)$$

$$CH_3CO\text{—}\underset{\displaystyle COOH}{\overset{\displaystyle HO}{\overset{|}{\underset{|}{C}}}}\text{—}CH_3 \xrightarrow{E_2} CO_2 + CH_3COCHOH\text{—}CH_3 \qquad (41)$$

Except for a few bacteria in which AMC is the end product of carbohydrate metabolism, AMC formation seems to be an insignificant reaction in most organisms. It has been discussed in detail because the studies on the

[147] E. Juni, *J. Biol. Chem.* **195**, 715, 727 (1952).

mechanism of its formation have provided the best clues towards understanding the nature of the primary product formed upon decarboxylation of pyruvate.

b. Products at the Acid Level. The oxidative decarboxylation of pyruvic acid occurs in all animal tissues and in many bacteria. It is a reaction of primary physiological significance, since it generally leads to the formation of an activated acetyl compound, acetyl-CoA, which feeds into the tricarboxylic acid cycle to provide the main source of energy to the organism. The pyruvic oxidases which have so far been studied seem to perform the oxidation and decarboxylation of pyruvate simultaneously. It has not yet been possible to break this reaction into individual steps.

The work of Lipmann[4] on the pyruvic oxidase system of *Lactobacillus delbrueckii* serves as the basis of present-day knowledge of pyruvate oxidation. The partially purified enzyme system for the oxidative decarboxylation of pyruvate required one or more enzymes, TPP, FAD, a divalent cation, and either inorganic phosphate or arsenate. No decarboxylation of pyruvate occurred unless both phosphate and FAD were present in addition to TPP. This indicated that the decarboxylation was intimately associated with the dehydrogenation of pyruvate, FAD accepting the hydrogens. An analysis of the phosphate requirement led to the isolation of acetyl phosphate. This significant contribution to the problem of "active acetate" has culminated recently in the work on acetyl-CoA. The absolute requirement for phosphate or arsenate by this pyruvic oxidase indicates that it is necessary to remove the acetyl portion from the coenzyme with phosphate in order to free the coenzyme to act catalytically. This study, then, clearly exemplifies the problems associated with the oxidative decarboxylation reaction in general. These are: (1) the apparent necessity for the simultaneous oxidation and decarboxylation giving rise to an active acetyl, and (2) the removal of the acetyl group to allow the system to act catalytically. The removal of the acetyl group may be accomplished hydrolytically yielding acetate, phosphorylitically yielding acetyl phosphate, or by transfer to CoA yielding acetyl-CoA.

The other pyruvic acid oxidases which have recently been described are additional examples which illustrate these same principles. They differ somewhat from each other either in the method of handling the hydrogens or in the ultimate fate of the 2-carbon residue. In *Clostridium butylicum*, the end products are acetyl phosphate, CO_2, and H_2. In these reactions, the hydrogen removed from pyruvate is not used to reduce oxygen to the state of water but instead is liberated as hydrogen gas. Similarly, in the pyruvate dismutation reaction, extensively studied by Korkes and his co-workers,[147a] acetyl-CoA or acetyl phosphate represents the fate of the C_2 fragment,

[147a] S. Korkes, A. del Campillo, I. C. Gunsalus, and S. Ochoa, *J. Biol. Chem.* **193**, 721 (1952).

while the hydrogens are used to reduce an additional molecule of pyruvate to lactate.

The pyruvic oxidase of heart muscle recently studied by Schweet *et al.*[144] is still another special case. In the absence of added CoA, the acetyl moiety is hydrolytically cleaved to acetate; TPP is the only required cofactor for this reaction; in the presence of CoA, the end product is acetyl-CoA. The disposition of the hydrogens in this reaction is not yet clearly indicated. It is reported that the enzyme is relatively free of pyridine nucleotides, as well as flavin derivatives. There is some evidence that H_2O_2 may be produced as a product of the reaction with oxygen. The authors reach the conclusions that their enzyme contains an auto-oxidizable electron carrier.

In general, then, the oxidative decarboxylation of pyruvate may be visualized as beginning with a decarboxylation of pyruvate leading to an acetaldehyde-TPP complex in a manner similar to the non-oxidative reactions discussed in a previous section. In some cases this acetaldehyde-TPP complex may react without further change to form acyloins, as in Schweet's experiments. In most instances the oxidation may be presumed to occur on the acetaldehyde-TPP complex and may lead to an "acyl-TPP" complex. The latter point is not yet clear, since there is no evidence of the existence of an acyl-TPP complex. The fact that the pyruvic oxidase of Schweet *et al.* leads, in the presence of TPP, to the formation of acetate suggests that the acetaldehyde-TPP complex may be oxidized to the acyl level and subsequently cleaved hydrolytically. The pyruvate oxidation factor (POF) almost certainly plays a role in this oxidation, but its exact function is not known (see Section IX, 3). Recently Reed and DeBusk[148] have reported the isolation and preparation of lipothiamide, the amide of thiamine and α-lipoic acid (POF). This compound appears to be part of the coenzyme for oxidative decarboxylation of α-keto acids. From what has already been discussed concerning TPP, it seems likely that lipothiamide may be the coenzyme which is capable of performing the dual function of a simultaneous oxidation and decarboxylation. This, however, remains yet to be demonstrated.

5. Methods of Assay

a. Chemical. The oxidation of thiamine by alkaline ferricyanide to the strongly fluorescent thiochrome was first used by Jansen[149] as the basis for a fluorometric method for the determination of thiamine. The thiamine phosphates are oxidized under the same conditions to the corresponding thiochrome phosphates. Thiochrome itself is readily extracted by isobutanol; the phosphorylated derivatives are not. Phosphatases convert the bound forms of thiamine to free thiamine, and it is therefore possible

[148] L. J. Reed and B. G. DeBusk, *J. Am. Chem. Soc.* **74,** 3457 (1952).
[149] B. C. P. Jansen, *Rec. trav. chim.* **55,** 1046 (1936).

to determine in this manner the free thiamine and the sum of thiamine monophosphate and TPP. This method is not satisfactory for the estimation of TPP as such, since thiamine monophosphate is included along with TPP.

b. Enzymatic. The enzymatic method is the only specific method for TPP. Alkaline-washed yeast carboxylase serves as the apoenzyme,[5] and the decarboxylation of pyruvate upon addition of the TPP solution is measured manometrically for CO_2 evolution or by determining the acetaldehyde formed and then compared with a curve prepared with known concentrations of TPP. Many factors can complicate this assay procedure.

6. The TPP Content of Tissues

Thiamine is found in nature in several forms—as free thiamine, the monophosphate, TPP, and probably as thiamine bound by an —S—S— linkage to protein, and as the disulfide of TPP. The recent report of Reed and DeBusk[148] indicates still another naturally occurring bound form of thiamine, lipothiamine, the amide of thiamine and α-lipoic acid. Many unsolved problems still remain. In the animal, although TPP constitutes 90% of the total thiamine content in some tissues, in others only 80% occurs as this form; muscle may contain more than 50% as free thiamine. The total thiamine content of tissues varies from 10 μg. per gram wet weight for liver to less than 1 μg. for muscle and brain.

7. The Metabolism of TPP

Although some early reports presented evidence for the formation of TPP from thiamine and inorganic phosphate apparently by the action of non-specific phosphatases, it is felt at present that ATP usually serves as the source of the phosphate groups of TPP. Many tissues have been shown to convert thiamine to the pyrophosphate under aerobic conditions. The early reports in this field have been summarized by Ochoa.[139] Weil-Malherbe[150] was the first to study the phosphorylation in cell-free extracts (1942). An extract of dried brewer's yeast was shown to form TPP from thiamine and thiamine monophosphate in the presence of ATP. The free thiamine yielded TPP at a more rapid rate than the monophosphate, and this was interpreted as indicating that the monophosphate must first be dephosphorylated to thiamine. Other workers have demonstrated thiaminokinase in extracts of dog liver[151] and rat liver.[152] Recently Steyn-Parvé[153] made a detailed study of a partially purified thiaminokinase system of yeast and concluded that the preparation can phosphorylate both free

[150] H. Weil-Malherbe, *Biochem. J.* **33,** 1997 (1939).
[151] Nguyen-Van-Thoai and L. Chevillard, *Bull. soc. chim. biol.* **31,** 204 (1949).
[152] H. Nielsen and F. Leuthardt, *Helv. Physiol. et Pharmacol. Acta* **8,** 32c (1950).
[153] E. P. Steyn-Parvé, *Biochim. et Biophys. Acta* **8,** 310 (1952).

thiamine and the monophosphate directly. The exact mechanisms of these reactions are not yet understood, but it has been shown that ATP, magnesium, and small concentrations of orthophosphate are required by this system.

VII. Coenzyme A

1. Introduction

Coenzyme A (CoA) has recently emerged as an important coenzyme in the transfer of acyl groups for a variety of biosynthetic reactions leading to the formation of carbon-to-carbon, carbon-to-nitrogen, and carbon-to-phosphorus bonds.

The presence of pantothenic acid in this coenzyme[154] and the fact that most, if not all, cellular bound pantothenate is in the form of the coenzyme explain the metabolic function of this vitamin.

Coenzyme A was discovered by Lipmann[155] in 1945 as necessary for the acetylation of sulfanilamide by pigeon liver extracts. Shortly thereafter[156] this same coenzyme was identified as the activator of choline acetylation earlier observed by Nachmansohn and Berman,[157] as well as Feldberg and Mann.[158] Subsequently, work from Lipmann's laboratory,[159] as well as from other laboratories, extended the role of CoA to a large variety of transacetylation reactions, i.e., acetylation of aromatic amines,[155, 160] synthesis of acetylcholine,[157] of citrate,[161, 162] of acetoacetate,[163] of fatty acids,[164] of sterols,[165] and of phospholipids.[166]

2. Isolation

The first isolation of CoA was carried out by using hog liver as a source.[167] After a series of heavy metal fractionation steps, material about 30% pure

[154] F. Lipmann, N. O. Kaplan, G. D. Novelli, L. C. Tuttle, and B. M. Guirard, *J. Biol. Chem.* **167**, 869 (1947).
[155] F. Lipmann, *J. Biol. Chem.* **160**, 173 (1946).
[156] F. Lipmann and N. O. Kaplan, *J. Biol. Chem.* **162**, 743 (1946).
[157] D. Nachmansohn and M. Berman, *J. Biol. Chem.* **165**, 551 (1946).
[158] W. Feldberg and T. Mann, *J. Physiol.* (*London*) **104**, 411 (1946).
[159] F. Lipmann, *in* Sebrell, Jr., and Harris, The Vitamins, Academic Press, New York, Vol. I, in press.
[160] R. C. Millican, S. M. Rosenthal, and H. Tabor, *J. Pharmacol. and Exptl. Therap.* **97**, 4 (1949).
[161] J. R. Stern and S. Ochoa, *J. Biol. Chem.* **179**, 491 (1949).
[162] G. D. Novelli and F. Lipmann, *J. Biol. Chem.* **182**, 213 (1950).
[163] M. Soodak and F. Lipmann, *J. Biol. Chem.* **175**, 999 (1948).
[164] H. A. Barker, *in* McElroy and Glass, Phosphorus Metabolism, The Johns Hopkins Press, Baltimore, 1951, Vol. I, p. 204.
[165] H. Klein, *Federation Proc.* **10**, 209 (1951).
[166] A. Kornberg and W. E. Pricer, Jr., *J. Am. Chem. Soc.* **74**, 1616 (1952).
[167] F. Lipmann and N. O. Kaplan, *Federation Proc.* **5**, 145 (1946).

was obtained.[154] Recently a microbiological fermentation has been found to be a good source of the coenzyme.[168] By a series of steps involving adsorption and elution from charcoal and chromatography on a charcoal column, the preparation of material estimated to be 50% pure has been reported. Starting with this material and using a Zn-HCl reduction step, a precipitation by Hg and passage over a resin material of about 95% purity has been obtained.[169] Green and his co-workers[170] have devised a similar method and have successfully applied this to the preparation of CoA from yeast. This group uses a charcoal adsorption and elution step, followed by reduction with glutathione and precipitation as the copper salt. After decomposition with H_2S, the glutathione is removed on a resin which gives them a product about 80% pure.

More recently, Stadtman and Kornberg[171] have made use of the low cross-linked resin Dowex-1, introduced by Cohn for the separation of nucelotides from nucleic acid, to produce high-potency CoA in a single step.

3. Chemical Composition and Structure

The complete structure of CoA is now established. The coenzyme is a pantothenic acid derivative in which the vitamin is doubly bound and is liberated by the combined action of two enzymes, an intestinal phosphatase and an unidentified peptidase from avian liver.[154, 169, 172] The latter enzyme has recently been found in fresh extracts of hog kidney.[173]

The best preparations of CoA give analyses which indicate that there is present, for each mole of pantothenate, 1 mole each of adenine, ribose, and β-mercaptoethylamine, and 3 moles of phosphate.[169] When these elements are combined with the elimination of 5 moles of water, a calculated molecular weight of 767 is obtained, which is in good agreement with the observed molecular weight obtained by the Northrop Diffusion Method of 800 ± 50.[174]

The structure of CoA is depicted in Fig. 10. The evidence for this structure of CoA has recently been reviewed[175, 176] and is too lengthy to be con-

[168] W. H. DeVries, W. M. Govier, J. S. Evans, J. D. Gregory, G. D. Novelli, M. Soodak, and F. Lipmann, *J. Am. Chem. Soc.* **72,** 4838 (1950).

[169] J. D. Gregory, G. D. Novelli, and F. Lipmann, *J. Am. Chem. Soc.* **74,** 854 (1952).

[170] H. Beinert, R. W., von Korff, D. E. Green, D. A. Buyske, R. E. Handschumacher, H. Higgens, and F. M. Strong, *J. Am. Chem. Soc.* **74,**

[171] E. R. Stadtman and A. Kornberg, unpublished method.

[172] G. D. Novelli, N. O. Kaplan, and F. Lipmann, *J. Biol. Chem.* **177,** 97 (1949).

[173] L. Levintow, unpublished observations.

[174] G. D. Novelli, R. M. Flynn, and F. Lipmann, *J. Biol. Chem.* **177,** 493 (1949).

[175] G. D. Novelli, *in* McElroy and Glass, Phosphorus Metabolism, The Johns Hopkins Press, Baltimore, 1951, Vol. I, p. 414.

[176] G. D. Novelli, Oak Ridge Symposium on Microbial Metabolism, April, 1952, *Suppl. to J. Cellular Comp. Physiol.* **41:** Supp. 1: 67 (1953).

sidered here in detail. There is one point regarding the structure on which there is still some doubt, namely, the position of the phosphomonoester on the adenosine moiety.

Kaplan and Shuster[177] and Wang, Shuster, and Kaplan[178] have recently obtained evidence suggesting that the phosphomonoester in CoA is esterified at a different position from the monophosphoester phosphate group of TPN. Here again the question of whether the phosphate is attached to the 2′- or 3′-hydroxyl of ribose must await a decision as to whether adenylic acid "a" and "b" are the 2′- and 3′-esters, as is presently thought. The phosphate in CoA is esterified at the same position as it is in adenylic acid "b." Regarding the position of attachment of the pyrophosphate bridge, Baddiley has presented evidence[179] suggesting that the pyrophos-

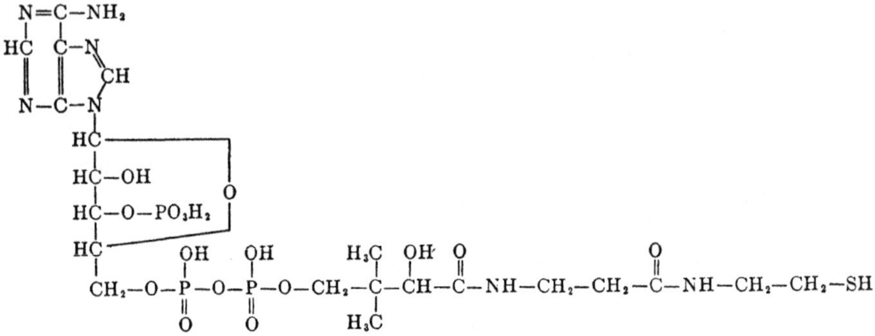

FIG. 10

phate bridge is attached to the 4′ hydroxyl of pentothenate. This has been confirmed by a chemical synthesis of 4′ phosphopantethine and the demonstration of its conversion to CA by enzymes present in pigeon liver extracts.[180] Furthermore, 4′ phosphopantethine is active in stimulating the growth of *Acetobacter suboxydans*[181] in a manner similar to the product of enzymatic hydrolysis of CoA by snake venom or potato nucleotide pyrophosphatase.[182, 183]

177 L. Shuster and N. O. Kaplan, *Federation Proc.* **11,** 286 (1952).
178 T. P. Wang, L. Shuster, and N. O. Kaplan, *J. Am. Chem. Soc.* **74,** 3204 (1952).
179 J. Baddiley and E. M. Thain, *J. Chem. Soc.* **1951,** 2253.
180 J. Baddiley, E. M. Thain, G. D. Novelli, and F. Lipmann, Nature **171:** 76 (1953).
181 L. Levintow and G. D. Novelli (unpublished).
182 G. D. Novelli, J. D. Gregory, R. M. Flynn, and F. J. Schmetz, Jr., *Federation Proc.* **10,** 229 (1951).
183 G. D. Novelli, N. O. Kaplan, and F. Lipmann, *Federation Proc.* **9,** 209 (1950).

4. Methods of Assay

CoA is usually measured by using the assay developed by Kaplan and Lipmann.[184] This method uses a crude bicarbonate extract of acetone-dried pigeon liver, which is converted to an apoenzyme by aging for 4 hr. at room temperature, during which time bound CoA is destroyed. Upon the addition of acetate, ATP, sulfanilamide (or other aromatic amine) and CoA, the acetylation of the aromatic amine is found to be dependent upon the amount of CoA added. This method measures from 0.25 to 2.0 units of CoA.

The original method has been subjected to several modifications. One modification utilizes a colored aromatic amine[185] and takes advantage of the change in color upon acetylation. The advantage of this method lies in the fact that the time and manipulations required for the development of the sulfanilamide color reaction of the original method are avoided.

It was found during the study of the original assay method that the crude piegon liver extract contained an enzyme capable of resynthesizing various split products of CoA[181] to the intact coenzyme in the presence of ATP. Thus the so-called "crude" assay responded not only to CoA but to some of its degradation products as well. By fractionating the pigeon liver extract, between 40 and 70% saturation with ammonium sulfate an enzyme extract is obtained which is still capable of acetylating sulfanilamide but is no longer able to synthesize the coenzyme from its split products. This procedure thus yields a more discriminating assay.

After the discovery that acetylphosphate could be activated in the presence of CoA by certain microbial extracts for various synthetic purposes,[162, 186] a new assay for CoA was developed which proved to be most specific since it no longer involved the use of ATP. The arsenolytic decomposition of acetyl phosphate in extracts of *Clostridium kluyveri*, a reaction discovered by Stadtman and Baker,[187] was shown to require CoA. A study of this reaction showed that this rate of decomposition of acetyl phosphate is directly proportional to the CoA concentration.[188] Since there is a convenient colorimetric method for acetyl phosphate,[189] the reaction is very easy to follow. The details of this reaction are described in a recent paper.[188]

[184] N. O. Kaplan and F. Lipmann, *J. Biol. Chem.* **174,** 37 (1948).

[185] R. E. Handschumacher, G. C. Mueller, and F. M. Strong, *J. Biol. Chem.* **189,** 335 (1951).

[186] E. R. Stadtman and H. A. Barker, *J. Biol. Chem.* **180,** 1117 (1949).

[187] E. R. Stadtman and H. A. Barker, *J. Biol. Chem.* **184,** 769 (1950).

[188] E. R. Stadtman, G. D. Novelli, and F. Lipmann, *J. Biol. Chem.* **191,** 365 (1951).

[189] F. Lipmann and L. C. Tuttle, *J. Biol. Chem.* **158,** 505 (1945).

Recently, Green[190] and his co-workers have described a new assay for CoA which is based on a different principle from the previous assays. It has been reported that CoA is required for the oxidation of α-ketoglutarate.[191] Since this reaction also requires the presence of DPN, a spectrophotometric method in which the reactants are so adjusted that the reduction of DPN is dependent upon CoA concentration is possible. This method is rapid and very sensitive.

Although several microorganisms have been described which will respond to CoA,[192, 193] none of these has been found to be specific for the coenzyme; i.e., they will also respond to various degradation products or to the free vitamin.

5. Mechanism of Action

Barker[164] has recently reviewed the metabolism of "active acetate" and has discussed the role of CoA in the various transformations of acetate. Only the more general developments regarding the mechanism of action of this coenzyme will be discussed.

As the study of CoA developed, it became apparent that the coenzyme was involved in reactions whereby acetate was activated by ATP and subsequently transferred to various acetyl acceptors. In pigeon liver extracts it was shown that acetate could be activated by ATP in the presence of CoA to acetylate sulfanilamide,[155] PABA,[194] histamine,[160] glucosamine,[193] to synthesize acetoacetic acid[163] and citrate.[161, 162] Acetyl phosphate, which has been demonstrated to be a product of pyruvate metabolism in several bacteria[4, 195, 196] and could theoretically be considered to be an intermediate in these reactions, was found to be unable to replace acetate and ATP in animal tissues. Eventually it was shown that there is present in certain bacteria an enzyme, phosphotransacetylase, which could convert acetyl phosphate to a reactive product which was thought to be acetyl-CoA.[188, 194] The isolation of acetyl-CoA from yeast extract by Lynen and Reichert[197] confirmed the idea that acetyl-CA is the reactive 2-carbon unit in these reactions. Stadtman has demonstrated that acetyl-CoA is indeed the product of the action of phosphotransacetylase.[198] Lipmann has recently

[190] D. E. Green, H. Beinert, D. Goldman, R. W. von Korff, and S. Mii, *Federation Proc.* **11,** 222 (1952).

[191] D. E. Green and H. Beinert, *in* McElroy and Glass, Phosphorus Metabolism, The Johns Hopkins Press, Baltimore, 1952, p. 330 (1951).

[192] E. Delwiche, private communication.

[193] T. C. Chou and M. Soodak, *J. Biol. Chem.* **196,** 105 (1952).

[194] T. C. Chou and F. Lipmann, *J. Biol. Chem.* **196,** 89 (1952).

[195] H. Koepsall and M. Johnson, *J. Biol. Chem.* **145,** 379 (1942).

[196] E. R. Stadtman and H. A. Barker, *J. Biol. Chem.* **180,** 1095 (1949).

[197] F. Lynen and E. Richert, *Angew Chem.* **63,** 47 (1951).

[198] E. R. Stadtman, *J. Biol. Chem.* **196,** 535 (1952).

discussed the metabolic function of CoA in some detail[159] and has summarized the formation of acetyl-CoA as well as its utilization in various metabolic systems. The following figure taken from his paper illustrates the mechanism of action of CoA as an acyl carrier.

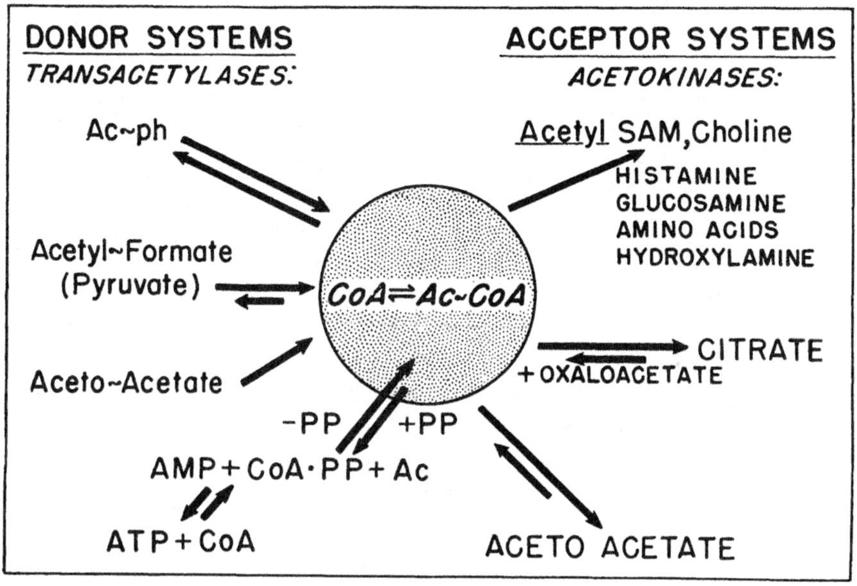

FIG. 11

6. Formation of Acetyl-CoA

After Lynen and Reichert[197] had isolated acetyl-CoA, they showed that the acetyl group is carried on CoA as a thioester of the following form.

$$R—NH \cdot CH_2 \cdot CH_2 \cdot S—\overset{\overset{\displaystyle O}{\|}}{C}—CH_3$$

Stadtman,[198] from equilibrium measurements of the phosphotransacetylase reaction, i.e.,

$$\text{Acetyl phosphate} + \text{CoA} \rightleftharpoons \text{Acetyl—CoA} + \text{phosphate} \qquad (42)$$

has shown that the acetyl thioester has a bond energy of approximately 10,000 cal. and is therefore an energy-rich compound. This explains the activation of acetate and its ready participation in a variety of condensation reactions.

Metabolically, acetyl-CoA may be formed in several ways. It can be derived in the dissimilation of pyruvate[147a, 199] by the oxidation of acetalde-

[199] H. Chantrenne and F. Lipmann, *J. Biol. Chem.* **187**, 757 (1950).

hyde,[200] by the oxidation of fatty acids,[164] and in the activation of acetate by ATP,[194] as well as from acetyl phosphate by phosphotransacetylase.[198]

a. Formation from Pyruvate. (1) Chantrenne and Lipmann[199] showed that the reversible exchange of formate with the carboxyl of pyruvate by extracts of *Escherichia coli* was CoA-dependent. They formulated the reaction as follows:

$$CH_3\overset{\overset{O}{\|}}{C}—COOH + CoA \rightleftharpoons CH_3\overset{\overset{O}{\|}}{C}—CoA + HCOOH \qquad (43)$$

In the virtual absence of phosphate the acetyl group of acetyl-CoA could be transferred to sulfanilamide.

(2) Korkes and co-workers[147a, 201] have demonstrated the requirement for CoA of the dismutation of pyruvate by extracts of *E. coli* and *Streptococcus faecalis*. In addition to several enzymes and CoA, the system requires DPN, cocarboxylase Mg^{++} or Mn^{++}, and phosphate. Two molecules of pyruvate are converted by this system to 1 molecule of acetyl phosphate, 1 molecule of lactate and 1 molecule of CO_2.

The reaction is formulated as follows:

$$\text{Pyruvate} + \text{phosphate} + \text{DPN} \rightleftharpoons \text{Acetyl phosphate} + CO_2 + \text{DPNH}_2 \qquad (44)$$

$$\text{Pyruvate} + \text{DPNH}_2 \rightleftharpoons \text{Lactate} + \text{DPN} \qquad (45)$$

$$\text{Sum } 2 \text{ Pyruvate} \rightleftharpoons \text{Lactate} + \text{acetyl phosphate} + CO_2 \qquad (46)$$

Acetyl phosphate is formed here due to the presence of phosphotransacetylase in these extracts. In the absence of inorganic phosphate the reaction is greatly diminished and no acetyl phosphate is formed. However, if in the absence of phosphate the system is coupled with crystalline-condensing enzyme and oxaloacetate, a rapid synthesis of citrate is observed. This indicates that acetyl-CoA is formed as an intermediate; in the absence of acetyl acceptors, such as phosphate or oxaloacetate, the reaction is limited by the CoA, since free CoA cannot be regenerated. In the presence of appropriate acetyl acceptors, free CoA is regenerated and allowed to act catalytically.

The reaction can be schematized as follows:

$$\text{Pyruvate} + \text{DPN} + \text{CoA} \rightleftharpoons \text{Acetyl-CoA} + CO_2 + \text{DPNH}_2 \qquad (47)$$

$$\text{Pyruvate} + \text{DPNH}_2 \rightleftharpoons \text{Lactate} + \text{DPN} \qquad (48)$$

$$\text{Acetyl—CoA} + X \rightleftharpoons \text{CoA} + \text{acetyl—X} \qquad (49)$$

$$\text{Sum } 2 \text{ Pyruvate} + X \rightleftharpoons \text{Lactate} + \text{acetyl—X} + CO_2 \qquad (50)$$

[200] R. M. Burton, *Federation Proc.* **11**, 193 (1952).

[201] S. Korkes, *in* McElroy and Glass, Phosphorus Metabolism, The Johns Hopkins Press, Baltimore, 1951, p. 259.

(3) Schweet *et al.*[144, 145] have reported the partial purification of a soluble pyruvic oxidase from pigeon breast muscle that converts pyruvate to to acetate and CO_2. The preparation contains flavin and protogen (pyruvate oxidation factor). Cocarboxylase and Mg^{++} are required for full activity. It is reported that CoA, DPN, or TPN are not required.

This preparation differs from the previously considered pyruvate system of bacteria in that free acetate is a product of the reaction and acetyl-CoA is not an intermediate. However, an activated acetyl derivative is certainly involved, since it is reported that, upon the addition of CoA and a fraction from pigeon liver extracts, the acetylation of sulfanilamide could be effected. Also the addition of CoA and bacterial phosphotransacetylase led to the formation of acetyl phosphate. These observations are summarized in the diagram.

$$\text{Pyruvate} \xrightarrow[\text{TPP}]{\text{pyruvic oxidase}} \text{Acetyl—X} + CO_2 + H_2O \tag{51}$$

$$\text{Acetyl—X} + \begin{array}{l} \text{HOH} \rightarrow \text{Acetate} \\ \text{CoA} \rightarrow \text{Acetyl—CoA} \end{array} \tag{52}$$

In the primary oxidation of pyruvate, it is visualized that pyruvate reacts with cocarboxylase, giving a 2-carbon complex at the level of oxidation of acetaldehyde with the liberation of CO_2. This is considered in greater detail in the section on cocarboxylase. The oxidase oxidizes this complex to an acetyl derivative which is not acetyl-CoA. The hypothetic acetyl derivative may subsequently undergo hydrolytic cleavage to free acetate or may be transferred to CoA, giving rise to acetyl-CoA. Korkes *et al.*[202] have described a pyruvate dismutation reaction present in heart muscle which requires at least two enzymes, CoA and DPN. The initial reaction is presumed to yield, per mole of pyruvate, 1 mole each of acetyl-CoA, CO_2, and reduced pyridine nucleotide. In this case the primary product appears to be acetyl-CoA.

b. Oxidation of Acetaldehyde. Acetyl-CoA may be formed by the oxidation of acetaldehyde by extracts from *Cl. kluyveri*[200] or from *E. coli.*[203] The enzyme requires DPN for activity. In the presence of acetyl acceptors with their appropriate enzymes the acetyl group may be transferred as usual. In the presence of phosphate and phosphotransacetylase, acetyl phosphate is formed.

c. Formation from Fatty Acid Oxidation. Barker[164] has discussed the role of CoA in fatty acid oxidation. In extracts of *Cl. kluyveri*, acetyl phosphate is formed from the oxidation of butyrate. Kennedy and Barker have

[202] S. Korkes, A. del Campillo, and S. Ochoa, *J. Biol. Chem.* **195**, 541 (1952).

[203] G. D. Pinchot, E. Racker, *in* McElroy and Glass, Phosphorus Metabolism, The Johns Hopkins Press, Baltimore, 1951, p. 366.

demonstrated the CoA-dependence of fatty acid oxidation in these extracts. It would appear that acetyl-CoA is necessary to activate fatty acid oxidation.

Recently Drysdale[204] has described a soluble enzyme system from rat liver mitochondria which oxidizes fatty acids to acetoacetate. CoA is required for this system. Likewise Green *et al.*[205] have observed a CoA-dependent fatty acid oxidation in a soluble preparation from heart muscle. It seems likely, therefore, that the fatty acids undergo β oxidation while attached to CoA as an acyl-CoA and that in the complete oxidation of fatty acid acetyl-CoA derived from the β oxidation is fed into the tricarboxylic acid cycle by means of the condensing enzyme of Stern and Ochoa.

d. Formation from Acetyl Phosphate. The formation of acetyl-CoA from acetyl phosphate by phosphotransacetylase has already been discussed.

e. Formation from Acetate and ATP.* Nachmansohn and Machado[206] first observed the activation of acetate by ATP in brain homogenates. In the discovery of CoA, Lipmann[155] showed that CoA was a necessary component for this activation. A reaction between acetate and ATP in *E. coli* extracts leading to the formation of acetyl phosphate has been shown not to require CoA.[189] The acetate-ATP reaction in liver extracts as well as in yeast extract did not give rise to acetyl phosphate,[194, 162] nor could these tissues utilize acetyl phosphate. However, the acetate-ATP reaction in these tissues, in the presence of hydroxylamine, leads to the formation of acethydroxamic acid. Since it is known that acetyl-CoA will react chemically with hydroxylamine to give acethydroxamic acid, this observation is now explained.

Lynen[197] postulated the formation of acetyl-CoA from acetate and ATP as follows:

$$\text{ATP} + \text{CoA} \rightleftharpoons \text{CoA—P} + \text{ADP} \tag{53}$$

$$\text{CoA—P} + \text{acetate} \rightleftharpoons \text{Acetyl—CoA} + \text{P} \tag{54}$$

This scheme avoids the formation of acetyl phosphate as an intermediate and is therefore consistent with the observations previously discussed. Recently Lipmann *et al.*[207, 208] have demonstrated that the activation of acetate by ATP in both liver and yeast extracts occurs by the intermediate

[204] G. R. Drysdale, *Federation Proc.* **11,** 204 (1952).

[205] D. E. Green, H. Beinert, D. Goldman, R. W. von Korff, and S. Mii, *Federation Proc.* **11,** 222 (1952).

[206] D. Nachmansohn and A. L. Machado, *J. Neurophysiol.* **6,** 397 (1943).

[207] S. Black, M. E. Jones, R. M. Flynn, and F. Lipmann, *Federation Proc.* **11,** 189 (1952).

[208] F. Lipmann, M. E. Jones, S. Black, and R. M. Flynn, *J. Am. Chem. Soc.* **74,** 2384 (1952).

* See Addendum at end of chapter.

formation of CoA pyrophosphate. The reaction is formulated as follows:

$$\text{ATP} + \text{CoA} \rightleftharpoons \text{CoA—P—P} + \text{AMP} \quad (55)$$

$$\text{CoA—P—P} + \text{acetate} \rightleftharpoons \text{Acetyl—CoA} + \text{P—P} \quad (56)$$

The participation of pyrophosphate in this reaction suggests that inorganic pyrophosphate may play a more important role in metabolic transformations than has hitherto been considered.

7. The Formation of Acyl-CoA's Other Than Acetyl-CoA

Recently evidence has been accumulated for the participation of CoA in reactions in which acetate does not seem to be involved. Kaufman[209] has reported on the isolation from pig heart of a soluble system, which in the presence of ammonium ions causes the dismutation of α-ketoglutarate to succinate, glutamate, and CO_2. The system is dependent upon CoA, DPN, and orthophosphate. The sequence of reaction is postulated as follows:

$$\alpha\text{-Ketoglutarate} + \text{DPN} + \text{CoA} \rightarrow \text{Succinyl—CoA} + \text{DPNH}_2 + \text{CO}_2 \quad (57)$$

$$\text{DPNH}_2 + \alpha\text{-ketoglutarate} + \text{NH}_3 \rightleftharpoons \text{Glutamate} + \text{DPN} \quad (58)$$

$$\text{Succinyl—CoA} + \text{P} + \text{ADP} \rightarrow \text{Succinate} + \text{ATP} + \text{CoA} \quad (59)$$

Thus succinyl CoA is a postulated intermediate in this reaction.

Sanadi and Littlefield[210] have reported on a soluble α-ketoglutarate oxidase system which is likewise CoA-dependent. This system has previously been referred to, since it was used as an assay for CoA. In the presence of the oxidase, a supernatant factor, DPN, and CoA, α-ketoglutarate is oxidized to succinate and CO_2. The reaction is visualized as follows:

$$\alpha\text{-Ketoglutarate} + \text{CoA} + \text{DPN} \rightarrow \text{Succinyl-CoA} + \text{DPNH}_2 + \text{CO}_2 \quad (60)$$

$$\text{Succinyl-CoA} + \text{Sn factor} \rightarrow \text{Succinate} + \text{CoA} \quad (61)$$

The formation of succinyl-CoA in this reaction has been established by coupling this system with a pigeon liver extract which succinylated sulfanilamide.

Novelli[211] has observed the formation of succinyl-CoA in extracts of *E. coli* by reaction of succinate, ATP and CoA.

These systems, then, indicate that, in addition to acetyl-CoA, it is possible that succinyl-CoA also exists as a reactive acyl intermediate. This raises the possibility that CoA may be a more general acyl carrier than was

209 S. Kaufman, *in* McElroy and Glass, Phosphorus Metabolism, The Johns Hopkins Press, Baltimore, 1951, Vol. I, p. 370.

210 D. R. Sanadi and J. W. Littlefield, *Federation Proc.* **11,** 280 (1952).

211 G. D. Novelli, unpublished observations.

previously suspected. One is reminded of the observation of Chantrenne that CoA is involved in the synthesis of hippuric acid. Here glycine is condensed with benzoic acid in the presence of ATP and CoA. It is possible that benzoyl-CoA may exist as an intermediate in this reaction. Recently Stadtman[212] obtained evidence for the formation of propionyl- and butyryl-CoA in *Cl. kluyveri* extracts.

Stadtman and Barker[213] had observed an apparent transphosphorylation between acetyl phosphate and propionate. The reaction is described by the following equation:

$$\text{Acetyl phosphate} + \text{propionate} \rightleftarrows \text{Acetate} + \text{propionyl phosphate} \quad (62)$$

A closer examination of this reaction revealed the following sequence:

$$\text{Acetyl phosphate} + \text{CoA} \rightleftarrows \text{Acetyl—CoA} + \text{P} \quad (63)$$

$$\text{Acetyl—CoA} + \text{propionate} \rightleftarrows \text{Propionyl—CoA} + \text{acetate} \quad (64)$$

$$\text{Propionyl—CoA} + \text{P} \rightleftarrows \text{Propionyl phosphate} + \text{CoA} \quad (65)$$

Reaction 63 is the well-known phosphotransacetylase reaction, and reaction 65 could be called a phosphotranspropionylase. Reaction 64 which interconverts acyl-CoA has been shown to occur with butyrate to give butyryl-CoA. This may be a general reaction for the formation of a large variety of acyl-CoA's.

Kornberg[166] has recently demonstrated the CoA-dependence of phospholipid synthesis in rat liver. Stearyl-CoA appears to be an intermediate in this reaction and is presumably formed by reaction of stearic acid, ATP, and CoA in a similar manner to Lipmann's formation of acetyl-CoA from acetate + ATP.

In summary of the metabolism of CoA we can visualize four general mechanisms for the formation of acyl-CoA:

$$\text{R—COOH} + \text{ATP} + \text{CoA} \rightleftharpoons \text{R—CO—CoA} + \text{AMP} + \text{PP} \quad (66)$$

$$\text{R—CO—COOH} + \text{CoA} + \text{X} \rightleftharpoons \text{R—CO—CoA} + \text{XH}_2 + \text{CO}_2 \quad (67)$$

$$\text{R—CO—CoA} + \text{R}'\text{—COOH} \rightleftharpoons \text{R}'\text{—CO—CoA} + \text{RCOOH} \quad (68)$$

$$\overset{\text{H}}{\text{RC}} = \text{O} + \text{CoA} + \text{X} \rightleftharpoons \text{R—CO—CoA} + \text{XH}_2 \quad (69)$$

In reaction 66 we postulate the activation of any carboxylic acid by ATP, going through the intermediate formation of CoA pyrophosphate to give the corresponding acyl-CoA. In reaction 67 is postulated the oxidative decarboxylation of any α-keto acid to yield CO_2 and the next lower acyl-CoA. This general type reaction would also apply to the formation of various

[212] E. R. Stadtman, *Federation Proc.* **11,** 291 (1952).
[213] E. R. Stadtman and H. A. Barker, *J. Biol. Chem.* **184,** 769 (1950).

acyl-CoA's through the β oxidation of fatty acids. Reaction 68 would be a general transfer reaction for the secondary formation of a variety of acyl-CoA species This is the transacylation reaction which Stadtman has observed with propionate and butyrate. Reaction 69 represents the formation of an acyl-CoA through the oxidation of an aldehyde.

8. Other Functions of CoA

By studying the synthesis of fat and sterols in pantothenate-deficient yeast as well as deficient rats, Klein[165] has observed that the formation of total lipid and sterol is markedly depressed by withholding the vitamin. A direct correlation between synthetic ability of the tissues and CoA content was made. These studies forecast a role of CoA in one or more of the enzymatic steps leading to the synthesis of these compounds.

Shemin[214] has also implicated CoA, presumably via succinyl-CoA in the synthesis of the pyrrole ring of the porphyrins.

VIII. Pyridoxal Phosphate

1. Introduction

This coenzyme occupies a unique position, in that it is involved in at least four types of reactions which all appear to be quite different. These are: the decarboxylation of amino acids, transamination, and the synthesis and cleavage of tryptophan. However, all these reactions, as we shall see, have one thing in common: they involve an α-amino acid and are concerned with either the amino group or the carbon atom adjacent to the amino group.

2. Historical

Pyridoxine (vitamin B_6) was first recognized as an essential factor in animal nutrition.[215] Later it was found to have growth-promoting ability for various strains of bacteria,[216] yeast,[217] molds, fungi, and plants.[218] Snell,[219] in a series of studies with bacteria, noted variations in response to certain natural extracts as source of B_6 among several bacteria. Out of this work came the appreciation that vitamin B_6 existed in several different forms to which various test organisms responded differently. The vitamin B_6 complex is now known to consist of the following compounds: pyridoxine, pyridoxal, pyridoxamine, and the phosphorylated forms of the two latter substances.

[214] D. Shemin and S. Kumin, *Federation Proc.* **11,** 285 (1952).
[215] P. György, *Nature* **133,** 478 (1934).
[216] E. F. Moller, *Z. physiol. Chem.* **254,** 285 (1938).
[217] A. S. Schultz, L. Atkin, and C. N. Frey, *J. Am. Chem. Soc.* **61,** 1931 (1939).
[218] W. J. Robbins and M. B. Schmidt, *Proc. Natl. Acad. Sci. U. S.* **25,** 1 (1939).
[219] E. E. Snell, B. M. Guirard, and R. J. Williams, *J. Biol. Chem.* **143,** 519 (1942).

The first observations which led to the discovery of the coenzyme function of pyridoxal came when Gale[220] noted that the decarboxylation of tyrosine and certain other amino acids by bacterial cells was catalyzed by an enzyme containing a dissociable cofactor. Gale and Epps[221] could not replace the cofactor with any known coenzyme and were able to prepare a concentrate of their cofactor which had 15,000 times the activity of yeast extract. In the meantime, Bellamy and Gunsalus[222] noted that the decarboxylation of tyrosine by resting cells was greatly influenced by the growth medium. Analysis of this effect led them to the finding that pyridoxine-deficient cells had a greatly decreased ability to decarboxylate tyrosine, and this ability was restored to normal by adding pyridoxal to the resting cells while pyridoxine or pyridoxamine were inactive.[223] It was subsequently shown that pyridoxal must be phosphorylated by cells before it is active.[224] A synthetic pyridoxal phosphate was prepared and shown to be active with purified amino acid decarboxylase. Thus pyridoxal phosphate became known as codecarboxylase.

The discovery of the cotransaminase function of vitamin B_6 goes back to Snell's nutritional observations.[219] It was early noted that the vitamin activity of pyridoxine many times depended upon its being autoclaved with amino acids. An investigation into the chemical change accompanying autoclaving led to the discovery of pyridoxal and pyridoxamine and a chemical method for their interconversion. The reaction as formulated by Snell[225] is a chemical transamination of the amino group of an amino acid, say glutamic acid, to pyridoxal, forming thereby pyridoxamine and the keto acid corresponding to the amino acid, thus, in this case, α-ketoglutarate; i.e.,

$$\text{Pyridoxal} + \text{glutamate} \rightleftharpoons \text{Pyridoxamine} + \text{ketoglutarate} \qquad (70)$$

Since an enzymatic transamination was already known,[226] the similarity between the chemical and the enzymatic reaction led to the suggestion that vitamin B_6 might be a coenzyme for enzymatic transamination. Schlenk and Snell[227] tested the hypothesis by measuring the transaminase activity of tissues from vitamin-B_6-deficient rats. They found that deficient tissues had a much lower transaminase ability than normal tissues. Subse-

[220] E. F. Gale, *Biochem. J.* **34,** 392 (1940).
[221] E. F. Gale and H. M. R. Epps, *Biochem. J.* **38,** 250 (1944).
[222] W. D. Bellamy and I. C. Gunsalus, *J. Bact.* **46,** 573 (1943).
[223] I. C. Gunsalus and W. D. Bellamy, *J. Biol. Chem.* **155,** 375 (1944).
[224] W. W. Umbreit, W. D. Bellamy, and I. C. Gunsalus, *Arch. Biochem.* **7,** 185 (1945).
[225] E. E. Snell, *J. Am. Chem. Soc.* **67,** 194 (1945).
[226] A. E. Braunstein and G. M. Kritzmann, *Enzymologia* **2,** 129 (1937).
[227] F. Schlenk and E. E. Snell, *J. Biol. Chem.* **157,** 425 (1945).

quently, a completely resolved transaminase apoenzyme was restored to full activity by the addition of pyridoxal phosphate.[228]

Pyridoxal phosphate has been established as a coenzyme in two reactions involving tryptophan. An enzyme has been isolated from *Neurospora* which catalyzes a synthesis of tryptophane from serine and indole.[229] This reaction requires pyridoxal phosphate. An enzyme has been isolated from *E. coli* which causes the decomposition of tryptophan to pyruvic acid, indole, and ammonia;[230] here too pyridoxal phosphate is a necessary cofactor.

3. Distribution and Method of Assay

Vitamin B_6 occurs in a wide variety of biological tissues. The vitamin occurs predominantly in the bound form as the phosphate of pyridoxal or pyridoxamine. In almost all tissues the pyridoxal phosphate is the major form with the exception of liver, where pyridoxamine phosphate appears to be in great excess over the aldehyde form.[231]

The best method for the assay of pyridoxal phosphate is the use of tyrosine decarboxylase as described by Gunsalus, Bellamy, and Umbreit.[13] The enzyme is prepared from a dried powder of cells of *S. faecalis* R. which has been grown deficient in vitamin B_6 by growth ln a vitamin-B_6-free alanine-rich medium. Thus, the decarboxylase is obtained almost completely resolved. This is a convenient preparation, since such a powder is stable for long periods and since the resolution of transaminases, decarboxylases, and tryptophanases isolated from tissues is a rather difficult task. The assay is performed manometrically by measuring the rate of CO_2 liberation from tyrosine by the dried powder in the presence of pyridoxal phosphate. The rate of CO_2 evolution is a function of the concentration of pyridoxal phosphate.

4. Structure of the Coenzyme

Pyridoxal phosphate was first synthesized by Gunsalus, Bellamy, and Umbreit in 1944,[13] by treating pyridoxal with phosphoryl chloride. Although pyridoxal phosphate has perhaps one of the simplest structures of the coenzymes, it was not until very recently that its structure was definitely established.[232] The report of the first synthesis of pyridoxal phosphate did not indicate its structure. Some of their spectral data led to the suggestion that the phosphate is not esterified in position 3.

[228] D. E. Green, L. F. Leloir, and V. Nocito, *J. Biol. Chem.* **161,** 559 (1945).
[229] W. W. Umbreit, W. A. Wood, and I. C. Gunsalus, *J. Biol. Chem.*
[230] W. A. Wood, I. C. Gunsalus, and W. W. Umbreit, *J. Biol. Chem.* **170,** 313 (1947).
[231] J. C. Rabinowitz and E. E. Snell, *J. Biol. Chem.* **169,** 643 (1947).
[232] D. Heyl, E. Luz, A. Harris, and K. Folkers, *J. Am. Chem. Soc.* **73,** 3430, 3436 (1951).

Karrer and Viscontini[233] synthesized the acetal of pyridoxal-3-phosphate and claimed that it had codecarboxylase activity but could not reactivate a transaminase system. In the meantime, Gunsalus[234] presented evidence that the active synthetic product was *not* the 3-phosphate (and by elimination must be the 5-phosphate). This discrepancy was resolved when the two products were compared simultaneously on the same test system.[235] The activity of the 3-phosphate was so low compared to the active pyridoxal phosphate that the 3-phosphate was definitely ruled out as the active coenzyme. In a recent series of papers,[232] the structure of pyridoxal phosphate is definitely established as pyridoxal-5-phosphate and that of pyridoxamine-phosphate as the 5 phosphate.

Pyridoxal phosphate Pyridoxamine phosphate

FIG. 12

5. MECHANISM OF ACTION

The chemical transamination of pyridoxal to pyridoxamine, discovered by Snell,[225] not only led to the discovery of the role of vitamin B_6 in enzymatic transamination but was suggested as a possible mechanism of action. (See Schlenk and Snell.[227])

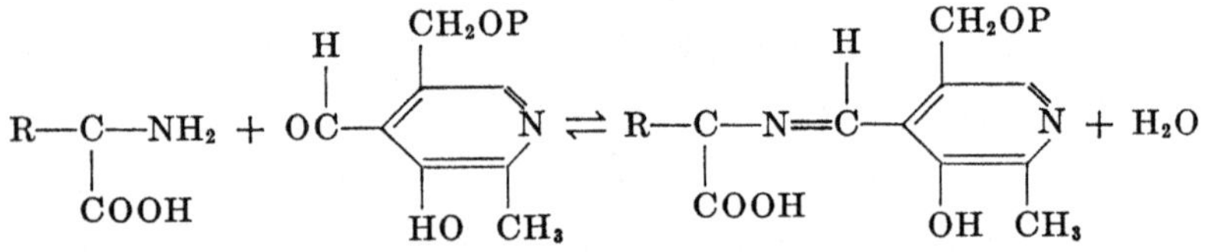
(71)

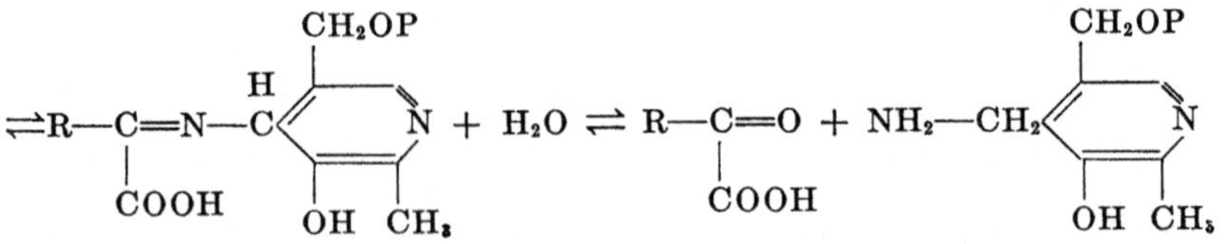
(72)

It was visualized that pyridoxal reacted with the amino group of an amino acid to form an intermediate Schiff base which was then hydrolyzed to the keto acid and pyridoxamine, equation 71.

[233] P. Karrer and M. Viscontini, *Helv. Chim. Acta* **30**, 52 (1947).
[234] I. C. Gunsalus and W. W. Umbreti, *Abstracts* 110th Meeting, American Chemical Society, 1944, p. 34b.
[235] W. W. Umbreit and I. C. Gunsalus, *J. Biol. Chem.* **179**, 279 (1949).

Starting with pyridoxamine and the keto acid the reverse reaction would occur, as given in equation 72.

Very recently Snell and Metzler[236] reinvestigated the chemical transformation and found that, starting with pyridoxamine and α-ketoglutarate and running the reaction at 100°, an equilibrium is reached in which 46% of the pyridoxamine is converted to pyridoxal. The same equilibrium point was reached in the reverse direction. They found that this reaction is metal catalyzed; iron, aluminum, and copper ions increased the rate about twentyfold, with copper being the most effective ion.

This model chemical reaction as an explanation for enzymatic transamination was further strengthened when Snell found that the phosphates of pyridoxal and pyridoxamine would undergo the same chemical transformations. In this case the equilibrium in much further toward pyridoxamine phosphate.

If this were indeed the mechanism of action of pyridoxal phosphate, then one would think that either pyridoxal or pyridoxamine phosphate would be active in catalyzing the enzymatic transamination. In one case recently with a highly purified glutamic aspartic transaminase only pyridoxal phosphate was capable of reconstituting the system.

Thus the mechanism of action of pyridoxal phosphate in transamination must be left open. There is as yet no clue to its mechanism of action in decarboxylation and in the tryptophanase reaction.

6. Reactions Catalyzed by Pyridoxal Phosphate

a. Amino Acid Decarboxylation.

$$R{-}\underset{\displaystyle H}{\overset{\displaystyle NH_2}{\underset{|}{\overset{|}{C}}}}{-}COOH \rightarrow CO_2 + R{-}CH_2{-}NH_2 \tag{73}$$

Amino acid decarboxylases shown to require pyridoxal phosphate are the following: (1) tyrosine, (2) arginine, (3) lysine, (4) ornithine, (5) glutamic acid and (6) dopa (3,4-dihydroxyphenylalanine).

These amino acids are decarboxylated to the corresponding amines with the liberation of CO_2.

b. Transamination. Braunstein and Kritzman[226] discovered this reaction. They reported that a large variety of keto acid analogues of monocarboxylic amino acids could act as amino acceptor from glutamic or aspartic acid. Cohen[236a] challenged this view and showed that the glutamic-aspartic

[236] E. E. Snell and A. Metzler, *in* McElroy and Glass, Phosphorus Metabolism The Johns Hopkins Press, Baltimore, 1951, Vol. I, p. 442.

[236a] P. P. Cohen, A Symposium on Respiratory Enzymes, University of Wisconsin Press, Madison, 1942, p. 210.

transaminase was by far the most active. Recently Feldman and Gunsalus[237] have demonstrated a wide variety of transaminases in bacteria. Cammarata and Cohen[238] have made similar observations. Feldman and Gunsalus[237] report that in bacteria these transaminases produce glutamic acid from α-ketoglutaric acid and the following amino donors: aspartate, alanine, valine, leucine, norleucine, tryptophan, tyrosine, phenylalanine, and methionine. Isoleucine, histidine, lycine, glycine, and threonine were much less effective amino donors.

The type reaction for transamination is given below:

$$\underset{\displaystyle NH_2}{R—\underset{|}{CH}}—COOH + R'—\underset{\displaystyle O}{\underset{\|}{C}}—COOH \rightleftharpoons R\underset{\displaystyle O}{\underset{\|}{C}}—COOH + R'—\underset{\displaystyle NH_2}{\underset{|}{CH}}—COOH \qquad (74)$$

In this reaction the coenzyme, pyridoxal phosphate, is considered to undergo the transformation given in equations 71 and 72.

c. Reactions Involving Tryptophan. Pyridoxal phosphate acts as a coenzyme for two reactions involving tryptophan. One of these reactions is the synthesis of tryptophan from indole and serine, and the other is a nonoxidative degradation of tryptophan to indole, pyruvic acid, and ammonia.

(1) *Synthesis of Tryptophan from Indole and Serine.* Tatum and Bonner[239] reported that a mutant of *Neurospora crassa* could utilize indole in place of tryptophan. Umbreit, Wood, and Gunsalus,[229] by using extracts from this mutant, showed that, with pyridoxal phosphate as a coenzyme, tryptophan was synthesized from serine and indole via the following reaction:

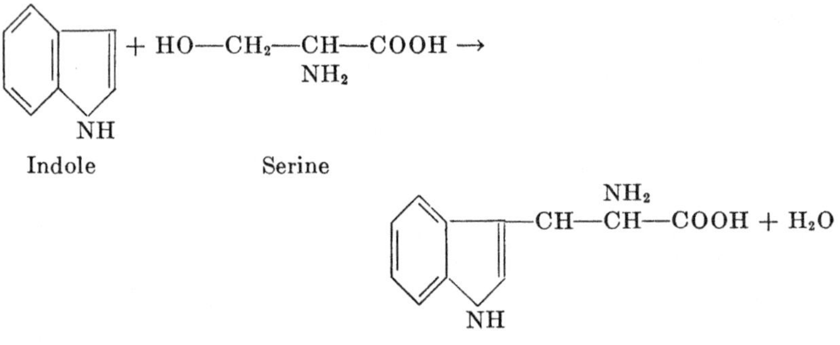

(75)

The reverse reaction, i.e., the breakdown of tryptophan to indole and serine, was not demonstrated with this extract.

(2) *The Degradation of Tryptophan.* Wood, Gunsalus, and Umbreit[230] studied the tryptophanase reaction in extracts of *E. coli*. This enzyme was

[237] L. I. Feldman and I. C. Gunsalus, *J. Biol. Chem.* **187,** 821 (1950).
[238] P. S. Cammarata and P. P. Cohen, *J. Biol. Chem.* **187,** 439 (1950).
[239] E. L. Tatum and D. Bonner, *Proc. Natl. Acad. Sci. U. S.* **30,** 30 (1944).

shown to require pyridoxal phosphate as cofactor and catalyzed the breakdown of tryptophan to indole, pyruvic acid, and ammonia.

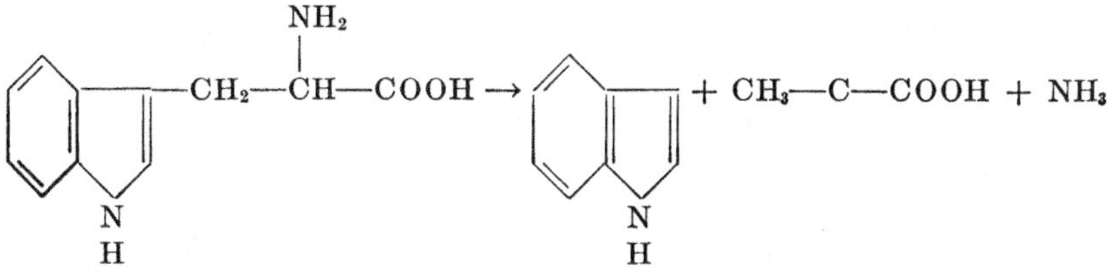

(76)

Tryptophan Indole Pyruvic acid Ammonia

The mechanism of this reaction is not known, since both alanine and serine, two possible intermediates, were shown not to be attacked by the enzyme.

IX. Miscellaneous

1. Uridine Diphosphate Glucose

a. Introduction. The nature of the interconversion of glucose and galactose has long been sought. The whole lactose molecule in milk can be derived exclusively from glucose. It is fair to assume that the galactose of the mucoproteins and the galactolipids are also derived from glucose. The conversion of glucose to galactose involves an isomerization at C_4 (Walden inversion). The enzyme system known as "galactowaldenase" or "galactoisomerase" has been found to equilibrate galactose-1-phosphate and glucose-1-phosphate in the presence of its coenzyme, uridine diphosphate glucose (UDP-Glucose).

b. Historical. Kosterlitz,[240, 241] in 1937, demonstrated the accumulation of α-D-galactose-1-phosphate in the livers of rabbits fed galactose. He also showed that this galactose phosphate could be fermented by cell-free extracts of galactose-adapted yeast and proposed the following scheme to account for the metabolism of galactose in these yeasts:

$$\text{ATP} + \text{galactose} \rightarrow \text{Galactose-1-P} + \text{ADP} \quad (77)$$

$$\text{Galactose-1-P} \rightleftharpoons \text{Glucose-1-P} \quad (78)$$

$$\text{Glucose-1-P} \rightleftharpoons \text{Glucose-6-P} \quad (79)$$

The work of Leloir and his group provided the facts in support of the scheme. Cell-free extracts of a galactose-fermenting *Saccharomyces fragilis* were studied. A galactokinase[242] was found which carries out reaction 77.

[240] H. W. Kosterlitz, *Biochem. J.* **37,** 318 (1943).
[241] H. W. Kosterlitz, *Biochem. J.* **37,** 322 (1943).
[242] R. E. Trucco, R. Caputto, L. F. Leloir, and N. Mittelman, *Arch. Biochem.* **18,** 137 (1948).

The phosphoglucomutase enzyme catalyzes reaction 79. Reaction 78, which involves an inversion (Walden inversion) at C_4 of the hexose, was shown to require a heat-stable cofactor.[243] This coenzyme for the galactowaldenase system was isolated and its structure determined.[244, 245] The coenzyme is uridinidiphosphoglucose (UDPGlucose).

c. Structure and Chemistry. The coenzyme consists of uridine-5-phosphate and glucose-1-phosphate in pyrophosphate linkage.[245] The catalytic activity of UDPGlucose is destroyed by heating with weak acid or alkali. Treating the coenzyme with 0.01 *N* acid at 100° for 5 to 10 min. yields free glucose and uridine diphosphate. The intact coenzyme is non-reducing. Further heating with 1 *N* acid for 15 min. at 100° releases 1 mole of inorganic phosphate to give uridine-5-phosphate. The presence of the pyrophos-

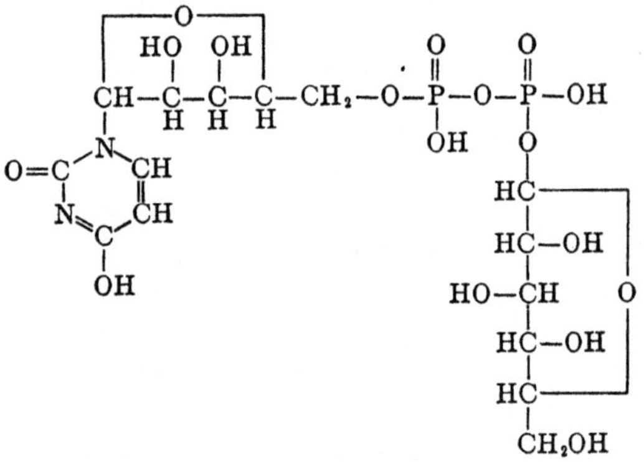

Uridine diphosphate glucose (UDPG)

FIG. 13

phate bridge was first deduced from titration data and confirmed by the use of Kornberg's pyrophosphatase which results in the formation of glucose-1-phosphate and the uridine-5-phosphate.

Heating UDPGlucose with 0.01 *N* alkali for a few minutes also completely destroys coenzyme activity. This reaction was explained by the use of paper chromatography. When the coenzyme is chromatographed by an ammonia-solvent system, two products are formed.[246] One is uridine-5-phosphate, and the other is a non-reducing doubly esterified phosphoric acid derivative, of glucose. It is probably the 1,2-cyclic phosphate, since

[243] R. Caputto, L. F. Leloir, R. E. Trucco, C. E. Cardini, and A. C. Paladini, *J. Biol. Chem.* **179**, 497 (1949).

[244] C. E. Cardini, A. C. Paladini, R. Caputto, and L. F. Leloir, *Nature* **165**, 191 (1950).

[245] R. Caputto, L. F. Leloir, C. E. Cardini, and A. C. Paladini, *J. Biol. Chem.* **184**, 333 (1950).

[246] A. C. Paladini and L. F. Leloir, *Biochem. J.* **51**, 426 (1952).

upon acid hydrolysis glucose-1-phosphate and glucose-2-phosphate are formed. FAD is split by alkali in an analogous manner.[62]

The ultraviolet absorption spectrum of the coenzyme corresponds to that of uridine. Bromine destroys the uracil moiety of the coenzyme and effects the destruction of biological activity.

The coenzyme is isolated from an alcoholic extract of toluene-treated baker's yeast. The procedure involves a selective mercury precipitation, charcoal chromatography, and the removal of ninhydrin-positive material by a cation exchange resin. The barium salt of the coenzyme is water insoluble.

d. Mechanism of Action. Crude extracts of the galactose-fermenting yeast contain an enzyme which catalyzes the interconversion of UDPGlucose to UDPGalactose.[247] No free galactose or galactose phosphate appear during the course of the reaction. The enzyme therefore carries out the Walden inversion at C_4 on the coenzyme molecule itself. The mechanism of the inversion is still not known. The over-all conversion of galactose-1-phosphate to glucose-1-phosphate probably occurs by a three-step reaction:

$$\text{Galactose-1-phosphate} + \text{UTP} \rightleftharpoons \text{UDPGalactose} + \text{pyrophosphate} \qquad (80)$$

$$\text{UDPGalactose} \rightleftharpoons \text{UDPGlucose} \qquad (81)$$

$$\text{UDPGlucose} + \text{pyrophosphate} \rightleftharpoons \text{Glucose-1-phosphate} + \text{UTP} \qquad (82)$$

UTP is the triphosphate of uridine and has been demonstrated by Kornberg.[17] Trucco[248] has shown that yeast extracts can best synthesize UDPGlucose when UDP, ATP, and glucose-1-phosphate are all present.

e. Function. Recent *in vivo* studies with isotopically labeled hexoses have demonstrated the conversion of galactose to glycogen in the liver,[249] and that of glucose to the lactose in milk,[250] without a breaking of the hexose skeleton. The fact that UDPGlucose is widely distributed in animal tissues suggests that the galactowaldenase system is involved in these conversions. However, no galactowaldenase activity has yet been demonstrated in non-galactose-fermenting yeast, which also contains some of the coenzyme, or in animal tissues, [251] except for the mammary gland.[252] This has suggested the possibility that the coenzyme may be involved in another function besides that of mediating the glucose-galactose transformation.

Calvin[252a] has made the interesting suggestion that UDPGlucose and UDPGalactose may function as hexose carriers in the synthesis of polysaccharide.

[247] L. F. Leloir, *Arch. Biochem. and Biophys.* **33,** 186 (1951).
[248] R. E. Trucco, *Arch. Biochem. and Biophys.* **34,** 482 (1951).
[249] Y. J. Topper and D. Stetten, Jr., *J. Biol. Chem.* **193,** 149 (1951).
[250] T. H. French, G. Popjak, and F. H. Malpress, *Nature* **169,** 71 (1952).
[251] R. L. Garner and G. F. Grannis, *Science* **114,** 501 (1951).
[252] R. Caputto and R. E. Trucco, *Nature* **169,** 1061 (1952).
[252a] *In* Phosphorus Metabolism, The Johns Hopkins Press, Baltimore, 1952, Vol. II, p. 440.

Park[253] has reported the isolation from penicillin-treated *Staphylococcus aureus* of some compounds containing what is probably uridine diphosphate-N-acetylglucosaminuronic acid. The latter may be a coenzyme of some kind, although no evidence for this is at hand.

f. Enzymatic Estimation and Distribution[245] of the Coenzyme. Extracts of galactose-fermenting *Saccharomyces fragilis*, as mentioned above, carry out the following transformations:

$$\text{Galactose-1-phosphate} \rightleftharpoons \text{glucose-1-phosphate} \rightarrow \text{glucose-6-phosphate}$$

By adding glucosediphosphate, the coenzyme for the second reaction in excess, the over-all reaction depends on the galactose-glucose transformation and, therefore, the UDPGlucose concentration. The rate of the over-all reaction serves as a measure of coenzyme content.

The UDPGlucose content in micromoles per gram of weight for tissues of the rat are as follows: 0.2 to 0.3 for kidney, brain, and muscle, and 0.1 to 0.2 for liver.

2. Glutathione

a. Introduction. The widespread occurrence of glutathione in all actively growing plant and animal tissues indicates that it is a substance of some metabolic importance. Because it readily undergoes oxidation and reduction, it has generally been assumed to act as a regulator of the oxido-reduction potential of the cell. Reduced glutathione plays a specific role as coenzyme for the glyoxalase system. Oxidized glutathione specifically accepts hydrogens from reduced TPN.

b. Historical. Glutathione was first isolated by Hopkins[254] in 1921 and was thought to be a dipeptide of glutamic acid and cysteine. A reinvestigation of the problem by Hopkins led to the elucidation of its structure as a tripeptide,[255] glutamyl-cysteinyl-glycine. Tissue extracts were found to contain systems capable of reducing oxidized glutathione (GSSG) and of rapidly oxidizing reduced glutathione (GSH) in the presence of oxygen. Glutathione has, therefore, generally been assumed to act as a hydrogen donor and acceptor, possibly in the control of the oxidation-reduction potential of cellular reactions, and possibly functioning in hydrogen transport. A specific funicton for GSH as coenzyme for the widely occurring glyoxalase system was demonstrated by Lohmann[256] in 1932. Recently an enzyme has been described in both plant[257, 258] and animal[259] tissues which mediates

[253] J. T. Park, *J. Biol. Chem.* **194,** 877, 885, 897 (1952).
[254] F. G. Hopkins, *Biochem. J.* **15,** 286 (1921).
[255] F. G. Hopkins, *J. Biol. Chem.* **84,** 269 (1929).
[256] K. Lohmann, *Biochem. Z.* **254,** 332 (1932).
[257] L. W. Mapson and D. R. Goddard, *Biochem. J.* **49,** 592 (1951).
[258] E. E. Conn and B. Vennesland, *J. Biol. Chem.* **192,** 17 (1951).
[259] T. W. Rall and A. L. Lehninger, *J. Biol. Chem.* **194,** 119 (1952).

specifically the reduction of GSSG by reduced TPN. This enzyme is called glutathione reductase.

c. Structure and Chemistry. GSH is the tripeptide γ-glutamyl-cysteinyl-glycine: The sulfhydryl group of the cysteine moiety can undergo oxidation to the disulfide form according to the following equation:

$$2GSH + A \rightleftharpoons GSSG + AH_2 \tag{83}$$

This oxidation is readily effected by oxygen at neutral or alkaline pH in the presence of traces of copper, iron, or cytochrome c. GSSG is reduced by thiol compounds like cysteine, thiol acids, H_2S, and BAL.

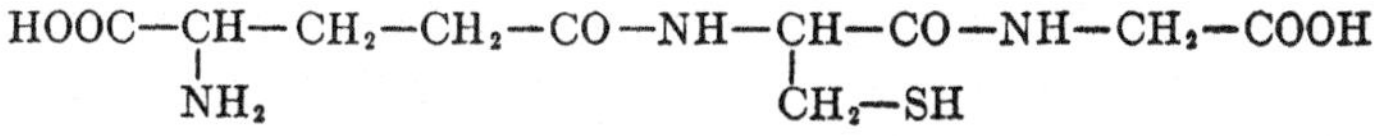

Glutathione

FIG. 14

GSH is generally detected by the nitroprusside test for free thiol groups and is quantitatively estimated by an iodine titration procedure which converts it to GSSG. This latter reaction may be utilized in preparing GSSG. GSH, like other thiols, has been used *in vitro* to maintain those proteins requiring an SH— group for activity in the reduced state, and GSSG can maintain those requiring an —S—S— group in that state. This is thought to be one of the biological functions of glutathione.

GSH is precipitated rather specifically in 0.5 *N* sulfuric acid by cuprous ions,[255] and this fact serves as the basis for the isolation of glutathione from natural sources.

d. Mechanism of Action of GSH in the Glyoxase System. The glyoxalase system carries out the conversion of methyl glyoxal to D-lactic acid. Racker[260] and Crook, and Law[261] have independently elucidated the essential steps in this conversion. The more detailed work by Racker has confirmed and extended older knowledge as to the mechanism of the over-all reaction. The two enzymes which are involved have been extensively purified. "Glyoxalase I" catalyzes the formation of the intermediate from methyl glyoxal and GSH, and "glyoxalase II," a hydrolytic enzyme, decomposes the intermediate to give lactate and regenerate the GSH. The scheme Racker proposes is the following:

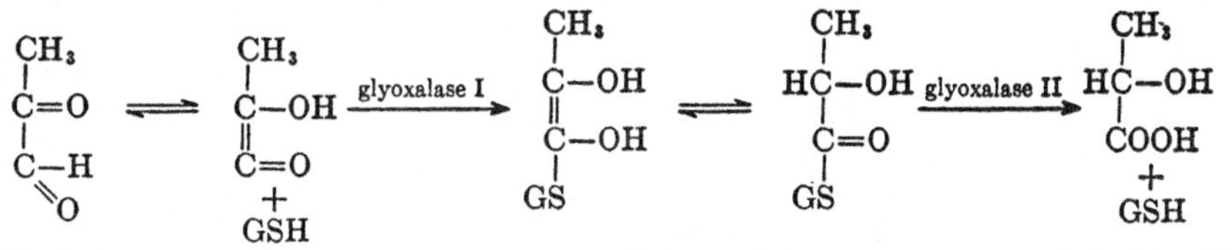

Methyl glyoxal — Intermediate — D-lactic acid

FIG. 15

The formula of the intermediate may be recognized as that of a thioester formed from lactic acid and GSH. Its ultraviolet absorption spectrum is characterized by a maximum at 235 mμ. This maximum absorption as well as its stability properties are characteristics manifested by other thioesters such as acetyl-CoA. The intermediate should also yield the hydroxamic acid derivative of lactic acid upon treatment with neutral hydroxylamine. The fact that it is a thioester suggests, by analogy with acetyl-CoA, that it is a high-energy compound.

e. Function. The glyoxalase system is present in significant amounts in most tissues and parallels the GSH content of these tissues. This fact suggests an important metabolic role for this system. The fact that the lactyl-glutathione intermediate is a high-energy compound adds to the possible significance of this system. As Racker points out, other aldehyde-acid transformations may take place via a similar mechanism involving high-energy compounds.

It has been known for some time that there exists a close correlation between the ascorbic acid and glutathione content of plant tissues. Plant tissues are also known to contain dehydroascorbic reductase, which catalyzes the transfer of hydrogens from GSH to dehydroascorbic acid, as well as ascorbic acid oxidase. Mapson and Goddard[257] have, therefore, provided a possible scheme for hydrogen transport which does not involve the cytochrome system. The hydrogens may take the following pathway:

$$\text{Substrate} + \text{TPN} \rightleftharpoons \text{TPNH}_2 + \text{oxidized substrate} \quad (84)$$

$$\text{TPNH}_2 + \text{GSSG} \rightleftharpoons 2\text{GSH} + \text{TPN} \quad (85)$$

$$2\text{GSH} + \text{dehydroascorbate} \rightleftharpoons \text{Ascorbate} + \text{GSSG} \quad (86)$$

$$\text{Ascorbate} + \text{oxygen} \rightleftharpoons \text{Dehydroascorbate} \quad (87)$$

Glutathione has also been implicated in other metabolic roles. The γ-glutamyl group of glutathione undergoes transpeptidation to other amino acids.[262] The significance of this reaction is still not known. Glutathione is thought to be involved in "detoxification" reactions, since all three of its amino acids have been found to occur in combination with certain toxic substances. In some bacteria glutathione has been demonstrated to play a role as a growth factor.[263] Glutathione reductase occurs in both plant and animal tissues. The exact significance of this enzyme is not known. Glutathione is known to exist in tissues almost exclusively in the reduced form, and this reductase may be one enzyme involved in maintaining this condition.

---

[260] E. Racker, *J. Biol. Chem.* **190,** 685 (1951).

[261] E. M. Crook and K. Law, *Biochem. J.* **46,** xxxvi (1950).

[262] G. S. Hanes, F. J. R. Hird, and F. A. Isherwood, *Nature* **166,** 288 (1950).

[263] H. D. Slade, G. A. Knox, and W. C. Slamp, *J. Bact.* **62,** 669 (1951).

f. Assay for Glutathione and Tissue Content. The iodometric technique for the determination of the GSH content in tissues is non-specific, since it includes other oxidizable substances, particularly ascorbic acid. The glyoxalase system serves as a specific reagent for the quantitative microestimation of GSH. An acetone-dried yeast preparation washed free of GSH is the source of the glyoxalase system. The conversion of methyl glyoxal to D-lactate is proportional to the GSH in the test solution and may be followed manometrically[264] in bicarbonate buffer, or by determining the disappearance of methyl glyoxal.[265] All the glutathione in blood and tissues is in the reduced form.[266] The GSH content of the various tissues of the rat range from 40 to 200 mg. per 100 g. of tissues.

g. Metabolism of the Coenzyme. When Dakin and Dudley, in 1913, first demonstrated the glyoxalase system and its widespread occurrence in tissues, they also observed the "antiglyoxalase" effect of certain tissues.[267] Schroeder and Woodward[268] and their co-workers later showed that this inhibition of the glyoxalase reaction was due to an enzyme system which hydrolyzes glutathione. This problem has recently been investigated by Binkley and associates[269] who have demonstrated that the breakdown of the GSH occurs in two steps. One enzyme hydrolyzes GSH to glutamic acid and cysteinylglycine; the second completes the degradation to cysteine and glycine. Bloch and his co-workers[270] have demonstrated the de novo synthesis of GSH in pigeon liver extracts from the constituent amino acids in the presence of ATP. The synthesis has been shown to take place in two stages. One enzyme catalyzes the formation of glutamylcysteine, and the second, the reaction of the latter with glycine to yield GSH. No coenzyme requirement for these reactions has been found.

3. Pyruvate Oxidation Factor

a. Introduction. Until very recently the status of the coenzyme form of the pyruvate oxidation factor (POF) was very obscure. However it now appears that this factor is lipothiamide pyrophosphate.[270a]

b. Historical. An unidentified growth factor for the protozoan *Tetra-*

[264] G. E. Woodward, *J. Biol. Chem.* **109,** 1 (1935).
[265] E. F. Schroeder and G. E. Woodward, *J. Biol. Chem.* **129,** 283 (1939).
[266] J. S. Dohan and G. E. Woodward, *J. Biol. Chem.* **129,** 393 (1939).
[267] H. D. Dakin and H. W. Dudley, *J. Biol. Chem.* **15,** 463 (1913).
[268] E. F. Schroeder, M. P. Munro, and L. Weil, *J. Biol. Chem.* **110,** 181 (1935); E. F. Schroeder and G. E. Woodward, *ibid.* **120,** 209 (1937); G. E. Woodward and F. E. Reinhart, *ibid.* **145,** 471 (1942).
[269] C. K. Olson and F. Binkley, *J. Biol. Chem.* **186,** 731 (1950); F. Binkley and C. K. Olson, *ibid.* **188,** 451 (1951).
[270] S. Yanari, J. Snoke, and K. Bloch, *Federation Proc.* **11,** 315 (1952).
[270a] L. J. Reed and B. G. De Busk, *J. Am. Chem. Soc.* **74,** 3964 (1952).

hymena geleii was first described by Kidder and Dewey.[271, 272] Several substances were shown to produce this effect[273] and were called the "protogens."[274]

Somewhat later O'Kane and Gunsalus[275, 276] observed that, when *Streptococcus faecalis* was grown in a synthetic medium, the harvested cells were unable to oxidize pyruvic acid unless supplemented with yeast extract or other natural materials. The activity of yeast extract could not be replaced by any of the known coenzymes or vitamins. The name pyruvate oxidation factor (POF) was therefore used to designate the biological activity of this factor.

The relationship of POF to protogen and other factors became clear when Snell and Broquist[277] demonstrated that POF as well as protogen could replace the acetate factor of Guirard, Snell, and Williams.[278] Gunsalus and Barker[279] showed that POF could replace the growth factor required by *Butyribacterium retgerii* and Lytle and O'Kane[280] have demonstrated that POF is able to replace reticulogen for the growth of *Streptococcus cremoris*. These nutritional studies indicated that all these growth factors were either identical or very closely related. This fact was subsequently borne out by the crystallization and synthesis of the protogens (thioctic acid) and of POF (lipoic acid).

c. Methods of Assay. Gunsalus and co-workers[281] have developed a method of assay for POF which is based on the oxidation of pyruvate by *S. faecalis* cells harvested from a synthetic medium. Such cells are unable to oxidize pyruvate unless supplemented with materials containing POF. The rate of respiration when preparations of POF are added to such cells is taken as a measure of POF activity and is compared to the activity given by a standard preparation of yeast extract.

Kidder and Dewey[271] as well as Stokstad *et al.*[273] have used the stimulation of growth of *Tetrahymena geleii* as an assay for protogen.

d. Chemistry. By careful analysis of protogen preparations and of preparations of POF it was shown that these natural materials exist in several

[271] G. W. Kidder and V. C. Dewey, *Biol. Bull.* **87,** 121 (1944).
[272] G. W. Kidder and V. C. Dewey, *Arch. Biochem.* **8,** 293 (1945).
[273] E. L. R. Stokstad, C. E. Hoffman, M. A. Regan, D. Fordham, and T. H. Jukes, *Arch. Biochem.* **20,** 75 (1949).
[274] E. L. Patterson, J. A. Brockman, Jr., F. P. Day, J. V. Pierce, M. E. Macchi, C. E. Hoffman, C. T. U. Fong, E. L. R. Stokstad, and T. H. Jukes, *J. Am. Chem. Soc.* **73,** 5919 (1951).
[275] D. J. O'Kane and I. C. Gunsalus, *J. Bact.* **54,** 20 (1947).
[276] D. J. O'Kane and I. C. Gunsalus, *J. Bact.* **56,** 499 (1948).
[277] E. E. Snell and H. P. Broquist, *Arch. Biochem.* **23,** 326 (1949).
[278] B. M. Guirard, E. E. Snell, and R. J. Williams, *Arch. Biochem.* **9,** 381 (1946).
[279] I. C. Gunsalus and H. A. Barker, unpublished, see ref. 10.
[280] V. L. Lytle and D. J. O'Kane, *J. Bact.* **61,** 240 (1951).
[281] I. C. Gunsalus, M. I. Dolin, and L. Struglia, *J. Biol. Chem.* **195,** 849 (1952).

forms. Patterson *et al.*[274] reported the crystallization of protogen B and showed that it contained 30% sulfur. Reed *et al.*[282] reported on the crystallization of α-lipoic acid and showed that it had POF activity. Upon studying its chemical nature they[283] reached the conclusion that α-lipoic acid is a cyclic disulfide with the empirical formula $C_8H_{14}S_2O_2$. In the preparation of bioautographs α-lipoic acid is partially converted to a more acidic substance, which suggests that sulfide is oxidized to sulfoxide.

Subsequently Brockman *et al.*[284, 285] reported on the synthesis of *dl*-thioctic acid which had biological activity corresponding to the protogens and the lipoic acids. Hornberg and his group[286] synthesized α-lipoic acid and showed that it had POF activity. They suggested that the structure of α-lipoic acid is most probably one of the optical isomers of the cyclic disulfide derived from 5,8-dithioloctanoic acid.[287]

The similarity of the results of the group working on the protogens with those of the group working with the lipoic acids further emphasizes the probable identity of these factors.

The results reported in an earlier section, showing the interconvertability of the protogens and the lipoic acids as growth factors for several microorganisms, seems to indicate that the materials are in reality nutritional factors (or vitamins) for the organisms concerned. Although no direct participation of POF in an enzyme system has been demonstrated, the results of Gunsalus and his group showing a requirement for POF in the oxidation of pyruvate suggest the function of POF or a derivative as a coenzyme in this reaction. It is quite possible that the forms of POF so far obtained are not the coenzyme form and that in the assay of POF activity the added materials are converted to the coenzyme form before they exert their activity.

Bound forms of POF do exist in nature and appear to be closely associated with enzymes involved in pyruvate metabolism. Korkes *et al.*[147a] although able to separate pyruvate dismutation into two enzymatic steps, were not able to remove POF from their enzymes. Schweet[144] reported that pyruvic oxidase isolated from pigeon breast muscle, although homogenous in the ultracentrifuge, still contains POF activity. It is possible that these

[282] L. J. Reed, B. G. DeBusk, I. C. Gunsalus, and C. S. Hornberger, Jr., *Science* **114,** 93 (1951).

[283] L. J. Reed, B. G. DeBusk, I. C. Gunsalus, and G. H. F. Schnakenberg, *J. Am. Chem. Soc.* **73,** 5920 (1951).

[284] J. A. Brockman, Jr., E. L. R. Stokstad, E. L. Patterson, J. V. Pierce, M. Macchi, and F. P. Day, *J. Am. Chem. Soc.* **74,** 1868 (1952).

[285] M. W. Biellock, J. A. Brockman, Jr., E. L. Patterson, J. V. Pierce, and E. L. R. Stokstad, *J. Am. Chem. Soc.* **74,** 1868 (1952).

[286] C. S. Hornberg, Jr., R. F. Heitmiller, I. C. Gunsalus, G. H. F. Schnakenberg, and L. J. Reed, *J. Am. Chem. Soc.* **74,** 2382 (1952).

[287] L. J. Reed, Q. F. Soper, G. H. F. Schnakenberg, S. F. Kern, H. Boaz, and I. C. Gunsalus, *J. Am. Chem. Soc.* **74,** 2384 (1952).

forms of POF which appear to be bound to protein may in fact be a tightly bound coenzyme form of POF.

In view of the fact that lipoic acid is a cyclic disulfide, it is tempting to speculate on its function as a 2-carbon carrier in pyruvate oxidation through a thioester in a fashion similar to the CoA function in this reaction. Similarly the disulfide nature of this factor invites speculation on its possible role as an oxidation-reduction coenzyme by going through a sulfhydryl form. It is a curious fact that three of the factors involved in pyruvate oxidation, thiamine pyrophosphate, CoA, and POF, all contain sulfur. In the first two of these factors, the sulfur appears to play an important role in their mechanism of action. By analogy one would suppose that the disulfide grouping of POF will be prominent in its mechanism of action.

Recently Reed and DeBusk[148] have isolated from natural materials a bound form of lipoic acid which seems to be somewhat closer to the coenzyme form. Analysis of this compound as well as its preparation from synthetic materials indicates that it is the amide of thiamine and lipoic acid (lipothiamide). This important discovery may eventually cast some light upon the mechanism of action of lipoic acid. It is significant that lipothiamide can restore pyruvate oxidation to deficient cells, and the authors conclude that lipothiamide may be part of the coenzyme for the oxidative decarboxylation of α-keto acids.

Even more recently, Reed and DeBusk[270a] have observed that lipothiamide pyrophosphate is required for pyruvate dismutation and α-ketoglutarate oxidation by soluble enzyme preparations from an *E. coli* mutant. They suggest that the initial step in the oxidative decarboxylation of α-keto acids is the formation of an acyl lipothiamide pyrophosphate complex. In a subsequent step, they assume that the acyl group is transferred to CoA. These observations cast considerable light on the mechanism of the oxidative decarboxylation of α-keto acids and certainly may explain some of the difficulties already discussed in the section on TPP. However it remains to be seen whethei lipothiamide pyrophosphate carries the acyl group as a thioester as in CoA, and whether it acts exclusively as an acyl carrier and not also as a hydrogen carrier. One point which is still not clear is whether the oxidation and decarboxylation occur simultaneously or in an orderly sequence.

4. Biotin

The vitamin, biotin, has been shown to influence a variety of enzymatic reactions. Although a bound form of biotin, biocytin,[288] has been obtained in crystalline form, no conclusive proof of the existence of a coenzyme form

[288] L. D. Wright, E. L. Crisson, H. R. Skeggs, T. R. Wood, R. L. Peck, D. E. Wolf, and K. Folkers, *J. Am. Chem. Soc.* **72**, 1048 (1950).

of biotin has been presented. Lichstein[289] has recently reviewed all the evidence which suggests the participation of biotin in certain enzymatic reactions and the controversial points are thoroughly considered.

Biotin has been implicated as having some kind of influence in the following enzyme systems: oxaloacetic decarboxylase,[290, 291, 292] aspartic acid, serine, and threonine deaminases,[293] succinic acid decarboxylase,[294] and the synthesis of citrulline.[295] Much of the evidence has been obtained in nutrition studies, and it is not clear whether biotin affects the enzymatic reaction itself, or whether it may be involved in the synthesis of the enzyme or even in some secondary reaction.

The close relationship of biotin deficiency to the behavior of these enzyme systems certainly suggests that, by analogy to the other members of the vitamin B group, biotin may indeed be a coenzyme in some enzymatic reaction. But this point must remain speculative until more concrete evidence for the direct participation of biotin or a biotin derivative in an enzymatic reaction is presented.

Addendum. Recent experiments have now shown[296] that CoA-P-P is not an intermediate in the acetate-ATP reaction. The reaction sequence is as follows:

$$\text{ATP} + \text{E} \rightleftharpoons \text{E-AMP} + \text{P-P} \tag{88}$$

$$\text{E-AMP} + \text{CoA} \rightleftharpoons \text{E-CoA} + \text{AMP} \tag{89}$$

$$\text{E-CoA} + \text{Acetate} \rightleftharpoons \text{Acetyl CoA} + \text{E} \tag{90}$$

This places the acetate-ATP reaction in the class of the "intermediate activation of the enzyme" as depicted in equation 32. If, however, for equation 26 we consider CoA to represent CoX, then the general type of transfer of either the phosphate or pyrophosphate group of ATP will apply. This appears to be a likely possibility in the succinate-ATP reaction as shown by equation 59.

289 H. C. Lichstein, *Vitamins and Hormones*, **9,** 27 (1951).

290 J. L. Stokes, A. Larsen, and M. Gunness, *J. Bact.* **54,** 219 (1947b).

291 H. A. Lardy, R. L. Potter, and C. A. Elvehjem, *J. Biol. Chem.* **169,** 451 (1947).

292 S. Ochoa, A. Mehler, M. L. Blanchard T. H. Jukes, C. E. Hoffman, and M. Regan, *J. Biol. Chem.* **170,** 413 (1947).

293 H. C. Lichstein and W. W. Umbreit, *J. Biol. Chem.* **170,** 423 (1947).

294 E. A. Delwiche, *J. Bact.* **59,** 439 (1950).

295 P. MacLeod, S. Grisolia, P. P. Cohen, and H. A. Lardy, *J. Biol. Chem.* **180,** 1003 (1949).

296 M. E. Jones, F. Lipmann, H. Hiltz, F. Lynen., *J. Am. Chem. Soc.*, in press.

CHAPTER 20

Iron Metabolism

H. F. W. KIRKPATRICK

Department of Clinical Investigation, The London Clinic, London, England

I. General Metabolism and Functions of Iron Compounds in the Body

Metabolism of iron differs from that of other nutritional elements in its divergence from the usual pattern of absorption, utilization, and excretion. The capacity of the body to excrete iron is so limited as to appear negligible, and in compensation absorption is closely regulated according to current body requirements. Within the body, priority in the supply of iron is accorded to cells and tissues for the production of the cytochromes, peroxidase and catalases, respiratory enzymes essential to cell metabolism, and of myoglobin, the oxygen storage compound of muscle, so that a deficiency of these substances is rarely observed. The rest of the body iron is concerned with the production and maintenance of the circulating hemoglobin, the chief function of which is to transport oxygen from the lungs to the tissues where it is utilized in cellular oxidative processes. Hemoglobin is synthesized in the bone marrow and travels in the blood stream entirely confined within the erythrocytes. The latter after an average life of 120 days undergo deterioration and are finally engulfed by the spleen and Kupffer cells of the liver where breakdown of hemoglobin is completed. From this point the pathway of the released iron appears to be divided, some of the iron going into storage in the liver and spleen, while a large part is transported to the bone marrow for immediate utilization in hemoglobin synthesis, thus completing the cycle. The small iron losses which occur normally are replaced by release of iron from the storage depots situated in the liver, the spleen, and the bone marrow in which iron is stored as ferritin, storage iron being maintained by absorption of dietary iron. Transportation of iron in the blood stream is an essential function of the serum or plasma, in which iron travels in inorganic ferric form reversibly bound to a specific globulin termed siderophillin.

The total amount of iron in the human body has been generally estimated at between 4 and 5 g., of which 65 to 75% is combined in complex forms with porphyrins and the rest is in inorganic form bound to proteins. The exact distribution of the various forms of iron present in the body is not known, but the following figures are approximations for the more important compounds,[1] the iron being expressed as percentage of total body iron: hemoglobin, 60 to 70; myoglobin, 3 to 5; hemoprotein enzymes, less than 1; ferritin, 15; serum iron, 0.1.

II. Absorption

Absorption of iron does not occur uniformly along the gastrointestinal tract, the capacity for iron absorption showing a gradient throughout the entire tract,[2] absorption normally taking place almost entirely in the region of the duodenum.[1] The area of absorption is extended under therapeutic administration of iron, but although total absorption is increased the efficiency falls off rapidly with increasing dose and there is a definite early limit to the amount that can be absorbed.

Absorption takes place in the form of ferrous ions, and the term "available" has been applied to the iron present in foods which is readily converted to this form. Its reaction with α,α'-dipyridyl has been employed to determine its proportion, but this procedure has not achieved any great biological significance and has fallen into disuse. In its wider sense availability depends upon a number of factors, the more well-defined of which are indicated below. In addition to these factors absorption of iron is de-

[1] S. Granick, *Bull. N. Y. Acad. Med.* **25,** 403 (1949).

[2] W. B. Stewart, C. L. Yuile, H. A. Claiborne, R. T. Snowman, and G. H. Whipple, *J. Exptl. Med.* **92,** 375 (1950).

pendent upon a regulatory mechanism which controls the passage of iron across the mucosal cells of the intestinal tract in accordance with the needs of the body.

1. Factors Affecting Absorption

a. Gastric Digestion. Production of ferrous ions from the iron of foodstuffs preparatory to absorption appears to be an essential function of the stomach. Reduction of iron by mixtures of certain foods with pepsin and hydrochloric acid observed *in vitro*[3] has been found to occur equally well in the stomach of normal human beings.[4] Thus breads, meats, and fruits give reduction of ferric iron to ferrous as high as 50 to 90%. Proteins and their digestion products and ascorbic acid all appear to be concerned in the reaction, and it is probable also that hydrochloric acid is essential for production of ferric ions from the colloidal ferric hydroxide of foods, and that achlorhydria, or even hypochlorhydria, has an adverse effect upon absorption.

b. Ascorbic Acid. The existence of a definite relationship between iron metabolism and ascorbic acid under physiological conditions is doubtful,[5] but there is considerable evidence that oral administration of ascorbic acid during the feeding of iron salts can increase the efficiency of absorption, apparently by reason of its reducing action.[6-9]

c. Phosphates. These compounds cause a decrease in iron absorption, an effect which may be related to the calcium-phosphorus ratio in the diet.[10] Where this ratio is low, excess phosphate combines with iron rendering it insoluble and not available for absorption. The influence of this factor has recently received further emphasis by the finding that the amount of iron deposited in the liver of rats is inversely related to the phosphorus content of the diet.[11]

d. Phytic Acid. Both ferrous and ferric phytates are even less soluble than the phosphates, and the presence of phytate in the diet consequently has been stated to have an adverse effect upon absorption.[12] Some doubt has been cast upon the effect by more recent work in which retention of iron

[3] S. L. Tompsett, *Biochem. J.* **34,** 961 (1940).
[4] O. Bergeim and E. R. Kirch, *J. Biol. Chem.* **177,** 59 (1949).
[5] L. E. Totterman, *Acta Med. Scand.* **134,** *Suppl.* 230, 153 (1949).
[6] C. V. Moore, V. Minnich, and J. Welch, *J. Clin. Invest.* **18,** 543 (1939).
[7] C. V. Moore, W. R. Arrowsmith, J. Welch, and V. Minnich, *J. Clin. Invest.* **18,** 553 (1939).
[8] J. F. Powell, *Quart. J. Med.* **13,** 19 (1944).
[9] I. C. Prina and L. Barbieri, *Boll. soc. ital. biol. sper.* **24,** 1312 (1948).
[10] L. S. P. Davidson, H. W. Fullerton, J. W. Howie, J. M. Croll, J. B. Orr, and W. Godden, *Brit. Med. J.* **1,** 685 (1933).
[11] D. M. Hegsted, C. A. Finch, and J. D. Kinney, *J. Exptl. Med.* **90,** 147 (1949).
[12] R. A. McCance and E. M. Widdowson, *Lancet,* **ii,** 126 (1943).

was found to be the same with both high and low phytate phosphorus diets.[13]

e. Pyridoxine. In pyridoxine deficiency in swine, absorption of iron continues without utilization, resulting in high serum iron values and hemosiderosis of liver and spleen.[14] In the same condition in rats continued absorption is demonstrated by a significant increase in total body iron.[15] The suggested explanation of these findings is that the regulatory mechanism controlling absorption breaks down in pyridoxine deficiency, and as pyridoxine is known to be concerned in protein metabolism it is possible that the defect occurs in the ferritin-apoferritin system.

2. Regulatory Mechanism of Absorption

Present theories with regard to the manner in which iron is absorbed are based upon the principle originally put forward by McCance and Widdowson[16] that there is a negligible excretion of iron by the intestines and that absorption is controlled by body requirements. Relationship between absorption and body needs has been demonstrated by feeding radioactive iron to dogs. Normal dogs absorbed only a small amount, whereas dogs rendered anemic by chronic bleeding absorbed up to twenty times this amount.[17] The observation of a latent period or lag in response to iron therapy in acute anemia[18] suggested the existence of a critical level of iron in the mucosal cells above which absorption is blocked, and this has been linked with ferritin by the finding of a significant increase in the ferritin content of the duodenal mucosa of guinea pigs after feeding ferrous iron.[19] A similar increase occurs in the intestinal mucosa of the horse.[20]

Some explanation of the one-way absorption of iron was afforded by the mucosal block theory in which discharge of iron into the blood stream by the mucosal cells was assumed to be controlled by the concentration of iron in the serum.[21] Serious objections to this concept were provided by the finding of high iron absorption in untreated pernicious anemia[22] and in pyridoxine deficiency,[14, 15] in which conditions serum iron is high. Uptake

[13] A. R. P. Walker, F. W. Fox, and J. T. Irving, *Biochem. J.* **42,** 452 (1948).

[14] G. E. Cartwright, M. M. Wintrobe, and S. Humphreys, *J. Biol. Chem.* **153,** 171 (1944).

[15] C. J. Gubler, G. E. Cartwright, and M. M. Wintrobe, *J. Biol. Chem.* **178,** 989 (1949).

[16] R. A. McCance and E. M. Widdowson, *Lancet* **ii,** 680 (1937).

[17] P. F. Hahn, W. F. Bale, E. O. Lawrence, and G. H. Whipple. *J. Exptl. Med.* **69,** 739 (1939).

[18] P. F. Hahn, W. F. Bale, J. F. Ross, W. M. Balfour, and G. H. Whipple, *J. Exptl. Med.* **78,** 169 (1943).

[19] S. Granick, *J. Biol. Chem.* **164,** 737 (1946).

[20] B. W. Gabrio and K. I. Salomon, *Proc. Soc. Exptl. Biol. Med.* **75,** 124 (1950).

[21] S. Granick, *Ann. N. Y. Acad. Sci.* **48,** 657 (1947).

[22] R. Dubach, S. T. E. Callendar, and C. V. Moore, *Blood* **3,** 526 (1948).

of iron thus appears to be independent of the level of serum iron, and an alternative hypothesis has been advanced. The suggested regulatory mechanism[1] may be summarized as follows: When absorbed into the mucosal cell ferrous iron is oxidized to the ferric condition and attaches itself to apoferritin to form ferritin. At the blood stream end of the cell the ferric iron is converted by cellular-reducing substances to the ferrous condition, detached from ferritin, and passed into the blood stream where it is autoxidized and attached to siderophillin for transport. Two regulatory mechanisms are assumed to exist: (1) absorption of iron into the cell is controlled by the content of ferritin, becoming blocked when the cell is physiologically saturated with iron; and (2) release of iron into the blood stream is dependent upon the oxygen level of the blood which is determined by its hemoglobin level. Where the latter is low, conditions are set up favoring the formation of ferrous iron and its transfer to the blood stream.

III. Excretion

1. Urine and Feces

Urinary excretion of iron occurs in normal persons at a fairly constant rate with an average of about 1 mg. per day. This amount is increased for about 24 hr. after intravenous injection, but the total amount excreted represents only 1 or 2% of that injected.[23, 24] The inability of the body to excrete iron by this route is well illustrated in hemolytic anemia and in hemolysis induced by acetylphenylhydrazine during treatment of polycythemia when less than 0.5% of the large amount of iron liberated in the body by destruction of red cells is excreted in urine and feces.[25, 26]

2. Bile

Knowledge with regard to excretion of iron by this route is rather limited, but experimental observations suggest that there is a regular daily excretion of iron in the bile, comparatively small in amount, which may be increased in certain conditions. In dogs the daily excretion varies between 0.1 and 0.2 mg. and is little affected by oral or parenteral iron, but it increases to ten times this amount when hemolysis is induced with phenylhydrazine.[27] In the latter condition, however, the increased excretion represents only 3% of the amount of iron calculated to be released from hemoglobin.

3. Skin

The statement that dermal excretion of iron amounts to 6.5 mg. per day under conditions of minimal sweating[28] has not been confirmed. In dogs no

[23] R. A. McCance and E. M. Widdowson, *J. Physiol. (London)* **94,** 148 (1938).

[24] P. F. Hahn, W. F. Bale, R. A. Hettig, M. D. Kamen, and G. H. Whipple, *J. Exptl. Med.* **70,** 443 (1939).

[25] R. A. McCance and E. M. Widdowson, *Quart. J. Med.* **30,** 277 (1937).

[26] R. A. McCance and E. M. Widdowson, *Nature* **152,** 326 (1943).

[27] W. B. Hawkins and P. F. Hahn, *J. Exptl. Med.* **80,** 31 (1944).

[28] H. H. Mitchell and T. S. Hamilton, *J. Biol. Chem.* **178,** 345 (1949).

significant excretion of radioactive iron occurs by way of the skin after giving iron orally or intravenously,[29] and the results of other experiments with human subjects suggest that the iron content of sweat is dependent upon its cell content and that dermal excretion of iron is due primarily to desquamation and not to sweating.[30]

IV. Iron in Blood

1. Hemoglobin

The primary function of hemoglobin and a general outline of its metabolic pathway so far as it concerns iron metabolism are given in the first section of this chapter. It is a conjugated protein containing 0.34% of iron, of molecular weight 68,000, in which the metalloporphyrin hem is in complex combination with the protein globin. The chemical structure of hem was established as ferroprotoporphyrin 9, type III, and its synthesis accomplished in 1929.[31] It consists of four substituted methylpyrrole units bound together by methene bridges with an iron atom situated in the center of the porphyrin lattice thus formed, bound to the nitrogen atom. The structure of the globin molecule has been deduced from X-ray analysis[32] to consist of four disk-like layers superimposed upon one another, each layer being made up of folded polypeptide chains. The hem molecules, which may also be considered of disk-like shape, although very much smaller than the globin units, are attached to the surface of the globin molecule, four hems to one globin, and lie parallel to one another with their flat surfaces normal to the polypeptide chains. Union is postulated to be brought about firstly by electrostatic attraction between ionized propionic acid groups of hem and surface basic groups of globin, and secondly by coordination of iron with a nitrogen containing group of the globin, probably imidazole from histidine. The latter union stabilizes the iron in the ferrous condition and allows it to be reversibly bonded to oxygen, thus giving hemoglobin its essential function of oxygen carrier.[33]

Biosynthesis of hemoglobin is assumed to take place in two stages in the bone marrow, globin being first built up in the differentiating stem cell,[34] while synthesis of hem and its attachment to globin occurs during later stages of development of the red cell. Knowledge concerning the biosynthesis of hem has been largely clarified by definite evidence that

[29] W. B. Stewart, R. T. Snowman, C. T. Yuile, and G. H. Whipple, *Proc. Soc. Exptl. Biol. Med.* **73,** 473 (1950).

[30] W. S. Adams, A. Leslie, and M. H. Levin, *Proc. Soc. Exptl. Biol. Med.* **74,** 46 (1950).

[31] H. Fischer and K. Zeile, *Ann.* **468,** 98 (1929).

[32] M. F. Perutz, *Proc. Roy. Soc.* (*London*) **A195,** 474 (1949).

[33] C. Rimington, *Lancet* **ii,** 551 (1951).

[34] B. Thorell, *Acta Med. Scand. Suppl.* 200 (1947).

[35] D. Shemin and D. Rittenberg, *J. Biol. Chem.* **166,** 621, 627 (1946).

glycine[35] and acetate[36] are utilized by the body as starting compounds. It is probable, therefore, that these are first condensed into pyrrole ring units, four of which unite to give protoporphyrin into which the iron enters to form hem. The exact steps by which this process is accomplished, however, are still a subject for speculation.[37]

The quantity of hemoglobin in blood shows considerable variation between normal healthy individuals and in addition is generally higher in men than in women, the probable limits expressed in grams of hemoglobin per 100 ml. blood being 14 to 18 and 12 to 16, respectively. Infants have a comparatively high blood hemoglobin at birth and during the first week of life when it averages about 19, but this figure then drops steadily to about 12 at 3 months. The difference between the sexes does not arise until puberty and is no doubt attributable to the loss from menstruation. The daily destruction and replacement of hemoglobin is of the order of 7 to 8 g., equivalent to a daily turnover of about 25 mg. of iron, an amount which necessitates the replacement of between 2 and 3 million red cells per second.

2. Serum Iron

It is now generally recognized that the transport of iron within the body is a function of the serum or plasma. Within the last few years the manner in which this iron travels in the blood stream has been elucidated, and it has been established that serum iron is quantitatively bound to a specific protein which has been isolated, crystallized, and characterized.[38, 39, 40] The protein is a β_1-globulin occurring in Fraction IV-7 (Cohn) of molecular weight 90,000, each molecule of which is capable of binding two atoms of iron, the complex being salmon-pink in color. This protein is now generally referred to as siderophillin, although the term transferrin has also been employed.[41]

Combination of iron with siderophillin occurs in the presence of carbon dioxide, the complex containing one CO_2 molecule to each iron atom, the iron being in the ferric state.[42] The reaction is dependent upon pH, being at a maximum in neutral solution (pH > 6.5) becoming less with decreasing pH until the latter is less than 5 when the whole of the iron becomes dialyzable. The iron is also directly removed from the complex when converted to the ferrous state by a reducing agent.

[36] K. Bloch and D. Rittenberg, *J. Biol. Chem.* **159,** 45 (1945).
[37] A. Neuberger, H. M. Muir, and C. H. Gray, *Nature* **165,** 948 (1950).
[38] A. L. Schade and L. Caroline, *Science* **104,** 340 (1946).
[39] C. B. Laurell and B. Ingelman, *Acta Chem. Scand.* **1,** 770 (1947).
[40] D. M. Sargenor, B. A. Koechlin, and L. E. Strong, *J. Clin. Invest.* **28,** 73 (1949).
[41] C. B. Laurell, *Blood* **6,** 183 (1951).
[42] A. L. Schade, R. W. Reinhart, and H. Levy, *Arch. Biochem.* **20,** 170 (1949).

It is not certain whether siderophillin serves as a true carrier of iron in the same manner as hemoglobin is a carrier of oxygen, but a consideration of the available evidence[41] suggests that the data largely support the carrier theory. It is probable, therefore, that iron leaves the blood stream in an ionized form and not as an iron-globulin complex.

The iron-binding property of siderophillin is specific to this particular protein and is not exhibited by any of the other proteins of serum. It follows, therefore, that the capacity of serum for transport of iron is limited by its content of siderophillin. Intravenous injection of iron produces a rise in the serum iron level to the saturation value only, any excess rapidly leaving the blood stream to enter the tissues where toxic effects may ensue.[43] The iron-binding capacity of serum has been measured by a number of investigators,[43-46] the results being expressed in micrograms of iron per 100 ml. of serum. The mean results for normal subjects have shown general agreement and vary from 300 to 360. The figures for the actual amount of serum iron present in the same subjects were 106 to 127 for males and 94 to 123 for females, so that the degree of saturation normally lies between 30 and 40%. Values occurring in various pathological conditions have also been recorded by the same workers, but the exact significance of the iron-binding capacity and degree of saturation is still obscure. The suggestion that the degree of saturation may be significant in affecting alimentary absorption[44] has not received any confirmation from later work.[47]

Variations in the serum iron observed pathologically are explicable on the assumption that the serum level indicates the iron available for hemoglobin synthesis.[48] In anemia due to iron deficiency arising from low intake or absorption, loss from hemorrhage, or excessive uptake of iron by the tissues as in the anemia of infection, the serum iron is low. It is also low when hemoglobin synthesis is accelerated, as during treatment of pernicious anemia. When hemoglobin synthesis is retarded in the face of adequate stores of iron, as in untreated pernicious anemia, aplastic anemia, and pyridoxine deficiency, the serum iron is high. In hemolytic anemia accelerated hemoglobin synthesis is coincident with accelerated breakdown of hemoglobin and the serum iron may fluctuate according to which of these processes predominates.

[43] C. G. Holmberg and C. B. Laurell, *Acta Physiol. Scand.* **10,** 307 (1945).

[44] C. B. Laurell, *Acta Physiol. Scand. Suppl.* 46 (1947).

[45] E. R. Rath and C. A. Finch, *J. Clin. Invest.* **28,** 79 (1949).

[46] G. E. Cartwright and M. M. Wintrobe, *J. Clin. Invest.* **28,** 86 (1949).

[47] C. L. Yuile, J. W. Hayden, J. A. Bush, H. Tesluk, and W. B. Stewart, *J. Exptl. Med.* **92,** 367 (1950).

[48] G. E. Cartwright, C. M. Huguley, H. Ashenbrucker, J. Fay, and M. M. Wintrobe, *Blood* **3,** 501 (1948).

V. Iron in Tissues

1. Ferritin

Ferritin was isolated by Laufberger in 1937[49] from horse spleen, which is the richest source, by addition of cadmium sulfate to aqueous extracts when ferritin readily crystallizes out. It is a brown, crystalline substance containing 17 to 23% of iron, the proportion not being absolutely constant, which is removed by the action of a reducing agent in acid solution leaving the iron-free protein apoferritin.[50] The latter is a colorless protein of the globulin class with a molecular weight of about 460,000, which crystallizes on addition of cadmium sulfate in the same manner as ferritin.

Ferritin has been shown to contain a small amount of phosphorus which is not contained in the apoferritin molecule, and the composition of the iron compound in ferritin has been suggested to approximate to the formula $(FeOOH)_8 \cdot (FeO \cdot OPO_3H_2)$, or one phosphate group to nine atoms of iron.[51] Examination by X-ray analysis[52] has indicated the structure of the ferritin crystal to be essentially that of the apoferritin crystal, and the former has been visualized[1] as consisting of ellipsoid apoferritin molecules, which are possibly bound together by cadmium atoms, with small clusters of ferric hydroxide phosphate attached to their surfaces and contained in the spaces between them. Attempts to synthesize ferritin from apoferritin and an iron compound have been unsuccessful, and it is believed that the formation is essentially a catalyzed process. This supposition is supported by the finding that the magnetic susceptibility of ferritin iron is quite characteristic, differing from all other physiological forms of iron as well as from iron in ferric hydroxide precipitated *in vitro* by chemical methods.[53]

The formation of apoferritin in the body appears to be remarkable in that the presence of apoferritin itself has not been demonstrated in any of the cells of tissues associated with ferritin production, viz., the liver, the spleen, bone marrow, and intestinal mucosa. The explanation may be either that synthesis and breakdown of apoferritin is a continuous process in which breakdown is prevented only by the presence of iron,[1] or that synthesis of apoferritin is stimulated by the presence of iron with breakdown occurring immediately upon removal of iron.[53]

The storage function of ferritin has been amply demonstrated by experiments with radioactive iron,[54] the results further indicating that iron newly absorbed from the alimentary tract is stored mainly in the liver,

[49] M. Laufberger, *Bull. soc. chim. biol.* **19,** 1575 (1937).
[50] S. Granick and L. Michaelis, *J. Biol. Chem.* **147,** 91 (1943).
[51] A. Rothan, *J. Biol. Chem.* **152,** 679 (1944).
[52] I. Frankuchen, *J. Biol. Chem.* **150,** 57 (1943).
[53] L. Michaelis, *Advances in Protein Chem.* **3,** 59 (1947).
[54] P. F. Hahn, S. Granick, W. F. Bale, and L. Michaelis, *J. Biol. Chem.* **150,** 407 (1943).

whereas iron released from hemoglobin during destruction of red cells may be stored in both the liver and the spleen. In addition to its storage function and its role in regulating absorption, ferritin has recently acquired further physiological significance by being identified with a vasopressor principle present in the liver which is concerned with the regulation of peripheral circulation.[55] Although not present in normal human serum, ferritin is detectable by immunochemical methods in the blood of patients suffering from hemorrhagic shock, hepatic cirrhosis, heart failure, toxemia of pregnancy, essential hypertension, and in the edema fluid of nephrosis and congestive heart failure.[56, 57, 58] It has further been shown to possess antidiuretic properties.[59]

2. Hemosiderin

This is a brown, granular pigment present in various tissues throughout the body, the iron in which is microscopically stainable in contrast to that of ferritin which cannot be viewed by histochemical means. It is not a normal storage form of iron as it is present only when there are ample supplies of ferritin, although it can be mobilized for hemoglobin synthesis when required, as the results of histochemical examination of the sternal marrow[60] have shown that in cases of iron deficiency anemia hemosiderin is absent from the marrow. This method has been employed for the estimation of the status of storage iron as an indication of the need for iron therapy, the latter being of benefit only when hemosiderin iron is absent from the marrow.

The iron in hemosiderin, which may be as high as 35% by weight, is similar in magnetic properties to that in ferritin, and it is possible that hemosiderin represents a further stage of iron deposition in tissues beyond the formation of ferritin.[1] On this basis the formation of hemosiderin may be regarded as ensuing from absorption of iron by the tissues at a rate greater than that at which the appropriate amount of apoferritin can be synthesized, with the result that the accumulating ferric hydroxide-phosphate units polymerize into large clusters instead of small ones for attachment to apoferritin.

Large deposits of hemosiderin occur in certain pathological conditions. In hemochromatosis deposition occurs spontaneously in tissues throughout

[55] A. Mazur and E. Shorr, *J. Biol. Chem.* **176,** 771 (1948).

[56] E. Shorr, B. W. Zweifach, and R. F. Furchgott, *Science* **102,** 489 (1945).

[57] E. Shorr, B. W. Zweifach, R. F. Furchgott, and S. Baez, *Trans. Assoc. Am. Physicians* **60,** 28 (1947).

[58] S. Baez, B. W. Zweifach, A. Mazur, and E. Shorr, *Proc. Soc. Exptl. Biol. Med.* **64,** 154 (1947).

[59] S. Baez, A. Mazur, E. Shorr, D. Metz, I. Litt, and R. Frenkel, *Am. J. Physiol.* **162,** 198 (1950).

[60] C. E. Rath and C. A. Finch, *J. Lab. Clin. Med.* **33,** 81 (1948).

the body, giving rise to dark pigmentation and damage to the tissues concerned; similar effects occur chiefly in the liver and the spleen in hemosiderosis resulting from multiple blood transfusions in the treatment of hemolytic and aplastic anemias.

VI. Iron Deficiency

1. Etiology

The commonest causes of iron deficiency are either nutritional or hemorrhagic. The former may arise from low intake or absorption and is rarely seen in the adult man but is fairly common in women. This sexual difference is due to the higher requirements of the latter to offset the loss of iron through menstruation or the extra demands of the fetus in pregnancy. The daily iron loss averages about 1 mg. in men and 2 mg. in women, and daily intakes of 5 to 10 mg. for men and 15 mg. for women are generally regarded as the minimum requirements for maintaining a normal blood picture. Growing children have the same requirements as women. In pregnancy fetal requirements are balanced in the early period by the saving of the loss from menstruation, and later by an increase in the efficiency of absorption,[61] but even so the minimal requirement, especially during the last 3 months, is probably higher than in the non-pregnant woman and is better regarded as 20 mg. per day.

Iron deficiency arising from hemorrhage may be due to a variety of common disorders, chronic or acute, such as gastric and duodenal ulcers, hemorrhoids, menorrhagia, and uterine hemorrhage at parturition.

2. Therapy

Treatment of iron deficiency by dietary means is a long and doubtful process, and recourse is invariably made to iron administration. Such treatment is likely to be of benefit only in anemia due to nutritional causes and blood loss, for in the anemia of infection or inflammation it is indicated only after the condition has subsided. In other anemias arising from disordered metabolism of iron it is likely to be harmful. The important point to be noted with regard to oral administration is that efficiency of absorption falls off with increasing dosage, so that although large doses are necessary for the best clinical results there is a limit beyond which all the iron merely goes to waste. For this reason it is usual to prescribe a daily dosage not greater than 25 grains of ferrous sulfate, absorption being aided by simultaneous administration of ascorbic acid.

Preparations of saccharated iron oxide suitable for intravenous injection are now available, and as a means of by-passing the barrier of regulated absorption and getting quick results their use is increasing, but as this

[61] P. F. Hahn, *Federation Proc.* **7**, 493 (1948).

method is potentially lethal its application is necessarily subject to circumspection. The obvious danger attending the use of intravenous iron is the development of hemosiderosis, but reports so far available, especially the study made by Andersson,[62] suggest that this possibility is rather more remote than anticipated. A total dosage by this route of 25 to 40 mg. of elemental iron for each 1% deficit of hemoglobin with a limit of 100 mg. for a single dose has given good results with very few toxic reactions,[63-66] being especially useful in the anemia of pregnancy.

[62] N. S. E. Andersson, *Acta Med. Scand.* **138,** *Suppl.* 241 (1950).

[63] D. L. Harrigan, J. F. Mueller, and R. W. Vilter, *J. Lab. Clin. Med.* **36,** 422 (1950).

[64] R. J. G. Sinclair and J. J. R. Duthie, *Brit. Med. J.* **ii,** 1257 (1950).

[65] J. M. Scott and A. D. T. Govan, *Lancet* **i,** 367 (1951).

[66] R. G. Holly, *Blood* **6,** 1159 (1951).

CHAPTER 21

Calcium and Phosphorus Metabolism

H. F. W. KIRKPATRICK AND J. DOUGLAS ROBERTSON

Department of Clinical Investigation, The London Clinic, London, England

I. Functions and Distribution

1. Functions

a. Calcium. The chief function of calcium in the body is concerned with the formation and maintenance of the skeletal structure. The important physiological actions of calcium, however, appear to provide some justification for the view held in some quarters that this is in reality a secondary function to the provision of an adequate store directed to the maintenance of a constant level of calcium ions in blood, lymph, and tissues. Calcium ions are an essential factor in nerve conductivity and in muscle contraction, variations in concentration from the limited optimal range having profound effects upon these processes. Thus, a low concentration produces hyperirritability and tetany, and a high concentration causes depressions of nerve conductivity and muscle rigor. Calcium ions play a part in the maintenance of normal rhythm of heart beat and also are necessary for the natural coagulation of blood, although there is no evidence that the coagulation time is ever affected by lack of calcium in pathological conditions.

b. Phosphorus. The major part of the body phosphorus is intimately associated with calcium in skeletal metabolism, but a higher proportion than is the case with calcium is concerned in other vital processes. By their function in phosphorylation, phosphates form an intermediate stage in the metabolism of fats and carbohydrates and are also used by the body in building up the more permanent organic phosphates, including catalysts, which are essential to cell structure and metabolism. Phosphates provide the energy-rich bonds in compounds such as adenosinetriphosphate which are important in muscle contraction, and they are utilized in the formation of the phosphatides of nerve tissue, the nucleoproteins of cell nuclei, and phosphoproteins such as casein of milk.

2. Distribution

Of the total calcium in the body, 99% is contained in bones and teeth, and if the tendons and ligaments are included the complete skeletal structure contains 99.5% of the body's calcium.[1] The remaining 0.5% is distributed among the blood, the lymph, and the soft tissues in which the concentration varies somewhat but in general that of the tissues and lymph approximates to the diffusible calcium concentration of plasma.[2] Variable estimates of the distribution of phosphorus suggest that rather more than 85% of the total phosphorus is combined with calcium in the skeleton, about 10% is in muscle tissue, rather less than 1% in brain, and about 0.3% in blood, the remainder being in other tissues and tissue fluids. The

[1] H. H. Mitchell, T. S. Hamilton, F. R. Steggerda, and H. W. Bean, *J. Biol. Chem.* **158,** 625 (1945).

[2] See C. L. A. Schmidt and D. M. Greenberg, *Physiol. Revs.* **15,** 297 (1935).

difficulties of obtaining accurate figures for distribution in humans are obvious, and very few analyses are available.

II. Calcium and Phosphorus in Blood

1. Calcium

In blood, calcium is confined to the plasma or serum, the red blood cells being essentially calcium-free. The concentration in serum, expressed in milligrams per cent, averages 10.4 and normally lies between 9.9 and 11.2[3] but extreme variations in health may be found between 9.0 and 11.4. Infants appear generally to have a higher serum calcium level than the adult average, the mean figure being 11.2,[4] but no variation has been detected in old age.[3]

It appears to be generally accepted that about 50 to 60% of the total serum calcium is dialyzable and therefore presumably in ionized form while the remainder is non-diffusible and is bound to protein, but any exact relationship between calcium and protein in serum remains obscure. The original suggestion that this relationship could be expressed by the simple mass law equation[5]

$$\frac{[\text{Ca}^{++}]\,[\text{Protein}^{--}]}{[\text{Ca proteinate}]} = K$$

received support from further data,[6] but a somewhat different relationship was arrived at by Greenberg and Tufts,[7] who suggested the following equation:

$$\text{Total Ca} = \frac{[\text{Ca}^{++}]\,[\text{Total protein}]}{[\text{Ca}^{++}] + K} + \text{Ca}^{++}$$

In normal sera the value of $[\text{Ca}^{++}]$ may be regarded as constant and the expression becomes Total Ca $= m$ Total protein $+ b$. This equation was found to be not applicable to sera from patients with hyperglobulinemia, and this led to an examination of the calcium-binding capacity of different globulin fractions,[8] the results of which showed considerable variation. A significant difference in the calcium-binding capacity of purified albumins from different species has also been found,[9] the results further suggesting that at least two albumins exist in normal sera which differ in calcium-binding power. The finding of such differences in the capacity of the various

[3] J. D. Robertson, *Lancet* **ii**, 97 (1941).
[4] A. T. Shohl, Mineral Metabolism, Reinhold Publishing Corp., New York, 1939.
[5] F. C. McLean and A. B. Hastings, *J. Biol. Chem.* **107**, 337 (1934); **108**, 285.
[6] E. G. Weir and A. B. Hastings, *J. Biol. Chem.* **114**, 397 (1936).
[7] D. M. Greenberg and E. V. Tufts, *J. Biol. Chem.* **119**, xxxix (1937).
[8] N. Drinker, A. A. Green, and A. B. Hastings, *J. Biol. Chem.* **131**, 641 (1939).
[9] N. H. Martin and D. J. Perkins, *Biochem. J.* **47**, 323 (1950).

protein fractions of serum for combining with calcium is in harmony with the conclusion of Ludewig and his associates,[10] formed from ultracentrifuge studies of the calcium-protein relationship, that the application of the simple mass law to human serum is unreliable. It would appear that the only positive statement to be made with regard to this relationship is that it is a reversible equilibrium and that the complex may be regarded as a reserve of calcium ions for physiological use.

The level of the calcium in blood is largely maintained by the action of the parathyroid glands which are stimulated to hypersecretion of their hormone by low serum levels and generally to hyposecretion by high levels, the effect being to bring about a return of calcium concentration to normal. The actions of this and other factors influencing blood calcium are discussed later in this chapter.

2. Phosphorus

The diversity of phosphorus compounds in the body is reflected in the blood in which, in addition to free inorganic phosphate, phosphoric esters, phospholipids, nucleotides, and a small amount of nucleic acid circulate. The presence of such complex compounds serves to emphasize the fact that although most of the body's phosphorus is associated with calcium in skeletal metabolism the inorganic phosphate of the blood, which is of chief significance in this respect, is in dynamic equilibrium with other vital processes beyond the range of this chapter, disorders in which may also affect phosphorus metabolism. The inorganic phosphate of the blood is normally present in completely ionized form, distributed equally between the cells and the plasma, the concentration in normal adults, in milligrams per cent, usually lying between 2 and 4, whereas in infants and children it is rather higher, being normally between 3 and 5. Increase in the calcium or phosphate concentration in blood to a high level, e.g., by intravenous injection of phosphate, leads to the formation of a colloidal calcium phosphate complex[11] which is stated to be phagocytized in the liver spleen and bone marrow.[12]

The level of inorganic phosphate in blood is controlled by a balance of various factors such as phosphatase activity, parathyroid hormone, and vitamin D.

III. Skeletal Calcium and Phosphorus

Bone consists essentially of an organic matrix laid down by osteoblasts upon which mineral matter, chiefly calcium and phosphate, is deposited, the mature bone consisting of a hard outer cortex and an inner area termed

[10] S. Ludewig, S. Chanutin, and A. V. Masket, *J. Biol. Chem.* **143,** 753 (1942).
[11] F. C. McLean and M. A. Hinrichs, *Am. J. Physiol.* **121,** 580 (1938).
[12] I. Gersh, *Am. J. Physiol.* **121,** 589 (1938).

the trabeculae. From a metabolic viewpoint the former area is relatively immobile and, except where there is a very severe demand for calcium and phosphorus, plays no important part in calcium and phosphorus interchange going on within the body. The trabeculae, on the other hand, are in a constant state of mobility and function as immediate storage depots in and from which calcium and phosphorus are stored or released as the need arises. Deposition of calcium and phosphate in bone occurs in the form of a double salt, the formula of which is usually written $CaCO_3 \cdot 3Ca_3(PO_4)_2$, some of the calcium being replaceable by magnesium. This salt is related both in composition and in X-ray characteristics to the apatite series of minerals.

Deposition and resorption of bone salt is a continuous process which is not yet fully understood although it is certain that a major participant in calcification is the phosphatase of bone which acts by accelerating the hydrolysis of organic phosphates to produce local concentrations of phosphate, thus bringing about precipitation. Calcification, however, is by no means as simple as this and is postulated by Logan and Taylor[13] to proceed stepwise, the first step consisting of aggregation of calcium and phosphate ions in almost equal numbers followed by loss of phosphate from the aggregate with formation of the apatite crystal lattice which then grows by removal of calcium and phosphate ions from solution. It is assumed that the carbonate moiety of bone salt is taken up in the last stage of formation by an exchange of ions between the liquid phase and the precipitate arising from changes in the composition of the former with which the latter has to be brought into equilibrium. Replacement of calcium by other cations may also occur at this stage. Logan[14] concludes that the critical factor in the initiation of deposition may be the ionic product $(Ca^{++})(HPO_4^{--})$ rather than $(Ca^{++})^3(PO_4^{---})^2$, although the resulting precipitate is $Ca_3(PO_4)_2$. $CaHPO_4$ has not been detected in normal bones, but it has been found to the extent of about 23% in the bones of rats fed a high-phosphorus low-calcium diet without vitamin D.[15]

Calcification proceeds in the adult as well as during growth, and an adequate supply of calcium and phosphorus is thus essential throughout life to a healthy skeletal structure, including teeth in which mineral turnover proceeds after growth at a reduced rate. Certain factors, therefore, which influence absorption or retention of these elements can bring about demineralization of bone in some conditions. Their effects are discussed in the following sections.

[13] M. A. Logan and H. L. Taylor, *J. Biol. Chem.* **119,** 293 (1937); **125,** 377, 391 (1938).

[14] M. A. Logan, *Physiol. Revs.* **20,** 522 (1940).

[15] A. Hirschmann, A. E. Sobel, B. Kramer, and I. Frankuchen, *J. Biol. Chem.* **171,** 285 (1947).

IV. Factors Influencing Absorption

1. Calcium-Phosphorus Ratio

In an adequate, well-balanced diet the ratio of calcium to phosphorus is of little significance from the nutritional viewpoint, but in less-balanced diets this ratio assumes some importance. At the pH of the intestine calcium phosphate is difficultly soluble, and excessive amounts of calcium will render most of the phosphate insoluble and thus not available for absorption; a high ratio of phosphate to calcium has a similar effect upon the calcium. This effect is enhanced when there is a deficiency of vitamin D, the incidence of rickets being directly related both to the actual amount ingested and to the calcium-phosphorus ratio.

In infants the ideal calcium-phosphorus ratio is 2:1, which represents the proportion of these minerals in mother's milk. With increasing age this ratio is not so important and may fall below 1.

2. Phytic Acid

The anticalcifying effect of cereals was first noted by Mellanby[16] and was ultimately traced to the presence of phytin, or inositol hexaphosphoric ester. This substance constitutes 50 to 80% of the phosphate of cereals and is hydrolyzed only to the extent of 37 to 64%[17] in the alimentary tract. Such phosphate is thus largely not available for absorption, and as in addition it forms with calcium an insoluble calcium phytate, a proportion of the ingested calcium is also rendered non-available. Some investigators[17, 18, 19] have concluded that the anticalcifying effect of cereals is due to a lowering of the available phosphate, whereas others[20, 21] have attributed it to the low calcium-phosphate ratio and have found that the addition of calcium lactate or carbonate to the diet counteracts the rachitic effect. Other experiments[22] have led to the suggestion that the effect of phytin can diminish naturally if it forms a constant item of diet. In three normal subjects an abrupt change from their usual diet to one high in phytate phosphorus brought about an immediate negative calcium balance which, however, in time was restored to equilibrium and the calcium deficit made good. Phosphorus retention was at first lowered in two subjects by the addition of the phytate but was later improved still further by a reduction in dietary phytate. These results were suggested to be due to hydrolysis of calcium phytate at a level of the alimentary tract high enough to permit subsequent

[16] E. Mellanby, *Brit. Med. J.* **11,** 849 (1922).
[17] R. A. McCance and E. M. Widdowson, *Biochem. J.* **29,** 2694 (1935).
[18] H. M. Bruce and R. K. Callow, *Biochem. J.* **28,** 517 (1934).
[19] J. T. Lowe and H. Steenbock, *Biochem. J.* **30,** 1991 (1936).
[20] N. Palmer and J. C. Mottram, *Biochem. J.* **33,** 512 (1939).
[21] N. Palmer, *Biochem. J.* **33,** 853 (1939).
[22] A. R. P. Walker, F. W. Fox, and J. T. Irving, *Biochem. J.* **42,** 452 (1948).

absorption, and also to a gradual adjustment of the body to the lowered calcium absorption. Such findings afford a useful illustration of the effect of phytic acid in the diet and demonstrate the efforts of nature to overcome the difficulties imposed by it, but it is obviously undesirable that the body should be forced to operate at a low level with possible detriment to normal calcification, and the practice of adding calcium carbonate to flour and other cereal foods to neutralize their anticalcifying effect is now generally accepted as physiologically sound.

3. Vitamin D

It has been established experimentally that vitamin D increases absorption of calcium from the intestine[23-26] and that although phosphorus absorption is not influenced directly an increase occurs as a result of the higher absorption of calcium. Experiments with radioactive calcium[27] have shown that in rats absorption occurs most rapidly during the first 2 to 4 hr. after administration and is not influenced by vitamn D, but that after this period absorption occurs only in those rats receiving this factor. The evidence indicated that the increase in efficiency of absorption produced by vitamin D was due to an extension of conditions favoring absorption from the proximal two-thirds of the small intestine to the distal one-third of the small intestine and to the large intestine. Regarding the mode of action of vitamin D it is a general finding that administration causes a lowering of intestinal pH. It has further been shown by Steenbock *et al.*[28] that there is a coincident increase in urinary pH which they liken to the postprandial alkaline tide induced by gastric secretion. These authors point out, however, that it cannot be stated whether the vitamin stimulates gastric secretion of acid or causes a depletion of base in the intestines, and that the possibility also remains that the increase in calcium absorption may itself bring about a depletion of base in the intestine, in which case the effect upon pH must be considered as a secondary one. The mode of action thus remains obscure and, can be clarified only by further experimental work.

4. Fat Absorption

When fat splitting and absorption is normal the ingestion of fats has in general been found to aid calcium absorption. In conditions in which there is excessive fat excretion, however, such as sprue or idiopathic steatorrhea, calcium is lost in the feces as calcium soaps. It has also been suggested

[23] R. Nicolaysen, *Biochem. J.* **31,** 105, 107, 122, 323 (1937).
[24] F. Albright and H. W. Sulkowitch, *J. Clin. Invest.* **17,** 305 (1938).
[25] D. M. Greenberg, *J. Biol. Chem.* **157,** 99 (1945).
[26] B. B. Migicovsky and A. R. G. Emslie, *Arch. Biochem.* **20,** 325 (1949).
[27] H. E. Harrison and H. C. Harrison, *J. Biol. Chem.* **188,** 83 (1951).
[28] H. Steenbock, S. A. Bellin, and W. G. Wiest, *J. Biol. Chem.* **193,** 843 (1951).

that in these conditions loss of vitamin D in the feces contributes to the inefficiency of calcium absorption.[29]

5. Citrates

Addition of citrate to a rachitic diet has been observed to render it non-rachitic[30] and to aid calcification.[31] It is suggested that citrates lower the pH of the intestinal tract, form calcium citrate which is relatively soluble, and aid deposition subsequent to absorption by raising the pH of tissue fluids.

6. Protein Intake

Addition of protein to a basal diet has been shown to increase calcium absorption and retention in men,[32] and addition of peptone to the diet increases calcium absorption in rats.[33] The effect upon absorption is presumed to be due to soluble calcium compounds formed with the amino acids produced by digestion of the protein. Better retention may possibly be due to the creation of a favorable nitrogen balance which assists in the laying down of matrix necessary for deposition.

7. Oxalic Acid

Absorption of calcium may be prevented by the formation of insoluble calcium oxalate where oxalic acid forms a significant proportion of intake. The most notable example is spinach which contains sufficient oxalic acid to render all its calcuim non-available with some to spare for other calcium present in the diet. There is some doubt, however, that this factor has any major effect upon nutrition in normal circumstances.

8. Gastric Hydrochloric Acid

It is generally accepted that a normal secretion of hydrochloric acid by the stomach is necessary for optimal absorption of calcium and phosphate, better absorption being due to a lowering of the pH of the gastrointestinal tract. The presence of hypochlorhydria or achlorhydria therefore exerts an adverse influence upon calcium and phosphate absorption.

9. Magnesium

Malcolm[34] has shown that magnesium caused increased loss of calcium in adult animals and hinders its deposition in young growing animals. There

[29] S. H. Bassett, E. H. Keutmann, H. Z. Hyde, and H. E. Van Alstine, *J. Clin. Invest.* **18,** 121 (1939).
[30] A. T. Shohl, *J. Nutrition* **14,** 69 (1937).
[31] M. L. Hathaway and F. L. Meyer, *J. Nutrition* **17,** 419 (1939).
[32] R. A. McCance, E. M. Widdowson, and H. Lehmann, *Biochem. J.* **36,** 686 (1942).
[33] T. C. Hall and H. Lehmann, *Biochem. J.* **38,** 117 (1944).
[34] J. Malcolm, *J. Physiol. (London)* **32,** 183 (1905).

seems little doubt that magnesium acts antagonistically to calcium utilization.[35]

V. Excretion

1. Calcium

Calcium is excreted in both feces and urine, about 80% of the total amount excreted being in feces and 20% in the urine. Fecal calcium consists of unabsorbed calcium together with calcium which has been absorbed and re-excreted. It has been generally assumed that the large intestine is the major site of excretion into the intestines, but recent experiments with radioactive calcium have demonstrated that in the rat all segments of the intestinal tract are active in this function with the small intestine predominating.[36] Calcium is excreted in urine mainly as chlorides and phosphates, and in feces as phosphate, carbonate, and soaps.

2. Phosphorus

In contrast to calcium, excretion of phosphorus occurs mainly in the urine, about two-thirds of the total amount excreted being lost by this route, almost entirely as phosphates of the various cations, the remainder being excreted in feces chiefly as calcium phosphate. Fecal phosphorus is composed of unabsorbed together with re-excreted phosphate, the latter having been estimated to average 25 to 30% of the total fecal phosphorus.[37]

VI. Factors Affecting Utilization of Absorbed Calcium and Phosphorus

The inclusion of factors under this heading has been made with some reservation, as it is obvious that conditions which induce profound disturbance of any vital metabolism may have indirect influence upon others. We have, therefore, included only those factors which appear to have a well-defined effect, either direct or indirect, upon calcium and phosphorus metabolism.

1. Parathyroid Hormone

This hormone is secreted by the parathyroid glands, which are four in number and lie on the posterior surface of the thyroid gland. The exact manner in which the hormone exerts its action upon calcium and phosphorus metabolism has not been proved satisfactorily, but the most immediate effect of administration of parathyroid extract which has been generally observed is an increased excretion of phosphorus in the urine, which is accompanied by a lowering of the serum inorganic phosphate. The increase in the urinary phosphate excretion has been ascribed by Harrison and Har-

[35] J. D. Robertson, *Chemistry & Industry* **62,** 222 (1943).
[36] H. D. Wallace, R. L. Shirley, and G. K. Davis, *J. Nutrition* **43,** 469 (1951).
[37] K. Kjerulf-Jensen, *Acta Physiol. Scand.* **3,** 1 (1941).

rison[38] to a lowering of the rate of reabsorption of phosphate by the renal tubules. This finding was not supported by Fay *et al.*,[39] but Tweedy *et al.*[40] found a prompt increase in urinary excretion of radiophosphorus by thyroid-parathyroidectomized rats in response to parathyroid extract, whereas the latter had no effect upon the retention or excretion of radiophosphorus in bilateral nephrectomized rats. It was concluded, therefore, that the parathyroid hormone exerts a direct action upon the kidney.

This initial effect of the hormone upon the urinary and serum phosphate is followed by an increase in the concentration of calcium in the serum and an increased urinary excretion of calcium, although the latter is not of sufficient magnitude to prevent hypercalcemia. Continued administration of the hormone brings about impairment of renal function so that urinary excretion of phosphate decreases and serum inorganic phosphate becomes increased above normal, with the result that calcium phosphate is precipitated in the form of renal calculi.

The effect of parathyroid hormone upon bone is considered by Albright[41] to be secondary to the increase in urinary phosphate excretion, the depletion of serum phosphate causing accelerated resorption of phosphate from bone which is inevitably accompanied by increased calcium resorption. This author distinguishes between hyperparathyroidism with bone disease and a like condition without bone disease. In the former there is an increase in the resorption of bone with a compensatory increase in bone formation, the osteoblasts and osteoclasts being numerically increased and the phosphatase activity of the serum raised. The latter condition shows normal bone metabolism, and the increased calcium excretion is postulated to come entirely from increased calcium intake and absorption, whereas serum phosphatase activity is normal.

Albright's hypothesis, however, is regarded as an over-simplification by Milne,[41a] who draws attention to the fact that the increase in urinary excretion of phosphorus is insufficient to account for the total loss of phosphate if that of the extracellular fluid, with which the plasma phosphate is in dynamic epuilibrium, is taken into account. Milne concludes that the renal effect of the hormone is of primary significance in the hypoparathyroid state, but its chief action in the normal subject is to stimulate osteoplastic resorption of bone.

Thus, although it appears certain that parathyroid hormone can exert a direct action upon both bone and kidney, considerable divergence of opinion exists as to which effect is primary. Further work will no doubt

[38] H. E. Harrison and H. C. Harrison, *Am. J. Physiol.* **134,** 781 (1941).
[39] M. Fay, V. G. Behrmann, and D. M. Buck, *Am. J. Physiol.* **136,** 716 (1942).
[40] W. R. Tweedy, M. E. Chilcote, and M. C. Patras, *J. Biol. Chem.* **168,** 597 (1947).
[41] F. Albright, *Recent Progr. in Hormone Research* **1,** 293 (1947).
[41a] M. D. Milne, *Clin. Sci.* **10,** 471 (1951).

be directed to other possibilities such as the existence of more than one hormone and the effect of changes in calcium-binding power of serum albumin.[41b]

2. Vitamin D

The beneficial effect of vitamin D upon calcium absorption has already been noted and is usually stressed as the major action of this vitamin upon calcium and phosphorus metabolism. There is considerable evidence, however, of further effect upon this metabolism within the body, but its precise nature is still a subject for speculation.

It has been demonstrated in hypoparathyroidism that administration of vitamin D has the effect of increasing urinary excretion of phosphate and lowering the level of the serum inorganic phosphate.[42, 43] This action is considered by Albright and his associates as a primary action of the vitamin, but as the apparent effect is usually quite different it is assumed that a balancing of actions occurs when vitamin D and parathyroid hormone are present together. In rickets and osteomalacia resulting from deficiency of vitamin D the usual findings are a normal serum calcium and a low serum inorganic phosphate. These are explained by a primary reduction of calcium absorption due to lack of vitamin D which causes a low serum calcium; the latter stimulates the parathyroids to increase their secretion, leading commonly to hyperplasia, restoring the calcium level to normal, but producing a coincident fall in the inorganic phosphate. When calcemic doses of vitamin D are administered, absorption of calcium from the intestine is increased, causing the serum calcium to rise, the need for hypersecretion of hormone ceases, and with depression of this activity the serum inorganic phosphate rises towards normal, while the low parathyroid secretion causes urinary phosphate excretion to fall. Toxic doses of vitamin D cause a very rapid absorption of calcium from the intestine which increases the serum calcium well above the normal level and leads to secondary hypoparathyroidism so that a rise in the serum phosphate level occurs. The physiological action of vitamin D upon phosphorus metabolism is thus assumed to be completely overshadowed by that of the parathyroid hormone.

There is some evidence that vitamin D also has a direct action upon the bone cells effecting an increased deposition of bone salt. In a study of the effect of adding vitamin D to a rachitic low-phosphorus diet,[44] it was observed that deposition of bone in rats was resumed at the expense of tissue phosphorus and that tissue growth was suspended. Deposition of radio-

[41b] *Lancet* **1**, 79 (1953).
[42] F. Albright and H. W. Sulkowitch, *J. Clin. Invest.* **17**, 305 (1938).
[43] F. Albright, H. W. Sulkowitch, and E. Bloomberg, *J. Clin. Invest.* **18**, 165 (1939).
[44] H. Schneider and H. Steenbock, *J. Biol. Chem.* **128**, 159 (1939).

phosphorus in the skeleton of the rat has been found to be increased by 25 to 50% by administration of vitamin D,[45] and tracer experiments with calcium and strontium have demonstrated that deposition in bone of these elements is approximately doubled in rachitic rats treated with vitamin D.[46]

3. Thyroid

It has been generally established that thyroid dysfunction leads to a disturbance of mineral metabolism. In hyperthyroidism an increased excretion of calcium and phosphorus is constantly found which is diminished by iodine medication and is still further diminished after subtotal thyroidectomy.[47] The high urinary calcium excretion also returns to normal after about 14 days' treatment with the antithyroid drug thiouracil.[48] A high fecal excretion of calcium has also been reported but is not a clearly established feature of hyperthyroidism. Long-standing untreated hyperthyroidism is accompanied by an osteoporosis, the cause of which is still a matter of differing opinion.

Robertson[49] suggests that the thyroid hormone exerts a direct action upon the kidney which has the effect of lowering the renal threshold for calcium, causing an increase in urinary excretion of calcium leading to decalcification. Albright[41] is of the opinion that the osteoporosis is due primarily to the negative nitrogen balance usually associated with this condition which results in an insufficient supply of organic material for matrix formation, and in support of this hypothesis he refers to the report that administration of testosterone propionate to a case of hyperthyroidism to reduce the negative nitrogen balance also reduced the urinary excretion of calcium.[50] Green and Lyall[48] report that the results of their experiments in six cases of severe hyperthyroidism show that grossly increased excretion of nitrogen and phosphorus is not a necessary accompaniment of a higher excretion of calcium, nor is this true in reverse. These authors appear to be attracted to a new version of the theory that the acidosis of hyperthyroidism may be the primary cause of the loss of calcium, and they criticize the rejection of an explanation upon this line by Albright *et al.*[51]

In hypothyroidism calcium and phosphorus excretions are generally decreased, becoming normal with thyroid medication.[48] No evidence of denser bone formation has resulted from radiographic studies of this condition.[52]

[45] W. E. Cohn and D. M. Greenberg, *J. Biol. Chem.* **130,** 625 (1939).
[46] D. M. Greenberg, *J. Biol. Chem.* **157,** 99 (1945).
[47] G. E. Beaumont, E. C. Dodds, and J. D. Robertson, *J. Endocrinol.* **2,** 237 (1940).
[48] J. Green and A. Lyall, *Lancet* **1,** 828 (1951).
[49] J. D. Robertson, *Lancet* **1,** 672 (1942).
[50] L. W. Kinsell, S. Hertz, and E. C. Reifenstein, *J. Clin. Invest.* **23,** 880 (1944).
[51] F. Albright, W. Bauer, and J. C. Aub, *J. Clin. Invest.* **10,** 187 (1931).
[52] R. Golden and H. Abbott, *Am. J. Roentgenol. Radium Therapy* **30,** 641 (1933).

4. Steroid Hormones

A review of the effect of sex hormones upon the skeletal system of both oviparous animals and mammals by Gardner and Pfeiffer[53] has emphasized the difference in response to these substances in different species and the difficulties of generalization even in individual species. These authors conclude that the steroid hormones have a part in the control of skeletal growth in that large amounts, in particular of estrogens, inhibit longitudinal osseous growth, whereas small amounts, particularly of androgen, may accelerate it. They suggest that estrogenic action upon calcification probably consists in a stimulation of osteoblastic activity.

Chief interest in the action of the steroid hormones upon bone metabolism in man has centered upon senile and postmenopausal osteoporosis. An investigation of the effect of estradiol benzoate upon the calcium and phosphorus metabolism in three cases of postmenopausal osteoporosis gave evidence of a considerable retention of calcium and phosphorus, the urinary and fecal excretions being reduced.[54] This work was later extended,[55] certain estrogens, androgens, and progesterone being used in a study of the effect of these substances upon various types of osteoporosis. It was observed that estradiol and diethylstilbestrol cause a decrease in the excretions of calcium and phosphorus, as also do testosterone and methyltestosterone, whereas progesterone is without effect. In addition it was found that the combined action of estrogen and androgen upon calcium metabolism is greater than either alone. In Cushing's syndrome estrogen appears to have some beneficial action upon calcium balance but androgen is very much more effective.

Estrogen therapy with dienestrol has been shown to increase the number and thickness of bone trabeculae, previously atrophic, with an increase in the number of osteoblasts and abnormally wide osteoid seams in osteoporosis complicating Paget's disease.[56] The effect of dienestrol has also been studied in postmenopausal osteoporosis and the beneficial effect appears to be maximal only when the intakes of calcium and phosphorus are both about twice the accepted optimal requirements.[57]

In explanation of the effect of steroid hormones upon bone formation, Albright[58] suggests that this effect is determined by their influence upon protein anabolism. Those which promote this, androgens and estrogens,

[53] W. U. Gardner and C. A. Pfeiffer, *Physiol. Revs.* **23,** 139 (1943).

[54] F. Albright, E. Bloomberg, and P. H. Smith, *Trans. Assoc. Am. Phys.* **55,** 298 (1940).

[55] E. C. Reifenstein and F. Albright, *J. Clin. Invest.* **26,** 24 (1947).

[56] M. S. Sherman, *J. Bone & Joint Surg.* **30A,** 915 (1948).

[57] I. A. Anderson, *Quart. J. Med.* **18,** 67 (1949).

[58] F. Albright, *Ann. Internat. Med.* **27,** 861 (1947).

stimulate the osteoblasts to lay down organic matrix and thus bring about higher retention and deposition of calcium and phosphorus, whereas those inhibiting it, "sugar-active" corticoids, depress osteoblastic activity.

VII. Requirements for Health

1. Calcium

The daily requirement of calcium for maintenance of health has been computed from balance experiments conducted by many independent observers. In normal healthy adults equilibrium is reached between intake and output when the former is in the region of 0.55 g. daily, but as it is desirable that the intake of calcium should be above the minimum required for balance a margin of 50% is allowed for safety, which brings the requirement of adults to 0.8 g. daily.

Children have a higher calcium requirement than adults, owing to demands for skeletal growth, and their minimum requirement, assessed in the same manner as that of adults, varies with age and in adolescence with sex. The Recommended Dietary Allowances of the Food Nutrition Board (1945) give the following scale for children: up to and including 9 years of age, 1.0 g.; 10 to 12 years, 1.2 g.; girls 13 to 15 years, 1.3 g.; girls 16 to 20 years, 1.0 g.; boys 13 to 20 years, 1.4 g.

In pregnancy and lactation the calcium intake should be at least 1.5 and 2.5 g. daily, respectively. As about two-thirds of the total calcium in the fetal skeleton is deposited during the last three months of pregnancy the higher intake of calcium should begin by the fifth month and should continue until lactation is over. These higher requirements are necessitated by the apparent absence of any conserving mechanism for calcium during this period of reproduction, and the calcium necessary for building the fetal skeleton and for milk production has to be provided by the mother in the face of her usual absorption and excretion.

It is notable that the general tendency in recent years has been to raise the requirement figures to higher levels, especially those for children and for pregnant and lactating women. This tendency is due in part to a recognition of the variations which occur both in absorption and utilization between individuals, and also to the observation of benefit to growth, weight, general health, and maternal efficiency in subjects receiving higher calcium diets. The figures quoted above therefore are now largely regarded as minimal rather than optimal requirements.

2. Phosphorus

In comparison with calcium, experimental work on phosphorus balance has been scanty. It appears to have been generally assumed that there is little chance of a deficiency of phosphorus arising from any diet which satisfies human requirements of other nutrients, and there is indeed good justification for such an assumption. Phosphorus is widely distributed in food substances and in addition is concentrated in the seeds of plants rather than the leaves, whereas the reverse is true of calcium, so that the prominence of flesh foods and grain products in human diets renders man subject less to phosphorus deficiency and more to calcium deficiency than animals.

A study of the phosphorus maintenance requirement of adults made by Sherman[59] showed that balance was achieved with an intake of 0.88 g. daily, so that the safe daily allowance is computed to be 1.32 g. Children should be allowed about half as much again as adults, and in pregnancy and lactation the intake should be twice the ordinary figure.

VIII. Calcium and Phosphorus Metabolism in Disease

Abnormal calcium and phosphorus metabolism may be due to the following causes.

1. Failure of Absorption or Lack of Available Calcium and Phosphorus

a. Rickets. Rickets is due to a vitamin deficiency in children. The ratio of calcium to phosphorus in bone remains constant, indicating that the type of salt laid down is normal, although the amount is insufficient. In the blood the calcium is normal, the phosphate low, and the phosphatase high. It has been noted that if the product of calcium and phosphate in the blood in milligram per cent is 40 or over rickets did not develop whereas a product of under 30 led to rickets.

b. The Steatorrheas. This group is meant to include sprue, coeliac disease, and the like. It is indirectly a vitamin D deficiency. The primary factors appear to be a fatty diarrhea and the formation of large amounts of insoluble calcium. In the blood both serum calcium and phosphate are low. Tetany is frequently present.

c. Osteomalacia or Vitamin D Deficiency in Adults. The picture differs from rickets in that the calcium loss from the bone is greater than phosphorus. The disease is commoner in women than in men and occurs most frequently in India and China. Absence of sunlight and/or foods lacking in calcium and vitamin D are the main etiological factors. The condition is accentuated by repeated pregnancies with their toll on the skeletal calcium. The blood shows a low calcium and phosphorus, and tetany is common.

d. Osteoporosis Due to Other Causes: (1) Bedridden patients because of reduced activity (2) Basophilic adenoma of the pituitary (3) Suprareno-

[59] H. C. Sherman, *J. Biol. Chem.* **41,** 173 (1920).

cortical tumors (4) Paget's disease where the changes may be confined to one bone (5) Osteogenesis imperfecta, a hereditary disease characterized by blue sclerotics and deafness; (6) Multiple myeloma.

2. An Increased Excretion of Calcium and Phosphorus Secondary to Diseases of the Thyroid and Parathyroid Glands

These conditions have already been discussed in Section VI of this chapter.

CHAPTER 22

Trace Elements*

E. J. UNDERWOOD

Institute of Agriculture, University of Western Australia, Crawley, Western Australia

* Material available up to the end of 1951 only was considered in this chapter.

I. Introduction

A very considerable number of elements has been shown to occur in a wide range of animal tissues and fluids in such minute amounts that they can appropriately be described as "traces." Numerous studies of biological materials from widely separated sources have established the fact that copper, manganese, zinc, iodine, cobalt, nickel, aluminum, chromium, tin, silicon, titanium, lead rubidium, lithium, molybdenum, arsenic, fluorine, bromine, barium, and strontium are commonly present in low concentrations in blood, milk, and tissues of higher animals, and other metals, such as silver, gold, boron, cadmium, and cerium are occasionally present. No doubt others will be detected as more-refined analytical methods are developed.

Demonstration of a physiological role for many of these trace elements, if any such exists, has lagged far behind their mere detection in the living organism. Of the twenty elements listed in the first-mentioned group only the first five, copper, manganese, zinc, iodine, and cobalt, have been definitely established as essential for animal life. Several others, including fluorine, bromine, arsenic, nickel, molybdenum, barium, and strontium have had physiological significance ascribed to them, based on evidence which, although not conclusive, is certainly suggestive. This includes the recent highly interesting finding of Rygh,[1] by the use of specially purified diets, that barium and strontium are essential for the growth and well-being of rats and guinea pigs and that strontium is necessary for proper calcification of bones and teeth in these animals. No such evidence exists for the remaining elements.

[1] O. Rygh, *Bull. soc. chim. biol.* **31**, 1052 (1949).

It may well be that animal tissues have developed a passive tolerance to the presence of some of these elements, owing to their more or less constant occurrence in the soils, the foods, the waters, and the atmosphere with which the organism is continuously associated. Even a consistent concentration of certain elements in particular organs or tissues, however teleogically tempting, cannot be taken as evidence that the element in question is there for some definite purpose, since such a concentration may merely mean a higher tolerance, or a greater chemical affinity, of one tissue or organ compared with that of another. Nevertheless, it seems probable, as more highly purified diets are produced and techniques improve, that at least some of these "accidental" trace elements will be found to have specific physiological functions in the animal body. It should not be forgotten that at one time copper, manganese, and zinc were also regarded as merely interesting contaminants.

Recognition of the fact that certain trace minerals are essential components of the diet of some or all of the higher animals has come mostly from two types of investigations. These were at first independent and distinct but have recently tended to merge. In this respect the history of the development of our knowledge of the trace minerals is strikingly similar to that of the vitamins. The first type is the investigation of certain naturally occurring diseases of man and animals which respond to treatment with minute doses of particular minerals. The demonstrations that iodine deficiency is the cause of endemic goiter of man and of farm stock and that cobalt deficiency is the cause of a group of wasting diseases of grazing sheep and cattle are examples of this type of investigation. They constitute one of the best illustrations, in the whole field of biological endeavor, of the way in which a thorough scientific attack on a naturally occurring "field" problem can lead to results of fundamental interest and importance.

The second of these types of investigation is that of the pure scientist interested in studying the effects on animals of highly purified or specially constituted diets deliberately designed to have an abnormally low content of a particular "trace" mineral. In this way the essentiality of copper, manganese, and zinc in the nutrition of small laboratory animals was first demonstrated. At first these studies were confined to small animals because of such obvious advantages as rapid growth and reproduction, small food consumption, and relative ease of control of environmental conditions. Advances in other branches of nutrition, notably in that of the vitamins, have enormously increased the effectiveness of this type of approach since the early days of mineral nutrition research. The purified diet or partially purified diet technique is now being applied, with some success, to large stock for the study of mineral deficiencies and toxicities which formerly could be studied only in special and restricted areas. Moreover, studies of naturally occurring deficiencies have provided new stimuli for the further investigation of the mode of action of the trace minerals in small laboratory animals by disclosing additional and often unexpected functions within the body. In this connection it is only necessary to mention the relation of copper to the development of nervous tissue in the fetal lamb and of manganese to bone formation in the chick embryo. The former relationship was established during investigation of the naturally occurring condition of enzootic ataxia in lambs, and the latter during investigation of the problem of nutritional chondrodystrophy and perosis of poultry.

The trace elements are frequently considered by nutrition workers as a group, much as the vitamins are. But, like the vitamins, they are not physiologically related and have little in common with each other, except their ability to function in minute quantities. Even in this respect they differ markedly in their quantitative requirements by animals. The minimum requirements of copper and manganese, for instance, are of the order of one hundred times those of iodine and cobalt, and zinc occurs in animal tissues and fluids in amounts so large compared with those of the elements just mentioned, that it hardly deserves the term "trace" element at all. Nor do physical and chemical similarities between elements imply similar physiological properties. Living cells are extraordinarily specific in their mineral requirements. It is only necessary to compare cobalt and nickel, in this respect, and iodine and bromine to make this point clear. Nevertheless, as a group, the trace elements must act as catalysts involved in hormone or enzyme systems, either as constituent parts of the molecules of hormones, enzymes, or coenzymes, or as enzyme activators. In fact, as Green[2] has said, "...enzymic catalysis is the only rational explanation of how a trace of some substance can produce profound biological effects." Many examples of trace minerals functioning as integral parts of enzyme molecules or in activating enzymic processes are now known, but we are far from any real understanding of their mode of action in the living body. The development of greatly improved methods of estimating accurately the minute amounts of the trace elements involved, such as has occurred recently in the case of iodine, and the increasing availability and use of suitable radioactive isotopes of these elements will, it is hoped, help considerably in the achievement of this aim.

One of the difficulties with which all writers on the subject of the trace elements have to contend is that of differentiating between their physiological, pharmacological, and toxicological aspects. The present writer has deliberately restricted his considerations to the physiological roles of these elements. Selenium and lead are therefore not dealt with at all, since, up to the present, they are of biological interest only through their toxic effects. Arsenic and fluorine, on the other hand, are briefly discussed because physiological properties have been ascribed to them, although they

[2] D. E. Green, *Advances in Enzymol.* **1,** 177 (1941).

are best known for their toxic properties. Fortunately for most of the trace elements there is a wide margin between physiological and toxic levels. The former can therefore be considered exclusively, or almost exclusively, without appearing unduly arbitrary and without serious loss of effectiveness.

II. Copper

Although it is more than a century since copper was first associated with living tissues, it was not until 1928[3] that it was conclusively shown to be an essential component of the diet of mammals. It is probable that Bucholz[4] (1816) and Miessner[5] (1817) were the earliest workers to show that copper is a constituent of plants, and Boutigny[6] (1833) of animals. During the next fifty years many investigations established the fact that copper is consistently present in plant and animal tissues, but most of the actual values obtained are suspect because of faulty methods of analysis. Moreover, its occurrence was considered to be merely accidental. The first suggestion that copper fulfills a vital role in the higher forms of life came in 1920 from several French investigators,[7] who worked with plants.

Long before this, considerable scientific effort had centered upon a physiological role for copper in the marine blue-bloods—the gastropods and arthropods. Harless[8] detected this element in snails in 1847 and showed that it existed in combination with the blood proteins. Fredericq[9] in 1878 established the fact that the copper-containing pigment (hemocyanin) in the blood of the octupus behaved as a respiratory pigment. Hemocyanins from different marine species are now known to vary in composition and copper content, but they all unite with oxygen in a definite ratio to copper. One atom of oxygen is taken up for each atom of copper in the molecule.

Another copper compound of early biological interest is the colored pigment, turacin, which is found in the feathers of the South African bird, touraco. It was first described by Church[10] in 1869, who later found it to contain 7.0% Cu and to be analogous to hematin. It is now known to belong to a derivative of the normal porphyrin pigments which are generally found in animals and plants.[11] The touraco is apparently the only animal

[3] E. B. Hart, H. Steenbock, J. Waddell, and C. A. Elvehjem, *J. Biol. Chem.* **77,** 797 (1928).

[4] C. F. Bucholz, *Rept. Pharm.* **2,** 253 (1816).

[5] W. Messner, *Ann. chim. et phys.* **4,** 106 (1817).

[6] d'E. J. Boutigny, *Chim. med.* **9,** 147 (1833).

[7] L. Maquenne and E. Domousby, *Compt. rend.* **170,** 87 (1920); B. Guerithalt, *ibid.* **171,** 196 (1920); G. Bertrand, *Bull. soc. sci. hyg. aliment.* **8,** 49 (1920).

[8] E. Harless, *Arch. Anat. u. Physiol.* 148, (1847).

[9] L. Fredericq, *Arch. zool. exptl. et gén.* **7,** 535 (1878).

[10] A. W. Church, *Phil. Trans.* **159,** 627 (1869); *Proc. Roy. Soc.* (*London*) **51,** 399 (1892).

[11] C. Rimington, *Proc. Roy. Soc.* (*London*) **B127,** 106 (1939).

species in which copper occurs combined with porphyrin. The reason for its occurrence in these birds is unknown.

1. Copper Content of Animal Tissues and Organs

The literature dealing with the distribution of copper throughout the animal body is voluminous. Data are available for the copper content of the principal tissues and organs of rat, rabbit, cat, dog, guinea pig, sheep, ox, horse, domestic fowl, and man. Copper occurs in varying concentrations in every tissue and organ of the body, but individual variability is too great to allow definite values to be assigned to particular organs. Nevertheless, the distribution of copper throughout the body is similar in all species so far examined, and certain organs and tissues consistently carry much higher concentrations of copper than others (see Table 1). The endocrine glands (pituitary, thyroid, and thymus) are examples of organs with very low copper content, whereas the liver, the kidneys, the heart, the brain, and the hair are usually among the highest in copper concentration. Relatively high values for the spleen and the lungs have also frequently been reported. The liver has received more attention than any other organ because it acts as a storage organ for copper and reflects the copper status of the animal, but it is apparent from the work of Tompsett,[12] Rusoff,[13] and others that the bones and muscles may contain much larger total stores of this element.

The age and the level of copper intake of the animal are the two principal factors affecting the copper content of the liver. In all species, including man, considerable storage of copper occurs in the liver during intrauterine life, but the extent of this storage and the time at which maximum concentration exists vary in different species. In man, rat, rabbit, guinea pig, dog, and ox the maximum concentrations generally occur at or near birth, but in pig and chick the peak occurs somewhat earlier in embryonic life.[15, 16] In early infancy, i.e., during the suckling period, the concentration of copper in the liver declines in most species. The sheep appears to be a notable exception in that liver values tend to rise after birth and normally reach exceptionally high levels at maturity. In the rat, also, the peak does not occur until 10 to 15 days after birth, although there is a decline thereafter.[17] The liver copper stores of the newborn cannot be greatly increased by raising the copper intake of the mother to levels well above normal. Neither can the copper content of eggs be raised by feeding extra copper to

[12] S. L. Tompsett, *Biochem. J.* **29**, 480 (1935).
[13] L. L. Rusoff, *Florida Agr. Expt. Sta. Tech. Bull.* 356 (1941).
[15] V. A. Wilkerson, *J. Biol. Chem.* **104**, 541 (1934).
[16] W. D. McFarlane and H. I. Milne, *J. Biol. Chem.* **107**, 309 (1934).
[17] S. Bruckmann and S. G. Zondek, *Nature* **146**, 30 (1940).

TABLE 1

COPPER CONTENT OF ORGANS OF DIFFERENT SPECIES (CUNNINGHAM[14])

(Parts per million on the dry basis)

Species	Description	Liver	Heart	Lungs	Spleen	Kidney	Pancreas	Brain	Flesh	Skin	Hair
Human	Adult*	24.9	—	—	5.2	17.5	4.3	17.5	—	—	
Bovine	Adult†	77.0	15.6	5.3	2.9	19.7	3.8	—	—	—	
Bovine	Newborn	470.0	14.8	4.9	4.8	15.7	8.5	—	4.8	—	
Bovine	Fetus	262.8	10.4	3.6	5.4	8.5	—	—	2.9	2.1	
Sheep	Adult	236.6	17.9	9.6	5.0	17.8	7.7	—	—		
Horse	Adult†	14.8	17.6	6.8	3.2	28.9	—	—	—		
Pig	Adult†	41.3	14.9	5.3	6.0	21.1	—	—	—		
Pig	Few days old	232.8	12.8	3.4	6.8	14.7	—	—	—		
Dog	9 days old adult†	98.2	17.4	6.2	—	14.2	—	8.5	—	9.9	22.7
Cat		25.3	14.4	3.8	5.2	10.1	—	14.6	2.3	4.2	11.9
Guinea pig	Adult	17.0	21.2	9.5	—	19.9	—	—	—		—
Rabbit	Adult	9.2	22.3	8.1	—	13.7	—	—	—		—
Rat	About 90 g. weight†	10.0	27.8	9.5	8.1	22.6	—	10.2	3.8	7.3	14.8
Badger	Adult†	21.7	12.8	5.6	3.0	9.4	—	10.8	—	3.2	—
Domestic fowl	Adult†	12.4	14.9	2.4	—	11.7	—	—	—	—	feathers 4.9
Average		103.6	16.8	6.0	5.0	16.4	—	—	—	—	—
Range		9.2–470	10.4–27.8	2.4–9.5	2.9–8.1	8.5–28.9	—	—	—	—	—

[14] I. J. Cunningham, *Biochem. J.* **25**, 1267 (1931).

* Average of 3 series of analyses.

† Average of 2 series of analyses.

the hen.[18] But these stores are markedly and significantly reduced by subnormal copper intakes by the mother. This has been demonstrated for several species and is well illustrated by the data of Bennetts and Beck.[19] These workers found the liver copper of five ataxic lambs from copper-deficient ewes to range from 4 to 8 p.p.m. Cu on the dry basis, whereas five normal lambs from healthy ewes gave values ranging from 120 to 350 p.p.m.

The functional significance of these congenital copper deposits is still not completely understood. The suggestion that copper is accumulated in the fetal liver because this is the chief site of blood formation during embryonic life is an attractive one, but it must be realized that the hematopoietic activity of the liver is at its highest during early fetal life and has nearly disappeared at birth when, in most species, the copper concentration is highest. The further claim that the liver copper stores of the newborn are provided to compensate for the low copper content of the maternal milk also seems obvious enough and is certainly supported by the fall in concentration of liver copper during the suckling period which occurs in most species. There is evidence, however, at least for the rat, the rabbit, the guinea pig,[17] and the sheep,[20] that the total *amounts* of copper in the liver actually increase during suckling.

The marked reduction in the copper content of the liver of the newborn resulting from subnormal copper intakes by the mother is paralleled by the effects of inadequate copper intakes in older animals. Low liver copper levels have been found in rats and pigs suffering from milk anemia[21] and in sheep and cattle grazing copper-deficient pastures.[19, 22, 23] When diets already adequately supplied with copper are supplemented by this element, greatly increased liver copper concentrations result.[24, 24a] These findings show clearly the great storage capacity of the liver for copper and make the level of copper in the liver the most reliable indicator of the copper status of animals at present known. When due regard is paid to the important factors of age and species, liver copper estimations can be used with confidence in the diagnosis of either copper deficiency or chronic or subacute copper poisoning. Data illustrating many of the statements made in the last three paragraphs are given in Table 2.

[18] C. A. Elvehjem, A. R. Kemmerer, E. B. Hart, and J. G. Halpin, *J. Biol. Chem.* **85,** 89 (1929).

[19] H. W. Bennetts and A. B. Beck, *Australia Council Sci. Ind. Research Bull.* 147 (1942).

[20] E. I. McDougall, *J. Agr. Sci.* **37,** 337 (1947).

[21] M. O. Schultze, C. A. Elvehjem, and E. B. Hart, *J. Biol. Chem.* **116,** 93 (1936).

[22] I. J. Cunningham, *New Zealand J. Sci. Technol.* **27A,** 381 (1946).

[23] H. R. Marston, H. J. Lee, and I. W. McDonald, *J. Agr. Sci.* **38,** 222 (1948).

[24] C. W. Lindow, W. H. Peterson, and H. Steenbock, *J. Biol. Chem.* **84,** 419 (1929).

[24a] I. J. Cunningham, *New Zealand J. Sci. Technol.* **27A,** 372 (1946).

TABLE 2

THE INFLUENCE OF AGE AND COPPER INTAKE UPON THE CONCENTRATION OF COPPER IN THE LIVER*

(Parts per million on the dry basis)

Species	Age and Treatment	Copper concentration	Reference
Rat	Mature; normal diet	11	24
Rat	Mature; stock diet plus copper	213	24
Rat	Young; Cu-deficient diet	3	21
Rat	Newborn; normal diet	58	25
Rat	3 weeks old; normal diet	20	25
Rat	2 months old; normal diet	9	25
Rat	Mature; normal diet	34	25
Rabbit	Newborn; normal diet	37	25
Rabbit	3 weeks old; normal diet	12	25
Rabbit	2 months old; normal diet	24	25
Rabbit	Mature; normal diet	23	25
Guinea pig	Newborn; normal diet	67	25
Guinea pig	3 weeks old; normal diet	21	25
Guinea pig	2 months old; normal diet	19	25
Guinea pig	Mature; normal diet	23	25
Pig	Newborn; normal diet		
Pig	Adult; normal diet		
Man	Newborn (0–7 weeks); normal diet	230	25a
Man	Adult; normal diet	22	12
Man	Adult; normal diet	35	25a
Sheep	Newborn; normal diet	168	22
Sheep	3 months old; normal diet	284	22
Sheep	Mature (aged); normal diet	599	22
Sheep	Newborn; Cu-deficient diet	13	22
Sheep	3 months old; Cu-deficient diet	16	22
Sheep	Mature (aged); Cu-deficient diet	27	22
Cattle	Newborn; normal diet	381	22
Cattle	Yearlings; normal diet	67	22
Cattle	Adult cows; normal diet	200	22
Cattle	Newborn; Cu-deficient diet	55†	22
Cattle	9–12 months old; Cu-deficient diet	16	22
Cattle	Adult cows; Cu-deficient diet	11	22

* A selection of data from the literature are included. In all cases the figures quoted are the mean of a number of samples.

† Very much lower levels have been obtained in Western Australia for Cu-deficient calves.

[25] E. J. Lorenzen and S. E. Smith, *J. Nutrition* **33,** 143 (1947).

[25a] S. Bruckmann and S. G. Zondek, *Biochem. J.* **33,** 1845 (1939).

2. Copper in Blood

The blood is a formidable rival to the liver as a subject for investigation of the physiology of copper. The normal range of concentration of copper in whole blood, serum, and plasma has been determined for many species, and the main factors influencing this concentration have been delineated. Less is known about the forms in which copper exists in blood. Most, if not all, appears to be in a non-dialyzable form combined with protein, but the protein fraction or fractions with which it is combined remain to be determined. Keiderling[26] provides some evidence that it occurs as a β-globulin compound, and Mann and Keilin[27] have isolated a copper-protein compound, hemocuprein, from the red cells of cattle. Hemocuprein is a bluish pigment containing 0.34% Cu in the cupric form. It has a molecular weight of 35,000 and two copper atoms per protein molecule.[27] Its function is unknown.

In man,[28, 29] sheep, and cattle,[28, 30] the evidence now seems conclusive that copper is normally equally divided between cells and plasma. This is fortunate because it means that in these species values for whole blood or for serum or plasma can be directly compared—they should in fact be identical for any particular blood. This statement needs some qualification, because it is not yet known whether the copper concentrations in cells and plasma are equally influenced by those factors which affect the copper content of whole blood. There is already evidence, to be discussed later, that in human pregnancy the copper content of the red cells remains normal, whereas that of the serum rises markedly.[29] It should be noted also that in birds there is a striking concentration of copper in the nucleated red blood cells.[31]

The normal range of concentration of copper in the blood of healthy animals is wide but very similar in all mammalian species studied (see Table 3). Not only are there considerable variations from individual to individual, but the same animal is subject to significant diurnal and day-to-day fluctuations for no known reasons. These normal levels are, however, markedly reduced by a low copper diet. Subnormal blood copper concentrations have been demonstrated in rats and pigs suffering from experi-

[26] W. Keiderling, *Klin. Wochschr.* **28,** 460 (1950).

[27] T. Mann and D. Keilin, *Proc. Roy. Soc. (London)* **B126,** 303 (1938).

[28] S. L. Tompsett, *Biochem. J.* **28,** 1544 (1934).

[29] H. Rottger, *Arch. Gynäkol.* **177,** 650 (1950).

[30] A. Eden and H. H. Green, *J. Comp. Pathol. Therap.* **52,** 301 (1939).

[31] O. Warburg and H. A. Krebs, *Biochem. Z.* **190,** 143 (1927); A. Locke, D. D. Robash, and L. E. Shenn, *J. Infectious Diseases* **51,** 54 (1934).

TABLE 3

THE COPPER CONTENT OF THE BLOOD OF DIFFERENT SPECIES

(Microgram per milliliter of whole blood* or serum†)

Species	Treatment	Copper concentration		Reference
		Mean	Range	
Man	Healthy, mature	1.14*		31a
Man	Healthy, mature	1.06†	0.70–1.40†	31b
Man	Healthy, female	1.23 ± 0.16†	—	31c
Man	Healthy, male	1.10 ± 0.12†	—	31c
Man	Pregnant, at delivery	2.69 ± 0.49†	—	31d
Man	Healthy, female	1.00†	—	29
Man	Late pregnancy	2.80†	—	29
Sheep	Healthy, mature	—	0.8–2.0*	31e
Sheep	Healthy, mature	—	0.4–1.6*	31f
Sheep	Healthy, mature	0.91*	—	22
Cow	Healthy, mature	—	0.7–1.7*	31f
Cow	Healthy, mature	0.93*	—	22
Cow	Copper deficient	0.53*	—	22
Sheep	Copper deficient	0.47*	—	22
Sheep	Copper deficient	—	0.1–0.5*	19
Pig	Healthy, mature	—	1.6–1.8*	31a
Pig	Healthy, young	—	1.5–1.7*	32
Guinea pig	Healthy, mature	0.70*	0.4–1.1*	33

mental milk anemia,[32] in prolonged anemia of children,[34] and in sheep and cattle grazing on copper-deficient pastures.[19, 22, 35, 36] Values as low as 0.1 and 0.2 μg. Cu per milliliter of blood have been reported in sheep and cattle under such conditions. Normal hematopoiesis cannot be sustained at these low levels, and anemia inevitably ensues if such low levels are maintained for any length of time. The quantity 0.2 μg. Cu per milliliter has also been suggested as the minimal level at which normal hematopoiesis can take place in the pig.[32] On the other hand, the addition of copper to diets already well

[31a] R. A. Kehoe, I. Cholak, and R. V. Storey, *J. Nutrition* **19,** 579 (1940).

[31b] L. Heilmeyer, W. Keiderling, and G. Stüwe, Kupfer und Eisen als körpereigene Wirkstoffe und ihre Bedeutung beim Krankheitsgeschehen, Gustav Fischer, Jena, 1941.

[31c] A. L. Nielsen, *Acta Med. Scand.* **118,** 87 (1944).

[31d] A. L. Nielsen, *Acta Med. Scand.* **118,** 92 (1944).

[31e] H. E. Albiston, L. B. Bull, A. T. Dick and J. C. Keast, *Australian Vet. J.* **16,** 233 (1940).

[31f] A. B. Beck, Australian *J. Exptl. Biol. Med. Sci.* **19,** 249 (1941).

[32] M. O. Schultze, C. A. Elvehjem, and E. B. Hart, *J. Biol. Chem.* **116,** 107 (1936).

[33] A. B. Beck, unpublished data (1951).

[34] E. Lesné and S. B. Briskas, *Acta Paediat.* **22,** 123 (1937), quoted by M. O. Schultze, *Physiol. Revs.* **20,** 37 (1940).

[35] H. R. Marston and H. J. Lee, *J. Agr. Sci.* **38,** 229 (1948).

[36] A. Eden, *J. Comp. Pathol. Therap.* **52,** 429 (1939).

supplied with this element does not necessarily result in consistently higher blood copper levels. Massive doses or injections, however, may produce significant but transitory increases.[25, 30] Blood copper values do not therefore provide quite as satisfactory an indication of the copper status of animals as do liver coppers. It is probable that the liver copper stores have a buffering effect on the blood copper levels. Nevertheless, blood copper determinations give extremely valuable supplementary information and have, of course, the distinct advantage of ready accessibility from the living animal.

Evidence of the effect of anemia resulting from severe hemorrhage on the copper content of blood is conflicting, even in the same species. Some workers have reported falls, whereas others have reported substantial rises. In the sheep Dick[37] could detect no change, even after the loss of large volumes of blood resulting in a fairly severe anemia.

Many workers have shown that the serum copper of women rises significantly during pregnancy and falls rapidly to normal levels after parturition. Nielsen[31c, 31d] studied 31 pregnant women and found that the serum copper usually began to increase from the third month of pregnancy. At delivery the average level in these women was 2.69 ± 0.48 μg. Cu per milliliter of serum, compared with the normal non-pregnant level for women which is given by this worker as 1.23 ± 0.16 μg. Cu per milliliter of serum.[31c, 31d] Blood taken from the umbilical cord of 20 infants gave the very much lower average figure of 0.56 μg. Cu per milliliter. These findings have recently been confirmed and extended by Rottger,[29] who found the serum copper of women to rise in pregnancy from 1.00 μg. to 2.80 μg. per milliliter. The concentration in the red cells of these women, however, remained constant throughout pregnancy at the normal level of 1.00 μg. The concentration of copper in the red cells of newborn infants was found to be very similar to that of the mother, namely 1.20 μg. Cu per milliliter, but, as in Nielsen's studies, the serum copper of the infants was markedly lower (0.54 μg.). No really satisfactory explanation of these changes has yet been put forward, although it has been suggested[38] that the high blood coppers of pregnancy have the function of a hemopoietic stimulus related to the physiological anemia of pregnancy in women. This problem deserves much further study in other species, especially as it has been established that in the normal, healthy sheep there are no changes in the copper content of whole blood characteristic of pregnancy.[31f, 39, 40] The fetal blood of the lamb, like that of the human fetus, is significantly lower in copper than the blood of its mother but there is a rapid post-partum rise to higher values in the

[37] A. T. Dick, *Australian J. Exptl. Biol. Med. Sci.* **17,** 271 (1939).
[38] A. Sachs, V. E. Levine, and A. A. Fabian, *Arch. Internal. Med.* **55,** 227 (1935).
[39] A. Eden, *Biochem. J.* **35,** 813 (1941).
[40] E. I. McDougall, *J. Agr. Sci.* **37,** 329 (1947).

first 1 to 2 weeks of life, which fall again within about 10 weeks to levels very similar to that of ewes.[40]

3. Copper in Milk

A considerable number of highly variable values for the copper content of milk, particularly cow's milk, exists in the literature. Without doubt some of this variability is due to the different methods of analysis employed, but it is also due to the different stages of lactation at which the samples have been taken. The importance of this latter point is either insufficiently realized by most investigators or not appreciated at all. The copper content of milk is influenced by (1) the species, (2) the stage of lactation, and (3) the level of copper intake. There is also considerable individual variation. This, of course, refers to milk taken directly from the animal. Treatment after milking, e.g., pasteurizing, drying, or holding in metal containers, usually results in a variable and often substantial contamination with copper.[41]

Certain species differences unquestionably exist but more data are needed, in which factors (1) and (2) above are strictly controlled, before this point can be clearly and generally established. Rat's milk appears to be very much richer in copper than that of any other species so far examined. Cox and Mueller[42] obtained the extremely high value of 7 mg. Cu per liter for pooled samples of rat's milk taken from four different stages of lactation, compared with 0.6 mg. for cow's milk and a quoted figure of 0.5 mg. for human milk. It appears also that human milk is slightly richer in copper than either cow's, ewe's, goat's, or mare's milk at similar stages of lactation.[43, 44, 45] Significant differences in the copper content of the milk of these latter species have not yet been established, and it is doubtful if they exist.

In all species studied colostrum is several times richer in copper than is normal milk. Also, in the early stages of lactation the copper content of the normal milk of cows, ewes, and women is significantly higher than it is in the later stages of lactation. There is suggestive, but not definite, evidence that a similar fall during lactation occurs in the milk of the goat and the mare. Beck[43] found the milk of normal ewes to fall progressively from 0.20 to 0.64 mg. Cu per liter in early lactation to 0.04 to 0.16 mg. several months later. Similar falls were observed in the milk of some cows but not in others. Most of the values lay between 0.05 and 0.20 mg. This

[41] W. L. Davies, The Chemistry of Milk, Chapman and Hall, London, 1936, p. 224.

[42] W. J. Cox and A. J. Mueller, *J. Nutrition* **13,** 249 (1937).

[43] A. B. Beck, *Australian J. Exptl. Biol. Med. Sci.* **19,** 145 (1941).

[44] E. Lesné, P. Ziziné, and S. B. Briskas, *Rev. pathol. comp. hyg. gén.* **36,** 1369 (1936).

[45] A. Brock and L. K. Wolff, *Acta Brevia. Neerl. and Physiol. Pharmacol. Microbiol.* **5,** 80 (1935).

worker also found the milk of women to fall steadily from 0.62 to 0.89 mg. Cu per liter in the first few weeks of lactation to 0.15 to 0.17 mg. six to seven months later.[46] Lesne and coworkers[44] studied the milk of five women and found a fall in copper content during the lactation period in each case but the figures were somewhat higher throughout than those of Beck. In four of these women the values fell from 1.05 to 0.80 mg. Cu per per liter in the first few weeks to 0.60 to 0.26 mg. at the end of lactation. In the fifth woman the "true" milk fell only from 1.07 to 0.90 mg. Cu per liter at the end of lactation. These workers make the further claim that there is a seasonal variation in the copper content of cow's and goat's milk related to the type of diet consumed.

From the limited evidence available it appears that the copper content of milk cannot be raised above normal levels by adding copper to diets already adequately supplied with this element.[47] This contrasts with the position with manganese and zinc. Under conditions of deficient copper intake, however, both ewes and cows produce milk much below normal in copper content. Beck[43] found the milk of ewes and cows grazing on pastures acutely deficient in copper (1 to 3 p.p.m. Cu on dry matter) to fall as low as 0.01 to 0.02 mg. Cu per liter. In all cases the concentration of copper in the blood was found to be higher than that in the milk. This, again, is in contrast to the position with manganese, zinc, and most other minerals.

4. Copper in Relation To Blood Formation

The classical work of Hart and his associates at Wisconsin on nutritional anemia produced by exclusive feeding of milk established beyond doubt that copper, in addition to iron, is necessary for blood formation in the rat. As little as 5 μg. Cu per day was found sufficient to initiate and maintain a steady formation of hemoglobin and erythrocytes. This finding was subsequently confirmed in many laboratories, and claims were made that other elements were similarly effective. The development of the work on milk anemia and the validity of these claims have been fully discussed by Elvehjem.[48] At present it can be stated unequivocally that the only metallic elements with any direct effect on the formation of hemoglobin and erythrocytes are iron, copper, and cobalt. In milk anemia cobalt is not involved.[49]

The hematopoietic effect of copper has been demonstrated for all species so far studied. These include rats, rabbits, chickens, pigs, dogs, sheep, goats, cattle, and man. The mechanism of this hematopoietic effect is not yet

[46] A. B. Beck, unpublished data.

[47] W. Liebscher, *Proc. 11th World's Dairy Congress*, Berlin, **1,** 130 (1937); B. H. Thomas, *Iowa State Coll. Agr. Mech. Arts Agr. Expt. Sta. Ann. Rept.* p. 87 (1937).

[48] C. A. Elvehjem, *Physiol. Revs.* **15,** 471 (1935).

[49] E. J. Underwood and C. A. Elvehjem, *J. Biol. Chem.* **124,** 419 (1938).

fully understood, although certain important facts are well established. First, copper is definitely not a constituent of the hemoglobin molecule, nor does it appear to be a necessary component of the structural material of the circulating erythrocytes. Second, copper is not necessary for the absorption and storage of iron in tissues, but it does facilitate, or is essential for, the utilization of iron by the blood-forming organs and for the mobilization of iron from the tissues. The entrance of therapeutic copper into the bone marrow of copper-deficient rats, in amounts too small to be detected by present chemical methods, has been demonstrated by the use of radioactive copper.[50] The stage of erythrogenesis at which copper exerts its action and the question as to whether copper is necessary for the formation of red cells only and not for the production of hemoglobin cannot yet be answered. Answers to these questions are most likely to come from detailed histological studies of the bone marrow correlated with chemical investigations of the blood carried out at varying copper intakes. Of value in this connection is the nature of the anemia which occurs as a result of uncomplicated copper deficiency. In the rat and the rabbit this is of the hypochromic, microcytic type,[51] whereas in the pig it appears to be of the hypochromic, normocytic type.[52] The position in sheep and cattle is not yet clearly established, although the anemia is certainly hypochromic and probably slightly macrocytic. These findings, together with the fact that some reticulocyte response occurs in milk anemia on feeding copper alone (but not iron alone), imply that copper acts as a catalyst both for the incorporation of iron into hemoglobin and for erythropoiesis and is not restricted specifically to either function.

5. Copper in Relation to Cell Respiration

In addition to its influence on blood formation copper has been shown to be concerned in many other functions within the body. A number of oxidases which play a part in the oxidation-reduction processes of the cell are known to be copper-protein compounds. These include the polyphenol oxidases of mushrooms and potatoes and ascorbic acid oxidase. Cytochrome oxidase is also usually considered a copper compound, although this is not certain.[53] It is generally believed that the copper ion is the reversibly attached prosthetic group of the enzyme whose activity is associated with a valency change of the copper. An adequate intake of copper has been shown to be essential for the formation and maintenance of normal cytochrome oxidase and catalase activity of certain tissues in the rat.[54]

[50] M. O. Schultze and S. J. Simmons, *J. Biol. Chem.* **142,** 97 (1942).
[51] S. E. Smith and M. Medlicott, *Am. J. Physiol.* **141,** 354 (1944); S. E. Smith, M. Medlicott, and G. H. Ellis, *ibid.* **142,** 179 (1944).
[52] T. S. Hamilton, S. E. Hunt, and W. E. Carroll, *J. Agri. Research* **47,** 543 (1933).
[53] G. Steensholt, *Acta Physiol. Scand.* **14,** 335 (1947); *C. A.* **42,** 4623 (1948).
[54] M. O. Schultze, *J. Biol. Chem.* **129,** 729 (1939); **137,** 727 (1940); **138,** 219 (1941).

The cytochrome oxidase activity of liver, heart, and bone marrow is greatly reduced under conditions of copper deficiency.[54] The catalase activity of liver and kidney is also decreased, but that of the heart is increased.[54] On the other hand, Marston and his colleagues[55] could find no such decrease in the cytochrome oxidase status of the tissues of copper deficient sheep. Much further study of this type is needed for the larger animals, particularly in relation to the naturally occurring copper-deficiency diseases of sheep and cattle. Only in this way does it seem likely that the different functional disturbances which occur in these animals, in addition to anemia, will be explained.

6. Copper Deficiency in Sheep and Cattle

Copper-deficiency diseases of grazing sheep and cattle have been shown to occur in various parts of the world. The history of their discovery and details of the conditions under which they occur have been given in detail by Russell[56] up to 1944 and will not be presented here. In general, three main lines of evidence have served to establish copper deficiency as the cause of the maladies: (1) sheep and cattle fail to thrive in certain areas, and anemia and specific symptoms of disease occur, all of which are overcome by copper supplements; (2) the liver and blood of affected stock contain subnormal amounts of copper; and (3) the herbage on which the stock graze, and in most cases the soils on which the herbage grows, contain subnormal amounts of copper.

In the "coast disease" of sheep in southern Australia and in "saltsick" of cattle in Florida the position is complicated by the existence of severe cobalt deficiency in the same areas. Administration of cobalt alone in these cases results in the onset of symptoms of uncomplicated copper deficiency.[57] "Swayback," the name given to a demyelinating disease of lambs which occurs in certain restricted areas of Great Britain, can be prevented by feeding copper to the pregnant ewes and is associated with a subnormal copper status of the blood and livers of both the lambs and their mothers. It can hardly be a simple copper deficiency, however, because the affected pastures contain normal amounts of copper. Apparently, abnormal amounts of some other element or elements in the herbage inhibit the absorption of copper.[58] On the other hand, the similar disease, enzootic ataxia, of lambs in Australia is established as an uncomplicated copper deficiency.[19]

There are marked differences between sheep and cattle in the symptoms of copper deficiency, and both these species exhibit metabolic disturbances which have not so far been reported in copper deficiency of laboratory ani-

[55] H. R. Marston, *Proc. Symposium on Fibrous Proteins*, Leeds, (1946).
[56] F. C. Russell, *Imp. Bur. Animal Nutrition, Tech. Commun.* 15 (1944).
[57] H. R. Marston, H. J. Lee, and I. W. McDonald, *J. Agr. Sci.* **38**, 216, 222 (1948).
[58] J. R. M. Innes and G. D. Shearer, *J. Comp. Pathol. Therap.* **53**, 1 (1940).

mals or man. Anemia appears to be a common occurrence in all species but it is not necessarily a dominant factor in the disease conditions of sheep and cattle. Apparently the very low blood copper levels which develop under these conditions limit blood formation, as in milk anemia of the rat and the pig, but in sheep and cattle other body processes may be more seriously affected. For instance, lesions in the wool and depigmentation are the first and most sensitive indications of copper deficiency in the sheep. They occur when no other symptoms are evident. Under conditions of acute copper deficiency the liver copper reserves and the blood copper levels of pregnant ewes become so seriously lowered that, in addition to loss of physical condition and moderate to severe anemia in late pregnancy, the developing embryo is affected by gross lesions (demyelination) of the central nervous system. Ths lamb is usually either born dead or affected with ataxia of the hind limbs. Myelin formation is apparently not affected by similar conditions in the pregnant cow, since ataxia has never been found in the newborn of this species, even when grazing on the most severely copper-deficient pastures known. On the other hand, Bennetts and co-workers[59] have reported the occurrence, in cattle grazing such pastures (which contain only 1 to 3 p.p.m. Cu on the dry basis), of a peculiar disease known locally as "falling disease." The sudden deaths which characterize this disease have been shown to be the terminal manifestation of severe and uncomplicated copper deficiency in which the primary lesion is starvation atrophy of the myocardium with replacement fibrosis. Anemia is rarely severe in this condition and usually only seasonal in occurrence. Further evidence of species differences is given by the frequent occurrence in many parts of the world of diarrhea or "scours" in cattle grazing copper-deficient pastures.[21, 60] Scouring is not a regular symptom in sheep grazing such pastures.

The minimum copper requirements of sheep and cattle cannot be stated with great accuracy, although pastures containing 6 to 8 p.p.m. on the dry basis are generally considered to be adequate for satisfactory growth, health, and reproduction. For normal keratinization of the wool of sheep and for the maintenance of liver copper reserves somewhat higher levels have been suggested.[57] Marston and co-workers[57] suggest that a total intake of 10 mg. Cu per day is essential for the wool sheep, judging from the results of long-term experiments in which the effects were compared of 1 mg., 5 mg. and 10 mg. per day as copper sulfate administered thrice weekly as a drench to sheep grazing on deficient pastures. These pastures were calculated to supply about 2 mg. Cu daily, but it should be noted that they also

[59] H. W. Bennetts et al., *Australian Vet. J.* **15,** 152 (1939); **17,** 85 (1941); **18,** 50 (1942); **24,** 237 (1948).

[60] B. Sjollema, *Biochem. Z.* **267,** 151 (1933).

contained above normal levels of molybdenum, which could raise the copper requirement.

7. Copper in Relation to the Growth and Pigmentation of Hair and Wool

Copper deficiency in the rat, the cat and the rabbit leads to achromotrichia, in addition to anemia. In the cat and the rabbit alopecia also occurs, and dermatosis has been reported in the rabbit.[61] The syndrome of achromotrichia, alopecia, and dermatosis, according to Smith and Ellis,[61] is a more sensitive indication of copper deficiency in the rabbit than is anemia. An adequate provision of copper is also necessary for normal pigmentation in black-wooled sheep,[62] and the level of copper intake, up to certain optimal values, profoundly influences the quantity and particularly the quality of wool grown.[19] The wool lesion is the first symptom of copper deficiency to appear in this animal and is, as pointed out earlier, the most specific and sensitive index of copper deficiency known.

The most obvious change in the character of the wool produced by copper-deficient sheep is loss of the characteristic crimp. This effect is much more noticeable in the Merino than in the British breeds, which do not normally produce wool with a marked crimp. The crimp becomes progressively less distinct as the animal's reserves of copper are used up, changing from normally crimped wool to straight, "stringy" or "steely" wool. A spectacular restoration of the crimp occurs when copper supplements are given. Further, the physical properties of the wool of copper-deficient sheep differ markedly from those of normal wool. The tensile strength is less and the elastic properties abnormal.[57]

Possible mechanisms involved in these profound changes have been well discussed by Marston,[55] who presents interesting evidence implicating copper in the keratinization process. Purified keratin from copper-deficient wool has been shown by this worker to be significantly lower (10 to 15%) in cystine than keratin similarly prepared from normal wool. The acid-combining power of the deficient keratin is also reduced. Apparently the chemical mechanisms involved in the production of normally keratinized wool, particularly the conversion of prekeratin to keratin, are dependent upon adequate concentrations of copper-containing catalysts, but this has yet to be established. In regard to pigmentation it should be mentioned that tyrosinase, the enzyme thought to be concerned in melanin formation, is a copper proteinate.[63]

[61] F. J. Gorter, *Z. Vitaminforsch.* **4**, 277 (1935); S. E. Smith and G. H. Ellis, *Arch. Biochem.* **15**, 81 (1947).

[62] H. J. Lee and G. R. Moule, *Australian Vet. J.* **23**, 303 (1947).

[63] F. Kubowitz, *Biochem. Z.* **296**, 443 (1938).

8. Copper in the Nutrition of Man

The value of copper supplements, with and without iron, in the treatment of anemias of infancy and of secondary anemias of adults has been extensively studied with results which are far from uniform. This is hardly surprising in view of the difficulties with humans of controlling such factors as previous dietary history, extent of tissue storage, and contamination of food and medicaments. Several investigators have found iron alone to be just as effective as iron and copper. A few cases have been reported in which copper therapy has been successful when iron therapy has failed, and a number of workers have found that iron and copper together give results in the treatment of nutritional anemia of infants superior to those of iron alone. A great deal of the literature dealing with this subject has been reviewed by Hutchinson,[64] who reports results of his own, demonstrating the value of copper in addition to iron. One of the most striking demonstrations of the value of copper is that of Usher *et al.*,[65] who found that 1 to 2 mg. Cu per day as a supplement to iron was markedly superior to iron only in producing and maintaining hemoglobin levels in a prophylactic experiment with a large group of infants. A number of investigators[66] have also shown that many cases of chronic idiopathic anemia of adults, refractory to iron, respond to copper therapy. It is clear that copper is necessary for blood formation in man, as in animals, although copper therapy is not always necessary in the treatment of human nutritional anemias. Nevertheless, the number of cases in which copper has been found necessary for maximum response are so numerous that it would seem wise to include some copper in all iron prescriptions, at least for the anemias of infancy and childhood.

The minimum copper requirements of the human adult are not known but they appear to be lower, per unit of body weight, than those of sheep or cattle. Acute copper deficiency must be very rare in humans and has never been demonstrated. Attempts to determine requirements by means of balance experiments have given extremely variable results. Tompsett's work indicates a minimum requirement as low as 0.6 mg. Cu daily,[28] whereas De's work indicates 2.0 mg.[67] Estimates of the average daily intakes of copper from ordinary diets also vary greatly. Tompsett and others[28, 68] give values of 1.5 to 2.6 mg. Cu for adults. On the other hand,

[64] J. H. Hutchinson, *Quart. J. Med.* [*N. S.*] **7,** 397 (1938).

[65] S. J. Usher, P. N. MacDermott, and E. Lozinski, *Am. J. Diseases Children* **18,** 642 (1935).

[66] E. S. Mills, *Am. J. Med. Sci.* **182,** 554 (1931).

[67] H. N. De, *Indian J. Med. Research* **37,** 301 (1949).

[68] R. M. Leverton and E. S. Binkley, *J. Nutrition* **27,** 43 (1944).

De[67] found the mean daily intake of a group of Indian male adults to be 4.5 mg. Cu on rice diets and 5.8 mg. Cu on wheat diets. Growing children have been estimated to retain 0.06 to 0.10 mg. Cu per kilogram body weight, indicating a minimum requirement of 1.2 to 2.0 mg. Cu daily for an average 4 to 6-year-old boy weighing 20 kg.[69] This amount would not always be easy to obtain from ordinary diets,[68a] at the level of consumption usual for such children, unless the estimates of normal adult intakes made by Tompsett and others are too low.

Estimates of requirements based upon balance experiments are hazardous with all the trace elements, because the amounts retained by the body normally represent such a small fraction of the amounts ingested that they fall within the experimental error of the chemical estimations involved. This is certainly true of copper and is a factor which has to be considered in addition to those inherent in all balance studies with minerals, namely, the influence of variable body stores of the element, and the difficulty, in fact the impossibility, of determining what proportion of the fecal excretion has passed through the tract unabsorbed and what proportion has been absorbed and subsequently excreted via the bile or other secretions or through the intestinal wall. There is evidence that the bile is an important path of excretion for copper and that urinary excretion is extremely small in all animal species studied. In the human species the amount of copper excreted in the urine has varied, in different investigations, from 0 to 0.7 mg. daily and from 0 to 15% of the amount ingested in the food.[26] In a very recent study of eleven normal adults, Porter[70] could detect no urinary copper in five of these subjects, in two the total daily excretion was less than 3.5 μg., and in four it ranged from 3.5 to 14.7 μg. Even the highest of these figures is almost certainly less than 1% of the amount taken in in the food. Raising the copper intake either by ingestion or injection does not greatly increase the amount excreted in the urine, but the amount retained depends very largely on the copper status of the animal's tissues, particularly the liver. Retention of copper in the body depends therefore upon two main factors, the level of intake and the copper status of the animal. The factors influencing absorption of copper are little understood and are, in fact, difficult to assess by any direct means, but there is no doubt that the ratio of copper to other minerals in the diet can be important. The significance of molybdenum in this connection is discussed in the following section.

[68a] S. W. Monier-Williams, Trace Elements in Food, Chapman and Hall, London, 1949, p. 30.

[69] T. P. Chou and W. H. Adolph, *Biochem. J.* **29**, 476 (1935); F. I. Scoular, *J. Nutrition* **16**, 437 (1938).

[70] H. Porter, *Arch. Biochem. Biophys.* **31**, 262 (1951).

9. Copper and Molybdenum

A relationship between copper and molybdenum in animals has been established as a result of recent work with sheep, cattle, and rats. The disease of cattle characterized by persistent scouring and known locally as "teart," which occurs in certain restricted areas in England, was shown by Ferguson and co-workers[71] to be caused by abnormally high intakes of molybdenum from the pastures (20 to 100 p.p.m. Mo on the dry basis) and to be prevented or cured by dosing with massive quantities of copper sulfate (1 to 2 g. daily). Intravenous injections of much smaller doses of copper sulfate (100 to 300 mg.) are also effective.[72] It must be emphasized that teart is not a molybdenum-induced copper deficiency, since the copper concentrations in the blood and liver of affected animals remain within normal limits. Neither does the copper therapy reduce the uptake of molybdenum, as judged by the levels of molybdenum in the blood and liver of treated animals, although it controls the scouring within a matter of days. Moreover, Green[72] reports an experiment with hypocupremic ewes in which sodium molybdate was fed, as a supplement to a normal ration, for 15 months in daily doses raised each quarter on the scale 14, 56, 112 and 224 mg. Mo. Liver and blood copper values rose in all cases to normal levels, in spite of the high molybdenum intake which brought about exceedingly high concentrations of molybdenum in the blood and liver of the animals.

The results of this latter experiment contrast with those of Dick and Bull[73] who added 10 mg. and 100 mg. Mo in the form of ammonium molybdate daily for 6 months to the diet of sheep receiving a normal diet and a normal diet plus additional copper. A marked and significant reduction in the copper concentration of the liver was obtained in the molybdenum-treated sheep with or without added copper. These findings indicate that molybdenum can limit the amount of copper assimilated and stored in the liver and suggest the possibility of increased molybdenum content of pastures being responsible for copper deficiency in sheep in some areas and accentuating it in others. Conversely, in areas where the copper content of the pastures is normal or high (15 p.p.m. Cu or more), abnormally low levels of molybdenum in the pastures (0.1 p.p.m. Mo or less) favor the development of a high copper status in sheep and may lead to chronic copper poisoning.[74] Confirmatory evidence of a reciprocal antagonism between copper and molybdenum has been obtained in New

[71] W. S. Ferguson, A. H. Lewis, and S. J. Watson, *Nature* **141,** 553 (1938); *Jealott's Hill Agr. Research Bull.* 1 (1940); *J. Agr. Sci.* **33,** 44, 116 (1943).

[72] H. H. Green, *Spec. Conf. in Agr. (Australia)* (1949).

[73] A. T. Dick and L. B. Bull, *Australian Vet. J.* **21,** 70 (1946).

[74] L. B. Bull, *Spec. Conf. in Agr. (Australia)* 1949.

Zealand by Cunningham.[75] Several experiments by this worker have shown that increased molybdenum intake either naturally from molybdenum-high pastures or from added molybdate reduces the copper content of bovine livers and blood. Data from one of these experiments are given in Table 4. Cunningham[75] claims further that molybdenum rather than copper is the significant factor in the scouring of cattle on copper-deficient peat lands since this symptom occurs where the pastures are moderately low in copper and moderately high in molybdenum but not where the pastures are deficient in copper but contain normal low amounts of molybdenum.

Copper is also effective in overcoming or reducing the effects of molybdenum toxicity in the rat.[76, 77] Comar and associates[76] found that newly

TABLE 4

COPPER AND MOLYBDENUM IN THE LIVERS AND BLOOD OF HEIFERS (CUNNINGHAM[75])

	Date sample taken (day/month/year)	Untreated group		Mo-dosed group*	
		Cu	Mo	Cu	Mo
Blood, g./ml.	5/11/46	0.9	0.06	0.4	0.6
Liver, p.p.m. in dry liver	19/2/47	96	3.8	10.1	5.4

* The animals of this group were drenched thrice weekly with a solution of ammonium molybdate supplying the following amounts of molybdate per dose: 1/8/45 to 4/11/45, 300 mg.; 5/11/45 to 7/3/46, 600 mg.; 8/3/46 to 29/4/46, 1200 mg ; 1/5/46 to 19/2/47, 600 mg.

weaned rats on a diet containing 2 p.p.m. Cu and 1 p.p.m. Mo grew well and appeared normal in every way. The addition of 80 p.p.m. Mo to this diet slowed growth and induced a high mortality. A further addition of 35 p.p.m. Cu as copper sulfate resulted in growth equivalent to that of the controls receiving no additional molybdenum and eliminated the mortality. Neilands and co-workers[77] were able to overcome completely the toxicity to rats of a diet containing 40 mg. Mo per 100 g. by the addition of 5% whole liver substance and very largely by the addition of 2 mg. Cu, as copper sulfate per 100 g., although the diet already contained 7.73 mg. Cu per 100 g. Using Mo^{99} as a tracer these workers showed that the distribution of molybdenum between stomach, intestines, feces, and urine was not affected by the copper supplement. The ability of copper to overcome molybdenum toxicity in the rat is suggested further by the work of

[75] I. J. Cunningham, *Proc. Intern. Vet. Congr. 14th Congr. London* (1949).
[76] C. L. Comar, L. Sniger, and G. K. Davis, *J. Biol. Chem.* **180,** 913 (1949).
[77] J. B. Neilands, F. M. Strong, and C. A. Elvehjem, *J. Biol. Chem.* **172,** 431 (1948).

Gray and Ellis,[78] but with the diets and mineral ratios used by these workers it was much less obvious.

III. Cobalt

The first record of cobalt's occurring in living tissues appears to be that of Legrip[79] (1841), who demonstrated its presence in the plant *Lathyrus odoratus*. It was more than half a century later before the general distribution of this element in biological material began to be recognized. Bertrand and co-workers[80] found that minute amounts of cobalt occur consistently in plants and in most of the organs of man and other mammals, and of the birds, fish, and marine animals which they examined. The widespread occurrence of cobalt in plant and animal tissues has since been confirmed by many workers.

No real indication of a possible function for cobalt was obtained from these early studies although the consistency with which cobalt occurs in the pancreas in higher concentration than in other mammalian organs led Bertrand and Macheboeuf[80] to suggest that it might be concerned in the elaboration and activity of insulin. Little support for this suggestion has come from subsequent work with cobalt, including the very recent studies of Ulrich and Copp[81] with radioactive cobalt. These workers could find no significant difference between the cobalt content of the pancreas of normal and alloxan diabetic rats and no localization of cobalt within the islet cells. Neither was the first attempt to demonstrate an essential role for cobalt in nutrition very convincing. Bertrand and Nakamura[82] found that mice fed a highly deficient ration survived longer when cobalt and nickel supplements were added. The diets fed were so deficient in other factors, however, that even with the cobalt and nickel supplements the mice only survived about three weeks.

Evidence that cobalt is biologically active was first demonstrated by Waltner and Waltner.[83] These workers showed that the administration of relatively large amounts of cobalt to the rat stimulates hematopoiesis and produces polycythemia. Many workers have subsequently confirmed this finding, but it must be emphasized that the amounts of cobalt necessary to induce polycythemia are considerably greater than the amounts every likely to be present in normal diets. The polycythemic action of

[78] L. F. Gray and G. H. Ellis, *J. Nutrition* **40,** 441 (1950).

[79] J. Legrip, *J. chim. med.* [2] **7,** 120 (1841), quoted by W. E. Brenchley, *Ann. Applied Biol.* **25,** 671 (1938).

[80] G. Bertrand and M. Mokragnatz, *Compt. rend.* **175,** 458 (1922); *Bull. soc. chim.* **37,** 554 (1925); G. Bertrand and M. Macheboeuf, *Compt. rend.* **180,** 1380, 1993 (1925); **182,** 1504; **183,** 5 (1926).

[81] F. Ulrich and D. H. Copp, *Arch. Biochem. Biophys.* **31,** 148 (1951).

[82] G. Bertrand and H. Nakamura, *Compt. rend.* **185,** 321 (1927); **186,** 1480 (1928).

[83] K. Waltner and K. Waltner, *Klin. Wochschr.* **8,** 313 (1929).

cobalt cannot therefore be construed as evidence of a function for this element in normal metabolic processes. Nevertheless, Waltner and Waltner's work stimulated further investigation of a possible physiological role for cobalt and was directly responsible for this element's being given an early trial in the treatment of "coast disease" of sheep. This is of great interest because "coast disease" was the first of the naturally occurring wasting diseases of sheep and cattle to be shown to respond to cobalt therapy.[84] The subsequent demonstration by workers in many parts of the world that these diseases are the result of cobalt deficiency, or of a dual deficiency of cobalt and copper, gave an enormous stimulus to work on the mode of action of cobalt in ruminants and its distribution in plant and animal tissues and organs. A further stimulus came from the important finding of Smith[85] and of Rickes *et al.*[86] in 1948 that the anti-pernicious anemia factor (vitamin B_{12}) is a cobalt compound.

1. Cobalt Deficiency in Sheep and Cattle

For very many years it had been recognized by farmers that certain localized areas in Australia, New Zealand, Europe, North America, and Africa were unsatisfactory for the raising of sheep and cattle, in spite of ample pasture of satisfactory quality as judged by conventional standards. Animals confined to these "affected" pastures showed a marked loss of appetite, emaciation, debility, progressive anemia, and finally death. Nonruminant species grazing under the same conditions remained perfectly healthy. The history of the various investigations, which led to the abandonment of the iron-deficiency hypothesis which was first used to explain their etiology and to the establishment of cobalt deficiency as their cause, has been reviewed by Marston,[87] Underwood,[88] and Russell[56] up to 1944 and will not be repeated here. As in the case of the naturally occurring copper-deficiency diseases, three main lines of evidence served to establish the position: (1) prevention and cure of the disease conditions by feeding minute cobalt supplements; (2) subnormal levels of cobalt in the liver of affected animals; and (3) subnormal concentrations of cobalt in the pasture and soils of the disease-producing areas.

As little as 0.07 p.p.m. Co in the dry matter of the diet is generally considered to be sufficient to provide the full requirements of sheep and cattle, although 0.1 p.p.m. is suggested by Stewart[89] as a safer minimum.

[84] E. W. Lines, *J. Council Sci. Ind. Research* **8,** 117 (1935).

[85] E. L. Smith, *Nature* **162,** 144 (1948).

[86] E. L. Rickes, N. G. Brink, F. R. Koniusky, T. R. Wood, and K. Folkers, *Science* **108,** 134 (1948).

[87] H. R. Marston, *Ann. Rev. Biochem.* **8,** 557 (1939).

[88] E. J. Underwood, *Nutrition Abstracts & Revs.* **9,** 515 (1940).

[89] J. Stewart, *Spec. Conf. in Agr.* (*Australia*) (1949).

"Healthy" pastures vary considerably in cobalt content but mostly lie within the range 0.1 to 0.3 p.p.m. Deficient pastures generally contain only 0.03 to 0.04 p.p.m., but figures as low as 0.01 p.p.m. have been reported. Sheep grazing such pastures have been observed to respond to the feeding of cobalt salts supplying as little as 0.05 mg. Co daily,[90] but for optimum growth and health a minimum of 0.1 mg. per day, fed either daily, thrice weekly, or weekly, is necessary.[90, 91, 92] Drenching with much larger doses of cobalt at longer intervals (monthly) will prevent obvious signs of deficiency in sheep,[93] but for optimum results frequent and regular treatment is necessary. This is due to the very limited capacity for storage of cobalt in the animal body, to which later reference will be made. The minimum supplements required for the normal growth and health of cattle grazing typical deficient pastures lie between 0.3 and 1.0 mg. Co per day, where these supplements are administered daily,[90] but considerably larger amounts than these minimum levels are usually supplied in practice. Annual or biennial treatment of the pastures with cobalt salts has also proved an efficient and effective form of treatment in many areas.[94, 95]

A striking feature of cobalt deficiency in both sheep and cattle is the absence of any symptoms which can be regarded as specific for this element. Varying degrees of anemia and aplasia of the bone marrow occur fairly consistently, but the appearance of a cobalt-deficient animal is that of extreme emaciation, indistinguishable from that of a starved animal. This is consistent with the fact that the first and dominant effect of lack of cobalt is failure of appetite. Wasting of the musculature (marasmus) and debility are a natural consequence of inadequate food consumption and usually precede any considerable degree of anemia. Similarly, the first measurable response to feeding cobalt salts is a rapid response in appetite with consequent weight gains, whereas improvement in the blood picture, though it may be equally dramatic, usually occurs somewhat later. Since anemia does not normally occur in animals suffering from ordinary starvation, it is clear that, in the ruminant, cobalt is concerned in some way with both the normal physiological controls of appetite and with hematopoiesis. On present evidence it looks as if this is not a direct effect on the host animal but an indirect one through the ruminal microflora. This question is discussed in the section on the mode of action of cobalt.

[90] J. F. Filmer and E. J. Underwood, *Australian Vet. J.* **13,** 57 (1937).

[91] H. O. Askew, *New Zealand J. Sci. Technol.* **20A,** 192 (1936).

[92] H. R. Marston, *Spec. Conf. in Agr.* (*Australia*) (1949).

[93] J. F. Filmer, *New Zealand J. Agr.* **63,** 287 (1941).

[94] H. O. Askew, *New Zealand J. Sci. Technol.* **28A,** 37 (1946).

[95] R. C. Rossiter, D. H. Curnow, and E. J. Underwood, *J. Australian Inst. Agr. Sci.* **14,** 9 (1948).

2. Cobalt Content of Animal Tissues and Organs

Data on the cobalt content of animal organs, obtained by reliable modern methods of analysis, are so few that worth-while species comparisons are impossible. Only in the rat and the sheep has the distribution of cobalt throughout the body been extensively studied. Bertrand and Machebouef's earlier work on human tissues is of interest, but their levels are suspiciously high, judged by more recent work on other species with improved methods. However, their finding that the concentration of cobalt in the pancreas is high, compared with other organs examined, has been confirmed by several workers for the rat and the sheep. Concentration in the spleen is also frequently high in normal healthy animals, although under conditions of relatively massive cobalt supplementation the liver and kidneys usually contain the highest concentrations. Ward *et al.*[96] found the concentration of cobalt in the tissues of normal newborn calves to range from 0.10 to 0.22 p.p.m. on the dry basis, except for the skin and hair which contained 1.0 p.p.m. and the urine which contained 0.36 p.p.m. Data on the cobalt content of bone marrow from healthy and cobalt-deficient animals are unfortunately almost non-existent.

In contrast to the above, the literature on the cobalt content of liver is fairly extensive, particularly for the sheep. The age of the animal does not appear greatly to affect the concentration of cobalt in the liver but it is markedly influenced by the level of intake in the diet. For these reasons liver cobalt determinations are extremely valuable, although not infallible, aids in diagnosing cobalt deficiency. Thus Underwood and Harvey[97] found the mean concentration of cobalt in the livers of a group of sheep suffering from cobalt deficiency to be 0.06 p.p.m. on the dry basis, compared with a figure of 0.28 p.p.m. for a group of healthy sheep. Askew and Dixon[98] found similar differences but somewhat lower values. Marston *et al.*[23] obtained a mean of 0.09 p.p.m. Co for the livers of a group of "coasty" sheep and 0.34 p.p.m. for a similar group of healthy, cobalt-treated animals. McNaught,[99] from an extensive study of healthy and "bush-sick" animals in New Zealand, suggests that 0.04 to 0.06 p.p.m. Co, or less, in the dry matter of the livers of sheep and cattle indicate cobalt deficiency, and values of 0.08 to 0.12 p.p.m., or more, indicate cobalt sufficiency.

McNaught's work is also of value because it shows that, unlike copper, cobalt does not accumulate in the fetal liver. In sheep and cattle it appears also that, under conditions of steady, adequate intake, the cobalt concen-

[96] G. M. Ward, E. J. Berne, H. D. Webster, C. W. Duncan, and C. F. Huffman, *J. Animal Sci.* **8**, 632 (1949).

[97] E. J. Underwood and R. J. Harvey, *Australian Vet. J.* **14**, 183 (1938).

[98] H. O. Askew and J. K. Dixon, *New Zealand J. Sci. Technol.* **18**, 707 (1937).

[99] K. J. McNaught, *New Zealand J. Sci. Technol.* **30A**, 26 (1948).

tration in the liver rises slightly from birth to weaning and is normally somewhat higher in the mature than in the newborn animal.

Many workers have drawn attention to the liver as a storage organ for cobalt, but it is not sufficiently realized that the storage capacity of this organ is very limited. In fact, substantial stores of cobalt have not been demonstrated in any of the organs or tissues of the body. This is no doubt related to the fact that cobalt is very poorly absorbed from the gut, at least of the ruminant.[100, 101]

Using orally administered radioactive cobalt, Comar *et al.*[101] found only 0.25% to be retained in the tissues (mostly the liver) of a bullock after several days, most of it being accounted for in the feces. Even when cobalt was placed directly in the rumen, none was detected in the blood and within 7 days only 1% remained in the rumen. On the other hand, Copp and Greenberg[102] found a single dose of radioactive cobalt given by stomach tube to be 50% absorbed by the rat and most of this to be rapidly excreted in the urine. The position with injected cobalt is somewhat different. Excretion takes place mostly in the urine, and percentage retention by the tissues is very low. Nevertheless, very much higher levels occur in the liver than when the cobalt is ingested. Ray *et al.*[103] found about ten times more cobalt in the livers of injected sheep than in sheep receiving the same amount of cobalt orally. In view of the fact that injected cobalt is largely ineffective, it is evident that liver cobalt is not a certain criterion of cobalt adequacy.

The cobalt content of the blood of sheep and cattle is so low that precise determination is extremely difficult. The few figures that are available suggest that there is a distinct tendency for the levels in the blood to follow that in the liver and therefore to reflect the cobalt status of the animal. Thus Askew and Dixon[98] found the blood for four affected sheep to contain 0.02, 0.02, 0.02, and 0.01 p.p.m. Co on the dry basis, compared with 0.05 and 0.04 p.p.m. for two similar healthy cobalt-treated animals. Keener *et al.*[104] also report a mean level of 0.04 p.p.m. Co for the blood serum of their cobalt-treated sheep and claim a mean figure of 0.004 p.p.m. for their cobalt-deficient animals.

3. Cobalt Content of Milk

The cow appears to be the only species for which figures are available for the cobalt content of the milk. Underwood and Elvehjem[49] and Ahmad

[100] H. O. Askew and S. W. Josland, *New Zealand J. Sci. Technol.* **18,** 888 (1937).

[101] C. L. Comar, G. K. Davis, and R. F. Taylor, *Arch. Biochem.* **9,** 149 (1946); C. L. Comar, G. K. Davis, R. F. Taylor, C. F. Huffman, and R. Ely, *J. Nutrition* **32,** 61 (1946).

[102] D. H. Copp and D. M. Greenberg, *Proc. Natl. Acad. Sci. U. S.* **27,** 153 (1941).

[103] S. N. Ray, W. C. Weir, A. L. Pope, G. Bohstedt, and P. H. Phillips, *J. Animal Sci.* **7,** 3 (1948).

[104] H. A. Keener, G. P. Percival, and K. S. Morrow, *J. Animal Sci.* **7,** 16 (1948).

and McCollum[104a] give values of the order of 10 to 20 μg. Co per liter, but more recent investigations, in which improved methods have been used, suggest strongly that these values are much too high. Thus Paulais[104b] found 0.4 to 0.7 μg. Co per liter and Archibald[104c] gives the normal range as 0.2 to 1.1 μg. The effect of cobalt deficiency on the cobalt content of milk is unknown, but it can be increased four-fold by feeding cobalt supplements to cows on normal rations.[104c]

4. Cobalt in the Nutrition of Man and Other Non-Ruminants

Sheep, cattle, and goats are the only species in which cobalt deficiency has been observed. Horses and pigs develop normally in the naturally occurring cobalt-deficiency areas, and rats and rabbits remain free from any evidence of cobalt deficiency on synthetic rations in which the cobalt concentrations have been reduced well below those which will induce symptoms of cobalt deficiency in ruminants. Houk *et al.* [105] have shown this for rats on a ration claimed to contain only 0.003 p.p.m. Co, and Thompson and Ellis[106] for rabbits on a ration claimed to contain only 0.0024 p.p.m. Co. It is apparent that if these species require cobalt at all it must be in extraordinarily small amounts.

In the human species work on cobalt has so far been restricted to its possible significance in the treatment of anemia. Underwood[107] found cobalt to be an invariable contaminant of iron salts and suggested that it might play some part in the treatment of those secondary anemias of man in which massive doses of iron compounds are necessary. What little evidence there is does not support this suggestion,[108] but far more critical experimentation is necessary before cobalt can completely be dismissed as a factor in the treatment of human secondary anemias. In infants suffering from nutritrional anemia Kato[109] reports an accelerated formation of red cells and hemoglobin when iron treatment was supplemented with relatively large amounts of cobalt, and Wilberg[110] claims good results from similar large (50 mg.) injections of cobalt in posthemorrhagic anemias and anemias due to infection or tumors. In milk anemia of rats, on the other hand, addition of moderate amounts of cobalt to the normal iron and copper supplements has no accelerating effect on hematopoiesis.[49]

The discovery that vitamin B_{12} is a cobalt compound[85, 86] presents the

[104a] B. Ahmad and E. V. McCollum, *Am. J. Hygiene* **29,** 24 (1939).
[104b] R. Paulais, *Ann. pharm. franç.* **4,** 110 (1946).
[104c] J. G. Archibald, *J. Dairy Sci.* **30,** 293 (1947).
[105] A. E. Houk, A. W. Thomas, and H. C. Sherman, *J. Nutrition* **31,** 106 (1946).
[106] J. F. Thompson and G. H. Ellis, *J. Nutrition* **34,** 121 (1947).
[107] E. J. Underwood, *Proc. Soc. Exptl. Biol. Med.* **36,** 296 (1937).
[108] C. R. Baxter, *Brit. Med. J.* **1,** 354 (1939); E. Cronin, *ibid.* **1,** 643 (1939).
[109] K. Kato, *J. Pediat.* **11,** 385 (1937).
[110] C. Wilberg, *Münch. med. Wochenschr.* **92,** 1373 (1950); *Nutrition Abstracts & Revs.* **20,** 1038 (1951).

whole problem of the cobalt requirements of man and other non-ruminant species in a fresh light. Since a supply of this vitamin in the diet is known to be necessary for growth or erythropoiesis, or both, in a number of species (man, rat, mouse, pig, chick, fox, and mink), a supply of cobalt in the diet must also be necessary. Whether these species require cobalt only in the form of vitamin B_{12} is not yet known. If they do, the amounts are extraordinarily small and far below the level of any other trace mineral. For instance, for the successful treatment of pernicious anemia in man, 1 to 2 μg. of vitamin B_{12} per day when injected and 5 μg. per day given orally with normal gastric juice have been found adequate. As vitamin B_{12} contains 4.5% Co, this represents, at the 5-μg. level, a daily intake of only about 0.2 μg. Co or about one five-hundredth of the daily requirement (0.1 mg.) of a sheep. Estimations of the cobalt content of human foods[104a, 111] show that all ordinary diets supply many times this amount of cobalt. The problem for these species seems therefore to be one of sources of vitamin B_{12} rather than of cobalt. This is outside the scope of this article and in any case has recently been discussed by Smith,[112] but it is of interest to note that non-ruminant herbivores, like ruminants, appear to be dependent upon the synthetic activities of the micro-organisms of their digestive tract because significant amounts of vitamin B_{12} do not occur in the higher plants.[112]

5. Polycythemic Action of Cobalt

Since the pioneer experiment of Waltner and Waltner,[83] which demonstrated that the feeding of cobalt causes polycythemia in rats, many investigators have confirmed this effect with rats and showed that it occurs also in mice, guinea pigs, dogs, frogs, pigs, rabbits, and chickens. Feeding or injecting equivalent levels of cobalt to ruminants does not induce polycythemia,[113, 114] but toxic effects, notably loss of appetite and depression of growth and body weight, occur at somewhat lower dose levels than those required to produce similar toxic effects in non-ruminants.[115] The polycythemia develops only when the diet is otherwise adequate for rapid blood formation. It is a true polycythemia resulting from hyperplasia of the bone marrow and not from a decrease in blood volume.

The polycythemic action of cobalt may be depressed or enhanced by various means which apparently do not act in the same way in different species. In the dog the simultaneous or separate feeding of liver and of choline or the injection of liver extracts or ascorbic acid reduces or

111 N. D. Sylvester and L. H. Lampitt, *J. Soc. Chem. Ind.* (*London*) **59,** 57 (1940); C. Hurwitz and K. C. Beeson, *Food Research* **9,** 348 (1944).

112 E. L. Smith, *Nutrition Abstracts & Revs.* **20,** 795 (1951).

113 S. W. Josland, *New Zealand J. Sci. Technol.* **19,** 31 (1937).

114 R. P. Geyer, E. W. Rupel, and E. B. Hart, *J. Dairy Sci.* **28,** 291 (1945).

115 R. E. Ely, K. M. Dunn, and C. F. Huffman, *J. Animal Sci.* **7,** 239 (1948).

nullifies the effect of cobalt.[116] In the rat on a milk diet choline has no such depressing effect, and whole liver powder, liver extract, and a P.A. liver concentrate have been found to enhance cobalt polycythemia.[117] The toxicity of high levels of orally ingested cobalt in the rat and in the calf is reduced, however, by concurrent or separate injections of methionine, cystine, cysteine, or histidine.[118] These effects cannot be paralleled by raising the intake of these amino acids through increased dietary protein levels or through increased consumption of proteins, such as casein, which supply equivalent amounts of methionine.[115] It appears that the action of these amino acids in reducing cobalt toxicity is due to their capacity to form coordination compounds with cobalt which are relatively non-toxic. Several workers have shown that cobalt forms coordination complexes with certain amino acids,[119] and the complex of cobalt and cysteine formed *in vitro* has been found to be relatively non-toxic in the rat.[120]

The way in which cobalt exerts its stimulating effect on the bone marrow is not yet understood. Barron and Barron[121] suggest that cobalt inhibits the respiration of immature red cells, causing their discharge into the circulation as imperfect, non-respiring cells and their replacement in the bone marrow by new cells. Griffith *et al.*[118] suggest that the formation of cobalt coordination complexes with compounds in the cell such as glutathione interferes with cellular oxidation which provides the necessary stimulus to hematopoiesis. Davis[116] suggests that cobalt stimulates erythropoiesis by interfering with a respiratory function of ascorbic acid. Bucciero and Orten[122] have shown that the mechanism of cobalt polycythemia is not one of lowering the O_2-capacity of the hemoglobin or of producing a methemoglobin possibly containing cobalt rather than iron in the hemoglobin molecule.

The failure of cobalt to induce polycythemia in ruminants is unexplained, although it has been suggested that this is due to the particularly low absorption and retention of cobalt by these species.

6. Mode of Action of Cobalt

Recent work in several laboratories has established that cobalt exerts its beneficial effects primarily, but not necessarily wholly, through its

[116] J. E. Davis, *Am. J. Physiol.* **127,** 322 (1939).

[117] H. D. Anderson, E. J. Underwood, and C. A. Elvehjem, *Am. J. Physiol.* **130,** 373 (1940); L. H. Marshall, *ibid.* **114,** 194 (1935).

[118] W. H. Griffith, C. L. Paveck, and D. J. Mulford, *J. Nutrition* **23,** 603 (1942).

[119] D. Burk, A. L. Schlade, M. L. Hesselback, and C. E. Fischer, *Federation Proc.* **5,** 126 (1946).

[120] L. Michaelis and S. Yamaguchi, *J. Biol. Chem.* **83,** 367 (1929).

[121] A. G. Barron and E. S. G. Barron, *Proc. Soc. Exptl. Biol. Med.* **35,** 407 (1937).

[122] M. C. Bucciero and J. M. Orten, *Blood* **4,** 395 (1949).

action on the ruminal microflora. In other words, the cobalt requirement of the ruminant is largely an indirect one, supplying the needs of rumen bacteria for this element, thus enabling them to produce one or more essential nutritional factors for their host. Many lines of evidence conspire

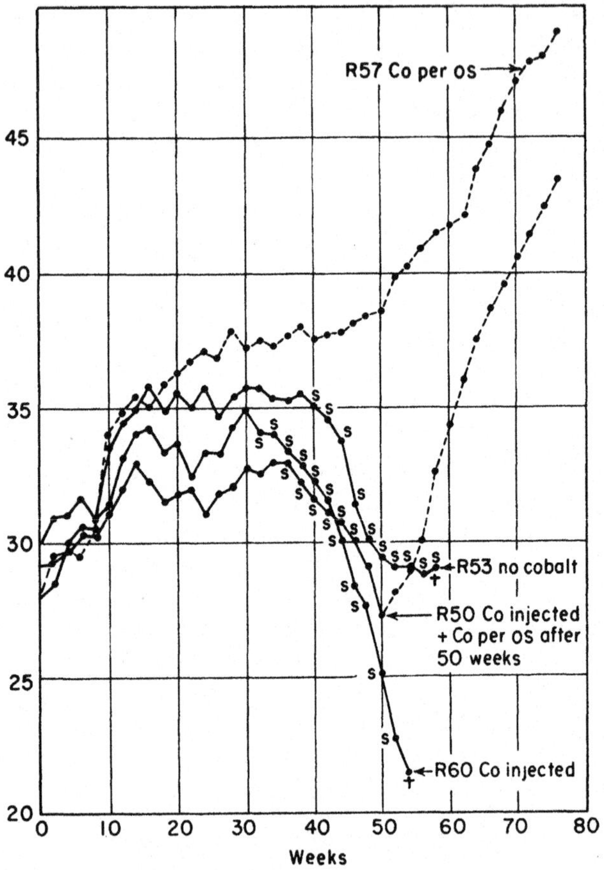

FIG. 1. Typical responses of ewes on cobalt-deficient pastures. R53, untreated; R57, 1 mg. Co daily *per os*; R60, 1 mg. Co daily injected; R50, 1 mg. Co daily injected plus 1 mg. Co daily injected after 50 weeks when the deficiency symptoms were in a terminal stage. S denotes recognizable deficiency systems. (Marston and Lee.[123])

to support this hypothesis. First, ruminants are the only species in which cobalt deficiency has been demonstrated. Second, cobalt must be ingested if it is to be completely effective (see Fig. 1), in spite of the fact that injected cobalt can increase the concentration of cobalt in the liver and blood tenfold.[103, 123, 124] This cobalt does not appear, however, in the rumen,

[123] H. R. Marston and H. J. Lee, *Nature* **164,** 529 (1949).
[124] D. E. Becker, S. E. Smith, and J. K. Loosli, *Science* **110,** 71 (1949).

the reticulum, or the omasum,[101] except where very large amounts are injected, and even then only in extremely small amounts.[125] Third, in cobalt-deficient animals there is a marked diminution in the numbers of bacteria in the rumen and significant changes in the types of organism present.[126] These bacterial changes are not the result of lowered feed intake *per se*, and the numbers can be raised to normal levels by cobalt feeding but not by injection.[126] Fourth, the cobalt present in the rumen contents is highly concentrated in the microbial fractions,[127] and both the total amount and the amount contained in the microflora are greatly reduced in the cobalt-deficient animal.[127] These latter findings are difficult to evaluate but certainly suggest that the rumen bacteria have a real requirement for cobalt for their own growth and synthetic purposes. Radiocobalt fed to sheep has been found to be incorporated in the fecal vitamin B_{12}, no doubt by the action of the ruminal microorganisms.[128] Many microorganisms have been shown to produce increased yields of vitamin B_{12} by the addition of traces (about 1 p.p.m.) of cobalt to the culture medium.[129] Supplementing the ration of ewes with cobalt, although this ration apparently supplied ample cobalt, has been shown to produce a highly significant increase in the vitamin B_{12} content of their milk, presumably as a result of increased synthesis by the rumen microorganisms.[130]

The two principal ways in which the rumen bacteria sssist their host is by (1) the fermentative breakdown of cellulose to organic acids (particularly acetic, proprionic, and butyric) which are readily absorbed from the rumen and constitute the main source of energy to ruminants and (2) the synthesis of B vitamins in amounts large enough to render the host independent of exogenous sources of these nutrients. There appears to be no impairment of the first of these activities in cobalt-deficient sheep, judging by the results of Becker and Smith,[131] who actually found significantly higher digestion coefficients for crude fiber in deficient than in cobalt-treated animals. How far this is a reflection of the lowered feed intake of the deficient sheep is unknown and can be resolved only by comparisons at the same levels of feed consumption.

The capacity of the rumen bacteria of cobalt-deficient animals to synthesize the animal's vitamin requirements, especially vitamin B_{12} is more interesting and has attracted much more attention. The blood of cobalt-deficient lambs has been shown to be below normal in the vitamin B_6

[125] H. A. Keener, R. R. Baldwin, and G. P. Percival, *J. Animal Sci.* **10,** 428 (1951).

[126] L. S. Gall, S. E. Smith, D. E. Becker, C. N. Stark, and J. K. Loosli, *Science* **109,** 468 (1949).

[127] J. Tosie and R. L. Mitchell, *Nature* **162,** 502 (1948).

[128] P. H. Abelson and H. H. Darby, *Science* **110,** 566 (1949).

[129] D. Hendlin and M. L. Ruger, *Science* **111,** 541 (1950).

[130] A. E. Harper, R. M. Richard, and R. A. Collins, *Arch. Biochem. Biophys.* **31,** 328 (1951).

[131] D. E. Becker and S. E. Smith, *J. Nutrition* **43,** 87 (1951).

group and in nicotinic acid.[132] Supplementing the diet of these animals with folic acid, thiamine, and pyridoxine administered separately and orally has no measurable beneficial effect,[103] but a complete B supplement given daily by mouth to two lambs for 7 weeks is reported to have produced "a complete reversal of all symptoms of cobalt deficiency."[133] This important finding needs confirmation.

The position with respect to vitamin B_{12} is now becoming much more clear. The rumen contents of normal animals are an exceedingly rich source of this nutrient, but the concentration is very greatly reduced in cobalt deficiency.[132] Nevertheless, injections of B_{12}, in doses of the same order as those used successfully in the treatment of pernicious anemia in man, have proved ineffective in overcoming the symptoms of cobalt deficiency in sheep.[123, 124, 131] However, Smith, Koch, and Turk[133a] have recently produced evidence that the failure of these earlier attempts was due to the small size of the doses used. These workers found that larger doses of this vitamin, 150 μg. or more injected daily, resulted in favorable responses in appetite, body weight, and hemoglobin levels in cobalt-deficient lambs. Support for this finding with respect to B_{12} is provided by the same workers who report curative effects from injections of an anti-pernicious anemia liver extract and a high correlation between the vitamin B_{12} activity of fractions of this extract and their curative effects. It seems clear from these results that at least one of the functions of cobalt in the ruminant is the synthesis of vitamin B_{12} by the ruminal flora and that cobalt deficiency in these animals is at least partially a vitamin B_{12} deficiency. Further the vitamin B_{12} content of liver may, after all, provide an explanation of the much earlier findings of Filmer and Underwood[134, 135] that whole liver, fresh or dried, can be highly curative of cobalt deficiency in calves and lambs, although it contains insufficient cobalt to account for its curative action. The question as to whether the ruminant has exceptionally high requirements for vitamin B_{12}, or whether man has exceptionally low requirements, compared with other animal species, must remain unanswered until further experimentation has disclosed the minimum effective doses for a range of species.

IV. Nickel

Nickel has not so far been shown to play any part in the nutrition of plants or animals, nor have any diseases in plants or animals been shown to be due to nickel deficiency. The close similarity in the physical and chemical

[132] W. H. Hale, A. L. Pope, P. H. Phillips, and G. Bohstedt, *J. Animal Sci.* **8**, 621 (1949).

[133] W. H. Hale, A. L. Pope, P. H. Phillips, and G. Bohstedt, *J. Animal Sci.* **9**, 484 (1950).

[133a] S. E. Smith, B. A. Koch, and K. L. Turk, *J. Nutrition* **44**, 455 (1951).

[134] J. F. Filmer, *Australian Vet. J.* **9**, 163 (1933).

[135] J. F. Filmer and E. J. Underwood, *Australian Vet. J.* **13**, 57 (1937).

properties of nickel and cobalt is thus not paralleled by their physiological properties. Filmer and Underwood[90] obtained results which indicated that nickel might partially, but only to a limited extent, replace cobalt in the treatment of enzootic marasmus, when suboptimal levels of cobalt are fed. Somewhat similar results were obtained by Dixon[136] with salts of nickel in the treatment of "Morton Mains" disease of lambs in New Zealand but nickel supplements have been found of no benefit in any of the wasting diseases of sheep and cattle without cobalt or when adequate cobalt has been fed.

The recognition of nickel as a plant constituent dates back nearly a hundred years, but it was more than fifty years later before its general distribution in plant and animal tissues became recognized. Bertrand and co-workers[80] and McHargue[137] made a considerable number of analyses of a wide range of plant and animal tissues and organs. They showed that nickel generally occurs in plants in higher concentrations than cobalt but in animal tissues the reverse is more usual. Confirmation of the fact that nickel is normally much more abundant in plants than cobalt is provided by the more recent work of Mitchell[138] using spectrographic methods. The mean nickel content of the dry matter of ten Scottish pastures was found to be 1.6 p.p.m. (range 0.52 to 3.68) compared with a mean of 0.09 p.p.m. (range 0.02 to 0.22) for the cobalt content of the same pastures. These differences are a reflection both of a higher uptake of nickel by plants from the soil and of higher nickel content of the soil.

Data on the nickel content of the tissues of man and higher animals are limited. The liver and pancreas appear usually to contain the highest concentrations, as is the case with cobalt, but a great deal more information is needed by the use of modern methods of analysis. In 1929 Martini[139] reported nickel to be a normal constituent of bones, and a suggestion has been made that nickel is associated with the activity of insulin.[80] Salts of nickel, in common with those of certain other metals, sustain the action of insulin when added in small amounts (0.05 to 4 mg. per 100 units) but in larger amounts inhibit its action.[140] Also a crystalline nickel insulinate can be prepared,[141] but there is no evidence that nickel is necessary for the normal functioning or production of insulin within the body.

V. Manganese

Interest in a physiological role for manganese began when Bertrand[142] in 1897 formulated his now disproven theory that this element is the active

[136] J. K. Dixon, *New Zealand J. Sci. Technol.* **19,** 326 (1937).
[137] J. S. McHargue, *J. Agr. Research* **30,** 193 (1925).
[138] R. L. Mitchell, *Soil Sci.* **60,** 63 (1945).
[139] A. Martini, *Mikrochemie* **7,** 235 (1929).
[140] H. Schwab, *Compt. rend.* **207,** 409 (1938).
[141] A. M. Fisher and D. A. Scott, *Biochem. J.* **29,** 1055 (1935).
[142] G. Bertrand, *Bull. soc. chim.* **17,** 753 (1897).

constituent of the oxidases. Later Bertrand and Medigreceanu[143] demonstrated that manganese occurs constantly in the tissues and organs of plants and animals, including man. Subsequent work has provided ample confirmation of this finding and shown clearly that, in contrast to iron, plant tissues are normally very much richer in this element than animal tissues. The relative constancy of the manganese content of most tissues led to a number of early, and largely abortive, attempts to demonstrate its essentiality in the nutrition of animals.[82, 144] In 1931 convincing evidence was obtained.[145] Improvement in the growth of mice was shown to follow additions of manganese to a milk diet supplemented with iron and copper. Many workers have since demonstrated that manganese is an essential component of the diet of the mouse, the rat, the rabbit, and the chick, not only for growth but for normal bone development and reproduction. Its importance as an activator of certain enzymes, notably liver arginase and bone phosphatase, is also established.

1. Manganese Content of Animal Tissues and Organs

Most animal tissues contain very low concentrations of manganese (1 to 2 p.p.m. on dry basis) and species differences are small. The concentrations in the animal body are therefore much lower than copper. They are also very much less variable. Substantial stores of manganese do not occur anywhere in the body in any species so far investigated, although the liver, the muscles, and the bones contain the largest amounts. Supplementing normal diets with additional manganese increases the amounts and concentrations to a small extent, but marked reduction in manganese content, particularly of the liver and bones, occurs under conditions of subnormal manganese intakes. This has been shown for the rat,[146] the rabbit,[147] the chick,[148] and the pig.[149] For instance, Skinner *et al.*[145] found that the bones of rats on low manganese diets contained 0.08 mg. Mn per 100 g. dry bone, compared with normal values of 0.22 mg. Mn per 100 g. Very similar figures, namely 0.06 and 0.20 mg. Mn per 100 g. dry bone, are reported for chicks on manganese-deficient and manganese-supplemented diets, respectively.[148] These values for normal rats and chicks are appreciably lower than those found by Kehoe *et al.*[150] for normal humans, namely 0.17 to 0.30 g. Mn per 100 g. of *fresh* bone.

[143] G. Bertrand and F. Medigreceanu, *Ann. inst. Pasteur* **27**, 282 (1913).
[144] J. S. McHargue, *Am. J. Physiol.* **77**, 245 (1926).
[145] J. T. Skinner, W. H. Peterson, and H. Steenbock, *J. Biol. Chem.* **90**, 65 (1931): E. Orent-Keiles and E. V. McCollum, *ibid.* **92**, 651 (1931).
[146] C. W. Lindow, W. H. Peterson, and H. Steenbock, *J. Biol. Chem.* **84**, 419 (1929).
[147] G. H. Ellis, S. E. Smith, and E. M. Gates, *J. Nutrition* **34**, 21 (1947).
[148] W. D. Gallup and L. C. Norris, *Science* **87**, 18 (1938).
[149] S. R. Johnson, *J. Animal Sci.* **2**, 14 (1943); R. H. Grummer, D. G., Bentley, P. H. Phillips, and G. Bohstedt, *ibid.* **9**, 170 (1950).
[150] R. A. Kehoe, I. Cholak, and R. V. Storey, *J. Nutrition* **19**, 579 (1940).

Data on the manganese content of blood are scanty, but it is apparent that the concentrations are normally extremely low. Normal human blood, according to Kehoe *et al.*[150] contains 0.12 to 0.18 μg per milliliter, about two-thirds of which occurs in the corpuscles. Bentley and Phillips[151] give substantially lower levels for cow's blood (0.05 to 0.07 μg Mn per milliliter). Investigation of the position in other species, at varying manganese intakes, is badly needed, but what little evidence is available suggests that subnormal manganese intakes do not reduce the manganese content of the blood to the same extent that they reduce the manganese content of the bones and the liver. Extraordinary increases in the serum manganese levels of cows turned out to pasture in certain "lactation tetany" areas in England were reported by Blakemore and co-workers.[152] Twenty days after the cows were turned out to pasture, the manganese content of the blood serum of six cows rose from a very low mean of 0.01 μg per milliliter to an extremely high mean of 1.5 μg. This increase was associated with abnormally high manganese contents of the pastures (540 to 1320 p.p.m. Mn on dry basis). No such remarkable increases have been reported elsewhere.

The concentration of manganese in the liver is little affected by the age of the animal and no reserve store of manganese is provided in the newborn of the rat, the rabbit, the guinea pig, the pig, or man.[25, 153, 154] Human livers from individuals of all ages have been found to be extremely constant in manganese content (6 to 8 p.p.m. on dry basis). The levels are slightly higher in cattle and slightly lower in the rat, the rabbit, and the guinea pig.[25, 151] There is a small but significant increase in these latter species during the suckling period.[25] As soon as solid food begins to be consumed, the amounts of manganese in the liver, but not the concentrations, increase substantially. In mammals it appears that almost all the manganese in the liver occurs in the arginase extract,[155] as the prosthetic group of that enzyme.

The absence of reserve stores of manganese in the liver of the newborn is in great contrast to the position with iron and copper, a point of considerable interest in view of the fact that milk is relatively just as low in manganese, i.e., compared with the amounts in most foods, as it is in iron and copper. In fact, milk has been employed as the principal item in practically all diets used for the production of manganese deficiency in mammals. Normal cow's milk contains 0.02 to 0.03 μg. Mn per milliliter,[156, 157] but, in contrast to the position with iron and copper, this

[151] O. G. Bentley and P. H. Phillips, *J. Dairy Sci.* **34,** 396 (1951).

[152] F. Blakemore, J. A. Nicholson, and J. Stewart, *Vet. Record* **49,** 415 (1937).

[153] J. H. Sheldon, *Brit. Med. J.* **ii,** 869 (1932).

[154] G. Bruckmann and S. G. Zondek, *Biochem. J.* **33,** 1845 (1939).

[155] S. Edlbacher and H. Bauer, *Naturwissenschaften* **26,** 268 (1938); S. Edlbacher and H. Pinösch, *Z. physiol. Chem.* **250,** 244 (1937).

[156] J. G. Archibald and J. G. Lindquist, *J. Dairy Sci.* **26,** 325 (1943).

[157] M. Sato and K. Murata, *J. Dairy Sci.* **15,** 461 (1932).

level can be increased substantially by feeding additional manganese.[156] It is curious that the mammary gland readily permits manganese to pass its barriers and yet successfully prevents iron and copper from doing the same. It should be noted, however, that colostrum is many times richer in manganese than normal milk.[157] From the limited data available it appears also that both ewe's and mare's milk are somewhat higher in manganese than cow's milk. Sato and Murata[157] give mean figures of 0.04 μg. Mn per milliliter for mare's milk and 0.05 μg. Mn per milliliter for ewe's milk.

2. Manganese in Mammalian Nutrition

The symptoms of manganese deficiency are now fairly well recognized for the mouse, the rat, and the rabbit, but the minimum requirements of this element are not yet definitely known, either for these species or for any other mammals. Mice, rats, and rabbits are unable to make normal growth on milk rations containing 0.1 to 0.2 p.p.m. Mn and on synthetic rations containing 0.2 to 0.3 p.p.m. Mn on the dry basis, unless additional manganese is provided.[158] Their requirement is therefore greater than 0.3 p.p.m.—how much greater remains to be determined. In the pig the minimum requirement for growth appears to be less than 0.5 p.p.m.,[149] although this needs confirmation. The requirement for reproduction is claimed to be very much higher. Smith and Ellis[159] have made an attempt to determine the minimum requirement of the rabbit, but further experimentation, in which larger groups and smaller manganese supplements are used, is needed to answer this question satisfactorily. The levels of calcium and phosphorus in the diet are also important since high calcium-phosphorus ratios enhance symptoms of manganese deficiency, presumably by depressing manganese absorption.

Rats, mice, and rabbits fed these manganese-deficient diets show abnormal bone development in most cases, but it is not an outstanding symptom as with manganese-deficient chicks. Shortening and bowing of the forelegs occurs in rats and rabbits[158] and lameness and crooked legs have been associated with manganese deficiency in pigs.[160] These gross symptoms are the result of a significant lowering of length, weight (but not volume), density, breaking strength, and ash content of the long bones.[158, 159] The calcium and phosphorus contents of the ash of the bones remain normal but the manganese content, as indicated earlier, is greatly reduced. In an endeavor to throw some light on the mode of action of manganese in bone formation, several workers have investigated the "alkaline" phosphatase

[158] L. W. Wachtel, C. A. Elvehjem, and E. B. Hart, *Am. J. Physiol.* **140,** 72 (1943); M. E. Shils and E. V. McCollum, *J. Nutrition* **26,** 1 (1943); S. E. Smith, M. Medlicott, and G. H. Ellis, *Arch. Biochem.* **4,** 281 (1944).

[159] S. E. Smith and G. H. Ellis, *J. Nutrition* **34,** 33 (1947).

[160] T. B. Keith, R. C. Miller, W. T. Thorp, and M. A. McClarty, *J. Animal Sci.* **1,** 120 (1942).

activity of the bones under conditions of manganese deficiency. In the rat and the rabbit, as well as in the chick, a highly significant reduction has been reported, indicating that manganese is concerned in bone formation through its influence on bone phosphatase activity.[158, 159]

The demonstration that manganese activates liver arginase[155] led naturally to the measurement of arginase activity in manganese-deficient animals. In the rat and the rabbit there is a significant reduction in liver arginase activity, but it is not entirely clear whether only the activation of the enzyme is lowered or whether the formation and concentration of the protein part of the enzyme are also reduced.[158, 159] The weight of evidence supports the former, but it should be emphasized that this has only been demonstrated *in vitro*. It is not known if arginase activity is also reduced *in vivo* and, if so, what effect it has on the animal. The only evidence known to bear on this question is that nitrogen excretion is not affected in manganese-deficient rats,[161] nor is there any accentuation of deficiency symptoms through increasing the burden on the urea-forming system by feeding ammonium citrate.[158]

Defective ovulation in female rats, testicular degeneration and sterility in male rats, and early mortality of young rats were among the earliest observations of the effects of manganese deficiency on milk diets. Claims that deficient lactation was the cause of the mortality in the young was elegantly disproved by Daniels and Everson[162] by placing the young from normal rats with manganese-deficient mothers. These young developed normally. More recently Shils and McCollum[158] have confirmed the fact that manganese-deficient females show no loss of ability to suckle normal young and have established further that there is no lack of maternal interest in these young. It seems highly probable that most of the conflicting results which have been obtained on the effects of manganese deficiency on reproduction in mice and rats can be explained in terms of the varying degrees of deficiency produced by different workers. Shils and McCollum suggest that there are three distinct stage of manganese deficiency in the female. In the least severe stage the female gives birth to viable young which develop symptoms of incoordination and paralysis. (Nervous symptoms of this type have also been observed in manganese-deficient chicks.) In the second and more severe stage non-viable young are born which die shortly after birth. In the third and most severe stage of deficiency, which has so far been developed only in the second generation of females on manganese-low diets, the estrus cycle is affected and sterility results. The mechanism of manganese action in reproduction is as yet unexplained.

Manganese deficiency under natural conditions in grazing sheep, cattle, or horses has never been recorded and, in fact, is hardly likely to occur in

[161] P. D. Boyer, J. H. Shaw, and P. H. Phillips, *J. Biol. Chem.* **143,** 417 (1942).
[162] A. L. Daniels and G. J. Everson, *J. Nutrition* **9,** 191 (1935).

view of the high manganese content of most pasture plants (50 to 150 p.p.m. Mn on dry basis). Even the vegetative parts of cereals growing on soils which give a marked response in crop yield and in manganese content to added manganese usually contain 10 to 15 p.p.m. Mn on the dry basis.[163] This is well above the level of requirement of laboratory mammals, but Bentley and Phillips[151] have obtained suggestive evidence that 10 p.p.m. Mn in the ration is marginal or borderline for optimal reproductive performance in dairy cows. A number of samples of hay examined in their experiments were unusually low in manganese (<20 p.p.m. Mn) and some contained less than 10 p.p.m. Abnormally high intakes of manganese from certain pastures may possibly be of some significance to grazing stock. The work of Blakemore and co-workers[152] on the relation of high manganese pastures to "lactation tetany" in cows has already been mentioned. In addition, Carlstrom and Hjarre[164] suggest that susceptibility to the virus of infectious anemia in horses on certain pastures in Sweden is associated with the high manganese content of these pastures.

Manganese has not yet been shown to be required by the human organism, although it would be extraordinary if it were not. Everson and Daniels,[165] from the results of balance experiments, recommend that the diet of children should contain 0.2 to 0.3 mg. Mn per kilogram body weight. This would amount to 3 to 5 mg. daily for a child weighing 35 lb., which seems rather high and would not always be easy to obtain from an ordinary child's diet high in dairy products. The estimate of 4.6 mg. Mn as the amount required daily to keep an adult male in manganese balance[166] would, on the other hand, easily be covered by most diets. Estimates made by the writer indicate an average intake of about 6 to 8 mg. Mn daily by adults on typical Australian diets. An average intake of 7 mg. Mn daily was calculated by Monier-Williams[68a] for adults on an English winter diet, but it should be noted that no less than 3.3 mg. of this came from tea which is particularly rich in manganese (150 to 900 p.p.m.). No other ordinary item in the diet approaches tea in manganese content.

Manganese resembles iron and cobalt in that it is poorly absorbed and retained in the body and largely excreted in the feces. Kent and McCance[167] found less than 0.1 mg. Mn in the urine of men consuming diets supplying 8 to 9 mg. Mn daily. Similar results have been obtained in the rat by the use of radioactive manganese both ingested and injected.[168] Studies with radioative manganese have shown further that the liver is particularly

[163] G. W. Leeper, *Proc. Roy. Soc. Victoria* **47,** 225 (1934).
[164] B. Carlstrom and A. Hjarre, *Skand. Vet.* **28,** 517 (1938).
[165] G. J. Everson and A. L. Daniels, *J. Nutrition* **8,** 497 (1934).
[166] K. P. Basu and M. C. Malakar, *J. Indian Chem. Soc.* **17,** 317 (1940).
[167] N. L. Kent and R. A. McCance, *Biochem. J.* **35,** 877 (1941).
[168] D. M. Greenberg and W. M. Campbell, *Proc. Natl. Acad. Sci. U. S.* **26,** 448 (1940).

active in manganese metabolism. From 50 to 75% of the intestinal excretion of such "labeled" manganese comes from the bile.[169]

3. Manganese in Avian Nutrition

Evidence that manganese is required by poultry was first reported by Wilgus, Norris, and Heuser[170] from a study of a means of prevention of perosis, a leg disorder of birds characterized by gross enlargement and malformation of the tibiometatarsal joint with displacement ("slipping") of the gastrocnemius or Achilles tendon from its condyles. Later Lyons and Insko[171] demonstrated that the condition known as nutritional chondrodystrophy, characterized by the development of a high proportion of deformed embryos with thick shortened legs and wings and globular heads, was also due to manganese deficiency in the diet of the hens. This results in the production of eggs containing insufficient manganese to allow normal hatchability and development of the embryo. Feeding additional manganese to the hen or injection of 0.03 mg. Mn directly into the eggs restores hatchability to normal and completely prevents the chondrodystrophy. Since these original findings extensive investigations by many workers have demonstrated that manganese is involved in much more than the prevention of perosis and chondrodystrophy, both of which are associated with abnormal bone development. Manganese deficiency in poultry results also in nervous symptoms (ataxia), inferior growth of chicks, failure to maintain body weight of mature birds, lowered egg production, decreased hatchability, reduced egg-shell breaking strength and egg-shell ash, and lower concentrations of manganese in the egg, embryo, bones, and liver.[172] The physiological mechanisms involved in these conditions, apart from bone formation, remain largely obscure.

The minimum manganese requirement of birds for the prevention of any of the above symptoms of deficiency depends upon three principal factors. These are (1) the breed, or even strain, of bird, (2) the quantity of calcium, phosphorus, and iron in the diet, and (3) the source, i.e., the form, of manganese supplied. The incidence of perosis may also be influenced by the level in the diet of several organic nutrients, of which choline[173] and inositol[174] appear the most important. This aspect of perosis has been well reviewed by Jukes[175] to 1941.

[169] J. T. Skinner and J. S. McHargue, *Am. J. Physiol.* **145,** 566 (1946).

[170] H. S. Wilgus, L. C. Norris, and G. F. Heuser, *Science* **84,** 252 (1936).

[171] M. Lyons and W. M. Insko, *Kentucky Agr. Expt. Sta. Bull.* 371 (1937).

[172] H. S. Wilgus, L. C. Norris, and G. F. Heuser, *Proc. World's Poultry Congr. 7th Congr. Cleveland, Ohio* p. 171 (1939).

[173] A. G. Hogan, *J. Nutrition* **21,** 327 (1941).

[174] A. C. Wiese, B. C. Johnson, C. A. Elvehjem, and E. B. Hart, *Science* **88,** 383 (1938).

[175] T. H. Jukes, *J. Nutrition* **22,** 315 (1941).

The lighter breeds have a somewhat lower manganese requirement than the heavier breeds. In White Leghorn chicks 30 p.p.m. Mn in the diet has been found sufficient for the prevention of perosis and for normal growth and 50 p.p.m. for New Hampshires.[176] Other workers report a minimum of 35 p.p.m. for their White Leghorns[177] and 41 p.p.m. for Barred Rocks.[178] A higher requirement for Barred Rock hens than for White Leghorns for egg production, hatchability, and prevention of chondrodystrophy has also been reported.[179] Considerable individual variability in inherent ability to assimilate or metabolize manganese exists.

Excess calcium and phosphorus in the diet increases the manganese requirement of birds by directly affecting its availablity. This is not due to the removal of manganese from solution as insoluble hydroxide or phosphate. In fact, manganese phosphate, $Mn_3(PO_4)_2$, is no less available than many other manganese compounds, as judged by its capacity to prevent perosis. It appears that the mechanism is rather through a reduction in soluble manganese through adsorption by solid mineral.[180, 181] Such adsorption of manganese by carbonates and phosphates, particularly the latter, have been shown in *in vitro* studies[180, 181] at acid reactions similar to those in the absorptive areas of the intestine. The amount of manganese remaining in solution is dependent upon the amount originally present, which explains how extra manganese in the ration tends to overcome the effect of excess mineral.

Although differences exist in the availability of different chemical combinations of manganese, they are not of great practical significance because all the normal sources of this element (oxide, carbonate, sulfate, chloride and permanganate) appear equally valuable.[180] However, one carbonate ore—rhodochrosite—and a silicate ore—rhodonite—have been found to be relatively unavailable.[176, 178]

The very high requirement of birds for manganese, compared with that of mammals, even under the most favorable dietary conditions, calls for some comment. To some extent this is undoubtedly due to lower absorption from the gut. Injection of manganese in quantities equivalent to 6 to 10 p.p.m. of the diet is completely effective in preventing perosis. Four to five times these quantities are needed in diets containing ordinary levels of calcium, phosphorus, and iron.[174, 178, 181] It appears, therefore, that only 20 to 25% of the manganese in the diet is ordinarily available. Poor ab-

[176] W. D. Gallup and L. C. Norris, *J. Biol. Chem.* **117,** 36 (1937); *Poultry Sci.* **18,** 76 (1939).

[177] W. M. Insko, M. Lyons, and J. H. Martin, *Poultry Sci.* **17,** 12 (1938).

[178] P. J. Schaible, S. L. Bandemer, and J. A. Davidson, *Mich. State Coll. Agr., Agr. Expt. Sta. Tech. Bull.* **159** (1938).

[179] W. N. Golding, P. J. Schaible, and J. A. Davidson, *Poultry Sci.* **19,** 263 (1940).

[180] P. J. Schaible and S. L. Bandemer, *Poultry Sci.* **21,** 8 (1942).

[181] H. S. Wilgus and A. R. Patton, *J. Nutrition* **18,** 35 (1939).

sorption of ingested manganese and practically complete elimination through the feces has been demonstrated in chicks in studies utilizing radioactive manganese.[182] It is doubtful, however, if low absorption accounts for all the difference between birds and mammals in manganese requirement. It seems highly likely that the former species have also a greater requirement for absorbed manganese, although in what way and for what purposes remains to be determined. It is significant in this respect that abnormal bone development is a much more prominent symptom of manganese deficiency in the chick than in the mammal, although in both cases the retardation of bone development is associated with a reduced level of phosphatase activity. In manganese deficient chicks Weise and associates[174] found the blood phosphatase levels to be depressed from 15.9 to 51.3 units per 100 ml. to 2.1 to 3.1 units and the bone phosphatase levels from 8.5 to 10.0 unit per gram to 3.6 to 7.7 units.

VI. Zinc

Zinc is one of the most recent of the trace minerals shown to be essential for the growth and well-being of the higher forms of animal life. Long before this, it was known to be widely distributed in plant and animal tissues and organs. Early attempts, by Bertrand[183] and McHargue[144] and their co-workers, to demonstrate an essential function for zinc in the nutrition of animals were not completely successful, as in the case of copper and manganese, and for the same reasons. The rations used were so deficient in other nutrients, notably vitamins, that even with added zinc the experimental mice and rats only survived for a few weeks. Todd, Elvehjem, and Hart[184] were more successful with their diets and in 1934 provided definite evidence that zinc is necessary for growth in the rat. Diets lower in zinc content and better in other respects were subsequently produced by the Wisconsin workers and others and ample evidence that this element is essential in the diet of both rats and mice is now available. Direct experimental evidence of a zinc requirement by other animal species has not yet been produced, but it has been obvious that there must be a requirement, however small, since the first demonstration by Keilin and Mann[185] in 1939 that zinc constitutes the prosthetic group of the enzyme carbonic anhydrase.

1. Zinc Content of Animal Tissues and Organs

Zinc occurs in the body of land animals in very much greater quantities than copper or manganese. It resembles copper to some extent in its rela-

[182] M. S. Mohamed and D. M. Greenberg, *Proc. Soc. Exptl. Biol. Med.* **54,** 197 (1943).
[183] G. Bertrand and R. Benson, *Compt. rend.* **175,** 289 (1922).
[184] W. R. Todd, C. A. Elvehjem, and E. B. Hart, *Am. J. Physiol.* **107,** 146 (1934).
[185] D. Keilin and T. Mann, *Nature* **144,** 442 (1939).

tive distribution in the different organs. In all tissues examined the zinc content is substantially higher than the copper content but the zinc-copper ratio is lower in the brain and higher in the muscles than in most other organs. Species differences appear to be small, judging by the limited amount of data for species other than the human. Special attention has been given to zinc in the human body since the extensive pioneer work of Lutz[186] and of Drinker and Collier,[187] who were interested particularly in this element as a possible industrial health hazard. This aspect of zinc

TABLE 5

THE ZINC CONTENT OF TISSUES

(Parts per million of *fresh* tissue)

	Rat			Cat			Man
	Normal		Added Zn	Normal		Added Zn	Normal
	Lutz	Drinker	Drinker	Lutz	Drinker	Drinker	Lutz
Blood	6.7	7.0	15.0	4.6	2.0	10.0	5.2
Gastro-intestinal tract	15.1	45.0	60.0	19.1	25.0	50.0	11.3
Pancreas		17.3	19.0	24.8	27.0	330.0	12.4
Liver	20.7	47.0	50.0	41.1	55.0	340.0	54.9
Kidney	14.4	50.0	110.0	14.1	14.0	220.0	35.0
Spleen	36.3	228.0	250.0	12.6	33.0	35.0	11.3
Lung	23.6	48.0	60.0	14.5	28.0	30.0	6.8
Testes	14.9	82.0	—	8.5	—	—	308.9
Brain and spinal cord	13.4	53.0	40.0	9.4	20.0	20.0	8.3
Muscle	13.6	38.0	30.0	21.1	22.0	25.0	30.2
Hide	33.8	36.0	50.0	39.1	30.0	60.0	—
Hair	—	—	—	224.1	—	—	163.0
Bone	178.3	—	—	125.9	227.0	260.0	100.8

has been well reviewed up to 1945 by Hegsted, McKibbin, and Drinker.[188] Table 5 gives some of the results of Lutz and Drinker and Collier for the zinc content of animal organs.

The following mean figures extracted from the more recent work of Eggleton[189] on twenty-six Chinese subjects agree fairly well with those of Table 5 and with others in the literature: liver, 245 p.p.m. Zn on dry basis; muscles, 226; kidneys, 186; pancreas, 135; heart, 100; adrenals, 82; spleen,

[186] R. E. Lutz, *J. Ind. Hyg.* **8,** 177 (1926).
[187] R. E. Drinker and E. S. Collier, *J. Ind. Hyg.* **8,** 257 (1926).
[188] D. M. Hegsted, J. M. McKebbin, and C. I. C. Drinker, *U. S. Public Health Rept. Suppl.* **179** (1945).
[189] W. G. E. Eggleton, *Biochem. J.* **34,** 991 (1940).

72; lungs, 67; cerebellum, 55; and cerebrum, 43. Individual variability was considerable in the liver, the muscles, and the kidneys but small in the other organs examined by Eggleton. This worker has shown further that the epidermal structures of the body are characteristically rich in in zinc[190] and several others have found high concentration (100 to 250 p.p.m.) in the bones and teeth.[191, 192, 193] Zinc resembles lead in its tendency to accumulate in the bones. In Table 6 are presented the mean levels of zinc in the epidermal structures of a group of healthy individuals and of a group of beriberi suffers. The much lower levels found in the latter group are regarded by Eggleton as a reflection of a low intake of zinc in beriberi-producing diets, due to a high positive correlation between the zinc and thiamine contents of foods. They imply that subnormal intakes of zinc

TABLE 6

ZINC CONTENT OF HUMAN EPIDERMAL STRUCTURES (EGGLETON[190])

(Parts per million of fat-free dry matter)

	No. of Samples	Healthy adults		No. of Samples	Beriberi sufferers	
		Mean	Range		Mean	Range
Head hair	13	255	84–444	7	173	110–229
Pubic hair	8	197	71–342	4	83	45–115
Fingernails	13	195	121–260	13	88	22–184
Toenails	11	198	96–340	6	90	31–153
Epidermis	1	97	—	—	—	—
Skin	8	26	12–55	7	13	6–21

result in subnormal levels of zinc in the tissues. There is no direct evidence for this in humans, but in rats on zinc deficient diets the total zinc content of the whole body is markedly below normal.[194] Adding zinc to normal diets, on the other hand, produces only slight increases in zinc concentration in the tissues.[191] The claim of Scott and Fisher[195] that the zinc content of the pancreas of diabetics is significantly lower than that of non-diabetics is not confirmed by the work of Eisenbrand and Sienz.[196] These latter workers could find little difference when their results were expressed on the basis of fat-free pancreas tissue. Their actual mean figures were 30.6 μg.

[190] W. G. E. Eggleton, *Biochem. J.* **33,** 403 (1939); *Chinese J. Physiol.* **13,** 339 (1938).

[191] K. R. Drinker, P. K. Thompson, and M. Marsh, *Am. J. Physiol.* **80,** 31 (1927).

[192] D. B. Cruickshank, *Brit. Dental J.* **61,** 530 (1936); **63,** 395 (1937); **68,** 257 (1940).

[193] R. E. R. Grimmett, I. G. McIntosh, E. M. Wall, and C. S. M. Hopkerk, *New Zealand J. Agr.* **54,** 216 (1937).

[194] F. E. Stirn, C. A. Elvehjem, and E. B. Hart, *J. Biol. Chem.* **109,** 347 (1935).

[195] D. A. Scott and A. M. Fisher, *J. Clin. Invest.* **17,** 725 (1938).

[196] J. Eisenbrand and M. Sienz, *Z. physiol. Chem.* **268,** 1 (1941).

Zn per gram for twenty-seven non-diabetics and 25.3 μg. per gram twenty-one diabetics.

2. Zinc Content of Blood

Figures for the zinc content of the blood of species other than human are too few to permit proper species comparisons, but the limited data suggest that mammalian differences are small. Values reported for whole human blood range from 3 to 9 μg. Zn per milliliter, with a high proportion lying between 5 and 7 μg. A higher mean value for normal human blood is given by Vallee and Gibson[197] in a recent study carried out by means of an improved dithizone method, accurate to 1 μg. They obtained the following mean concentrations: whole blood, 8.8 ± 2.0 μg. Zn per milliliter; plasma, 3.0 ± 1.6 μg. per milliliter; packed erythrocytes, 14.4 ± 2.7 μg. per milliliter; erythrocytes, $1.34 \times 10^{-3} \pm 0.2$ μg. per million cells; leucocytes, $3.2 \pm 1.3 \times 10^{-2}$ μg. per million cells. Calculations showed that 75% of the total zinc in the blood was present in the erythrocytes, 22% in the plasma, and 3% in the leucocytes. The individual leucocyte, however, contained twenty-five times as much zinc as the individual erythrocyte. Neither the role of zinc in leucocytes nor the nature of the zinc compound or compounds present is yet known, but the concentration of zinc in the peripheral leucocytes of patients with chronic myelocytic, lymphocytic, and monocytic leukemia is very greatly reduced and cannot be raised by injections of stable zinc gluconate.[198] A rise to normal levels occurs, however, in clinical remission and under therapy with X-rays or urethane, accompanying the falling leucocyte count.[199] The significance of this fall in zinc concentration in leucemia cannot be evaluated at present but it suggests that a study of the concentration of copper and other trace elements in leukemic cells would be of great interest.

The whole of the zinc in erythrocytes can be accounted for as carbonic anhydrase,[199] and the carbonic anhydrase activity of blood is confined to the erythrocytes.[200] There is none in the plasma or leucocytes.[200] This suggests that the carbonic anhydrase activity of blood might be low in anemia and high in polycythemia and that zinc determinations might provide a useful means of estimating the *amount* of the enzyme in red blood cells. Both these points have been convincingly answered by Vallee and his associates.[201, 202] They have shown that in almost all patients with anemias,

[197] B. L. Vallee and J. G. Gibson, *J. Biol. Chem.* **176,** 445 (1948).

[198] J. G. Gibson, R. L. Vallee, R. G. Fluharty, and J. E. Nelson, *Proc. Intern. Cancer Res. Congr., 4th Congr.* St. Louis, p. 104 (1947).

[199] E. Hove, C. A. Elvehjem, and E. B. Hart, *J. Biol. Chem.* **136,** 425 (1940).

[200] H. D. Lewis and M. D. Altschule, *Blood* **4,** 442 (1949).

[201] B. L. Vallee and J. G. Gibson, *Blood* **4,** 455 (1949).

[202] B. L. Vallee, H. D. Lewis, M. D. Altschule, and J. G. Gibson, *Blood* **4,** 467 (1949).

other than pernicious anemia, both zinc and carbonic anhydrase activity are lowered in parallel fashion, so that the decreases are proportional to the decreases in hematocrit, hemoglobin levels, and erythrocyte counts. The zinc and enzyme values per unit of R.B.C. remain in the normal range (Fig. 2). Patients with pernicious anemia, on the other hand, were found to show no decrease in absolute values for zinc and carbonic anhydrase activity in spite of marked lowering of hematocrit, hemoglobin levels, and erythrocyte counts. The zinc and enzyme values per unit of R.B.C. in this disease are therefore significantly elevated above normal (Fig. 3). This occurs even when the zinc concentration is calculated per million

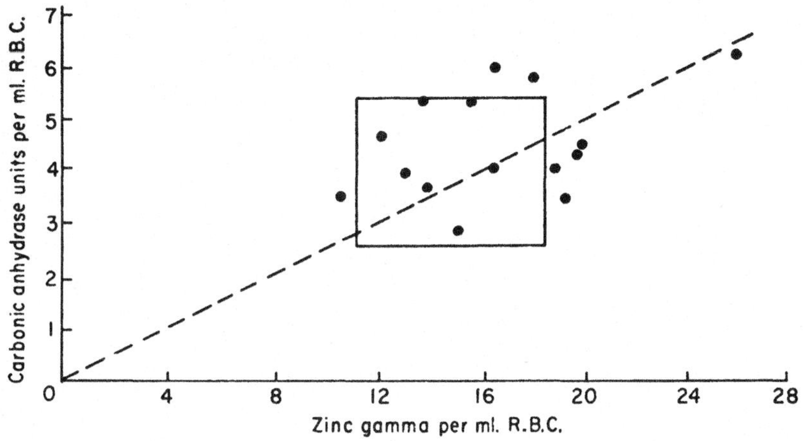

FIG. 2. The relationship of zinc in micrograms per cubic centimeter of erythrocytes to carbonic anhydrase in U units per cubic centimeter of erythrocytes in anemias other than pernicious anemia.[201] The rectangle denotes the normal range. The line connecting with the zero point is drawn through the normal mean of both parameters.

cells so that the increased average red cell size in pernicious anemia is eliminated as a contributing factor. The significance of these interesting findings is impossible to evaluate at present but, as Vallee and Gibson[201] have pointed out, they do indicate that "the haemoglobin and carbonic anhydrase systems are structurally discrete though functionally related." They indicate further that erythrocyte zinc concentration might be used as a relatively simple index of the carbonic anhydrase content in studying various phases of respiratory physiology.

3. Zinc Content of Milk

The zinc content of the milk of only three species (cow, ewe, and human being) has been studied with any thoroughness. The results show clearly that milk is relatively rich in zinc, i.e., compared with its content of copper or manganese. In fact, most of the values are about ten times higher than copper and about fifty times higher than manganese. Species differences either do not exist or are very small. Some workers have found human milk to be lower in zinc than cow's milk;[203] others have found it to be higher.[204] Sato and Murata[205] were unable to detect significant differences between human, ewe's, and cow's milk. These latter workers found a high proportion of the samples of each species to lie between 3 and 4 μg. Zn per milliliter. A mean figure of 3.9 μg. was obtained by Archibald[206] for normal cow's milk. Feeding a zinc supplement to the cows consistently raised the zinc content of the milk to a mean of 5.1 μg. Whether the converse holds, i.e., whether subnormal intakes of zinc result in subnormal levels of zinc

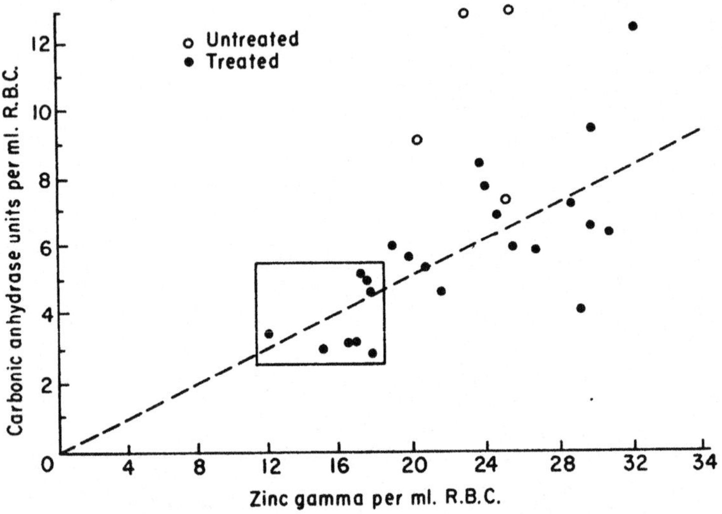

FIG. 3. The relationship of zinc in micrograms per cubic centimeter of erythrocytes to carbonic anhydrase in U units per cubic centimeter of erythrocytes in pernicious anemia.[201] The rectangle denotes the normal range. The line connecting with the zero point is drawn through the normal mean of both parameters.

in the milk, is not yet known. There is no evidence that a fall in the zinc content of milk takes place throughout lactation, as occurs in the case of copper, but all workers agree that the zinc content of colostrum, in all species studied, is about three to four times that of later milk.

4. Zinc Deficiency in Animals

Zinc deficiency has not yet been observed in man or in farm stock, either under naturally occurring or experimental conditions. It is difficult to

[203] E. Rost, *Med. Klin.* (*Munich*) **17,** 123 (1921).

[204] A. Brock and L. K. Wolff, *Acta Brevia Neerl. and Physiol. Pharmacol. Microbiol.* **5,** 80 (1935).

[205] M. Sato and K. Murata, *J. Dairy Sci.* **15,** 451 (1932).

[206] J. G. Archibald, *J. Dairy Sci.* **27,** 257 (1944).

imagine that it would ever arise in grazing animals, since the zinc content of pasture plants and forages usually lies within the range 30 to 100 p.p.m. Zn on the dry basis. Even when growing on soils in which the plants respond to zinic applications in both yield and composition, values below 10 p.p.m. rarely occur. The symptoms of zinc deficiency must therefore be considered in the rat and mouse.

In both these species the only clinical evidences of deficiency are impaired growth, progressive emaciation, and alopecia. These effects were obtained with diets estimated to contain 1.6 p.p.m. Zn or less in rats[184, 207] and as little as 0.3 p.p.m. Zn in mice.[208] In this latter work the deficiency was so severe that 18% of zinc-deficient mice died within 8 weeks, although 100% of the controls survived. Specific pathological changes and a reduced activity of several enzyme systems also occur, but their relationship to the symptoms of zinc deficiency are by no means clear. The histological changes accompanying zinc deficiency in rats, as found by Follis *et al.*,[209] are as follows: extreme parakeratosis of the esophagus with a thick layer of keratinized cells, hyperkeratinization of the skin with thickening of the epidermis, and loss of hair follicles but not of sebaceous glands. Vascularization of the cornea and leucocyte infiltration, reminiscent of riboflavin deficiency, occur in some animals.

Highly purified preparations of at least three enzymes have been shown to contain zinc, namely carbonic anhydrase,[185] uricase[210] and kidney phosphatase.[211] Figures for the zinc content of carbonic anhydrase range from 0.16 to 0.35%. There is some evidence that the lower figure, actually 0.2% Zn, is correct for human carbonic ahnydrase and that the higher figure is for the enzyme prepared from ox and sheep blood.[212] Carbonic anhydrase is widely distributed throughout the animal body and is present in relatively high concentration in the red blood cells, the gastric mucosa, and the renal cortex. It catalyzes the reaction $H_2O + CO_2 \rightleftharpoons H_2CO_3$, facilitating the conversion of CO_2 derived from the tissues to bicarbonate in the blood and the breakdown of bicarbonate to release CO_2 in the lungs. This reaction could not proceed at a rate sufficient to maintain life were it not for the presence of carbonic anhydrase in the red cells, but the minimum amounts compatible with normal respiratory function are not yet known. The carbonic anhydrase activity of the blood of zinc-deficient rats was not significantly reduced under the experimental conditions imposed by Hove and associates.[199] Neither was the liver uricase of such animals re-

[207] E. Hove, C. A. Elvehjem, and E. B. Hart, *Am. J. Physiol.* **124,** 750 (1938).

[208] H. G. Day and B. E. Skidmore, *J. Nutrition* **33,** 27 (1947).

[209] R. H. Follis, H. G. Day, and E. V. McCollum, *J. Nutrition* **22,** 223 (1941).

[210] C. G. Holmberg, *Biochem. J.* **33,** 1901 (1933).

[211] R. Cloetens, *Biochem. Z.* **308,** 37 (1941).

[212] D. A. Scott and J. R. Mendive, *J. Biol. Chem.* **139,** 690 (1944); **140,** 445 (1944).

duced, although there was a rise in the plasma uric acid.[213] In zinc-deficient rats and mice, however, a reduction in intestinal phosphatase activity, without any measurable impairment of carbonhydrate metabolism,[214] and a marked reduction in liver and kidney catalase activity[207] have been demonstrated. It is apparent that zinc is essential for the proper functioning of a number of enzyme systems within the body, but it is equally apparent that it must be involved in other functions, not necessarily related to either carbonic anhydrase, uricase, or even intestinal phosphatase, because significant retardation of growth, emaciation, and alopecia occur in zinc deficiency before there is a measurable reduction in the activity of these enzymes.

5. Zinc and Insulin

Despite numerous experiments indicating that the pancreas is involved in zinc metabolism and that the addition of zinc to insulin solution causes a delay in its physiological action and prolongs the hypoglycemia, there is still no conclusive evidence that this element plays any part in the normal production or action of insulin *in vivo*. Insulin, as ordinarily prepared, contains zinc, and when the ash content of crystalline insulin is lowered the hormone will not again crystallize unless salts of zinc, or of nickel, cobalt, or cadmium, are added. The amounts of these metals which combine with the insulin are fixed and roughly proportional to the atomic weights of the elements. This suggests that they are not merely contaminants but are chemically combined constituents of the insulin molecule. Cohn and co-workers[215] crystallized two pure amorphous preparations of insulin with salts of radioactive zinc and found 0.31% and 0.36% Zn in the two crystalline products formed. As indicated earlier, however, the zinc content of the pancreas of diabetics is not significantly lower than that of non-diabetics, although the insulin content is markedly lower. This, of course, does not prove that zinc plays no part in insulin production, but it seems to rule out lack of zinc in the diabetic pancreases as the reason for the reduced insulin production.

Studies with radioactive zinc have shown that the pancreas, at least in dogs and mice, is one of the most active organs of the body in terms of turnover of zinc.[216] A considerable proportion of such zinc is excreted in

[213] L. W. Wachtel, E. Hove, C. A. Elvehjem, and E. B. Hart, *J. Biol. Chem.* **138,** 361 (1941).

[214] E. Hove, C. A. Elvehjem, and E. B. Hart, *Am. J. Physiol.* **119,** 768 (1937).

[215] E. S. Cohn, J. D. Ferry, J. J. Livingood, and M. H. Blanchard, *Science* **90,** 183 (1939).

[216] G. E. Sheline, I. L. Chaikoff, H. B. Jones, and M. L. Montgomery, *J. Biol. Chem.* **147,** 409 (1943).

the pancreatic juice of the dog, whereas only a minute proportion is excreted in the bile.[217]

6. Zinc and Human Nutrition

No suggestion of zinc deficiency in human nutrition has been made, apart from that of Eggleton,[190] who suggests that zinc deficiency may be a factor in the beriberi syndrone. It is also faintly suggested by the finding of Stevenson[218] that the blood of premature infants has a low carbonic anhydrase activity. Preschool children have been shown to retain dietary zinc in amounts of 0.3 mg. per kilogram body weight per day by one worker[219] and of 0.4 to 0.6 mg. per kilogram per day by others.[220] This corresponds to an intake of 7 to 14 mg. Zn daily for a 50-lb. child. The average daily intake of zinc from normal, well-balanced diets by adults, according to Eggleton, is about 12 mg., or about 6 mg. from the diets of the poorer class of people in South China. The first figure corresponds well with the amounts of zinc which several workers have shown to be excreted daily in the feces and urine. Practically all food-ingested zinc is eliminated in the feces. This has been shown with normal zinc in balance studies in human subjects[221, 222] and in studies with dogs and mice in which radiozinc was employed.[216] The minute amounts of zinc normally appearing in the urine (0.3 to 0.9 mg. daily) do not vary with the intake of zinc, even when the plasma zinc level is raised following zinc injections. Apparently the kidney has no capacity to excrete zinc. The trace which does appear in the urine is possibly only an end product of a metabolic function of the kidney itself, as McCance and Widdowson have suggested.[222]

VII. Iodine

Iodine is unique among the trace elements, not only because of its very long and interesting history in relation to a disease of man and his domestic animals (endemic goiter) but also because it functions as an indispensable constituent of a hormone, thyroxine of the thyroid gland, which exercises control over the rate of energy metabolism and influences many other functions of the body. No other functions for iodine in the nutrition of the higher animals are known, and for no other trace element is there similar clear and unequivocal evidence of essentiality for the proper functioning of an endocrine gland *in vivo*. The significance of a properly functioning thyroid,

[217] M. L. Montgomery, G. E. Sheline, and I. L. Chaikoff, *J. Exptl. Med.* **78,** 151 (1943).

[218] S. S. Stevenson, *J. Clin. Invest.* **22,** 463 (1943).

[219] F. I. Scoular, *J. Nutrition* **17,** 103 (1939).

[220] A. Stern, M. Nalder, and I. G. Macy, *J. Nutrition* 21 *Suppl.* p. 8 (1941).

[221] K. R. Drinker, J. W. Fehnel, and M. Marsh, *J. Biol. Chem.* **72,** 375 (1927).

[222] R. A. McCance and E. M. Widdowson, *Biochem. J.* **36,** 692 (1942).

and therefore of adequate dietary iodine, is related to much more than the regulation of energy metabolism, although this appears to be primary. It vitally concerns such important physiological processes as physical and mental growth and development, reproduction, lactation, and egg production.

A vast literature, far greater than that of all the other trace elements combined, exists on the physiology of iodine. It is obviously impossible to deal adequately with all phases of this element if this chapter is to be kept within reasonable bounds. Such aspects as the historical development of our knowledge of the role of iodine, the pathology of the thryoid gland, and the extremely interesting modern work with artificially iodinated proteins are, therefore, reluctantly omitted. Nevertheless, three events of outstanding historical importance must at least be mentioned. These are (1) the demonstration by Baumann,[223] in 1895, that iodine is a normal constituent of the thyroid gland and that the amount in the thyroid diminishes in endemic goiter, (2) the isolation, in crystalline form, of the active constituent of this gland by Kendall[224] in 1915, which he found to contain 60% iodine and named thyroxine, and (3) the working-out of the structure and the synthesis of thyroxine by Harington and Barger[225] in 1927.

It is well to point out at this stage that the terms thyroxine and thyroid hormone are by no means synonymous. Thyroxine is prepared from the thyroid only by fairly drastic hydrolysis. Thyroid hormone is a generic term indicating any substance which will relieve human myxedema and whose administration causes a specific sequence of biological events. A number of these substances are known, of which thyroxine is one. They vary in their physical and chemical constitution, but all contain a single common denominator which is essential for physiological activity. This is now known to be a chemical grouping named thyronine[226] which consists of a diphenyl ether combination with a phenolic hydroxyl group at one end and a substituted alanine chain at the other. All thyro-active substances contain the thryonine nucleus, and all those prepared from natural sources or with a substantial degree of activity contain iodine substituted in the inner aromatic ring. Thyroxine contains iodine substituted in the 3,5 positions in both rings. Removal of the two outer iodine atoms (3′, 5′) results in a loss of over 90% of the activity of thyroxine.[227] When all four iodine atoms are removed, no measurable activity remains. These facts indicate the essential nature of both thyronine and iodine, although strictly

[223] E. Baumann, *Z. physiol. Chem.* **22,** 1 (1896–1897).
[224] E. C. Kendall, *J. Am. Med. Assoc.* **64,** 2042 (1915).
[225] C. R. Harington and G. Barger, *Biochem. J.* **21,** 169 (1927).
[226] C. R. Harington, *Biochem. J.* **20,** 300 (1926).
[227] A. Anderson, C. R. Harington, and D. M. Lyon, *Lancet* **2,** 1081 (1933).

iodine should not be classed with thyronine in this way, since both tetrabromthyronine and tetrachlorothyronine show weak thyroidal activity.

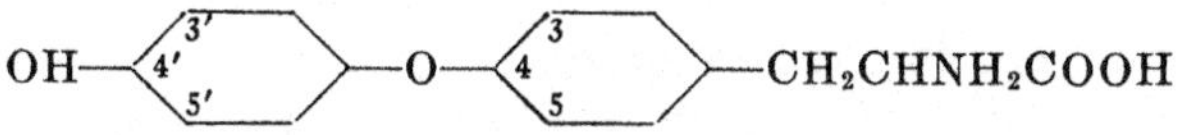

Thyronine [4-(4′-Hydroxyphenoxy)phenylalanine]

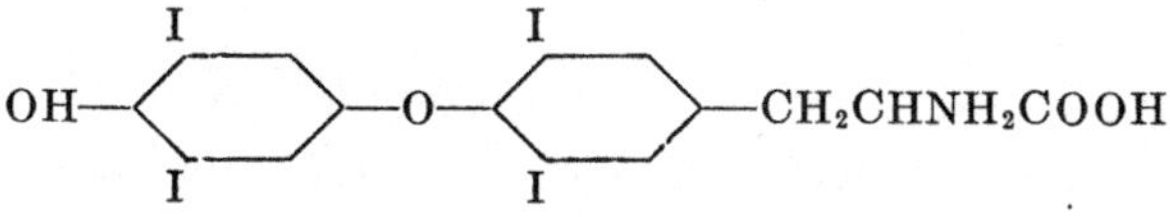

Thyroxine [Tetraiodothyronine]

1. Iodine Content of Tissues and Organs

Iodine has been shown to occur in every cell of the body, but the total amount in the whole body is extremely small. In the human adult it usually ranges from 20 to 50 mg.[228] A high proportion of this total, variously estimated to be 20 to 40%, is concentrated in the thyroid gland—a tremendous concentration in view of the fact that the mass of this gland is normally only one five-hundredth of the whole body. The other great iodine reserve of the body exists in the skeletal muscles. The concentration of iodine in the muscles is less than one-thousandth of that in the thyroid, but the total muscle iodine is not far from equal to the total thyroid iodine. The concentration in the ovaries is about three to four times that of the muscles and is higher than in any tissue of the body except the thyroid, not only in the human body but in all species studied.[228, 229] There is evidence of a cyclic change in ovarian iodine in association with ovarian activity, and of lower values before puberty and after the menopause in women, but much of this work needs to be repeated with improved modern methods.

The above figures refer to total iodine which includes inorganic iodide and organically bound iodine. The concentrations of inorganic iodide in the tissues are extremely low, of the order of 1 to 2 μg.%, and are in equilibrium with the circulating iodide of the body fluids.[230] It seems that iodide ions, like chloride ions, permeate practically all tissues and are distributed in extracellular fluids much like chlorides, as described by Peters.[231] The concentrations of inorganic iodide in both tissues and fluids are very greatly increased during the therapeutic administration of iodide, both at the prophylactic levels necessary for the treatment of simple goiter and at the much higher levels used in exophthalmic goiter.[230] The concentrations

[228] A. Sturm and B. Buchholz, *Arch. klin. Med.* **161,** 227 (1928).
[229] E. Maurer and H. Dugrue, *Biochem. Z.* **193,** 356 (1928).
[230] W. T. Salter, in The Hormones, Academic Press, New York, 1950, Vol. II.
[231] J. P. Peters, *Yale J. Biol. and Med.* **13,** 739 (1941).

of organically bound iodine in the tissues are also normally small (about 5 μg.% in muscle). Much remains to be learned about this fraction. Its solubility is different from that of thyroxine added to tissue extracts, and its distribution is not uniform between the muscle protein fractions of Szent-Györgyi (myosin and actin).[232] However, the amounts in the tissues decrease in hypothyroidism and increase in hyperthyroidism.[232] It seems probable that it consists largely of thyroxine loosely bound to protein.

2. Iodine in the Thyroid Gland

The concentration of total iodine in the thyroid varies widely with the iodine intake, with the activity of the gland, with age, and with the individiual, but there is no conclusive evidence that it varies significantly with either sex or species, except that seafish have thyroids very much richer and rats have thyroids slightly lower in iodine than most mammalian species. The concentration in the normal, healthy thyroid of mammalian species ranges usually from 0.2 to 0.5% I on the dry basis. The total amount in the normal adult human thyroid is about 8 mg. In endemic goiter this amount may be reduced to as low as 1 mg., but the concentration is more affected than the total content because of the characteristic compensatory hyperplastic changes which take place in the gland in this condition. Marine[233] showed long ago that, when the iodine concentration in the thyroid falls below 0.1% hyperplastic changes are regularly found, so that 0.1% may be regarded as somewhere near the minimum effective level in the gland. Marine's claim has been substantially confirmed in a recent study with sheep and pigs made by Andrews and associates.[234] These workers found a significant inverse correlation between iodine concentration and height of the thyroid epithelium in these species. Sheep's thyroids showing moderate and marked hyperplasia contained, respectively, 0.04% and 0.01% I on the dry basis. Pigs' thyroids showing only slight hyperplasia contained 0.11% I. Feeding stabilized iodized salt to these animals completely eliminated the hyperplasia of the follicular epithelium and increased the iodine content of the glands so that very few members of either the sheep or swine herds contained less than 0.2% and none less than 0.12% I on the dry basis.

Iodine is believed to exist in the thyroid in the following forms: inorganic iodide, thyroxine, diiodotyrosine, polypeptides containing thyroxine, diiodotyrosine, and thyroglobulin. Monoiodotyrosine may also exist in thyroid tissue. According to Salter,[230] thyroxine and diiodotyrosine do not exist in the free state in significant concentrations, and inorganic iodide

[232] W. T. Salter and McA. W. Johnson, *J. Clin. Endocrinol.* **8**, 924 (1948).
[233] D. Marine, *Physiol. Revs.* **2**, 521 (1922).
[234] F. N. Andrews, C. L. Shrewsbury, C Harper, C. M. Vestal, and L. P. Doyle, *J. Animal Sci.* **7**, 298 (1948).

constitutes only about one-tenth of the total iodine present in the normal gland. Recent evidence suggests that the inorganic iodide of the thyroid is of two types—the free and the bound. The free iodide is in equilibrium with the iodide of the circulating plasma and, in turn, with the bound iodide which is held in loose combination by a colloidal system which cannot pass cell membranes.[235] Thyroglobulin, the thyroid "colloid," is the storage form of the thyroid hormone and contains a very high proportion of the total iodine of the normal gland. It is a very large molecule, comparable in size to the gamma globulin of blood plasma, but it is not yet clear whether it is a chemical entity in the sense that the crystalline blood proteins are entities. Thyroxine and diiodotyrosine have been shown to be bound in peptide linkage as an inherent part of the polypeptide chain of the protein, but the amounts and proportions of these amino acids and therefore the iodine content of thyroglobulin vary with the supply of iodine available to the gland. It is clear, therefore, that the native protein of the gland must be regarded simply as a storage form and that the thyroid serves several functions with respect to iodine. It converts inorganic iodine to a higher state of oxidation, through diiodotryosine to the physiologically active thyroxine; it serves as a reservoir for thyroid hormone, which it fixes and stores as thyroglobulin; it regulates the release of this stored hormone under the control of the pituitary; and it traps with very great efficiency the inorganic iodide of the body to which the tissue and circulating "hormonal" iodine reverts in the course of metabolism. Much remains to be learned of the mechanisms, enzymic and otherwise, by which the thyroid is able to perform these functions. This has been very fully discussed by Salter[230] in a recent review and need not be repeated here.

3. Blood Iodine

Values for the total iodine content of the whole blood of normal humans have been reported ranging from 3 to 30 μg. I per 100 ml., with a high proportion of these values lying beteen 8 and 12 μg. per 100 ml.[236] Other animal species contain similar amounts of total iodine in their blood. When iodine-rich diets, high in marine fish or seaweeds, are being consumed or when iodides are being administered, higher values are usual. In fact, therapeutic doses (e.g., 1 g. of NaI) may raise the level temporarily by several hundred per cent. The establishment of these norms led to repeated attempts to use total blood iodine determinations in the clinical diagnosis of thyroid diseases and in studying thyroid function. The results were disappointing mainly because of the extremely tedious and unsatisfactory nature of the chemical methods available but also because total blood

[235] W. T. Salter, R. E. Cortell, and E. A. McKay, *J. Pharmacol. Exptl. Therap.* **85,** 310 (1945).

[236] C. B. Davis, G. M. Curtis, and V. V. Cole, *J. Lab. Clin. Med.* **19,** 818 (1934).

iodine includes both inorganic iodide and protein-bound iodine. It is this latter fraction, properly determined, which gives a truer measure of circulating thyroid hormone. Ionized (inorganic) iodine is normally very low in blood (1 to 2 μg. per 100 ml.) and remains approximately constant in different states of thyroid activity. It increases markedly, however, when exogenous iodine enters the organism and gives a "false" increase in protein-bound iodine which can lead to diagnostic confusion.[237] The development of exceedingly sensitive and relatively quick and simple catalytic methods of determining the protein-bound iodine in as little as 0.5 to 3.0 ml. of serum or plasma[238, 239] has revolutionized the position. These methods make use of the power of iodide to catalyze the reduction of ceric to cerous ions by arsenious acid. Serum or plasma is now very largely used in the determination of protein-bound iodine for clinical purposes, but reports on the distribution of iodine between the red cells and plasma are conflicting.[240]

The protein-bound iodine of normal human blood serum or plasma, frequently referred to as the serum-precipitable iodine (S.P.I.), has been found to be much less variable than the total blood iodine of earlier workers. The limits of normality have repeatedly been established as 3 to 8 μg. per 100 ml., with a mean value lying between 5 and 6 μg. per 100 ml.[240, 241] Slightly lower norms (3 to 4 μg. I per 100 ml.) exist in the rat, the mouse, the dog, and the domestic fowl.[242] In the human, S.P.I. does not appear to be significantly influenced by age or sex, and menstrual changes are small and inconsistent. It is, however, significantly raised above normal in pregnancy and in hyperthyroidism and reduced below normal in hypothyroidism. Heinemann and others[241, 243] have shown that, as early as the third week in human pregnancy, S.P.I. values rise sharply to concentrations in the upper range of normal or even to levels which, outside of pregnancy, are characteristic of hyperthyroidism (Fig. 4). They have shown further that, if such a rise does not occur, the pregnancy is likely to end in abortion in the first four months. This rise in S.P.I. is not attended by any clinical evidence of excessive activity of the thyroid and occurs long before the increase in B.M.R. of pregnancy. It would be of great interest to determine

[237] W. T. Salter, G. Karandihar, and P. Block, *J. Clin. Endocrinol.* **9,** 1080 (1949).

[238] A. C. Connor, G. M. Curtis, and R. E. Swenson, *J. Clin. Endocrinol.* **9,** 1185 (1949).

[239] W. T. Salter and I. Rosenblum, *J. Endocrinol.* **7,** 180 (1950).

[240] R. L. Rapport and G. M. Curtis, *J. Clin. Endocrinol.* **10,** 735 (1950).

[241] B. L. Hallman, P. K. Bondy, and M. A. Hagewood, *Arch. Internal Med.* **87,** 817 (1951).

[242] A. Taurog and L. Chaikoff, *J. Biol. Chem.* **163,** 313 (1946).

[243] M. Heinemann, C. E. Johnson, and E. B. Man, *J. Clin. Invest.* **27,** 91 (1948); E. B. Man, M. Heinemann, C. E. Johnson, D. C. Leary, and J. P. Peters, *ibid.* **30,** 157 (1951).

if pregnancy is accompanied by similar rises in S.P.I. in other animal species.

A relationship between level of thyroid activity and S.P.I. level has been demonstrated by many workers.[230, 240] The fall in hypothyroidism and the rise in hyperthyroidism (Fig. 5) occur even when these conditions are not associated with corresponding deviations in the B.M.R. Such changes in the S.P.I. levels do not normally occur when hypometabolism and hypermetabolism are not of thyroid origin. It is obvious, therefore, that S.P.I.

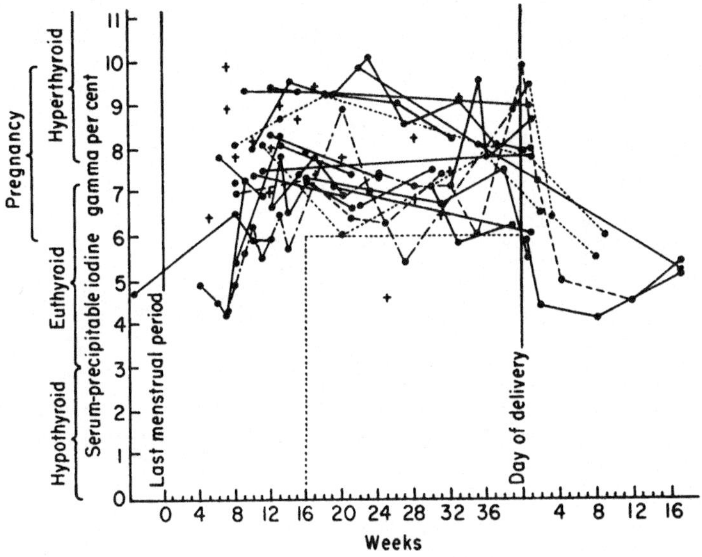

FIG. 4. Serum-precipitable iodine in the course of normal pregnancy.[243] Crosses represent single observations. Circles connected by lines represent repeated observations.

determinations are a valuable aid in clinical diagnosis of thyroid dysfunction, although, as Rapport and Curtis[240] are careful to point out, "diagnosis of thyroid dysfunction cannot be made from it alone, any more than a diagnosis of acute appendicitis can be made from the total and differential leucocyte count."

In spite of the great value and wide clinical use of S.P.I. as an index of thyroid activity, its exact chemical nature is unknown. Present knowledge can be summarized somewhat as follows:

1. The major portion of S.P.I. is loosely bound to those proteins within the tranditional crude albumin fraction of the serum.

2. S.P.I. contains more than one iodine-containing compound but consists largely of (*a*) a thyroxine-like moiety, subject to fluctuations with

thyroid activity, and (*b*) a much smaller diiodotyrosine-like moiety which contributes little to changes in S.P.I.

3. The circulating thyroid hormone consists of thyroxine loosely attached to plasma protein.

4. S.P.I. provides a very good measure of the level of circulating thyroid hormone, even under conditions of iodine therapy, if suitable precuations are taken to avoid "spurious" elevations.[237]

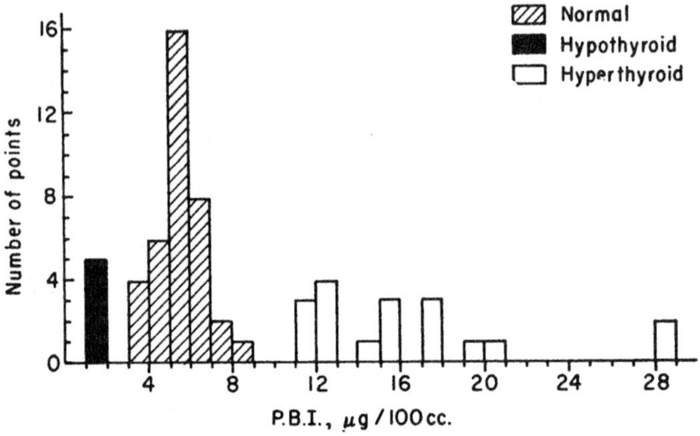

FIG. 5. Distribution of protein-bound iodine in normal subjects and in patients whose thyroid disease was confirmed by a satisfactory response to appropriate treatment.[241]

4. IODINE IN MILK

Available data indicate that the milk of all species normally contains small and highly variable amounts of iodine. Values ranging from 0 to more than 100 μg. per 100 ml. have been reported. The concentration depends to a large extent on the level of intake of iodine in the diet and may be raised far beyond "normal" levels by iodide supplements. It is difficult to state just what is a "normal" level, but 3 to 7 μg. per 100 ml. is a good representation of a large number of investigations of milk from cows on healthy, unsupplemented diets in various parts of the world. Thus Orr and Leitch[244] found that adding 0.18 g. I as KI daily to the diet of cows raised the iodine content of the milk from a "normal" 4 to 7 μg. per 100 ml. to 33 μg. per 100 ml. Similarly, Blom[245] found that a supplement of 0.1 g. KI daily raised the iodine content of cow's milk from a "normal" 2 to 7 μg. per 100 ml. to 51 to 107 μg. per 100 ml. The percentage recovery in milk of added dietary iodine is low and, according to Hanford and as-

[244] J. B. Orr and I. Leitch, *J. Roy. Agr. Soc. Engl.* **87,** 43 (1926).
[245] I. J. B. Blom, *Onderstepoort J. Vet. Sci. Animal Ind.* **2,** 139 (1934).

sociates,[246] of the same magnitude irrespective of the level or form of intake. It has been shown that milk iodine exists almost entirely in the skimmed milk fraction and not in the fat and is precipitated with the milk proteins, probably by simple adsorption. But it is not yet clear whether it occurs in organic or inorganic form, although no thyroxine-like fraction can be isolated from milk after hydrolysis.[230]

The normal iodine content of goat's, ewe's, and human milk appears to be of the same order, although slightly higher than that of cow's milk, but the values reported are so variable that it is virtually impossible to make valid comparisons. Values of 5 to 24 μg. I per 100 ml. are quoted by Salter[230] for human colostrum, and 8 to 4 μg. I per 100 ml. for human milk after lactation has been established. Therapeutic doses of iodide of 1 g. or more given to the mother can result in milk containing several milligrams per cent I.[230] In the milk of all species studied the bulk of evidence suggests that the iodine content of colostrum is lower than that of true milk, which contrasts greatly with the position in all other trace elements studied.

5. Iodine Metabolism

During the last century a good deal has been learned about the distribution of iodine throughout the body and its relation to thyroid function. These aspects have been considered in preceding sections. In this section some of the findings on the metabolism of iodine made by means of radioiodine will be discussed. Radioiodine has proved an exceptionally useful tool in the study of iodine metabolism and thyroid function because (1) it permits the use of physiological quantities of material under physiological conditions, (2) it "tags" a given dose with respect to time, and (3) it can be measured in body fluids and tissues in intact animals with satisfactory accuracy—a point of considerable importance in view of the difficulties inherent in the determination of the minute amounts of stable iodine present. At first the only radioiodine available was I^{128} which has a half-life of only 26 min. Some fine work was done with this isotope, but the advent of I^{131} with a half-life of 8 days has greatly facilitated investigations. These have involved the use of inorganic iodides containing radioiodine and of calorigenically active materials containing radioiodine in organic combination. The former can be regarded as taking part in an anabolic phase and the latter in a catabolic phase of iodine metabolism. Studies have been made with normal individuals and with patients suffering from various thyroid disorders. Much of this work has been reviewed by Keating and Albert[247] and by Salter.[230]

[246] Z. M. Hanford, G. C. Supplee, and L. T. Wilson, *J. Dairy Sci.* **17,** 771 (1934).
[247] F. R. Keating, Jr. and A. Albert, *Recent Progr. in Hormone Research* **4,** 429 (1949).

A number of differences in the metabolism of radioiodine present in inorganic iodides and in physiologically active organic compounds have been disclosed. Inorganic iodide is rapidly and probably completely absorbed from the small intestine at an exponential rate which varies more or less in direct proportion to the level of thyroid activity.[247] Ordinarily very little (about 2%) of it appears in the feces.[247] Hypothyroid patients excrete over 85% in the urine in 5 days as compared with 65% in normal individuals in 2 days and less for hyperthyroid subjects. These values represent the position when very small doses are given and are believed to reflect the natural state of iodide metabolism in man. Absorption of physiologically active organic iodine compounds, on the other hand, is measurably slower and varies somewhat with the metabolic level of the subject; i.e., it is faster in the normal than in the hypothyroid individual. Moreover, a substantial proportion of the dose (11 to 60%) appears in the feces, all of which is organically bound.[247] It should be noted, however, that the fecal iodine after inorganic iodide has been administered is also organically bound. The iodine which is excreted in the urine after the administration of organically bound iodine is largely inorganic iodide, indicating that the major pathway of organic iodine metabolism is deiodination. The small fraction of urinary iodine which is in organic combination has the solubility characteristics of diiodotyrosine. No significant fraction behaving like thyroxine has yet been demonstrated in urine. This indicates that, in addition to deiodination, there is a cleavage of thyroxine into diiodotyrosine.

Absorbed inorganic iodide disappears from the blood at an exponential rate which is the sum of the individual rates of removal by the thyroid, the kidneys, and other tissues. Each of these takes place at rates proportional to its concentration in blood.[247] An idea of the relative importance of these three pathways can be gained from the fact that in normal human adults the thyroid accumulates iodide at the rate of about 2.5% per hour, the kidneys excrete it at the rate of about 6% per hour, and the rest of the tissues of the body at about 1 to 2% per hour.[247] The marked concentrating power of the thyroid, the iodide "trap" to which earlier reference has been made, varies with the activity of the gland. In exophthalmic goiter the rate of accumulation by the thyroid approximates 20% per hour, or eight times the rate of the euthyroid individual. Although the tracer is accumulated faster and to a larger degree by the thyroid in hyperthyroidism, it is soon lost as the hyperactive gland expends its synthesized hormone. The appearance and disappearance of tracer iodide in the thyroid therefore affords a useful means of appraising the activity of the gland. Calculations of the "thyroidal iodide clearance," defined as the volume of plasma cleared of its iodide content per minute, have been made. Salter[230] states that "in round numbers the 'average' normal thyroid clears 10 ml.

per minute, whereas the 'average' exophthalmic goiter clears 130." The sensitivity of this index can be gauged from the fact that the maximal accumulation in Graves' disease rarely exceeds three times the euthyroid value.

6. Minimum Iodine Requirements

Attempts to establish the minimum iodine requirements of animals have been made by various means, none of which have been entirely satisfactory. Comparisons of the estimated amounts of iodine ingested and excreted in goitrous and in non-goitrous areas have given useful data but can be misleading because of the great fluctuations in intake from food and water and the difficulty of assessing borderline states of deficiency. Balance experiments have been valuable, but in addition to the difficulties inherent in all balance experiments with trace minerals (see p. 444) there are special difficulties with iodine. Excretion occurs to a variable extent through the skin and lungs, as well as in the feces and urine. These must be taken into account. Further, the efficiency with which the thyroid salvages iodine from the breakdown of iodine compounds in the body and the amount of iodine released by the destruction of tissue, particularly the muscles, are important factors determining the net iodine balance.

A more precise, but still somewhat arbitrary, means of assessing minimum iodine requirements is through the use of iodine-low diets to which graded amounts of iodine are added. The minimum requirements are taken as those just necessary to either (1) maintain a concentration of 0.1% I in the moisture-free thyroid gland or (2) prevent any significant enlargement of this gland. Levine and associates[248] have used these criteria and set the minimum requirement of the rat as 0.9 μg. I per day by the first and 1 to 2 μg. I per day by the second criterion. If these two measures are taken as limits, it can be calculated that the rat requires 20 to 40 μg. I daily per 1000 food calories consumed. On the same basis the average human adult consuming 3000 calories daily would require 60 to 120 μg. I per day. This range includes most of the estimates made from balance experiments. Thus Cole and Curtis[249] found that four adults remained in equilibrium or positive balance when the intake of iodine was 39 to 162 μg. per day, and Scheffer[250] showed that equilibrium could be maintained in a normal subject on intakes of 54 to 155 μg. per day.

Whether the criteria of Levine and associates can justifiably be applied to species of animals other than the rat and human being is not known. These workers suggest that "until precise data are obtained for different species of animals . . . 20–40 μg./1000 calories of the ration be consid-

[248] H. Levine, R. E. Remington, and H. von Kolnitz, *J. Nutrition* **6,** 347 (1933).
[249] V. V. Cole and G. M. Curtis, *J. Nutrition* **10,** 493 (1935).
[250] L. Scheffer, *Biochem. Z.* **259,** 11 (1933).

ered as the minimum requirements for farm animals." Unfortunately, no such "precise data" are yet available. The fact that goiter does not commonly occur among livestock in many areas where it occurs to a mild but definite degree in the human population suggests a lower iodine requirement for farm stock; but when the very different dietary habits of these species and the range in iodine content of various foodstuffs are taken into consideration such a deduction becomes dubious, if not untenable. A further point of some importance is that in adult rats and human beings energy intake and heat production are nearly the same, whereas in farm animals gross energy intake frequently greatly exceeds heat production. The requirement for iodine is more properly related to heat production than to energy intake, if it is assumed that the destruction of thyroxine is proportional to the metabolic rate it induces. Mitchell and McClure[251] have calculated the minimum iodine requirements of various classes of farm animals by applying Levine's standards to estimates of heat production as follows:

Animal	Body weight, lb.	Heat production, cal.	Iodine requirement, g./day
Chicken	5	225	4.5–9
Sheep	110	2,500	50–100
Pig	150	4,000	80–160
Cow in milk (40 lb. daily)	1,000	20,000	400–800

It is worth noting that these estimates of requirements compare extremely well with the minimum consumption figures given by Orr and Leitch[252] for these species in non-goitrous areas, except for the cow. Apparently the cow ordinarily consumes very much more iodine than it requires.

The iodine requirements of the normal, healthy, resting adult may be greatly increased in various functional activities and disturbances. Very strenuous physical exercise and fever or infection increase the demand. During pregnancy and lactation there is also a considerable increase in iodine requirement, at least in the human species. Further investigation of this point for farm stock is badly needed, although it is perhaps significant that the birth of dead or weak and hairless foals, calves, lambs, and piglets is the most obvious sign of iodine deficiency in these species. Further, the requirements for growth are significantly greater than those of maintenance in all species studied. Growing babies need about 22 to 44 μg. I daily, and older children somewhat more.[230] Numerous surveys in goiter areas in different countries have disclosed a much higher incidence of visible goiter in growing children than in adults in the same region.

The nature of the rest of the diet cannot be excluded entirely from a

[251] H. H. Mitchell and F. J. McClure, *Bull. Natl. Research Council U. S.* 99 (1937).

[252] J. B. Orr and I. Leitch, *Med. Research Council Brit. Special Rept. Ser.* 123 (1929).

consideration of iodine requirements. Little is known of the forms of iodine in foods and their relation to iodine uptake, but provided that there is not a marked excess of elements such as arsenic and fluorine, for which there is some evidence of iodine antagonism (see pp. 498–499[102, 104]), or a high consumption of certain goitrogenic substances present in some foods (discussed below) it can be stated with confidence that iodine content is the only dietary factor ordinarily influencing thyroid function.

7. Goitrogenic Substances in Food

Innumerable experiments have established beyond doubt that simple or endemic goiter is associated with a deficient intake of iodine in the food and water, and that the use of iodized salt, or other sources of supplemental iodine, is accompanied by a marked decrease in the incidence of such goiter in the human and stock populations. Nevertheless, cases of non-toxic thyroid enlargement occur sporadically where the iodine ingestion is well above the estimated requirements. No really satisfactory

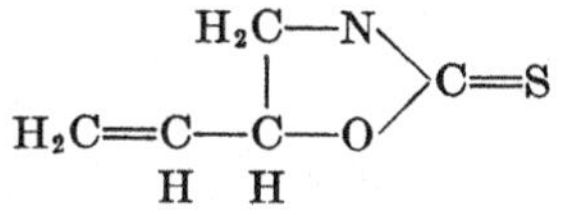

Fig. 6. L-5-Vinyl-2-thiooxazolidone, the anti-thyroid compound of the cabbage family.[253]

explanation of these cases can yet be given, although in some instances they can be related to the consumption of unusually large amounts of certain foods which have been shown to be goitrogenic. Extensive experiments with rats and with humans have particularly implicated cabbage, cauliflower, broccoli, Brussel sprouts, white and yellow turnips or "swedes" (rutabagas), and Brassica seeds.[253] Soybeans have also been shown to be definitely goitrogenic in the rat and the chick[254] and several vegetables and fruits, such as strawberries, pears, and carrots have been found to possess a certain degree of antithyroid activity in man.[255]

It is doubtful if the goitrogenic activity of all these foods is due to the same chemical compound or compounds or to the same mechanism within the animal body. Astwood and his associates[253] have wholly or predominantly accounted for the antithyroid activity of Brassica seeds by the presence in these seeds, in a combined inactive form, of a compound which they have isolated and characterized as L-5-vinyl-2-thiooxazolidone (Fig. 6) the antithyroid compound of the cabbage family.[25]

[253] M. A. Greer, M. G. Ettlinger, and E. B. Astwood, *J. Clin. Endocrinol.* **9,** 1069 (1949).

[254] H. S. Wilgus, F. X. Gassner, A. R. Patton, and R. G. Gustavson, *J. Nutrition* **22,** 43 (1941).

[255] M. A. Greer and E. B. Astwood, *Endocrinology* **43,** 105 (1948).

This compound has been demonstrated in the edible portions of yellow and white turnips, in addition to Brassica seeds, but not in the edible portions of cabbage.[253] This raises a number of interesting points. Thiooxazolidone, like the antithyroid drugs thiouracil and propyl-thiouracil which are used as effective therapeutic agents in the treatment of the hyperthyroidism of exophthalmic goiter, is goitrogenic because it inhibits thyroxine synthesis by the thyroid gland. Administration of iodine exerts a very incomplete inhibition of the antithyroid action of these substances. Cabbage goiter, on the other hand, can be completely inhibited by iodine feeding. Soybean goiter in chicks and rats can also be prevented by iodine feeding.[254] These facts suggest a similarity between the goitrogenic action of cabbage and soybeans and that of thiocyanates, which can also be inhibited completely by iodine administration. Thiocyanates apparently do not limit thyroxine synthesis within the thyroid cells but are believed to limit the capacity of the thyroid to concentrate iodide ions.

The above findings make it clear that iodine content is not always the only dietary factor influencing thyroid function and iodine requirements. It must be emphasized, however, that goiter is unlikely to result if these goitrogenic foods are consumed in normal quantity and still less likely if they are consumed when cooked in the ordinary way. Greer *et al.*[253] have shown that cooking vegetables containing thiooxazolidone, either whole or in chunks, prevents any goitrogenesis, presumably by destroying the enzyme which liberates this substance from its inactive precursor. Only where the diet is composed almost entirely of goitrogenic foods is the amount of antithyroid material ingested likley to be sufficient to inhibit thyroxine synthesis and cause goiter. Some extremely interesting examples of such eating habits, forced upon certain communities in Europe during the recent war, which were accompanied by an increased incidence of simple goiter have been described.[253]

8. Sources of Iodine

By far the major source of iodine to the animal body is the food. Potable waters generally contain extremely small amounts of iodine, of the order 0.2 to 2 μg. per liter.[256] Lower concentrations than 0.2 μg. per liter are common in goitrous areas, and records are available of particular deep-well waters with extraordinarily high iodine contents. Normally, however, the water supply does not contribute important amounts of iodine. The low iodine content of potable waters in goitrous areas is not in itself a significant factor determining the incidence of goiter but is a reflection of a low iodine content of rocks and soils and hence of the vegetables and other plant foods grown in these soils. Sea water is not particularly rich in iodine, in spite of

[256] A. T. Shohl, Mineral Metabolism, American Chemical Society, Monograph 82, 1939.

popular belief to the contrary. Orr and Leitch[252] give a figure of 17 to 18 μg. per liter, and Goldschmidt[257] a figure of 50 μg. per liter. This iodine, however, is greatly concentrated in various forms of marine life, particularly seaweeds. Where seaweeds are consumed as vegetables, as is the case with certain oriental peoples, they constitute an extremely rich source of dietary iodine. Figures as high as 0.2% I on the fresh basis have been recorded for some seaweeds. Marine plankton is also very rich in iodine, but the contribution of this iodine to human and stock populations is largely indirect i.e., through acting as a source of nutrient to fish. Remington and co-workers[258] have shown that the effect of marine plankton and seawater, thrown up in the spray, upon the iodine content of soil and crops is negligible, except for a very narrow coastal strip. The iodine contents of seafish and of shellfish are high but so variable that it is difficult to present average figures which have much meaning. Monier-Williams[68a] claims that 400 μg. I per kilogram for seafish and 900 μg. I per kilogram for shellfish are reasonable representative figures for these species. The liver oils of many seafish are exceptionally rich in iodine. Heddle and Brawn[259] obtained levels as high as several hundred parts per million.

All plants contain iodine in amounts ranging from 2 μg. per kilogram or less to several hundred times this level. The concentration in land plants used either as human or stock food depends far more upon the available iodine in the soil than upon species. In fact, an outstanding feature of the iodine content of plants is the wide variation found in the same species grown in different localities. Adding potassium or sodium iodides, or certain manures containing iodine, to soils can markedly increase the iodine content of the crops or pastures grown thereon. An idea of the extent of the variation in the iodine content of different plant materials used as human food can be obtained from the following figures given by Orr and Leitch:[252] lettuce, 50 to 620 μg. per kilogram; wheat, 1 to 64 μg. per kilogram; watercress, 190 to 450 μg. per kilogram; and nuts, 15 to 200 μg. per kilogram. Monier-Williams[68a] states that the "average" figures for whole cereals taken from various parts of the world can be taken as 25 μg. per kilogram for vegetables, 30 μg. per kilogram; and for fruit, 10 μg. per kilogram. Animal products, apart from seafish and shellfish which have been mentioned earlier, generally contain somewhat lower levels of iodine than green vegetables. According to Hercus and Roberts,[260] who conducted a comprehensive survey of the iodine content of foodstuffs raised in goitrous and non-goitrous districts in New Zealand, the foods which contain most

[257] V. M. Goldschmidt, *J. Chem. Soc.* **1937,** 655.

[258] R. E. Remington, F. B. Culp, and H. von Kolnitz, *J. Am. Chem. Soc.* **51,** 2492 (1929).

[259] R. D. Heddle and J. S. Brawn, *Can. J. Research* **18B,** 386 (1940).

[260] C. E. Hercus and K. C. Roberts, *J. Hyg.* **26,** 49 (1927).

iodine are, in order of merit, fish, green vegetables, whole cereals, milk, meat, and root vegetables. In a normal, well-balanced diet, in which large quantities of fish are not consumed, the greatest contribution to the iodine intake is made by vegetables and milk, However, there is no evidence that, in endemic areas, sufficient iodine can be obtained from these or other dietary sources (except seafish) to protect all age groups from goiter. To do this additional iodine must be supplied, either directly through iodized table salt, tablets, or confectionery, by addition of iodine to the water supplies, or by iodide fertilization of vegetables and other crops. The introduction of iodized table salt, containing sodium or potassium iodide in proportions ranging from 1:50,000 in the United States to 1:250,000 in New Zealand, is generally considered the most satisfactory.

VIII. Bromine

1. Distribution and Sources of Bromine

Bromine is widely distributed in rocks and soils and especially in waters and is constantly present in plant and animal tissues. Whether its presence in living tissues is merely fortuitous, owing to passive ingestion from soils and foods, or whether it has some physiological function is still undecided.

Study of plant and animal materials shows that the food regularly supplies considerable, but highly variable, quantities of bromine to the human body. Common salt can be a particularly fruitful source. One sample of salt examined by Dixon[261] contained no less than 1 mg. Br per gram Cl. In general ten to one hundred times as much bromine as iodine occurs in foods and water. Thus Damiens and Blaignan[262] give the bromine content of bread as 0.9 to 6.1 p.p.m. Ford and associates[263] found 2.4 to 7.7 p.p.m. Br in untreated white flours. Winnek and Smith[264] give values of 5.2 and 7.9 p.p.m. Br for two samples of white flour but very much higher values for milk powder (40 to 42 p.p.m.) and egg albumin (94 p.p.m.). Neufeld[265] analyzed a large number of marine and land plants and found the former to be much richer in bromine than the land plants commonly used as food. The materials lowest in bromine which Neufeld examined were the cereal grains. These were found to contain 1 to 11 p.p.m. Br on the dry basis.

Animal tissues are normally fifty to one hundred times richer in bromine than they are in iodine, with the exception of the thyroid gland for which

[261] T. F. Dixon, *Biochem. J.* **28,** 86 (1935).

[262] M. A. Damiens and S. von Blaignan, *Compt. rend.* **193,** 1460 (1931); **194,** 2077 (1932).

[263] W. P. Ford, D. W. Kent-Jones, A. M. Maiden, and R. C. Spalding, *J. Soc. Chem. Ind.* (*London*) **59,** 177 (1940).

[264] P. S. Winnek and A. H. Smith, *J. Biol. Chem.* **119,** 93 (1937); **121,** 345 (1937).

[265] A. H. Neufeld, *Can. J. Research* **B14,** 160 (1936).

the reverse holds. Figures available for the blood and various organs of man and of the rat and the pig[261, 264, 266] indicate that differences between these species are small and that bromine is not especially concentrated in any organ or tissue so far examined. The levels in the various tissues are markedly influenced by the level of intake of bromine, either in the diet or as added bromide, and by the bromine-chlorine ratio of the diet (see Table 7). There is an interchange between bromide and chloride in the body fluids and tissues so that the administration of bromide results in the replacement by bromide of a proportion of the chloride of the body, thus increasing the bromine concentrations present.[267] Conversely the administration of large amounts of chloride accelerates the elimination of bromide, thus reducing the bromine concentrations present.[268] Dosing with

TABLE 7

EFFECT OF BROMINE INTAKE AND BR:CL RATIO ON THE BROMINE CONTENT OF THE TISSUES OF RATS (WINNEK AND SMITH[264])

(Parts per million on dry basis)

Tissues	Stock diet, 20 p.p.m. Br, Br:Cl 0.009:1	Synthetic diet plus KBr, 20 p.p.m. Br, Br:Cl 0.021:1	Synthetic diet <0.5 p.p.m. Br
Whole blood	117–130	336–503	2.8–14
Hair and skin	41–47	90–170	3.5–5.4
Liver	26–29	39–130	2.2–3.5
Muscle	13–17	41–62	2.9–6.2
Kidney	65–70	220–230	7–20
Spleen	54–56	160–190	8–32
Brain	25–26	82–89	3.3–21
Young at birth	85–105	320–410	7.3–14

chloride has been recommended as a treatment for bromide intoxication for this reason.

The only attempt to produce a bromine low ration and to test its effects on animals is, so far as is known, that of Winneck and Smith.[264] These workers prepared a specially purified diet containing less than 0.5 p.p.m. Br. On this diet rats made reasonable growth over a period of 11 weeks, showed no signs of ill-health, and produced apparently normal young. A similar group receiving the same basal diet plus 20 p.p.m. Br as KBr made no better growth and and did not differ in appearance or reproduction records from the control animals. The amount of bromine in the young rats at birth was, however, very much lower in the unsupplemented group. It is interesting to speculate upon the possibility of a bromine deficiency

[266] H. Ucko, *Biochem. J.* **30,** 992 (1936).

[267] M. F. Mason, *J. Biol. Chem.* **113,** 61 (1936).

[268] P. W. Preu, J. Romano, and W. T. Brown, *New Eng. J. Med.* **214,** 56 (1936), quoted in ref. 264.

arising in the second generation on such a bromine-low diet, when the bromine stores might be still further depleted. For the present it can be stated that either bromine is non-essential or that it is required by the rat at less than 0.5 p.p.m. on the diet.

2. Bromine and Mental Disease

A great stimulus to investigation of the physiology of bromine was given by the work of Zondek and Bier[269] in which the claim was made that the blood of depressive psychotics contains subnormal concentrations of bromine. A further impetus to such work was provided by the claim of Bernhardt and Ucko[270] that the pituitary gland contains fifteen to twenty times the bromine content of blood and that a bromine-containing, hormone-like organic compound is stored in the pituitary and circulates in the blood, causing sleep and a sedative effect. Support for these claims was given by several subsequent investigations. The development of improved methods for the determination of bromine has provided abundant evidence which completely invalidates the earlier work. The Pincussen and Roman method used by these earlier workers was shown to be unreliable and subject to gross errors. The bromine content of the pituitary is not significantly higher than that of blood,[261, 266] there is no evidence for the existence of an organic bromine compound in blood,[261, 267, 271] and the bromine content of blood is not significantly below normal in manic-depressive states.[261, 272] Thus Dixon found the mean bromine content of ten normal human bloods to be 0.73 mg. per 100 ml. (range 0.39 to 1.36) and of twelve manic-depressive psychotic patients to be 0.77 mg. per 100 ml. (range 0.72 to 1.72). Using a very delicate micromethod, Conway and Flood[273] obtained a lower mean figure for normal human blood, namely 0.37 mg. bromine 100 ml.

3. Bromine and the Thyroid

Although it provides a very obvious field for investigation, a possible relationship between bromine and iodine metabolism has attracted very few workers. In fact, it is not even definitely established that the normal thyroid contains more bromine than any other tisuse in the body. Perlman and associates[274] studied the distribution of administered radioactive bromine and concluded from their data that the thyroid showed a greater uptake of the labeled bromine than any of the other tissues tested and that this uptake could not be explained on the basis of simple diffusion from

[269] H. Zondek and A. Bier, *Klin. Wochschr.* **11,** 633, 759 (1932); **12,** 55 (1933).
[270] H. Bernhardt and H. Ucko, *Biochem. Z.* **170,** 459 (1926).
[271] J. H. Quastel and E. D. Yates, *Biochem. J.* **28,** 1530 (1934).
[272] P. Chatagnon and C. Chatagnon, *Compt. rend.* **202,** 1119 (1936).
[273] E. J. Conway and J. C. Flood, *Biochem. J.* **30,** 716 (1936).
[274] I. Perlman, M. E. Morton, and I. L. Chaikoff, *Am. J. Physiol.* **34,** 107 (1941).

the blood serum. Baumann and co-workers[275] cound find no more bromine in the normal thyroid of rabbits than in the blood, but glands which were hyperplastic due to a relative or absolute deficiency of iodine were found to be significantly richer in bromine than the blood. These workers argue that thyroid tissue can distinguish only imperfectly between bromine and iodine, so that in the absence of iodine it seizes some bromine in its place. When iodine is supplied to animals with hyperplastic thyroids they quickly lose the bromine which was accumulated in a futile effort to overcome the lack of iodine. Apparently the bromine accumulated in this way cannot be utilized by the thyroid to synthesize a hormone. No inhibition of the development of iodine-deficient goiters in rats could be obtained by Richards *et al.*[276] by the administration of large amounts of bromine as NaBr, although KI was completely effective. It seems that the thyroid is incapable of the brominating process. The bromine analogue of thyroxine (tetrabromothyronine), however, has been shown by several workers to have, in several animal species, the same qualitative effect as thyroxine, although in much lower degree. In the treatment of human myxedema tetrabromothyronine is one-fifteenth to one-sixtieth as active as thyroxine in the prevention of the development of goiters.[277] Since the corresponding chlorine analogue of thyroxine (tetrachlorothyronine) also has some thyroxine-like activity, although of a still lower order, it is clear, as was pointed out in the section on iodine, that the physiological action of thyroxine is not dependent upon the presence of iodine in the thyronine nucleus.

IX. Fluorine

For many years interest in fluorine as a trace element has been concentrated upon its toxic effects in man and his domestic animals. At the present time fluorine is particularly associated with intense activity in the field of preventive dentistry. The latter aspect is dealt with below, but the effects of fluorine intoxication have been so well discussed in many excellent reviews[278-281] that they will not be considered, except incidentally.

Long before the present interest in fluorine in relation to the prevention of dental caries the view was held that it was an essential element in animal nutrition especially concerned in the formation of the enamel of teeth. This view no doubt arose from its constant presence in the bones and teeth, as well as in other organs and tissues, notably the thyroid and epidermal

[275] E. J. Baumann, D. B. Sprinson, and D. Marine, *Endocrinology* **28**, 793 (1941).
[276] C. E. Richards, R. O. Brady, and D. S. Riggs, *J. Clin. Endocrinol.* **9**, 1107 (1949).
[277] J. Lerman and C. R. Harington, *J. Clin. Endocrinol.* **9**, 1099 (1949).
[278] F. J. McClure, *Physiol. Revs.* **13**, 277 (1933).
[279] F. DeEds, *Medicine* **12**, 1 (1933).
[280] A. W. Peirce, *Nutrition Abstracts & Revs.* **9**, 253 (1939).
[281] K. Robolm, Fluorine Intoxication—A Clinical, Hygienic Study, Copenhagen, 1937.

structures. This was shown by Gay-Lussac and Berthollet as long ago as 1805 and has been amply confirmed since that time. So far attempts to demonstrate an impairment of growth, health, tooth structure, or reproduction by means of fluorine-low diets, have entirely failed.[282, 283] Evans and Phillips[284] found that a mineralized milk diet containing 0.1 to 0.2 p.p.m. F and supplying only about 2 μg. F daily was adequate for the growth, general well-being, and reproduction of rats through five generations. The bones were strong, smooth, evenly and well calcified and there was no depletion of fluorine stores or increase in demand for fluorine throughout the five generations. Moreover, additional fluorine from 0.1 to 20.0 p.p.m. produced no measurable improvement in the rats, although bleaching of the teeth occurred at a level of 10 p.p.m. and over. It is clear that if fluorine is required at all by the rat it must be at a level below 2 μg. per day.

1. Fluorine and Dental Caries

The first suggestion of a relationship between the fluoride content of domestic waters and the incidence of caries came as a result of the remarkable series of epidemiological investigations begun in 1908 in Colorado and followed by studies of the effects of changed water supply and of fluorine ingestion by laboratory animals, which demonstrated the now well-established fact that the presence of excessive quantities of fluorine in the drinking water causes mottled enamel. Incidental to these studies it was noted that dental caries incidence was highest in communities whose water supply was free from fluorides and approached a minimum in communities whose water supplies contained approximately 1.0 to 1.5 p.p.m. F. No significant mottling occurs at these concentrations of fluorine, which have come to be regarded as optimal.[285] Caries inhibition has been most effectively demonstrated for 12- to 14-year-old children continuously exposed to fluoride waters, but the evidence is by no means conclusive for older individuals whose exposure to fluoride waters followed their formative tooth life.[286-289] All the investigations made so far emphasize that the major beneficial effect occurs when the fluorine is ingested during the period of calcification and maturation of the teeth. These investigations include the first three years of the extraordinarily thorough and impressive ten-year Newburgh-

[282] G. R. Sharpless and E. V. McCollum, *J. Nutrition* **6,** 163 (1933).
[283] P. H. Phillips, E. B. Hart, and G. Bohstedt, *J. Biol. Chem.* **105,** 123 (1934).
[284] R. S. Evans and P. H. Phillips, *J. Nutrition* **18,** 353 (1939).
[285] Dental Caries and Fluorine, American Association for the Advancement of Science, Washington, D. C., 111 pp., 1946.
[286] H. T. Dean, P. Jay, F. A. Arnold, and E. Elvolve, *U. S. Public Health Repts.* **56,** 761 (1941).
[287] C. F. Deatherage, *J. Dental Research* **22,** 129, 173 (1943).
[288] R. Weaver, *Brit. Dental J.* **76,** 29 (1944); **77,** 185 (1944).
[289] F. S. McKay, *Am. J. Pub. Health* **38,** 828 (1938).

Kingston study,[290] in which the effect of adding sodium fluoride to maintain a level of 1.0 to 1.2 p.p.m. F to the water supply of one city (Newburgh) is being compared with the neighboring control city (Kingston), which has a fluorine-free water supply. Erupted teeth appear to benefit less from exposure to fluoride-bearing waters.

A mass of experimental results with laboratory animals supports the human caries-fluorine relationship. These have been summarized by McClure[291] as follows:

(i) Induced experimental caries in small animals is fluoride-inhibited.

(ii) Dental tissues contain fluorine which may be classified as primary fluorine acquired during formative tooth life, secondary fluorine acquired after tooth eruption, and adsorbed secondary enamel fluorine acquired possibly by local oral enamel surface adsorption.

(iii) Reduced acid solubility and surface changes in the enamel and dentine are attributed to fluorine reactions on dental tissues *in vitro*.

(iv) Fluoride may affect oral bacterial activity and exert antienzymatic effects possibly associated with dental caries aetiology.

(v) The ameloblasts (enamel-forming cells) are extremely sensitive to fluoride.

The mechanism of the anticaries action of fluorine is obscure. An early finding that the enamel of carious teeth contains less fluorine than enamel of sound teeth[292] is not supported by later work.[293, 294] McClure[294] examined the dentine and enamel of several hundred sound and carious teeth, which showed no evidence of fluorosis, from nearly 100 individuals and obtained the following results:

Sound teeth:	Enamel	$0.0102 \pm 0.004\%$ F	Dentin	$0.0241 \pm 0.001\%$ F
Carious teeth:	Enamel	$0.0098 \pm 0.003\%$ F	Dentin	$0.0225 \pm 0.0007\%$ F

Moreover, the best analytical procedures available have failed to reveal an increase in the fluorine content of topically treated teeth, although most of the studies of topical applications of fluoride to tooth surfaces have demonstrated caries-preventive effects. The most likely explanation seems to be that fluorine influences the physical and biochemical properties of teeth and particularly of enamel surfaces *in vivo*, perhaps by the surface absorption of minute amounts of fluorine by the OH-apatite of the enamel with the production of a protective layer of acid-resisting fluorapatite. Studies with radioactive fluorine have shown that enamel, dentine, bone, and OH-apatite absorb fluorine according to the Freundlich adsorption isotherm.[295] That physicochemical changes do occur which confer acid-

[290] D. B. Ast and I. McCaffrey, *Am. J. Public Health* **40,** 716 (1950).

[291] F. J. McClure, *Ann. Rev. Biochem.* **18,** 335 (1949).

[292] W. D. Armstrong and P. J. Brekhus, *J. Dental Research* **17,** 27 (1938).

[293] T. Ockerse, *J. Dental Research* **22,** 441 (1943).

[294] F. J. McClure, *J. Dental Research* **27,** 287 (1948).

[295] J. F. Volker, H. C. Hodge, H. G. Wilson, and S. N. Van Voorkis, *J. Biochem.* **134,** 543 (1940).

resisting properties upon teeth is supported by the finding that fluorine in concentrations as low as 1.0 p.p.m. in the drinking fluids is effective in reducing the erosion produced *in vivo* on rats' molar teeth by citrate and lactate drinking fluids.[296]

A further possibility, which cannot be entirely overlooked, is that the presence of fluorides in the fluids of the oral cavity alters bacterial metabolism and inhibits enzyme activity. In support of this explanation is the fact that iodoacetic acid, also an active antienzymatic substance, inhibits caries when added to food,[297, 298] and that injected fluoride, which does not reach the oral cavity, is not caries-inhibitory.[299]

2. Sources of Fluorine

Water-borne fluorine is the major fluorine exposure in most human population groups apart from industrial sources, which explains the emphasis upon water supplies in the public health studies on mottled enamel and caries prevention. Drinking water containing 1 p.p.m. F supplies about 1 to 2 mg. F daily under average conditions, which may be compared with an estimated 0.2 to 0.3 mg. F ingested daily from the food in an average American adult diet, i.e., exclusive of drinking water.[300] Very few foods, with the exception of fish and fish products, contain more than 1 p.p.m. F, and most of them less than 0.5 p.p.m. Seafish may contain relatively large amounts, of the order of 5 to 10 p.p.m. F. Another dietary item of significance in relation to fluorine intake is tea. Fluroine contents of 100 p.p.m. or more in tea, particularly China tea, are not uncommon.[301, 302, 303] About two-thirds of this goes into the infusion. Under English and Australian dietary habits this would imply a daily intake of fluorine from tea as high as 1 mg. for adults but very much less for children, a factor which must be taken into account in considering total dietary intakes.

The fluorine content of edible foods grown in areas where the local water is above normal in fluorine content, or where phosphatic fertilizers containing fluorine are used, does not appear to be appreciably greater than normal, although small increments of fluorine may be added to certain foods when they are cooked in fluoride waters.[304] The fluorine content of

[296] J. S. Restarski, R. A. Gortner, Jr., and C. M. McCay, *J. Am. Dental Assoc.* **32,** 668 (1945).
[297] B. F. Miller, *Proc. Soc. Exptl. Biol. Med.* **39,** 389 (1938).
[298] F. J. McClure and F. A. Arnold, *J. Dental Research* **20,** 97 (1941).
[299] F. A. Arnold and F. J. McClure, *J. Dental Research* **20,** 457 (1941).
[300] F. J. McClure, *Am. J. Diseases Children* **66,** 362 (1943).
[301] E. Reid, *Chinese J. Physiol.* **10,** 259 (1936).
[302] H. C. Lockwood, *Analyst* **62,** 775 (1937).
[303] M. Lawrenz and H. H. Mitchell, *J. Nutrition* **22,** 621 (1941).
[304] H. V. Smith, M. C. Smith, and M. Vavich, *Arizona Agr. Expt. Sta. Rept.* 77, 6 (1945).

milk is also apparently not significantly increased above the normal levels of 0.05 to 0.25 p.p.m. by the addition of fluorides to the cow's food or drinking water.[305, 306] The addition of fluorine as rock phosphate to the ration of laying hens, on the other hand, produces a marked increase in the fluorine content of the egg.[307] No doubt similar increases would also occur from the consumption of fluoride waters by the hens.

A program of deliberately raising the fluorine content of water supplies to a level of 1.0 to 1.5 p.p.m. with the aim of reducing caries incidence raises the important possibility of the development of a cumulative toxic fluorosis. Fluorine accumulates slowly in the bones with age under normal conditions and rapidly at high intakes, but at the moderate levels associated with caries prevention the toxic hazard appears to be very small. Balance studies with human subjects show that when the quantities of fluorine ingested from food and water do not exceed 4 to 5 mg. daily, which is appreciably higher than normal, very little is retained by the body.[308] Absorption is high but excretion is also high, roughly three-quarters being excreted in the urine and one-quarter in the sweat.[308] At higher levels of intake more is absorbed and excreted but more is retained. There is also evidence that the nature of the diet also influences the position. Intakes of calcium above certain minimum levels in particular reduces the absorption of dietary fluorine.[303]

3. Fluorine and Enzyme Action

The capacity of fluorides to inhibit enzyme action is well recognized and was briefly referred to in connection with the mechanism of the caries-preventive action. A comprehensive review of this subject up to 1945 has been prepared by Borei.[309] Those enzymes which require metals such as calcium, magnesium, manganese, iron, zinc, and copper as catalysts are particularly affected, owing to the formation of fluorine metal complexes. Fluorine may also combine with the prosthetic groups of phosphoproteins, thus inhibiting activity by the formation of fluorophosphoprotein complexes. Enolase has been shown to be inactivated by the formation of magnesium fluorophosphate.[310] The lipase of ricin, succinodehydrase, acetylcholine esterase, and esterases in general are inhibited by fluorine, according to Dufait and Massart.[311] An effect on bone phosphatase is probably the

[305] P. H. Phillips, E. B. Hart, and G. Bohstedt, *Wisconsin Agr. Expt. Sta. Bull.* 430 (1935).

[306] J. F. McClendon and W. C. Foster, *Federation Proc.* **5**, 67 (1946).

[307] P. H. Phillips, J. G. Halpin, and E. B. Hart, *J. Nutrition* **10**, 93 (1935).

[308] F. J. McClure, H. H. Mitchell, I. S. Hamilton, and C. A. Kinser, *J. Ind. Hyg. Toxicol.* **27**, 159 (1945).

[309] H. Borei, *Arkiv Kemi, Mineral. Geol.*, **A20**, No. 8, 215 pp. (1945).

[310] O. Warburg and W. Christian, *Biochem. Z.* **310**, 384 (1942).

[311] R. Dufait and L. Massart, *Enzymologia* **7**, 337 (1939).

explanation of the interesting finding that extremely low concentrations of fluoride have a pronounced inhibiting effect on calcification *in vitro*.[312] It is likely also that the preservative action of fluorides is related to their effect on enzyme action and microbial respiration due to the formation of complexes with those metals which constitute an essential part of certain enzyme systems.

4. Fluorine and Thyroid Activity

The chemical similarity between fluorine and iodine and the marked capacity of the thyroid gland to absorb fluorine during fluoride administration[313] has led to a number of studies of a possible fluorine-iodine antagonism and of the influence of fluorine upon the basal metabolism of animals and upon human thyrotoxicosis. The results are confusing and in some ways contradictory. Goldemberg[314] found oral administration or injection of NaF to bring about a reduction of basal metabolism in the normal rat or rabbit and in patients suffering from thyrotoxicosis. Similar results were obtained in humans by Gorlitzer[315] by the use of warm-water baths containing HF and by Raveno[316] by the use of alterante I and NaF feedings. In marked contrast to these results are the findings of Phillips and coworkers.[317] These workers administered NaF, both orally and by injection over a wide range of doses including those used by Goldemberg, to normal rats and guinea pigs and to these animals rendered hyperthyroid by the administration of desiccated thyroid. The NaF was found to be without effect on the basal metabolic rate of the normal animal but produced a slight rise during the period of desiccated thyroid sensitivity. In so far as thyrotoxicosis and hyperthyroidism from the feeding of thyroid are related conditions, it would seem that fluoride administration is actually contraindicated.

X. Arsenic

Arsenic has been known for many years to be a regular "trace" constituent of plant and animal tissues, but whether it is an essential component of living cells or whether its presence is due merely to passive ingestion by plants from the soil and by animals from plants is still undecided. Small amounts of arsenic in addition to iron have been used in the treatment of human anemias for centuries, though convincing evidence of its value is

[312] R. Robinson and A. H. Rosenheim, *Biochem. J.* **28,** 684 (1934).

[313] C. Y. Chang, P. H. Phillips, E. B. Hart, and G. Bohstedt, *J. Dairy Sci.* **17,** 695 (1934).

[314] L. Goldemberg, *Semana méd.* (*Buenos Aires*) **360,** 1639 (1932); *J. physiol. et pathol. gén.* **28,** 556 (1930).

[315] V. Gorlitzer, *Med. Klin.* (*Munich*) **28,** 717 (1932).

[316] W. S. Raveno, *J. Michigan State Med. Soc.* **33,** 359 (1934).

[317] P. H. Phillips, H. E. English, and E. B. Hart, *Am. J. Physiol.* **113,** 441 (1935); P. H. Phillips, *ibid.* **117,** 155 (1936).

entirely lacking. There are indications, however, that arsenic may play some part in physiological processes. This is quite apart from its toxic effects, which are well known and need not detain us here.

1. Arsenic and Growth

Hove and co-workers[318] attempted, unsuccessfully, to induce arsenic deficiency in rats. No improvement in growth, hemoglobin levels, or the number and fragility of the red cells was observed when 1 μg. or 5 μg. As per day was added to a mineralized milk diet supplying 2 μg. As daily. It was concluded that "if the rat requires As, this requirement must be less than 2 μg. per day." About 80% of the arsenic in the blood was found by these workers to be concentrated in the red cells, and in anemia there was a fall in the amounts in both cells and plasma. Guthmann and Grass[319] had earlier made the intriguing discovery that the arsenic content of the blood of women rises significantly during menstruation and pregnancy. Normal levels of 64 μg. As per 100 ml. compared with almost 100 μg. during menstruation and as high as 220 μg. in the fifth and sixth months of pregnancy were obtained. Levels above normal were also reported for the blood of cancer patients. These findings, which need confirmation, led the authors to conclude that arsenic is directly related to tissue growth and cell proliferation. A stimulation of the growth of tissue cultures[320] and of the rates of growth and metamorphosis of tadpoles[321] by minute amounts of arsenic has been reported.

2. Arsenic and Thyroid Function

Interesting evidence of antagonism between arsenic and iodine and arsenic and selenium have also been reported. It seems likely that this antagonism is related, in both cases, to the action of arsenic in combining with sulfhydril groups, thereby interfering with certain oxidation-reduction enzyme systems. There are numerous reports of an antagonism between arsenic and thyroid.[322, 323] Sharpless and Metzger,[324] in experiments with rats, showed that arsenic can act as a positive goitrogenic substance, that it can increase iodine requirements, and that iodine can reduce the goitrogenic action as well as the toxicity of small amounts of arsenic. These workers point out that arsenic, in non-toxic amounts, has an insignificant action and suggest that in man a goitrogenic effect from arsenic can be anticipated only where the iodine intake is low and the arsenic intake sufficient to be slightly toxic.

[318] E. Hove, C. A. Elvehjem, and E. B. Hart, *Am. J. Physiol.* **124,** 205 (1938).
[319] H. Guthmann and H. Grass, *Arch. Gynäkol.* **152,** 127 (1932).
[320] K. Sonjo, *Folia Pharmacol. Japon.* **151** (1934).
[321] J. Godonniche and G. Dastugue, *Bull. soc. chim. biol.* **16,** 253 (1934).
[322] E. Hesse, *Klin. Wochschr.* **12,** 1060 (1933), quoted in ref. 324.
[323] M. Scott, *Trans. Intern. Goitre Conf. 3rd Conf.* p. 34 (1938).
[324] G. R. Sharpless and M. Metzger, *J. Nutrition* **21,** 341 (1941).

3. Arsenic and Selenium Poisoning

In 1938, Moxon[325] reported the alleviation of selenium poisoning in rats by arsenite. Since then arsenic has been successfully used to counteract selenium poisoning in pigs, dogs, chicks, and cattle.[326, 327] Sodium arsenite and arsenate are equally effective, but arsenic in the form of certain organic compounds (neoarsphenamine and sulfarsphenamine) is less effective, and arsenic sulfides (AsS_2 and AsS_3) are ineffective.[326] Five parts per million of arsenic as sodium arsenite in the drinking water completely prevents the symptoms of selenium poisoning on "natural" seleniferous diets or diets made seleniferous by additions of sodium selenite or the selenium analogue of cystine.[326] Ten parts per million in the ration gives slightly better results than 5 p.p.m. in the drinking water. It is important to note that these levels of arsenic produced no symptoms of arsenic poisoning, either when fed with control rations or with seleniferous rations. Moreover, they are effective in counteracting or preventing the symptoms of chronic selenium poisoning on rations containing very much higher levels of selenium than those which will, without arsenic, induce some signs of selenium toxicity.[326] The mode of action of arsenic in antagonizing selenium has not been thoroughly investigated. It does not act by reducing selenium absorption, since the selenium content of the blood and tissues of arsenic-treated animals is not significantly different from that of untreated animals showing evidence of selenosis.[328] Recent work provides strong evidence that an important aspect of its action is the release of the inhibition of succinic dehydrogenase brought about by selenium.[329] Klug and associates[329] have shown that the liver succinic dehydrogenase levels of rats fed seleniferous diets are reduced and that the inclusion of arsenic in such diets restores the enzyme values to normal after a relatively short period of depression. Whether the toxic action of selenium in the animal body is due wholly to the inactivation of succinic dehydrogenase and whether the action of arsenic in alleviating selenium poisoning is due wholly to its capacity to counteract this particular action of selenium remain to be determined.

XI. Boron

Although boron is one of the earliest of the trace minerals shown to be essential for plant life there is no certain evidence of its essentiality for animals. As a regular constituent of plant tissues, however, it is an in-

[325] A. L. Moxon, *Science* **88,** 81 (1938).

[326] A. L. Moxon and M. Rhian, *Physiol. Revs.* **23,** 305 (1943).

[327] A. L. Moxon, M. Rhian, H. D. Anderson, and O. E. Olson, *J. Animal Sci.* **3,** 299 (1944).

[328] M. Rhian and A. L. Moxon, *J. Pharmacol. Exptl. Therap.* **78,** 249 (1943).

[329] H. L. Klug, A. L. Moxon, D. F. Petersen, and V. R. Potter, *Arch. Biochem.* **28,** 253 (1950).

evitable component of human and animal diets and takes on additional significance in human physiology from its use, as boric acid, as a food preservative.

Several unsuccessful attempts have been made to produce boron deficiency in rats by the use of highly purified synthetic diets containing 0.15 to 0.16 p.p.m. B on the dry basis.[330, 331, 332] The rats on these diets, which supplied as little as 0.6 μg. B per rat per day, grew normally, reproduced, and reared young which appeared normal. Boron supplements produced no measurable improvement in the animals. It is apparent that if boron is needed by the growing rat it must be at a level below 0.15 p.p.m. of the dry ration. A possible relationship between boron and potassium has been investigated by Skinner and McHargue[333] and by Follis.[334] Potassium-deficient rats were found by the former workers to survive longer when boron supplements were added to the rations, and after 21 days the rats receiving the boron had more glycogen in their livers and better stores of body fat than their non-boron receiving controls. Follis in a somewhat similar study found that boron additions had no effect on the growth rate or survival time of potassium-deficient rats or on the heart and kidney lesions associated with potassium deficiency. Further studies are required on this question, in which the food consumption of the test and control animals are carefully controlled.

The boron content of a wide range of human foods has been investigated, since the pioneer work of Bertrand and Agulhon[335] in 1912. Fruits and vegetables are normally very much richer in boron (5 to 15 p.p.m. B)[336] than cereal grains (1 to 2 p.p.m.)[337] and constitute the principal source in the average human diet. Flesh foods and dairy produce are, in most cases, lower in boron content than cereal grains. Cow's milk contains 0.5 to 1.0 p.p.m. B[330] and appears to vary little with breed or stage of lactation. Supplementing the cow's diet with boric acid or borax, however, results in a fivefold to tenfold increase in the boron content of the milk.[330, 338] Thus Owen[338] found that the addition of about 20 g. borax daily to the ration increased the boron content of the milk from 0.7 to over 3 p.p.m. Feeding additional boron to hens results also in a marked increase in the boron content of the eggs.

[330] E. Hove, C. A. Elvehjem, and E. B. Hart, *Am. J. Physiol.* **127,** 689 (1939).
[331] J. D. Teresi, E. Hove, C. A. Elvehjem, and E. B. Hart, *Am. J. Physiol.* **140,** 513 (1944).
[332] E. Orent-Keiles, *Proc. Soc. Exptl. Biol. Med.* **44,** 199 (1941).
[333] J. T. Skinner and J. S. McHargue, *Am. J. Physiol.* **143,** 385 (1945).
[334] R. H. Follis, Jr., *Am. J. Physiol.* **150,** 520 (1947).
[335] G. Bertrand and H. Agulhon, *Compt. rend.* **155,** 248 (1912).
[336] W. B. Brown, *Analyst* **61,** 671 (1936).
[337] G. Bertrand and G. De Wall, *Compt. rend.* **202,** 605 (1936).
[338] E. C. Owen, *J. Dairy Research* **13,** 243 (1944).

The intake of boron from the food by human adults ranges from 10 to 20 mg. per day according to the data of Kent and McCance.[339] The higher levels are associated with the consumption of large amounts of fruit. This food boron is rapidly and almost completely absorbed and excreted, mostly in the urine. Boric acid taken in by mouth is also rapidly excreted in the urine.[339] The position is very similar in the cow.[338]

XII. Aluminum

Aluminum is one of the most abundant elements in the earth's crust and is present in relatively high concentrations in most soils and in atmospheric dust, but it is poorly absorbed and rarely occurs in plant and animal tissues in more than minute amounts. The use of aluminum cooking utensils and baking powders containing aluminum salts undoubtedly increases the amounts of aluminum ingested by humans, but the amounts are not large and in any case are so poorly absorbed that they do not normally constitute the slightest health hazard. This aspect of aluminum has been extensively investigated and so thoroughly reviewed[340, 341, 342] that it will not be discussed here. Suffice to say that several investigators have shown that ten times the quantity of aluminum likely to be ingested daily in this way can be tolerated by humans without deleterious effects. Very much larger intakes of aluminum than these have been shown to produce gastrointestinal irritation and colic[342] and to produce rickets by interference with the absorption of phosphate,[343, 344] but the amounts required are far beyond those ever likely to occur under ordinary conditions of living. Moreover, therapeutic doses of aluminum hydroxide gel used in the treatment of peptic ulcer are unlikely, according to Street,[345] to reduce phosphorus absorption to dangerously low levels, because of the lower solubility of this form of aluminum compared with the sulfate.

There is no evidence that aluminum is an essential element in the diet of animals. Hove and co-workers[346] attempted to produce an aluminum-deficient diet for rats but were unsuccessful. They concluded from the results of their experiments that "if Al is required by the rat the requirement can be met by as little as 1 μg. daily." Balance experiments with children and adults do not suggest that aluminum is essential in the human

[339] N. L. Kent and R. A. McCance, *Biochem. J.* **35,** 837 (1941).
[340] J. H. Burn, *Analyst* **57,** 428 (1932).
[341] G. D. Beal, G. J. Cox, R. B. Unangst, and H. B. Wigman, *Ind. Eng. Chem.* **24,** 405 (1932).
[342] J. Wurher, *Arch. Hyg.* **112,** 98 (1934).
[343] G. J. Cox, J. L. Dodds, H. B. Wigman, and F. J. Murphy, *J. Biol. Chem.* **92,** xi (*Proc.*) (1931).
[344] H. J. Deobald and C. A. Elvehjem, *Am. J. Physiol.* **111,** 118 (1935).
[345] H. R. Street, *J. Nutrition* **24,** 111 (1942).
[346] E. Hove, C. A. Elvehjem, and E. B. Hart, *Am. J. Physiol.* **123,** 640 (1938).

diet.[150, 347] In fact, Scoular[347] obtained negative balances in more than half the studies made with normal preschool boys, and several investigators have actually found slightly more aluminum in the feces than was ingested in the food. Thus Kehoe and associates[150] found the mean intake of aluminum from the food and beverages of a normal adult American to be 36.4 mg. daily over 28 days and the mean daily excretion in the feces during the same period to be 41.9 mg. The urine of this individual contained a mean of 0.05 mg. Al per liter, which indicates that aluminum is absorbed only with diffiulty. Similar low absorption and predominant excretion via the feces have been demonstrated for the rat, the rabbit, and the dog, even when substantial additions of aluminum are made to the diet.

The occurrence of aluminum in animal tissues, blood, and urine has been the subject of controversy. The majority of investigators now agree that aluminum is regularly present in very small amounts, although there is less agreement on the magnitude of these amounts. The following mean figures taken from the work of Kehoe *et al.*[150] on normal human tissues agree well with those of Myers and Mull[348] for man, dog, and rat and those of Wührer[349] for dog: kidney, 0.42 p.p.m. fresh tissue; heart, 0.56; brain, 0.04; liver, 1.6; spleen, 1.3; lung, 59.4; muscle, 0.15; long bone, 5.0; rib bone, 2.4; stomach, 0.73; intestines, 0.87. Their figure for normal whole human blood (0.013 mg. Al per 1C0 ml.—mainly in the plasma) is very considerably lower, however, than any other recorded values, except those of Lewis.[350] The relatively high figure for lung tissue is worthy of comment. Similar figures have been obtained by several workers, indicating presumably an accumulation of aluminum from atmospheric dust.

XIII. Silicon

Silicon resembles aluminum in its high concentration in soils and atmospheric dust and its relatively low concentrations in animal tissues. It occurs, however, in comparatively large amounts in many plants, particularly gramineous species. Silicon is apparently essential in the nutrition of the higher plants[351, 352] and for the reproduction of marine plankton,[353] but whether it is an essential component of the diet of animals has not yet been decided.

The development of modern micromethods of analysis has shown that much of the early work on the distribution of silicon in the animal body

[347] F. I. Scoular, *J. Nutrition* **17,** 393 (1939).
[348] V. C. Myers and J. W. Mull, *J. Biol. Chem.* **78,** 605, 625 (1928).
[349] J. Würher, *Biochem. Z.* **265,** 169 (1933).
[350] S. J. Lewis, *Biochem. J.* **25,** 2162 (1931).
[351] A. L. Somner, *Univ. Calif. Pubs. Agr. Sci.* **5,** 57 (1926).
[352] C. B. Lipman, *Soil Sci.* **45,** 189 (1938).
[353] O. Richter, *Verhandl. Ges. Deut. Naturforsch. Ärzte* **2,** 249 (1904), quoted in ref. 354.

gave results entirely too high.[354] However, even fetal tissue contains appreciable quantities of silica, the normal range being 2 to 20 mg. Si per 100 g. dry tissue, compared with 5 to 100 mg. Si per 100 g. for normal adult human tissues.[355] The highest levels in the adult are found in the lungs, presumably owing to the inhalation of atmospheric dust, as in the case of aluminum, and the lowest are found in the muscles. In the fetus, on the other hand, the lungs are the lowest in silica and the muscles the highest. The blood of man and other species contains an average of 0.5 mg. Si per 100 ml. This level is not increased appreciably by the inhalation of silica dust, by ingestion in the food, or even by injection.[354, 355] Apparently the body has a very low renal threshold for silicates because there is considerable absorption and excretion in the urine and the amounts excreted in this way can be markedly increased by raising the level of silicates in the diet. This contrasts greatly with the position for aluminum.

Herbivorous animals ingest very large amounts of silicon in their diets and excrete ten to thirty times as much in their urine as carnivorous species.[355] A metabolism of large amounts of silicon by ruminants was also demonstrated many years ago by Forbes and Beegle.[356]

A relationship between blood silicic acid and the parathyroid glands and serum calcium has been claimed[357] but could not be confirmed by King and co-workers.[356] However, these latter workers suggest that silicon may play some part in the maintenance of acid-base equilibrium in the animal body.

[354] E. J. King and T. H. Belt, *Physiol. Revs.* **18,** 329 (1938).
[355] E. J. King, H. Stantial, and M. Dolan, *Biochem. J.* **27,** 1002, 1007 (1933).
[356] E. B. Forbes and F. M. Beegle, *Ohio Agr. Expt. Sta. Bull.* No. 295 (1916).
[357] J. Gursching and H. Kraut, *Arch. exptl. Path. Pharmakol.* **167,** 146 (1932).

CHAPTER 23

Application to Human Nutrition

GRACE A. GOLDSMITH

Tulane University, School of Medicine, New Orleans, Louisiana

I. Historical Background—Development of the Science of Human Nutrition

Knowledge of human nutrition has stemmed from many diverse fields and from all parts of the world. Man was the subject of the earliest scientific studies in nutrition many centuries ago, when it was observed that a normal man varied little in size from year to year, despite ingestion of large amounts of food. Physiologists appreciated that "the weight added by the ingestion of food and drink was lost in the urine, the feces and the 'insensible perspiration.' "[1] From the early part of the eighteenth century, astute observations of practicing physicians linked scurvy to consumption of a faulty diet consisting usually of biscuits and salt meat. It was also known that recovery followed administration of green vegetables and juices of oranges, limes, and lemons.[2] Pellagra, which was first described in the eighteenth century by Casal, Frapolli, and others, was likewise considered to be related to consumption of a monotonous diet of poor quality.[3]

[1] G. Lusk, The Elements of the Science of Nutrition, 2nd ed. rev., W. B. Saunders Co., Philadelphia, 1909, p. 17.

[2] Kramer, *in* Funk, The Vitamines, authorized translation from 2nd German ed. by Dubin, Williams & Wilkins Co., Baltimore, 1922, p. 27.

[3] (*a*). G. Casal, *in* Vitamins: A Survey of Present Knowledge, Medical Research Council, Special Report Series No. 167, His Majesty's Stationery Office, London, 1932, p. 176; (*b*) F. Frapolli, *in* The Vitamins, A Symposium, American Medical Association, Chicago, 1939, p. 308.

Unfortunately these clinical observations failed to stimulate specific investigation in this field.

The modern era of nutrition began in the late eighteenth century with the brilliant research of the French chemist Lavoisier and his contemporaries. The discovery of the significance of oxygen in combustion and the observation that life processes are those of oxidation with resultant elimination of heat laid the foundation for the emergence of nutrition as a science. Lavoisier and his associates[4] conducted experiments on animal heat and respiration and deveoped an apparatus for measuring the metabolism of man. The influence of food, work, and temperature on heat production was determined.

Rapid advances in chemistry in the nineteenth century added much to nutritional knowledge. Foods were shown to contain proteins, carbohydrates, fats, and inorganic salts rather than a single nutrient substance. As late as 1832, however, William Beaumont[5] in his studies of digestion indicated his belief in a single nutritive "element." Liebig,[6] the great German chemist, showed that it was not carbon and hydrogen which were burned in the body but protein, carbohydrate, and fat. Liebig's contributions stimulated the research of Voit, who, with his students at Munich, established such concepts as nitrogenous equilibrium, specific dynamic action of food stuffs, and relationship of basal metabolism to body surface area (see Chapter 1).

These early metabolic studies were continued into this century by Atwater, Benedict, Lusk, and many others. The continuity of this research is strikingly illustrated in the dedication of DuBois' book on basal metabolism: "To Graham Lusk, pupil of Voit, pupil of Liebig, pupil of Gay-Lussac, pupil of Berthollet and La Place, pupils of Lavoisier."[7]

From the study of energy, metabolic research extended to finer problems. It was shown that carbohydrate could be converted to fat and that part of the protein molecule could be converted to carbohydrate.

By 1900 the caloric value of foods had been carefully determined. Between 1890 and 1900 efforts were made to study food requirements by statistical methods. Dietary studies were conducted in several parts of the world, diets being evaluated for energy and protein according to standards developed by Voit and Atwater. Data collected by Atwater in the United States represent, as stated by McCollum, "an epoch in development of knowledge of human nutrition."[8]

[4] A. L. Lavoisier, *in* Lusk, ref. 1, p. 18.

[5] W. Beaumont, *in* McCollum, The Newer Knowledge of Nutrition, 2nd ed., Macmillan, New York, 1922, p. 40.

[6] J. von Liebig, *in* Lusk, ref. 1, p. 19.

[7] E. F. DuBois, Basal Metabolism in Health and Disease, Lea and Febiger, Philadelphia, 1924 [p. iii].

[8] E. V. McCollum, ref. 5, p. 50.

Another valuable contribution was publication of tables of the chemical composition of an extensive list of human foods by Atwater in this country and by Voit in Germany.[9] Until about 1900 there was no fear that diets would be unsatisfactory if protein and energy were adequately supplied. Soon after this, the investigations of Fischer and Kossel[10] showed that the composition of protein from different sources varied greatly, and the concept of quality of protein was introduced.

Recognition of vitamins and formulation of the concept of deficiency diseases developed from two different lines of research: one, animal experiments with simplified and purified diets; the other, epidemiological and public health investigations. Forster[11] in 1873 found that dogs fed meat (after preparation of Liebig's meat extract), fat, sugar, and starch died earlier than did starving animals and manifested tremor, weakness, and a peculiar gait. Similar findings were noted in pigeons. In the school of Bunge at Basal, Lunin,[12] in 1881, and Socin,[13] ten years later, found, in the course of experiments designed to study inorganic elements in nutrition, that animals fed the known constituents of milk, i.e., proteins, carbohydrates, salts, and fats, failed to survive. They concluded that a natural food such as milk must contain small amounts of unknown substances essential to life.

In 1884, it was demonstrated in the Japanese Navy, under the direction of Admiral Takaki,[14] that beriberi was due to a faulty diet. Addition of meat, bread, fruit, and vegetables to a diet which had consisted largely of polished rice almost eliminated beriberi from the Japanese Navy. Eijkman,[15] a military doctor in the Dutch East Indies, concluded in 1897 that beriberi resulted from consumption of decorticated rice. His conclusion was based on evidence from two sources:(1) experimental production of a disease resembling beriberi, polyneuritis gallinarum, in pigeons fed a diet of polished rice and prevention of the disease with whole rice; and (2) statistical findings in an epidemiologic study by Vorderman in prisoners in Java, which showed that the incidence of beriberi was associated with ingestion of white rice. Eijkman believed that rice was overrich in starch which acted as a poison to nerve cells and that the outer layer of the grain provided an antidote. Grijns[16] in 1901 interpreted the findings differently,

[9] (*a*) W. O. Atwater, *in* McCollum, ref. 5, p. 3; (*b*) C. v. Voit, *in* McCollum, ref. 5, p. 3.

[10] (*a*) E. Fischer, *in* McCollum, ref. 5, p. 4; (*b*) A. Kossel, *in* McCollum, ref. 5, p. 4.

[11] J. Forster, *in* Funk, ref. 2, p. 21.

[12] N. Lunin, *in* Funk, ref. 2, p. 21.

[13] C. A. Socin, *in* Funk, ref. 2, p. 22.

[14] K. Takaki, *in* Funk, ref. 2, p. 29.

[15] C. Eijkman, *in* Vitamins, ref. 3*a*, p. 13.

[16] G. Grijns, *in* Funk, ref. 2, p. 13.

concluding that beriberi in birds or man was due to absence from the food of an essential nutrient found in the cortical parts of rice. Thus Grijns was the first to perceive clearly the concept of a deficiency disease.

Eijkman's work made it possible to turn from human to animal experiments in studying beriberi. Production of experimental scurvy in guinea pigs by Holst and Frölich[17] in 1907 provided another non-human experimental tool. Many investigators in the early twentieth century presented evidence for unknown essential substances in foods such as milk, yeast, and fats. Evidence was based largely on studies in animals fed "purified" diets and on clinical observations of the association of scurvy and rickets, beriberi and pellagra with dietary factors. Outstanding investigators included Pickelwaring, Hopkins, Muller, Stepp, McCollum, Osborne, and Mendel.

The term "vitamine" was suggested by the Polish chemist, Casimir Funk,[18] in 1912 in a paper on the etiology of deficiency diseases.

He indicated that vitamins appeared to be essential for growth and for general health and suggested that beriberi, scurvy, and pellagra were deficiency diseases.

In 1913, McCollum and Davis[19] and Osborne and Mendel[20] published papers indicating that certain fats contained an essential nutritional factor. In 1915, McCollum and Davis[21] published important papers which indicated that at least two accessory food factors were needed for normal growth, one of which was found in butter fat, the other in milk sugar. The names proposed for these factors were fat-soluble A and water-soluble B, thus initiating the vitamin alphabet. In 1919–1920 the antiscorbutic factor was designated water-soluble C by Drummond.[22] In 1918, E. Mellanby[23] discovered the antirachitic vitamin in certain fats. Thus the early era in human nutrition emphasized energy metabolism and investigation of the energy-producing foodstuffs, followed by discovery of accessory food factors or vitamins and formulation of the concept of deficiency disease. In these investigations physicians, physiologists, chemists, and biologists all played a role.

From these beginnings, investigation in the past thirty years has been unbelievably rapid. Vitamins have been isolated, identified, and synthesized. Deficiency syndromes have been clearly delineated and studied clinically and pathologically. Many nutrient requirements have been established. The role of essential nutrients in metabolic processes has been

[17] A. Holst and T. Frölich, *in* Funk, ref. 2, p. 128.
[18] C. Funk, *J. State Med.* **20,** 341 (1912).
[19] E. V. McCollum and M. Davis, *J. Biol. Chem.*, **15,** 167 (1913).
[20] T. B. Osborne and L. B. Mendel, *J. Biol. Chem.* **15,** 311 (1913).
[21] E. V. McCollum and M. Davis, *J. Biol. Chem.* **23,** 181, 231 (1915).
[22] J. C. Drummond, *Biochem. J.* **13,** 77 (1919).
[23] E. Mellanby, *J. Physiol.* **52,** xi (1918–1919).

widely studied, and enzymes, coenzymes, systems of energy transfer, and nutrient interrelationships have received much attention. Radioactive isotopes have been used in tracing metabolic pathways in both animals and man. Investigation of the nutrition of bacteria, fungi, single-celled animals, insects, etc., have added information relative to human nutrition. It is only because of contributions by investigators in nearly every field of science that knowledge of human nutrition has progressed to the present high level.

II. Energy Metabolism in Man

1. Energy Requirement

Estimation of human energy requirement is neither simple nor exact. Consideration must be given first to needs in a basal state and second to needs resulting from physical activity. Basal requirements are related to age, sex, bodily size, activity of the thyroid gland, and environmental temperature. The influence of each of these factors has been established with reasonable accuracy and applied in estimation of the energy requirements. The relationship between basal metabolism and surface area of the body, demonstrated in man as well as in numerous animal species, is widely used in estimating basal caloric needs. The energy required in various occupational and non-occupational activities is less well documented. Some of the earlier findings have been summarized by DuBois and Chambers.[24] Unconscious tensing of muscles in anxiety may increase heat production 10 to 20%, moderate activity 20 to 50%, while heavy exercise may lead to a three- to tenfold increase in metabolism. Recent important contributions to knowledge of energy expenditure are those by Lehmann and associates,[25] who studied industrial workers engaged in 107 different activities, and those conducted at Columbia University in cooperation with the Bureau of Human Nutrition and Home Economics of the U. S. Department of Agriculture on energy needs of children.[26]

In formulation of diets to meet energy requirements, proteins and carbohydrates are considered to yield 4 cal. per gram, and fats, 9 cal. per gram. The specific dynamic action of foodstuffs must also be considered and is usually calculated as 6% of the total caloric value of the food over a 24-hr. period.[24]

In spite of the difficulties in determining exact caloric requirements,

[24] E. F. DuBois and W. H. Chambers, Calories in Medical Practice, *in* Handbook of Nutrition, A Symposium, American Medical Association, Chicago, 1943, pp. 55ff.

[25] G. Lehmann, E. A. Müller, and H. Spitzer, *Arbeitsphysiol.* **14,** 166 (1949–1950).

[26] (*a*) E. M. Thompson, M. E. R. Bal, E. M. McIntosh, G. MacLeod, and C. M. Taylor, *J. Nutrition* **44,** 275 (1951); (*b*) C. M. Taylor, O. F. Pye, M. Schafer, and S. Wing, *ibid.* **44,** 295 (1951).

attempts have been made to devise standards which may be useful in planning and evaluating diets of individuals and population groups. Caloric allowances for the growth period and allowances for the maintenance of health in adults have been recommended by the Food and Nutrition Board of the National Research Council, U.S.A.,[27] by the Canadian Council on Nutrition,[28] and by the Committee on Nutrition of the British Medical Association.[29] In addition, the committee on calorie requirements of the Food and Agriculture Organization of the United Nations[30] has suggested standards which may be applied in determining caloric needs of population groups which vary in size, age, sex distribution, and geographical location. This international collaboration is an important step in estimating food supplies needed by various countries and should assist in the over-all planning for more adequate production and more equitable distribution of world food supplies. Thus from Lavoisier's early experiments on respiration, animal heat, and calorimetry has evolved information of great importance in planning one of the most fundamental aspects of a world public health program.

Caloric requirements in disease have also been the subject of study. There is an increase in basal metabolism of about 13% for each degree (centigrade) rise in body temperature. Accordingly, all febrile illnesses increase energy needs. Basal metabolism is increased in hyperthyroidism and is often elevated in other endocrinopathies, in certain blood dyscrasias and in heart or lung diseases accompanied by severe dypsnea. Caloric undernutrition is common in these conditions which are associated with an increased energy requirement. It is also frequently encountered in illnesses in which anorexia is a prominent finding and in conditions in which calories are lost in the stool or urine, such as diarrheal diseases or diabetes mellitus.

2. Energy Transfer

The complicated metabolic processes by which energy transfer is accomplished have been described in Chapter 16. Knowledge of the important role of adenosine triphosphate (ATP) in energy mechanisms, of relationships of the breakdown of foodstuffs to generation of ATP, and the concept of high-energy bonds are of importance in human nutrition in explaining, at least in part, the manner by which "metabolic energy is harnessed

[27] Recommended Dietary Allowances, National Research Council, Food and Nutrition Board, Reprint and Circular Series No. 129, Washington, 1948.

[28] Canadian Bulletin on Nutrition, Council on Nutrition, Ottawa, Vol. 2, No. 1, 1950.

[29] Report of Committee on Nutrition, British Medical Association, British Medical House, London, 1950.

[30] Calorie Requirements, Report of Committee on Calorie Requirements, United Nations Food and Agriculture Organization, Nutritional Studies No. 5, Columbia University Press, New York, 1950.

TABLE 1

RECOMMENDED DAILY DIETARY ALLOWANCES[a]

(Food and Nutrition Board, National Research Council; revised 1948)

	Calories[b]	Protein, gm.	Calcium, gm.	Iron, mg.	Vitamin A,[c] I.U.	Thiamine,[d] mg.	Riboflavin,[d] mg.	Niacin (Nicotinic acid),[d] mg.	Ascorbic acid, mg.	Vitamin D, I.U.
Man (154 lb., 70 kg.)										
Sedentary	2400	70	1.0	12[e]	5000	1.2	1.8	12	75	[f]
Physically active	3000	70	1.0	12[e]	5000	1.5	1.8	15	75	[f]
With heavy work	4500	70	1.0	12[e]	5000	1.8	1.8	18	75	[f]
Woman (123 lb., 56 kg.)										
Sedentary	2000	60	1.0	12	5000	1.0	1.5	10	70	[f]
Moderately active	2400	60	1.0	12	5000	1.2	1.5	12	70	[f]
Very active	3000	60	1.0	12	5000	1.5	1.5	15	70	[f]
Pregnancy (latter half)	2400[g]	85	1.5	15	6000	1.5	2.5	15	100	400
Lactation	3000	100	2.0	15	8000	1.5	3.0	15	150	400
Children up to 12 yr.[h]										
Under 1 yr.[i]	110/2.2 lb. (1 kg.)	3.5/2.2 lb. (1 kg.)	1.0	6	1500	0.4	0.6	4	30	400
1–3 yr. (27 lb., 12 kg.)	1200	40	1.0	7	2000	0.6	0.9	6	35	400
4–6 yr. (42 lb., 19 kg.)	1600	50	1.0	8	2500	0.8	1.2	8	50	400
7–9 yr. (58 lb., 26 kg.)	2000	60	1.0	10	3500	1.0	1.5	10	60	400
10–12 yr. (78 lb., 35 kg.)	2500	70	1.2	12	4500	1.2	1.8	12	75	400
Children over 12 yr.[h]										
Girls, 13–15 yr. (108 lb., 49 kg.)	2600	80	1.3	15	5000	1.3	2.0	13	80	400
16–20 yr. (122 lb., 55 kg.)	2400	75	1.0	15	5000	1.2	1.8	12	80	400
Boys, 13–15 yr. (108 lb., 49 kg.)	3200	85	1.4	15	5000	1.5	2.0	15	90	400
16–20 yr. (141 lb., 64 kg.)	3800	100	1.4	15	6000	1.7	2.5	17	100	400

[a] Objectives toward which to aim in planning practical dietaries. The recommended allowances can be attained with a good variety of common foods which will also provide other minerals and vitamins for which requirements are less well known.

[b] Calorie allowances must be adjusted up or down to meet specific needs. The calorie values in the table are therefore not applicable to all individuals but rather represent group averages. The proper calorie allowance is that which over an extended period will maintain body weight or rate of growth at the level most conducive to well-being.

[c] The allowance depends on the relative amounts of vitamin A and carotene. The allowances of the table are based on the premise that approximately two-thirds of the vitamin A value of the average diet in this country is contributed by carotene and that carotene has half or less than half the value of vitamin A.

[d] For adults (except pregnant and lactating women) receiving diets supplying 2000 cal. or less, such as reducing diets, the allowances of thiamine and niacin may be 1 mg. and 10 mg., respectively. The fact that figures are given for different calorie levels for thiamine and niacin does not imply that we can estimate the requirement of these factors within 500 cal., but they are added merely for simplicity of calculation. In the present revision, riboflavin allowances are based on body weight rather than caloric levels. Other members of the B complex also are required, though no values can be given. Foods supplying adequate thiamine, riboflavin, and niacin will tend to supply sufficient of the remaining B vitamins.

[e] There is evidence that the male adult needs relatively little iron. The need will usually be provided for if the diet is satisfactory in other respects.

[f] The need for supplemental vitamin D by vigorous adults leading a normal life seems to be minimum. For persons working at night and for nuns and others whose habits shield them from the sunlight, as well as for elderly persons, the ingestion of small amounts of vitamin D is desirable.

[g] During the latter part of pregnancy the calorie allowance should increase to approximately 20% above the preceding level. The value of 2400 cal. represents the allowanc for pregnant, sedentary women.

[h] Allowances for children are based on the needs for the middle year in each group (as 2, 5, 8, etc.) and are for moderate activity and for average weight at the middle year of the age group.

[i] Needs for infants increase from month to month with size and activity. The amounts given are for approximately 6 to 8 months. The dietary requirements for some of the nutrients such as protein and calcium are less if derived largely from human milk.

Further recommendations

Fat. There is available little information concerning the human requirement for fat. Fat allowances must be based at present more on food habits than on physiological requirements. While a requirement for certain unsaturated fatty acids (the linoleic and arachidonic acids of natural fats) has been amply demonstrated with experimental animals, the human need for these fatty acids is not known. In spite of the paucity of information on this subject there are several factors which make it desirable (1) that fat be included in the diet to the extent of at least 20 to 25% of the total calories and (2) that the fat intake include essential unsaturated fatty acids to the extent of at least 1% of the total calories. At higher levels of energy expenditure, e.g., for a very active person consuming 4500 cal. and for children and for adolescent persons, it is desirable that 30 to 35% of the total calories be derived from fat. Since foodstuffs such as meat, milk, cheese, nuts, etc., contribute fat to the diet, it is necessary to use separated or "visible" fats such as butter, oleomargarine, lard, or shortenings to supply only one-third to one-half the amounts indicated.

Iodine. The requirement for iodine is small, probably about 0.002 to 0.004 mg. daily for each kilogram of body weight, or a total of 0.15 to 0.30 mg. daily for the adult. This need is met by the regular use of iodized salt; its use is especially imporant in *adolescence* and *pregnancy*.

Water. A suitable allowance of water for adults is 2.5 l. daily in most instances. An ordinary standard for diverse persons is 1 ml. for each calorie of food. Most of this quantity is contained in prepared foods. At work or in hot weather, requirements may reach 5 to 13 l. daily. Water should be allowed *ad libitum*, since sensations of thirst usually serve as adequate guides to intake except for infants and sick persons.

Salt. The needs for salt and for water are closely interrelated. A liberal allowance of sodium chloride for the adult is 5 g. daily, except for some persons who sweat profusely. The average normal intake of salt is 10 to 15 g. daily, an amount which meets the salt requirements for a water intake up to 4 l. daily. When sweating is excessive, one additional gram of salt should be consumed for each liter of water in excess of 4 l. daily. With heavy work or in hot climates 20 to 30 g. daily may be consumed with meals and in drinking water. Even then, most persons do not need more salt than usually occurs in prepared foods. It has been shown that after acclimatization persons produce sweat that contains only about 0.5 g. to the liter in contrast with a content of 2 to 3 g. for sweat of the unacclimatized person. Consequently, after acclimatization, need for increase of salt beyond that of ordinary food disappears.

Phosphorus. Available evidence indicates that the phosphorus allowances should be at least equal to those for calcium in the diets of children and of women during the latter part of pregnancy and during lactation. In the case of other adults the phosphorus allowances should be approximately 1.5 times those for calcium. In general it is safe to assume that, if the calcium and protein needs are met through common foods, the phosphorus requirement also will be covered, because the common foods richest in calcium and protein are also the best sources of phosphorus.

Copper. The requirement for copper for adults is about 1 to 2 mg. daily. Infants and children require approximately 0.05 mg. for each kilogram of body weight. The requirement for copper is approximately one-tenth that for iron. A good diet normally will supply sufficient copper.

Vitamin K. The requirement for vitamin K usually is satisfied by any good diet except for the infant in utero and for the first few days after birth. Supplemental vitamin K is recommended during the last month of pregnancy. When it has not been given in this manner, it is recommended for the mother preceding delivery or for the baby immediately after birth.

Folic Acid. Evidence for recognizing folic acid (pteroylglutamic acid, vitamin B_c, *L. casei* factor or vitamin M) as an essential human nutrient is presented in the text. The quantitative requirement cannot be closely estimated from evidence now available.

in a form useful to the living body" (Chapter 16). The concept of the dynamic state of constituents of living cells, formulated by Schoenheimer[31] and documented by many data from isotopic tracer studies, also has an impact on human nutrition. Not only is there a continuous expenditure of energy to maintain this dynamic equilibrium, but all the essential nutrient substances must be available continuously for energy transfer to occur.

A number of the vitamins and minerals participate in enzyme systems, and a constant supply of certain amino acids which the body cannot manufacture is essential for the formation of proteins. These findings emphasize the importance of the quality of the diet even when the prime consideration is meeting energy requirements.

Another important concept, recently demonstrated, is the interrelationship of protein, fat, and carbohydrate oxidative mechanisms and the final common pathway through the Krebs citric acid cycle (Chapter 16).

The 2-carbon unit or activated acetate, which is the acetylated form of coenzyme A, the pantothenic acid-containing enzyme, has been shown to occupy a central position in many converging pathways. The role of mitochondria and other essential cellular units in oxidative metabolism, which have been elucidated recently, link cytology and biochemistry closely together and may lead to closer correlation between pathologic anatomical changes demonstrable in tissue section and biochemical abnormalities found in body fluids and tissues (Chapter 12).

3. Evaluation of Caloric Nutrition

Clinical evaluation of caloric nutrition is inexact. In most instances, the body weight is used as the criterion and the observed weight is compared with the weight for persons of similar height, age, and sex as given in some type of standard table. Unfortunately such standards usually represent arithmetic averages of findings in selected population groups and do not even include standard deviation or range of values. No consideration is given to the relative proportion of the several components of total body mass i.e., muscles, adipose tissues, bones, and extracellular fluid. Genetic variations and environmental circumstances are other factors which are seldom taken into account.

Brozek and Keys[32] have recently reviewed quantitative methods for evaluation of leanness and fatness in man. The most promising indirect method is measurement of the thickness of skin folds with specially constructed calipers. In adult males, there is good correlation between findings

[31] R. Schoenheimer, The Dynamic State of Body Constituents, Harvard University Monographs in Medicine and Public Health No. 3, Harvard University Press, Cambridge, Mass., 1942.

[32] J. Brozek and A. Keys, *Nutrition Abstracts & Revs.* **20**, 247 (1950); *Brit. J. Nutrition* **5**, 194 (1951).

with this method and estimation of total body fat from determination of the specific gravity of the body. Other indirect methods include estimation of body fat from the water content of the body and determination of the thickness of subcutaneous tissues from X-ray photographs.

Caloric balance can be maintained over a fairly wide range of body weight and caloric intake in a given individual. A continued loss beyond 10 % in weight may be considered representative of caloric deficit. Weighing a patient at frequent intervals is the best single method of determining caloric balance, although fluctuations may represent changes in fluid rather than in metabolic mass, especially in acutely ill patients.

4. Biochemical and Physiologic Changes in Starvation

Keys and associates[33] have reviewed knowledge of the biology of human starvation in detail in a recent volume. Many of the physiologic changes which occur appear to be useful or adaptive to survival; undoubtedly multiple interrelated factors are involved. Outstanding findings in starvation include loss of weight, marked weakness, mental and emotional changes which are reflected in behavior, circulatory changes including bradycardia, hypotension, reduced peripheral circulation and decreased venous pressure, reduction in body temperature, edema, and a decrease in the general reactivity of the organism to various stimuli. There is a high incidence of amenorrhea and sterility. The growth and development of children is retarded. Nitrogen balance is negative, although this becomes less marked as starvation proceeds. When protein deficiency is severe, edema and anemia are prominent findings. Signs of vitamin deficiency occur only if the diet has been deficient in these factors. Basal oxygen consumption declines, and this decrease applies to actively metabolizing tissue, as well as to the organism as a whole. Whether this decline is due to fuel exhaustion, a decrease in oxidative enzymes, reduced circulation, low body temperature, or other factors is unknown. The level of blood sugar during fasting may be low, and in starvation experiments in human volunteers blood pyruvate rose to abnormally high levels after exercise. Although glycogen stores are reduced early in starvation, they are not totally depleted. The causes of starvation or famine edema are not completely understood. Undoubtedly many factors are contributory. Although hypoproteinemia with lowered colloid osmotic pressure influences edema formation, it does not control it; nor can edema be explained by increased capillary permeability to colloids or by elevated venous or capillary pressure. It has been suggested that edema represents simply a persistence in the body of its normal complement of extracellular fluid.[33] The progressive loss of

[33] A Keys, J. Brozek, A. Henschel, O. Mickelsen, and H. L. Taylor, The Biology of Human Starvation, Vols. 1 and 2, University of Minnesota Press, Minneapolis, 1950.

cellular bulk produces a relative increase in hydration of the body as a whole. Edema appeared only when thiocyanate space was in excess of the normal proportion of body weight by about 10%.

In starvation, the degree of weight loss is an important factor in prognosis; if this is more than 40%, the outlook is very serious, even in previously healthy young adults.

5. Caloric Overnutrition—Obesity

Caloric overnutrition is an extremely important problem in many parts of the world. The many hazards of obesity have been well documented by Armstrong.[34] Obesity is associated with an increase in mortality rate and an increase in the incidence of a number of diseases, namely, hypertension, atherosclerosis, renal disease, diabetes mellitus, gall bladder disease, and even neoplasia. Possible relationships between fat metabolism and diseases of the blood vessels will be discussed in a subsequent section. In most instances, obesity is the result of simple overeating, the caloric intake being greater than the energy expenditure. It is related only rarely to disease of the endocrine glands.

III. Protein Nutrition

1. Protein Requirement

In the late nineteenth century, Voit and Atwater[35] suggested, on the basis of habitual protein intake of certain population groups, that an intake of 118 to 120 g. of protein daily was a desirable level. Early in the twentieth century, Chittenden[36] published studies which indicated that the health and efficiency of young men could be maintained with diets which furnished about 44 to 53 g. of protein daily (for men weighing 70 kg.). Investigations of Fischer and Kossel[10] showed that the composition of proteins from different sources varied greatly, and it was demonstrated subsequently that certain proteins were capable of supporting growth in animals but others were not. Since proteins were known to be hydrolyzed to amino acids in the digestive process, it appeared that the biological value of proteins was dependent on their amino acid composition. From studies of nutritionally inadequate proteins and from supplementation experiments, the concept of essential amino acids arose. The essential amino acids, that is, those which cannot be synthesized by the tissues, have been determined for a number of animal species, including man.

[34] (*a*) D. B. Armstrong, L. I. Dublin, G. M. Wheatley, and H. H. Marks, *J. Am. Med. Assoc.* **147,** 1007 (1951); (*b*) D. B. Armstrong, L. I. Dublin, E. C. Bannett, and H. H. Marks, *Postgraduate Med.* **10,** 407 (1951).

[35] C. v. Voit and W. O. Atwater, *in* McCollum, ref. 5, p. 51.

[36] R. H. Chittenden, *in* Sherman, Chemistry of Food and Nutrition, 6th ed., Macmillan, New York, 1941, p. 204.

a. Amino acid requirements. Rose[37] demonstrated in metabolic balance studies in adult males on purified diets that eight amino acids are required for maintenance of nitrogen equilibrium, namely, isoleucine, leucine, lysine, methionine, phenylalanine, threonine, tryptophan, and valine. A table of minimal and recommended intakes is given in Chapter 5 (p. 129). The diet must also furnish sufficient nitrogen for synthesis of the non-essential amino acids. The eight amino acids essential in human nutrition are also required for growth by the dog, the rat, the mouse, and the chick, but each species requires one or more additional amino acids. Histidine and arginine, which are essential for dog, rat, and chick, do not appear to be essential for man, although it has been reported that a dietary deficiency of arginine results in decreased spermatogenesis in man (Chapter 5).

A few studies of the amino acid requirement of human infants have been carried out; in addition to the eight amino acids needed by adults, cystine and tyrosine may be required under certain conditions.

There is some evidence that in rats and mice a mixture of essential amino acids may not support optimal growth and that some polypeptide fraction may be required. Whether this is true in man is unknown.

Recent studies indicate that all the amino acids needed for the formation of the protein molecule must be available at the same time and that the proportion of essential amino acids to each other remains relatively constant at various levels of protein intake. Thus, the balance of amino acids in the diet is another factor requiring consideration.[38] The distribution of proteins in the three daily meals may also influence protein requirement and utilization.[38e]

b. Total protein requirement. The protein needs of man are dependent on age, sex, body size, the previous supply of protein, the energy value of the diet, and a number of other factors. The body tends to adjust protein metabolism to protein supply, and carbohydrate and fat exert a protein-sparing action. When the diet is practically devoid of protein, the nitrogen excretion of an adult male weighing 70 kg. has been found to be approximately 3 g. a day, which represents about 20 g. of protein. Recently, patients with hypertension have been maintained for many months on rice diets which furnish 20 to 27 g. of protein. Approximate nitrogen balance is usually reached after 3 to 6 months.[39] Certainly, larger amounts of protein would appear to be desirable under most circumstances to provide a safe

[37] W. C. Rose, *Federation Proc.* **8,** 546 (1949).

[38] (*a*) E. Geiger, *Science* **111,** 594 (1950); (*b*) H. Borsook, *Physiol. Revs.* **30,** 206 (1950); (*c*) R. W. Wissler, L. E. Frazier, and R. E. Slayton, *Proc. Soc. Exptl. Biol. Med.* **72,** 589 (1949); (*d*) H. J. Almquist, *ibid.* **72,** 179 (1949); (*e*) R. M. Leverton and M. R. Gram, *J. Nutrition* **39,** 57 (1949).

[39] (*a*) V. P. Dole, L. K. Dahl, G. C. Cotzias, H. A. Eder, and M. E. Krebs, *J. Clin. Invest.* **29,** 1189 (1950); (*b*) E. Peschel and R. L. Peschel, *ibid.* **29,** 455 (1950).

allowance for good nutrition and health. From findings in more than one hundred nitrogen balance experiments conducted by a number of independent investigators, Sherman[40] calculated that the average protein requirement for the maintenance of equilibrium was 44 g. for an adult of either sex weighing 70 kg., or ⅔ g. per kilogram of body weight per day. Some of these experiments were of relatively short duration, and Sherman suggested that the best data would indicate an average need not far from 0.5 g. per kilogram of body weight. Rose and associates[41] showed that in healthy young men nitrogen balance could be maintained on a level of nitrogen consumption of 7.02 g. daily when more than 95 % of the nitrogen of the diet was in the form of a mixture of the ten amino acids found to be indispensable for animals. This represents about 44 g. of protein daily.

The Food and Nutrition Board of the National Research Council[27] recommends as a desirable allowance of protein for adults 1 g. per day for each kilogram of body weight. This factor of safety is designed to cover individual differences in requirement as well as variations in the nutritive efficiency of protein in different diets. Protein is needed in increased amounts in the latter half of pregnancy and during lactation.[42] In children, nitrogen balance studies indicate that positive balances may be obtained with the following intakes of protein per kilogram of body weight: infancy, 4 to 3.5 g.; early childhood, 3 to 2.5 g.; late childhood and adolescence, 2 to 1.5 g. The amounts required depend on the size of the child and the quality of the protein.

The numerous factors which influence protein utilization have been reviewed recently in a symposium; data obtained in nitrogen balance experiments represent the summation of many variables which must be considered in interpretation.[43]

2. Metabolism of Protein

In the digestive tract, proteins are hydrolyzed to peptides and amino acids (Chapter 17) which are absorbed into the portal circulation and carried to the liver and from there to the tissues by the systemic blood. Several fates await the amino acids. (1) They may be deaminated, the nitrogenous fraction being largely converted to urea and the remainder of

[40] H. C. Sherman, ref. 36, p. 204.

[41] (*a*) W. C. Rose, W. J. Haines, and J. E. Johnson, *J. Biol. Chem.* **146,** 683 (1942); (*b*) W. C. Rose, W. J. Haines, J. E. Johnson, and D. T. Warner, *ibid.* **148,** 457 (1943).

[42] (*a*) I. G. Macy, H. A. Hunscher, B. Nims, and S. S. McCosh, *J. Biol. Chem.* **86,** 17 (1930); (*b*) C. K. Shukers, I. G. Macy, E. Donelson, B. Nims, and H. A. Hunscher, *J. Nutrition* **4,** 399 (1931); (*c*) C. M. Coons, Studies in Metabolism in Pregnancy, *Oklahoma Agr. Expt. Sta. Bull.* 223, Stillwater, 1935.

[43] (*a*) P. P. Swanson, *Federation Proc.* **10,** 660 (1951); (*b*) E. Geiger, *ibid.* **10,** 670 (1951); (*c*) J. B. Allison, *ibid.* **10,** 676 (1951).

the molecule transformed into dextrose or fatty acids and utilized for energy or stored as fat. As previously noted, carbohydrates, protein, and fat meet in a common metabolic pathway in which the 2-carbon unit or coenzyme A occupies a central position. (2) They may be used in the synthesis of cellular proteins of any of the tissues, or of special proteins such as hemoglobin, albumin, globulin, fibrinogen, etc., or in the formation of hormones, which are protein or protein derivatives, or in the synthesis of enzymes (Chapter 6). Studies of Schoenheimer[31] demonstrated that tissue proteins are labile compounds which are undergoing continual breakdown and synthesis. Cannon[44] has recently discussed this dynamic equilibrium and the mechanisms which maintain it. Sprinson and Rittenberg[45] estimated that the average half-life of the total protein in man is about 80 days, the rate of replacement of protein nitrogen being about 1 g. per day. The rate of amino acid exchange varies in different tissues; the turnover in liver and plasma accounts for 41 % of the total exchange in man. Kinsell and associates[46] reported that plasma protein synthesis and degradation proceed at a constant rate in normal persons. In patients with liver disease the rate of protein formation is diminished, and in idiopathic hypoproteinemia the rate of protein degradation is increased.

3. Amino Acids

a. Blood levels and excretion of amino acids. Numerous studies of the concentrations of amino acids in blood and of the excretion of these substances in urine and feces have been carried out in man by means of microbiologic and chromatographic methods. In general, the fasting amino acid content of plasma does not fluctuate significantly, despite considerable variation in the quantity of amino acids ingested.[47] Considerable variation in urinary excretion of amino acids has been observed, and the rate of excretion of the several amino acids is not directly correlated with the respective amounts ingested nor with the blood levels of these acids.[47a, b] Abnormal urinary excretion of amino acids has been reported in liver disease,[48] but not in peptic ulcer,[49] chronic ulcerative colitis,[50] or following

[44] P. R. Cannon, *Am. J. Clin. Path.* **19,** 99 (1949).

[45] D. B. Sprinson and D. Rittenberg, *J. Biol. Chem.* **180,** 715 (1949).

[46] L. W. Kinsell, S. Margen, H. Tarver, J. M. Frantz, E. K. Flanagan, M. E. Hutchin, G. D. Michaels, and D. P. McCallie, *J. Clin. Invest.* **29,** 238 (1950).

[47] (*a*) B. F. Steele, M. S. Reynolds, and C. A. Baumann, *J. Nutrition* **40,** 145 (1950); (*b*) J. B. Kirsner, A. L. Sheffner, and W. L. Palmer, *J. Clin. Invest.* **28,** 716 (1949); (*c*) E. S. Nasset and R. H. Tully, III, *J. Nutrition* **44,** 477 (1951).

[48] M. S. Dunn, S. Akawaie, H. L. Yeh, and H. Martin, *J. Clin. Invest.* **29,** 302 (1950).

[49] J. B. Kirsner, A. L. Sheffner, W. L. Palmer, and K. Sterling, *J. Clin. Invest.* **29,** 869 (1950).

[50] J. B. Kirsner, A. L. Sheffner, and W. L. Palmer, *J. Clin. Invest.* **29,** 874 (1950).

surgical procedures when nitrogen balance is negative. After radiation injury, an increase in amino acid excretion was noted in some subjects.[51] Albanese,[52] who has been studying amino acid deficiencies in man, expressed the view that changes in amino acid levels in urine and blood may be useful in detecting certain of these deficiencies. He also suggested that in protein-depleted states, a diet poor in one essential amino acid may be more deleterious than one poor in protein.

b. Specific amino acids in human nutrition. Knowledge of the metabolic role of individual amino acids is increasing. Tryptophan has been shown to be a precursor of niacin in neurospora and in many species of animals, including man.[53] Administration of tryptophan to human subjects leads to an increase in the excretion of N^1-methylnicotinamide (N^1-Me), the pyridone of N^1-Me, and quinolinic acid.[54] Studies in neurospora and in the rat have elucidated some of the steps in the conversion of tryptophan to niacin (Chapter 5). This conversion appears to take place in the tissues and not in the intestinal tract.[55] Recent findings in man suggest that quinolinic acid is not in the main pathway of conversion, although some increase in excretion of niacin metabolites followed administration of this substance.[54b] Pyridoxine is involved in the conversion mechanism and, in deficiency of this vitamin, conversion is impaired.[56] The finding that the amino acid tryptophan can be converted to niacin in the body explains, in part, the role of corn diets in the pathogenesis of pellagra. Not only is corn low in niacin, but the main protein, zein, is low in tryptophan (see p. 560).

Knowledge of metabolic pathways of phenylalanine and tyrosine has been obtained by study of certain inborn errors of metabolism in man (see Chapter 5). Of particular interest in human nutrition is the relationship of ascorbic acid and folic acid in the metabolism of these two amino acids. In premature infants or in persons with scurvy, the feeding of high protein diets or the administration of tyrosine leads to hydroxyphenyluria. Both ascorbic acid and folic acid (large doses) will prevent the excretion of abnormal quantities of hydroxyphenyl derivatives.[57] Recently, Sealock

[51] L. H. Hempelmann, H. Lisco, and J. G. Hoffman, *Ann. Internal Med.* **36,** 279 (1952).

[52] A. A. Albanese, *J. Clin. Nutrition* **1,** 44 (1952).

[53] H. P. Sarett and G. A. Goldsmith, *J. Biol. Chem.* **177,** 461 (1949).

[54] (*a*) G. A. Goldsmith, H. P. Sarett, U. D. Register, and J. Gibbens, *J. Clin. Invest.* **31,** 533 (1952); (*b*) H. P. Sarett, *J. Biol. Chem.* **193,** 627 (1951).

[55] H. P. Sarett, *J. Nutrition* **47,** 275 (1952).

[56] S. Snydermann, R. Carretero, and L. E. Holt, Jr., *Federation Proc.* **9,** 371 (1950).

[57] (*a*) C. W. Woodruff, M. E. Cherrington, A. K. Stockell, and W. J. Darby, *J. Biol. Chem.* **178,** 861 (1949); (*b*) J. E. Morris, E. R. Harpur, and A. Goldbloom, *J. Clin. Invest.* **29,** 325 (1950); (*c*) C. W. Woodruff, *J. Lab. Clin. Med.* **36,** 640 (1950); (*d*) W. F. Rogers, Jr., and F. Gardner, *J. Clin. Invest.* **28,** 806 (1949).

and Goodland[58] demonstrated that ascrobic acid is a necessary coenzyme in the metabolic oxidation of tyrosine.

Although investigation of methionine metabolism in animals has been extensive, little is known about the application of these findings to man. Relationships between vitamin B_{12}, folic acid, choline, and methionine have been observed in a number of animal species. All these nutrients seem to participate in transmethylation. The importance of methyl groups in the prevention of fatty livers and cirrhosis in animals fed diets high in fat, or low in protein and low in fat, is well known. The methyl groups are presumably of similar importance in man. Many patients with cirrhosis give a dietary history indicative of deficient intake of protein and of vitamins of the B complex. Many investigators have reported improvement in such patients when a high protein diet was instituted. In subjects with alcoholic fatty cirrhosis, Eckhardt and colleagues[59] have shown that administration of choline will cause moderate reduction in liver fat and that provision of a diet adequate in protein will be followed by a decrease in fat and a return of the size and appearance of cells to normal. Cayer and Cornatzer[60] have shown an increase in phospholipid turnover in untreated patients with chronic hepatitis after a large single dose of choline or methionine. Ralli and co-workers[61] reported that administration of liver extract, in conjunction with conventional therapy, increased survival rate in patients with cirrhosis. It is conceivable that vitamin B_{12} may have played a role in the improvement noted. Studies of Kirsner[62] suggest that amino acids, with the possible exception of methionine, are metabolized adequately in subjects with cirrhosis although less efficiently than normal. Concentrations of methionine in plasma were greatly increased in acute severe hepatitis.

Gross and associates[63] found an increased loss of fat in the stools in patients with cirrhosis and postulated defective formation of bile salts as the cause. Serum concentrations of sodium, potassium, calcium, and phosphorus have been found to be lower than normal, and blood pyruvic acid levels elevated, in patients with decompensated alcoholic cirrhosis.[64] When an adequate diet was instituted, these findings reverted to normal. The low levels of calcium and phosphorus may have resulted from an increase

[58] R. R. Sealock and R. L. Goodland, *Science* **114,** 645 (1951).

[59] R. D. Eckhardt, N. Zamcheck, R. L. Sidman, G. J. Gabuzda, Jr., and C. S. Davidson, *J. Clin. Invest.* **29,** 227 (1950).

[60] D. Cayer and W. E. Cornatzer, *Southern Med. J.* **42,** 669 (1949).

[61] E. P. Ralli, S. H. Leslie, G. H. Stueck, Jr., H. E. Shorr, J. S. Robson, D. H. Clarke, and B. Laken, *Medicine* **28,** 301 (1949).

[62] J. B. Kirsner, A. L. Sheffner, W. L. Palmer, and O. Bergeim, *J. Lab. Clin. Med.* **36,** 735 (1950).

[63] J. B. Gross, M. W. Comfort, E. E. Wollaeger, and M. H. Power, *Gastroenterology* **16,** 140 (1950); *Proc. Staff Meetings Mayo Clinic* **26,** 9 (1951).

[64] D. S. Amatuzio, F. Stutzman, N. Shrifter, and S. Nesbitt, *J. Lab. Clin. Med.* **39,** 26 (1952).

in fat in the stools. In hepatic coma, marked elevations of pyruvic acid have been observed in both blood and spinal fluid which were not affected by thiamine administration. It was suggested that failure of the liver to assimilate pyruvic acid to form the carboxylic acids of the Krebs cycle might explain these findings or that they might be due to a partial deficiency of cocarboxylase. Failure of mechanisms of cellular aerobic metabolism might perhaps explain the neurologic and renal manifestations of hepatic coma.

c. Amino acids in therapy. Protein hydrolyzates and mixtures of pure amino acids have been used in the therapy of many pathologic states. Normal protein requirement can be satisfied and nitrogen equilibrium maintained by intravenous administration of properly prepared hydrolyzates or solutions of amino acids. In disease, it may be impossible to fulfill the requirement by intravenous feeding. If hydrolyzates are supplemented by individual amino acids in large amounts, the resulting imbalance may cause a loss of body nitrogen. Nausea and vomiting which may follow administration of amino acid preparations appear to be related to the free glutamic acid content of the mixture.[65]

4. Protein Deficiency Syndromes

The probable importance of protein in the pathogenesis and treatment of cirrhosis of the liver has already been mentioned. Protein deficiency is usually associated with caloric deficiency and occurs in a variety of conditions other than simple starvation. Deficiency is common in prolonged febrile illnesses, thyrotoxicosis, and other hypermetabolic states; in diseases associated with loss of protein from the body, such as nephrosis, effusion into serous cavities, weeping wounds, etc.; in conditions interfering with digestion and absorption; and in metabolic diseases which interfere with utilization. Following trauma such as fractures, burns, and operative procedures, a profound negative nitrogen balance may occur, persisting for days.[66] Nitrogen loss may amount to as much as 0.6 g. per kilogram of body weight daily. In chronically starved individuals, the loss following injury is less than in normal subjects. It is, at times, impossible to produce a positive nitrogen balance even with a large protein intake, but if serious depletion has occurred such an attempt may be life-saving.

The clinical syndrome of protein deficiency includes the following findings: loss of weight, loss of subcutaneous fat, wasting of muscles, pigment changes in the skin, hypotension, bradycardia, weakness, and edema. The basal metabolic rate falls, anemia is common, and a decrease in serum proteins, especially of the albumin fraction, may be noted. Hemoconcentration and a decrease in blood volume are frequently observed.

[65] S. Levey, J. E. Harroun, and C. J. Smyth, *J. Lab. Clin. Med.* **34,** 1238 (1949).
[66] R. Elman, *Ann. Surgery* **120,** 350 (1944).

Procedures for evaluating the state of protein nutrition are far from exact. Estimation of the percentage of body weight which has been lost is a good index of the severity of protein deficiency, although edema may mask findings. Determination of the concentration of serum proteins, especially of the albumin fraction, is informative as to the extent of protein depletion and is of assistance in planning and evaluating therapy.

In calculating needs for replacing body stores of protein, it is assumed, on the basis of animal experiments, that each gram decrease in total circulating plasma albumin represents a loss of 30 g. of tissue protein.[67] Rehabilitation of patients with severe protein deficiency may require long periods of time and with continuous loss of protein from the body may be impossible. The slow recovery is readily appreciated in view of some of the considerations noted above.

Kwashiorkor,[68] a syndrome observed in infants and young children in Africa and in certain other parts of the world, appears to be due to deficiency of foods containing animal protein, or vegetable protein of high biological value. The daily food supply in areas where Kwashiorkor is common is very low in methionine. Characteristic findings in this syndrome include retarded growth, alteration in skin and hair pigmentation, edema, fatty infiltration with cellular necrosis and fibrosis of the liver, and a high mortality in the absence of proper dietary treatment. Skim milk powder has been shown to be effective therapy.

IV. Carbohydrates in Nutrition

Carbohydrates furnish 50 to 60% of the total calories in the average American diet and a still larger percentage in diets in other parts of the world. Since carbohydrate may be synthesized in the body from certain amino acids and fatty acids, there is no true dietary requirement. Much of our knowledge of carbohydrate metabolism has been gleaned from studies in animals or from *in vitro* experiments, but clinical investigations indicate that most of the concepts are applicable to man. Since carbohydrate metabolism is discussed in detail in Chapter 4, only a brief résumé of pertinent findings will be given here.

1. Digestion and Absorption of Carbohydrates

In the digestive process, carbohydrates are broken down to monosaccharides which are absorbed from the small intestine either by diffusion or by a specific mechanism involving phosphorylation. Glucose, galactose, and fructose are absorbed by both processes. Many factors influence ab-

[67] L. A. Sachar, A. Horvitz, and R. Elman, *J. Exptl. Med.* **75,** 453 (1942).

[68] J. F. Brock and M. Autret, Kwashiorkor in Africa, United Nations Food and Agriculture Organization, Nutritional Studies No. 8, Columbia University Press, New York, 1952.

sorption: the food mixture in the intestine, the condition of the intestinal mucous membrane, the intake of vitamins of the B complex, and the function of the endocrine glands, especially the thyroid, the anterior pituitary, and the adrenal cortex.

2. Intermediate Metabolism and Functions of Carbohydrates

Carbohydrate is utilized chiefly for energy and is more efficient than protein or fat in providing fuel for muscular exercise. It is stored in the liver and muscles as glycogen, or it may be converted to protein or fat. Carbohydrate is present in the blood and extracellular fluids as glucose, all hexoses being converted to glucose in the liver. The formation of carbohydrate from non-carbohydrate precursors, gluconeogenesis, occurs in the liver and probably also in the kidney.

The concentration of glucose in the blood is maintained at a relatively constant level by homeostatic processes which include enzymatic, nervous, and hormonal components (Chapter 4). A balance is preserved between glucose production and glycogenolysis, on one hand, and glucose utilization, on the other. If the concentration of glucose in the blood falls below a critical level, serious damage may result. Carbohydrate is essential for normal functioning of the heart and nervous system, tissues of the latter using only glucose to satisfy energy requirements. Hypoglycemia may precipitate severe myocardial injury and even death, in older individuals with arteriosclerosis. Prolonged episodes of hypoglycemia may induce permanent damage to the brain.

Glycogen stores in the liver vary with the type of diet ingested; they are highest when the diet contains large amounts of carbohydrate, intermediate when the diet is rich in protein, and lowest when the diet is high in fat. A diet high in carbohydrate ensures adequate glycogen storage which assists in protection of the liver against injury. Carbohydrate exerts a detoxifying action in the liver; e.g., acetyl groups derived from carbohydrate are used in acetylation of numerous compounds, including sulfa drugs, and glycuronic acid derived from carbohydrate combines with compounds containing phenolic-hydroxyl groups. The latter mechanism may be important in regulating the metabolism of steroid hormones and in preventing accumulation of excessive amounts of the sex hormones.

Carbohydrate has an antiketogenic action in that it inhibits the breakdown of fatty acids in the liver. It exerts a sparing action on protein by decreasing the rate of deamination of amino acids in the liver.

The intermediary metabolism of carbohydrate is discussed in Chapter 4, to which the reader is referred for details of the enzymatic conversion of hexoses, the formation of glycogen, and the Embden-Meyerhof scheme of glycolysis. Terminal oxidation of carbohydrate is also discussed; the oxidation and reactions of pyruvic acid, the monophosphate shunt, and the tricarboxylic acid cycle are illustrated diagrammatically.

It might be emphasized here that a number of the coenzymes concerned in the chemical reactions involving carbohydrate are formed from vitamins of the B complex. Coenzymes containing thiamine, niacinamide, riboflavin, pyridoxine, and pantothenic acid are vital in various steps in the degradation and synthesis of carbohydrate. An insufficient supply of any one of these factors will prevent metabolism from proceeding normally. This knowledge is applied in diagnosing deficiency of at least one of the B vitamins. In thiamine deficiency, pyruvic acid, an intermediate in the breakdown of glucose, accumulates in the blood and tissues. Determination of the concentration of pyruvic acid in blood and of the lactate-pyruvate ratio is used in evaluating thiamine nutrition.

Diets high in carbohydrate require a generous supply of vitamins of the B group. When sugar and highly milled cereals, from which most of the B vitamins have been removed, are the chief sources of carbohydrate, vitamin B complex deficiency may result. In order to provide an adequate intake of these vitamins for large population groups who use refined cereals as the staple article of food, a program of enrichment has been developed. Vitamins of the B complex have been added to white flour, polished rice or corn meal in amounts equal to those present in whole grain. Use of these enriched products has resulted in improvement in the nutritional status of populations in certain areas of the world, for example, in Newfoundland and in the Philippines.[69]

The hormonal regulation of carbohydrate metabolism is considered in Chapter 4, including the roles of insulin and of the adrenal and pituitary hormones. This knowledge, and speculations based thereon, have proved most useful in the understanding and therapy of diabetes mellitus and in the management of diseases associated with adrenal or pituitary dysfunction. The extensive use of adrenocorticotropic hormone (ACTH) and cortisone in clinical medicine during the past few years has greatly enhanced our knowledge of the part played by these factors in carbohydrate metabolism in man.

3. Abnormalities of Carbohydrate Metabolism

In clinical medicine, the tests most often used to detect abnormalities of carbohydrate metabolism are determination of glucose excretion in the urine and estimation of the glucose concentration in the blood during fasting and after administration of a large dose of glucose (glucose tolerance

[69] (*a*) G. A. Goldsmith, W. J. Darby, R. C. Steinkamp, A. S. Beam, and E. McDevitt, *J. Nutrition* **40,** 41 (1950); (*b*) W. R. Aykroyd, N. Jolliffe, O. H. Lowry, P. E. Moore, W. H. Sebrell, R. E. Shank, F. F. Tisdall, R. M. Wilder, and P. C. Zamecnik, *Can. Med. Assoc. J.* **60,** 329 (1949); (*c*) J. Salcedo, Jr., E. O. Carrasco, R. F. Jose, and R. C. Valenzuela, *J. Nutrition* **36,** 561 (1948); (*d*) H. B. Burch, J. Salcedo, Jr., E. O. Carrasco, C. L. Intengan, and A. B. Caldwell, *ibid.* **42,** 9 (1950).

test). The small amounts of glucose in urine in normal persons, less than 0.1 %, are not detectable by the usual clinical procedures. An increase in glucose excretion and an abnormally high concentration of glucose in blood in the post-absorptive state are presumptive evidence of diabetes mellitus but occur in other conditions. The glucose tolerance test is of assistance in establishing a diagnosis. In diabetes mellitus, the blood sugar rises to abnormally high levels and remains elevated for 3 hr. or more, after a large dose of glucose. The blood sugar response to oral administration of glucose is dependent in part on the previous diet; when the diet is high in fat, the elevation of blood sugar is greater and more prolonged than when the diet is high in protein; when the diet is high in carbohydrate, the rise is least and of shortest duration. A diabetic type of glucose tolerance test may be found in association with hyperactivity of the pituitary gland or of the adrenal cortex, in thiamine deficiency and during starvation. Abnormally rapid absorption of glucose is found in hyperthyroidism, while abnormally slow absorption occurs in hypothyroidism, Simmonds' disease (hypopituitarism), and Addison's disease (adrenal insufficiency), in the steatorrheas, including sprue, and in nutritional macrocytic anemia. In these conditions a "flat" glucose tolerance test is characteristic.

An excessive fall in blood sugar several hours after a meal, especially one which is high in sugar, is observed in certain persons who are otherwise normal. Hypoglycemia is also observed in hyperinsulinism.

4. Metabolism of Fructose and Invert Sugar

Recent studies of the metabolism of fructose in normal and diabetic subjects indicate that fructose is assimilated more rapidly than glucose and its metabolism appears to be independent of the action of insulin.[70] Solutions of invert sugar have been used in the treatment of non-diabetic subjects who must be nourished by the parenteral route. Advantages claimed for invert sugar as a source of calories, in comparison to glucose, are more rapid assimilation by the body and smaller loss of sugar in the urine. Actually, the caloric increment is not great.

V. Fats in Human Nutrition

Fat metabolism in man has been the subject of increasing interest and investigation. In the past few years, attention has been directed particularly to the role of fat in the pathogenesis of atherosclerosis.

[70] (*a*) A. I. Mendeloff and T. E. Weichselbaum, *J. Lab. Clin. Med.* **38,** 929 (1951); (*b*) T. E. Weichselbaum and W. H. Daughaday, *ibid.* **38,** 958 (1951); (*c*) J. W. Craig, W. R. Drucker, M. Miller, J. E. Owens, and H. Woodward, Jr., *ibid.* **38,** 800 (1951); (*d*) M. Miller, W. R. Drucker, J. E. Owens, and H. Woodward, Jr., *ibid.* **38,** 931 (1951); (*e*) M. Miller, W. R. Drucker, J. E. Owens, J. W. Craig, and H. Woodward, Jr., *J. Clin. Invest.* **31,** 115 (1952).

1. Fat Requirement

Fat is needed in human nutrition for growth and replacement of tissues, for certain lipid secretions, and as a source of energy. The optimal level of fat in the diet is not known, and since fat may be formed from carbohydrate or protein, no definite requirement can be formulated. Fat increases the palatability of the diet and also reduces the bulk, since the energy value of fat is more than twice that of an equivalent amount of protein or carbohydrate. Dietary fat also acts as a carrier for the fat-soluble vitamins.

In animals, certain unsaturated fatty acids, such as linolenic or linoleic acids, are essential dietary components. Whether these compounds can be synthesized by man has not been determined, but since they are widely present in fats ordinarily consumed, deficiency seems an unlikely possibility. In dogs, there is evidence for a need of fat above that necessary to relieve specific effects of essential fatty acid deficiency (Chapter 7). This may be true in other species, including man.

2. Digestion and Absorption of Fats

The energy lipids, chiefly triglycerides, are hydrolyzed by pancreatic lipase in the small intestine to fatty acids and glycerol. According to Fraser (Chapter 7), hydrolysis may be only partial and fats can be absorbed both as fatty acids and as glycerides. Bile salts are of great importance in the emulsifying system which effects absorption. Fine fat particles, either particulate glycerides or fatty acids, enter the cell border and pass into the chyle. Fatty acids of less than 10-carbon length, or glycerides containing short-chain fatty acids, may pass to the liver by way of the portal vein or may be metabolized in the intestinal wall (Chapter 7). There is some controversy as to whether or not unsplit fat can enter the lacteals directly. It is postulated by some that only fatty acids can be absorbed and that glycerides are resynthesized in the intestinal mucosa by a process which involves phosphorylation. Dietary phospholipid is not hydrolyzed and passes into the intestinal wall as such. Cholesterol is freely absorbed only in the presence of fat. The lecithin which leaves the intestine is formed from part of the glyceride (Chapter 7).

Absorption of fat is seriously impaired in man in the absence of bile from the intestinal tract, for example, in obstructive jaundice, although splitting of fat to fatty acids and glycerol proceeds normally. In sprue, idiopathic steatorrhea, and the celiac syndrome, fat is also poorly absorbed, although there is no decrease in available bile and the enzymatic breakdown to fatty acids is unimpaired. It has been suggested that a defective phosphorylating mechanism, due to deficiency of one or more vitamins of the B complex, may account for poor absorption in the sprue syndrome. Folic acid may be one of the factors involved.

3. Blood Levels of Lipids

The lipids in the blood during fasting consist largely of phospholipids, cholesterol, and cholesterol esters and almost no triglycerides. After a fat meal, glycerides in the blood increase 300 to 500 mg. per 100 ml. Maximum hyperlipemia is observed 2½ to 3 hr. after ingestion of a moderate amount of fat and the fasting level is attained by the fifth hour (Chapter 7). Hyperlipemia depends upon the rate of removal of fat from the blood as well as on the rate of intake. Administration of heparin has been shown to prevent normal alimentary lipemia. This action may be related to the beneficial effect of heparin reported by Jones and associates[71] in patients with atherosclerosis, namely, reduction in the level of certain lipoproteins in the blood.

Endogenous hyperlipemia occurs in uncontrolled diabetes and, at times, in starvation, pregnancy, lactation, severe anemia, nephrosis, and after hemorrhage. The lipid changes appear to be associated with a decrease in plasma protein concentration or in oxygen-carrying capacity of blood (Chapter 7).

The normal level of cholesterol in blood is in the neighborhood of 200 mg. per 100 ml., of which over 60 % is esterified. An increase in cholesterol level occurs with age. The concentration is not significantly related to the intake of cholesterol over a range of 250 to 800 mg. daily but is influenced by the amount of fat in the diet, as will be discussed subsequently.[72] The phospholipid level in the blood is about 220 mg. per 100 ml. and increases with age in a manner similar to that noted for cholesterol. The amount of phospholipid in relation to that of cholesterol in blood may be a factor in maintaining normal lipid dispersion.

The lipids in plasma are associated closely with protein, mainly with the alpha and beta globulin fractions, and recent investigation has indicated that lipoprotein complexes are of great metabolic importance. They appear to play a role in antigen-antibody reactions, and certain lipoproteins may be implicated in the development of atherosclerosis.

4. Intermediate Metabolism of Fats

Ekstein[73] has summarized the fate of absorbed fatty acids as follows: incorporation into more complex lipids, conversion into other fatty acids, transformation into carbohydrate, use in the production of milk fat, excretion through the gut, deposition in fat depots, or oxidation to carbon di-

[71] H. B. Jones, J. W. Gofman, F. T. Lindgren, T. P. Lyon, D. M. Graham, B. Strisower, and A. V. Nichols, *Am. J. Med.* **11**, 358 (1951).

[72] (*a*) A. Keys, *Science* **112**, 79 (1950); (*b*) A. Keys, *Circulation* **5**, 115 (1952).

[73] H. C. Ekstein, Fat in in Nutrition, *in* Handbook of Nutrition, A Symposium, 2nd ed., The Blakiston Company, Philadelphia, 1951, p. 23.

oxide and water with the liberation of energy. The metabolic processes involved in these pathways are discussed in Chapter 7.

Cholesterol can be synthesized in the human body from 2-carbon fragments, and it has been estimated that 1.5 to 2 g. may be formed daily.[74] Cholesterol is closely related to other steroid compounds, including the bile acids, vitamin D_3, and the adrenal cortical and sex hormones, and may be a precursor of these compounds. The half-life of cholesterol in man has been estimated to be 8 days.

Phospholipids can be synthesized in many tissues, including liver, kidney, muscle, and brain, and the turnover in these tissues has been measured using P^{32} (Chapter 7).

Lipids are important components of cell structure and are found in cytoplasm, microsomes, mitochondria, and cell membranes usually closely bound to protein.

5. Storage of Fat–Obesity

Adipose tissue contains triglycerides of palmitic and oleic acids. The metabolism of cells in which fat is deposited is more active than is commonly appreciated. These cells synthesize, assimilate, store, and mobilize glycerides and glycogen. Studies with labeled fat indicate that mobilization of fat from depots is constantly taking place. The amount of fat stored in these depots is dependent upon the intake of energy-producing foods in relation to energy expenditure. Excessive storage of fat results in obesity, with its related problems. These have been discussed in the section on caloric overnutrition (p. 516).

6. Utilization of Fat for Energy

Fatty acids are used for energy in the liver and many other tissues. This is a continuous process, since the respiratory quotient is always somewhat less than 1. An increase in the demand of fat for energy occurs when there is a shortage of carbohydrate. This is of particular importance in clinical medicine during starvation and in diabetes mellitus. Recent work with isotopes indicates that long-chain fatty acids are degraded by rapid fragmentation of the whole chain to form 2-carbon and, occasionally, 4-carbon units. Two-carbon units may enter the tricarboxylic acid cycle, may be synthesized into fatty acids or cholestorol, or may condense to form acetoacetic acid. Acetoacetic acid is not broken down in the liver but may be used in the tissues. The formation of this compound is greatly increased if there is interference with passage of 2-carbon units into the tricarboxylic acid cycle, as in diabetes mellitus and starvation. In these conditions, the

[74] R. G. Gould, *Am. J. Med.* **11**, 209 (1951).

concentration of acetoacetic acid increases in the blood and tissues, and when it reaches 20 mg. per 100 ml. ketonuria results. Much of the acetoacetic acid may be converted to β-hydroxybutyric acid and acetone. Acidosis due to excessive formation of ketone bodies is a serious hazard in diabetes mellitus. Also, in diabetes, fatty acid synthesis in the liver is defective while formation of cholesterol may be increased. The proper management of diabetes with diet and insulin will prevent ketosis and assist in reversal of the pathologic processes discussed above.

7. Fat and Liver Disease

The majority of the fat in liver is phospholipid, the remainder being glycerides and cholesterol. Phospholipid turnover is greater in the liver than elsewhere, and the liver is the main source of plasma phospholipids. Although the lipid concentration in normal liver is about 4%, no fat is visible in the cells with the usual staining techniques. In certain pathologic conditions, lipids accumulate in the liver and fat may be visualized in the cells following staining. Such accumulation must be due to either excessive synthesis or defective removal. In man, as in many animals, fatty livers appear to be associated with a deficiency of choline or methionine. Accumulation of fat in the liver cells is followed by fibrosis and cirrhosis.

Cirrhosis of the liver in man is often found in chronic alcoholism and is probably due to dietary deficiency. In active fatty alcoholic cirrhosis, choline administration has been shown to lead to a decrease in liver fat.[59] An increase in the rate of phospholipid turnover, following administration of 10 g. of choline or methionine, has been demonstrated in patients with cirrhosis who had evidence of fatty infiltration of the liver as shown by biopsy.[75] In animals, vitamin B_{12} and folic acid are intimately related to choline and methionine metabolism and are important in the prevention of fatty livers under certain conditions.[76] Whether these vitamins are related to accumulation of fat in the liver and cirrhosis in man remains to be ascertained. The value of high protein diets in the prevention and treatment of experimental dietary cirrhosis in animals is well established; there is much evidence that such is also true in man[76] (see also p. 521).

[75] (*a*) W. E. Cornatzer and D. Cayer, *J. Clin. Invest.* **29,** 534 (1950); (*b*) *ibid.* **29,** 542 (1950); (*c*) D. Cayer and W. E. Cornatzer, *Gastroenterology* **18,** 79 (1951); (*d*) W. E. Cornatzer, D. Cayer, and W. A. Lambeth, Jr., *J. Lab. Clin. Med.* **38,** 705 (1951); (*e*) D. Cayer, *in* Conference on Liver Injury, Transactions of the 10th Conference, New York, Josiah Macy, Jr. Foundation, 1951, p. 91; (*f*) Editorial, *J. Am. Med. Assoc.* **144,** 1566 (1950).

[76] (*a*) G. A. Goldsmith, W. G. Unglaub, and J. Gibbens, *Arch. Internal Med.* **90,** 526f (1952); (*b*) G. A. Goldsmith and J. Gibbens, *Arch. Internal Med.* **88,** 100f (1951).

8. Lipids and Atherosclerosis

Experimental atherosclerosis has been produced by feeding cholesterol to rabbits and rats, by administration of diethylstilbestrol which induces endogenous hypercholesteremia in chicks, and by administration of an anti-thyroid drug in association with a high cholesterol intake in dogs.[77] The relationship of cholesterol to atheromatous deposits in the blood vessels of man remains speculative. Diets used in the induction of atherosclerosis in animals contain much more cholesterol than is found in human diets. Furthermore, cholesterol is synthesized in the human body, and the serum level seems to be largely independent of the cholesterol intake. However, serum cholesterol concentration appears to be influenced by the fat content of the diet, decreasing when the diet is low in fat and increasing when fat is abundant.[72b] The following evidence suggests a relationship of cholesterol to the pathogenesis of human atherosclerosis: (1) early and extensive atherosclerosis occurs in persons with diabetes mellitus and nephrosis, diseases which are associated with hypercholesteremia; (2) serum cholesterol levels tend to be high in coronary atherosclerosis; and (3) cholesterol concentration in the blood and the incidence of atherosclerosis show a somewhat parallel increase with age. It has been postulated, from a genetic analysis of families with xanthomatosis, that early coronary atherosclerosis may be a hereditary disturbance of lipid metabolism manifested by elevation of serum cholesterol and transmitted as a dominant trait.[78] It has been suggested, also, that the ratio of the concentration of phospholipid to that of cholesterol is more important than the precise level of these lipids in serum. Phospholipid concentration increases with age in a manner similar to that noted for cholesterol. In patients with coronary atherosclerosis, the increase in phospholipid concentration with age has been found to be less than in normal subjects.[79]

Chylomicrons, which are large fat particles found in the blood after a fat meal, have also been implicated in the development of human atherosclerosis. These particles tend to be present in greater number, and to disappear from the blood more slowly, in old than in young subjects.[80] A qualitative difference in the physical size of lipid particles has been reported between atherosclerotic and non-atherosclerotic subjects.[81]

When oxylated blood from persons with myocardial infarction, vascular disease, and diabetes mellitus was pulsated for 72 hr., against strips of

[77] L. M. Katz, *Circulation* **5**, 101 (1952).

[78] D. Adlersberg, *Am. J. Med.* **11**, 600 (1951).

[79] (*a*) M. M. Gertler, S. M. Garn, and J. Lerman, *Circulation* **2**, 205 (1950); (*b*) M. M. Gertler, S. M. Garn, and E. F. Bland, *ibid.* **2**, 517 (1950).

[80] H. Necheles, *Am. J. Digestive Diseases* **18**, 229 (1951).

[81] W. J. Zinn and G. C. Griffith, *Am. J. Med. Sci.* **220**, 597 (1950).

normal aorta obtained at autopsy, examination with polarized light microscopy showed birefringent particulate material deposited in the aortic wall in 75% of tests; the changes resembled early atheromas. Blood from normal persons pulsated in a similar manner induced changes in only 5% of tests.[82]

Lipo-proteins in human serum have been studied by an ultracentrifugal flotation technique by Gofman and associates.[83] A strong correlation was reported between atherosclerosis and the concentration of certain of these lipoproteins, which are classified according to the rate at which migration occurs in terms of Svedberg flotation (Sf) units. Concentration of Sf 12–20 molecules was above normal in subjects with diabetes mellitus, hypothyroidism, hypertension, nephrosis, and myocardial infarction. The recurrence rate of myocardial infarction was significantly greater when levels of Sf 12–20 lipoproteins were found to be 100 mg. per 100 ml. than when levels were 50 mg. per 100 ml. Diets low in fat and cholesterol were associated with a decrease in Sf 12–20 lipoprotein concentration and a lower incidence of recurrent coronary occlusion. Jones and associates[71] reported that administration of heparin was followed by changes in lipoprotein patterns in human sera and was associated with relief of anginal pain in some subjects. Deficiency of heparin, or of some similar substance, was postulated as a causative factor in disturbances of lipid metabolism. As noted previously, heparin has been found to prevent normal alimentary lipemia, and this action may be related to the above observations.

Much more investigation must be undertaken before the role of cholesterol, chylomicrons, or Sf 12–20 lipoproteins in the development of human atherosclerosis can be ascertained. At present, the only practical dietary advice which may prove useful in either the prevention or treatment of atherosclerosis is avoidance of excessive caloric intake and of diets high in fat. Choline and inositol have not been shown to influence serum cholesterol concentration or atherosclerosis in animals or man.

9. Lipids and Gall Bladder Disease

Relationships between lipid metabolism and disease of the gall bladder, particularly cholelithiasis, have been suggested for many years but precise information is scanty. Bile contains cholesterol, and it is recognized that the gall bladder wall can absorb lipids. Deposits of cholesterol, in crystalline form, in the wall and lumen of the gall bladder are common and increase with advancing age. Factors which influence this deposition and subsequent stone formation are only partly understood. Cholesterol levels are

[82] S. M. Evans, H. K. Ihrig, J. A. Means, W. Zeit, and E. E. Haushalter, *J. Lab. Clin. Med.* **38,** 807 (1951).

[83] J. W. Gofman, H. B. Jones, T. P. Lyon, F. Lindgren, B. Strisower, D. Colman, and V. Herring, *Circulation* **5,** 119 (1952).

known to rise during pregnancy, and gall bladder disease is common in women who have had repeated pregnancies. The frequent occurrence of gall bladder disease in other conditions associated with hypercholesteremia is well known.

10. Fats in Therapeutic Nutrition

Fat is an important source of calories and, when oral feeding is contraindicated, a preparation of fat which could be given intravenously would be of great therapeutic value, since solutions of dextrose and amino acids can furnish only a limited supply of energy. Attempts have been made to prepare stable fat emulsions for intravenous use but, as yet, most emulsions develop thermogenicity after standing. Particle size and the emulsifying agent appear to be of great importance. The widespread use of fat in parenteral feeding must await further technicological development. However, it has been demonstrated that fat emulsions can be given to humans without causing demonstrable pathology; the fat disappears rapidly from the blood stream and appears to be utilized.

VI. Mineral Nutrition

At least thirteen mineral elements have been shown to be essential in human nutrition. These include calcium, magnesium, sodium, potassium, phosphorus, sulfur, and chlorine (principal mineral elements), and iron, copper, iodine, manganese, cobalt, and zinc (trace elements). It is uncertain as to whether fluorine, aluminum, and boron are essential nutrients.

Minerals are widely distributed in foods and, for many of them, there is little danger that deficiency will occur. Only the minerals which are of practical importance in clinical nutrition will be discussed in this section.

1. Water and Electrolytes

a. Water. Water has been recognized as indispensable since ancient times; man can live only a few days without it, the survival time being dependent largely upon external environmental temperature. The requirement for water is about 1 ml. for each calorie of food or about 2.5 l. daily for an average adult living in the temperate climate.[27] This water is usually supplied as follows: 1200 ml. as liquids, 1000 ml. from water in solid foods, and 300 ml. as water of oxidation. During fasting and under conditions of limited activity, a minimum intake of 0.8 l. of water is required. This is about 50% of the essential need, the remainder being furnished by oxidation and release of cellular water incident to protein utilization (Chapter 3). Requirement may be greatly increased when there is excessive loss of fluid, as in profuse sweating, vomiting, diarrhea, or polyuria.

The amount of water lost from the body in 24 hr. is approximately 2.5

l.—500 ml. from the skin as insensible perspiration, 500 ml. in the expired air, 100 ml. in the intestinal discharge, and 1500 ml. excreted by the kidney.

The kidney is the organ primarily involved in maintenance of the volume and composition of body fluids. Water and the normal constituents of body fluids are conserved or excreted according to physiologic need. The normal kidney is capable of excreting the catabolites formed in 24 hr. in as little as 500 ml. of urine.

Insensible loss of water from the skin is greater in the infant than in the adult. The urine volume per kilogram of body weight is also greater in the infant, owing to the higher metabolic rate and the resultant increase in catabolites which must be excreted. Since ability to concentrate urine is limited in the infant, depletion of body water can occur more rapidly and be more serious than in the adult.

The total amount of water in the body is about 60 % of total body weight, or 71 % of the weight of the fat-free body; of this, intracellular water amounts to 50 %, extracellular water 15 %, and plasma 5 %. Methods for determining the volume of plasma, of extracellular fluid, and of total body water are discussed in Chapter 3.

Exchange of fluid between the various compartments of the body is constant and extensive. In health, balance is maintained through a number of regulatory mechanisms. Water secreted into the gastrointestinal tract each day, as saliva, gastric, pancreatic, and intestinal juice and bile, may amount to as much as 8000 ml.; of this, all but 100 ml. is normally reabsorbed. The role of the kidney in the regulatory mechanism has already been mentioned. Hormones are also of great importance.

Excessive loss of water may occur under a number of circumstances, including any condition associated with high fever, vomiting, diarrhea or polyuria, and high external environmental temperature. As much as 14 l. may be lost daily in the sweat or urine, and as much as 8 l. from the gastrointestinal tract. Evaluation of the extent of dehydration is not easy. Frequent determinations of the volume of packed erythrocytes (hematocrit) and of plasma protein concentration or specific gravity are of considerable assistance. Measurements of changes in body weight are also useful in estimating the degree of water depletion and in following the response to therapy.

Black, McCance, and Young[84] studied water depletion in human volunteers. Intense thirst was the initial and most prominent symptom. Other findings were dryness of the mouth, difficulty in swallowing, oliguria, loss of weight proportional to water deficit, some weakness, a change in temperament, confusion, and hallucinations. Plasma volume decreased slightly; serum sodium increased and serum potassium decreased. In profound

[84] D. A. K. Black, R. A. McCance, and W. F. Young, *J. Physiol.* **102,** 406 (1944).

dehydration, prostration and collapse occur, and anemia, azotemia, and serious electrolytic disturbances are observed.

b. Sodium. The minimal requirement of salt is about 1 g. daily. The average intake in this country is 10 to 15 g. daily, which meets the requirements for water intake up to 4 l. A liberal allowance for an adult would be 5 g., except under conditions of profuse sweating, in which instance 1 g. of salt should be ingested for each liter of water in excess of 4 l. In hot environments, prior to acclimatization, sweat may contain 2 to 3 g. of salt per liter; after acclimatization, 0.5 g. per liter.

Sodium and chloride comprise the bulk of the electrolytes in plasma and interstitial fluid. Sodium constitutes 90% of the total base of the plasma, the normal concentration being 140 meq. per liter. The normal concentration of chloride is 104 meq. per liter. The sodium ion plays an important role in the maintenance of acid-base equilibrium and in the maintenance of osmotic pressure, which depends largely on total base. Cations in blood, other than sodium, are calcium, potassium, and magnesium; anions, other than chloride, are bicarbonate, protein, and small amounts of organic acid. The pH is usually regulated by the relative amounts of chloride and bicarbonate. Acidosis and alkalosis are encountered in many diseases of man, but these problems belong in the field of clinical medicine rather than nutrition and will not be discussed here.

The amount of salt excreted by the kidney is dependent on the amount ingested in the absence of significant sweating. When the diet is very low in salt, almost no sodium or chloride is excreted, owing to the efficient regulatory mechanism controlling reabsorption by the renal tubules. This activity is controlled in part by adrenal cortical hormones. In adrenal insufficiency, large amounts of salt are lost in the urine; many of the features of Addison's disease are attributable to salt depletion. In hyperfunction of the adrenal cortex, salt is retained and there is an associated retention of water.

The influence of sodium chloride on water retention is illustrated in many medical conditions. Diets low in salt assist in the control of edema in acute hemorrhagic nephritis, nephrosis, congestive heart failure, hypoproteinemia, and cirrhosis of the liver. Recently diets very low in salt (sodium content, 200 to 300 mg.) have been used in the therapy of hypertension, with beneficial results in a percentage of cases. The role of sodium in disease states has been reviewed recently by Danowski.[85]

When diets very low in salt are used for prolonged periods, signs of salt depletion may make their appearance. The signs are gradual in onset and include weakness, excessive fatigue, lassitude, apathy, anorexia, nausea, and muscle cramps.[86] Thirst is absent. Urine volume is normal until late.

[85] T. S. Danowski, *Am. J. Med.* **10,** 468 (1951).

[86] H. L. Marriot, *Brit. Med. J.* **1,** 246, 328 (1947).

Salt is absent from the urine, and the concentrations of sodium and chloride in the plasma are diminished. Plasma volume is decreased. There is hemoconcentration, increased blood viscosity, and increased blood urea nitrogen. The blood pressure falls, and peripheral circulatory collapse is the mode of death.

Combined water and salt depletion is a syndrome commonly encountered in clinical medicine. Manifestations are those of both sodium and water deficiency and include anorexia, nausea, weakness, muscle cramps, peripheral vascular collapse (sodium deficiency), thirst, and oliguria (water deficiency). Ingestion of water alone increases the severity of symptoms. "Heat cramps," which occur in persons working in very hot environments, are an example of this syndrome. Deficiency may be prevented by the addition of salt to drinking water to make a final concentration of 0.1%.

c. Potassium. The minimum daily requirement of potassium is probably similar to that of sodium. A normal person can ingest as much as 6 g. of potassium daily without danger, but this amount may be toxic in subjects with renal damage or adrenal insufficiency. A balance between the intake and output of potassium is maintained by the kidney in a manner resembling that for sodium, except during fasting and after damage to tissues when this ion is released from the cell and is excreted in the urine.

Potassium is found largely in the intracellular compartment, the concentration in extracellular fluid being only 5% of that within the cells. Studies with K^{42} have indicated a constant exchange of potassium between intra- and extracellular phases. However, concentration of potassium in serum does not necessarily reflect concentration within the cells. In normal subjects, the average concentration of potassium in serum was found to be 4.18 meq. per liter by Elliott and Holley.[87] Deane and Smith,[88] using K^{42}, found the concentration of intracellular potassium to range from 96 to 125 meq. per liter with a mean of 112 meq.

Potassium has an important role in metabolism as evidenced by the close relationship between cell growth and breakdown and potassium accumulation and loss (see Chapter 3). The excitability of nerve tissue and the transmission of impulses is profoundly influenced by the concentration of potassium in the cells and extracellular fluids. Potassium plays an important role in the contractility of all types of muscle. Paralysis of striated muscle occurs when the level of serum potassium is less than 2.05 to 2.56 meq. per liter.[89] Smooth muscle is also affected in potassium deficiency, and paralytic ileus may result. The activity of cardiac muscle is affected

[87] H. C. Elliott, Jr., and H. L. Holley, *Am. J. Clin. Path.* **21,** 831 (1951).

[88] N. Deane and H. W. Smith, *J. Clin. Invest.* **31,** 197 (1952).

[89] H. E. Martin, T. B. Reynolds, E. N. Snyder, C. J. Berne, R. E. Homann, Jr., H. A. Edmondson, N. Blatherwick, I. Fields, M. Wertman, and L. Westover, *J. Am. Med. Assoc.* **147,** 24 (1951).

by potassium concentration. This is reflected in the electrocardiogram which shows changes in both hypo- and hyperpotassemia. Potassium is involved in certain enzyme reactions. Greisheimer states that "its transport across cell boundaries against existing ionic concentration gradients seems linked with the carbohydrate phosphorylation cycle."[90]

Primary dietary deficiency of potassium is not observed, but depletion secondary to some pathologic condition is often encountered in clinical medicine. Common causes are intravenous alimentation without including potassium in the mixture, increased loss of potassium in the urine due to altered renal tubular function as in alkalosis, shock, and renal disease, particularly the lower nephron syndrome, and following administration of diuretics or adrenal cortical hormones. Potassium loss may be excessive in gastrointestinal diseases when intestinal secretions are lost in large amounts. Negative nitrogen balance is associated with loss of potassium which is released from cells when protein is broken down. Potassium deficiency is common in diabetic acidosis during insulin therapy. When glucose metabolism is restored, potassium shifts from the extracellular compartment into the cells, thus reducing the concentration in serum and interstitial fluid.

Signs of potassium deficiency are primarily those of decreased muscular irritability and disturbances of conduction and contractility in the heart muscle. Muscular weakness or paralysis are characteristic and may be accompanied by lethargy and coma. Cardiac dilatation, hypertension, congestive failure, and cardiac arrest may occur. Death may result from respiratory or cardiac failure or from paralytic ileus.

Darrow[91] has reviewed the findings in hyperpotassemia. These include, mental confusion, numbness and tingling of the extremities, weakness, pallor, cold skin, disturbances in cardiac rhythm, and peripheral collapse. This condition may be the result of extensive tissue breakdown, terminal renal failure, adrenal insufficiency, advanced dehydration, or the administration of excessive amounts of potassium.

2. Calcium

Ten years ago Macy[92] stated that "neither the most satisfactory level of calcium intake nor the optimal retention of calcium at any physiologic age or stage of man's development is known." This is still true today. The daily recommended allowances of the Food and Nutrition Board are 1.0 g. for adults, 1.5 g. during pregnancy, and 2.0 g. during lactation. The recommendation for infants and children is 1 g. daily until adolescence, when 1.2 to 1.4 g. is suggested. These allowances are based largely on findings in

[90] E. M. Greisheimer, *J. Am. Med. Women's Assoc.* **12,** 448 (1952).

[91] D. C. Darrow, *New Engl. J. Med.* **242,** 978, 1014 (1950).

[92] I. G. Macy, Principal Mineral Elements in Nutrition, *in* ref. 73, p. 99.

metabolic balance studies and, in adults, are thought to cover the needs of all but one in one hundred persons.[27] Many persons can undoubtedly adjust satisfactorily to lower intakes. Unfortunately, no procedure for detecting potential calcium deficiency is available, nor are there definitive criteria for estimating the adequacy of calcium nutrition. Long-term studies in rats in Sherman's laboratory indicate beneficial results from an abundant calcium intake.

Ninety-nine per cent of the calcium is in the bones and teeth. The calcium in bones serves not only as a structural component but as a reserve store which can be mobilized for maintenance of serum calcium concentration. The small amount of calcium in serum and in other body fluids has important functions, particularly in neuromuscular excitability and probably also in the blood-clotting mechanism.

Absorption of calcium is influenced by many factors. The maintenance of normal gastric acidity facilitates absorption, and vitamin D enhances it. Absorption is increased by protein, lactose, and citrates in the diet. Absorption is decreased by the presence of phytates (in cereals), oxalates, and fatty acids, since in each instance insoluble calcium salts are formed. The ratio of calcium to phosphorus in the diet influences absorption. Calcium enters the gastrointestinal tract not only in food but as a component of digestive secretions; as much as 0.3 to 0.8 g. daily may be secreted into the gastrointestinal tract. Reabsorption of this calcium is influenced by the factors just enumerated.

Calcium is absorbed directly into plasma where a concentration of 9 to 11 mg. per 100 ml. is maintained. Calcium is present in blood in both ionized (50 to 60%) and non-ionized form, the latter bound to protein (Chapter 21). The concentration of protein-bound calcium, and hence of total blood calcium, varies with the concentration of serum proteins. The ionized calcium fraction is the one responsible for prevention of tetany.

The process of deposition of calcium in bone is not fully understood, but alkaline phosphatase, vitamin D, vitamin C, and probably other factors are of great importance. Removal of mineral from bone and maintenance of normal serum calcium concentration is dependent largely on the activity of the parathyroid hormone. Albright considers that the primary effect of the parathyroid hormone is an increase in phosphorus excretion (Chapter 21). This is followed by a decrease in serum phosphorus and resorption of phosphorus and calcium from bone. Other hormones also influence utilization of calcium and phosphorus (Chapter 21). In hyperthyroidism, excretion of calcium and phosphorus is increased and osteoporosis may develop. The steroid hormones, e.g., estradiol and testosterone, decrease calcium and phosphorus excretion and may therefore be useful in the therapy of postmenopausal and senile osteoporosis.

Approximately 80% of calcium is excreted in the feces, and 20% in the urine (Chapter 21). Excretion of calcium in the urine is influenced by many factors including the calcium intake, the skeletal size, the acid-base balance, and "endogenous factors" which are presumably the resultant of various body hormones.[93] Urinary calcium excretion varies widely in normal subjects at any given level of intake; maximum excretion may be ten times minimal.

Calcium deficiency is manifested by tetany with characteristic carpopedal spasm and, at times, laryngospasm and convulsive seizures. When tetany is present, the concentration of serum calcium is usually less than 8 mg. per 100 ml. Tetany occurs in osteomalacia and often accompanies rickets, sprue, the celiac syndrome, and other steatorrheas.

3. Phosphorus

Phosphorus is abundantly supplied in foods, and deficiency in man is an unlikely possibility. The daily allowance of phosphorus should probably be at least equal to that of calcium for children and for women during pregnancy, and perhaps 1.5 times that of calcium for other adults.[27]

Phosphorus is more readily absorbed from the intestinal tract than calcium. Approximately two-thirds of the phosphorus excreted is found in the feces, one-third in the urine. Between 10 and 20% of phosphorus is found in tissues other than bone, and this phosphorus appears to have metabolic priority.[93] The mechanisms which regulate deposition and release of phosphorus from bone are the same as those for calcium. Phosphorus in soft tissues plays a very vital role in many metabolic processes. The importance of adenosine triphosphate (ATP) in energy transfer systems has been discussed previously (p. 203, Chapter 16). Phosphorylation appears to be essential for the absorption of a number of nutrients, e.g., fatty acids and glucose. The phosphate radical is bound to proteins, fatty acids, carbohydrates, and enzymes. Phosphate is the chief inorganic anion of intracellular fluid, and phosphates of extracellular fluid participate in acid-base regulation. The inorganic phosphate level of blood ranges from 2 to 4 mg. per 100 ml. in adults and 3 to 5 mg. per 100 ml. in children.

Phosphorus metabolism, like that of calcium, is disturbed in rickets, osteomalacia, and diseases of the parathyroid and thyroid glands. The steroid hormones likewise influence metabolism of this element (Chapter 21).

4. Iodine

Iodine is essential for the formation of the thyroid hormone which is an important regulator of energy metabolism. In iodine deficiency, simple or so-called endemic goiter develops. The human requirement of iodine is

[93] G. Stearns, Human Requirement of Calcium, Phosphorus and Magnesium, *in* ref. 73, pp. 71ff.

about 1 μg. per kilogram of body weight daily, for basal needs, and 2 μg. additional for maintenance of normal metabolic activity.[94] The requirement for an adult weighing 70 kg. would thus be about 200 μg. More iodine is needed during the period of growth and during pregnancy. The Food and Nutrition Board suggests a daily allowance of 0.002 to 0.004 mg. per kilogram of body weight. This requirement may be met by the regular use of iodized salt.[27]

Recent studies with radioactive iodine have added much to our knowledge of iodine metabolism. Iodine is rapidly and probably completely absorbed from the small intestine at an exponential rate which varies with the level of thyroid activity (Chapter 22).

Iodine is taken up by the thyroid gland where it is converted to diiodotyrosine and thyroxine. The thyroid hormone is stored in the form of thyroglobulin. In normal persons, the rate of iodine uptake by the thyroid is proportional to the blood concentration and averages about 2.5% per hour. In hyperthyroidism, the uptake may reach 20% per hour. The release of thyroid hormone from the gland is under pituitary regulation.

The adult human body contains 20 to 50 mg. of iodine, of which 20 to 40% is found in the thyroid. The normal adult thyroid contains about 8 to 10 mg. of iodine. In endemic goiter, the gland may contain less than 1 mg. (Chapter 22). The ovaries contain a higher concentration of iodine than any tissue other than the thyroid.

Iodine is found in the blood as "inorganic" and protein-bound iodine, the latter fraction probably representing the circulating thyroid hormone. The total iodine content of normal whole blood is about 8 to 12 μg. per 100 ml. (range 3 to 30 μg.); protein-bound iodine varies from 3 to 8 μg. per 100 ml., with a mean of 5 to 6 μg. (Chapter 22). The level of protein-bound iodine is increased in pregnancy and in hyperthyroidism and decreased in hypothyroidism.

The urinary excretion of iodine is influenced by the amount in the diet, the activity of the thyroid gland, and probably other factors.[95] The kidney removes iodine from the blood at a rate of about 6% per hour (Chapter 22). Urinary excretion in non-goitrogenous regions has ranged from 72 to 343 μg. daily; in goitrogenous regions, from 27 to 64 μg. daily.[95]

Iodine is essential for normal growth, since it is a requisite part of the thyroid hormone. In cretinism, or congenital hypothyroidism, which is observed in goitrogenous areas, growth may be stimulated by administration of dessicated thyroid. Iodine also plays a role in human and animal fertility and in lactation.[96] Simple goiter and cretinism are still prevalent in many areas of the world; prophylaxis could be obtained through the universal use of iodized salt.

[94] G. M. Curtis and M. B. Fertman, Iodine in Nutrition, *in* ref. 73, p. 133.
[95] G. M. Curtis and M. B. Fertman, Iodine in Nutrition, *in* ref. 73, p. 122.
[96] G. M. Curtis and M. B. Fertman, Iodine in Nutrition, *in* ref. 73, p. 130.

5. Fluorine

Whether fluorine is an essential human nutrient is unknown. Fluorine is of importance to human health because of its influence on the incidence of dental caries. It has been shown that the incidence of caries is low in children who have received water containing fluorine in amounts of 1 to 1.5 p.p.m. during the period of calcification and maturation of teeth. A high incidence of dental caries has been observed in children who have received water containing less than 1 p.p.m. of fluorine. When water contains more than 2.5 p.p.m. of fluorine, mottling of the enamel of the teeth is observed. In a number of communities in the United States, addition of fluorine to the water supply (1 p.p.m.) has resulted in a decrease in caries incidence. Topical application of fluorine to teeth has also been found beneficial.

There appears to be little danger of toxicity when the fluorine content of water is 1 p.p.m. This concentration should supply about 1 to 2 mg. of fluorine daily. The average American diet contains only 0.2 to 0.3 mg. of fluorine. Balance studies in man indicate that, if the intake is not over 4 to 5 mg. daily, very little fluorine is retained (Chapter 22). Fluoride intakes of more than 5 mg. daily for many years appear to be necessary for intoxication to occur.

6. Iron

Iron is essential for the formation of hemoglobin, for the production of certain essential respiratory enzymes, i.e., cytochromes, peroxidases, and catalases, and for the formation of myoglobin. The latter two functions have priority in the body.

Absorption of iron is never very efficient and is influenced by many factors. Hydrochloric acid, pepsin, proteins and their digestion products, and ascorbic acid appear to be concerned in the reduction of iron to the ferrous state in which form it is absorbed. Absorption takes place largely in the duodenum (Chapter 20). Iron absorption may be increased by administration of ascorbic acid and decreased by phosphates and perhaps by phytates.

Studies with radioactive iron have indicated that absorption of iron salts is dependent on the need for iron. Normal persons absorb only small amounts, whereas persons with iron deficiency take up much larger quantities. In growing children, absorption appears to be related to the requirement for this element.[97] In the latter part of pregnancy, absorption is greatly increased, presumably due to the need of the fetus for iron.[98]

[97] W. J. Darby, P. F. Hahn, M. M. Kaser, R. C. Steinkamp, P. M. Densen, and M. B. Cook, *J. Nutrition* **33,** 107 (1947).

[98] P. F. Hahn, E. L. Carothers, W. J. Darby, M. Martin, C. W. Sheppard, R. O. Cannon, A. S. Beam, P. M. Densen, J. C. Peterson, and G. S. McClelland, *Am. J. Obstet. Gynecol.* **61,** 477 (1951).

Recent studies of Moore and Dubach[99] have shown that normal subjects absorb about 10% of iron from foodstuffs. Ascorbic acid enhanced absorption, but administration of hydrochloric acid and antacids were without effect. Subjects with iron-deficiency anemia showed no increased efficiency of iron absorption from foods. This finding is in contrast to studies of absorption of inorganic iron salts, in which subjects with iron deficiency showed greater absorption than normal subjects.

It has been postulated that iron absorption is regulated by an iron-containing protein, ferritin, in the intestinal mucosa.[100] Iron is thought to attach itself to the iron-free form of this protein, apoferritin, to produce ferritin. Iron is released from ferritin at the blood stream end of the mucosal cell (Chapter 20).

Iron is transported in the serum in ferric form attached to a specific globulin, siderophillin (Chapter 20). The normal level of serum iron is about 100 μg. per 100 ml. (range 80 to 180 μg.); the maximum iron binding capacity of serum is 300 to 360 μg. per 100 ml. In iron-deficiency anemia, serum iron is usually less than 50 μg. per 100 ml. Serum iron is low in the anemia associated with infections and is high in untreated pernicious anemia. Recent investigation suggests that the hypoferremia of infection may be mediated, at least in part, by the adrenal cortex.[101] Similar hypoferremia has been produced by a number of agents which produce varying degrees of stress and by administration of adrenal cortical hormones. Studies of the rate of disappearance of radioactive iron from plasma have indicated that the turnover rate of Fe^{59} furnishes a reliable index of the rate of erythrocyte formation.[102]

The total amount of iron in the body is 4 to 5 g., a large percentage of this being present in hemoglobin. Iron is stored in the form of ferritin and hemosiderin in the liver, spleen, and bone marrow. Evaluation of the extent of iron stores may be made by estimating the amount of hemosiderin in bone marrow. The daily destruction and replacement of hemoglobin is about 7 to 8 g., which is the equivalent of a daily turnover of about 25 mg. of iron (Chapter 20).

Very little iron is excreted from the body; only about 0.1 mg. daily is found in the urine and about 1 mg. daily in the feces.

[99] C. V. Moore and R. Dubach, *Trans. Am. Assoc. Physicians* **64,** 245 (1951).

[100] S. Granick, *J. Biol. Chem.* **164,** 737 (1946).

[101] (*a*) G. E. Cartwright, L. P. Hamilton, C. J. Gubler, N. M. Fellows, H. Ashenbrucker, and M. M. Wintrobe, *J. Clin. Invest.* **30,** 161 (1951); (*b*) *Nutrition Revs.* **9,** 250 (1951).

[102] (*a*) R. L. Huff, T. G. Hennessy, R. E. Austin, J. F. Garcia, B. M. Roberts, and J. H. Lawrence, *J. Clin. Invest.* **29,** 1041 (1950); (*b*) R. L. Huff, P. J. Elmlinger, J. F. Garcia, J. M. Oda, M. C. Cockrell, and J. H. Lawrence, *ibid.* **30,** 1512 (1951); (*c*) L. R. Wasserman, I. A. Rashkoff, D. Leavitt, J. Mayer, and S. Port, *ibid.* **31,**, 32 (1952).

In view of the conservation of iron by the body, estimation of requirement is difficult. Heath and Patek[103] have calculated, from estimates of the iron required to build body tissue and to replace loss, that the total amount needed by males from birth to 21 years is 3.148 g., and by females from birth to 47 years, 12.222 g. These figures do not represent dietary requirement. Moore and Dubach[99] have suggested that the normal adult male must assimilate 0.5 to 1 mg. of iron daily to maintain balance, the adult female about 1 to 2 mg. daily. If only 10% of the iron in food is absorbed, the diet should furnish 10 to 15 mg. of iron daily. Recommended allowances of the Food and Nutrition Board are of this order (Table 1).

Iron-deficiency anemia may be nutritional in origin or the result of loss of blood, especially chronic repeated hemorrhage. Dietary deficiency of iron occurs during periods of growth, particularly in infancy and adolescence, and during pregnancy and lactation. The infant at birth has a certain amount of stored iron, the quantity being related to the maternal supply during pregnancy.

Iron-deficiency anemia is characterized by a proportionately greater decrease in hemoglobin than in erythrocyte count. The cells are hypochromic, either normal or small in size; the color index and mean corpuscular hemoglobin concentration are low. The bone marrow shows an increase in the number of normoblasts. Clinical findings include easy fatigability, anemia, pallor, and, in children, retardation of growth. In some subjects, especially in women, dysphagia, glossitis, and angular stomatitis may be present. Koilonychia, spoon-shaped finger nails, may be observed in anemia of great chronicity.

Iron may be toxic, particularly if given intravenously, since there is no significant excretory pathway. The total dose of intravenous iron should not exceed 25 to 40 mg. of elemental iron for each 1% deficit in hemoglobin, and not more than 100 mg. should be given as a single dose (Chapter 20).

7. Copper

No definite evidence for the occurrence of copper deficiency in man has been presented, and human requirement is unknown. The diet usually furnishes 2 to 4 mg. of copper daily.[104] Studies in man indicate that 2 mg. daily will maintain an adult in balance.[104] The Food and Nutrition Board suggests a daily allowance of 1 to 2 mg. of copper for an adult, and 0.05 mg. per kilogram of body weight for infants and children. The copper content of whole blood is between 90 and 150 μg. per 100 ml., equally divided between cells and plasma.[104] There is little evidence that the administration of copper is of assistance in the therapy of hypochromic microcytic anemia.

[103] C. W. Heath and A. J. Patek, Jr., *Medicine* **16**, 267 (1937).

[104] W. J. Darby, Iron and Copper, ref. 73, p. 108.

8. Cobalt

Cobalt has been shown to be a constituent of vitamin B_{12}; whether man needs cobalt only in this form is unknown (Chapter 22). Cobaltous chloride has been administered in an attempt to influence anemias associated with chronic infection and advanced neoplasia. Although stimulation of erythropoiesis has been reported in some patients, findings are inconclusive.

VII. Fat-Soluble Vitamins

1. Vitamin A

a. Absorption and storage of vitamin A. Vitamin A, an unsaturated alcohol, is found solely in animal tissues, while the provitamin, carotene, is present in large amounts in green and yellow vegetables and fruits. In man, the major portion of the dietary supply of vitamin A is furnished by carotene which is converted to vitamin A ester, probably in the intestinal wall,[105] from which area it passes via the lymphatics to the blood stream and finally to the liver. The amount of carotene biologically available from different food sources varies greatly; in certain vegetables, the percentage absorption ranged from 34 to 69% as tested by the rat growth bioassay method.[106] At present, the availability of carotene in foods is being investigated in human subjects in several laboratories. The conversion of carotene to vitamin A may be defective in diabetes mellitus and in hypothyroidism; in these diseases evidence of both vitamin A deficiency and hypercarotenemia is common.

Absorption of vitamin A appears to be facilitated by the simultaneous absorption of fat. Poor absorption has been demonstrated in the steatorrheas, such as sprue and celiac disease, and in infectious hepatitis.[107] Aqueous emulsions of vitamin A have been found to be more adequately absorbed than oily preparations, particularly in conditions associated with poor absorption of fat.[107b, 108] Liquid petrolatum, administered with meals, interferes considerably with absorption of carotene and to a less extent with that of vitamin A.

Vitamin A is stored chiefly in the liver, the median level in normal subjects being 320 I.U. per gram of liver, according to Moore (Chapter 8). Glover and Morton[109] have suggested that vitamin A may be stored in two sites in the liver, the Kupffer cells and the true storage cells, and that

[105] S. K. Kon and S. Y. Thompson, *Brit. J. Nutrition* **5,** 114 (1951).

[106] E. C. Callison, E. Orent-Keiles, R. Frenchman, and E. G. Zook, *J. Nutrition* **37,** 139 (1949).

[107] (*a*) W. J. Darby, E. Jones, H. F. Warden, and M. M. Kaser, *J. Nutrition* **34,** 645 (1947); (*b*) T. Moore and I. M. Sharman, *Brit. J. Nutrition* **5,** 119 (1951).

[108] (*a*) J. M. Lewis and S. Q. Cohlan, *Med. Clinics N. Amer.* **34,** 413 (1950); (*b*) J. M. Lewis, S. Q. Cohlan, and A. Messina, *Pediatrics* **5,** 425 (1950).

[109] J. Glover and R. A. Morton, *Biochem. J.* **43,** xii (1948).

only the latter contribute to the level of functional vitamin A in the blood. They also postulate that the amounts reaching the Kupffer cells increase with the rate at which vitamin A enters the blood stream. If these hypotheses are correct, there may be an optimal rate of ingestion of vitamin A.

b. Blood levels of carotene and vitamin A. The level of carotene in the blood is related to recent dietary intake and, in subjects receiving adequate diets, ranges from 50 to 150 μg. per 100 ml., often being above 100 μg. The concentration of vitamin A in the blood in normal subjects in the post-absorptive state is in the neighborhood of 30 μg. (100 I.U.) per 100 ml. Mean levels as high as 48 μg. (160 I.U.) per 100 ml. have been reported in certain studies.[107b] Knowledge of factors which control mobilization of vitamin A from the liver is limited. The blood level decreases in a number of febrile illnesses and returns to normal concomitant with recovery without vitamin A having been administered. It is obvious that low levels do not necessarily imply exhaustion of liver reserves. Plasma vitamin A is decreased in advanced cirrhosis of the liver, in the steatorrheas, and in uncomplicated vitamin A deficiency, and in these instances vitamin A reserves in the liver are reduced.[110]

Vitamin A is not found in normal human urine but may be present in the urine of patients with pneumonia, obstructive jaundice, or chronic nephritis. An abnormality in vitamin A metabolism has been reported in the nephrotic syndrome; high plasma levels follow administration of vitamin A, perhaps owing to failure of utilization or storage by the liver.[107b]

c. Metabolic functions of vitamin A. Vitamin A forms an integral part of rhodopsin, which is necessary for vision in dim light. When the dark-adapted retina is exposed to light, rhodopsin is changed through several steps to retinene, which is vitamin A aldehyde plus protein. Rhodopsin may be regenerated from retinene plus "opsin" (a retinal protein), or from vitamin A, opsin, and enzymes present in the retina and pigmented layers of the eye which are activated by cozymase (see Chapter 8).

A second function of vitamin A is the maintenance of normal epithelium, but the precise role that this vitamin plays in the metabolic processes of epithelial cells is unknown. The changes which occur in vitamin A deficiency have been studied extensively by Wolbach and associates and consist essentially of atrophy of the epithelium with reparative proliferation of the basal cells and differentiation into stratified keratinizing epithelium (Chapter 13) regardless of the original type of epithelial cell. This keratinizing metaplasia has been observed in many animal species, including man.

Vitamin A has been found to influence bone growth in certain animal species and, perhaps secondarily, to cause degeneration of nervous tissue

[110] G. A. Goldsmith, *Federation Proc.* **8**, 553 (1949).

as a result of pressure (see Chapter 13). In animals, vitamin A deficiency during pregnancy has resulted in a number of congenital malformations in the offspring. Little is known about application of these aspects of vitamin A metabolism to man. In children who received excessive amounts of vitamin A, abnormalities of bone have been observed, including hyperostoses and periosteal elevations. Other findings in hypervitaminosis A include anorexia, loss of weight, low-grade fever, hepatomegaly, splenomegaly, sparseness of hair, and hypoplastic anemia.[111] The serum level of vitamin A is elevated. It has been postulated that the syndrome may be related to hepatic dysfunction.

d. Vitamin A deficiency. The outstanding manifestations of vitamin A deficiency in man are night blindness and xerophthalmia. In the latter, keratinization of the cornea and conjunctiva are followed frequently by infection, panophthalmitis, and blindness. Xerophthalmia and night blindness are lesions common to many animal species. Bitot's spots, triangular thickenings of the conjunctiva with accumulation of white foam-like epithelial cells situated lateral to the cornea, occur in vitamin A deficiency but are rarely observed in temperate climates. They must be differentiated from other common lesions, unrelated to malnutrition, such as pingueculae.[112]

It seems probable that xerosis of the skin with hyperkeratinization, so-called toad skin or phrynoderma, also represents vitamin A deficiency in man. The relationship of vitamin A to mild skin changes such as localized areas of keratinization of the hair follicles, follicular hyperkeratosis, is less well documented. Vitamin A deficiency may be responsible for the lesions in some instances but certainly not in all. Unequivocal follicular keratotic changes have not been produced experimentally in man.[113]

In monkeys and rats, continuous cornification of the vaginal smear is evident in vitamin A deficiency. The influence of vitamin A on the vaginal mucosa in humans is uncertain, although administration of this vitamin has brought about improvement of senile vaginitis (Chapter 13). Keratinization of the uterine epithelium in children with vitamin A deficiency has been reported.

Epithelial changes in the urogenital system in animals have been associated with an increase in the incidence of urinary calculi. There is no precise information linking vitamin A deficiency to formation of renal calculi in man.

[111] (*a*) M. B. Sulzberger and M. P. Lazar, *J. Am. Med. Assoc.* **146,** 788 (1951); (*b*) I. E. Rineberg and R. J. Gross, *ibid.* **146,** 1222 (1951); (*c*) G. Bair, *ibid.* **146,** 1573 (1951).

[112] Z. A. Leitner, *Brit. Med. J.* **1,** 1110 (1951).

[113] E. M. Hume and R. A. Krebs, Vitamin A Requirement of Human Adults: An Experimental Study of Vitamin A Deprivation in Man, Medical Research Council Special Report Series 264, His Majesty's Stationery Office, London, 1949.

Changes in the epithelium of the respiratory tract with blockage of the small bronchioles, bronchiectasis, and atelectasis has been reported in human infants.[114] Keratinizing metaplasia has also been noted in the epithelium of the urinary tract, accessory sinuses, and pancreas.

Diagnosis of vitamin A deficiency is dependent on a history of an inadequate diet and the presence of abnormal dark adaptation and epithelial changes. Carotene levels in blood may corroborate the dietary history, while low levels of vitamin A in blood suggest deficiency, if other causes of such decrease can be ruled out.

e. Vitamin A requirement. Vitamin A requirement in animals appears to be related to body weight. The requirement for prevention of night blindness in six species (cattle, sheep, swine, horse, rabbit, and dog) and for prevention of cornification of the vaginal epithelium in the rat was found to be about 20 I.U. per kilogram of body weight daily if furnished by preformed vitamin A (cod-liver oil).[115, 115a] If carotene in food was used to supply vitamin A, almost five times as much was needed. Evidence also indicated that about three times the minimal vitamin A level and five times the minimal carotene level was needed for significant storage of vitamin A and for normal reproduction.[115] A recent study of experimental vitamin A deficiency in adult men indicated a minimal protective dose of 1300 I.U. of vitamin A daily or approximately 20 I.U. per kilogram of body weight.[113] Earlier studies[27] suggested a requirement of 25 to 55 I.U. per kilogram of body weight per day to permit normal dark adaptation. The Committee on Nutrition of the British Medical Association[29] recommended 2500 I.U. of vitamin A daily for adults as a satisfactory allowance for maintenance of good nutrition, and three times this amount, or 7500 I.U. daily, if carotene in food was the sole source of vitamin A. In a mixed diet supplying two-thirds of the vitamin A as carotene, 5000 I.U. daily was suggested as adequate. The amount proposed in the Recommended Dietary Allowances of the Food and Nutrition Board of the National Research Council in this country is 5000 I.U. daily for the normal adult receiving a mixed diet.[27] Allowances for children and for pregnancy and lactation are indicated in Table 1. In the Canadian Dietary Standard,[28] a

[114] (*a*) K. D. Blackfan and S. B. Wolbach, *J. Pediat.* **3,** 679 (1933); (*b*) L. K. Sweet and H. J. K'Ang, *Am. J. Diseases Children* **50,** 699 (1935).

[115] H. B. Guilbert, C. E. Howell, and G. H. Hart, *J. Nutrition* **19,** 91 (1940).

[115a] In these studies, it was estimated that 0.21 μg. of vitamin A approximated 1 I.U., and hence vitamin A requirement was considered to be 4 μg. per kilogram of body weight. The new international standard for vitamin A defines one unit as equal to 0.34 μg. of vitamin A acetate or 0.30 μg. of the alcohol.[116] The vitamin A requirement would therefore be 6 μg. per kilogram of body weight.

[116] E. M. Hume, *Brit. J. Nutrition* **5,** 104 (1951).

vitamin A intake of 72 I.U. of carotene per kilogram of body weight per day is recommended. Although this amount may prevent night blindness or skin lesions, it seems a meager supply for storage and perhaps also for growth and other functions, in view of the data obtained in animal studies.

2. Vitamin D

Vitamin D consists of a group of sterol derivatives produced by ultraviolet irradiation of the corresponding precursors. The two important members of this group are vitamin D_2 or calciferol, which is activated ergosterol, and vitamin D_3, which is activated 7-dehydrocholesterol. The latter is the compound formed in the skin on exposure to sunshine or ultraviolet rays.

a. Absorption, storage, and functions of vitamin D. The amounts of vitamin D in the human diet are very small, and the quantity formed in the skin by exposure to sunshine is often inadequate, especially in the winter months in this country. Medicinal preparations are accordingly widely used to satisfy requirements. Since vitamin D is fat-soluble, it is absorbed with the fat of the diet and absorption is impaired in conditions which decrease fat absorption, such as the absence of bile from the intestinal tract. Vitamin D is stored in the liver and to a small extent in skin, brain, spleen, and bones.

The functions of vitamin D have not been completely elucidated. There is good evidence that this vitamin increases absorption of calcium from the intestine, and it may increase phosphorus absorption as well. In addition, the excretion of phosphorus by the kidney is decreased by increasing resorption in the renal tubules. This effect may be related, in part at least, to increase in serum calcium, with resultant diminution in the activity of the parathyroid glands. In vitamin D deficiency, excretion of calcium and phosphorus in the stool is increased, excretion of calcium in the urine is decreased, and that of phosphorus increased. In addition to maintaining levels of calcium and phosphorus in the blood suitable for deposition of bone, vitamin D is thought to have a direct effect at the site of deposition.[117] Vitamin D may aid in the conversion of organic to inorganic phosphorus, and alkaline phosphatase may play a part in this picture. A recent study by Zetterstrom and associates,[118] using a water-soluble phosphorylated vitamin D_2, indicates that vitamin D activates alkaline phosphatases of bone, kidney, and intestine. This work, if substantiated, provides an enzymatic explanation of the role of vitamin D in the prevention of rickets. Alkaline phosphatases are thought to have a role in intestinal absorption, renal tubular reabsorption, and deposition of phosphate in bone.

b. Vitamin D requirement. Vitamin D is expressed quantitatively in terms of International Units, one unit being equivalent to the activity of

[117] P. C. Jeans, Vitamin D, *in* ref. 73, p. 205.

[118] R. Zetterstrom and M. Ljunggren, *Acta Chem. Scand.* **5**, 283, 343 (1951).

0.025 μg. of calciferol. If an infant receives an ample supply of calcium and phosphorus, as in a customary cow's milk formula, an intake of 90 to 100 I.U. of vitamin D daily leads to retention of 25 to 30% of the calcium in the diet, rickets is prevented, and linear growth is average. If 300 to 400 I.U. of vitamin D is given daily, retention of calcium increases to 35 to 40% of the intake and linear growth is greater than average. Larger amounts of vitamin D do not increase the percentage of calcium retained. Maximum calcium retention is associated with excellent skeletal growth and early dentition. If more than 1800 units of vitamin D are administered daily, appetite is decreased, food intake is smaller, calcium retention is less, and linear growth becomes retarded.[117] The recommended allowance for vitamin D in this country is 400 units daily from birth to 20 years of age.[27] Premature infants do not appear to require more vitamin D than normal infants. Although some children may not need supplementary vitamin D, evidence is good that the majority require it throughout the period of growth. The need in adolescence is probably universal and may be as great as that in infancy.

During pregnancy and lactation supplemental vitamin D would seem desirable in view of the increase in requirement of calcium and phosphorus and the effect of vitamin D on utilization of these elements. Approximately 400 units daily appears to be adequate.

c. Vitamin D deficiency. Vitamin D deficiency leads to poor retention of calcium and phosphorus and retarded skeletal growth. It causes rickets during the period of growth, especially in infancy and in early childhood, and osteomalacia in adult life. Not only human children but dogs and poultry are prone to rickets when the diet is deficient in vitamin D. The susceptibility to rickets in animals is influenced by dietary changes, such as the inclusion of cereals containing phytic acid, which decreases absorption of calcium. Phytate also appears to interfere with absorption of calcium from the human intestinal tract.[119] Mellanby[120] has shown that phytate may exercise an anticalcifying action in puppies, even when vitamin D is provided. Whether this finding is applicable to human nutrition is unknown.

Pathologic findings in rickets include changes in the bones and teeth, which are described in detail in Chapter 13, and hyperplasia of the parathyroid glands. Manifestations of rickets in human infants include restlessness, irritability, excessive sweating about the head and neck, digestive disturbances due to impaired intestinal motility, and often delayed dentition. Growth is retarded, the fontanels fail to close at the anticipated time, and chest deformities, the "rachitic rosary," square head, and changes in the long bones make their appearance. The last occur as a result of muscle

[119] *Nutrition Revs.* **8,** 59 (1950).

[120] E. Mellanby, *J. Physiol.* **109,** 488 (1949).

pull acting on imperfectly calcified bone; when the child is walking, the weight of the body increases the deformity. Tetany due to low serum calcium concentration is occasionally associated with rickets.

In the early diagnosis of rickets, determination of alkaline serum phosphatase is of great value. Normal levels in young children range from 5 to 15 Bodansky units. Elevation of serum phosphatase above 20 units is highly suggestive of rickets, since few other conditions influence phosphatase activity in this age group. Serial determinations of alkaline phosphatase in serum are helpful in following therapy. Except in very early rickets, changes in the bones are demonstrable by roentgenographic examination and healing may be followed by repeated roentgenograms. Changes are similar to those observed in the line test which is used in vitamin D assay in animals. The level of inorganic phosphorus in serum decreases in rickets to about 4 mg. or less and promptly returns to normal with treatment, even though recovery may not be complete.

Rickets which is refractory to treatment has been reported, especially in children over 3 years of age. The nature of the underlying defect is unknown, but there appears to be interference with the normal calcifying action of vitamin D. Massive doses of vitamin D may promote healing without correcting all of the metabolic defects.[121]

Osteomalacia is most often seen during pregnancy and lactation and is characterized by osteoporosis, deformity and fracture of bones, and tetany. The concentration of calcium in serum is low, often less than 7 mg. per 100 ml., and the level of alkaline phosphatase is elevated above the usual adult value of 3 to 5 units.

d. Vitamin D intoxication. Severe toxic reactions follow administration of massive doses of vitamin D for long periods of time. Serum calcium becomes elevated, and metastatic calcification develops. The cause of death is usually renal failure.

3. Vitamin E

It has not been demonstrated that vitamin E is an essential dietary substance for man. None of the pathologic changes of vitamin E deficiency in animals have been shown to have their counterpart in man.[122] Vitamin E has been administered as a therapeutic agent in a number of human diseases but without adequate demonstration of real benefit.[123]

4. Vitamin K

a. Absorption and requirement of vitamin K. Vitamin K is present in a large variety of foods and is synthesized by bacteria in the intestinal tract.

[121] S. Freeman and I. Dunsky, *Am. J. Diseases Children* **79**, 409, (1950).

[122] J. B. Youmans, *J. Am. Med. Assoc.* **144**, 34 (1950).

[123] Report of Council on Pharmacy and Chemistry, *J. Am. Med. Assoc.* **142**, 485 (1950).

Since naturally occurring forms of vitamin K are fat-soluble, absorption is related to that of fat and is impaired when bile is absent from the intestinal tract and in the steatorrheas. It is conceivable that prolonged administration of antibiotics may depress intestinal synthesis of vitamin K and lead to deficiency. The deficiency which occurs in newborn infants may be due largely to limited food intake or, perhaps, to sterility of the intestinal tract.

The requirement of vitamin K is unknown but is apparently minute and easily supplied by an average diet, in addition to the amount synthesized by intestinal bacteria, except in the newborn infant. The infant needs supplementary vitamin K during the first few days of life unless the mother has received the vitamin prior to delivery. A single oral dose of 1 mg. appears to be sufficient to prevent postnatal decrease in prothrombin.[124] There is recent evidence suggesting that vitamin K has functions other than formation of prothrombin and probably should be administered to the mother during the latter part of pregnancy. If the mother has not received vitamin K, a suitable dose for the infant at birth is 1 mg. as a single dose. Actually, much less would be sufficient as the daily requirement of the infant is approximately 1 μg.[124]

b. Functions of vitamin K. Vitamin K is essential for the formation of prothrombin, which is necessary for blood coagulation. Prothrombin is formed in the liver, but the exact role of vitamin K in this process is unknown. It has been postulated that vitamin K forms the prosthetic group of an enzyme responsible for the synthesis of prothrombin.[125] Vitamin K may have functions other than prothrombin formation. Bleeding in the newborn period has not always ceased with restoration of the prothrombin level of the blood to normal.[126] Clinical observations have indicated that certain tissue lesions may precede hemorrhage in infants deficient in prothrombin.[127] In chicks and infant rats, vascular and parenchymal lesions, which might lead to hemorrhage or tissue injury, have been found in the brain in vitamin K deficiency. A hemorrhagic condition has been observed in newborn dogs in which intracranial blood vessels were markedly engorged. This condition may be related to vitamin K deficiency. Although it has not been established that degenerative changes in blood vessels or brain substance occur in the human fetus as a result of vitamin K deficiency, evidence suggests this possibility. In a study of 1531 children in Oslo born to women who received vitamin K during the last three weeks of pregnancy not a single instance of cerebral hemorrhage was observed.[127a]

[124] R. L. Sells, S. A. Walker, and C. A. Owen, *Proc. Soc. Exptl. Biol. Med.* **47,** 441 (1941).

[125] (*a*) A. J. Quick and G. E. Collentine, *J. Lab. Clin. Med.* **36,** 976 (1950); (*b*) G. E. Collentine and A. J. Quick, *Am. J. Med. Sci.* **222,** 7 (1951).

[126] A. H. Parmalee, *J. Michigan State Med. Soc.* **42,** 455 (1943).

[127] (*a*) K. U. Toverud, *Acta Paediat.* **18,** 249 (1936); (*b*) Maternal Nutrition and Health, National Research Council, Bulletin 123, Washington, 1950.

Interesting findings were recently reported by Sørbye and associates[128] in a study of the factors concerned in the normal clotting mechanism. They isolated from the plasma of vitamin K-deficient chicks a protein factor, designated the "kappa factor" which reduced plasma prothrombin time in animals treated with bishydroxycoumarin (dicumarol) but which had no effect upon prothrombin time in vitamin K deficiency. A second factor, the "delta factor," was isolated from plasma of chicks treated with dicumarol which reduced prothrombin time of plasma of vitamin K-deficient chicks but not that of plasma from chicks receiving dicumarol. They postulated that at least four factors are necessary in the clotting mechanism: prothrombin proper, the kappa and delta factors, and the labile factor of fresh plasma (factor V). Deficiency of vitamin K apparently produced a defect in the delta factor, while administration of dicumarol produced a deficiency of the kappa factor.

c. Vitamin K deficiency. Hemorrhage is the outstanding manifestation of vitamin K deficiency and is observed in clinical medicine in the newborn infant and in patients with obstructive jaundice, biliary fistulae, steatorrhea, severe prolonged diarrheal disease, or pancreatic fibrosis. Vitamin K deficiency may also follow operations on the gastrointestinal tract in which large areas of absorptive surface have been removed. Diagnosis is dependent upon estimation of the prothrombin activity of blood.

Prothrombin time may be prolonged in situations other than vitamin K deficiency. Administration of dicumarol or of large doses of salicylates will be followed by decreased prothrombin activity. In severe liver disease, prothrombin formation is defective and administration of vitamin K is without benefit, which finding is used as a test of liver function.

VIII. Vitamins of the B Complex

The vitamin B complex contains a number of factors which are closely associated in their distribution in nature and have related functions in intermediate metabolism. Of the eleven factors which are available in pure form, five have been shown to be constituents of coenzymes, namely, thiamine, riboflavin, niacinamide, pyridoxine, and pantothenic acid. It seems likely that other B vitamins may be found to function in a similar manner. Two members of the B complex, choline and inositol, appear to have lipotropic activity, and two others, folic acid and vitamin B_{12}, have antianemic properties. Deficiency of vitamins of the B complex is one of the most frequently encountered syndromes of malnutrition in man.

In view of the common distribution of many of the B vitamins in foods, deficiency of several factors is more often observed than deficiency of a single vitamin. In view of the interrelationship of these vitamins in metabo-

[128] Ø. Sørbye, I. Kruse, and H. Dam, *Acta Chem. Scand.* **4**, 549, 831 (1950).

lism, the clinical manifestations of deficiency may be similar when any one of several factors is lacking in the diet. Not all the B vitamins have been shown to be essential in human nutrition. Attention will be directed particularly to those which are of established importance in man.

1. Thiamine

Thiamine, or vitamin B_1, is a water-soluble compound which is rapidly broken down by moist heat in neutral or alkaline solutions into its constituent pyrimidine and thiazole rings. The ready destructability of thiamine is important in human nutrition, since much may be lost in the preparation of food. Some of the biochemical methods used in evaluating thiamine nutrition are based on reactions with the thiazole and pyrimidine portions of the thiamine molecule. The thiochrome method is widely used in assaying biological materials for thiamine, while determination of the urinary excretion of pyramine (a pyrimidine-like compound) has been used to assist in assessment of nutritional status.

a. Absorption, storage, blood levels, and excretion of thiamine. Thiamine is readily absorbed from the small intestine and is probably phosphorylated in the intestinal mucosa (Chapter 9). Concentrations of thiamine in human tissue are not great, the following having been reported: heart muscle, 2 to 3 μg. per gram; skeletal muscle, 0.5 μg. per gram; and brain, liver, and kidney, 1 μg. per gram. These concentrations were approximately doubled shortly after thiamine therapy and were approximately halved by inadequate diets which were low in thiamine but which did not result in overt evidence of deficiency.[129] Thiamine is present in tissues in several forms: free, monophosphate, pyrophosphate and probably bound to protein by an S-S linkage, and as the disulfide of thiamine pyrophosphate and lipothiamine (Chapter 19).

Thiamine is excreted in the urine, the amount being dependent on dietary intake and the relative saturation of the tissue stores. Determination of thiamine excretion in the urine, especially after a test dose of thiamine has been administered, is one of the methods used in evaluating nutritive status relative to this vitamin. After intramuscular injection of 1 mg. of thiamine, persons who are adequately nourished excrete at least 100 μg. in the subsequent 4 hr., whereas patients with signs of thiamine deficiency usually excrete less than 50 μg. during this period.[130] Estimation of the concentration of thiamine in blood has also been used in nutritional appraisal. Mean

[129] J. W. Ferrebee, N. Weissman, D. Parker, and P. S. Owen, The Thiamin Content of Human Tissue, *in* The Role of Nutritional Deficiency in Nervous and Mental Disease, Williams and Wilkins Co., Baltimore, 1943, p. 42.

[130] (*a*) R. D. Williams, H. L. Mason, and R. M. Wilder, *J. Nutrition* **25,** 71 (1943); (*b*) R. D. Williams, H. L. Mason, M. H. Power, and R. M. Wilder, *Arch. Internal Med.* **71,** 38 (1943); (*c*) L. E. Holt, Jr., *Federation Proc.* **3,** 171 (1944).

levels in normal subjects have been found to be 4.7 μg. per 100 ml. in whole blood, 8.0 μg. % in erythrocytes, and 67.5 μg. % in leucocytes.[131] In middle-aged and older persons, the average concentration of thiamine was reported to be 3.4 μg. per 100 ml.[132] In beriberi, mean values of 3.2 μg. per 100 ml. have been observed.[69d]

Absorption of thiamine may be interfered with by ingestion of live yeast which competes with the host for available thiamine.

b. Funtions of thiamine. Thiamine functions in the tissues in the form of two coenzymes: thiamine pyrophosphate or cocarboxylase and lipothiamide, which is the amide of thiamine and lipoic acid. These coenzymes are important in the decarboxylation of α-keto acids and in the biosynthesis of certain acyloins. In decarboxylation of the oxidative type, lipothiamide appears to be the coenzyme. Substances formed include acetic acid, acetyl phosphate and acetyl-coenzyme A, the last being of great importance since it feeds into the tricarboxylic acid cycle to provide energy for the organism. (See Chapter 19).

In thiamine deficiency, pyruvate accumulates in the blood and tissues and there is a change in the lactate-pyruvate ratio. These findings have been used in detecting thiamine deficiency in man.[133] Horwitt[134] has proposed as a test of thiamine nutrition the simultaneous measurement of lactic acid, pyruvic acid, and glucose in the blood 5 min. after mild exercise and 60 min. after ingestion of glucose. The carbohydate index (C.I.) is determined from the formula

$$\mathrm{CI} = \frac{\left(L - \frac{G}{10}\right) + \left(15P - \frac{G}{10}\right)}{2}$$

where G is glucose, L is lactic acid, and P is pyruvic acid. An index greater than 15 is suggestive of thiamine deficiency.

c. Thiamine requirement. Numerous experimental studies in adult humans suggest that the minimum thiamine requirement is approximately 0.23 to 0.3 mg. per 1000 cal.[27,135] Although requirement is related to caloric consumption, there is some evidence that relatively less is needed at high caloric levels. The carbohydrate content of the diet influences requirement

[131] H. B. Burch, O. A. Bessey, R. H. Love, and O. H. Lowry, *J. Biol. Chem.* **198,** 477 (1952).

[132] E. Kirk and M. Chieffi, *J. Nutrition* **38,** 353 (1949).

[133] (*a*) E. Stotz and O. A. Bessey, *J. Biol. Chem.* **143,** 625 (1942); (*b*) G. A. Goldsmith, *Am. J. Med. Sci.* **215,** 182 (1948).

[134] M. K. Horwitt, Biochemical Observations, *in* Investigation of Human Requirements for B-Complex Vitamins, National Research Council Bulletin No. 116, Washington, June 1948, p. 12.

[135] K. Daum, W. W. Tuttle, and M. Wilson, *J. Am. Dietet. Assoc.* **25,** 398 (1949).

to some extent and fat exerts a sparing action. Requirement is increased during periods of active growth and in pregnancy and lactation. Few studies have been made of requirement in infancy. Deficiency has been prevented with intakes of 0.14 to 0.20 mg. daily[136] and with approximately 30 μg. per kilogram of body weight daily.[137] The Recommended Dietary Allowances for thiamine are shown in Table 1. For adults, the allowance is 0.5 mg. per 1000 cal., which is about 100% more than the minimal requirement. This factor of safety does not seem excessive, since thiamine stores of the body are never large and may be readily exhausted in diseases associated with elevated metabolism and probably in other stress situations.

Although some thiamine may be synthesized by intestinal bacteria, the amount available to the organism to supplement the dietary supply is probably not great.

d. Thiamine deficiency. Some of the biochemical changes in thiamine deficiency have been discussed above. Urinary excretion of thiamine falls to low levels, tissue stores are depleted, the level of thiamine in the blood falls slightly, and there is accumulation of pyruvic acid in the blood and tissues. In experimental thiamine deficiency in man, early clinical findings have included the following: fatigability, anorexia, apathy, nausea, epigastric pain, and certain mental and personality disturbances including irritability, vague fears, moodiness, depression, quarrelsomeness, and lack of ambition. Daum and associates[135] found that maximum work output and mechanical efficiency were the most sensitive physiologic indices of a decrease in dietary thiamine. Paresthesias of the lower extremities may be the first indication of polyneuritis, and electrocardiographic abnormalities often precede clinical evidence of heart involvement. Advanced thiamine deficiency or beriberi is characterized by polyneuritis, cardiovascular disturbances, and edema. The heart is enlarged, the right side being predominantly involved, and all the signs of congestive heart failure may develop. The heart failure is usually of the high output type, and the circulation time is relatively rapid. Sudden death from acute peripheral circulatory collapse may occur.

Manifestations of thiamine deficiency in animals resemble certain aspects of human beriberi. Electrocardiographic changes indicative of myocardial damage have been observed in cats, dogs, pigs, and monkeys, as well as in man. Bradycardia may occur in rats and in man. Histological lesions do not seem to be sufficient to account for either bradycardia or the

[136] L. E. Holt, Jr., R. L. Nemir, S. E. Snyderman, A. A. Albanese, K. C. Ketron, L. P. Guy, and R. Corretero, *J. Nutrition* **37,** 53 (1949).

[137] (*a*) E. M. Knott, S. C. Kleiger, and F. W. Schultz, *J. Pediatrics* **22,** 43 (1943); (*b*) E. M. Knott, S. C. Kleiger, and F. Torres-Bracamonte, *J. Nutrition* **25,** 49 (1943).

abnormalities of the electrocardiogram (Chapter 13). In man, mild scarring, hydropic degeneration, and fatty infiltration have been observed in the heart.

Polyneuritis, with degenerative changes in the peripheral nerves, has been reported in thiamine deficiency in a number of animal species, but the diets were often deficient not only in thiamine but in other factors. Evidence is good, although not unequivocal, that thiamine deficiency is accompanied by neuritis in pigeons and probably in dogs and other mammals (Chapter 13). Degeneration of the peripheral nerves, demyelinization of nerve roots, and degeneration of tracts in the spinal cord have been reported in man. The polyneuritis of chronic alcoholism and that which may occur in pernicious vomiting of pregnancy are due to thiamine deficiency. Alcohol does not increase thiamine requirement; actually less thiamine may be needed for utilization of alcohol than for utilization of glucose.[138] Polyneuritis develops in the chronic alcoholic on the basis of dietary thiamine deficiency. The neuritis which accompanies diabetes mellitus is rarely, if ever, related to lack of thiamine.

2. Riboflavin

Riboflavin is a water-soluble, yellow pigment with green fluorescence, having the chemical formula 6,7-dimethyl-9-(D,1′-ribityl)-isoalloxazine. It is widely distributed in both the plant and animal kingdoms and appears to be an essential dietary factor for all species of animals, including man, and for some types of bacteria. Although chemical research on riboflavin dates back to 1879, the function and importance of this substance in animal and human nutrition was not apparent until the 1930's.[139] In 1932 Warburg and Christian obtained a yellow enzyme from yeast which they were able to split into a protein and pigment component (flavin), thus separating and identifying the prosthetic group of an enzyme for the first time.

Riboflavin is fairly stable to heat, especially in acid solutions, but it is extremely labile when exposed to light. As a result of this lability, considerable loss of riboflavin occurs in food such as milk.

a. Absorption, storage, and excretion of riboflavin. Riboflavin in its naturally occurring free form is readily absorbed from the small intestine. Some riboflavin is synthesized by the intestinal flora, but the extent to which this supply is available to the organism is uncertain. Experiments of Najjar and associates[140] suggest that some of this riboflavin may be

[138] R. E. Butler and H. P. Sarett, *J. Nutrition* **35**, 539 (1948).

[139] G. A. Goldsmith, Riboflavin Malnutrition, *in* Clinical Nutrition, Paul B. Hoeber, Inc., New York, 1950, p. 512.

[140] V. A. Najjar, G. A. Johns, G. C. Medairy, G. Fleishmann, and L. E. Holt, Jr., *J. Am. Med. Assoc.* **126**, 357 (1944).

absorbed. There are no special organs in which riboflavin is stored. Comparatively large amounts have been found in human liver (16 μg. per gram) and in human kidney (20 to 25 μg. per gram). Muscle contains about 2 to 3 μg. per gram.[141] Tissue saturation may be readily attained when the riboflavin intake is high, but reserve stores are not great and may be lost quickly. A relationship between the retention of riboflavin and the retention of protein has been demonstrated.

The concentration of riboflavin in the serum of well-nourished adults was found by Burch and associates[142] to be 0.8 μg. per 100 ml. for free riboflavin and 3.2 μg. per 100 ml. for total riboflavin. In the white cell-platelet layer, total riboflavin was 252 μg. per 100 ml.; in the red blood cells, 22 μg. per 100 ml.

Riboflavin is excreted in the urine, the output varying with the intake and the degree to which tissue stores are saturated. In normal persons, values range from 150 to 2000 μg. daily.

b. Riboflavin requirement. Experimental riboflavin deficiency has been induced in man when the diet contained less than 0.6 mg. daily.[143] No signs of deficiency were observed over a period of 288 days when the intake was 0.7 mg. daily.[144] Studies of urinary excretion have suggested that an intake of 1.1 to 1.6 mg. daily will provide adequate body stores.[145] The need for riboflavin is probably not related to caloric consumption but may be related to weight or some function thereof. In several species of animals, and in man, ingestion of 15 to 20 μg. of riboflavin per kilogram of body weight daily has been shown to prevent deficiency.[27] Long-term experiments in animals suggest that benefits result from amounts of riboflavin above minimum requirement. Accordingly, the recommended dietary allowances of riboflavin suggested by the Food and Nutrition Board of the National Research Council are 1.5 mg. daily for a 56-kg. woman and 1.8 mg. daily for a 70-kg. man (Table 1).

In animals, about twice as much riboflavin is needed for growth as for adult maintenance. This is probably true in man also, although evidence for this is meager. The recommended allowances provide about 60 μg. per kilogram of body weight per day, which probably represents a 100% factor of safety.

c. Functions of riboflavin. Riboflavin is present in tissues largely in bound

[141] F. W. Clements, *Vitamins and Hormones*, **4,** 91 (1946).

[142] H. B. Burch, O. A. Bessey, and O. H. Lowry, *J. Biol. Chem.* **175,** 457 (1948).

[143] (*a*) M. K. Horwitt, O. W. Hills, C. C. Harvey, E. Liebert, and D. L. Steinberg, *J. Nutrition* **39,** 357 (1949); (*b*) W. H. Sebrell and R. E. Butler, *Public Health Repts.* (U.S.) **53,** 2282 (1938).

[144] R. D. Williams, H. L. Mason, P. L. Cusick, and R. M. Wilder, *J. Nutrition* **25,** 361 (1943).

[145] M. K. Horwitt, C. C. Harvey, O. W. Hills, and E. Liebert, *J. Nutrition* **41,** 247 (1950).

form, flavin adenine dinucleotide (FAD) being the chief component, and flavin mononucleotide (FMN) making up most of the remainder (Chapter 19). Free riboflavin is found in milk and in urine and, in certain animal species, in the retina. Flavoproteins function as important enzymes in tissue respiration. They play a major role in the hydrogen transport system where they catalyze the oxidation of the reduced pyridine nucleotides (Chapter 18). Thus, they mediate hydrogen transport between coenzymes rather than between substrates. Flavoproteins also occur as "oxidases" which catalyze the direct oxidation of various substrates by oxygen (Chapter 18). Most of the flavoproteins contain FAD as their prosthetic group (Chapter 19). FMN and FAD undergo alternate oxidation and reduction in a manner similar to niacin-containing enzymes. A list of the flavoprotein enzymes and further discussion of their functions may be found in Chapters 18 and 19.

d. Riboflavin deficiency. Early symptoms of riboflavin deficiency in man include soreness and burning of the lips, mouth, and tongue, and visual complaints such as photophobia, lacrimination, and burning and itching of the eyes. The lips show redness and denudation along the line of closure with maceration, crusting, and fissures at the angles of the mouth.[143b] Cheilosis and angular stomatitis are not pathognomonic of riboflavin deficiency but have been observed, also, in experimental human niacin deficiency,[54a] following administration of desoxypyridoxine,[146] and in conditions of non-nutritional origin. Other findings in human riboflavin deficiency are dermatitis, glossitis, and lesions of the eye. The dermatitis, which is of seborrheic type, may involve the ears, the outer and inner canthi of the eyes, and the scrotum. The tongue may be purplish-red in color, the papillae may be hypertrophic or atrophic, and fissures may be present. Eye lesions include blepharitis, conjunctivitis, iritis, and superficial vascularization of the cornea.

Many of the findings in riboflavin deficiency in man are similar to those observed in various animal species[139] (see also Chapter 13). Skin changes occur in riboflavin deficiency in the rat, hamster, dog, pig, and monkey. Lesions of the lips and oral cavity have been noted in rats and young calves. Conjunctivitis, blepharitis, and corneal vascularization have been found in rats. Other changes resulting from riboflavin deficiency in animals, such as nerve degeneration, cataract formation, anemia, and sudden collapse, have not been encountered in man. In animals, the ability of the liver to inactivate estradiol is reduced. It has been suggested that a similar change may occur in human riboflavin deficiency.

Little is known as to the manner in which an inadequate supply of riboflavin produces the pathologic changes observed. Liver slices obtained

[146] J. F. Mueller and R. W. Vilter, *J. Clin. Invest.* **29,** 193 (1950).

from riboflavin-deficient animals show a reduction in the ability to oxidize D-amino acids. The activity of the enzymes xanthine oxidase and succinic dehydrogenase has been found to be decreased. In animals, where death was due to lack of riboflavin, the concentration of this vitamin in the liver, the kidney, and the heart was found to be one-third of normal.[141]

The urinary excretion of riboflavin is decreased in human deficiency, both in 24-hr. periods and after administration of test doses of riboflavin. Because of the close relationship between retention of riboflavin and retention of protein, excretion data must be interpreted with caution. In starvation and in conditions associated with negative nitrogen balance, riboflavin excretion is increased.

In animals, riboflavin deficiency during pregnancy has resulted in striking abnormalities in the embryo and, at times, in abortion. Whether congenital defects in man are ever related to an inadequate maternal supply of riboflavin is unknown.

3. Niacin (nicotinic Acid)

Niacin, the pellagra-preventive factor, is a unique vitamin in that one of the amino acids, tryptophan, serves as a precursor of this vitamin in many animal and plant species. It is interesting to note that niacinamide was shown to be part of two coenzymes prior to the discovery of its nutritional importance (Chapter 19).

a. Absorption, storage, and excretion of niacin. Niacin is readily absorbed from the intestinal tract, a fact easily demonstrable by the erythematous skin reaction which occurs 15 to 30 min. after administration. Since niacin is stable to heat, loss in food preparation is minimal. The storage of niacin is not great, and any excess of the vitamin is disposed of largely by a methylating process, in man, dog, rat, and pig, and by other pathways in the herbivora and polygastric species.[147] Niacin metabolites in human urine are chiefly N^1-methylnicotinamide (N^1-Me) and the pyridone of N^1-methylnicotinamide (pyridone), plus a small amount of nicotinic acid. Normal persons receiving a good diet excrete approximately 6 to 8 mg. of N^1-Me daily and 10 to 12 mg. of the pyridone. Niacin is present in tissues largely as coenzymes I and II, which are diphosphopyridine nucleotide (DPN) and triphosphopyridine nucleotide (TPN), respectively. The concentration of pyridine nucleotides in whole blood in normal subjects is about 36 μg. per milliliter, practically all of which is in the erythrocytes.

b. Metabolic functions of niacin. Niacin functions in the body as a component of the important coenzymes DPN and TPN, which are concerned in glycolysis and tissue respiration. These enzymes have been found in every type of cell which has been examined. The major functions of

[147] W. A. Perlzweig, F. Rosen, and P. B. Pearson, *J. Nutrition* **40,** 453 (1950).

DPN and TPN are removal of hydrogen from certain substrates, in cooperation with dehydrogenases, and transfer of hydrogen (or electrons) to another coenzyme in the hydrogen transport series, or to another substrate, which is correspondingly reduced (Chapter 19).

As noted above, the amino acid tryptophan can be converted to niacin in a number of plant and animal species and in man.[148] Administration of tryptophan to human beings is followed by an increase in the urinary excretion of N^1-Me, the pyridone of N^1-Me, quinolinic acid, and tryptophan.[53,149] The increase in excretion of niacin metabolites is approximately proportional to the amount of tryptophan administered. The chemical steps by which tryptophan is converted to niacin are not completely known, although much information has been obtained from studies in *Neurospora* and in rats.[150] Possible intermediate metabolites include kynurenine, 3-hydroxyanthranilic acid, and quinolinic acid. Studies in man indicate that quinolinic acid is formed from tryptophan but that it is probably not in the main pathway of conversion.[54b] Several years ago, it was suggested that tryptophan catalyzed niacin synthesis by intestinal bacteria rather than being converted to niacin in the tissues. Subsequently, this hypothesis was shown to be erroneous. In the enterectomized rat, tryptophan administration is followed by an increase in excretion of niacin compounds.[151] In man, the conversion of tryptophan to niacin is not affected by oral administration of streptomycin in amounts which largely inhibit growth of coliform organisms in the intestinal tract.[152] Incontrovertible evidence of the conversion of tryptophan to niacin was demonstrated by administering tryptophan labeled with C^{14} to rats and recovering niacin with the isotope in the carboxyl group in the urine.[153]

Administration of tryptophan is followed by an increase in the concentration of pyridine nucleotides in red blood cells; the rise following administration of 10 g. of tryptophan is similar to that following 1 g. of niacin.[154]

c. **Niacin requirement.** It is apparent that the niacin requirement can be established only in relation to the dietary intake of tryptophan, since this amino acid functions as a precursor of niacin. In view of this amino acid-vitamin relationship, data concerning human niacin requirement are

[148] H. P. Sarett and G. A. Goldsmith, *J. Biol. Chem.* **167,** 293 (1948).

[149] (*a*) H. P. Sarett and G. A. Goldsmith, *J. Biol. Chem.* **182,** 679 (1950); (*b*) W. A. Perlzweig, F. Rosen, H. Levitas, and J. Robinson, *J. Biol. Chem.* **167,** 511 (1947).

[150] D. M. Bonner and C. Yanofsky, *J. Nutrition* **44,** 603 (1951).

[151] L. M. Henderson and L. V. Hankes, *Proc. Soc. Exptl. Biol. Med.* **70,** 26 (1949).

[152] H. P. Sarett and G. A. Goldsmith, *Federation Proc.* **9,** 369 (1950).

[153] C. Heidelberger, E. P. Abraham, and S. Lepkovsky, *J. Biol. Chem.* **179,** 151 (1949).

[154] M. Duncan and H. P. Sarett, *J. Biol. Chem.* **193,** 317 (1951).

meager. Goldsmith and associates[54a] produced experimental pellagra in human subjects who received corn diets containing 190 mg. of tryptophan and 4.7 mg. of niacin daily for more than 50 days. In studies conducted by Horwitt and associates,[155] pellagra has not been observed in subjects who received diets containing 265 mg. of tryptophan and 5.8 mg. of niacin daily for more than a year.

Frazier and Friedemann[156] calculated the niacin and protein content of pellagra-producing and pellagra-preventing diets and concluded that, with marginal diets containing corn products, the minimum need of niacin was about 7.5 mg. daily. Recent studies of Goldsmith and associates[54a] indicate that, when the diet is rich in corn and furnishes approximately 200 mg. of tryptophan daily, the minimal niacin requirement of adult females is about 6 to 7 mg. per day. Requirement may be slighly less when the diet contains wheat rather than corn as the chief cereal.

The role of corn diets in the production of pellagra may be explained in part by the low tryptophan content of corn. Recent experiments in rats suggest another possible explanation. Some of the niacin in corn, and in certain other cereals, appears to be present in bound form which is unavailable to the organism, unless previously hydrolyzed by alkali.[157]

In dogs, the niacin requirement appears to be about ten times that of thiamine.[27] In rats, niacin is an essential nutrient only when the tryptophan content of the diet is low, and under these circumstances requirement is about ten times the thiamine need.[158] The recommended dietary allowances for niacin in this country are ten times the thiamine allowances (Table 1). In view of the above data, these allowances should provide a fair margin of safety.

d. Niacin deficiency—pellagra. (1) *Clinical Findings.* Niacin deficiency in man leads to pellagra, which is characterized by (1) dermatitis, particularly of skin surfaces exposed to light or subjected to trauma; (2) severe inflammation of mucous membranes manifested by glossitis, stomatitis, diarrhea, proctitis, and vaginitis; and (3) psychic changes including anxiety, irritability, depression, and, in advanced stages, delirium, hallucinations, disorientation, confusion, and stupor. Lesions of niacin deficiency in animals resemble those found in man. Blacktongue in dogs is characterized by changes in the skin and mucous membranes (Chapter 13). A condition re-

[155] M. K. Horwitt, personal communication.

[156] E. I. Frazier and T. E. Friedemann, *Quart. Bull. Northwestern Univ. Med. School* **20,** 24 (1946).

[157] (*a*) E. Kodicek, *Biochem. J.* **48,** viii (1951); (*b*) J. Laguna and K. J. Carpenter, *J. Nutrition* **45,** 21 (1951).

[158] (*a*) W. A. Krehl, P. S. Sarma, and C. A. Elvehjem, *J. Biol. Chem.* **162,** 403 (1946); (*b*) W. A. Krehl, L. M. Henderson, J. de la Huerga, and C. A. Elvehjem, *J. Biol. Chem.* **166,** 531 (1946).

sembling canine blacktongue has been observed in pigs. In chicks, poor growth and feathering, and inflammation of the mouth have been reported (Chapter 9).

In experimental pellagra, which was induced in three human subjects who received a "corn" diet which furnished 4.7 mg. of niacin and 190 mg. of tryptophan daily, findings included dermatitis, glossitis, stomatitis, bloody diarrhea, and mild mental disturbances.[54a] In addition, cheilosis, angular stomatitis, and dermatitis of the nasolabial folds were observed, lesions commonly found in riboflavin deficiency, yet the diet furnished 2 mg. of riboflavin daily. The occurrence of similar lesions in deficiency of niacin and riboflavin reflects the close metabolic relationship of these vitamins.

Macrocytic anemia may occur in niacin deficiency in both dogs and humans. In dogs, the anemia may be cured by niacin, except in repeated episodes of deficiency when folic acid is required for response. In humans, the anemia often responds to folic acid rather than to niacin.

Achlorhydria is of common occurrence in pellagra and free hydrochloric acid may reappear in the gastric juice following recovery. Amenorrhea has likewise been observed in both endemic and experimental pellagra.

(2) *Biochemical Findings.* In both experimental and endemic pellagra, urinary excretion of the pyridone of N^1-Me falls to non-detectable levels.[54a,159] In experimental pellagra, excretion of N^1-Me was in the range of 0.5 to 0.6 mg. daily at the time lesions first appeared and remained at that level.[54a] In endemic pellagra, N^1-Me excretion is usually low, both in the 24-hr. urine and in urine collected after the administration of a test dose of niacin.[149a] In one subject with pellagra, who showed azotemia and an extremely large output of nitrogen in the urine, excretion of N^1-Me was above normal and excretion of quinolinic acid was markedly elevated.

Administration of tryptophan, as well as of niacin, has been shown to bring about healing of the lesions of pellagra.[54a, 149a, 160] Abnormal and varied excretion of tryptophan and niacin metabolites has been noted after administration of tryptophan to subjects with endemic pellagra. Unusually high levels of quinolinic acid were noted in some instances,[149a] suggesting the possibility of an abnormality in the conversion of tryptophan to niacin. In experimental niacin deficiency, a gradual increase in excretion of niacin metabolites followed tryptophan administration.

4. Pyridoxine (Vitamin B_6)

a. Sources and functions of pyridoxine. Vitamin B_6 is a complex consisting of three closely related compounds: pyridoxine, pyridoxal, and pyridoxamine. Although these three forms of the vitamin are equally active

[159] W. I. M. Holman and D. J. de Lange, *Nature* **165,** 112 (1950).

[160] (*a*) R. W. Vilter, J. F. Mueller, and W. B. Bean, *J. Lab. Clin. Med.* **34,** 409 (1949); (*b*) W. B. Bean, M. Franklin, and K. Daum, *ibid.* **38,** 167 (1951).

when injected into animals, pyridoxine is the most active when given with food, presumably because it is used less well by intestinal microorganisms.[161] In the rat, vitamin B_6 which is synthesized by the intestinal flora appears to be available for metabolic needs. Whether this is true in man is unknown.

Fecal excretion of vitamin B_6 is fairly constant in animals and in man; average values for man are 0.8 mg. daily in adults and 0.2 mg. daily in infants.[162] Urinary excretion of vitamin B_6 varies with the intake and is largely in the form of an oxidation product, 4-pyridoxic acid.

In the tissues, vitamin B_6 occurs predominantly as the phosphate of pyridoxal or pyridoxamine, especially the former, except in the liver. Pyridoxal phosphate functions as a coenzyme in four types of reactions: decarboxylation of amino acids, transamination, and the synthesis and cleavage of tryptophan (Chapter 19). This coenzyme is necessary for the deamination of amino acids and for the formation of urea nitrogen.[163] It appears to be essential for the conversion of tryptophan to the pyridine coenzymes.[164] Pyridoxine may be related to fatty acid metabolism[165] and seems to be necessary for normal adrenal cortical function.[166]

In pyridoxine deficiency in both animals and man, an abnormal product of tryptophan metabolism, xanthurenic acid, appears in the urine after administration of tryptophan. This finding has been suggested as a test in evaluating pyridoxine nutrition.[167] Another biochemical test proposed for detection of vitamin B_6 deficiency is administration of alanine; in deficient animals, blood urea nitrogen remains elevated for more than 12 hr. and glutamic acid disappears slowly, suggesting impaired deamination.[168]

b. Pyridoxine deficiency. Recent studies have demonstrated that vitamin B_6 is an essential human nutrient although the exact requirement is uncertain. When a normal adult subject was given a purified diet deficient in vitamin B_6 for 54 days, mental depression, confusion, albuminuria, and

[161] E. E. Snell, B. M. Guirard, and R. J. Williams, *J. Biol. Chem.* **143,** 519 (1942).

[162] (*a*) H. Linkswiler and M. S. Reynolds, *J. Nutrition* **41,** 523 (1950); (*b*) P. Moller, *Acta Physiol. Scand.* **23,** 47 (1951).

[163] W. J. McGainty, E. W. McHenry, H. B. Van Wych, and G. L. Watt, *J. Biol. Chem.* **178,** 511 (1949).

[164] J. P. Kring, K. Ebisuzake, J. N. Williams, and C. A. Elvehjem, *J. Biol. Chem.* **195,** 591 (1952).

[165] (*a*) C. W. Carter and P. J. R. Phizackerley, *Biochem. J.* **49,** 222 (1951); (*b*) H. Sherman, *Vitamins and Hormones*, **8,** 55 (1950).

[166] R. B. Stebbins, *Endocrinology* **49,** 25 (1951).

[167] (*a*) S. Lepkovsky, E. Roboz, and A. J. Haagen-Smit, *J. Biol. Chem.* **149,** 195 (1943); (*b*) L. D. Greenberg, D. Bohr, H. McGrath, and J. F. Rinehart, *Arch. Biochem.* **21,** 237 (1949); (*c*) H. S. Glazer, J. F. Mueller, C. Thompson, V. R. Hawkins, and R. W. Vilter, *Arch. Biochem.* **33,** 243 (1951); (*d*) S. E. Snyderman, L. E. Holt, Jr., R. Carretero and K. Jacobs. L. Clin. Nutrition 1, 200 (1953).

[168] J. R. Beaton, R. M. Ballantyne, R. E. Lau, A. Steckley, and E. W. McHenry, *J. Biol. Chem.* **186,** 93 (1950).

a decrease in neutrophiles in the blood were noted.[169] The abnormal findings disappeared after pyridoxine administration. In two adults who received a diet low in vitamin B_6, excretion of xanthurenic acid after ingestion of tryptophan was greatly increased.[167b] In two human infants who were given a diet deficient in pyridoxine, definite evidence of deficiency was observed.[167d] Pyridoxic acid disappeared from the urine, and the ability to convert tryptophan to niacin was lost. The infants ceased to gain weight, and convulsions developed in one, hypochromic anemia in the other. All abnormalities were relieved by pyridoxine administration.

Pyridoxine deficiency has been induced by administration of desoxypyridoxine to adults receiving a diet low in B complex vitamins.[146] Seborrheic skin lesions developed about the eyes, nose, and mouth, and cheilosis, glossitis, and stomatitis were observed. Although these findings resemble those commonly seen in riboflavin and niacin deficiency, healing was dependent on administration of pyridoxine. The deficient subjects excreted large amounts of xanthurenic acid in the urine after a test dose of tryptophan, but ability to convert tryptophan to niacin was unimpaired.

Pyridoxine has been used empirically for a number of years in the treatment of nausea and vomiting of pregnancy, with presumably beneficial results. Recent findings suggest that pyridoxine metabolism may be altered during human pregnancy. Pregnant women have lower levels of urea nitrogen in the blood than do non-pregnant women. In women with hyperemesis gravidarum, levels are lower than those found during normal pregnancy and the response to a test dose of alanine is abnormal, i.e., blood urea nitrogen remains elevated for more than 12 hours.[163] It has also been shown that, after a test dose of tryptophan, pregnant women excrete abnormally large amounts of xanthurenic acid in the urine as compared to women who are not pregnant. This abnormality is corrected by pyridoxine.

Pyrodoxine has been reported to bring about healing of cheilosis in some subjects. In seborrheic dermatitis of the secca type, application of an ointment containing pyridoxine has been followed by improvement.[170]

Human requirement for vitamin B^6 can be estimated at present only by analogy from animal experiments. On the basis of studies of rat growth, about 1.5 mg. daily would approximate the human need.[171]

5. Pantothenic Acid

Pantothenic acid is a constituent of coenzyme A, which has an extremely important role in a large variety of transacetylation reactions (see Chapter

[169] W. W. Hawkins and J. Barsky, *Science* **108**, 284 (1948).

[170] A. W. Schreiner, W. Slinger, V. R. Hawkins, and R. W. Vilter, *J. Lab. Clin. Med.* **40**, 121 (1952).

[171] U. D. Register, U. J. Lewis, W. R. Ruegamer, and C. A. Elvehjem, *J. Nutrition* **40**, 281 (1950).

19). Although pantothenic acid and coenzyme A are undoubtedly important in nutrition in man as well as in animals, no evidence of human pantothenic acid deficiency has been recorded. The wide distribution of pantothenic acid in food may explain this failure to observe deficiency even on restricted diets. The human requirement is unknown but probably is not above 5 mg. daily.[172] In dogs, the need is similar to that for thiamine and riboflavin.

6. Choline

Choline is considered to be an important member of the vitamin B complex, since most animals, when given diets low in this compound, develop deficiency characterized by fatty livers and hemorrhagic lesions of the kidney. Choline is believed to function in at least three ways: (1) as an integral part of acetylcholine, (2) as a source of labile methyl groups, and (3) in stimulating the formation of phospholipids.

Little is known of choline requirement in either animals or man since the need is dependent, in part, on other sources of methyl groups in the diet, namely methionine and betaine. Elvehjem[173] has suggested that, in the light of animal studies, the human requirement is probably less than 500 mg. daily. This amount may be furnished by the average diet which contains about 250 to 600 mg. of choline.

Choline has been used in the therapy of liver disease in man, either alone or in conjunction with methionine and a diet high in protein and carbohydrate. The value of choline in the treatment of fatty livers and cirrhosis has not been determined with accuracy, since long-term human experiments are difficult to control and, in most instances, multiple therapy has been prescribed. It has been shown that administration of a single 10-g. dose of choline increases the rate of phospholipid turnover in subjects in whom fatty infiltration of the liver was observed on initial biopsy.[75b] When fat disappeared from the liver, this effect could not be demonstrated. In patients with active fatty alcoholic cirrhosis, choline administration brought about a moderate reduction in liver fat as shown histologically in biopsy specimens.[59] Provision of adequate protein in the diet was followed by much greater improvement in the liver cell.

In animals, choline, methionine, vitamin B_{12}, and folic acid have been shown to be interrelated in the prevention of fatty livers under certain dietary conditions.[174] Whether this is true in man is unknown. In patients

[172] C. A. Elvehjem, The Vitamin B Complex *in* ref. 73, p. 178.

[173] C. A. Elvehjem, The Vitamin B Complex, *in* ref. 73, p. 181.

[174] (*a*) M. M. Burns and J. M. McKibbin, *J. Nutrition* **44,** 487 (1951); (*b*) D. R. Strength, E. A. Schaefer, and W. D. Salmon, *ibid.* **45,** 329 (1951); (*c*) H. R. V. Arnstein and A. Neuberger, *Biochem. J.* **48,** ii (1951); (*d*) *Nutrition Revs.* **8,** 145, 269, 301 (1950); (*e*) A. E. Schaefer, W. D. Salmon, D. R. Strength, and D. H. Copeland, *J. Nutrition* **40,** 95 (1950).

with cirrhosis who were receiving the usual dietary therapy, improvement after administration of liver extract has been reported.[175]

The absorption and excretion of choline have been studied in normal subjects and in patients with hepatobiliary diseases.[176] After oral administration of 2 to 8 g. of choline base, only small amounts of choline were detected in the urine, while approximately two-thirds of the ingested choline nitrogen was excreted as trimethylamine and its oxide. Incubated stool dilutions were shown to transform choline to trimethylamine. After intravenous administration of choline, urinary trimethylamine excretion did not increase. These findings suggest that trimethylamine is formed from choline in the intestinal tract. Since trimethylamine is a substance without lipotropic activity, it would appear that oral therapy with choline could not be expected to produce markedly beneficial effects.

7. Inositol

Little is known about the role of inositol in nutrition. Inositol exhibits lipotropic activity in animals only when added to a hypolipotropic diet devoid of fat, and it exerts either no or slight activity when fat is added to the diet.[177] Inositol has been suggested as an adjuvant in the prevention and treatment of chronic liver disease in man. It has been reported that inositol administration leads to a decrease in the level of cholesterol and lipid phosphorus in the serum of patients with diabetes mellitus.[178] It has been postulated that inositol plays a role in creatine metabolism and therefore may be beneficial in the therapy of muscular dystrophy in conjunction with tocopherol.[179] The value of inositol in any of the above situations remains unproved.

8. Biotin

Biotin, the anti-egg white injury factor, is presumably needed in human nutrition, but requirement would be difficult to establish.[180] The intestinal bacteria synthesize this factor, and the average diet appears to supply

[175] E. P. Ralli, S. H. Leslie, G. H. Stueck, Jr., H. E. Shorr, J. S. Robson, D. H. Clarke, and B. Laken, *Medicine* **28,** 301 (1949).

[176] (*a*) J. de la Huerga and H. Popper, *J. Clin. Invest.* **30,** 463 (1951); (*b*) H. Popper in discussion at Conference on Liver Injury, *in* Transactions of the 10th Conference, Josiah Macy, Jr. Foundation, New York, 1951, p. 124; (*c*) J. de la Huerga, H. Popper, and F. Steigmann, *J. Lab. Clin. Med.* **38,** 904 (1951).

[177] (*a*) C. H. Best, J. H. Ridout, J. M. Patterson, and C. C. Lucas, *Biochem. J.* **48,** 448 (1951); (*b*) C. H. Best, C. C. Lucas, J. M. Patterson, and J. H. Ridout, *ibid.* **48,** 452 (1951).

[178] W. C. Felch and L. B. Dotti, *Proc. Soc. Exptl. Biol. Med.* **72,** 376 (1949).

[179] A. T. Milhorat and W. E. Bartels, *Science* **101,** 93 (1945).

[180] (*a*) T. W. Oppel, *Am. J. Med. Sci.* **204,** 856 (1942); (*b*) V. P. Sydenstricker, S. A. Singal, A. B. Briggs, N. M. DeVaughn, and H. Isbell, *J. Am. Med. Assoc.* **118,** 1199 (1942).

adequate quantities. Urinary excretion of biotin in man exceeds the amount supplied in the diet, while fecal excretion is much higher than the intake.[180a] Experimental human deficiency was induced by Sydenstricker and coworkers[180b] by feeding raw egg white. Characteristic findings were a dry, scaly dermatitis, anorexia, and muscle pains. It seems unlikely that spontaneous biotin deficiency will be observed in man.

p-Aminobenzoic Acid

There is no evidence that *p*-aminobenzoic acid must be supplied in the diet of human beings.

10. Folic Acid (Pteroylglutamic Acid) and Citrovorum Factor (Folinic Acid)

a. Folic acid requirement. Folic acid, a factor essential for growth and blood formation in several animal species, is also essential in man. The source of folic acid is both dietary and from synthesis by the intestinal flora. The latter source of supply may be important in man, as in the dog and rat, since experimental human deficiency has not been produced with diets low in folic acid. However, macrocytic anemia which responds to folic acid has been observed in human beings under a number of circumstances, e.g., in infancy, during pregnancy, in sprue, and in association with poor diets, so-called nutritional macrocytic anemia. The human requirement is unknown, but on the basis of animal requirements and of response of macrocytic anemia to therapy it may be in the vicinity of 0.1 to 0.2 mg. daily.[27] One of the difficulties in estimating the need for folic acid is the occurrence of this factor in nature in three forms: free, as pteroyltriglutamic acid, and as pteroylheptaglutamic acid. Information concerning amounts of these compounds in foods and their biologic availability is meager. The best food sources appear to be liver, deep-green leafy vegetables, other green vegetables, cauliflower, kidney, muscle meat, and wheat cereals. Loss during cooking and storage may be considerable.

Another member of the folic acid group of vitamins, citrovorum factor or folinic acid, is also found in natural materials, both in free and combined form. Citrovorum factor is believed to be a metabolically active form of folic acid and is formed in the body from folic acid. The citrovorum content of foods is largely unknown.

b. Functions of folic acid. Folic acid and citrovorum factor are believed to function by participation in the synthesis of compounds utilized in the formation of nucleoproteins, and in transmethylation processes.[181] According to Jukes (Chapter 10), folic acid is concerned with the incorporation

[181] (*a*) B. Kelley, *Federation Proc.* **10,** 206 (1951); (*b*) J. Lascelles, M. J. Cross, and D. D. Woods, *Biochem. J.* **49,** lxvi (1951); (*c*) J. N. Williams, Jr., *Proc. Soc. Exptl. Biol. Med.* **76,** 206 (1951).

of the "single-carbon unit" into the 2 and 8 positions of the purine ring and the 5-methyl group of thymine, with the reversible formation of serine from glycine and "formate", and with other reactions relating to the formation of choline, creatine, and histidine. It seems likely that citrovorum factor is the form in which folic acid participates in several enzyme systems, although these systems have not been isolated.

Folic acid exerts a hematologic effect in practically all types of human macrocytic anemia. This hematopoietic property is probably related to the formation or utilization of thymine and other pyrimidines and purines. Thymine, in large doses, stimulates blood regeneration in macrocytic anemia.[182] In patients with pernicious anemia treated with folic acid, neurologic lesions appear and hematologic relapse occurs if vitamin B_{12} is not administered. It has been postulated that folic acid exerts its effect in pernicious anemia by "mass action," pushing a chemical reaction through to completion in persons already deficient in vitamin B_{12} and further depleting body stores of this factor.

Aminopterin, a folic acid antagonist, has been administered to patients with leukemia and has brought about temporary remissions in about one-third of the subjects.[183] It has been suggested that aminopterin may block the synthesis of desoxyribonucleic acid by inhibiting incorporation of the "single-carbon" fragment into the methyl group of thymine and into the 2 and 8 positions of the purine ring (Chapter 10). It has been postulated, also, that aminopterin interferes with the conversion of folic acid to citrovorum factor. Toxic symptoms following aminopterin administration represent the syndrome of folic acid deficiency, i.e., glossitis, diarrhea, gastrointestinal lesions, and anemia. These manifestations may be reversed by citrovorum factor if it is administered promptly.[184]

(1) *Folic Acid—Ascorbic Acid Relationships.* Folic acid and ascorbic acid have some interesting metabolic relationships, among which is a role in tyrosine metabolism. Abnormal excretion of tyrosine metabolites occurs in human infants with scurvy, and in premature infants when the diet is high in this amino acid. Folic acid, in large doses, will prevent or relieve this abnormal excretion, as does ascorbic acid.[185] Folic acid exerts a similar action in scorbutic guinea pigs.[57a]

Megaloblastic anemia of infancy appears to be related to deficiency of ascorbic acid as well as of folic acid. This anemia has been observed chiefly

[182] T. D. Spies, W. B. Frommeyer, C. F. Vilter, and A. English, *Blood* **1,** 185 (1946).

[183] (*a*) W. Dameshek, *Blood* **4;** 168 (1949); (*b*) S. Farber, *ibid.* **4,** 160 (1949); (*c*) J. H. Burchenal, O. A. Karnofsky, C. M. Southam, W. P. L. Myers, L. F. Carver, H. W. Dargeon, Jr., and C. P. Rhoads, *Am. J. Med.* **7,** 420 (1949).

[184] (*a*) A. M. Earle, W. A. Reilly, and W. G. Lawson, *J. Pediat.* **39,** 560 (1951); (*b*) E. B. Schoenbach, E. M. Greenspan, and J. Colsky, *J. Am. Med. Assoc.* **144,** 1558 (1950).

[185] J. E. Morris, E. R. Harpur, and A. Goldbloom, *J. Clin. Invest.* **29,** 325 (1950).

in infants who received a powdered milk preparation low in vitamin C and has almost disappeared since ascorbic acid was added to this food. The anemia occurs during the same age period in which scurvy shows the highest incidence. Megaloblastic anemia of infancy responds to folic acid or to citrovorum factor.[186] May and associates[187] produced a comparable anemia in monkeys on a diet deficient in both ascorbic and folic acids. The anemia could be prevented by ascorbic acid alone and responded to treatment with folic acid but only partially to ascorbic acid.

In vitro studies indicate that folic acid may be converted to citrovorum factor by liver slices and that this conversion is aided by ascorbic acid.[188] Vitamin B_{12} may also have a role in the conversion of folic acid to citrovorum factor,[189] and both vitamin B_{12} and ascorbic acid may stimulate synthesis of folic acid in the body.[190] Relationships between vitamin B_{12} and folic acid are complex: both function in hematopoiesis, probably in nucleoprotein synthesis, and in the metabolism of labile methyl groups and choline.

Folic acid has a role in the reproductive process in the rat, the chick, and the monkey. No evidence is available as to a similar role in human nutrition.

c. Excretion of folic acid and citrovorum factor. The urinary excretion of folic acid and citrovorum factor by fourteen human subjects on normal diets was found to average 4.1 and 2.6 μg. daily, respectively. Excretion increased when the diet was high in purines.[191] After a 50-mg. oral dose of folic acid, 0.1 % was excreted in the urine as citrovorum factor in the subsequent 6 hr.[188b] Oral administration of ascorbic acid with folic acid increased excretion of citrovorum factor threefold as compared to administration of folic acid alone.[188] In human scurvy, only small amounts of citrovorum factor were found in the urine and excretion increased slowly after administration of folic acid. If ascorbic acid was given in large doses, only a slight rise in excretion of citrovorum factor was observed. Although the above interrelationships between ascorbic acid and folic acid require elucidation, ascorbic acid appears to have a role in the conversion of folic acid to citrovorum factor. Recent studies indicate that ascorbic acid ac-

[186] C. W. Woodruff, J. C. Peterson, and W. J. Darby, *Proc. Soc. Exptl. Biol. Med.* **77,** 16 (1951).

[187] C. D. May, R. D. Sundberg, F. Schaar, C. U. Lowe, and R. J. Salmon, *Am. J. Diseases Children* **82,** 282 (1951).

[188] (*a*) C. A. Nichol and A. D. Welch, *Proc. Soc. Exptl. Biol. Med.* **74,** 52 (1950); (*b*) H. P. Broquist, E. L. R. Stokstad, and T. H. Jukes, *J. Lab. Clin. Med.* **38,** 95 (1951).

[189] L. S. Dietrich, W. J. Monson, and C. A. Elvehjem, *Proc. Soc. Expt. Biol. Med.* **77,** 93 (1951).

[190] L. S. Dietrich, C. A. Nichol, W. J. Monson, and C. A. Elvehjem, *J. Biol. Chem.* **181,** 915 (1949).

[191] U. D. Register and H. P. Sarett, *Proc. Soc. Exptl. Biol. Med.* **77,** 837 (1951).

tivates an enzyme which liberates citrovorum factor from a bound form present in liver.[192] Perhaps ascorbic acid functions in a similar manner in the intestinal tract to release citrovorum factor from bound form in food.

It has been reported that some patients with pernicious anemia or steatorrhea excrete less folic acid after intravenous administration of small doses than do normal subjects. This might indicate an increase in the utilization of folic acid in these diseases.[193]

d. Folic acid deficiency. Folic acid deficiency in man is characterized by macrocytic anemia, glossitis, diarrhea, and impaired absorption from the gastrointestinal tract. Folic acid will stimulate hematologic response in practically all types of macrocytic anemia. Certain of these anemias appear to represent syndromes of folic acid deficiency, namely, macrocytic anemia of pregnancy and megaloblastic anemia of infancy. In these anemias, response always follows administration of folic acid but does not always occur after administration of vitamin B_{12}.[194] Some cases of sprue and nutritional macrocytic anemia likewise appear to be due to deficiency of folic acid, whereas others may be related to deficiency of vitamin B_{12} or of both factors.[195] Administration of folic acid is followed by healing of glossitis, subsidence of diarrhea, and improvement in absorption of glucose, fat, and fat-soluble vitamins from the intestinal tract, when these findings have been present.

Citrovorum factor stimulates hematologic improvement, similar to that induced by folic acid, in sprue, nutritional macrocytic anemia, macrocytic anemia of pregnancy, pernicious anemia and megaloblastic anemia of infancy.[193] Whether this factor is more active than folic acid has not been determined. Neither folic acid nor citrovorum factor will prevent hematologic relapse or the development of neurologic lesions in pernicious anemia.[196] Folic acid is helpful at times in the therapy of macrocytic anemia associated with cirrhosis of the liver.

11. Vitamin B_{12}

Vitamin B_{12}, the anti-pernicious anemia factor of liver, was originally isolated in 1948 by Rickes and associates[197] in this country and, independently, by E. Lester Smith in England.[198] This vitamin is now known to

[192] C. H. Hill and M. L. Scott, *J. Biol. Chem.* **196,** 195 (1952).

[193] G. A. Goldsmith, W. G. Unglaub, and J. Gibbens, *Arch. Internal Med.* **90,** 540ff (1952).

[194] *Nutrition Revs.* **8,** 260 (1950).

[195] G. A. Goldsmith, Proceedings of the Intern. Society of Hematology, Grune and Stratton, New York, 1951, p. 11.

[196] L. M. Meyer and W. C. L. Diefenbach, *Am. J. Clin. Path.* **21,** 1054 (1951).

[197] E. L. Rickes, N. G. Brink, F. R. Koniuszy, T. R. Wood, and K. Folkers, *Science* **107,** 396 (1948).

[198] (*a*) E. L. Smith and L. F. J. Parker, *Biochem. J.* **43,** *viii* (1948); (*b*) E. L. Smith, *Nature* **161,** 636 (1948).

include a group of closely related compounds with similar physiologic activities; vitamin B_{12} contains the cyan group; vitamin B_{12a}, B_{12b}, and B_{12d} (which are identical) contain the hydroxy group; and vitamin B_{12c} contains the nitrite group. Isolation of the anti-pernicious anemia factor was hampered by inability to produce a disease comparable to pernicious anemia in experimental animals and the consequent necessity of testing fractions of liver for activity in human subjects who had pernicious anemia in relapse. Shorb's finding[199] that certain lactobacilli apparently required the anti-pernicious anemia factor of liver for growth greatly facilitated research. At present, several microorganisms are used in assaying material for vitamin B_{12} including *L. leichmannii*, a mutant of *B. coli* and *Euglena gracilis*. Although vitamin B_{12} was originally isolated from liver, it has subsequently been obtained by fermentation from *Streptomyces* (*Actinomyces*) *griseus*.[200] Vitamin B_{12} has been found in nature in complex or bound forms which may be inactive either microbiologically or clinically. Excellent reviews of the chemistry, methods of assay, sources, functions, and clinical aspects of vitamin B_{12} have recently been published.[201]

a. Metabolic functions of vitamin B_{12}. (1) *Pathologic Changes in Pernicious Anemia.* In order to discuss the probable functions of vitamin B_{12} in human nutrition, it is necessary to review briefly the concepts of the pathogenesis of pernicious anemia, which represents the syndrome of vitamin B_{12} deficiency in man, and to outline the pathologic changes which occur in this condition. Pernicious anemia is not a simple dietary deficiency disease. Many years ago, Castle demonstrated that some factor which is present in normal gastric juice is absent from the gastric juice of patients with pernicious anemia. Administration of beef muscle with normal gastric juice stimulated blood regeneration in pernicious anemia. He postulated that an "intrinsic factor" in gastric juice combined with an "extrinsic factor" in food to form the anti-pernicious anemia factor of liver.

In pernicious anemia, the bone marrow is hyperplastic and contains many large young cells which are designated megaloblasts. The peripheral blood shows severe macrocytic anemia, leucopenia with a high percentage of old polymorphonuclear leucocytes, and thrombocytopenia. Erythrocyte destruction is excessive, as manifested by an increase in the level of bilirubin in serum and a rise in urobilin excretion in the urine. An abnormality of tyrosine metabolism is demonstrated by the excretion of certain phenolic compounds in the urine. Hydrochloric acid is absent from the gastric juice.

[199] M. S. Shorb, *Science* **107,** 397 (1948).

[200] E. L. Rickes, N. G. Brink, F. R. Koniuszy, T. R. Wood, and K. Folkers, *Science* **108,** 634 (1948).

[201] (*a*) E. L. Smith, *Nutrition Abstracts & Revs.* **20,** 795 (1950–1951); (*b*) C. C. Ungley, *ibid.* **21,** 1 (1951–1952).

The patient frequently has severe glossitis, gastrointestinal complaints, and may develop neurologic lesions which include subacute combined degeneration of the spinal cord and peripheral neuritis.

(2) *Effect of Vitamin B_{12} on Pathologic Changes in Pernicious Anemia.* In order to prove that vitamin B_{12} was the anti-pernicious anemia factor, it was necessary to show that this vitamin would influence the pathologic changes in pernicious anemia in the same manner as liver extract. It has been demonstrated that the parenteral administration of either vitamin B_{12} or liver extract, in adequate and comparable amounts, induces rapid return of the megaloblastic bone marrow to normal, a marked rise in reticulocytes (young red cells) in the blood, a gradual return of the erythrocyte, leucocyte, and platelet counts to normal, a decrease in bilirubin concentration in blood and urobilin excretion in urine to normal levels, disappearance of abnormal phenolic compounds from the urine, healing of glossitis, and improvement of lesions of the nervous system. The amount of vitamin B_{12} which will induce maximal response in pernicious anemia is in the neighborhood of 1 to 3 μg. daily.[195, 202] Vitamin B_{12} has also been shown to be effective in the maintenance of patients who have pernicious anemia in a normal condition, in a manner comparable to liver extract. The maintenance requirement of vitamin B_{12} ranges from less than 0.5 to 4 μg. daily,[195, 203] when administered parenterally at monthly or bimonthly intervals. The exact mode of action of vitamin B_{12} in erythropoiesis has not been elucidated. This vitamin may be involved in nucleoprotein synthesis as is folic acid. Hausmann[204] obtained a hematopoietic response, with reversal of megaloblastic bone marrow to normal, following administration of 2.0 to 2.8 g. of thymidine to two patients with pernicious anemia. These findings were interpreted as supporting the hypothesis that vitamin B_{12} exerts an enzyme influence on nuclear metabolism.

(3) *Vitamin B_{12} and "Extrinsic" Factor.* Vitamin B_{12} not only appears to be the anti-pernicious anemia factor of liver but also the "extrinsic" factor of food. Beef muscle, which is a good source of extrinsic factor, is not effective orally in the treatment of pernicious anemia, unless given in association with normal gastric juice. Beef muscle extract, administered parenterally to patients with pernicious anemia, has an activity equivalent to its vitamin B_{12} content.[205] Vitamin B_{12}, administered orally in amounts comparable to those found in beef muscle, is likewise not effective in the

[202] E. Jones and W. J. Darby, *Blood* **4,** 827 (1949).

[203] (*a*) J. F. Mueller, T. Jarrold, V. R. Hawkins, and R. W. Vilter, *Ohio Med. J.* **46,** 225 (1950); (*b*) C. C. Ungley, *Brit. Med. J.* **2,** 1370 (1949); (*c*) F. H. Bethell, *J. Am. Dietet. Assoc.* **26,** 89 (1950); (*d*) *Lancet* **1,** 353 (1950).

[204] K. Hausmann, *Lancet* **1,** 329 (1951).

[205] E. H. Morgan, B. E. Hall, and D. C. Campbell, *Proc. Staff Meetings Mayo Clinic* **24,** 594 (1949).

treatment of pernicious anemia, unless given in conjunction with normal gastric juice.[201b, 206] Oral doses of 25 to 400 μg. of vitamin B_{12} rarely stimulate a hematologic response; doses of 500 to 1000 μg. at times stimulate partial, and occasionally complete, remission. Oral doses of 3000 μg. may stimulate maximal hematologic improvement;[207] the median response in seventeen subjects was found by Ungley to be equal to that produced by injection of 20 to 40 μg.[201b]

(4) *Vitamin B_{12} and "Intrinsic" Factor.* The nature of the "intrinsic" factor in gastric juice has not been determined and its mode of action remains poorly understood. From data discussed above, intrinsic factor appears to facilitate the absorption of vitamin B_{12}. It may do this by protecting vitamin B_{12} from destruction by some toxic or inhibitory factor in the intestinal tract, either by combining with vitamin B_{12} or by neutralizing the toxic factor. Macrocytic anemia has been observed in animals after operative procedures which produced stagnant loops of intestine, and it has been noted in human beings with similar pathologic findings. The anemia disappears with relief of the stagnation. Intrinsic factor may prevent bacteria in the upper intestinal tract from utilizing vitamin B_{12}, thereby making it available for absorption. The bacterial flora of the upper intestine is abnormal in patients with pernicious anemia.[208] Perhaps intrinsic factor must combine with vitamin B_{12} before absorption can occur. Gastric juice contains a substance which will combine with vitamin B_{12}, rendering it microbiologically inactive, but it has not been proved that this substance is identical with intrinsic factor.[201a, 209] A number of other substances combine with vitamin B_{12}. In blood serum, vitamin B_{12} is present largely in combined form. There is some evidence that the absorbable form of vitamin B_{12} may be a complex.[210]

In marrow culture studies, Callender and Lajtha[211] found that vitamin B_{12} and normal gastric juice formed a thermolabile hematopoietic factor which ripened megaloblasts, whereas either substance alone was inactive. They also reported that normal serum contains a thermolabile factor which ripens megaloblasts in vitro, whereas serum from patients with untreated pernicious anemia had an inhibitory action. They postulated an extragastric as well as a gastric source of intrinsic factor. These findings are in

[206] G. A. Goldsmith, W. G. Unglaub, and J. Gibbens, *Arch. Internal Med.* **90,** 542ff (1952).

[207] W. G. Unglaub, H. L. Rosenthal, and G. A. Goldsmith, *Federation Proc.* **12,** 432 (1953).

[208] P. R. Burkholder, *Science* **114,** 478 (1951).

[209] G. B. J. Glass, L. J. Boyd, M. A. Rubinstein and C. S. Svigals, *Science* **115,** 101 (1952).

[210] F. H. Bethell, M. E. Swendseid, S. Miller, and A. A. Cintron-Rivera, *Ann. Internal Med.* 35, 518 (1951).

[211] S. T. Callender and L. G. Lajtha, *Blood* **6,** 1234 (1951).

contrast to those of Horrigan and associates,[212] who injected vitamin B_{12} into the bone marrow of one iliac crest and produced maturation of erythrocytes in this area, with no change in the marrow of the other iliac crest. It is obvious that vitamin B_{12}-intrinsic factor relationships must be clarified by future investigation.

(5) *Blood Levels and Excretion of Vitamin B_{12}.* The urinary excretion of vitamin B_{12} has been studied by several groups of investigators using different methods. Register and Sarett,[191] using *L. leichmannii* as the test organism, found the average excretion of fourteen subjects on normal diets to be 31 mμg. daily; daily excretion was 62 mμg. during fasting. Girdwood,[213] using *Euglena gracilis* as the test organism, reported mean excretion of vitamin B_{12} to be 126 mμg. daily in ten normal subjects, and 95 mμg. daily in twenty-two patients with pernicious anemia. Oral administration of vitamin B_{12} in amounts up to 10,000 μg. has resulted in very minute or no increase in urinary excretion.[214] A major portion of the ingested vitamin appears in the feces.[214b] Essentially the same amount of vitamin B_{12} has been found in the feces of patients with pernicious anemia as in normal subjects.[215] Extracts prepared from feces have been found to be hematopoietically active when injected into patients with pernicious anemia.

After intramuscular administration of vitamin B_{12}, Chesterman and associates[214c] found that the amount excreted in the urine was related to the size of the dose according to the formula: Excretion = Dose − 1.2 $\text{Dose}^{0.89}$. Experiments of Unglaub and associates[207] indicate that the relationship between dosage and excretion cannot be predicted with accuracy. Patients with pernicious anemia excrete less vitamin B_{12} than do normal subjects after comparable doses.[207, 214c]

Vitamin B_{12} activity in human sera has been measured by Ross[216] using *Euglena gracilis* as the test organism, and by Rosenthal and Sarett[217] using *L. leichmannii*. Vitamin B_{12} appears to be present in sera largely in combined form, probably bound to protein. In normal subjects, Ross[216] found levels of vitamin B_{12} activity of 0.35 to 0.75 mμg. per milliliter of serum; Rosenthal and Sarett[217] found levels of 0.08 to 0.42 mμg. per milliliter. Unglaub and associates[218] found serum activity to be less than 0.08 mμg.

[212] D. Horrigan, T. Jarrold, and R. W. Vilter, *J. Clin. Invest.* **30,** 31 (1951).

[213] R. H. Girdwood and K. M. Carmichael, *Brit. Med. J.* **2,** 1357 (1950).

[214] (*a*) C. L. Conley, J. R. Krevans, B. F. Chow, C. Barrow, and C. A. Lang, *J. Lab. Clin. Med.* **38,** 84 (1951); (*b*) B. F. Chow, *J. Nutrition* **43,** 323 (1951); (*c*) D. C. Chesterman, W. F. J. Cuthbertson, and H. F. Pegler, *Biochem. J.* **48,** li (1951).

[215] S. T. E. Callender and G. H. Spray, *Lancet* **1,** 1391 (1951).

[216] G. I. M. Ross, *Nature* **166,** 270 (1950).

[217] H. L. Rosenthal and H. P. Sarett, *J. Biol. Chem.* **199,** 433 (1952).

[218] W. G. Unglaub, H. L. Rosenthal, and G. A. Goldsmith, *Proc. South. Soc. Clin. Research,* Jan. 31, 1953, p. 33.

per milliliter in patients with macrocytic anemia. Serum levels increased following injections of vitamin B_{12}, the increase being proportional to the size of the dose.[207] A rise in serum vitamin B_{12} activity also followed oral administration of 3000 μg. of vitamin B_{12} in normal subjects and in patients with pernicious anemia.[207, 218] The rise was equal to that observed after parenteral administration of 10 to 50 μg. of the vitamin. An increase in serum activity was observed after oral administration of 500 and 1000 μg of vitamin B_{12} in one patient with pernicious anemia but not in several normal subjects.

No vitamin B_{12} was detected microbiologically in the tissues of a patient with untreated pernicious anemia who died of coronary thrombosis.[219] However, skin specimens obtained by biopsy from three patients with pernicious anemia in relapse showed vitamin B_{12} activity.

(6) *Vitamin B_{12} and Growth.* Vitamin B_{12} is essential for the growth of many animal species and is probably the principal but not necessarily the only component of the animal protein factor (Chapters 9 and 10). Preliminary reports indicate that, under certain conditions of diet and nutrition, vitamin B_{12} may promote growth in human beings.[214b, 220]

(7) *Other Functions of Vitamin B_{12}.* Relationships of vitamin B_{12} to methionine, choline, and folic acid metabolism have been discussed in Chapter 10. Whether vitamin B_{12} is related to transmethylation processes in man remains to be determined.

b. Clinical importance of vitamin B_{12}. The importance of vitamin B_{12} in the therapy of pernicious anemia has been discussed above. This vitamin is effective, also, in the treatment of certain other types of human macrocytic anemia. Satisfactory clinical and hematologic response has been obtained in sprue and nutritional macrocytic anemia after parenteral administration of vitamin B_{12},[206] and in some subjects after oral administration.[195] Gastric juice appears to potentiate the effect of oral vitamin B_{12} in some subjects.[221] Macrocytic anemia after gastrectomy responds to injections of vitamin B_{12} but folic acid may also be required.[222] Vitamin B_{12} is not as effective as folic acid in the treatment of macrocytic anemia of pregnancy or megaloblastic anemia of infancy; response occurs in some cases and not in others.[223] Administration of vitamin B_{12} is useful in anemia associated with fish tapeworm infestation. It is believed that the anemia results from competition between the worm and the host for vitamin B_{12}. In

[219] R. H. Girdwood, *Brit. J. Nutrition* **5,** *xi* (1951).

[220] (*a*) N. C. Wetzel, W. C. Fargo, I. H. Smith, and J. Helikson, *Science* **110,** 651 (1949); (*b*) N. C. Wetzel, H. H. Hopwood, M. E. Kuechle, and R. M. Grueninger, *J. Clin. Nutrition* **1,** 17 (1952).

[221] F. Diez-Rivas, R. M. Suarez, F. H. Morales, and E. P. Santiago, *Ann. Internal Med.* **36,** 583 (1951).

[222] N. S. Conway and H. Conway, *Brit. Med. J.* **1,** 158 (1951).

[223] (*a*) C. W. Woodruff and J. C. Peterson, *Postgrad. Med.* **10,** 189 (1951); (*b*) V. Ginsberg, J. Watson, and H. Lichtman, *J. Lab. and Clin. Med.* **36,** 328 (1950).

some cases of macrocytic anemia associated with cirrhosis of the liver, vitamin B_{12} may stimulate hematologic response.

Vitamin B_{12} has been used in the treatment of certain neurological conditions unrelated to pernicious anemia. Dramatic relief of pain in nutritional neuropathy was reported by Bean and associates.[224] Marked improvement has followed administration of massive doses of vitamin B_{12}, 1000 μg. daily, in trigeminal neuralgia. The neurological manifestations of diabetes mellitus have also improved following therapy with vitamin B_{12}.[225]

IX. Ascorbic Acid (Vitamin C)

The disease scurvy was recognized many centuries ago and satisfactory empiric treatment was devised several hundred years before ascorbic acid was isolated. The many studies which have been conducted since identification of this vitamin are reviewed in Chapter 11. Therefore, only findings of particular importance in human nutrition will be summarized in this section.

1. Absorption, Excretion, Blood Levels, and Storage of Ascorbic Acid

Since ascorbic acid is water-soluble and very easily oxidized, loss of this factor in the storage and preparation of food may be great. Destruction is more rapid in neutral and alkaline solutions than in acid solutions. In patients who have achlorhydria and in those who have received alkaline medication, ascorbic acid may be destroyed in the upper intestinal tract. The vitamin is absorbed readily, and the levels of ascorbic acid in plasma are related to recent dietary intake. Tissue saturation is achieved easily with large doses of ascorbic acid, and amounts in excess of this are excreted in the urine. It has been estimated that when the tissues of an adult human being are saturated the body contains about 5 g. of ascorbic acid.

The physiology of excretion of ascorbic acid, which is a threshold substance, has been reviewed by Smith.[226] The mechanism of reabsorption in the renal tubules is poorly understood. Average maximum tubular reabsorption was found by Ralli and associates to be 1.77 mg. per 100 ml. of tubular filtrate per minute (see Chapter 11).

Many studies of the concentration of ascorbic acid in blood and urine in relation to dietary intake have been carried out. In studies of population groups, a significant correlation between dietary intake and plasma level

[224] W. B. Bean, M. Franklin, and A. L. Sahs, *Am. J. Med. Sci.* **220,** 431 (1950).

[225] S. M. Sancetta, P. R. Ayres, and W. R. Scott, *Ann. Internal Med.* **35,** 1028 (1951).

[226] H. W. Smith, The Kidney, Structure and Function in Health and Disease, Oxford University Press, New York, 1951, pp. 136ff.

of ascorbic acid has been reported.[227] When the diet contains 75 to 100 mg. of ascorbic acid, plasma levels of 1.0 to 1.4 mg. per 100 ml. are observed. When the diet includes only 15 to 25 mg., plasma values range from 0.1 to 0.3 mg. per 100 ml. When vitamin C is essentially absent from the diet, plasma concentration falls rapidly to zero. Determination of plasma concentration is not a good measure of the vitamin C nutrition of individuals, since a zero level does not signify impending scurvy. Also, there is wide individual variation in plasma levels on comparable diets, and a decrease may occur in many pathological states.

The concentration of ascorbic acid in the white cell-platelet layer of blood is a much better index of tissue concentration and body stores of vitamin C then the level in plasma. In normal subjects whose tissues are nearly saturated, the white cell-platelet concentration is 25 to 30 mg. per 100 ml. The concentration falls gradually during depletion and reaches zero shortly before signs of scurvy make their appearance.[228] Peters and associates[228c] suggest that a white cell-platelet concentration of less than 2 mg. per 100 g., on repeated analyses, supports the diagnosis of scurvy.

The urinary excretion of ascorbic acid falls when the dietary intake is reduced, but it is a poor measure of tissue stores. Administration of test doses of ascorbic acid, followed by determination of the amount excreted in the urine or of the concentration in blood, has been widely used in nutritional evaluation.[227b] When the tissues are saturated with ascorbic acid, a larger percentage of the test dose is excreted in the urine, and the blood concentration reaches higher levels and remains elevated for longer periods, than when tissues are depleted.

2. Metabolic Functions of Ascorbic Acid

Although ascorbic acid has been shown to have many biologic functions, little is known of its chemical role in metabolic processes. Recently, Sealock and Goodland[58] and Knox and LeMay-Knox[229] have demonstrated that ascorbic acid is a coenzyme in the oxidation of tyrosine. This is "the only section of the vitamin's activity in which the molecular processes that depend on it are clearly understood" (Chapter 11). This explains the abnormal excretion of tyrosine metabolites in human scurvy and the return to

[227] (*a*) P. Putnam, D. F. Milam, R. K. Anderson, W. J. Darby, and P. A. Mead, *Milbank Mem. Fund Quart.* **27**, 355 (1950); (*b*) G. A. Goldsmith, *Federation Proc.* **8**, 561 (1949).

[228] (*a*) J. H. Crandon, C. C. Lund, and D. B. Dill, *New Engl. J. Med.* **223**, 353 (1940); (*b*) C. J. Farmer, *Federation Proc.* **3**, 179 (1944); (*c*) R. A. Peters, K. H. Coward, H. A. Krebs, L. W. Mapson, L. G. Parsons, B. S. Platt, J. C. Spence, and J. R. P. O'Brien, *Lancet* **1**, 853 (1948).

[229] (*a*) M. LeMay-Knox and W. E. Knox, *Biochem. J.* **48**, *xxii* (1951); (*b*) W. E. Knox and M. LeMay-Knox, *Biochem. J.* **49**, 686 (1951).

normal after administration of ascorbic acid. As previously noted, folic acid will also correct this metabolic defect. Relationships between ascorbic acid, folic acid, and vitamin B_{12}, particularly in megaloblastic anemia of infancy, have been discussed in a previous section (pp. 568 and 569). Ascorbic acid may release citrovorum factor from bound form or may assist in conversion of folic acid to citrovorum factor. The choline oxidase system may be concerned in relationships between ascorbic acid, folic acid, and vitamin B_{12} (Chapter 11).

It is generally agreed that connective tissue is defective in ascorbic acid deficiency but the exact nature of the defect is controversial. Ascorbic acid appears to be essential for the production of collagen. According to Wolbach,[230] all intercellular substances of supporting tissue have a substructure of collagen. Changes in bones, teeth, and gums in scurvy are believed to be secondary to this fundamental defect, i.e., failure to form normal connective tissue. The defect in capillaries in scurvy, which is manifested by hemorrhagic phenomena, is perhaps due to abnormal pericapillary connective tissue rather than to absence of an intercellular cement substance (Chapter 11). Defective wound healing has been demonstrated in ascorbic acid deficiency in man[228a] as well as in guinea pigs and is explicable on the same fundamental basis. In clinical medicine, close attention is directed to the ascorbic acid nutrition of subjects who are injured, or who are to undergo surgical operations, to ensure an adequate supply of vitamin C for tissue repair.

Ascorbic acid has a close relationship to adrenal cortical activity (Chapter 11). Stimulation of the adrenal cortex leads to depletion of ascorbic acid stores in the gland, but even in severe depletion cortical function is maintained. In conditions of stress and after administration of corticotropin, requirement and utilization of ascrobic acid is increased. It has been suggested that ascorbic acid is concerned in the oxidation of a precursor of some adrenal cortical steroid (Chapter 11). Pirani[231] postulated that ascorbic acid may have a nonspecific function related to cellular respiratory activity and metabolic rate.

The relationship between ascorbic acid and adrenal cortical activity suggests that ascorbic acid should be administered in conditions of stress, particularly after burns and other trauma, and to persons receiving prolonged therapy with corticotropin. Dao and Shank[232] reported that the decline in plasma ascorbic acid which follows surgical procedures can be prevented if 1 g. of ascorbic acid is given intravenously prior to operation. In patients who received this therapy, urinary excretion of steroids in-

[230] S. B. Wolbach, *Am. J. Path. Suppl.* **9**, 689 (1933).

[231] C. L. Pirani, *Metabolism* **1**, 197 (1952).

[232] T. L. Dao and R. E. Shank, *J. Lab. Clin. Med.* **38**, 802 (1951).

creased post-operatively. Klasson[233] found that administration of ascorbic acid to patients with burns reduced the time interval which must elapse before skin grafting could be undertaken.

Plasma ascorbic acid decreases in a number of infections, particularly those of long duration, such as rheumatic fever and tuberculosis. It seems logical to supplement the diet with ascorbic acid in these conditions even though the role of vitamin C in infection is not clear. This role may be related to adrenal cortical activity, to some function of ascorbic acid in phagocytosis, or perhaps to some general effect on tissue reaction to injury (Chapter 11). In hyperthyroidism, the requirement for ascorbic acid appears to be increased and supplementation with this vitamin would appear to be indicated.

Ascorbic acid has a number of other biologic functions. It exerts some effect on the hyaluronidase-hyaluronic acid reaction.[234] High concentration of ascorbic acid in the ovary is associated with secretion of the steroid hormone, progesterone. Ascorbic acid influences the reaction of small blood vessels to adrenaline. Whether ascorbic acid functions as a respiratory catalyst is still uncertain. It has been suggested that the movement of ascorbone into erythrocytes, the anterior chamber of the eye, and the central nerous system may indicate a role as an oxygen carrier (Chapter 11). Many of the biochemical changes in scorbutic guinea pigs seem to be related to oxidation-reduction processes. Further investigation will undoubtedly clarify the relationship of ascorbic acid to hormones and to enzyme systems and permit additional applications to human nutrition and clinical medicine.

The recent studies of Burns and associates[235] with radioactive L-ascorbic acid showed that 30% of a dose injected into guinea pigs was recovered in respiratory carbon dioxide during 24 hr. The urine contained 3 to 6% of total activity, 60% of which was found in calcium oxalate. Presumably ascorbic acid will be found to be a precursor of oxalate in man.

3. Ascorbic Acid Requirement

The minimal amount of ascorbic acid which will prevent scurvy in infants or adults is approximately 10 mg. daily. There is disagreement as to the amount of ascorbic acid which should be included in the diet for the maintenance of good nutrition. The recommended daily allowance in the dietary standard of the United States (Table 1) is 30 mg. for infants, 70 to 75 mg. for adults, 100 mg. during pregnancy, and 150 mg. during lacta-

[233] D. H. Klasson, *New York State J. Med.* **51,** 2388 (1951).

[234] E. Reppert, J. Donegan, and L. E. Hines, *Proc. Soc. Exptl. Biol. Med.* **77,** 318 (1951).

[235] J. J. Burns, H. B. Burch, and C. G. King, *Federation Proc.* **9,** 157 (1950).

tion.[27] The recommendation in the dietary standard of Great Britain is 10 mg. daily for infants, 20 mg. daily for adults, and 50 mg. during lactation.[29] The Canadian recommendation is 30 mg. daily for individuals of all age groups.[28]

The bases for the high allowances suggested in this country are several. In young growing guinea pigs, almost five times as much ascorbic acid is required for the development of normal tooth structure and for rapid healing of wounds than is needed for prevention of the classical lesions of scurvy; still more is required for maximum phosphatase activity in wounds.[27] Ascorbic acid is needed in amounts greater than those which will just prevent scurvy to maintain a tissue concentration of ascorbic acid in guinea pigs and primates comparable to that found in animals which synthesize this vitamin.[27] Ascorbic acid is needed in increased amounts during periods of stress. Infections are associated with a decrease in the plasma concentration and urinary excretion of ascorbic acid. Considerable evidence suggests a close metabolic relationship between ascorbic acid and adrenal cortical activity.

Kellie and Zilva[236] found, in one subject, that, when 30 mg. of ascorbic acid was ingested daily, almost none was excreted in the urine; this suggested complete utilization. When the dietary intake was 100 mg., about 50 mg. was "metabolized"; when intake was 50 mg. about 40 mg. was "metabolized." Dodds[237] studied plasma concentration and urinary excretion of ascorbic acid at several levels of intake (50, 75, and 100 mg. daily) and calculated that "utilization" of ascorbic acid was slightly less than 1 mg. per kilogram of body weight.

The amount of ascorbic acid necessary for the maintenance of tissue saturation in adults is in the neighborhood of 100 mg. or more daily. Although there is no evidence that saturation is necessary for the maintenance of normal nutrition, the above findings suggest that something above minimal is desirable. The exact level which will ensure adequate body stores and provide for emergency situations remains unknown.

4. Ascorbic Acid Deficiency

In experimental ascorbic acid deficiency in man,[228a, 228c] it was noted that the plasma concentration of ascorbic acid fell to zero in five to six weeks, while concentration in the white cell-platelet layer dropped to less than 1 mg. per 100 g. in about sixteen weeks. Shortly after this, clinical signs of deficiency appeared. These included hyperkeratotic papules surrounding hair follicles, exacerbation of acne, perifollicular hemorrhages and petechiae, tiny hemorrhages in the tips of the interdental papillae, and poor wound healing. In one study,[228c] the gums became purplish,

[236] A. E. Kellie and S. S. Zilva, *Biochem. J.* **33,** 153 (1939).

[237] M. L. Dodds, E. L. Price, and F. L. MacLeod, *J. Nutrition* **40,** 255 (1950).

spongy, and swollen in subjects who had previously shown periodontal disease. In another,[228a] X-ray films of the teeth showed interruptions of the lamina dura. In a third investigation,[228b] a saturation test in which blood and urine concentration were measured after giving 15 mg. of ascorbic acid per kilogram of body weight proved to be a reliable index of tissue depletion. The capillary fragility test was of no value in detecting deficiency.

In spontaneous scurvy in adults, findings similar to those noted in experimental deficiency are observed. When deficiency is severe, hemorrhagic phenomena are more marked and bleeding may occur into any tissue or from any mucous membrane. Anemia and edema are common findings. Severe lesions of the gums, atrophy of alveolar bone, and loosening of the teeth occur in advanced scurvy.

Scurvy in infants is characterized by hemorrhagic phenomena, including subperiostial hemorrhage, gum changes, and pathology of the bones. The lesions are described in detail in Chapter 13.

X. Trends in Human Nutrition Research and Future Possibilities

The preceding discussion summarizes a portion of the vast store of information which has accrued in the field of human nutrition over the years. Early investigation dealt with studies of energy metabolism and determination of the nutrients essential for man. Subsequently, attention was directed to delineation of specific deficiency syndromes. More recently, interest has centered on the role of nutrients in metabolic processes and on nutrient interrelationships. Concomitant with basic research, knowledge of nutrition has been applied in preventive and curative medicine and in public health. The only major deficiency disease in the world today in which the etiology remains to be clarified is kwashiorkor. If food supplies could be equitably distributed and if education could keep pace with advancing knowledge, it would be possible on the basis of present information to provide a reasonably adequate diet for all normal persons. Poverty, ignorance, lack of trained personnel, problems inherent in national productivity and in international trade relationships are some of the factors limiting application of current knowledge to the solution of world nutritional problems. Education and international cooperation should assist in solving these problems but, from their inherent nature, progress will be slow.

Future research should elucidate many intricacies of metabolism, not only in health but in disease. Relationships between hormones and nutrients appear to be a fruitful field for exploration. That certain nutrients have important functions in stress situations is beginning to be appreciated. Little precise information relative to nutritive requirements in disease has been obtained. When more data become available, nutritional therapy may be applied with greater accuracy in the management of many pathologic states. Recently, much attention has been directed to the role of nutrition in diseases which are primarily non-nutritional in origin, and in diseases in which etiologic relationships to nutrition are uncertain. A case in point is the extensive investigation of relationships between lipid metabolism and atherosclerosis. Perhaps nutrition will be found to be a factor in the pathogenesis of some of the so-called degenerative diseases, which increase in frequency with advancing age. Nutrition may play a role in the development of some types of human neoplasia, as has been found for certain tumors in animals.

As new insight into metabolic processes is gained through basic research, findings will be applied to man and should be of great assistance in providing abundant health to more persons, and in the prevention and treatment of diseases which have been resistant to therapeutic efforts in the past.

Author Index*

*Numbers in parentheses are footnote numbers and are inserted to enable the reader to locate a cross reference when the author's name does not appear at the point of reference in the text.

C

D

G

H

I

M

Q

R

T

U

V

Waltner, K., 448, 449, 454

Y

Z

Subject Index

D

H

I

M

Q

R

S

T

U

Printed in Dunstable, United Kingdom

91338425R00366